Computational Analysis of Biochemical Systems

Throughout the foreseeable future, research in medicine, molecular biology, and biotechnology will be dominated by scientific advancements resulting from the Human Genome Project. Knowledge of genomes will elucidate disease patterns and promises improved and personalized treatments. Food and drug development will be revolutionized by the focused manipulation of genes and biochemical processes. Researchers will begin to construct metabolic pathways from scratch.

A true understanding of genetic and metabolic function and design will crucially depend on mathematical and computational methods for analyzing biochemical systems. It will require new ways of thinking and novel approaches of integrative analysis.

This book teaches biochemists and molecular biologists in a hands-on fashion the use of modern computational methods for the analysis of complex biomedical systems; a modest mathematical background will suffice. The book begins with representations of biochemical systems, provides guidelines for setting up models, discusses in detail mathematical and computational methods of parameter estimation and model analysis, and ultimately connects to the modern literature with four detailed case studies. Every step is illustrated with examples and exercises, and is explored with the accompanying software, *PLAS*. A dedicated Web site featuring color illustrations, further exercises, research news, as well as resources and links also accompanies this book.

Eberhard O. Voit is Professor of Biometry in the Department of Biometry and Epidemiology at the Medical University of South Carolina.

PLAS was created by António E. N. Ferreira, Instructor in the Department of Chemistry and Biochemistry at the University of Lisbon, Portugal.

PLAS (Power Law Analysis and Simulation) Software License Agreement and Warranty

Copyright © 1996–2000 by António E. N. Ferreira. All rights reserved.

You must accept the following terms prior to using the Software on the accompanying CD for any purpose. If you do not accept these terms, you must not use this Software.

License Agreement

This Software may be freely used, distributed, or reproduced by any individual for educational, personal, and non-commercial use only. Any other use of this Software, including business or corporate use, is strictly prohibited without permission from the copyright holder. You may not attempt to reverse compile, modify, translate, or disassemble the Software in whole or in part. All of the components of the Software should be distributed or reproduced in the original archive form and should not be modified in any way. The Software cannot be included in another software package without the copyright holder's permission.

Warranty Disclaimer

THIS SOFTWARE IS PROVIDED "AS IS" WITHOUT WARRANTY OF ANY KIND. TO THE MAXIMUM EXTENT PERMITTED BY APPLICABLE LAW, THE AUTHOR, COPYRIGHT HOLDER, AND CAMBRIDGE UNIVERSITY PRESS FURTHER DISCLAIM ALL WARRANTIES, INCLUDING, WITHOUT LIMITATION, ANY IMPLIED WARRANTIES OF MERCHANTABILITY, FITNESS FOR A PARTICULAR PURPOSE, AND NONINFRINGEMENT. THE ENTIRE RISK ARISING OUT OF THE USE OR PERFORMANCE OF THE PRODUCT AND DOCUMENTATION REMAINS WITH RECIPIENT. TO THE MAXIMUM EXTENT PERMITTED BY APPLICABLE LAW, IN NO EVENT SHALL THE AUTHOR, COPYRIGHT HOLDER, OR CAMBRIDGE UNIVERSITY PRESS BE LIABLE FOR ANY CONSEQUENTIAL, INCIDENTAL, DIRECT, INDIRECT, SPECIAL, PUNITIVE, OR OTHER DAMAGES WHATSOEVER (INCLUDING, WITHOUT LIMITATION, DAMAGES FOR LOSS OF BUSINESS PROFITS, BUSINESS INTERRUPTION, LOSS OF BUSINESS INFORMATION, OR OTHER PECUNIARY LOSS) ARISING OUT OF THIS AGREEMENT OR THE USE OF OR INABILITY TO USE THE PRODUCT, EVEN IF THE COPYRIGHT HOLDER HAS BEEN ADVISED OF THE POSSIBILITY OF SUCH DAMAGES.

System Requirements for PLAS, Version 1.0

PLAS will run on any IBM PC or compatible system with an Intel Pentium or equivalent processor running at 100 MHz or faster, 32 MB of RAM, an 8X speed or faster CD-ROM drive (for CD installations only), a color graphics card and color monitor capable of displaying 640 × 480 pixel resolution, a mouse or other pointing device (optional but strongly recommended) and a Windows^R 32-bit operating system (Windows 95, Windows 98, Windows NT 4 or above, or Windows 2000). Installation requires 2 MB of hard disk space.

Computational Analysis of Biochemical Systems

A PRACTICAL GUIDE FOR BIOCHEMISTS AND MOLECULAR BIOLOGISTS

EBERHARD O. VOIT
Medical University of South Carolina

Software **PLAS**
by António Ferreira
Universidade de Lisboa

PUBLISHED BY THE PRESS SYNDICATE OF THE UNIVERSITY OF CAMBRIDGE
The Pitt Building, Trumpington Street, Cambridge, United Kingdom

CAMBRIDGE UNIVERSITY PRESS
The Edinburgh Building, Cambridge CB2 2RU, UK   http://www.cup.cam.ac.uk
40 West 20th Street, New York, NY 10011-4211, USA   http://www.cup.org
10 Stamford Road, Oakleigh, Melbourne 3166, Australia
Ruiz de Alarcón 13, 28014 Madrid, Spain

© Cambridge University Press 2000

This book is in copyright. Subject to statutory exception
and to the provisions of relevant collective licensing agreements,
no reproduction of any part may take place without
the written permission of Cambridge University Press.

First published 2000

Printed in the United States of America

Typefaces Sabon 10/13 pt. and Franklin Gothic   *System* LaTeX 2_ε [TB]

A catalog record for this book is available from the British Library.

Library of Congress Cataloging in Publication Data
Voit, Eberhard O.
    Computational analysis of biochemical systems : a practical guide for biochemists and molecular biologists / Eberhard O. Voit.
        p.   cm.
    Includes bibliographical references (p.   ).
    ISBN 0-521-78087-X (hardbound)
    1. Biological systems – Mathematical models.   2. Biochemistry – Mathematical models.
I. Title
    QH323.5.V65   2000
    570′.1′5118 – dc21
                                                                                00-021913

ISBN 0 521 78087 X hardback
ISBN 0 521 78579 0 paperback

PLAS (Power Law Analysis and Simulation) program (on CD-ROM accompanying this book) copyright © 1996–2000 by António Ferreira. All rights reserved.

Windows and Windows NT are registered trademarks of Microsoft Corporation.

To Ann, Walter, Richard, and Benedict,
who all encouraged me in their own ways.

Contents

	Preface	page IX
	Introduction	1
1	Graphical Representation of Biochemical Systems	11
2	Models of Biochemical Systems	37
3	From Maps to Equations	76
4	Computer Simulation	97
5	Parameter Estimation	143
6	Analytical Steady-State Evaluation	193
7	Sensitivity Analysis	222
8	Case Study 1 – Anaerobic Fermentation Pathway in *Saccharomyces cerevisiae*	260
9	Case Study 2 – Diagnosis and Refinement of a Model of the Tricarboxylic Acid Cycle in *Dictyostelium discoideum*	293
10	Case Study 3 – A Sequence of Models Describing Purine Metabolism	326
11	Case Study 4 – Algebraic Analysis of the Initial Steps of the Glycolytic–Glycogenolytic Pathway in Perfused Rat Liver	365
12	Epilogue – Canonical Modeling Beyond Biochemistry	399
	Appendix	413
	Hints and Partial Solutions	443
	References	469
	Author Index	499
	Subject Index	507

Preface

Entering a new millennium invites predictions. What will daily life be like, and where shall we be scientifically, philosophically, and socially, a hundred years from now? A thousand years from now? If the development of scientific knowledge during the twentieth century is any indication, we can expect a lot of excitement.

Three trends profoundly shaped biological thinking during the past half century and will continue unabated into the next millennium. First, the hard work and ingenuity of many researchers led to an explosion in factual knowledge about biology. Advanced methods of molecular biology brought forth an unprecedented wealth of detailed information about the most intricate component parts of organisms, and this information is now begging to be evaluated and integrated. Ten thousand strings of genetic code can be analyzed simultaneously with modern chip and microarray technologies; about three million genes have already been sequenced; and the complete human, mouse, and rat genomes will be known within the next few years. Second, the ready availability of personal computers made data management and computation accessible to most scientists. Huge amounts of information can be stored and retrieved with an efficiency that was unbelievable just a decade ago. Where it took many long minutes of mainframe computing in the 1970s to integrate even three or four moderately simple differential equations, systems of dozens of equations can now be evaluated very quickly with off-the-shelf personal computers. The third important development occurred in the scientific community at large, when the realization set in that the paradigm of reductionism is reaching its limits. It was recognized that even the complete hierarchical reduction of organisms to their fundamental components cannot always answer questions about function and purpose. An appreciation emerged for a new need of reconstructing organismic systems from the bottom up. It became evident that this reconstruction is an essential prerequisite for understanding the rationale for natural designs, for predicting their responses under untested conditions, and, eventually, for producing new pathways of medical or biotechnological significance with some confidence and reliability. It was realized that this reconstruction is only possible with the new mathematics of nonlinear systems analysis and modeling. The integration of knowledge with mathematical and computational means became a new frontier in biology.

The tasks at this new frontier require thinking beyond linear chains of causes and effects: thinking in terms of integrated functional entities; thinking in systems, networks, models. Over the past thirty years, a mathematical framework has evolved that deals with these issues. It is based on a number of theorems and on rigorous analytical and numerical methods, yet it has the advantage of being accessible also to the scientist who does not have an advanced mathematical background. The framework is called *canonical modeling*. This uncommon designation reflects the fact that the approach is based on a small number of principles that apply irrespective of the subject area. Canonical models can be constructed with a minimal amount of information, and the design processes and evaluation procedures follow a handful of guidelines. Canonical models always have the same structure, and this has facilitated the development of specific, very efficient computer algorithms, which are now part of an entirely interactive and intuitive software package, PLAS, that does not require any programming skills.

Throughout the thirty-year history of canonical modeling, great effort was expended to document and publicize new developments in the peer-reviewed literature, in dissertations, and at conferences. In spite of these efforts and of the relative ease with which canonical modeling can be learned, the new approach was embraced only by a limited group of scientists, most of whom had significant mathematical training. At a meeting a few years ago, some of the canonical modelers pondered this discrepancy and tried to pinpoint the possible hurdles that kept a larger audience from experiencing the fascination of canonical modeling. The main reason emerging from this soul searching was the lack of an up-to-date introductory text that would facilitate access to the technical literature and correct the wrong perception that canonical modeling was only for the mathematical whiz. I was asked to spearhead the project of writing such a text and accepted the challenge. This is the result.

As agreed, the book is written with the biochemist and molecular biologist in mind. It does not assume mathematical expertise beyond high school, but it requires an open mind and a willingness to learn new mathematical concepts. For those feeling uneasy about the underlying math, a crash course on some background material is provided in the Appendix. Techniques germane to the canonical modeling approach are introduced step by step in the first seven chapters. These technical chapters are followed by four detailed case studies, most of which are excerpted from the literature, reformulated, adapted, and annotated. It is my strong belief that studying these cases in detail and re-executing some of the analyses conducted by experts in the field will greatly benefit the reader's understanding. Beyond general strategies, modeling is not always straightforward in its nuances and sometimes has an almost artistic flavor. It is therefore of great value to retrace step by step the rationale for setting up a model and analyzing it. It is said that Beethoven copied entire Mozart symphonies note by note, just to see how the master had worked the details.

Several chapters are augmented with *Complements*, which present material that expands the horizon of the chapter but is not essential for later chapters. This material, combined with numerous references to the pertinent literature, may well be useful for further independent study. Every chapter also contains a set of exercises, which range from working numerical examples to thinking about a newly introduced concept and to related projects that are open-ended. In addition to the *Complements*, the *Epilogue* summarizes canonical modeling efforts that reach beyond biochemistry and molecular biology and may serve as an inspiration to branch out into other application areas.

Because modeling is a hands-on endeavor, the book comes with the computer program PLAS, which allows the reader to try out almost all of the concepts discussed in the book. PLAS is user-friendly and does not require any programming skills. The overhead of developing expertise in PLAS is minimal, and the learning curve has been steep for all who have worked with the program so far.

When I agreed to take on the book project, I did it under the condition of "communal help" from the leaders in canonical modeling. This help was solemnly pledged, and I have indeed received it generously throughout the process. Everyone involved in the early discussions contributed with published or unpublished material and was eager to help with input and feedback. It is very difficult to rank my appreciation to these individuals, so I will thank them in alphabetical order.

Marta Cascante was very helpful in the planning stage and discussed early versions of the material with me in great detail. She wrote the first draft on parameter estimation with methods developed by our colleagues in the related area of *metabolic control analysis*. Marta also provided the material for the last case study and for some of the smaller examples. Her contributions were of enormous help.

Raul Curto spent many long hours developing a series of models of purine metabolism. As is often the case, only the final model eventually appeared in the literature, thereby obscuring the hard work involved in the process of model development. I am very grateful to Raul for generously giving me all his background material and without hesitation allowing me to compose a case study from it.

António Ferreira, of course, is a main contributor to this project. His computer program PLAS may be the strongest part of it. It never ceases to amaze me with what fervor and patience António is willing to implement new ideas in PLAS or to transport the program to yet another platform. In addition to his computational efforts, António has given me uncounted comments and valuable feedback, especially for Chapter 4.

Rebecca Knapp gave me a lot of moral support and good suggestions for the subtitle.

Michael Savageau has always been and will always be "the one with the big picture." He has supported this project in multitudinous tangible and intangible ways. He impressed on me the importance of having the reader start with PLAS as early as possible and not following the standard procedure of doing all the theory first and finally getting to the fun (playing with PLAS) after many readers might already have put the book aside. He really liked my original 30-page introduction giving a historical perspective of biochemical endeavors beginning early B.C. but suggested throwing it out anyway in order to get going with the modeling. Only a good friend could have done that!

Albert Sorribas always reminded me that serious science and writing are not necessarily divorced from enjoying the process. I greatly appreciate his frank and constructive criticism, his help with Chapter 7, and his good sense of humor in person and through the Internet.

Néstor Torres kept me on track, reminding me regularly that I had promised to help him with another project after this book was done. He expended great effort checking several versions of the last case study and re-computing the model with several sets of parameter values. I could not have done that chapter without him.

Zhen Zhang helped me uncounted times to outfox my computer when it didn't behave.

Several of my students helped me by proofreading various versions of various chapters, giving me good suggestions and fixing "unusual grammatical constructions" in my English. I would like to express my gratitude in particular to Keith Bangerter, Russell Goodman, and Nicole White for making insightful comments.

While they were not directly involved with this book project, I wish to thank Drs. Alan Gross (in memoriam), Clinton Miller, Philip Rust, and Peter Sands, who never failed to encourage me in my work on canonical models. Last but definitely not least I thank my family here in Charleston, in Germany, and in Michigan for their unconditional support.

The publication of a book constitutes the completion of a big project. I consider it also a beginning. It is my hope that the readers will include many newcomers who develop an interest in canonical models and biochemical systems analysis and who will give me honest feedback and new impulses. On occasion I have been criticized for being too verbose. That doesn't bother me much. I remember a friend who wrote a 50-page dissertation draft on a very complicated, obscure topic of abstract mathematical logic. His professor, himself a brilliant mathematician, pleaded with my friend *please* to put a few sentences between the theorems, proofs, and corollaries, indicating which direction the argument was going. Obligingly, my friend wrote up the same mathematical material in 150 pages. Pulling him aside at graduation, the professor congratulated him, saying, "Now that I understand your dissertation, I really like the short version." I hope that some of my audience, having read the book, will really like the concise technical papers that are plentiful in the literature.

Eberhard O. Voit
Charleston, SC
2000

Introduction

Interest in biochemistry and metabolism is as old as mankind. By becoming *Homo sapiens*, man has developed a desire to understand life. Over the millennia, no other object of science has received as much attention as the human body, the food that keeps it going, and the medicines that restore its health after suffering from ailments. Indeed, most of our biochemical and pharmacological knowledge has its roots in centuries of trial and error that have yielded a remarkable understanding of drugs, dosages, and contraindications.

One anecdote that epitomizes the early human experience of expanding the body of pharmacological knowledge tells about the local shepherd who was accepted by the villagers as the expert pharmacist. A blacksmith presented to him with unexplained abdominal pain. The shepherd thought up a new herbal treatment and confidently promised the blacksmith he would not feel that pain much longer. Indeed, the blacksmith was up and running very quickly, and the satisfied shepherd recorded the efficacious new cure in his notebook. A few weeks later the shepherd was visited by a tailor with the same malady and gave the same advice, with more confidence than before. The tailor also did not feel the pain much longer: He died within a week. The shepherd dryly added to his prior record that "this treatment does not work for tailors" (Crummenerl, 1969: pp. 9–10).

While trial-and-error experiments have contributed much to our knowledge about metabolism, they do not provide insight into the underlying biochemical and physiological processes that ultimately prove one treatment effective and another not. Realizing this limitation in the last century, medical research began to shift from the former holistic approach to one of trying to characterize physiological and biochemical mechanisms. In view of the reliable predictions and explanations in physics and astronomy, it appeared only plausible to postulate that deterministic laws, similar to those in astronomy, would also govern metabolism. To develop such laws, the emphasis of biochemistry and physiology began to embrace controlled experimentation and quantification. Textbooks began to characterize chemical compounds according to their measurable physical and chemical properties, and blood pressure measurements and quantitative urine analysis became standard diagnostic tools of internal medicine (Leicester, 1974).

The wealth of new technologies, methods, and chemicals precipitated an unparalleled increase in physiological and biochemical knowledge. Science was so successful that it appeared to have no limits. Physical phenomena were described mathematically with great precision and predictive power, and it seemed just a question of time and intensive research until biological phenomena could be understood in the same way. In the late eighteenth century, Antoine Lavoisier, who is well known for his studies on oxygen and respiration, called upon the great mathematician Pierre Simon Laplace to attack some of his research problems (Leicester, 1974). This early form of biomathematical teamwork must have been quite successful. Lavoisier has been given credit for being the first to describe a biochemical phenomenon in the form of a mathematical equation:

must [juice] of grapes = carbonic acid + alcohol.

He explained the use of this equation in the following way:

> We may consider the substance submitted to fermentation, and the products resulting from that operation, as forming an algebraic equation, and, by successively supposing each of the elements in this equation unknown, we can calculate their values in succession and thus verify our experiments by calculation and our calculations by experiments reciprocally (Leicester, 1974: p. 139).

Out of these humble beginnings of employing mathematics in biochemistry eventually grew a concentrated effort to formulate physiological and biochemical phenomena with suitable algebraic functions. Mechanistic rate laws became part of the standard repertoire of the biochemist, and the characterization of rate law parameters, such as the maximum velocity and the Michaelis constant, became the subject of thousands of biochemical studies.

Reductionism

The second potent ingredient in the success of science in the eighteenth, nineteenth, and twentieth centuries was the strategy of *reductionism* (which Garfinkel (1985) traces back to Julius Caesar's motto "divide and conquer"). Its key rationale was that to understand an organism, one had to understand its organs, and to understand an organ, one would have to understand its cells and their subcellular components. In the pure philosophy of reductionism, science was to proceed down the hierarchy of organization until one had a firm understanding of the most fundamental elements and processes (cf. Simpson, 1975). These ideas were accompanied by the conviction that knowledge of the component parts would be rich enough to reconstruct an organism and its functionality and, hence, to yield an understanding of life. The approach of reductionism directly led to the present-day mainstream experimentation in the life sciences.

The first doubts about the unlimited power of reductionism came in the form of surprises at the beginning of the twentieth century: the mechanistic theories failed, even within the physical sciences where they had been most successful. Heisenberg proposed the uncertainty principle, according to which it was not possible to measure simultaneously the location and velocity of an electron. The discovery of radioactive decay highlighted a new dualism of determinism and randomness, since the decay overall followed a simple and precise mathematical law, while the decay of any given atom was entirely unpredictable. Random events were postulated by some, while Einstein maintained that "God does not cast dice." Confusing these complicated issues further were observations that components of systems behaved differently within the system than in isolation. It became evident that one had to pay attention to the interacting relationships among components, and these relationships were of such staggering complexity that simple mechanistic explanations had to be questioned. Eighty years later, many of these deep issues of science remain unresolved.

Why is reductionism insufficient? To put it simply, there is a tremendous difference between a living organism and a bottle containing all its chemical components. A birthday cake is more than flour, milk, eggs, butter, sugar, and candles. As Savageau (1991d) pointed out:

> Paradoxically, it is at the very height of its success that the weaknesses of this paradigm [of reductionism] are becoming apparent. We shall soon have the complete parts catalog of *E. coli*. Yet, by comparison, we still know little about the integrated system, what makes it a living cell, or how it will respond to novel environments, and to specific changes in its molecular constitution. Our knowledge is fragmented and descriptive; we have almost no understanding of the 'design principles' that govern the intact biological system. ... We need a radically different approach that is able to elucidate quantitative and qualitative features of complex integrated systems.

The intrinsic problem with biochemical systems *in vivo* is their enormous complexity. Considering that the human mind is apparently limited to simultaneously processing 7 ± 2 items of information (Miller, 1956), there is no way that we could intuitively keep track of all the changes occurring in such systems. Realistic biochemical systems are characterized by large numbers of components, and their governing processes are manifold, strongly coupled, and associative rather than additive. The functional characteristics and relationships within the subsystem of these systems are nonlinear, and the reactions take place far from thermodynamic equilibrium (e.g., von Bertalanffy, 1968: p. 19; Yates, 1977, 1978; Savageau, 1992b: p. 58). An organism like *E. coli* is rather simple in comparison with a higher animal or plant, yet it consists of thousands of biochemical components. The human brain contains on the order of 100 billion neuronal components and hundreds of trillions of interconnections (Koch and Laurent, 1999). The human body is composed of some five octillion atoms (Laszlo, 1972: p. 5). Just describing, not even analyzing, these components poses an enormous bookkeeping problem that can only be managed with mathematical means. Discussing issues of complexity, Goldenfeld and Kadanoff (1999) pointedly remarked: "Everything is simple and neat – except, of course, the world." Adding to the complexity of a typical biochemical or cellular system is that "it is not a machine (however complex) drawn to a well-defined design, but a machine that can and does constantly rebuild itself within a range of variable parameters" (Weng et al., 1999).

Nonlinearities in the governing processes cause two problems for any intuitive, non-mathematical approach. First, they create situations that lead our cause-and-effect way of thinking astray. Without the aid of mathematical analysis, we cannot reliably predict the effects of even the simplest regulatory mechanism in metabolism, the ubiquitous *feedback inhibition*. It is impossible to evaluate the advantages of inhibition of product formation over activation of its degradation, or the merits of repressible versus inducible gene circuits (e.g., Savageau, 1976, 1996). Synergistic responses of biochemical systems are commonplace, and it happens that integrated systems react in ways that seem totally counterintuitive. It is very difficult to find intuitive explanations for why systems in nature are designed the way they are, and we cannot reliably predict the correct system response without a rigorous quantitative approach (Savageau, 1992b). Since most constituents of biochemical systems play multiple roles, it is not sufficient to study their function in isolation and to expect them to behave similarly *in vivo*. We are trained to apportion big problems into manageable tasks, but when we follow this approach in the analysis of an organizationally complex system, we are at risk of missing some of its most interesting systemic properties.

Second, nonlinearities result in relationships between components and processes whose qualitative behavior depends on quantitative properties of the system: If some parameter is within a certain range, the system hovers around a normal state; if it is less than a critical lower threshold value, the system may exhibit sustained oscillations; and when the parameter exceeds some other threshold, the system may cease to function altogether. With numerous parameters having these characteristics, there is no possibility of understanding or predicting the effects of changes in some system components without a mathematical analysis (Savageau, 1992b). While the "lure of the linear" (Savageau, 1991e: p. 7) tempts us with all its theoretical underpinnings and a wide variety of powerful tools, nature is more often than not nonlinear, and it is necessary that we develop germane methods of capturing the essence of nonlinear phenomena.

Integration and Reconstruction

Summarizing the state of the art of reductionism, Simpson (1975) concluded that reduction to atomic and molecular levels alone was neither philosophically nor practically sufficient and that all levels of the biological hierarchy had to be studied if biological phenomena were to be explained. Yates (1977), a strong supporter of studying complexity and a proponent of an integrated approach to biology, wrote that even among true believers in reductionist biology, "there is a residual mystery after the reduction is accomplished." The failure of reductionism to yield a true understanding of biological phenomena led to the advent of a paradigm shift (Kuhn, 1962) in scientific thinking. It became evident to some forward-looking biologists that new laws had to be postulated – not laws describing parts, but laws capturing integrated systems.

One of the crucial beginnings of this paradigm shift in biology can actually be seen in the early work of Alfred Lotka and of Ludwig von Bertalanffy. Lotka (1924/1956) summarized his insights in a classic text that can be considered the first book on biomathematics. Appreciating the important role of systems in nature, Lotka formulated a *program of physical biology*, which was based on three principles: first, attention must be paid to phenomena displayed by the component aggregates in bulk; Lotka speaks of *bulk mechanics* or *macro-mechanics*. Second, one must study the behavior of the individual components; Lotka calls this approach *micro-mechanics*. And third, there is an inherent relationship between these two branches, which is the subject of *statistical mechanics*. The scope of the latter includes the treatment of the (micro-)dynamics of the components as well as the (macro-)dynamics of the living organism as a whole (Lotka 1924/1956: pp. 50–52).

Ludwig von Bertalanffy laid the foundation of modern systems science with writings beginning in the 1920s. In his classic *General Systems Theory* of 1968, he recounts the seminal ideas in the following way:

> In many phenomena in biology and also in the behavioral and social sciences, mathematical expressions and models are applicable. ... The structural similarity of such models and their isomorphism in different fields became apparent; and just those problems of order, organization, wholeness, teleology, etc., appeared central which were programmatically excluded in mechanistic science. This, then, was the idea of general system theory.

But as is typical for every paradigm shift (cf. Kuhn, 1962), the community of peers was not enthusiastic about changing the traditional and successful ways of doing science. Somewhat frustrated, von Bertalanffy noted in his 1968 book that nothing essentially new had been added to what he had proposed forty years earlier, mainly because the time had just not yet been ripe and biology was synonymous with laboratory work:

> The proposal of system theory was received incredulously as fantastic or presumptuous. Either – it was argued – it was *trivial* because the so-called isomorphisms were merely examples of the truism that mathematics can be applied to all sorts of things, and therefore carried no more weight than the 'discovery' that $2 + 2 = 4$ holds true for apples, dollars and galaxies alike; or it was *false* and *misleading* because superficial analogies – as in the famous simile of society as an 'organism' – camouflage actual differences and so lead to wrong and even morally objectionable conclusions. Or, again, it was philosophically and methodologically *unsound* because the alleged 'irreducibility' of higher levels to lower ones tended to impede analytical research whose success was obvious in various fields such as the reduction of chemistry to physical principles, or of life principles to molecular biology (von Bertalanffy, 1968: p. 14).

The situation improved slightly in the late forties, when three fundamentally novel concepts appeared: von Neumann and Morgenstern's game theory (1944), Wiener's cybernetics (1948), and Shannon's information theory (1948). In the decades to follow, the systems approach slowly became the "emerging contemporary view of organized complexity, one step beyond the Newtonian view of organized simplicity, and two steps beyond the classical world views of divinely ordered or imaginatively envisaged complexity" (Laszlo, 1972: pp. 12, 15).

The paradigm shift from reductionism to a systems approach of *reconstructionism* is still in progress (cf. Savageau, 1991cd, 1996; Kanehisa, 1998; Gallagher and Appenzeller, 1999). The emerging paradigm proposes that, in addition to traditional experimentation, the isolated parts must be put together again, and that intact systems and their interactions with the environment must be analyzed (Garfinkel, 1985;

Savageau, 1991cd). The paradigm of reconstructionism does not suggest abandoning well-controlled *in vitro* experiments. Systems approaches and modeling rely on good, hard data, and they are often inspired by hypotheses that arise from experimental observation. Similarly, mechanistic rate laws are not becoming obsolete. They have been the cornerstones of our traditional biochemical terminology and thinking and have proven successful. They provide a convenient language and convey an immediate meaning among experts. Nonetheless, the new paradigm acknowledges that mechanistic rate laws are not always sufficient and that they need to be complemented with an appropriate mathematical modeling framework that is capable of dealing with integrated systems (e.g., Yates, 1977, 1978; Savageau, 1991cd; Service, 1999).

With a paradigm shift from reductionism to reconstructionism unfolding before us, we must ask what the new challenges are and what types of strategies we must devise to address them. Three components seem to be essential for accepting the challenge: a valid means of problem *simplification*, a convenient *terminology*, and a convenient *mathematical representation* for large systems.

Simplification is a standard tool of theoretical research and not a strategy of avoiding reality. Much of our learning in all areas of life, be it geometry, history, or foreign languages, is based on simplified problems. Biochemistry books are full of graphical models, and six-foot models of DNA adorn many science labs. All these models are at best crude representations of reality, but they are still very useful, not *in spite* of their simplified nature but *because* of it. To demonstrate the staircase arrangement of bases, sugars, and phosphoric acid in the DNA molecule, it is unnecessary and inconvenient to consider super-helical turns and stabilizing histones, and therefore these features are (rightfully) omitted in models for novices. Employing simplifying models is nothing radically new.

The real question is not whether we should allow simplifications, but what types of simplifications are best suited for the analysis of organizationally complex systems. Some are provided by the hierarchical structure of nature itself (cf. Savageau, 1976: pp. 80–83). If a biochemical reaction is the focus of investigation, the laws of particle physics are still in effect, but it seems unnecessary and undesirable to represent the reaction based on a description of the motion of each contributing particle. Thus, one simplification consists of replacing the complexity at all sub-molecular levels with some average behavior. By the same token, we are often justified in ignoring processes at higher levels, such as the growth and development of the organism in which the biochemical reaction of interest takes place. When we are interested in the biochemical interactions between insulin, adrenaline, and glucose in diabetes, we might be justified in ignoring the cardiovascular implications of diabetes, knowing full well that they may have long-term effects that are of medical interest.

In many instances, the hierarchy in organizational levels corresponds to a hierarchy of time scales at which processes occur. For biological purposes one may differentiate the biomolecular, biochemical, developmental, and evolutionary time scales. Limiting an analysis to a single time scale provides a significant simplification, since processes occurring at much higher rates approach their steady states so fast that their dynamics are irrelevant, while processes at much lower rates don't change much and thus are essentially constant (Savageau, 1976; Heinrich et al., 1977).

Another simplification results from the spatial organization within the cell. Enzymes usually do not swim about, colliding with all kinds of proteins, until they incidentally find their substrate. They are confined to compartments, attached to tubules or surfaces, or exist in complexes that serve consecutive steps of metabolic pathways. These and other natural simplifications translate directly into significant mathematical simplifications that can and should be exploited in biochemical systems analysis (Savageau, 1976). We will discuss them later in this book in more detail.

The second ingredient of an integrated biochemical systems approach is a convenient terminology. Selecting a terminology is not just a cosmetic decision like the color of a car; it determines what types of questions can be asked and answered by the analysis. For instance, without the concept of inhibition constants it is very difficult to quantify regulatory patterns of biochemical systems. We will discuss questions of terminology in Chapters 1 and 2.

The third ingredient of an integrated biochemical systems approach is the selection of a convenient mathematical representation, a *model*. This may seem a surprising statement, since for most of our lives we were taught that mathematics is either right or wrong. While this is true in some sense, alternative mathematical approaches may greatly differ with respect to complexity, tractability, interpretability, and, again, the types of questions that can be addressed.

What can we expect from such models? Models help us recognize general patterns or relationships in nature that are otherwise masked by details. In doing so, they reduce uncertainty and facilitate the search for quantitative laws (cf. Carnap, 1966; Steinbuch, 1977: p. 30). Models can add information, because the simplified mathematical formulation can be compared with existing physical and mathematical knowledge, which is important in the detection of inconsistencies between the modeled experimental data and the commonly accepted body of knowledge (Steinbuch, 1977: p. 13). Once a model has revealed a pattern that underlies experimental data, it allows us to order and categorize these data by assigning them to classes that are defined by the pattern. Such classifications can help us objectively to distinguish between normal and abnormal phenomena, between healthy and sick.

Maybe the most important feature of a mathematical model is its ability to integrate collections of observations into an entity and connect them to a higher level of biological organization. In a biochemical model, system responses become explainable in terms of kinetic parameters; in a cell population model, the growth characteristics of the population may become explainable in terms of cell cycle phenomena. Explanations afforded by models include a better understanding of design principles in nature and answer questions like *Why is this process regulated in this particular way and not in an alternative fashion?* In this explanatory function, models enhance our understanding of the phenomenon, because they allow us to relate levels of organization and to predict responses at one level from information at another.

A good model can be used to predict phenomena under altered conditions and to execute simulations of what-if scenarios. For instance, a reliable metabolic model can help us screen or test the effects of potential drugs by simulating their interactions with metabolites and their side effects. Such simulations are fast, cheap, and precisely reproducible. In contrast to biological and medical experiments, they are also not impeded by issues of ethics. Simulations can be used to discard hypotheses and will eventually reduce the number of animal experiments. In the not-too-distant future, models will become a necessary prerequisite for the efficient design of pathways or even entire organisms. The technical side of genetic engineering is advancing fast, but to be successful, newly designed pathways and their control structures must be understood. Models will provide cheap and reliable screening tools for this purpose.

The Future

As Oliver Wendell Holmes once said, "it is not so important where we stand as in what direction we are going." (Incidentally – or maybe not – this quote could be interpreted as a non-mathematician's description of the concept of a differential equation.)

So, where *do* we stand and where *are* we going? Molecular biology, mathematics, and computer technology have created an intellectual environment that offers greater opportunities and challenges than ever before. Huge amounts of new, good data are published every day. In the fall of 1998, the U.S. National Library of Medicine contained more than ten million records from published articles and more than two and a half million DNA sequences (Boguski, 1998). The genomes of several bacteria, including the medically important *Haemophilus influenzae* (Fleischmann et al., 1995) and *Mycobacterium tuberculosis* (Cole et al., 1998), have been deciphered entirely. Insight into these genomes and the associated metabolic processes is very valuable, for instance, in better understanding the resistance of these organisms to current treatments and the developments of new drugs (Cole et al., 1998). In an unprecedented collaborative effort, about 600 scientists from around the world have sequenced the almost 6,000 genes of the first complete genome of an eukaryote, the baker's yeast *Saccharomyces cerevisiae* (Goffeau et al., 1996; see also supplement issue 6632S to *Nature*, Volume 387, 1997). The first genome of a multi-cellular organism, the nematode *Caenorhabditis elegans*, followed with about 19,000 genes in late 1998 (*C. elegans* Sequencing Consortium, 1998), and the genomes of mice and men will be available within the next few years (Collins et al., 1998; Battey et al., 1999). Parallel to the exponentially growing information base, the power of readily available personal computing has become greater than most of us can exploit, and it is steadily increasing. Meanwhile, new mathematical approaches and numerical algorithms on personal computers allow us to grind out mathematical solutions that could not have been obtained by any means ten or twenty years ago.

The convergence of biology and computer technology has formed a new field called *bioinformatics*, a term that did not even exist ten years ago (cf. Boguski, 1998). While the main focus of this field may currently be acquisition, storage, and retrieval of biological data (cf. Baxevanis and Ouelette, 1998), future bioinformaticians will increasingly emphasize the *interpretation* of the unbelievably large data stream coming from experimental labs around the world. Current strategies of *genomics* research will more and more be complemented with *comparative genomics* (Kanehisa, 1998) and *functional genomics* (Brownstein et al., 1998). They will be expanded to include *proteomes*, which are the complete sets of proteins that a living cell can synthesize (Kahn, 1995; Goffeau et al., 1996; Wilkins et al., 1996), and *proteomics*, which studies "expression profiles, time correlations, tissue-specific proteins (normal and abnormal), disease-related proteins and 'personal' genomes" (Thornton, 1998). *Pharmacogenomics* and intimate knowledge of biochemical systems will elucidate the role of nucleotide and enzyme polymorphisms in the inter-individual variability of responses to drugs. The result will be the genome-based development of entirely customized treatments and of drug dosages that vary from person to person. Eventually, *physiomics* will attempt to integrate knowledge about genes and proteins in an effort to reconstruct organisms or design new metabolic pathways (Boguski, pers. comm.). Brownstein et al. (1998) project the importance of connecting genomic data mining with biochemical interpretation in the following way:

> Biochemical pathways to which a particular transcript belongs could be identified or genes with which the transcript is thought to interact could be found. In a time-course experiment, sets of genes with similar temporal expression profiles could be sought. In the long run, the software could, indeed should, be made capable of pre-interpreting the data (using a biochemical knowledge base and a set of heuristics) and presenting the investigator with alternative hypotheses or explanations about its meaning. It is the only way that experiments involving tens of thousands of genes, of which a considerable fraction shows changes, can be managed.

The integration of the various aspects of genome research will eventually define the field of computational biology to include all steps from the exploration of whole genomes and singular nucleotide polymorphisms to the mathematical interpretation and design of biochemical and physiological systems. Modeling will play a crucial role at this new frontier. As Thornton (1998) puts it:

> The emphasis will therefore shift to understanding the principles and control of biological function and the interaction between molecules. Modeling cellular processes, such as signaling and metabolic pathways, will become increasingly important, especially as more proteomic data become available. Understanding and modeling function is essential to enable the rational design or modification of proteins or ligands for new functions. In my view, this is the greatest challenge for bioinformatics in the next millennium.

Indeed, the strategic plan for the Human Genome Project for the years 1998–2003 specifically acknowledges that progress in the field will eventually "need tools for modeling complex systems and interactions" (Collins et al., 1998). This plan also stresses – as one of seven goals – the interdisciplinary training of scientists in biology combined with mathematics, engineering, computer science, physics, chemistry, or social sciences (Collins et al., 1998).

This book may in some small way contribute to the outlined emphases. It only covers a tiny fraction of the entire spectrum of computational biology. In particular, it does not deal with the analysis of gene sequences, but rather concentrates on the second part of the research process, which integrates genomic and biochemical knowledge into functional entities. It equips the biologist with effective methods for developing and analyzing mathematical models that capture the essence of biochemical systems.

It would be a true milestone of science if this book could present a correct and "complete" mathematical theory (cf. von Weizsäcker, 1979: p. 193) of biochemistry and molecular biology, but we are still far from that point. Every day, a newly discovered phenomenon offers another glimpse into a wonder world of unbelievable complexity and catches us by surprise. We may not yet understand many of these phenomena, but we are equipped with more and better tools to deal with questions of complexity than ever before. These tools need to be tempered in the analysis of relevant systems. Their potential and limitations need to be explored. The tools need to be re-sharpened periodically or even recast into new and better tools. A new frontier of biology is challenging us. It is time to take up the challenge.

REFERENCES

[21–2], [32], [39], [42], [48], [54], [57], [64], [91], [102], [106], [119], [121], [144], [182], [184], [199], [208], [210], [216], [224], [249], [319], [335–7], [339], [344], [371–2], [383], [402], [412], [439], [474], [476–7], [485], [491–2], [516–7].

CHAPTER ONE

Graphical Representation of Biochemical Systems

One of the great challenges in biology and biochemistry is complexity. In the macroscopic world, this complexity is manifest in the multitude of individuals and species that interact within our ecosystems, but also in the physical and behavioral intricacy of each and every organism. In the microscopic world, the variety and sophistication even among the simplest unicellular organisms is just as awesome. A typical bacterium contains more than ten thousand proteins alone, and thousands of other constituents have evolved that are needed for life under ever varying conditions.

The focus of this book lies at yet a lower level of organization, the level of biochemical systems and metabolic pathways. At this level, the "species" are metabolites; the "players" are enzymes; the "interactions" are kinetic processes, enzyme-catalyzed reactions, and a variety of modulations of these reactions. Although this level constitutes merely one aspect of life, the complexity and sophistication of metabolic phenomena are still overwhelmingly great; they exceed our intuition and tax our most advanced mathematical and computational means.

No matter how simple a biochemical phenomenon may look at first, closer scrutiny usually reveals that it is in reality affected by many factors. A pathway that seems to be rather simple in a textbook is actually part of a complex network of metabolic processes. Factors and cofactors that are constant in the textbook example in reality are products and substrates of other pathways and subject to ongoing turnover. Modulators, which are neatly distinguished in verbal descriptions of the pathway, in reality are not independent of each other but interact in complex synergistic or antagonistic ways.

These and many other manifestations of the ubiquitous complexity of biochemical and metabolic systems mandate as the first and possibly most significant step in any modeling effort a decision on what to include in a model and what to exclude. A good first approach to this step is a graphical representation of what exactly we intend to model.

Maps

There are many ways to describe a biochemical pathway. In the simplest case of a single reaction or a sequence of a few reactions, a brief verbal description often contains all the information that needs to be conveyed. For example, the statement "5-phosphoribosyl-α-1-pyrophosphate (PP-ribose-P) is formed by transfer of the terminal pyrophosphate group of ATP to the carbon 1 of ribose-5-phosphate in a reaction catalyzed by PP-ribose-P synthetase" (Stanbury et al., 1983) definitely describes one of the initial steps of purine synthesis *de novo*. The compounds involved in the reaction are listed (ATP and ribose-5-phosphate), and the sentence also offers some explanation of the mechanism: "transfer of a pyrophosphate group." Further details are readily added: "The reaction requires the presence of magnesium (Mg^{2+}) and inorganic phosphate P_i" (Stanbury et al., 1983).

A variation on verbal description is a representation in terms of chemical structures. In the case of synthesis and degradation of PP-ribose-P, such a representation may look like Fig. 1.1.

This representation exhibits the chemical composition of the primary constituents that contribute to the reaction. On one hand, this representation shows more detail than the verbal description; for instance, the structure of the ribose ring is immediately evident. On the other hand, the types of reactions are not explicitly named, and one has to deduce them from a comparison of substrates and product. Secondary products are not always shown. For instance, the degradation of PP-ribose-P generates glutamic acid and PP_i, but this is not evident from Fig. 1.1.

A pathway with many reactions and branch points and with a high degree of regulation or modulation becomes problematic for a purely verbal description. Our language is a linear stream, and it runs into difficulties when we want to represent simultaneous events or multiple roles of a constituent. As an analogy, imagine an exclusively verbal description of a street map. We rather readily tell people over the phone how to find their way from A to B, but we would have to give up if we had to describe all paths from any point on the map to any other point. Biochemical systems are much more complicated than street maps. The street map merely compares with the flow of material through the network of biochemical reactions, but the control structure of the metabolic network is not included. For instance, a complete verbal description would have to include information on how inhibitors affect an enzyme-catalyzed process, and because this effect is not dichotomous – on or off – but concentration-dependent, we would soon literally run out of words describing

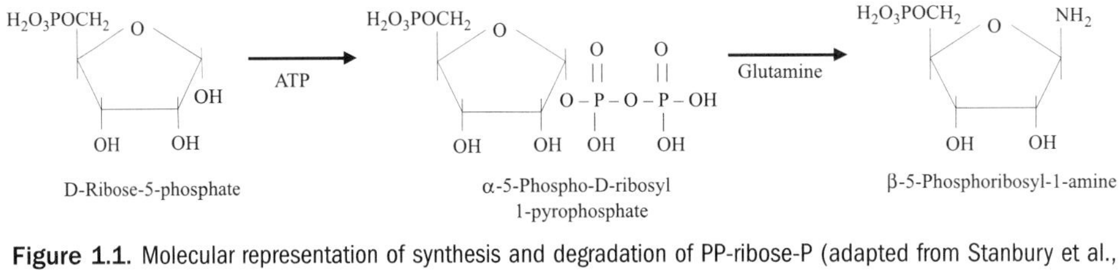

Figure 1.1. Molecular representation of synthesis and degradation of PP-ribose-P (adapted from Stanbury et al., 1983).

exactly the degree of inhibition in any given situation. To force the actual multi-dimensional complexity into language, branches of a pathway must be considered separately, and if a product exerts inhibition of a previous reaction, its two roles in the pathway require individual mention. If, in addition, constituents are recycled and used as modulators of other reactions of the same pathway, it becomes very difficult to develop an intuitive impression of the global structure of the biochemical system.

The graphical representation of a pathway in terms of chemical structures is two-dimensional and offers greater flexibility in visualizing a biochemical system than a verbal description. Explicitly or implicitly, all our analyses will use such a graphical representation as a starting point, and because of its importance, we will henceforth call a proper graphical representation of a biochemical system a *biochemical map*, a *metabolic map*, or simply a *map*. A map may depict the chemical structures of all metabolites, as exemplified in Fig. 1.1. Or the structures may be replaced with names or abbreviations, if the global features of a system are of primary interest, if the chemical composition of its constituents is less relevant, or if the map becomes too confusing.

Again using purine biosynthesis as an example, consider the very much simplified system representation in Fig. 1.2. A biochemist easily extracts the overall flow of material as well as the control structure that governs the system. As an exercise demonstrating the advantage of a map over a verbal description, try to describe the system to somebody else, without showing the diagram and without the other person drawing something like a map.

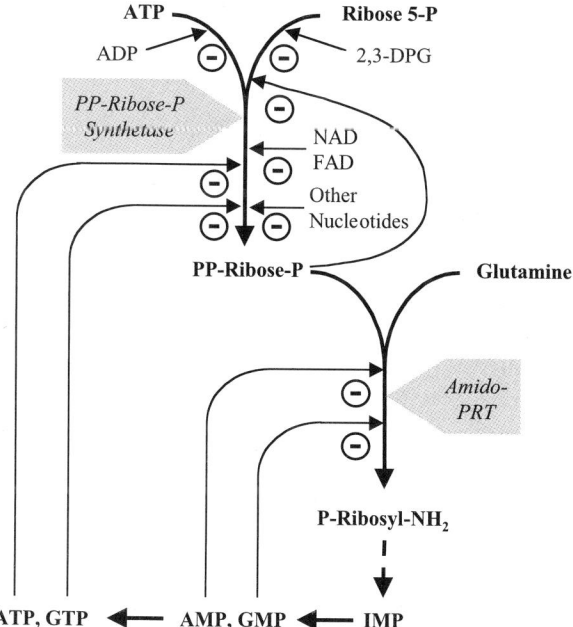

Figure 1.2. Purine biosynthesis is subject to several feedback controls (light arrows). Redrawn from Stanbury et al. (1983).

It is useful – and indeed mandatory – to list some ground rules for setting up a map. For our purposes, the map consists of three key elements:

- system components or *pools* of components,
- arrows that indicate *flow of material*, and
- arrows that indicate *flow of information* or *signals*.

By connecting pools with heavy arrows we indicate which system components can be transformed into others, and following these arrows within the map, we obtain an impression of the different routes through which material can be processed by the system. In complicated metabolic systems, several maps may be necessary to describe different aspects of a phenomenon: One map may represent carbon flow, another one nitrogen flow, and yet another may show how ATP molecules are employed to transfer chemical energy.

In contrast to heavy arrows that represent flow of material, we use a light, dashed, or differently colored arrows to indicate that a system component can affect or modulate a process in the system. An arrow of this type may represent, for instance, a feedback inhibition or the activation of a reaction.

The overall strategy of analyzing a biochemical system consists of a sequence of six steps:

1. List all components or pools of components that affect the system.
2. List all interactions between these components and all modulations by which components affect the system.
3. Arrange components, pools, interactions, and modulations in the form of a map.
4. Transcribe the map in terms of mathematical symbols and equations.
5. Analyze these equations.
6. Interpret the results.

In the remainder of this chapter, we discuss the first three steps; steps 4–6 will occupy the rest of the book.

The list of steps clearly signals the fundamental importance of setting up the map in a proper form. The map connects reality and mathematical analysis, and if this connection is faulty, the results are unreliable or even wrong.

When all components and interactions of a biochemical system are known, and when we strictly adhere to a few rules, it is usually not too difficult to construct a proper map. On the other hand, ill-defined components, using the wrong types of arrows, or confusing the flow of material with a regulatory influence often leads to incorrect conclusions. In some cases, one may detect those problems early on, for instance, when one tries to set up the model in the form of mathematical equations. However, once the equations are formulated – correctly or incorrectly – it is often very difficult to detect inconsistencies in the map or the equations until it is time to interpret the results.

Maps are thus the bridges between the real world and the model, and it is clearly very important to construct them correctly and to use the best available information about their components. In olden days – which in modern science means "until a couple years ago" – data on metabolic pathways were extracted from original scientific articles, reviews, books (such as Lehninger, 1970; Nishizuka, 1980, 1997; Schomburg, 1990; Lehninger et al., 1993; Stryer, 1995; Michal, 1998a), or the famous *Boehringer Metabolic Map* (see Michal, 1993, 1998b) that used to adorn every biochemistry department. Indeed, these sources still serve as "staple food" for metabolic models and will continue to do so throughout the foreseeable future.

Nonetheless, as in all parts of science, there is a clear trend toward information storage and retrieval in the form of electronic databases, most of which can be accessed freely, for academic use, through the Internet. While the emergence and disappearance of such databases exhibit a dynamic with an oftentimes very short half-life, some of the interesting currently available electronic options for metabolic data retrieval should be mentioned here. Many of these databases are annotated by experts in the field and linked to other sites, thereby facilitating explorations of connections between genes, proteins, metabolic pathways, and diseases. Some of the databases are listed in the annual Database Issue of the journal *Nucleic Acids Research* (see e.g., Burks, 1999), which itself is electronically available at *http://www.oup.co.uk/nar/Volume_27/Issue_01/summary/gkc105_gml.html*; another list can be found at *http://www.cc.um.edu.my/biocomp/pathwaydb.html*. Printed descriptions of many databases were collated in recent years in Volumes 24(1), 25(1), 26(1), and 27(1) of *Nucleic Acids Research*.

URL addresses of some of these databases are presented in alphabetical order in a separate section of the references. A recent guide to browsing the Net for biological databases was provided by Baxevanis and Ouellette (1998). This text focuses primarily on methods for finding and analyzing gene sequences, rather than on metabolic databases, but it is still a useful tool in our context, because the separation between gene expression and metabolism is becoming increasingly blurred and may disappear altogether in the not-too-distant future.

Of particular interest in the current context are databases that contain metabolic pathways and information about their components. Three relatively simple but very useful tools for exploration are Bairoch's (1993, 1999) *EC Enzyme Database*, Selkov and coworkers' (1996) *EMP Enzymology Database*, and Selkov and coworkers' (1997, 1998) *MPW Metabolic Pathway Database*. The *EC*, *EMP*, and *MPW* web sites let the user interactively explore features of enzymes and query combinations of enzymes, substrates, and – in *MPW* – coenzymes and pathway intermediates.

Several sites contain graphical pathway representations that are very similar to the maps we are using here. Clicking on an enzyme links automatically to an enzyme database and exhibits detailed information about enzyme classifications, substrates, products, cofactors, genes, pathways, and associated diseases (see also Hofestädt and Scholz, 1998). Clicking a metabolite links to information about its chemical structure as well as to alternate names, synthesizing and degrading pathways, and enzymes that use the metabolite as a substrate. A good example for this type of database is *KEGG* (*Kyoto Encyclopedia of Genes and Genomes*). KEGG is maintained by the Institute for Chemical Research at Kyoto University and contains numerous metabolic pathways, as well as databases for enzymes and gene maps (see Kanehisa, 1997; Ogata et al., 1998, 1999). A recent extension, *LIGAND* (Goto et al., 1998, 1999), connects KEGG with chemical databases and with *DBGET/LinkDB*, which contains genomic and disease information, along with pertinent literature. LIGAND furthermore contains a "path computation tool," *PathComp*, which – for a user-selected pair of a substrate and a product – computes reaction pathways that are known to exist in a given organism (see Goto et al., 1997; Fujibuchi et al., 1998).

EcoCyc is a web site that is philosophically similar to KEGG but specializes in the known metabolic pathways of *Escherichia coli* (e.g., Karp et al., 1996, 1999; see also Karp and Paley, 1996; Karp et al., 2000). EcoCyc allows the user to query metabolites, enzymes, and related genes. An interesting feature with respect to model design is the option of displaying more or fewer details of the pathway. The system provides summaries of the pathways, along with references, mentions *superpathways*, and shows locations of mapped genes. Information on genes and proteins of *E. coli* can also be found in *GenProtEC* (see Riley, 1998).

Clickable metabolic pathways can be explored in a similar fashion with *Boehringer Mannheim's Biochemical Pathways Module*, which is part of the *ExPASy Molecular Biology Server* of the Swiss Institute of Bioinformatics (SIB). This server also offers a plethora of links to relevant databases, proteomics tools, and other servers. *FlyNets* offers pathways of *Drosophila melanogaster* (FlyBase Consortium, 1999), and *UMBBD* (University of Minnesota Biocatalysis/Biodegradation Database) lets the user explore microbial biocatalytic reactions and biodegradation pathways primarily for xenobiotic, chemical compounds. The *Yeast Protein Database* (*YPD*) and the *Caenorhabditis elegans database* (*WormPD*) are interactive databases that give researchers instant access to the current literature on these organisms (Hodges et al., 1999).

Selkov and coworkers of the Computational Biology Group at the Mathematics and Computer Science Division at Argonne National Laboratory, with additions and support from Integrated Genomics Inc., Chicago, have been developing an outstanding database of detailed metabolic models (e.g., Selkov et al., 1997). The primary goal of this database, called *WIT* (What Is There?), is to help the user with "interactive metabolic reconstruction" from a combination of gene sequences, biochemical, and phenotypic data. WIT currently contains more than 2,900 diagrams depicting metabolic pathways, which are linked to companion information. An extension is the *Enzyme Reaction Mechanism* (ERM) database, which represents data on more than 90 mechanisms involving one or several types of substrates and different types of inhibition. In addition to the schematic representations, the site offers graphs, stoichiometric matrices, and mathematical formulations of rate laws in the tradition of Michaelis and Menten, Hill, Monod, and Wyman and Changeux. Another site, offering help with the analysis of enzyme reactions is maintained by the University of Washington Resource Facility for Kinetic Analysis.

In addition to these metabolic databases, numerous web sites offer access to gene and protein sequences and their analysis. Before we discuss some of these genomic databases, one might ask why genomic information is of interest for metabolic modeling. The answer is that some of the new gene technologies offer options for setting up metabolic maps that were unthinkable just a few years ago. Right now these options may still be too expensive in many cases, but affordability is probably only a matter of time.

With the new technologies, it is possible – at least in principle – to induce a *perturbation* in an organism, and simultaneously to measure the degree of expression of all potentially involved genes at a series of time points, until the cell has recovered from the perturbation. A perturbation may take very different forms. It may consist of the injection of some endogenous metabolite or toxicant, a drop or rise in temperature, or essentially any stress or stimulus that evokes a biochemical reaction. By interpreting gene expression in terms of altered enzyme activities, one can identify which pathway is "turned on" at a given point in time and trace how the functioning of all involved pathways is orchestrated. As police headquarters follows the positions of fleeing crooks and chasing police cars on electronic maps, the biochemist is able to follow signal and material flows in response to stimuli.

The identification of pathways that are turned on or off is highly relevant to the inclusion of variables and processes in a biochemical system model. In fact, it may become a major contributor to the design of the map, which in turn determines the repertoire of questions the model will be able to address. Furthermore, if the degree of gene expression can be measured, this information allows deductions about the relative strengths of responses and the relative sizes of fluxes, for instance, at branch points. It may also shed light on metabolites that regulate the processes following the perturbation.

One of the breakthroughs that have led to these novel approaches is the development of DNA chips and microarrays, which allow the researcher to monitor the expression levels of hundreds or even thousands of genes simultaneously (e.g., Schena et al., 1996; DeRisi et al., 1997; Fodor, 1997). A collection of articles describing these technological developments was recently published as a supplement to the journal *Nature Genetics* (Volume 21, January 1999), entitled *The Chipping Forecast*.

According to Corton et al. (1999), who give a good, brief introduction to the topic, a DNA chip is "a solid support containing silicon to which have been synthesized short DNAs of 20 to 30 nucleotides in length" (see also Lipshutz et al., 1999). These chips are used to measure the amounts of mRNAs expressed in a cell before and after some stimulus. They can also be used to identify mutations and genetic polymorphisms, which is of interest in this context if they correspond to different levels of enzyme activity. An example for such a situation is the well-documented polymorphism of the enzyme glucose-6-phosphate dehydrogenase, whose manifestations can lead to severe metabolic deficiencies and diseases (e.g., Stanbury et al., 1983; Martini and Ursini, 1996; Pandolfi et al., 1996; Jollow and McMillan, 1998).

A DNA microarray is "a solid support to which 100s or 1000s of cDNAs have been covalently attached at high density" (Corton et al., 1999). For its construction, mRNAs are isolated from cells or tissues of interest and converted with reverse transcriptase into individual copy DNAs (cDNAs) of about 1,000 to 2,000 base pairs. The mixture of cDNAs is cloned into bacterial plasmids, and these are sequenced, thus yielding a complete characterization of the DNA fragments. The fragments are subsequently amplified through the polymerase chain reaction, purified, and "spotted" by a robotic arrayer onto glass slides or nylon membranes (e.g., see Brown and Botstein, 1999; Corton et al., 1999). For the actual experiment, cells are subjected to some stimulus, stress, or insult, and mRNAs are isolated and converted with reverse transcriptase to cDNAs that contain either radioactive or fluorescent nucleotides. The labeled cDNAs are hybridized to the thousands of individual cDNAs on the microarray, each of which represent an expressed gene. Measuring either fluorescence or radioactivity, an imaging system estimates the degree of hybridization and, by comparison with controls, determines whether a gene is positively or negatively regulated. Based on similarities of their expression profiles, genes can be grouped into clusters that suggest their coordinated regulation (Duggan et al., 1999).

Details on technical aspects can be obtained from the web site of *Affymetrix*, which currently produces chips and arrays allowing the simultaneous expression monitoring of about 10,000 genes. Also, the web site of Patrick Brown's Laboratory at Stanford, which was instrumental in the development of these techniques, describes some of the procedural details of this technology and, among other features, offers information about the "yeast genome on a chip" (see also DeRisi et al., 1997; Lashkari et al., 1997; Wodicka et al., 1997). Related, up-to-date literature about microarray techniques can also be found in the *CBC Bibliography Database*.

A premier address for studying genomes themselves is the web site of the National Human Genome Research Institute, which provides an introduction to the field and numerous links to related sites. Augmenting the raw genomic information, Lawrence Berkeley National Laboratory has made it its goal "to add biological value to the raw data being generated by the Human Genome Project. As part of this aim, a variety of biological, computational, and expression array strategies are being developed and applied to the identification of genes and the determination of their function in the context of biology and human disease" (quoted from the web site).

Numerous other web sites are dedicated to gene and protein sequences. Among the databases for DNA sequences, *GenBank*, *EMBL*, and *DDBJ* stand out. GenBank is operated by the National Center for Biotechnology Information, which is part of the National Library of Medicine (for recent descriptions of GenBank, see Benson et al., 1998, 1999). Similar to GenBank, EMBL (European Molecular Biology Laboratory) maintains a site that permits searches of sequences, explorations of sequence alignments, as well as structure comparisons and predictions (e.g., Stoesser et al., 1997, 1999). DDBJ (DNA Data Bank of Japan) offers similar options and is linked to a number of other databases (Sugawara et al., 1999).

Protein sequences, along with descriptions of structure, function and post-translational modifications, can be found in *SWISS-PROT* (Geneva Bioinformatics; for descriptions see, e.g., Bairoch and Boeckmann, 1991; Bairoch and Apweiler, 1997ab, 1999), the *Prokaryotic Database*, and the *Protein Information Resource*, PIR (George et al., 1996ab; Barker et al., 1999), which is closely linked to *MIPS* (Mewes et al., 1999). Three-dimensional biomolecular structures are accessible through the Protein Data Bank (*PDB*); it was initiated in 1971 (for an early description see Bernstein et al., 1977) and is now maintained by the Research Collaboratory for Structural Bioinformatics (RCSB). Similar sites are operated by the European Bioinformatics Institute (*Macromolecular Structure Database: EBI/MSD*) and the Australian National Genomic Information Service (ANGIS). RasMol & Chime's *Molecular Visualization Freeware* shows pseudo-three-dimensional representations of macromolecules. Many of the sites on sequences and protein models are accessible from the *ExPASy Molecular Biology Server* of the Swiss Institute of Bioinformatics (SIB).

Variables and Parameters

Before we can effectively set up a map for a biochemical system and analyze it, we need to develop a streamlined terminology. At its heart is a convenient nomenclature for the constituents of the system, which can be identified with different types of names. Some of the alternatives are molecular formulas, such as $C_{11}H_{12}N_2O_2$; IUPAC names, such as *(R)-2-amino-3-(3-indolyl)propionic acid*; common names, such as *tryptophan*; mnemonic abbreviations, such as *Trp*; or symbolic names, such as X, A_4, or Y_{ij}. All of these types of names are legitimate and have their distinct advantages in the appropriate context. For systematic mathematical analyses, however, all the common names are clumsy in comparison with symbolic names with indices of the type X_i. In fact, the use of symbolic variable names and the assigning of mathematical values to variables is a milestone on our way to understanding how and why mathematical analyses of biochemical systems work.

Rudolph Carnap (1966: Chapter 11) lucidly explained the importance of a quantitative terminology in his treatise *Philosophical Foundations of Physics*:

> Quantitative concepts are not given by nature; they arise from our practice of applying numbers to natural phenomena. What are the advantages of doing this? If the quantitative magnitudes were supplied by nature, we would no more ask this question than we would ask: what are the advantages of colors? Nature might not have colors, but it is pleasant to find them in the world. They are simply there, a part of nature. We cannot do anything about it. The situation is not the same with respect to quantitative concepts. They are part of our language, not part of nature. It is *we* who introduce them; therefore, it is legitimate to ask *why* we introduce them. Why do we go to all the trouble of devising complicated rules and postulates in order to have magnitudes that can be measured on numerical scales? ...
>
> First of all – though this is only a minor advantage – there is an increase in the efficiency of our vocabulary. Before a quantitative concept is introduced, we have to use dozens of different qualitative terms with respect to that magnitude. Without the concept of temperature, for example, we have to speak of something as "very hot," "hot," "warm," "lukewarm," "cool," "cold," "very cold," and so on. These are all what we have called classificatory concepts. If we had a few hundred such terms, perhaps it would be necessary, for many everyday purposes, to introduce the quantitative concept of temperature. Instead of saying, "It is 95 degrees today," we would have a nice adjective that meant just this temperature, and for 100 degrees we would have another adjective, and so on.
>
> What would be wrong with this? For one thing, it would be exceedingly hard on our memory. We would not only have to know a great number of different adjectives, but we would also have to memorize their order, so we would know immediately whether a certain term was higher or lower on the scale than another. But, if we introduce the single concept of temperature, which correlates the states of a body with numbers, we have only one term to memorize. It is true, of course, that we must have previously memorized the numbers, but once we have done so, we can apply those numbers to any quantitative magnitude.

The major advantage ... is that quantitative concepts permit us to formulate quantitative laws. Such laws are enormously more powerful, both as ways to explain phenomena and as means for predicting new phenomena. Even with an enriched qualitative language, in which our memory is burdened with hundreds of qualifying adjectives, we would have great difficulty expressing even the simplest laws. ... A law expressed in a quantitative language is thus much shorter and simpler than the cumbersome expressions that would be required if we tried to express the same law in qualitative terms. Instead of one simple, compact equation, we would have dozens of "if–then" sentences, each pairing a predicate of one class with a predicate of another. ... Once we have the law in numerical form, we can employ that powerful part of deductive logic we call mathematics and, in that way, make predictions.

There is no need to abandon common names or molecular formulas. They remain very important because much of the final evaluation and presentation of our mathematical results will be done in the terminology of biochemistry. So, we will not do away with mnemonic names, but temporarily suspend their use. For the mathematical analysis, we code the relevant components of the system with symbolic definitions and record all these definitions in a dictionary of the form

X_1 = tryptophan,

X_2 = isoleucine,

X_3 = pool of all aromatic amino acids.

For the analysis of the system, we work exclusively with the symbolic names like X_1, and when all the mathematics is executed, simple translation back to common names will help us to interpret and communicate our results.

The use of variables with indices is the terminology of choice, because it allows us to talk succinctly about all or a few select constituents. For instance, it is cumbersome to list the masses of the first sixteen constituents of a pathway every time we want to talk about this collection, whereas it is easy to address $X_1, \ldots, X_{16}$. In symbolic form, one readily defines the sum of masses as a new variable, using the same streamlined notation. If our current system contains twenty variables altogether, we just add for convenience the variable X_{21} as

$$X_{21} = \sum_{j=1}^{16} X_j = X_1 + X_2 + \cdots + X_{16}$$

and update our dictionary by including X_{21} as the total mass of the first sixteen constituents.

Having established a homogeneous nomenclature for the components of the system, we now analyze the different roles that these components can play in the overall action of the model. As an illustration, consider the simplified pathway for the synthesis of serine from 3-phosphoglycerate depicted in Fig. 1.3. It consists of seven components that interact in an organized fashion, ultimately yielding serine. The representation in Fig. 1.3 indicates that there are components of different types. 3-Phosphoglycerate, 3-phosphohydroxypyruvate, phosphoserine, and serine are shown as metabolites, and they are connected by arrows that indicate a flow of material, whereas 3-phosphoglycerate dehydrogenase, phosphoserine transaminase, and

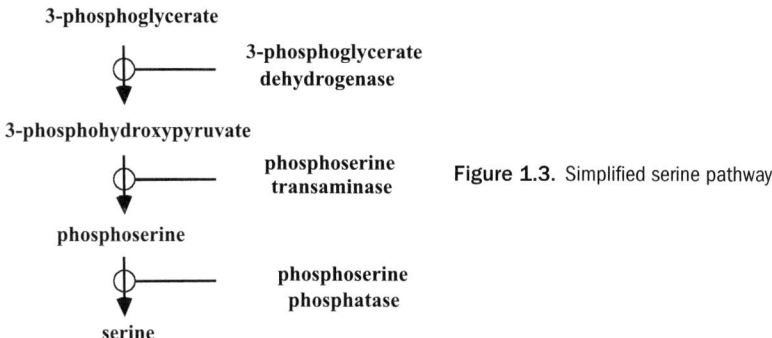

Figure 1.3. Simplified serine pathway.

phosphoserine phosphatase are shown as enzymes that catalyze the steps of the pathway. The enzymes are depicted here in a fashion that indicates that they are not involved in the pathway as sources of material but as facilitators that control conversions between metabolites.

The different roles of metabolites and enzymes need to be considered in the mathematical representation of the system. To accomplish this, we must ask ourselves what the essential differences between the enzymes and the metabolites in our example are. For instance, one could argue that the catalyzing enzymes themselves are metabolites in pathways that lead to their synthesis. Although this is certainly true, we must not lose sight of the specific pathway under investigation. Restricting our attention exclusively to our example, we find that the concentrations of metabolites may noticeably change over the duration of the experiment. In fact, if no 3-phosphoglycerate is supplied, the initial amount will be used up and no further reactions will take place. In contrast, the three participating enzymes are catalysts whose overall (free plus bound) concentrations remain fairly constant. Their effectiveness is more or less independent of their concentrations, but it is essentially determined by properties that are quantified through *parameters* like K_M or K_{eq}.

Thus, we can distinguish three types of entities that contribute to the dynamics of a system:

- *metabolites*, whose concentrations change significantly during an experiment;
- *enzymes* or *modifiers*, which do not change appreciably during the experiment; and
- *parameters*, such as K_M or K_{eq}, which have particular, constant values during the experiment.

Enzyme or modifier concentrations and parameters may have different values in different experiments, but for the duration of any one experiment they are considered constant.

To make these distinctions clear, and in order to avoid undesired connotations, we define three main entities of a biochemical system:

- *Dependent Variable:* A variable, representing a system component or a pool of components, whose value is affected by the system. Typically, the values of dependent variables change during an experiment.
- *Independent Variable:* A variable, representing a system component or a pool of components, that itself is unaffected by the system. Typically, independent variables

are constant during any given experiment or they change in a manner that is controlled by the experimenter.
- *Parameter:* An entity with a constant numerical value that quantifies a property of the system.

As an example, consider a biotechnological batch process in which yeast cells ferment sugars. Typical dependent variables are intermediates and products of glycolysis, such as pyruvate. A typical independent variable is the substrate fed to the cells by the experimenter. The parameters in this case are equilibrium constants and the controlled, constant temperature.

In the above example (Fig. 1.3), phosphoserine is a dependent variable whose concentration significantly depends on the dependent variable 3-phosphohydroxypyruvate. Phosphoserine phosphatase is to be considered an independent variable, because it is not affected by phosphoserine or 3-phosphohydroxypyruvate or any other system constituent. The identification of dependent and independent variables may change from one analysis to the next. For instance, one could study a slightly different experiment in which phosphoserine phosphatase itself is the product of some pathway of interest. In such an experiment, the concentration of phosphoserine phosphatase would change with time, and this metabolite would become a dependent variable. This implies that the terms *dependent* and *independent* are not absolute and may change with the focus of the analysis.

Several remarks are in order:

1. Typically, dependent variables are affected by independent variables and often by other dependent variables. In contrast, independent variables are never affected by dependent variables or by other independent variables.
2. Dependent variables are also called *system variables, state variables*, or *internal variables*. They are an integral part of the system and may change in value during the experiment.
3. In contrast to the dependent variables, the independent variables are not affected by the system. In many cases, these independent variables are outside the system and remain constant during an experiment, or the experimenter changes them in order to study responses by the system. However, independent variables are not always directly accessible. Enzymes are clearly "inside" the system, but if their activities do not change during an experiment, they are to be classified as independent variables. Independent variables are sometimes called *external variables, control variables, input variables*, or *input signals*.
4. The term *independent variable* is slightly different from general mathematical usage, in which the independent variable in a biochemical experiment would be *time*. Time is, strictly speaking, a variable (after all, its numerical value changes during the experiment), and because it is not influenced by the system or by the experimenter, it is an independent variable. Nonetheless, we shall use the term *independent variable* as defined above, because this definition is most familiar to biochemists and mathematical confusion about this terminology is unlikely. Thus, we have three types of variables: time, usually denoted by t; dependent variables, which represent the components of primary interest; and independent variables, which are not affected by other variables.

5. The distinction between dependent and independent variables must be made clear at the beginning of every analysis. It is sometimes questionable whether a variable changes during the experiment or is essentially constant. In such cases, a decision and subsequent definition are necessary. This decision is based on biochemical considerations and may be changed from one experiment to the next. In complicated systems, one might even perform analyses for both scenarios and compare the results. If in doubt, such a variable should be represented as a dependent variable.
6. The distinction between independent variables and parameters may sound obvious but it is an issue worth discussing. For instance, temperature and pH can be considered as either parameters or independent variables, depending on the viewpoint and questions asked. One could even imagine temperature as a dependent variable, for instance, in batch processes with microbial heat generation.
7. In most cases, we represent a dependent or independent variable with a capital X with one index, e.g., X_1, X_6, X_i, or X_j; we list the dependent variables first and the independent variables second. For parameters we will use lowercase Latin or Greek letters without an index or with one or two indices, e.g., α, β_i, g_{ij}, h_{12}, or $f_{i,n+1}$.

A typical symbolic representation for the above example is given in Fig. 1.4. The first four indexed variables represent the dependent variables ($X_1 = $ 3-phosphoglycerate, $X_2 = $ 3-phosphohydroxypyruvate, $X_3 = $ phosphoserine, and $X_4 = $ serine), and the catalyzing enzymes are represented as independent variables ($X_5 = $ 3-phosphoglycerate dehydrogenase, $X_6 = $ phosphoserine transaminase, and $X_7 = $ phosphoserine phosphatase). Parameters are not yet included explicitly but will become important when we establish equations describing the pathway.

Rules for Constructing a Proper Map

A proper map is constructed in a straightforward fashion and includes the listing of all relevant components, the listing of processes, and the graphical arrangement of components and processes.

1. List all components or pools of components that affect the system. This first step decides which components of the metabolic pathway or biochemical system

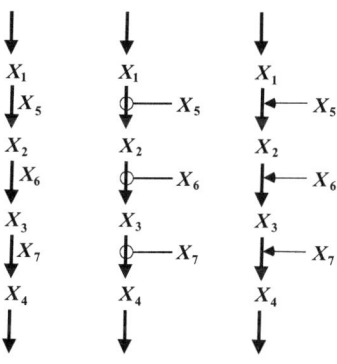

Figure 1.4. Alternative maps for a simple linear pathway of enzyme-catalyzed reactions. In each case, the flow of material between the (dependent) metabolites $X_1, \ldots, X_4$ is represented by heavy arrows, which represent reactions that are catalyzed by (independent) enzymes X_5, X_6, and X_7. The first representation simply associates an enzyme with a process. The second representation is advantageous when there is not enough space in the map or if the coding of arrows is confusing. The third representation follows directly the rules for setting up maps by indicating that the enzymes affect the conversions between metabolites; as a minor disadvantage, this map does not distinguish between the role of an enzyme and that of any other modulator.

will be considered in the model. This task sounds easier than it often is. When we analyze an *in vitro* system, the components are usually rather well defined, but for the analysis of a pathway *in vivo*, we have to accept compromises. On one hand, the mathematical representation of the biochemical system should be simple enough to allow analyses and provide new insights. Consequently, we want to keep the number of system components to a minimum. On the other hand, no constituent within a living organism is independent of its biochemical surroundings, but accounting for all metabolic details would require an overwhelming number of variables and equations. Thus, we have to rely on our biochemical expertise to decide which components are most relevant and should be included in a model and which components are secondary, so that – at least initially – they can be ignored.

Mathematically speaking, the decision about the inclusion or exclusion of components amounts to the definition of our model, and once this decision is made, the model becomes the reality that we analyze. It is very difficult to devise specific, generally applicable guidelines for which components a mathematical system description should incorporate. As a rule of thumb, one tries to define a system in such a way that the number of interactions within the system is maximized and the number of interactions between a system and its environment is minimized (see Fig. 1.5). We must be aware that this decision on inclusion or exclusion of components may be contentious. In fact, this decision is the source of many disagreements about the relevance and reliability of models.

The variables of the biochemical system are typically, but need not be, individual chemical substances or pools of substances. They can also indicate substances in different compartments, such as "glucose in the pancreas" and "glucose in the bloodstream," or they can be of an entirely different nature, for instance, representing effectors like temperature or pH if these change during the experiment.

2. List all interactions and modulations. Once all constituents are identified, we list the processes that directly affect any components of the system. These processes may be very complicated, but at this point there is no need to describe them quantitatively or mechanistically. The crucial issue is to identify which independent and dependent variables have an influence on which of the dependent variables in the system. We distinguish the *flow of material* between constituents, such as in the phosphorylation of glucose to glucose-6-phosphate, and the *modulation of a process*, such

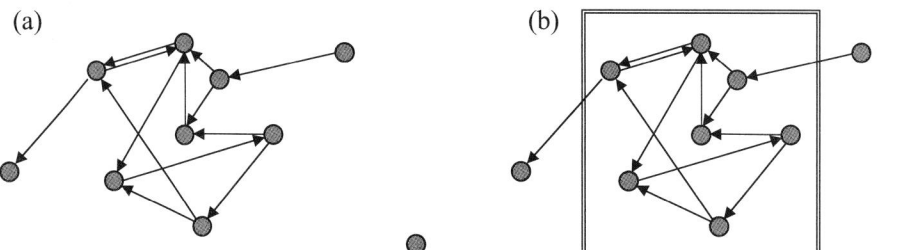

Figure 1.5. Delineation of a system by maximizing the number of interactions within the system and minimizing the number of interactions between the system and its environment.

as the competitive inhibition of an enzyme-catalyzed process by one of the other variables.

In many cases, a constituent may have several distinct roles. It may be the product of one reaction and the substrate for another reaction, and at the same time it may modulate one of the pathways modeled in the system. The list of interactions and modulations should take account of each of these roles separately. For instance, serine is the result of the hydrolysis of 3-phosphoserine; serine is also an inhibitor of the hydrolyzing enzyme, serine phosphatase; and of course, serine is a substrate or intermediate in several other pathways. If production and further use of serine are represented in a model, then serine has to appear at least three times in the list of interactions and modulations: as a product, as an inhibitor, and as a substrate.

Even though no detailed information about the contributing processes is needed at this point, it is prudent to start a notebook containing all available information that might be of use later. Such information includes *in vivo* concentrations of system constituents, *in vivo* or *in vitro* reaction rates, and details about the particular mechanisms of inhibition within the system. It should also include the source of each piece of information in the form of a complete reference.

3. Arrange components, pools, interactions, and modulations in the form of a map.

The map integrates components and interactions. The metabolites or pools are arranged as *nodes* in some convenient way, and arrows show the flow or transport of material and the modulation of processes.

We use heavy arrows for a flow or transport of material. These *flow arrows* originate at a node and enter into a node. For instance, if $X_1 =$ 3-phosphoserine is hydrolyzed to $X_2 =$ serine, a flow arrow originates at the node X_1 and enters the node X_2, as shown in Fig. 1.6.

This unidirectional flow arrow indicates an *irreversible reaction*. It may also represent a reversible reaction in which the back flux is negligible in comparison with the forward flux, a situation we may call *essentially irreversible*.

In some pathways the flow between X_1 and X_2 is *reversible* and neither of the directions can be ignored. Prominent examples are amphibolic pathways. If the net flux in such a system goes in only one direction (during an experiment of interest), we may use the same arrows as above in the essentially irreversible case. However, if flux reversal is possible, we indicate this with two regular arrows or half-arrows, as illustrated in Fig. 1.7.

In many pathways, material is channeled away from a metabolite in different directions. We call such a point in the pathway a (*divergence*) *branch point*; an example is shown in Fig. 1.8. Similarly, a metabolite may be generated from different sources. This situation is represented with a (*convergence*) *branch point*, as shown in Fig. 1.9.

At a divergence branch point, material may flow independently into any of the divergent pathways. This situation is to be distinguished from a splitting reaction in

$X_1 \longrightarrow X_2$ **Figure 1.6.** Representation of an unmodified process or reaction.

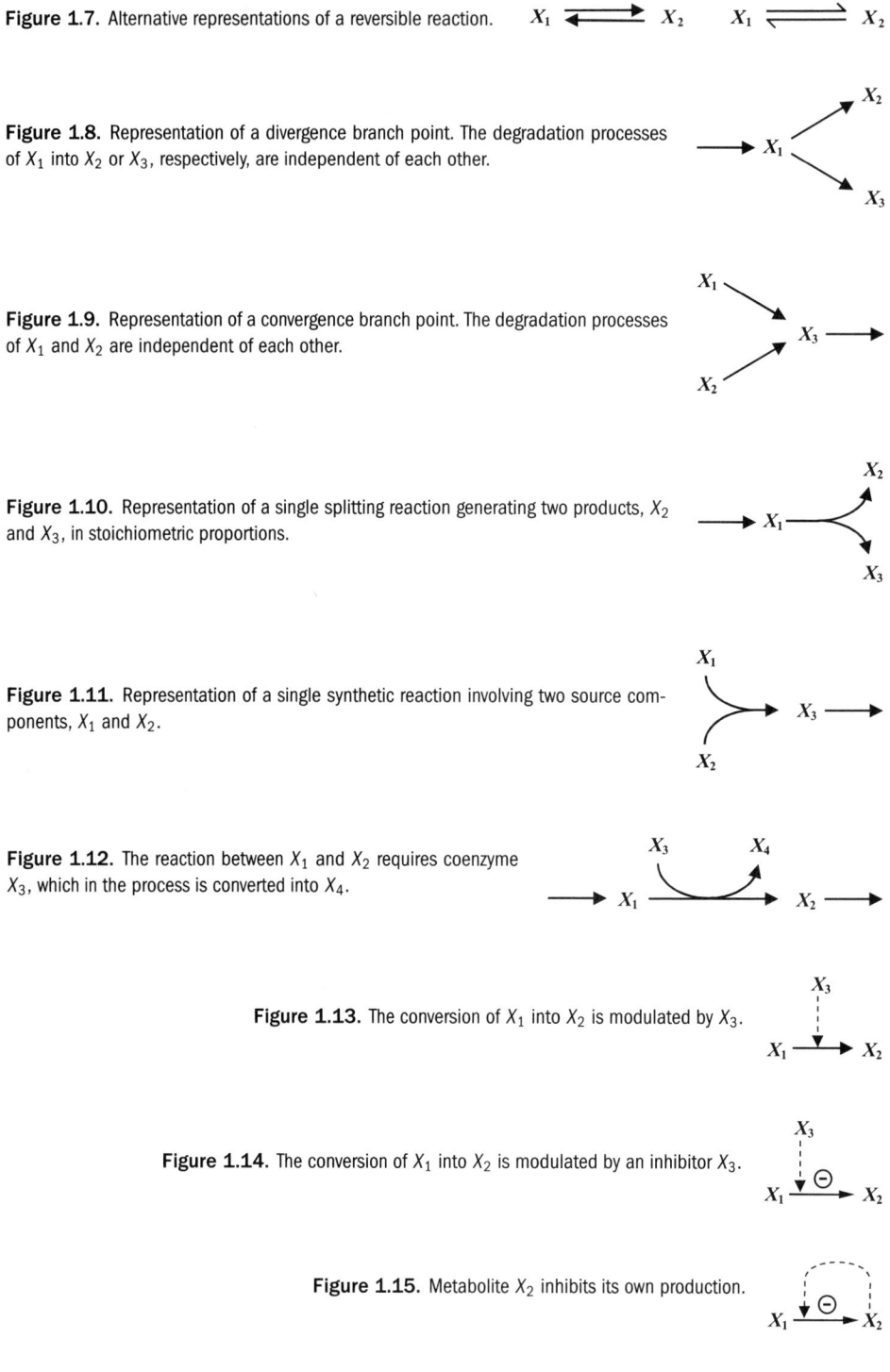

Figure 1.7. Alternative representations of a reversible reaction.

Figure 1.8. Representation of a divergence branch point. The degradation processes of X_1 into X_2 or X_3, respectively, are independent of each other.

Figure 1.9. Representation of a convergence branch point. The degradation processes of X_1 and X_2 are independent of each other.

Figure 1.10. Representation of a single splitting reaction generating two products, X_2 and X_3, in stoichiometric proportions.

Figure 1.11. Representation of a single synthetic reaction involving two source components, X_1 and X_2.

Figure 1.12. The reaction between X_1 and X_2 requires coenzyme X_3, which in the process is converted into X_4.

Figure 1.13. The conversion of X_1 into X_2 is modulated by X_3.

Figure 1.14. The conversion of X_1 into X_2 is modulated by an inhibitor X_3.

Figure 1.15. Metabolite X_2 inhibits its own production.

Figure 1.16. Metabolite X_7 inhibits an early step in its own production pathway.

which a metabolite is degraded into two parts, which are thus generated in stoichiometric proportions. We represent this splitting reaction with a double-headed arrow, as shown in Fig. 1.10. Analogous arguments hold for a convergence branch point. Here the branch point metabolite may be synthesized independently from each of the sources. In contrast, some metabolites are synthesized from two components, and both sources are required in stoichiometric proportions for the reaction to occur. We denote this situation with a double-tailed flux arrow, as shown in Fig. 1.11.

Many important reactions require coenzymes or second substrates that are converted from a high free energy form to a low free energy form, as in the case of ATP and ADP, or from a reduced to an oxidized form, as in the case of NADH and NAD^+. Such a reaction is illustrated in Fig. 1.12.

To distinguish modulating influences from flux of material we represent modulations with light or dashed *modulation arrows*. Modulation arrows originate at a metabolite or pool and terminate at a flow arrow. The point of origin indicates the system constituent that exerts the modulating influence, and the flow arrow to which the modulation arrow points indicates the modulated process. For instance, if X_3 modulates the transformation of X_1 into X_2, the corresponding map may look as shown in Fig. 1.13.

The modulation can be of any type. The modulator may be an enzyme or an inhibitor; it may be a cofactor, or it may represent effectors like temperature or pH. Typically, the modulator itself is a system component and as such is modeled as a dependent variable. A modulation arrow may also signify an influence that is constant during any given experiment but may be varied from one experiment to the next. In such a case the modulator is typically an independent variable.

If details about a modulation are known, the map may acknowledge them to some degree, but most of this information will only become relevant when we determine parameter values for the model equations (see Chapter 5). For instance, when we know that a modulator has an inhibiting effect, we may associate a minus sign with the arrowhead that represents this modulation and a plus sign with activating influences. For added emphasis, we sometimes frame the plus or minus sign with a box or circle. Figure 1.14 shows an example in which X_3 inhibits the transformation of X_1 into X_2; the inhibitory effect is indicated with a minus sign next to the head of the modulation arrow.

Of course, the modulator itself can be a product or intermediate of the pathway. Figure 1.15 could represent hydrolysis of 3-phosphoserine (X_1) to serine (X_2), a process that is inhibited by the end product serine. Similarly, a metabolite may inhibit a process that is located several steps upstream or even in a different pathway. Examples are given in Figs. 1.16 and 1.17.

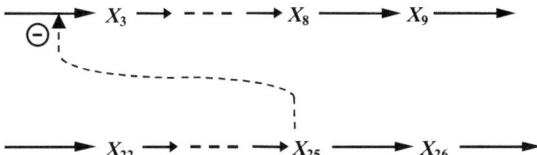

Figure 1.17. Metabolite X_{25} inhibits the synthesis of X_3 in another pathway.

Figure 1.18. X_4 activates a process downstream.

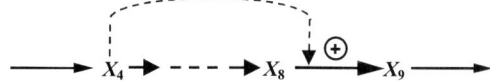

In an analogous manner, the map in Fig. 1.18 conveys that X_4 activates the transformation of X_8 into X_9, a reaction that takes place downstream from X_4 in the same pathway.

Examples of Proper Maps

The concepts of setting up a proper map are best illustrated with a few specific examples.

1. Glycolysis. The prominent metabolic role of glucose in mammalian cells is to provide potential chemical energy that is used to synthesize ATP. Glycolysis is very well understood and widely described in journals and textbooks, so we can design the map without presenting many details.

The first task is identification of dependent and independent variables and assignment of symbolic names. Recall that this step is somewhat subjective but, at the same time, crucial, because it includes an almost final decision about which components are considered relevant and subject to further analysis and which potential influences are assumed to be constant or are simply ignored. Our dictionary for the system is:

Dependent Variables:

X_1 = Glucose-1-phosphate
X_2 = Glucose-6-phosphate
X_3 = Fructose-6-phosphate
$X_4 = P_i$

Independent Variables:

X_5 = Glucose
X_6 = Phosphorylase
X_7 = Phosphoglucomutase
X_8 = Phosphoglucose isomerase
X_9 = Phosphofructokinase
X_{10} = Glucokinase
X_{11} = Glycogen

All components are arranged in the map and connected with flow arrows. This process is pretty straightforward, except that we have to decide which of the conversions are to be considered essentially irreversible and which need to be modeled as bi-directional. Figures 1.19 and 1.20 show proper maps of glycolysis in terms of biochemical names and mathematical symbols.

GRAPHICAL REPRESENTATION OF BIOCHEMICAL SYSTEMS

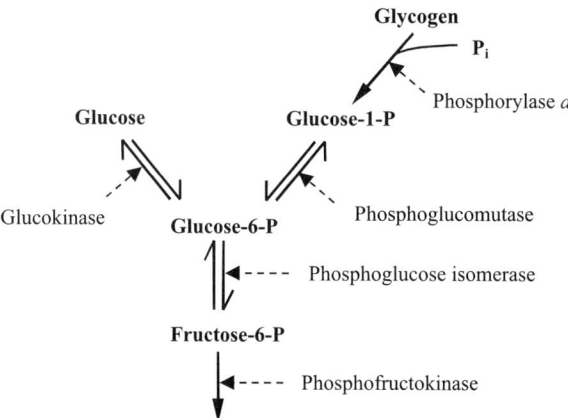

Figure 1.19. Map of glycolysis in biochemical terminology.

2. Activation of pancreatic zymogen. The activation of zymogen constitutes a cascade mechanism, which, in simplified form, can be represented as shown in Fig. 1.21. We begin again with the identification of relevant dependent and independent variables and the assignment of symbolic names. The result is the following dictionary:

Dependent Variables:
X_1 = Trypsin
X_2 = Carboxypeptidase
X_3 = Chymotrypsin
X_4 = Elastase

Independent Variables:
X_5 = Procarboxypeptidase
X_6 = Chymotrypsinogen

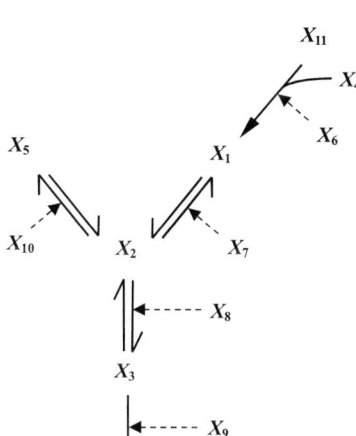

Figure 1.20. Map of glycolysis in mathematical symbols.

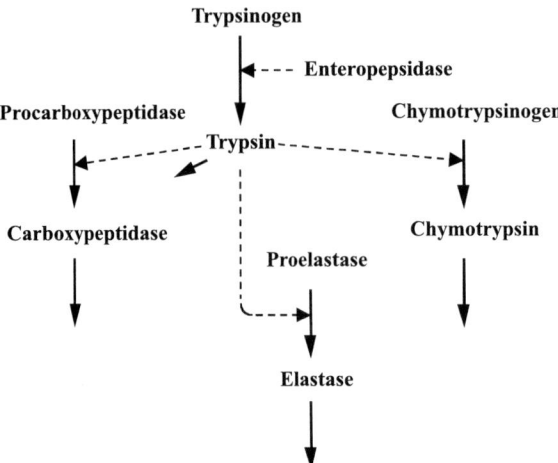

Figure 1.21. Activation of pancreatic zymogens in biochemical terms.

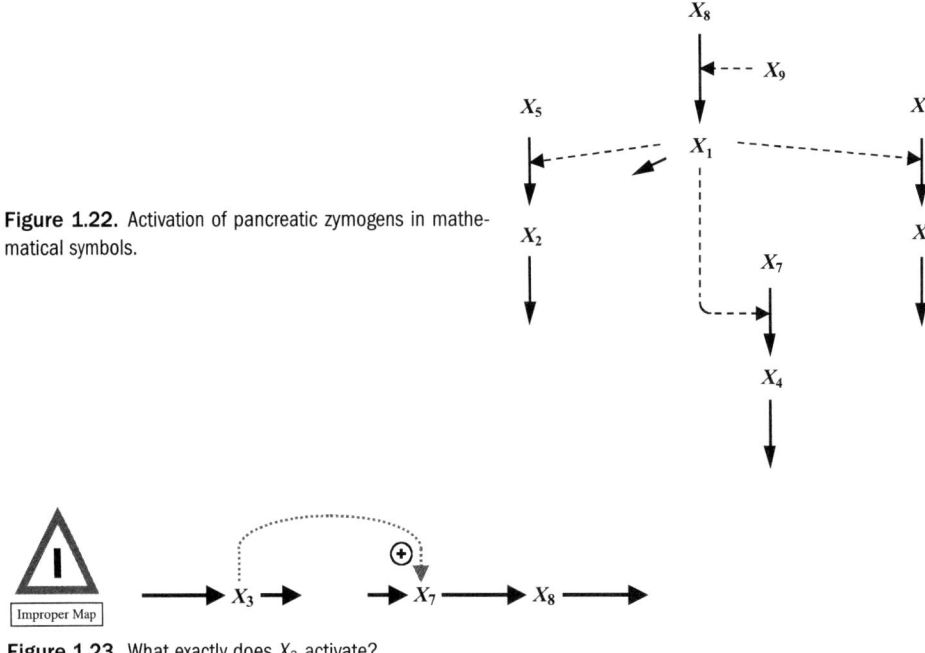

Figure 1.22. Activation of pancreatic zymogens in mathematical symbols.

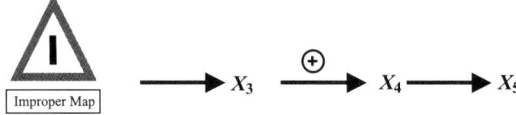

Figure 1.23. What exactly does X_3 activate?

Figure 1.24. What is positive about the conversion of X_3 into X_4?

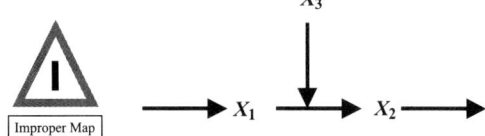

Figure 1.25. No material should flow out of X_3 into the process between X_1 and X_2.

GRAPHICAL REPRESENTATION OF BIOCHEMICAL SYSTEMS

X_7 = Proelastase,
X_8 = Trypsinogen
X_9 = Enteropeptidase

Proper maps for the system are shown in Figs. 1.21 and 1.22.

More complicated examples are presented in the case studies of later chapters.

Improper Arrow Diagrams

It is instructive to study a few arrow diagrams that do not constitute legitimate maps. Typically, errors arise when components or arrows are not defined precisely.

A vague statement like "X_3 affects X_7" may lead to a map like the one in Fig. 1.23, which may look correct at first glance but does not specify whether X_3 affects the production or degradation of X_7. Violating the formal requirements of setting up a map, the modulation arrow does not terminate at a flow arrow.

The middle arrow in Fig. 1.24 is improper because the plus sign is associated with a flow arrow. It is not clear whether material flows between X_3 and X_4 or whether X_3 or some other constituent activates the production and/or degradation of X_4.

If X_3 is an enzyme that catalyzes the production of X_2 from substrate X_1, then Fig. 1.25 is inappropriate, because a flow arrow is used to represent the role of X_3, when in fact there is no flow of material out of X_3.

Summary

The first step in the analysis of a metabolic pathway is the development of a proper map. This map contains all components of interest with their symbolic names. It shows with heavy arrows the material flow between components, and with light arrows which components modulate the flows into, between, and out of components.

It cannot be emphasized enough that the construction of a proper map is a key step, because all further analyses are based on it. With the acceptance of the map, important and almost final decisions have been made as to which components are relevant and which are being ignored. The strict distinction between flow arrows and modulation arrows helps the investigator to develop a clear picture of the processes within the system. It is essential for the development of the model equations.

The advantages of using symbolic variable names with indices instead of common biochemical names may not yet be obvious but will become clear in later chapters.

EXERCISES

1. Construct maps for the following scenarios and decide which components should be represented by independent or dependent variables, and which are best represented as parameters.

    1.1. The persistent form of hyperlysinemia involves a deficiency of the enzyme that converts lysine and α-ketoglutarate to saccharopine (from Stanbury et al., 1983: p. 439).

1.2. "Ornithine-δ-aminotransferase is a pyridoxal phosphate-requiring Ω-transaminase, which catalyzes the reversible conversion of ornithine and α-ketoglutarate to Δ^1-pyrroline-5-carboxylate and glutamate. Glutamate semialdehyde is the initial product formed by removal of the δ-amino group of ornithine; however, it cyclizes spontaneously to form Δ^1-pyrroline-5-carboxylate" (from Stanbury et al., 1983: p. 386).

1.3. "... Methylcobalamin (MeCbl) is a cofactor in the complex series of reactions by which homocysteine is remethylated to methionine. This reaction requires S-adenosylmethionine and N^5-methyltetrahydrofolate (Me-H$_4$folate), as well as the methyltransferase apoenzyme and methylcobalamin. The exact mechanism of homocysteine remethylation remains obscure but probably involves the following sequence: Me-H$_4$folate is converted to tetrahydrofolate (H$_4$folate) by transferring its methyl group to a cobalamin prosthetic group of the methyltransferase apoenzyme; in turn, the methyl group is transferred from MeCbl to homocysteine, leading to the formation of methionine" (from Stanbury et al., 1983: p. 480).

1.4. *Biosynthesis of isoleucine, methionine, and lysine.* The mother compound for these three amino acids is aspartate. Aspartate is converted into β-aspartylphosphate, a process that is inhibited by threonine and lysine. In a second step, β-aspartylphosphate is catalyzed to aspartate β-semialdehyde, which, in turn, is the substrate for two pathways. One eventually leads to lysine, the other to homoserine. Homoserine again constitutes a branch point. Homoserine kinase converts homoserine into phosphohomoserine, phosphohomoserine is converted into threonine, and threonine is the first substrate in an unbranched pathway leading to isoleucine. The end product, isoleucine, inhibits the synthesis of threonine from phosphohomoserine. Homoserine succinylase catalyzes the acyl transfer that converts homoserine into O-succinylhomoserine. The enzyme is subject to feedback inhibition by methionine, which is the end product of the unbranched pathway that converts O-succinylhomoserine in several steps into methionine.

2. Which of the following diagrams constitute proper maps? For proper maps, describe the reaction scheme in words. For improper maps, explain what is wrong and how the map could be amended.

2.1.
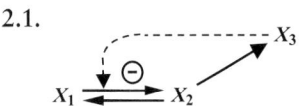

2.2.
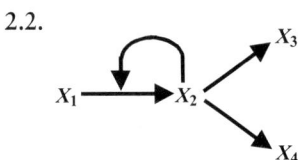

2.3.
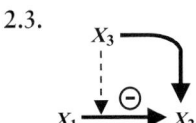

GRAPHICAL REPRESENTATION OF BIOCHEMICAL SYSTEMS

2.4.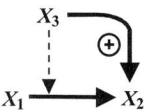

2.5.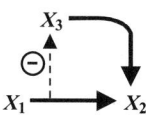

2.6.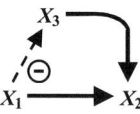

2.7.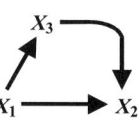

2.8.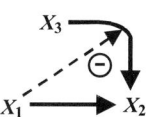

2.9.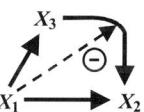

2.10.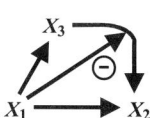

2.11.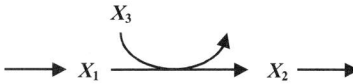

2.12.

2.13.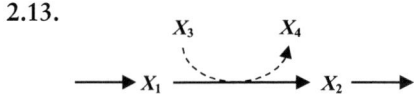

2.14.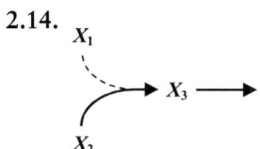

2.15.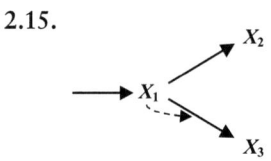

2.16.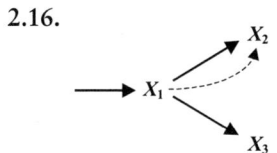

2.17.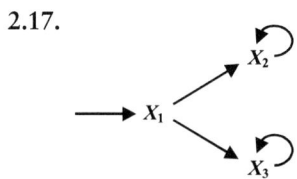

2.18.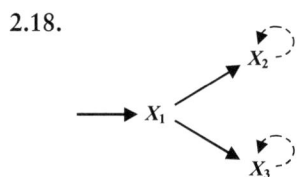

2.19.

GRAPHICAL REPRESENTATION OF BIOCHEMICAL SYSTEMS

2.20.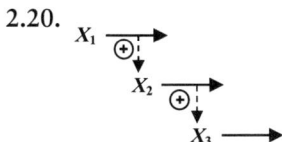

3. Compare the following systems, and explain the differences and similarities.

   3.1. (a)

   $\longrightarrow X_1 \Rightarrow \Rightarrow \Rightarrow X_8 \longrightarrow X_9 \longrightarrow$

   (b)

   $X_1 \Rightarrow \Rightarrow \Rightarrow X_8 \longrightarrow X_9 \longrightarrow$

   (X_1 independent)

   (c)

   $X_1 \Rightarrow \Rightarrow \Rightarrow X_8 \longrightarrow X_9 \longrightarrow$

   (X_1 dependent)

   3.2. (a)

   $\longrightarrow X_1 \Rightarrow \Rightarrow \Rightarrow X_8 \longrightarrow X_9 \longrightarrow$

   (b)

   $\longrightarrow X_1 \Rightarrow \Rightarrow \Rightarrow X_8 \longrightarrow X_9$

   (c)

   $X_1 \Rightarrow \Rightarrow \Rightarrow X_8 \longrightarrow X_9$

   3.3.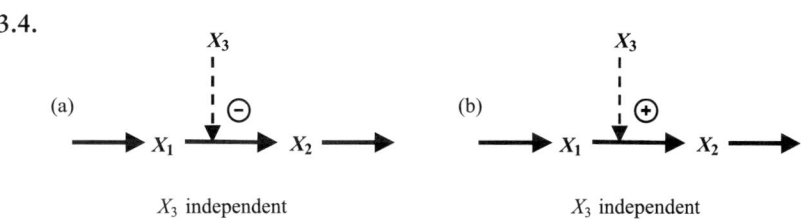

   3.4.

3.5.

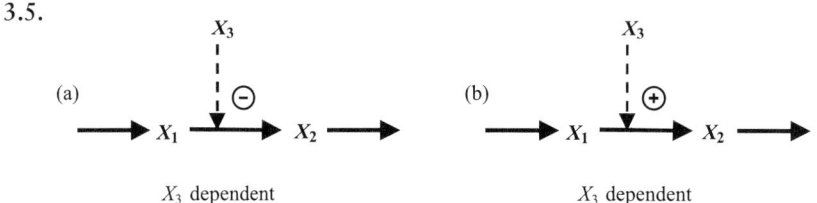

(a) X_3 dependent (b) X_3 dependent

REFERENCES

[8–13], [17], [22], [25–6], [28], [38], [40], [42], [61], [73], [77], [93–4], [98], [114–5], [122–4], [159], [161], [174], [183], [185–8], [209], [214–5], [221], [231], [243], [245–7], [260–1], [264–5], [276], [292], [361–2], [367–9], [399], [405–6], [408], [410], [495].

CHAPTER TWO

Models of Biochemical Systems

Once we have listed all components and interactions of a biochemical system and arranged them in a map, this map becomes the focus of our analysis. Aspects we have ignored, intentionally omitted, or forgotten in constructing the map are out of sight. Our task is now to translate the map into a mathematical structure that is appropriate and convenient to analyze. The problem we face is that there is *a priori* no prescription for what the "exact" mathematical representation even of the simplest reactions is. We may use a Michaelis–Menten rate law or one of the other traditional kinetic descriptions, but it is well known that all these formulations are based on simplifications and assumptions, such as the quasi-steady-state assumption that is fundamental to the Michaelis–Menten law.

Even though the traditional rate laws are simplifications of a more complex reality, they are nevertheless very useful. They are mathematical representations that capture particular aspects in a convenient form, while they ignore other aspects that are deemed less relevant. The discrepancy between loss of detail and usefulness alludes to a deeper truth: The quest for "the exact" mathematical representation of a biological phenomenon is ill posed.

For instance, one can argue that even if a mathematical formula expressed the rate of a reaction as a function of substrate and effectors in such a way that it precisely modeled all available data, it could not capture in detail all molecular interactions and the properties of all atoms and electrons involved. In addition, such a rate law would not include all influences that could possibly have other, minute effects. Of course, one would counter that these minute details and eventualities would unnecessarily complicate the picture, and that one has to draw the line somewhere. That is the key point. We must accept that there are no unique, exact mathematical descriptions of processes in nature. We have to search for *approximations* that capture all aspects of interest as accurately as feasible and at the same time allow us to gain insight from their analysis. If we pursue these thoughts further, we recognize that all "laws of nature" are approximations. The paradigms of natural laws, Newton's laws of physics, are in fact approximations that ignore issues addressed in Einstein's relativity theory. Once we realize that the search for "exact laws" is futile, the emphasis of biochemical systems analysis shifts to developing useful approximations.

The main criterion in the search for a useful approximation is the question of what the mathematical model should do for us. If we are interested in events at the atomic or molecular level, the variables should represent individual atoms, atomic groups, free radicals, molecules, or groups of molecules. The dynamics of the model then will elucidate interactions between molecules and, for instance, show how individual bonds or entire molecules are formed. This level of detail becomes overwhelming – and indeed unnecessary – when the analysis addresses a higher level, such as an entire enzyme-catalyzed pathway or the cooperation between organs. Shifts in focus are typically accompanied by new definitions of variables. At the pathway level, it is appropriate to represent metabolites, groups of metabolites, and enzymes as variables, and the dynamics of the model addresses flow of material from the initial substrate through all intermediates to the end product. At the organ level, the variables may represent main output products, such as insulin or adrenaline. Even though all processes at the atomic level are still responsible for the dynamics at these higher levels, the higher-level model is not able – and not intended – to represent such detail.

Our primary interest in biochemical systems analysis is not the mechanics of individual reactions but the masterly orchestrated cooperation between the many reactions and modulations that govern biochemical and metabolic systems. It is this level of systems of reactions where we have to focus our attention. In the search for appropriate mathematical representations, an obvious first candidate is the Michaelis–Menten rate law (Michaelis and Menten, 1913). This function, along with its derivatives and generalizations, such as the mechanisms of Hill (1910), Cleland (1967), and Koshland (e.g., see Koshland and Neet, 1968), has been tremendously successful in guiding biochemical research (for fuller treatises see, e.g., Savageau, 1976; Segel, 1991; Cornish-Bowden and Cárdenas, 1990; Heinrich and Schuster, 1996; Fell, 1997). However, it turns out that the apparently simple mathematical structure of rational functions becomes unwieldy if the pathway exceeds a fairly small size.

One could argue that computers have become so powerful that the size and mathematical complexity of a system shouldn't be a hindrance. In fact, even when computers had only a fraction of today's power, Garfinkel and others successfully developed simulation programs for fairly complex metabolic pathways (e.g., Chance et al., 1960; Garfinkel, 1968, 1980, 1985; Garfinkel et al., 1968, 1970, 1974; Roman and Garfinkel, 1978; Achs et al., 1991). Today, it appears, there should be no limit to what can be implemented in a modern computer algorithm. That may be so, but the problem with analyzing realistically big computer models is that the results of the analyses in themselves become too complex to permit real insights (for further discussion see Shiraishi and Savageau, 1992a).

In addition to problems of assessing results, the use of rational functions, such as the Michaelis–Menten rate law, as fundamental descriptions runs into other severe limitations. Tightly regulated pathways make it difficult to determine the mathematical form of all rate laws, and often, a fair number of assumptions are unavoidable. Furthermore, rate laws of this type may contain an enormous number of parameters when the pathway consists of a realistic number of metabolites and modulators. As an extreme example, Savageau (1976: p. 75) mentions the enzyme glutamine synthetase, which is affected by at least eight reactants and modifiers (cf. Woolfolk and Stadtman, 1967): Even under the restrictive assumption that none of the reactants or modifiers enters the rate law with a power higher than one, this rate law would consist of about 500 terms, and it would take at the order of 100 million experimental assays to establish it.

As an alternative to experimental assays, one could try to use computer simulations with representative sets of different parameter values. However, to explore a system with ten parameters, using ten different values for each parameter, one needs ten billion simulations, and yet it would be almost impossible to evaluate the validity and reliability of predictions based on such simulations.

A quite trivial example may demonstrate the insufficiency of computer simulations to provide general results. Suppose we want to test whether all numbers have an inverse, i.e., whether 1 divided by any real number again is a real number. The computer simulation approach would be randomly or systematically to select many numbers a, b, etc., and to compute the inverses $1/a$, $1/b$, etc. With all probability, each result will be a real number, even if we select very large or very small positive or negative numbers, and we are tempted to conclude that our hypothesis is correct. However, it is not, since zero has no real-number inverse. This example is rather simple, but it demonstrates the danger of making general predictions based on computer simulations. Biochemical systems are much harder to understand, and it can happen easily that simulations suggest a wrong conclusion without our noticing it.

Let us summarize the advantages and problems associated with the typical traditional approach. In cases where it is possible, the mechanistic approach yields insight at the molecular level in that it explains the role of the contributing system components and elucidates the dynamics of enzyme binding and dissociation. However, the mathematical analysis of mechanistic models requires factual or assumed knowledge of all involved mechanisms and parameter values. For small *in vitro* systems this information may be available, but *in vivo* it is usually not. Furthermore, the detailed inclusion of mechanisms in even moderately complicated models results in mathematical representations that are simply too hard to analyze. Computer simulations can remedy some of the evaluation problems, but by their very nature they can never establish generally valid results or predictions.

If mechanistic models cannot be used for larger systems, is there any alternative? The answer is yes. What we have to do, though, is shift our emphasis away from the individual mechanism toward an integrated cooperation of all system components. This shift is accompanied by a change in the type of questions that can be asked and answered. In mechanistic models, we may ask what the amount of a particular enzyme complex is at a given time, or what the value of a particularly interesting inhibition constant is. In contrast, when analyzing integrated system models we may ask how the temporal change in one variable affects the concentration of the other variables, how the system is affected by changes in modulation, or why nature has selected a particular mode of regulation over other possible modes. We may ask what types of perturbations an organism can tolerate or which components of a system are most sensitive to such perturbations.

Integrated models for biochemical systems must satisfy a number of requirements. They must:

- *Capture the essence of the system under realistic conditions.* The model should respond to various inputs the same way the actual biochemical system does. For instance, any increase in an inhibitor concentration should slow down or turn off the inhibited reaction. It would be ideal to have a model that responds correctly under all imaginable conditions, but this would be too restrictive a requirement. Instead, we require the model to react correctly under most *realistic* conditions (Savageau, 1976; Garfinkel, 1985). If systems *in vivo* are being investigated, then *in vivo* conditions should determine what "realistic" means. A case in point is the variation in concentrations *in vivo*. In an experimental setting, substrate, modulator, and enzyme concentrations can be varied almost arbitrarily. However, actual measurements *in vivo* (e.g., Geigy, 1960; Gerber et al., 1974; Wallace, 1986), as well as theoretical considerations (Savageau, 1976: pp. 80–83), strongly suggest that concentration ranges in intact organisms are very restricted; variations in excess of ten or twenty percent in many cases constitute a pathological condition. For example, the total protein and glucose contents in serum deviate under normal condition only within about 10% of their means (Geigy, 1960). This restricted variability in concentration implies that a mathematical model of a biochemical system *in vivo* should not be required to respond correctly to very large variations in concentrations. If a model correctly captures variations as they are encountered in intact organisms, then it may be considered appropriate.
- *Be qualitatively and quantitatively consistent with key observations.* The model should offer one-to-one relationships between standard experimental procedures and the mathematical representation. For instance, it is customary to study changes in flux as a response to changes in one of the metabolites. A good model should render it possible to execute the corresponding "mathematical experiment."
- *In principle, allow analyses of arbitrarily large systems.* The structure of models for biochemical systems should be independent of the size of the system. An example of a size-independent structure is a system of linear equations: The addition of a new variable increases the number of terms in each equation and requires a further equation, but the system is still linear, its structure is unchanged, and all methods of linear analysis still apply. In contrast, if we want to allow for an additional inhibitor in a mechanistic model, the structure of several rate laws is likely to change, and there are no clear recipes for implementing these changes.
- *Be generally applicable.* Many models require the investigator to know or assume the mathematical form for the rate laws of all processes of a system before the model can be analyzed. This requirement becomes very severe in biochemical systems of realistic size. A preferable modeling approach must allow the investigator to formulate model equations solely from the structure of the metabolic map, i.e., from a diagram that shows the flow of material and the existence of modulations, as we discussed in Chapter 1.
- *Be characterized by measurable quantities.* Although the general theory of science allows models to include components that have no directly measurable counterpart in the real world (Carnap, 1966), those components should be held to a minimum. Ideally, every variable and every parameter should have a uniquely defined role and meaning. A system component may represent a feature that traditionally is not

measured or cannot be measured with today's methods, but it should, in principle, be a measurable feature.
- *Allow simple translation of results back into biochemical language.* Once a map is translated into a mathematical representation, the analysis consists of procedures of applied mathematics and computer simulation, and typical results are expressed in mathematical terminology. A good modeling approach should render it easy to translate what the numerical or symbolic results mean in biochemical terms.
- *Have a mathematical form that is amenable to analytical and numerical evaluation.* All of the previous features were concerned with the appropriateness of a good mathematical representation. Almost as important is mathematical tractability. There is not much use for a mathematical model, however appropriate it may be biochemically, if one cannot analyze it.

No mathematical model is known to combine all desired features in an ideal fashion. In many cases, a model either accurately represents the majority of biochemical aspects and is mathematically very complex, or facilitates a great variety of mathematical analyses but shows deficiencies in comparison with the actual system. Among the best compromises known are models in which the interactions between components are represented as products of power functions. These functions may look rather unusual if one is familiar with traditional rate laws, and they may appear counterintuitive at first. However, power functions are very convenient mathematically and logistically, and the *Complements* section shows that they are backed up by strong theoretical support and justification.

Change

To understand a biochemical system, we need to know the dynamics of its constituents, that is, we need to determine functions of the type $X_i(t)$. The function $X_i(t)$ describes the status of any of the dependent variables discussed in the previous chapter at a time point t. If we have this information for all components $i = 1, \ldots, n$ and for all time points t of interest, we know everything about the system.

Experiments that aim to reveal the dynamics directly can be very difficult or even impossible to execute, and it is often easier to measure temporal *changes* in dependent variables. For instance, one often measures instantaneous product formation in response to changes in the exogenous substrate, inhibitor, or enzyme concentration. Most kinetic laws relate a reaction rate to concentrations, and a reaction rate is nothing but an instantaneous temporal rate of change in concentration of substrate or product. The availability of these common procedures raises the question of whether information about changes in concentrations is sufficient to deduce the dynamics of a biochemical system. Within some limitations, the answer is yes.

Let's explore the question with a simple, non-biochemical analogue: a car driving along a highway. For simplicity, let's assume the car starts at milestone 210 and maintains a constant speed of 50 mph for 12 min. Then, the car drives at 35 mph for 6 min, and finally, it drives at 55 mph for another 6 min. Straightforward calculation yields that during the first 12 min the car drove 10 miles, during the next 6 min it drove 3.5 miles, and during the last 6 min it drove 5.5 miles. So altogether the car drove 19 miles, and its final location is at milestone 229. Thus, we have computed the *location* of the car from information about its *speed* and its *initial location*. If the car changes its speed more often, the number of calculations increases, but the principle remains the same. Even if we push the situation to the limit and let the speed change continuously, we can imagine that, in essence, we could go through the same operations to compute the final location of the car. In this case, the sum is replaced by an integral, but evaluation of the integral by a computer is again done by a summation of many tiny steps.

Now let's look at the analogous situation in biochemical systems. Speed in the car example means the temporal rate of change of location. In a biochemical setting, it corresponds to a rate of change in concentration. The initial location of the car corresponds to the initial concentration. We conclude, by analogy, that given the initial concentration and all rates of change in concentration over the time period of interest, we can compute the concentration itself at any point of this time period. Except for some biochemically irrelevant cases, one can show mathematically that this conclusion is true.

Because of its great importance for all that follows, it is prudent to introduce a succinct terminology for the *(instantaneous) rate of change in a dependent variable*. One could use a Greek delta or define a function like "Change(X_i)". In fact, this "change function" is nothing else but the derivative of X_i with respect to time, or the slope of X_i, expressed as a function of time. Thus, we could use the common mathematical notation dX_i/dt. However, this notation is sometimes clumsy, and as long as the changes in X_i are changes with respect to time, it is much more convenient to introduce a notation that is widely used in applied mathematics, physics, and engineering; namely, a dot above the symbol. Thus, we define

$$\dot{X}_i(t) = \text{(instantaneous) rate of change in } X_i \text{ at time } t = \frac{dX_i}{dt}. \qquad (2.1)$$

It is usually clear that $\dot{X}_i(t)$ is a function of t. Therefore, the argument (t) is omitted, and we simply denote the (instantaneous) rate of change in X_i as $\dot{X}_i$.

In the car example, the speed was measured directly by the speedometer. In biochemistry, one often measures *rates* of reactions or *fluxes*, and these correspond directly to changes in concentrations or pool sizes. When we express such changes in terms of substrate concentrations, enzymes, factors, and products, we can write the relationship symbolically in form of an equation:

$$\begin{aligned}\dot{X}_i &= \text{function of substrate concentrations, enzymes, factors, and products} \\ &= f(S_1, S_2, \ldots, E_1, E_2, \ldots, F_1, F_2, \ldots, P_1, P_2, \ldots).\end{aligned} \qquad (2.2)$$

This type of equation, which contains one or more derivatives and one or more algebraic functions, is called a *differential equation*.

The well-known rate laws are in fact differential equations that express the rate of change in substrate or product concentration as functions of substrate and inhibitor concentrations. For instance, the Michaelis–Menten rate law for a substrate S and a product P asserts

$$v(S) = \frac{V_{max} S}{K_M + S} \qquad (2.3)$$

(Michaelis and Menten, 1913), and since the rate $v(S)$ is equivalent to the loss of substrate, we obtain

$$v(S) = -\dot{S}. \qquad (2.4)$$

This equation makes us realize that the Michaelis–Menten reaction actually constitutes a differential equation of the form

$$\dot{S} = -\frac{V_{max} S}{K_M + S}. \qquad (2.5)$$

This equation translates into two messages: (1) the rate of change in substrate concentration is a function of time; and (2) its magnitude depends on the current substrate concentration and on two parameters, V_{max} and K_M, and this dependence is described mathematically by the function on the right-hand side of the equation.

As in the car example, one is interested in the status of the concentration S at a desired point t, e.g., 20 min after the beginning of the experiment. Extracting $S(t)$ from the Eq. (2.5) is called *solving the differential equation*. In other words, the solution of the differential Eq. (2.5) is S expressed as a function of time. A plot of this temporal change in substrate is shown in Fig. 2.1. This type of representation of $S(t)$ is not often seen in biochemical texts, but it is important, because it actually shows the dynamics of the enzyme-catalyzed reaction over a period of time. The plot shows a gradual decrease without recovery, which confirms our intuition, because no substrate is added to the system, but substrate is being used up all the time. This general trend can directly be deduced from the differential equation, without the computation of an actual solution of the equation, since the right-hand side of Eq. (2.5) is always negative.

If we are interested in product formation as well, we can augment this equation with the second differential equation:

$$\dot{P} = \frac{V_{max} S}{K_M + S}. \qquad (2.6)$$

Equations (2.5) and (2.6) together constitute a small *system of differential equations*.

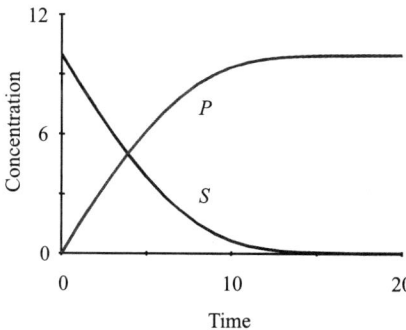

Figure 2.1. Temporal decrease in substrate concentration and increase in product concentration, according to Eqs. (2.5) and (2.6) with $V_{max} = 2$ and $K_M = 4$. Substrate concentration at time 0 is 10, and product concentration at time 0 is 0.

It is time to try out some of these issues first hand with PLAS. Suppose substrate X_1 is converted into product X_2 in a kinetic process of first order. The laws of elemental chemical kinetics teach us that the decrease in substrate concentration is given as

$$\dot{X}_1 = -kX_1, \tag{2.7}$$

where the dot on the left-hand side again represents the rate of change with time and k on the right-hand side the rate constant of the process. The minus sign indicates that the concentration is going down. The differential equation says that the change in X_1 is proportional to the current concentration X_1, and the speed of the process is characterized by the rate constant k.

The equation is implemented in PLAS as the file *first.plc*. When we open this file, we see that the left-hand side is coded as X1' and that the subscript on the right-hand side is written on the line: X1. The equation in PLAS reads

$$X1' = -k\ X1. \tag{2.8}$$

A few more pieces of information are required before a numerical solution can be obtained. One obviously is the numerical value of k, and another is the concentration at the beginning of the experiment, which in jargon is called the *initial value* or *initial condition*. To specify these in PLAS, the lines

$$\begin{aligned} k &= 0.2 \\ X1 &= 10 \end{aligned} \tag{2.9}$$

were added. Note that no units are entered. The rate constant is measured in inverse time units (e.g., per minute,) and these units are implicit in the time derivative X1'. If k is given in inverse minutes, then the solution is given minutes. Similarly, no unit is entered for the concentration, because this unit is given by the specification of the initial value X1 = 10.

It is also necessary to tell PLAS the start and finish times of the experiment, and the desired *step size*, which is the time period between recordings of results. In *first.plc* the specifications are

$$\begin{aligned} t0 &= 0 \\ hr &= 1 \\ tf &= 20 \end{aligned} \tag{2.10}$$

Clicking on the "solution button" initiates the numerical solution. Since this example is very simple, the solution is obtained almost instantly. The graph shows the decrease in substrate concentration (see Fig. 2.2), and the accompanying table, obtained from the *Results* menu, contains the results in a numerical form for $t = 0, 1, 2, \ldots 20$.

This numerical solution does not provide information about what kind of mathematical function $X_1(t)$ is. In this particular case, it is an exponential function, but numerical solvers don't know that. PLAS routinely approximates the solution with a numerical algorithm (see the *Complements* section), but, like any other algorithm, it cannot identify the underlying mathematical structure.

MODELS OF BIOCHEMICAL SYSTEMS

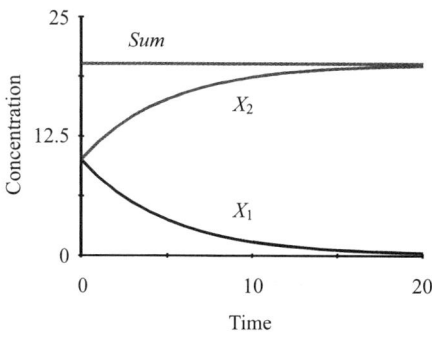

Figure 2.2. Dynamics of X_1 and X_2 according to Eqs. (2.7) through (2.13). As expected, the substrate concentration decreases from a value of 10 at time 0 to almost 0, while the product concentration increases from 10 at time 0 to almost 20. The sum remains constant at 20, which reflects that no material is entering or leaving the system.

It is informative to explore the simple equation in order to develop a feel for dynamic processes. For instance, we can initiate the process with X1 = 20, i.e., with twice the initial amount. Does it take twice as long for the process to arrive at a small concentration such as 0.1? How much substrate is lost during the first 5 time units? How does this loss compare with the loss in the first process? How do the relative losses compare? All of these questions can be answered from the comparison of the two solutions, one starting with 10 units, and the other with 20 (see Exercise 4).

Another set of questions can be answered when we change the rate constant k. How does the degradation speed compare if k is doubled or halved? What happens if we set $k = 0$? What does it mean if we replace $-k$ with a positive constant $+c$? The exercises address some of these questions, but you should explore more scenarios on your own.

A third set of parameters is concerned with the final time and the step size. Setting $tf = 40$ produces exactly the same solution as before but extends it up to 40 time units. Similarly, setting $hr = 4$ produces the same solution, but the solution is reported only at every other time point. The graph reflects the decreased number of output points, because it connects subsequent points with straight lines. At the breakpoints, the solution is rather accurate, but on the connecting line segments it is usually not. If solution points are desired here, the step size must be reduced.

There are infinitely many other questions that straightforward extensions of the simple model can answer. We consider just a few here, but you are strongly encouraged to come up with additional situations. First, let's look at X_2, which is the recipient of material lost in the degradation of X_1. The implementation consists of two steps. First, a differential equation needs to be formulated. In this case, that is not very difficult because all the material leaving the pool X_1 enters the pool X_2. Thus, the same term, but with a positive sign, appears on the right-hand side of the X_2 equation:

$$X2' = k\,X1 \tag{2.11}$$

Note that X_1 appears on the right-hand side, and not X_2. The formation of product does not depend on the current amount of product, but it does depend on the current amount of substrate.

The second step is setting an initial value for X_2. For instance, suppose 10 units of X_2 are already present when the experiment begins. To model this, we add to the

PLAS program the equation

$$X2 = 10 \qquad (2.12)$$

No additional parameters are required, because the rate of product formation is the same as the rate of substrate degradation. Also, the same parameters *t0, hr,* and *tf* are to be used for both equations. Clicking the solution button now produces two graphs, one showing the degradation of X_1 and the other the accumulation of product X_2 (Fig. 2.2). The associated table of numerical results shows time and both concentrations.

Is material conserved in the system? There are no leaks, and we should expect conservation of mass. The question is easily answered by adding to the PLAS program the equation

$$\text{Sum} = X1 + X2 \qquad (2.13)$$

(There is no necessity to call this new variable *Sum*; we could choose almost any name.) The solution now shows three variables, among which the variable *Sum* is constant at a value of 20 (Fig. 2.2). This number corresponds to the sum of initial values, that is, to the original mass in the system. This mass doesn't change during the experiment, but the distribution between the pools does.

We need not limit ourselves to differential equations whose right-hand sides contain only one variable. For instance, in elemental chemical kinetics, the equation describing the formation of product X_3 from a bi-substrate reaction of the two substrates X_1 and X_2 has a right-hand side that consists of the product of substrate concentrations and a rate constant:

$$\dot{X}_3 = k X_1 X_2. \qquad (2.14)$$

In general, the right-hand sides of the differential equations we need will contain some function V of all substrates and effectors:

$$\dot{X}_i = V(X_1, X_2, X_3, \ldots, X_n). \qquad (2.15)$$

The great biomathematician Alfred Lotka called equations of this type the "fundamental equations of *kinetics* of evolution, since they furnish expressions for the *velocities* of transformation and exhibit the relations between these velocities and the masses of the several components" (Lotka, 1924/1956: p. 51).

Before we specify what the function V may look like in a particular situation, it is useful to write down these equations to indicate which constituents of the system affect the change in a given component X_i. For a pathway of the type $X_1 \rightarrow X_2$, the function V describing the change in X_2 is a function of X_1, and substrate depletion and product formation take the form

$$\begin{aligned}\dot{X}_1 &= -V(X_1), \\ \dot{X}_2 &= V(X_1).\end{aligned} \qquad (2.16)$$

A simple example is the linear function discussed above.

Product formation in the irreversible bi-substrate reaction $X_1 + X_2 \to X_3$ is represented symbolically as

$$\dot{X}_3 = V(X_1, X_2). \tag{2.17}$$

If this reaction is modulated by X_4, then V is a function of all three contributors:

$$\dot{X}_3 = V(X_1, X_2, X_4). \tag{2.18}$$

For most systems, some information about V is available, but the exact form of the function is unknown. Later in this chapter, we will discuss how appropriate differential equations can be deduced from knowledge about the biochemical system under study.

In all these rate laws in the form of a differential equation, the symbol $\dot{X}_i$ signifies rate of change over time, while the function V is a function of concentrations. This is not inconsistent, since the concentrations themselves are changing with time and therefore are functions of time. Thus, V, as a function of functions of time, itself is a function of time. If it is important to note the time dependence explicitly, we can write the unabridged equation as

$$\dot{X}_i(t) = V(X_1(t), X_2(t), X_3(t), \ldots, X_n(t)), \tag{2.19}$$

but usually the time dependence is clear, and we omit the argument (t).

So far we have focused on one reaction, either the production of X_i or the depletion of X_i. In most biochemical pathways, of course, X_i will be the product of one or more reactions and the substrate for one or more other reactions. It is rather obvious that the change in product carries a positive sign whereas the change in substrate carries a negative sign: if we are limited to a single reaction, product accumulates while substrate is depleted. Combining production and depletion of a given metabolite, the total change is composed of two parts, one for production and one for depletion, and these parts are different because they reflect two entirely different sets of processes. Mathematically, they are readily put together, and the differential equation taking account of both production and depletion, in the most general terms, reads

$$\dot{X}_i = V_i^+(X_1, X_2, X_3, \ldots, X_n) - V_i^-(X_1, X_2, X_3, \ldots, X_n). \tag{2.20}$$

Note that we have embellished the production and depletion functions V with an index and a plus or a minus sign. This notation has terminological advantages over calling one function V and the other one W, because the index signals immediately the metabolite whose temporal changes are addressed, and the sign indicates whether the function represents production or depletion. For systems that consist of many dependent and independent variables and require a lot of bookkeeping, this streamlined index notation is the only feasible choice.

One could argue that many processes may contribute to the production and to the depletion of X_i and that, therefore, the right-hand side of the differential equation should consist of many functions. This may be true, but we don't have to worry about it, because this scenario is already included in our differential equation: Since V_i^+ and V_i^- are not yet specified, they can, in particular, be sums of functions that each represent one individual reaction. For instance, our formulation automatically includes an equation like

$$\dot{X}_i = V_{i1}^+(X_1, X_2, \ldots, X_n) + V_{i2}^+(X_1, X_2, \ldots, X_n) + V_{i3}^+(X_1, X_2, \ldots, X_n)$$
$$- V_{i1}^-(X_1, X_2, \ldots, X_n) - V_{i2}^-(X_1, X_2, \ldots, X_n) - V_{i3}^-(X_1, X_2, \ldots, X_n)$$
$$- V_{i4}^-(X_1, X_2, \ldots, X_n), \tag{2.21}$$

which consists of three production terms and four depletion terms, because we can define V_i^+ as the sum of V_{i1}^+, V_{i2}^+, and V_{i3}^+, and V_i^- as the sum of V_{i1}^-, V_{i2}^-, V_{i3}^-, and V_{i4}^-.

A pathway composed of many metabolites is represented by a collection of differential equations for the dependent variables. There is no need to write equations for independent variables, because these are fixed at one value throughout the experiment. Nevertheless, the independent variables are included in right-hand sides of the system equations to make their influence explicit and to facilitate later extensions of the system description. For n dependent variables and m independent variables the general differential equations for the entire system are

$$\dot{X}_1 = V_1^+(X_1, X_2, \ldots, X_n, X_{n+1}, \ldots, X_{n+m})$$
$$- V_1^-(X_1, X_2, \ldots, X_n, X_{n+1}, \ldots, X_{n+m}),$$
$$\dot{X}_2 = V_2^+(X_1, X_2, \ldots, X_n, X_{n+1}, \ldots, X_{n+m})$$
$$- V_2^-(X_1, X_2, \ldots, X_n, X_{n+1}, \ldots, X_{n+m}), \tag{2.22}$$
$$\vdots$$
$$\dot{X}_n = V_n^+(X_1, X_2, \ldots, X_n, X_{n+1}, \ldots, X_{n+m})$$
$$- V_n^-(X_1, X_2, \ldots, X_n, X_{n+1}, \ldots, X_{n+m}).$$

Making better use of the index notation, we can simply write

$$\dot{X}_i = V_i^+(X_1, X_2, \ldots, X_n, X_{n+1}, \ldots, X_{n+m})$$
$$- V_i^-(X_1, X_2, \ldots, X_n, X_{n+1}, \ldots, X_{n+m}) \quad \text{for} \quad i = 1, 2, \ldots, n \tag{2.23}$$

We call these n differential equations the *system equations*, or, collectively, the *system equation*, or the *system description*. Sometimes, they are also referred to as *Kirchhoff's node equations*, because they show some similarity to equations describing flow of currents in branched electric circuits (e.g., Hayt and Kemmerly, 1978). When we talk about the *behavior* or *response* of a biochemical system, we mean the collective change of all its constituents. We say we *understand the behavior of the system* when we understand the temporal behavior of all its dependent variables. In this form, the system description is very general, since we have not specified what the functions V_i^+ and V_i^- on the right-hand sides are. Identification of appropriate functions is the key to a successful system analysis.

(Note: Studying the changes in all components does not imply that these components are studied in isolation or that they are independent of each other. In particular, it does not imply that the superposition principle holds, which would allow us to study components of the system individually and then to superimpose all the results. On the contrary, each differential equation of the system depends on some or all dependent and independent variables and usually cannot be solved without the others.)

MODELS OF BIOCHEMICAL SYSTEMS

The example explored above in Eqs. (2.8) and (2.11) is a very simple special case of this equation, namely, $n = 2$, $m = 0$, $V_1^+ = 0$, $V_1^- = kX_1$, $V_2^+ = kX_1$, $V_2^- = 0$.

S-SYSTEMS

Without making any stringent mathematical assumptions, we have found that the general equation describing the temporal changes in a biochemical system can be formulated as

$$\dot{X}_i = V_i^+ - V_i^- \quad \text{for } i = 1, 2, \ldots, n, \tag{2.24}$$

where the functions $V_i^+ = V_i^+(X_1, X_2, \ldots, X_n, X_{n+1}, \ldots, X_{n+m})$ and $V_i^- = V_i^-(X_1, X_2, \ldots, X_n, X_{n+1}, \ldots, X_{n+m})$ are in general functions of all dependent variables $X_1, X_2, X_3, \ldots, X_n$ and independent variables $X_{n+1}, X_{n+2}, \ldots, X_{n+m}$.

In order not to get confused with all the indices, consider a system with two dependent and one independent variable. It consists of two equations:

$$\begin{aligned}\dot{X}_1 &= V_1^+(X_1, X_2, X_3) - V_1^-(X_1, X_2, X_3), \\ \dot{X}_2 &= V_2^+(X_1, X_2, X_3) - V_2^-(X_1, X_2, X_3).\end{aligned} \tag{2.25}$$

The functions on the right-hand side do not necessarily depend on all three variables, but the possibility is not excluded in the general formulation above. To be specific, suppose we analyze a simple conversion of X_1 into X_2 that is catalyzed by X_3. A proper map is shown in Fig. 2.3.

The details of translating a map into equations are the subject of Chapter 3, and it is sufficient here just to state the results. There is a constant influx into the system that replenishes the pool X_1; hence, the corresponding function V_1^+ is a constant, and we represent it with the symbol α. The degradation of X_1 depends on the concentration or pool size of X_1 itself and also on the enzyme X_3. Thus, the corresponding function V_1^- depends on X_1 and X_3, but not on X_2. The production of X_2 constitutes the same process as the degradation of X_1; in mathematical terms, $V_2^+ = V_1^-$. Finally, the degradation of X_2 depends only on its current concentration or pool size. With these specifications, the system description reads

$$\begin{aligned}\dot{X}_1 &= \alpha - V_1^-(X_1, X_3), \\ \dot{X}_2 &= V_1^-(X_1, X_3) - V_2^-(X_2).\end{aligned} \tag{2.26}$$

Although it is of some value, this system description cannot be analyzed much further until we determine what the functions V_1^- and V_2^- actually are. Nobody knows the answer to this question, but numerous considerations of the properties and dynamic reactions of biochemical systems suggest that a convenient mathematical representation for a process V_i^+ or V_i^- is given by a product of power-law functions of

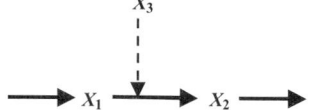

Figure 2.3. Conversion of X_1 into X_2, catalyzed by enzyme X_3.

those and only those variables that directly affect this process; this product is furthermore multiplied by a rate constant that determines the speed of the process (Savageau, 1969ab, 1970, 1972). There is no mathematical proof that these functions are the best possible description or come closest to our vague idea of an "exact" model. Nonetheless, there is a lot of circumstantial evidence that supports these types of functions (e.g., Savageau, 1976, 1995a; Voit, 1991; see also Voit and Yi, 1990 and Chapter 12: *Epilogue*). This evidence is summarized later in this chapter and in the *Complements*. Power-law representations for the processes $V_1^-(X_1, X_3)$ and $V_2^-(X_2)$ are

power-law representation of $V_1^-(X_1, X_3)$: $\beta X_1^a X_3^b$,

power-law representation of $V_2^-(X_2)$: γX_2^c.

The rate constants are commonly symbolized by lowercase Greek letters, and the powers by lowercase Latin letters. Note that the process $V_1^+ = \alpha$ adheres to the same mathematical form. However, since this process does not depend on any of the analyzed variables, the rate constant α is all that's left.

It doesn't take much imagination to see that we would fast run out of symbols for larger systems unless we use indexed notation. There is one rate constant for the production of each pool and one rate constant for its degradation. These are traditionally called α_i and β_i, where i signifies the pool in question. For each power, we use one subscript for the affected pool (X_i) and one subscript for the variable that is exerting the effect (X_j). In the production term, the power is called g, and in the degradation, it is called h. Thus, instead of the symbols a, b, and c in the above power-law representations, we use h_{11}, h_{13}, and h_{22}, respectively. The distinction in nomenclature between V_i^+ and V_i^- is made for reasons of convenience, intuition, and interpretation: The parameters α_i and g_{ij} always refer to production or synthesis, and the parameters β_i and h_{ij} always refer to degradation or loss.

With the use of indices, the size of a system no longer poses a notational problem. For a system with n dependent and m independent variables, we write (for $i = 1, 2, \ldots, n$)

$$V_i^+(X_1, X_2, \ldots, X_n, X_{n+1}, \ldots, X_{n+m}) = \alpha_i X_1^{g_{i1}} X_2^{g_{i2}} \cdots X_n^{g_{in}} X_{n+1}^{g_{i,n+1}} \cdots X_{n+m}^{g_{i,n+m}} \tag{2.27}$$

and

$$V_i^-(X_1, X_2, \ldots, X_n, X_{n+1}, \ldots, X_{n+m}) = \beta_i X_1^{h_{i1}} X_2^{h_{i2}} \cdots X_n^{h_{in}} X_{n+1}^{h_{i,n+1}} \cdots X_{n+m}^{h_{i,n+m}}. \tag{2.28}$$

The first feature to note about these functions is their homogeneous structure. Whether a dependent or an independent variable, a constituent is always formally treated in the same way, namely, as a factor in the product V_i^+ or V_i^-, raised to a power. This formulation is very different from conventional rate laws consisting of rational functions, in which substrates or modulators may appear several times in the same term. At first glance, it may appear that the rigid power-law equations (2.27) and (2.28) could not possibly suffice to represent all the complicated behaviors that we know to exist in biochemical systems. This intuition is wrong, though, since it has been shown that virtually any differentiable nonlinearity can be captured by these equations, including most complex oscillations and even chaos (e.g., Savageau and Voit, 1987).

Let's formulate in words what the definitions of V_i^+ and V_i^- say: For each dependent variable, the change in time is modeled as a difference of two functions, V_i^+ and V_i^-, which may depend on some or all dependent and independent variables. Each of these functions is defined as a product of a constant (α_i or β_i) and all the variables raised to individual powers (g_{ij} or h_{ij}, respectively). From the above example, it is clear that if a variable X_k does not affect the ith production function V_i^+, then the corresponding power, g_{ik}, is equal to zero. This effectively eliminates X_k from the production term, because $X_k^0 = 1$. The analogous statement is true for the degradation term.

For example, if X_4 does not affect the degradation of X_2, then the corresponding power h_{24} in the function V_2^- equals zero. If X_4 does not affect X_2 at all, then g_{24} in V_2^+ and h_{24} in V_2^- both equal zero. We will come back to these situations in the next chapter.

Substitution of the functions (2.27) and (2.28) in the general scheme (2.24) and using the usual notation $\prod$ for a product, our biochemical system representation becomes very succinct, taking the form

$$\dot{X}_i = \alpha_i \prod_{j=1}^{n+m} X_j^{g_{ij}} - \beta_i \prod_{j=1}^{n+m} X_j^{h_{ij}} \quad \text{for} \quad i = 1, 2, \ldots, n. \tag{2.29}$$

This system of equations is called an *S-system*, where the S refers to *synergism* and *saturation* of the investigated system. Synergism and saturation are two fundamental properties of biochemical and biological systems that are inherent in these equations but not in many other types, such as linear differential equations. Note that all S-system equations have the same mathematical form but differ in their parameters.

Before we discuss the role and interpretation of these parameters and their numerical values, we explore some examples of S-systems in PLAS. Note that the example of the previous section also falls into this category.

Parameters

Rate constants. α_i is the constant factor in the production term for X_i. It may be positive or zero, but cannot be negative. If α_i equals zero, the entire function V_i^+ vanishes, and there is no production of the metabolite X_i. The larger α_i is, the larger is the function V_i^+ and thus the production of X_i. In other words, α_i represents the rate with which the concentration or the pool X_i is enlarged, and it is therefore called the *rate constant* for the production of X_i. If we imagine for a moment that there is no degradation of X_i, then the function V_i^- is equal to zero and the right-hand side of Eq. (2.29) consists exclusively of the α term:

$$\dot{X}_i = \alpha_i \prod_{j=1}^{n+m} X_j^{g_{ij}}. \tag{2.30}$$

Not altering any of the exponents g_{ij}, but increasing or decreasing α_i, we provoke a concomitant change in $\dot{X}_i$, that is, in the speed with which X_i changes. This again shows that α_i is directly interpretable as a rate constant.

β_i also is a rate constant, but it refers to the degradation of X_i. Like α_i, β_i can be positive or zero, but not negative. If β_i is increased while everything else is fixed, V_i^- increases and $\dot{X}_i$ decreases. $\dot{X}_i$ even may become negative, indicating that more material X_i is used than replenished.

Since the total change in X_i is given by the difference between the α-term and the β-term, the concentration of X_i may go up (the α-term dominates) or down (the β-term dominates), or it may not change at all, if the two terms are in balance. Which term dominates, of course, not only depends on the rate constants α_i and β_i, but also on the other parameters, g_{ij} and h_{ij}, and on the current concentrations of all metabolites that are involved in V_i^+ and V_i^-. If all equations are balanced, we say that the system is in a *steady state*. Reformulated mathematically, a steady state is characterized by the condition that no metabolite is changing, i.e., that $\dot{X}_i = 0$ for all $i = 1, 2, \ldots, n$. We shall discuss in Chapter 6 how one can compute concentrations of all metabolites for which the biochemical system is in a steady state.

Example. It is very useful to develop some intuitive feel for the role of the parameters in S-systems. This is accomplished most easily with an example in PLAS. The file *Example1.plc* represents the following system with only one dependent and two independent variables. In this system, X_2 is supplied by the experimenter at a constant rate. It is converted into X_1 and subsequently degraded. The degradation is inhibited by the independent variable X_3. The map is shown in Fig. 2.4. It is immaterial for this example how the S-system equations are developed for this map. Rules for this procedure will be given in the next chapter. The only purpose of the example at this point is an exploration of the role of the rate constants.

Since there is only one dependent variable, the S-system contains only one equation. Upon numerical specification of all parameter values except the rate constants, the mathematical description may read

$$\dot{X}_1 = \alpha_1 X_2 - \beta_1 X_1^{0.5} X_3^{-1}, \qquad X_1(0) = 0.01,$$
$$X_2 = 2, \tag{2.31}$$
$$X_3 = 0.5.$$

Note that X_2 and X_3 are independent variables, which have constant values for the duration of each of the following "experiments."

For the first, *baseline* experiment, we set $\alpha_1 = \beta_1 = 1$; these values are actually stored in the PLAS file *Example1.plc*. Related cases are left as an exercise.

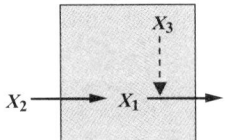

Figure 2.4. An externally supplied substrate X_2 is converted into X_1 and subsequently degraded. X_3 inhibits the degradation.

MODELS OF BIOCHEMICAL SYSTEMS

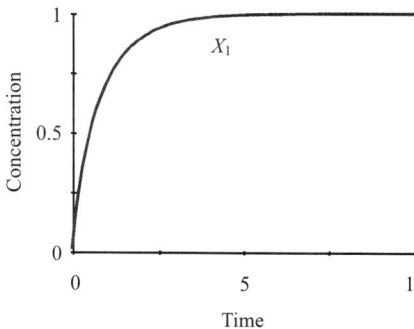

Figure 2.5. Dynamics of X_1, according to Eq. (2.31) and the PLAS implementation of *Example1.plc*.

Kinetic orders. In order to interpret the exponents g_{ij} and h_{ij}, recall some elemental chemical kinetics. In a bimolecular reaction of the type $X_1 + X_2 \rightarrow X_3$ the production of X_3 is equal to the product of a rate constant k and the two concentrations of X_1 and X_2:

$$\dot{X}_3 = k X_1 X_2. \tag{2.32}$$

If the reaction involves two molecules of X_1 and one molecule of X_2, then X_1 enters the equation with a power of 2:

$$\dot{X}_3 = k X_1^2 X_2, \tag{2.33}$$

and one says that *the kinetic order of the reaction with respect to the chemical species X_1 is 2* and that *the kinetic order of the reaction with respect to the chemical species X_2 is 1*. In elemental chemical kinetics, these powers are directly interpreted as the numbers of molecules involved in each individual reaction, and as a consequence, these powers assume values like 0, 1, or 2. Recent biochemical studies suggest that many enzyme-catalyzed reactions *in vivo* follow a similar type of rate law, but that the powers can be non-integer numbers. For instance, second order reactions occurring on surfaces have a kinetic order of 2.46 (Savageau, 1995a).

Based on the tradition of elemental chemical kinetics and on these newer biochemical studies, we call the exponents g_{ij} and h_{ij} in our system equation *kinetic orders* and allow them to assume not only integer values but any real numbers. The effect of a kinetic order on the change in a metabolite is easy to explore. Let's look at the example

$$\dot{X}_2 = 2 X_1 X_3^{-1} X_4^{1/2} - 0.2 X_2, \tag{2.34}$$

in which the change in X_2 depends on X_1 through X_4. The change in the concentration of X_2 results from the difference of its production, represented as $2 X_1 X_3^{-1} X_4^{1/2}$, and its degradation, represented as $0.2 X_2$. The degradation depends only on X_2, which is present with an exponent of 1, signaling an exponential function or, in chemical terms, a first-order reaction with rate constant 0.2. The production of X_2 depends on X_1, X_3, and X_4, which enter with different powers. With respect to X_1, the process is of first order. The kinetic orders with respect to X_3 and X_4 cannot be interpreted this way. However, we can see that the relative influence of X_4 on the production of X_2 is weaker than that of X_1, since X_4 enters the term only with the power 1/2. X_3 has a negative power, and this indicates an inhibiting effect, since an increase in X_3 leads to a diminution of the α-term and therefore to reduced production of X_2.

To match up our production term with the general functional description of V_i^+ in (2.27), we could include X_2 in the production term with a power of zero. It is easy to see that X_2 would still not affect V_2^+, because no matter what the positive value of X_2, the zero power would make it equal to 1, which in turn would have no effect on V_2^+. Thus, a power of zero reflects that the corresponding variable has no influence on the production or degradation in question, and the variable can be dropped from that term altogether.

Let's illustrate these findings with a numerical example in which the concentrations of X_1 through X_4 all have an arbitrary value of 100 of some unit. The production of X_2 is thus $2 \times 100 \times 0.01 \times 10 = 20$. If X_1 is increased by 10%, the production of X_2 becomes $2 \times 110 \times 0.01 \times 10 = 22$, which corresponds to an increase of 10%. If X_3 is increased by 10%, the production of X_2 becomes $2 \times 100 \times 0.00909 \times 10 = 18.2$, which corresponds to a *decrease* of about 9% and reflects the inhibitory effect of X_3. If X_4 is increased by 10%, the production of X_2 becomes $2 \times 100 \times 0.01 \times 110^{1/2} = 20.98$, which corresponds to an increase of about 5%. A change in X_2 does not affect the production of X_2, because X_2 is not a part of the production term; however, a change in X_2 does affect its degradation: If X_2 is increased by 10%, the degradation of X_2 becomes $0.2 \times 110 = 22$, which corresponds to an increase of 10% in degradation.

Example. This example can be considered an expansion of the previous example in which X_2 is now a dependent variable and its production is also under investigation. Specifically, X_2 is the product of a reaction that uses X_4 as substrate and is activated by X_1 and inhibited by X_3. The map is given in Fig. 2.6.

The single equation of the previous example is augmented with an equation describing the dynamics of X_2. With numerical specifications, the system is

$$\begin{aligned}
\dot{X}_1 &= 2X_2 - 1.2 X_1^{0.5} X_3^{-1}, & X_1(0) &= 2, \\
\dot{X}_2 &= 2 X_1^{0.1} X_3^{-1} X_4^{1/2} - 2X_2, & X_2(0) &= 0.1, \\
X_3 &= 0.5, \\
X_4 &= 1.
\end{aligned} \qquad (2.35)$$

This system is implemented in PLAS as *Example2.plc*. Note that if we had started from scratch, we would probably have come up with a set of equations in which X_1

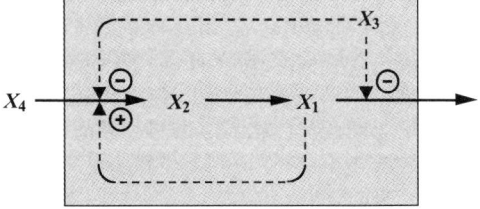

Figure 2.6. The map of Fig. 2.4 is expanded to include X_2 as a dependent variable, whose production is modulated by X_3, and an additional independent variable X_4.

MODELS OF BIOCHEMICAL SYSTEMS

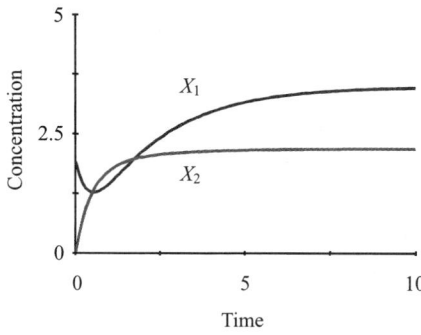

Figure 2.7. Dynamics of X_1 and X_2 in the example depicted in Fig. 2.6 and formulated in Eq. (2.35). X_1 initially undershoots but ultimately reaches a level higher than at the beginning of the experiment. X_2 shows a simple monotonic increase.

and X_2 are exchanged. Mathematically, this difference does not matter at all. The dynamics of the system is shown in Fig. 2.7.

Several changes may be made to explore the roles of the kinetic order parameters (see Exercise 5).

In general, the roles of the kinetic orders g_{ij} and h_{ij} can be summarized as follows: g_{ij} represents how the production of X_i, whose index matches the first index of the kinetic order, is influenced by the variable X_j, whose index matches the second index of the kinetic order. h_{ij} has the same role but refers to degradation instead of production. If the kinetic order g_{ij} equals zero, then the ith equation contains a production term that is independent of the metabolite X_j; the metabolite X_j can be dropped from the term. If the kinetic order h_{ij} equals zero, then the ith equation contains a degradation term that is independent the metabolite X_j; the metabolite X_j can be dropped from the term. Positive kinetic orders indicate activating influences, negative kinetic orders express inhibition. These interpretations are discussed in more detail in the following chapters.

Properties of S-System Models

Validity. S-system models are consistent with a number of specific features of biochemical systems as well as with observations that are common to other branches of biology. We have already mentioned the consistency of S-system models with the traditional laws of elemental chemical kinetics and the observation that biochemical reactions in heterogeneous media show fractal kinetic orders. It is difficult to find convincing measurements of kinetic processes *in vivo*, but power functions, as they are the basis for S-systems, appear to provide reasonably accurate rate laws (see Voit and Savageau, 1987, and Savageau, 1995b, for a discussion of this question).

A different type of observation that has been made time and again is that of *allometric relationships* between different parts of a growing organism. According to the etymology of the term, the parts of interest grow "at different scales." Specifically, one finds that the absolute growth of one part is not linearly related to the growth of the other part, but instead, that the *relative* growth of two parts very often is linearly related: If the first part grows by 5% over a certain period of time, then the other part grows by approximately the same percentage. This empirical rule goes back at least to Galileo in the seventeenth century, and it has found support by many investigators ever since (for references, see Savageau, 1979b, 1991e). Thompson (1917) and Huxley (1932) treated allometry in the context of growth, form, and development in great depth. Summarizing numerous chemical examples where the changes in different components are allometrically related, Needham (1942) suggested that there must be a "chemical groundplan" based on allometry. Allometric relationships are so commonplace that Adolph (1949) proposed that every new finding concerning growth should satisfy the allometric law because otherwise it would be inconsistent with the harmony among the component parts of the growing organism. Allometry also governs the relationships between individuals or groups within populations and societies (for some collections of allometric phenomena, see Batschelet, 1979; Savageau, 1979b; White, 1981; Pickard, 1983; Peschel and Mende, 1986).

Formulated mathematically, an allometric growth relationship between two variables X_1 and X_2 requires that the relative change $\dot{X}_1/X_1$ be a linear function of the relative change $\dot{X}_2/X_2$. Savageau (1979b) has shown that S-system models of growing organisms under rather unrestrictive conditions exhibit this type of allometry. Biochemical systems show linear relationships between relative changes in dependent and independent variables when the metabolite concentrations are close to steady state. It is shown in the *Complements* section that this relationship, substituted into the general system description we developed earlier in this chapter, suggests the S-system form as an appropriate model for biochemical systems.

Another interesting argument for the validity of the S-system form is its so-called *telescopic property* (cf. Savageau, 1979a, 1985a). It is often of interest to model a phenomenon at different levels. In a first analysis, one could be interested in the enzyme-catalyzed reactions that constitute a biochemical pathway. Then, in a second step, one might want to study interactions between organelles or cells, and finally, the focus could be the dynamics of a system with different organs.

As an example, suppose we are interested in the biochemical features of purine metabolism. At the lowest level, we could study the interconversion of the various adenylates, using metabolites like adenine, adenosine, adenyl succinate, AMP, ADP, and ATP as internal variables. At the next level, it might be desirable to study the dynamic interactions between adenylates, guanylates, and oxypurines, and we could approach this question by pooling all adenine derivatives and considering this pool as one variable, "adenylates." At yet another level, we might be interested in the synthesis and degradation of DNA and RNA, and see a need to pool all nucleosides and nucleotides.

This hierarchy of foci raises the question, *To what degree can the structure and results of lower-level models be incorporated into higher-level models?* In other words, if a single variable of the higher-level system constitutes an entire system at the lower level, how do the two models relate to each other? In linear models, where the right-hand sides of all differential equations consist of sums of variables, this substitution of a variable with a lower-level system does not change the linear structure of the higher-level model. However, as soon as the differential equations contain nonlinear terms, the structure is usually destroyed. A very rare exception is S-systems and related differential equations that are based on products of power functions; like linear models, S-systems preserve their structure when variables are exchanged for lower-level systems (Savageau, 1979a). Because one can focus on different levels of the same phenomenon and always use S-system models of the same type, one says that S-systems have the *telescopic* property.

Theoretical justification. No model can ever be proven mathematically to be correct. A model is based on abstractions and simplifications, which on one hand make the model easier to understand than the modeled reality, but on the other hand result in differences between model responses and reality. Depending on the type of abstractions and simplifications, the compromise between validity and simplification will turn out differently, and it is difficult, if not impossible, to rank competing models according to their overall quality.

In addition to its consistency with experimental observations, the S-system form is supported by different types of theoretical arguments.

- S-systems can be derived from the general system description through well-founded mathematical methods of approximation theory. This theory assures us that S-systems are valid representations if the concentrations of all variables remain fairly close to their steady-state values (cf. *Complements*). Very often this is the case, as we discussed before: Many *in vivo* systems are very well buffered against variation in concentrations. Experience with a variety of biological and non-biological phenomena has furthermore shown that the products of power functions on which the S-system equations are based are indeed valid representations. Often, a metabolite concentration can be varied 10 times or 100 times, and still the power-function representation is highly accurate. These ranges of variation are much wider than typically seen *in vivo*. The most prominent and ubiquitous mechanism that tends to keep concentrations within close limits is feedback inhibition, but other mechanisms, such as feedforward regulation and the physiological shortening of pathways (for references, see Savageau, 1976: p. 190) are also powerful means for buffering concentration variations *in vivo*. These observations and theoretical explanations further justify the use of S-system representations in biochemical systems analysis.
- A consequence of the particular approximation on which S-systems are based is that relative changes in metabolite concentrations are linearly related to relative changes in production and degradation rates. These types of relationships are of great importance and will be discussed in detail in Chapter 7. Not only do S-systems represent these phenomena correctly, but they are among the very few mathematical models that are able to capture them at all (see above discussion about allometry, and the *Complements* section).
- Every mathematical model can describe a certain repertoire of behaviors, but not others. For instance, monotonic functions cannot represent oscillations, and linear differential equations cannot describe saturation or chaos. To be able to model these types of responses is important, in that oscillations are abundant in biology, and chaos can be observed even in rather simple biochemical processes (e.g., Olsen and Degn, 1977; see also Yates, 1992; Zimmer, 1999). It is often not easy to determine what types of behaviors non-linear differential equations can capture and what is outside their reach. In the case of S-systems, it has been shown mathematically that virtually any phenomenon that can be formulated as a differential equation at all can also be formulated as an equivalent S-system (Savageau and Voit, 1987). This

demonstrates that one is not likely to encounter a biochemical system that cannot be analyzed with S-system methods.

Analytical convenience. S-system models have unique properties that make mathematical and numerical analyses of biochemical systems possible and often relatively simple.

- For S-system models, the translation of a biochemical map into rate equations follows straightforward recipes (see Chapter 3). This is possible because the differential equations always have the same homogeneous structure, and because all parameters have a well-defined meaning. In contrast, *ad hoc* models require a high degree of mathematical ingenuity and a lot of assumptions and simplifications that may be hard to justify. For instance, there are no generally applicable rules for devising the optimal mechanistic rate law for a highly regulated pathway *in vivo*. Furthermore, the complexity of *ad hoc* models grows immensely with the inclusion of additional metabolites and regulators, whereas the structure of the S-system differential equations always remains the same. Of course, the number of variables, equations, and parameters grows, but even for large biochemical systems it is still possible to set up the equations with the same simple recipes.
- All parameters of an S-system model have a clearly defined meaning. This is very important. In principle, all parameter values can be deduced from the biochemical system through experimental measurements (cf. Chapter 5). These experiments may be difficult to execute, or we may not yet have the techniques to perform them at all. But in contrast to "fudge factors" that are mathematically necessary but have no biochemical meaning and therefore never will be measurable, time and biochemical ingenuity will provide the technology necessary to estimate S-system parameters.
- The following chapters will make it clear that the particular mathematical form of the S-system differential equations offers outstanding features when it comes to mathematical and computer-aided analysis. No other general type of differential equations is known to provide biochemically relevant representations of the same mathematical convenience.
- The rigid structure of S-systems permits comparative analyses of alternative model structures that are conceptually analogous with control experiments in experimental sciences. These mathematically controlled analyses elucidate the relative importance of any of the system components and explain why nature has selected observed regulatory patterns over other possible designs (e.g., Savageau, 1985a; Irvine and Savageau, 1995ab; Hlavacek and Savageau, 1995, 1996, 1997). Even though this type of mathematical experimentation provides more insight than many other typical analyses, it is not often performed in other approaches to systems analysis, because mathematical models without a homogeneous structure are not well suited for such mathematically controlled experiments.

ALTERNATIVE MODELS

Of course, S-system models are not the only dynamical models of biochemical systems. Although they are usually excellent compromises with regard to accuracy and mathematical feasibility, other representations are very useful, and, given a different context or a different set of goals, they may be preferable to S-system models.

Three such classes of models are mentioned here. The first consists of *alternative power-law representations*. The philosophy behind these models is very similar to that leading to S-systems, but the approximation is executed in a slightly different way, yielding mathematically different representations that have their own advantages and disadvantages. The second class focuses on the *stoichiometry of metabolic networks*. Because the representation is limited to the topology of interactions, the resulting models are linear differential equations, which are more easily analyzed. The third class contains approaches that formulate metabolic or genetic networks as *formal languages*, in which metabolites, enzymes, and other constituents are words and *metabolic rules* govern which sentences are admissible and which are not.

Power-Function Representations

In the derivation of S-systems, we briefly mentioned the idea of allowing V_i^+ and V_i^- to consist of several functions each. For instance, the rate of change in X_i in such a situation could be

$$\dot{X}_i = V_{i1}^+ + V_{i2}^+ + V_{i3}^+ - V_{i1}^- - V_{i2}^- - V_{i3}^- - V_{i4}^-, \qquad (2.36)$$

where each function V_{ij}^+ or V_{ij}^- could depend on several or all of the dependent and independent variables $X_1, \ldots, X_{n+m}$. In this representation, the index i again represents the dependent variables, $i = 1, 2, \ldots, n$, whereas j symbolizes all processes associated with the variable of interest, and the number of these processes can be small or large, but is unrelated to n and m.

Even though this case can be considered a special case of the simple difference of two terms, we may ask what happens when we approximate each of the production and degradation functions separately. Following exactly the same rationale as before, that is, developing a product of power-law functions for each function V_{ij}^+ and V_{ij}^-, the resulting system description for n dependent and m independent variables has the form of a *generalized mass action* (GMA) system with positive, negative, or zero rate constants γ_{i1} and kinetic orders f_{ijl}. Its mathematical structure is

$$\dot{X}_i = \gamma_{i1} \prod_{j=1}^{n+m} X_j^{f_{ij1}} + \gamma_{i2} \prod_{j=1}^{n+m} X_j^{f_{ij2}} + \cdots + \gamma_{ik} \prod_{j=1}^{n+m} X_j^{f_{ijk}}, \quad i=1,2,\ldots,n. \quad (2.37)$$

This type of equation, also called a *multinomial system* (Peschel and Mende, 1986), is a legitimate system description, and many comparisons between this representation and corresponding S-systems have been performed in the context of biochemical systems analysis. These comparisons have elucidated the derivation, development, analytical and computational tractability (Savageau et al., 1987ab), biochemical validity (Sorribas and Savageau, 1989abc), accuracy (Voit and Savageau, 1987; Curto et al., 1998a), feasibility for optimization (Torres et al., 1997), and general mathematical properties (Savageau and Voit, 1987; Voit, 1991; Hernández-Bermejo and Fairén, 1997) of the two representations.

Both representations, S-systems and GMA systems, have advantages in their own right, and depending on the purpose of the analysis, one or the other power-law representation may be preferable. Whether one or the other model is closer to the true metabolic dynamics are an unanswered question, and one can easily imagine that no one model would always be superior. Furthermore, for analyses, numerical or algebraic, that remain in the close vicinity of a normal operating point, the two modeling strategies usually yield rather similar results.

We have seen in the examples developed so far that it is not difficult to develop S-system representations. Nonetheless, it is sometimes easier to identify one power function with each process, and this strategy leads to the GMA form. We will use GMA systems primarily for the first step in the estimation of parameter values for S-system models. The rationale for this procedure is the relative ease with which the GMA model is developed from existing biochemical rate laws, combined with the fact that the corresponding S-system can be computed directly from the GMA model.

A different approach to analyzing biochemical systems, called *metabolic control analysis*, is mathematically based on GMA systems, even though this is not always evident at first glance. The approach was originally proposed by Kacser and Burns (1973) and Heinrich and Rapoport (1974) and subsequently extended by a number of researchers [for recent reviews, see Cornish-Bowden and Cárdenas (1990), Heinrich and Schuster (1996), and Fell (1997), as well as the special volume 182(3) of the *Journal of Theoretical Biology* (1996), which was dedicated to the memory of the late Dr. Kacser]. Detailed comparisons of Metabolic Control Analysis and Biochemical Systems Theory (using S-systems or GMA systems) are readily available in the literature (e.g., Savageau et al., 1987ab; Savageau and Sorribas, 1989; Sorribas and Savageau, 1989a; Savageau, 1991a; Cascante et al., 1995b; Curto et al., 1995; Puigjaner et al., 1995; Sorribas et al., 1995); some crucial aspects are also discussed in Chapter 7.

There is a third power-law alternative, different from both S-systems and GMA systems. One could argue that the difference between V_i^+ and V_i^- is itself a function V_i which could be approximated by a single product of power-law functions. The general system description in this case would simply be

$$\dot{X}_i = V_i(X_1, \ldots, X_{n+m}). \tag{2.38}$$

Again developing the Taylor polynomial, we obtain the rate of change in X_i as

$$\dot{X}_i = \gamma_i \prod_{j=1}^{n+m} X_j^{f_{ij}}, \quad i = 1, 2, \ldots, n. \tag{2.39}$$

Systems consisting of these types of equations are known as *Riccati* systems (e.g., Peschel and Mende, 1986) or *half*-systems (Savageau and Voit, 1987). While mathematically very interesting, they are generally inconvenient for the analysis of biochemical systems.

Stoichiometric Network Models

In some cases, not enough kinetic information is available to estimate all the parameters of a model that consists of traditional rate laws or power-law representations.

MODELS OF BIOCHEMICAL SYSTEMS

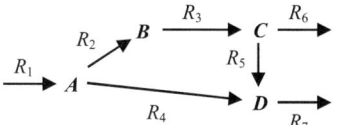

Figure 2.8. Stoichiometric network with metabolites A, B, C, D and reactions R_1 through R_7.

An alternative approach in this situation is focusing just on the topology of the biochemical systems and stating which metabolite may be converted into which other metabolite. The result is a simple "wiring diagram," which can be represented visually by a directed graph that is similar to a simplified map, or mathematically by a balance equation. Such an equation expresses the rate of change in concentrations in terms of a stoichiometric matrix $\mathbf{N}$ and a vector $\vec{V}$ of reaction rates (e.g., Gavalas, 1968; Heinrich et al., 1977). In typical notation, a balance equation reads

$$\frac{d\vec{S}}{dt} = \mathbf{N} \cdot \vec{V}, \tag{2.40}$$

where $\vec{S}$ is a vector containing all metabolite concentrations (for terminology and explanations of matrix equations, see Appendix).

The stoichiometric matrix $\mathbf{N}$ consists of one row for each metabolite involved and one column for each reaction. If a reaction leads toward a metabolite, the corresponding matrix element is $+1$, and if it leads away from a metabolite, the corresponding matrix element is -1. If a metabolite and a reaction are unrelated, the corresponding matrix element is zero. For instance, the graph in Fig. 2.8 leads to the stoichiometric matrix

$$\mathbf{N} = \begin{array}{c} \\ \\ \end{array} \begin{array}{cccccccc} R_1 & R_2 & R_3 & R_4 & R_5 & R_6 & R_7 & \\ \begin{pmatrix} 1 & -1 & 0 & -1 & 0 & 0 & 0 \\ 0 & 1 & -1 & 0 & 0 & 0 & 0 \\ 0 & 0 & 1 & 0 & -1 & -1 & 0 \\ 0 & 0 & 0 & 1 & 1 & 0 & -1 \end{pmatrix} & \begin{array}{c} A \\ B \\ C \\ D \end{array} \end{array}$$

where the metabolites A, B, C, D and the reactions R_1 through R_7 are explicitly identified for convenience.

The representation of the system dynamics in form of a matrix equation is very advantageous, since a lot of theory and a corresponding plethora of methods are available for analysis. For instance, considerations of the size and rank of the matrix give insights into conservation relationships between metabolites and into the steady-state conditions of the system, for which none of the metabolite concentrations are changing (e.g., Gavalas, 1968; Heinrich et al., 1977; Reder, 1988; Heinrich and Schuster, 1996, 1998).

Studying the steady-state conditions further, it is possible – to some degree – to predict all reaction rates within the system when only the input and output fluxes are known (cf. Vallino and Stephanopoulos, 1993; Varma and Palsson, 1995). Going beyond algebraic analysis, some groups of researchers have developed computer algorithms that, based on a metabolic database and principles of artificial intelligence, design possible metabolic pathways leading from a given substrate to a desired product (e.g., Seressiotis and Bailey, 1988; Mavrovouniotis et al., 1990b).

Finally, the linearity of stoichiometric network models permits optimization with standard methods of operations research (e.g., Luenberger, 1984). Fell and Small (1986) used this strategy to analyze lipogenesis in adipose tissue, and Savinell (1991) and Savinell and Palsson (1992a–d) optimized metabolic flux patterns in *Hybridoma* cells for biotechnological purposes. Also using a stoichiometric approach, but analyzing it in a different way, Kacser and Acerenza (1993) proposed a method of genetically manipulating organisms for optimized yield.

Networks as Languages

Faced with the enormous size of genetic and metabolic systems, several groups of researchers have set out to describe the functioning of these systems as *formal languages*. A formal language consists of two types of components (e.g., Maurer, 1969; Collado-Vides, 1989; Hofestädt and Scholz, 1998). First, there are *symbols* – taken from an *alphabet* – that correspond to words in a spoken language. In a *metabolic language*, the symbols represent substrates, products, enzymes, modulators, and other constituents of the system. In a *genetic language*, the words may be promoters, structural genes, or binding sites. Through smart coding, even concentrations of the metabolites can be modeled (cf. Hofestädt and Scholz, 1998). Secondly, there is a *grammar* that specifies how words may be lined up to form proper sentences. As in a spoken language, where a singular subject takes the singular form of the verb, *metabolic rules* or *genetic rules* determine which constituents can be matched with what types of processes. A typical question that can be answered in this type of language is: Is it possible that a given substance appears in a pathway of interest, and if so, how likely is that (Hofestädt, 1993)? A long-term goal of the formal language approach is to find uniform and universal principles governing gene regulation and expression (cf. Collado-Vides et al., 1998).

Formal definitions of genetic and metabolic systems may at first seem overly abstract and even somewhat stilted, but they have great appeal for two reasons. First, Hofestädt (1993) proposed that every metabolic pathway can be formalized by derivation from a genetic grammar, which gives the formal approach the generality needed for real-world applications. Secondly, the formal coding of processes as a rule set is an ideal basis for computer implementation. [The interested reader is encouraged to check out the computerized version of the *lac* operon in *Escherichia coli* (Hofestädt, 1993) and a computer-compatible representation of gene regulation (Collado-Vides et al., 1998; Hofestädt and Scholz, 1998).] Once a problem is coded as a computer algorithm, its size becomes less of an issue, and this is of utmost importance. To make optimal use of formal languages, their computer implementations can be designed in a fashion that they are compatible with some of the vast metabolic databases like KEGG (Goto et al., 1997), which were mentioned in Chapter 1 (cf. Hofestädt and Scholz, 1998).

An alternative to the rule-based formal language approach is the representation of genetic and metabolic interaction networks with methods of *Boolean algebra* (e.g., Mendelson, 1964) that reduce the networks to their principal features (cf. Kauffman, 1993; Somogyi and Sniegoski, 1996; Somogyi and Liang, 1998). The translation into a

Boolean model is based on four crucial steps (cf. Somogyi and Sniegoski, 1996): (i) it is assumed that the state of each gene or element can be reduced to either *on* (1) or *off* (0); (ii) the combinatorial control of gene expression can be reduced to a "wiring diagram"; (iii) the combinatorial nature of control follows Boolean rules; and (iv) all elements update their states synchronously. The resulting Boolean network follows the rules and theorems of *automata theory*, which was first proposed and studied by Turing (1936) and von Neumann (1951). A Boolean network is characterized by the number of its elements, each element's array of inputs, and each element's Boolean function or rule. The state of the network is given by the current values of all elements. A typical set of rules for transitions between a set of elements at subsequent time points $[(A, B, C) \Rightarrow (A', B', C')]$ may read:

$A' = B$,
$B' = A$ or C,
$C' = (A$ and $B)$ or $(B$ and $C)$ or $(A$ and $C)$

(Liang et al., 1998).

In the specific context of genetic networks, Somogyi and Sniegoski (1996) define *Input* as the *cis*-acting elements on the DNA that respond to a *trans* signal, and *Output* as a *trans*-acting element, which is a protein or enzymatic product encoded by a gene. Computations consist of linking appropriate *trans*- and *cis*-acting elements and evaluating the Boolean value of this linkage.

Somogyi and Liang (1998) generalized the approach by allowing the system variables to be any "biological parameters," whose alterations may have an effect on other biological parameters. Typical examples include the level of a particular mRNA, the amount, phosphorylation status, or location of a protein, and the presence of a cofactor or hormone. Allowing for this broader view raised the number of elements from about 100,000 in a pure genetic network to about 1,000,000. The regulatory interactions within the network result from a combination of a directed graph, showing which elements affect which other elements, and a set of logical functions. The dynamics of the network is given as a time sequence of states, which are determined by each element being either *on* or *off*. The dynamics of large networks can be very complex and exhibit different types of oscillations and chaos (e.g., Kauffman, 1993; Somogyi and Liang, 1998). Typical questions to be answered with the Boolean network representation address the temporal expression patterns of large numbers of genes, a systematic assessment of multiple gene expression, and the clustering of genes into related expression patterns (e.g., Wen et al., 1998).

The immense advantage of the Boolean approach is that networks and their evaluations are ideally suited for computer implementation and analysis, even if the networks are very large. It is not only possible to simulate time sequences of states that represent phenomena like cell cycle control (e.g., Somogyi and Liang, 1998), but it is possible – to some degree – to infer the architecture of a regulatory network through computer-aided reverse engineering (Liang et al., 1998).

An approach that is in some sense intermediate between Boolean networks, stoichiometric models, and quantitative network descriptions as we use them throughout this book is the representation of a biochemical system as a *Petri net* (e.g., Peterson, 1981; Reisig, 1985; Kohn and Lemieux, 1991; Reddy et al., 1996). A Petri net consists of two types of nodes, called *places* (or *chemnodes*) and *transitions* (or *relnodes*), and of directed connections, called *arcs*. Arcs connect places to transitions and transitions to places. Metabolites, enzymes, cofactors, and other constituents are represented as places, and individual reactions are represented as transitions. The places contain *tokens*, which represent the amount of metabolite in a given place. Different states of a metabolite (e.g., with respect to phosphorylation) are coded as separate places. Instead of using several parallel arcs between a place and a transition, the arcs may be weighted by non-negative integer numbers.

The dynamics of the Petri net is generated by *enabling* and *firing* transitions. A transition is enabled if the number of tokens in a place is greater than the weight of the connecting arc. An enabled transition can fire, thereby moving tokens from the input to the output place. The dynamics of a Petri net is readily described with matrix equations, which permit large network sizes and offer a rich repertoire of analytical methods. The linearity of the approach also allows zooming in and out: if some part of the biochemical system is not of interest or cannot be characterized quantitatively, it can be collapsed into a single transition; conversely, a transition can be expanded into a Petri (sub)-net. This feature is similar to the telescopic property of S-system models, which we discussed above. Questions that can be approached with this type of approach deal, for instance, with the accumulation of metabolites for biotechnological purposes and the prediction of deadlocks in the pathway structure (cf. Reddy et al., 1996).

EXERCISES

Many of the exercises in this and the following chapters do not have one single answer. They will often read: "Study the system...," "explore what happens if...," or "describe the effects of...." As in the wet lab, uncounted situations and scenarios can be analyzed, and it is often up to the researcher to decide how many experiments are needed to "understand" the phenomenon. Also as in the wet lab, it is strongly recommended to write down in detail which analyses were done, why they were done, and what the results were. It is amazing how fast details of a mathematical analysis are forgotten and how often one has to repeat an analysis because of incomplete records.

To develop a feel for the dynamics of systems, it is recommended always to predict the results before they are computed numerically. In simple systems, the expectations will often come true, but in more complicated systems, predictions are often difficult to make. If the predictions are wrong, try to explain why they were wrong.

1. Equations (2.5) and (2.6) describe the dynamics of an enzyme-catalyzed reaction that converts substrate into product. The system is implemented in PLAS as *MMRL.plc*. Change the numerical values of V_{max} and K_M, and study the effects on the dynamics. Describe the results in words.

2. For the same system *MMRL.plc* as in Exercise 1, change the initial values of S and P. Study the effects and summarize the results.

MODELS OF BIOCHEMICAL SYSTEMS

3. For the same system *MMRL.plc* as in Exercise 1, begin the experiment at initial time $t0 = 10$ and solve until $tf = 30$. Compare with the original results, and explain the differences and similarities.
4. Explore the first-order system in the PLAS program *first.plc*, as specified in Eqs. (2.7) through (2.13). Change the parameter k to different values and study the effects. How long does it take until 90% of X_1 is converted into X_2? Compare solutions with different values of *hr*.
5. (Project) Study the system in Eq. (2.35). Before you solve the system in PLAS, make predictions about the outcomes of the numerical analysis. How do the initial values of X_1 and X_2 affect the results? Change the values of X_3 and X_4. Change rate constants and kinetic orders. Keep a detailed logbook.

COMPLEMENTS

How Are Solutions of Differential Equations Obtained?

Canonical models in S-system or GMA form are sets of differential equations. Hundreds of books have been written about differential equations, and an important issue, of course, is the derivation of their solutions. While mathematical treatises naturally focus on differential equations for which solutions can be obtained with paper and pencil, the fraction of biochemically relevant equations that can be solved in this fashion is small. Nonetheless, it is useful to look at some properties of a simple such case before we discuss numerical solutions, which can be obtained when analytical solutions are unknown or even non-existent.

One of the simplest differential equations is

$$\dot{X}(t) = a X(t). \tag{2C.1}$$

This equation can be translated into words as follows: The instantaneous rate of change of the variable $X(t)$ is equal to the current value of X, multiplied by a constant a. The more X there is, the faster X will change. If the parameter a is positive, then X grows, and the more it grows the faster it grows. If the parameter is negative, then X decreases, and the decrease itself slows down, because the value of X becomes smaller and this value determines the speed of change, as is expressed in Eq. (2C.1).

The only function that satisfies this requirement is the exponential function, and the solution to Eq. (2C.1) is $X(t) = \exp(at + b)$. It is easily checked that this actually is a solution because the derivative of the function $X(t) = \exp(at + b)$ is $dX/dt = \dot{X}(t) = a \exp(at + b)$, which is equal to $a X(t)$, as required.

How can we find such a solution without guessing? In general, this is difficult and mostly even impossible. In our particular case, the solution is found by first dividing both sides by $X(t)$ and then integrating both sides. The result is

$$\int \frac{dX}{dt} \frac{1}{X} dt = \int a \, dt. \tag{2C.2}$$

Integral tables tell us that the integral on the left-hand side equals the logarithmic function plus a constant, whereas the right-hand side corresponds to at plus a constant. Thus, we obtain

$$\ln X + c_1 = at + c_2. \qquad (2C.3)$$

Exponentiating both sides, we obtain

$$\exp(\ln X) = \exp(at + c_2 - c_1), \qquad (2C.4)$$

which is equivalent to the exponential solution given above, with $b = c_2 - c_1$.

One may ask what we know about the constant b in the exponential solution. As it stands, any real number b can be chosen, and still the differential equation is satisfied. In other words, for every b we obtain a different solution in form of an exponential function. The constant b becomes uniquely determined as soon as we specify an *initial value* for $X(t)$, which is the value that $X(t)$ has at the beginning of the experiment. In the earlier car example, this initial value was mile 210 at the beginning ($t_0 = 0$) of our "experiment." Changing this initial value to mile 250, but not altering anything else in the example, the car's location at any time point of the observation period would be shifted by 40 miles, and at the end of the observation period the car would have reached mile 269 instead of mile 229.

Suppose that in the example of our exponential function the initial value is $X(0) = 1$, which is equivalent to saying that "$X(t) = \exp(at + b)$ at $t = 0$ must equal 1." Substitution of 0 for t yields $\exp(b) = 1$, and since we know that the exponential function has the value 1 only at 0, we conclude that b must equal 0. Thus, the only solution to the above problem, under consideration of the initial condition $X(0) = 1$, is $X(t) = \exp(at)$. If instead the initial value is 0.5 at time 10, i.e., $X(10) = 0.5$, then substitution in the equation $X(t) = \exp(at + b)$ yields $0.5 = \exp(10a + b)$ or, equivalently, $b = \ln 0.5 - 10a$. Let's say $a = 0.2$; then b must be $b = \ln 0.5 - 2 = -2.693147$, and we easily confirm that, with these values for a and b, $X(10) = \exp(0.2 \times 10 - 2.693147) = 0.5$, as required.

Given the analytical solution, values of the parameters, and an initial value, the value of X can be computed for any time point t. One simply substitutes t in the exponential function. Compare some values with the numerical solution obtained in PLAS. (Remember that the file *first.plc* deals with this system, except that we used $-k$ instead of a for the rate constant.)

As said before, in a practical situation of biochemical systems analysis, it is a truly exceptional case that one can solve the differential equations with such analytical methods. Therefore, we will not deal with analytical solutions of differential equations any further, but henceforth obtain solutions with numerical methods.

Even though it is very difficult, if not impossible, to compute X as a function of t from a differential equation in an analytical fashion, the theory of differential equations assures us that in biochemically relevant cases such a solution exists and that it is unique, as soon as one selects an initial value $X(t_0)$ at a start time t_0. Barring unusual circumstances, the numerical solution can be obtained with Euler's algorithm or with more modern algorithms that are based on similar principles but are more efficient.

MODELS OF BIOCHEMICAL SYSTEMS

Euler's method, by modern standards, is rather inefficient but demonstrates most clearly how numerical algorithms work. Suppose we analyze a rate law that is described by the differential equation

$$\dot{X}_1 = X_1^2 - X_1^3 \tag{2C.5}$$

and the initial concentration of X_1 at time $t = 0$ is $X_1(0) = 0.1$. From this initial value, we easily compute what the instantaneous change at this point is, since the differential equation tells us that this change is equal to $X_1^2 - X_1^3$. Substituting 0.1 for X_1, we compute the instantaneous rate of change as

$$\dot{X}_1(0) = 0.1^2 - 0.1^3 = 0.009. \tag{2C.6}$$

Thus, X_1 increases slightly, since the rate of change (i.e., its slope) is positive but has a small magnitude of 0.009. Equipped with the original value of X_1 and the slope at time 0, we can project the development of X_1. For instance, we can project a value of X_1 at time 0.005. This projection is given as

$$\text{New } X_1\text{-value} = \text{Old } X_1\text{-value} + \text{slope} \times \text{amount stepped forward.} \tag{2C.7}$$

In our example,

$$X_1(0.005) \approx 0.1 + 0.009 \times 0.005 = 0.100045. \tag{2C.8}$$

Since the differential equation is also valid at $t = 0.005$, we can substitute the freshly computed value of $X_1(0.005)$ into the differential equation and compute the slope at this time point:

$$\dot{X}_1(0.005) = 0.100045^2 - 0.100045^3 = 0.009007651. \tag{2C.9}$$

It is easy to imagine how the numerical solution inches forward in this fashion until the solution of X_1 over the desired period of time is completely computed. That is exactly what Euler's method does. We must note that every time the method steps forward by 0.005, it typically makes a slight error, since the slope changes continuously, but we pretend it remains the same between subsequent time points, such as $t = 0$ and $t = 0.005$. These errors are small if we don't step forward too fast. How far we are allowed to move forward with each step is a very complicated question that has received enormous attention from the mathematical discipline of numerical analysis. Figure 2C.1 shows a generic example in which the solutions steps forward with a step length of 0.01.

More modern algorithms are often based on similar principles, but they allow for bigger steps without compromising accuracy. These bigger steps are achieved, for instance, by approximating the right-hand side of the differential equation with a polynomial. PLAS and its precursor ESSYNS use such a strategy in a very efficient manner (Irvine and Savageau, 1990; Ferreira, 1992).

Why Do the Recipes for Setting Up S-System Equations Work?

The translation of a biochemical map into an S-system representation follows simple rules. The procedure for the ith equation is to include in the α_i-term a rate constant

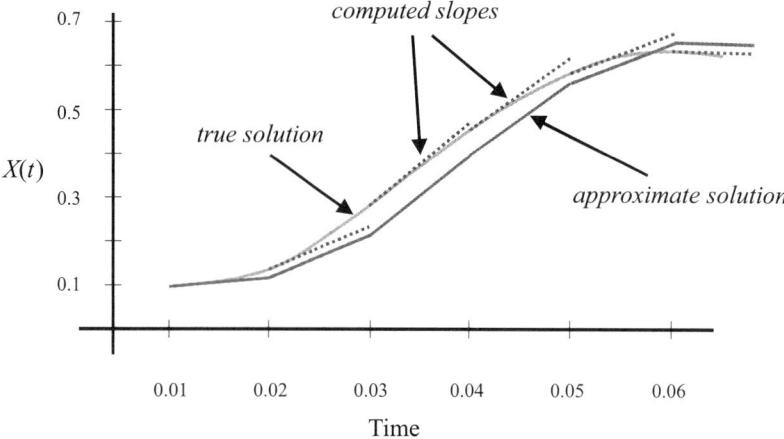

Figure 2C.1. The numerical solution begins at the specified initial point [$t = 0.01$, $X(t) = 0.1$], computes the slope from the differential equation, and steps forward in a linear fashion to the time point $t = 0.02$. The slope at this new time point is again computed from the differential equation and concatenated to the former projection. The result is an approximate solution whose quality depends on the differential equation and the step size.

and all variables that contribute to the production of X_i, raised to a power; the degradation term is constructed analogously. In this section, we explain why this procedure works (see also Savageau, 1969ab; Voit and Savageau, 1982b; Voit, 1991).

Let's look at the dynamics of metabolite X_i in the very general system representation

$$\dot{X}_i = V_i^+(X_1, X_2, \ldots, X_n) - V_i^-(X_1, X_2, \ldots, X_n), \tag{2C.10}$$

where we have omitted independent variables for the sake of simplicity but without losing generality. *A priori*, we have no specific information about the functions V_i^+ and V_i^-. They could be very complicated, and we only require that they have positive values and can be differentiated. We do know from biochemical findings that it is often more appropriate to study *relative* effects in responses to *relative* changes in metabolites rather than to study *absolute* effects. The relative change is dimensionless and automatically allows for the widely diverse concentrations of metabolites, enzymes, and effectors that are typical *in vivo* and even *in vitro*.

The relative rate of change in a variable $X(t)$ can be defined as the absolute rate of change in relation to the present concentration, i.e., as $\dot{X}(t)/X(t)$. This relative rate of change is mathematically equal to the rate of change in the logarithm of $X(t)$. It is immaterial which base we choose for the logarithmic function; often biochemists use logarithms to the base 10, but we will use the natural logarithm to the base e. When we thus define $Y(t) = \ln X(t)$, the relative change in $X(t)$ can be formulated as

$$\begin{aligned}
&\text{relative rate of change in } X(t) \\
&= \frac{\dot{X}(t)}{X(t)} = \frac{d}{dt} \ln \dot{X}(t) = \dot{Y}(t) \\
&= \text{absolute rate of change in } Y(t) = \ln X(t).
\end{aligned} \tag{2C.11}$$

MODELS OF BIOCHEMICAL SYSTEMS

Since the relative rate of change in a quantity is the absolute rate of change in its logarithm, it is reasonable to study the logarithm of V_i^+ as a function of the logarithms of X_1 through X_n and to do the same with V_i^-.

Let's begin with a very simple example in which V_i^+ depends only on one variable, say X_k, and define

$$W = \ln V_i^+ \qquad (2C.12)$$

and

$$Y = \ln X_k. \qquad (2C.13)$$

W is a function of Y, since V_i^+ is a function of X_k; however, in realistic situations we don't know what this function looks like, and there is probably no way to find it out experimentally. A solution to the problem comes from the famous theorem of Taylor, which states that whatever the function $W(Y)$ might be, it can be written as a polynomial, the so-called *Taylor polynomial* of the form

$$W(Y) = b + a_1 Y + a_2 Y^2 + a_3 Y^3 + \cdots \qquad (2C.14)$$

(cf. Appendix). For our purposes, it is convenient to write this polynomial as the sum of a linear function plus of a collection of higher-order terms (h.o.t.), such as $a_2 Y^2, a_3 Y^3$:

$$W(Y) = b + mY + \text{h.o.t.} \qquad (2C.15)$$

Taylor showed that one can ignore the higher-order terms as long as Y does not deviate too much from one chosen value, the so-called *operating point*. This means that (i) exactly at the operating point, the polynomial without h.o.t. is exact, (ii) close to this point, the polynomial without h.o.t. is still very similar to the original function (which we often do not even know), and (iii) nothing really can be said about the polynomial without h.o.t. if Y is far away from the operating point. What defines "close" or "far away" depends on the specific case and cannot easily be specified.

With the caveat of not deviating too far from the operating point, we can thus write

$$W(Y) \approx b + mY. \qquad (2C.16)$$

The same theorem holds when V_i^+ depends on the variables X_1 through X_n. The only difference is that W now is a function of all $Y_i = \ln X_i$:

$$W(Y_1, Y_2, Y_3, \ldots, Y_n) \approx B + m_1 Y_1 + m_2 Y_2 + m_3 Y_3 + \cdots + m_n Y_n, \qquad (2C.17)$$

and, again, this representation of the original function V_i^+ is valid as long as the system does not deviate too much from one point of our choice.

Of course, we are primarily interested in the metabolite concentrations themselves, rather than in their logarithms. This transition is easily accomplished when we exponentiate both sides:

$$\exp[W(Y_1, Y_2, Y_3, \ldots, Y_n)] \approx \exp[B + m_1 Y_1 + m_2 Y_2 + m_3 Y_3 + \cdots + m_n Y_n]. \tag{2C.18}$$

Recalling $W = \ln V_i^+$ and $Y_i = \ln X_i$, we obtain $\exp(W) = V_i^+$ and $\exp(Y_i) = X_i$, and (2C.18) becomes

$$V_i^+(X_1, X_2, \ldots, X_n) \approx \exp(B) X_1^{m_1} X_2^{m_2} X_3^{m_3} \cdots X_n^{m_n} \tag{2C.19}$$

Simply renaming $\alpha_i = \exp(B)$ and $g_{ij} = m_j$, the right-hand side of this equation is the same as the production term of the ith S-system equation, and we obtain

$$V_i^+ \approx \alpha_i \prod_{j=1}^{n} X_j^{g_{ij}}. \tag{2C.20}$$

The same arguments hold for the degradation term V_i^-. If a flux, say V_i^+, does not depend on one of the dependent or independent variables (say it is unaffected by X_8), then W does not depend on Y_8. Mathematically, this means $m_8 = 0$ and consequently $g_{i8} = 0$.

Let's recapitulate what we have done and what the result is. We begin with the very general system representation (2C.10) in which the rate of change in each metabolite is given as the difference of an influx V_i^+ and an efflux V_i^-. These fluxes are unknown smooth functions of some or all variables in the system. No matter what these functions might be, Taylor's theorem allows us to replace each of them with the product of a rate constant and power functions of exactly those variables on which the production and degradation depend. Variables that do not affect production or degradation also do not appear in the corresponding α-term or β-term. For instance, if the original V_i^+ depends on X_2, X_3, X_5, and X_8, then the α_i-term that results from this procedure is

$$V_i^+ = \alpha_i X_2^{g_{i2}} X_3^{g_{i3}} X_5^{g_{i5}} X_8^{g_{i8}}. \tag{2C.21}$$

If instead V_i^- depends on $X_k, X_{k+1}, \ldots, X_{k+p}$, then the corresponding degradation β_i-term of the ith S-system equation reads

$$V_i^- = \beta_i \prod_{j=k}^{k+p} X_j^{h_{ij}}. \tag{2C.22}$$

Because the result of this procedure always is a product of power functions of the contributing variables only, the translation of a biochemical map into this type of system description is not only possible but very simple.

Further Justification of the S-System Description

S-system models indeed are amazing constructs for the representation of biochemical systems. If one assumes strict allometry between constituents of a biochemical system,

MODELS OF BIOCHEMICAL SYSTEMS

then the S-system form follows almost necessarily from first principles of systems analysis (cf. Voit, 1992d). Consider again the general system description

$$\dot{X}_i = V_i^+(X_1, X_2, \ldots, X_n) - V_i^-(X_1, X_2, \ldots, X_n), \tag{2C.23}$$

in which the functions $V_i^+(X_1, X_2, \ldots, X_n)$ and $V_i^-(X_1, X_2, \ldots, X_n)$ are not yet specified. The law of allometry asserts that the relative change in one component is linearly related to relative changes in other components. In our terminology, the law of allometry can thus be formulated as

$$\frac{\dot{X}_i}{X_i} = a_{ij} \frac{\dot{X}_j}{X_j}, \tag{2C.24}$$

where a_{ij} is some number. This relationship is a linear function of the type $f(z) = mz + b$, with $f = (\dot{X}_i/X_i)$, $z = (\dot{X}_j/X_j)$, $m = a_{ij}$, and $b = 0$. The constant term b is necessarily equal to zero, because if everything else in the system remains unchanged and there is no change in X_j, then there should be no change in X_i. In other words, if $\dot{X}_j$ equals zero, then $\dot{X}_i$ should be zero, and that is only the case if Eq. (2C.24) does not include a constant term.

Recalling that the relative rate of change in a variable is equal to the absolute rate of change in its logarithm [cf. Eq. (2C.11)], we can reformulate Eq. (3C.2) directly as

$$\dot{Y}_i = a_{ij} \dot{Y}_j, \tag{2C.25}$$

where we again define $Y_i = \ln X_i$ and $Y_j = \ln X_j$. Applying standard mathematics of integration, one can see that Eq. (2C.25) implies the linear relationship

$$Y_i = a_{ij} Y_j + b_{ij} \tag{2C.26}$$

between Y_i and Y_j. This is most easily checked if we differentiate (2C.26) with respect to t, which directly returns (2C.25). The linear relationship between the logarithms of variables is the cornerstone of allometry.

Now consider a system with many variables and perfect allometry, which requires that the relative changes between any two variables be linearly related. In this case, Eqs. (2C.24), (2C.25) are satisfied for each pair of variables Y_i and Y_j, and if we sum the relationships between the ith variable and all other variables, the result that corresponds to Eq. (2C.25) is

$$nY_i = \sum_{j=1}^{n} (a_{ij} Y_j + b_{ij}). \tag{2C.27}$$

For simplicity of notation, this equation includes the trivial allometry $Y_i = a_i Y_i$ with $a_{ii} = 1$. Now subtract nY_i from both sides and define $\tilde{a}_{ii} = a_{ii} - n = 1 - n$ for all i and $\tilde{a}_{ij} = a_{ij}$ for all i and j that are different from each other. Furthermore, define b_i as the sum of all constant terms b_{ij}. The result is the linear equation

$$0 = \sum_{j=1}^{n} \tilde{a}_{ij} Y_j + b_i. \tag{2C.28}$$

Now recall that $Y_j = \ln X_j$, and exponentiate both sides. The result is

$$1 = \exp(b_i) \cdot \prod_{j=1}^{n} X_j^{\tilde{a}_{ij}}, \qquad (2C.29)$$

or

$$0 = \exp(b_i) \cdot \prod_{j=1}^{n} X_j^{\tilde{a}_{ij}} - 1. \qquad (2C.30)$$

Equation (2C.30) remains satisfied when we multiply both sides by a number or even a function Φ_i that depends on X_1 through X_n:

$$0 = \Phi_i(X_1, \ldots, X_n) \exp(b_i) \cdot \prod_{j=1}^{n} X_j^{\tilde{a}_{ij}} - \Phi_i(X_1, \ldots, X_n). \qquad (2C.31)$$

Let's recall where Eq. (2C.31) comes from: It is a constraint that the system variables X_1 through X_n have to satisfy under the assumption of complete allometry.

Now consider a biochemical system in which all production and degradation processes are in balance. There is convincing experimental evidence that, close to such a *steady state*, relative changes in metabolite concentrations are linearly related. Restricting our theoretical considerations to systems at steady state, Eq. (2C.31) can be interpreted as the ith steady-state equation of a general biochemical model in which

$$V_i^+ = \Phi_i(X_1, \ldots, X_n) \exp(b_i) \cdot \prod_{j=1}^{n} X_j^{\tilde{a}_{ij}} \qquad (2C.32)$$

and

$$V_i^- = \Phi_i(X_1, \ldots, X_n). \qquad (2C.33)$$

The dynamical system that possesses this steady state is

$$\dot{X}_i = \Phi_i(X_1, \ldots, X_n) \exp(b_i) \cdot \prod_{j=1}^{n} X_j^{\tilde{a}_{ij}} - \Phi_i(X_1, \ldots, X_n), \qquad i = 1, 2, \ldots, n \qquad (2C.34)$$

In comparison with the general system equation (2C.23), the system in (2C.34) is forced to respond to changes at steady state in an allometric fashion. We have not yet specified what mathematical form the functions V_i^+, V_i^-, and Φ_i should take, but we know that they must have the same structure, because they describe biochemical processes of the same type. For instance, if we look at a linear pathway of the type $\cdots \to X_1 \to X_2 \to X_3 \to \cdots$, then the degradation of X_1, as represented by V_1^-, is the same as the synthesis of X_2, which is represented by V_2^+. Thus, V_i^+ and V_i^- may differ in their variables and parameters, but their mathematical structure must be the same.

For the two terms on the right-hand side of (2C.34) to have the same structure, $\Phi_i(X_1, \ldots, X_n)$ must have a form comparable to the product of power-law terms that appears in the first term of the right-hand sides of (2C.32) and (2C.34). Thus, we define

$$\Phi_i(X_1, \ldots, X_n) = \beta_i \prod_{j=1}^{n} X_j^{h_{ij}}, \tag{2C.35}$$

where β_i is a positive number and the exponents h_{ij} are real numbers. Substituting Eq. (2C.35) into Eq. (2C.34), we obtain

$$\dot{X}_i = \beta_i \prod_{j=1}^{n} X_j^{h_{ij}} \exp(b_i) \cdot \prod_{j=1}^{n} X_j^{\tilde{a}_{ij}} - \beta_i \prod_{j=1}^{n} X_j^{h_{ij}}, \tag{2C.36}$$

which simplifies to

$$\dot{X}_i = \alpha_i \prod_{j=1}^{n} X_j^{g_{ij}} - \beta_i \prod_{j=1}^{n} X_j^{h_{ij}} \tag{2C.37}$$

when we rename $\alpha_i = \beta_i \exp(b_i)$ and $g_{ij} = \tilde{a}_{ij} + h_{ij}$. This is in fact a system equation in S-system form. This result is quite amazing, because it is based merely on two conditions: that the general system description (2C.23) is valid, and that the system, close to steady state, exhibits strict allometry.

Relative Changes in Flux

Many kinetic experiments suggest that, close to a steady state, relative changes in a metabolite concentration provoke proportional relative changes in flux. S-system representations of biochemical systems respond in the same way. To see this, we have to identify fluxes in the system equation. For metabolite X_i, the incoming flux is given as the α-term of the ith equation, because this term by definition encompasses all processes that produce, synthesize, or augment X_i. Similarly, the β_i-term represents the outgoing flux. If incoming and outgoing fluxes are equivalent, the net change in the concentration of the ith metabolite is zero, i.e., the metabolite is in a steady state. At the steady state, where most flux experiments are performed, we can thus use the ith α-term or the ith β-term to characterize the flux through X_i.

The flux F_i through metabolite X_i at steady state is thus

$$F_i = \alpha_i \prod_{j=1}^{n} X_j^{g_{ij}} = \beta_i \prod_{j=1}^{n} X_j^{h_{ij}} \tag{2C.38}$$

where all X_j are to be interpreted as metabolite concentrations at steady state.

The relative rate of change of a metabolite is given again as $\dot{X}_j / X_j$ and the relative rate of change of a flux is given analogously as $\dot{F}_i / F_i$. To express mathematically how a relative change in X_j leads to a relative change in F_i, we define again $Y_j = \ln X_j$

and $G_i = \ln F_i$. Application of the chain rule yields

$$\frac{dG_i}{dY_j} = \frac{dG_i}{dt} \bigg/ \frac{dY_j}{dt}$$

$$= \frac{\dot{F}_i}{F_i} \bigg/ \frac{\dot{X}_j}{X_j}$$

$$= \frac{dF_i/dt}{F_i} \cdot \frac{dX_j/dt}{X_j}$$

$$= \frac{dF_i}{dX_j} \frac{X_j}{F_i}$$

$$= g_{ij} F_i X_j^{-1} \frac{X_j}{F_i}$$

$$= g_{ij}. \tag{2C.39}$$

This equation says that relative changes in metabolites and in fluxes are linearly related, as observed. The result can be paraphrased as follows: A one-percent change in X_j evokes a g_{ij}-percent change in the flux through X_i at steady state. An immediate consequence of this result is that all relative changes in fluxes can be read off the S-system equations without any further analysis: They are given as the kinetic orders g_{ij} and h_{ij}. In later chapters we shall discuss the implications of this important finding in greater detail.

One should note that the linear relationship between relative changes in metabolites and fluxes is not necessarily realized by other nonlinear models. For example, suppose that X_i is produced from X_j via an enzyme-catalyzed reaction, described by a Michaelis–Menten rate law:

$$\dot{X}_i = v = \frac{V_{max} X_j}{K_m + X_j}. \tag{2C.40}$$

If the substrate X_j is allometrically replaced with a power-law function $X_j = \alpha R^g$, the resulting rate law is

$$v = \frac{\tilde{V}_{max} R^g}{K_m + R^g}, \tag{2C.41}$$

which has the form of a Hill equation (Hill, 1910). This form is qualitatively different from a Michaelis–Menten rate law in that it is S-shaped and has a zero slope at zero, whereas the Michaelis–Menten law is simply saturated with a slope of 1 at 0.

With computations similar to those above, one computes the relative change in a flux in response to a change in a metabolite as

$$\frac{\text{relative rate of change in } F_i}{\text{relative rate of change in } X_j} = \frac{V_{max}(K_m + X_j) - V_{max} X_j}{(K_m + X_j)^2} \frac{X_j(K_m + X_j)}{V_{max} X_j}$$

$$= \frac{V_{max} K_m}{V_{max}(K_M + X_j)}$$

$$= \frac{K_m}{K_m + X_j}. \tag{2C.42}$$

MODELS OF BIOCHEMICAL SYSTEMS

The computation shows that in this type of Michaelis–Menten model the relative change in a flux is not constant but a hyperbolic function of X_j.

EXERCISES

C.1 Consider the differential equation $\dot{x} = -x^2$, $x(0) = 1$.

    1.1. Use the Euler algorithm to solve the equation numerically. Use as step sizes 0.1 and 0.5.

    1.2. Use PLAS to solve the equation.

    1.3. Show that $x(t) = 1/(t+1)$ is the analytical solution of this differential equation.

    1.4. Compare your numerical solutions with this analytical solution.

    1.5. What happens in part 1.1 if one replaces the initial condition with $x(0) = 0$? From your result, deduce (guess) the analytical solution in this case.

C.2 Use the Euler algorithm to compute the substrate degradation over time which is given as the solution of the Michaelis–Menten equation

$$\dot{S} = -\frac{V_{max} S}{K_M + S},$$

and determine the substrate concentration as a function of time. Begin at $t = 0$, and advance by 0.1 with each step, thus computing the solution at time points $0.1, 0.2, \ldots, 1$.

REFERENCES

[1–2], [20], [42], [44], [49], [53], [55–6], [60], [66], [68], [87–9], [104–9], [111–13], [122], [139], [144–7], [149], [154–7], [160–63], [167], [177–8], [189], [201], [204], [219], [224], [226], [234], [237], [241], [244], [253], [269], [279–80], [282–3], [285–6], [289], [294], [310–15], [319], [321–22], [329], [333], [342–43], [350], [352–56], [366], [370], [374], [387], [391], [394–96], [410–11], [423], [427–28], [432], [439], [443], [463], [466], [472], [475], [478], [484], [489], [496], [518], [524].

CHAPTER THREE

From Maps to Equations

One of the outstanding features of the S-system approach is the ease with which models are developed from a metabolic map. In traditional models of biochemical systems, all mechanisms need to be known in detail, before one can begin to set up equations. For instance, in the framework of Michaelis–Menten reactions, one must find out not only which metabolite or effector inhibits a particular reaction, but also what type of inhibition is present – whether it is competitive, uncompetitive, allosteric, or some mixture of these idealized types of inhibition. Depending on the alleged mechanism, the system equations take different forms, and it is not easy to explore how significant uncertainties or misclassifications in the choice of mechanisms are. In contrast, S-system and GMA models are constructed directly and exclusively from the biochemical map; mechanistic aspects may affect the parameter values, but do not influence the structure of the equations. As a consequence, uncertainties or the lack of knowledge about particular details result in uncertainties only in the numerical, not the structural, properties of the equations.

The previous chapter already mentioned the relationships between maps and equations, and by exploring the roles of rate constants and kinetic orders we gained some intuitive familiarity with S-system and GMA equations. This chapter provides general and specific rules for the translation of a map into GMA and S-systems. The construction of the model equations begins with a listing of the dependent and independent variables and the map of the actual pathway. With the exception of a few special scenarios, which are discussed later, the system equations are set up with two simple recipes.

1. Enumerate dependent and independent variables. If it has not been done during the development of the map for the metabolic pathway (cf. Chapter 1), it is first necessary to decide which of the variables involved are dependent and which independent. Only the dependent variables will be represented by system equations; the independent variables will be assigned constant values that may change from one experiment to the next but not during one experiment. In typical cases, intermediate metabolites and products are dependent variables, whereas externally supplied substrates and enzymes are independent variables. Of course, we have to be careful with this simplistic classification: for instance, enzymes themselves are the products of metabolic pathways and in that connection are to be considered dependent variables.

Suppose there are n dependent and m independent variables, and they are numbered in such a way that the first n indices refer to dependent variables and the remaining m indices to independent variables. The S-system then will consist of n differential equations, for $\dot{X}_1$ through $\dot{X}_n$, which contain the variables X_1 through X_{n+m}.

2. Write equations for all dependent variables. There is one differential equation for each dependent variable X_i. In the case of an S-system, the form of all these equations is exactly the same: $\dot{X}_i$ is equal to the difference between two terms, one for production or accumulation (the so-called α-*term*) and one for degradation or clearance (the so-called β-*term*). Each term is the product of a rate constant, α_i or β_i, which is positive or zero, and all dependent and independent variables that *directly* affect production or degradation, respectively. Each variable X_j is raised to an individual kinetic order, g_{ij} or h_{ij}. The index of the rate constant refers to the dependent variable X_i whose change is being described – in other words, to the number of the equation, i. The first index of the kinetic orders again is i, and the second index refers to the dependent or independent variable X_j that affects the production or degradation of X_i.

Thus, the α-term and β-term of an S-system equation always read

$$\alpha_i X_1^{g_{i1}} X_2^{g_{i2}} \cdots X_n^{g_{in}} X_{n+1}^{g_{i,n+1}} \cdots X_{n+m}^{g_{i,n+m}} \tag{3.1}$$

and

$$\beta_i X_1^{h_{i1}} X_2^{h_{i2}} \cdots X_n^{h_{in}} X_{n+1}^{h_{i,n+1}} \cdots X_{n+m}^{h_{i,n+m}}, \tag{3.2}$$

and using the notation for a product, as described in Chapter 2, each S-system equation (for $i =, 1, 2, \ldots, n$) has the form

$$\dot{X}_i = \alpha_i \prod_{j=1}^{n+m} X_j^{g_{ij}} - \beta_i \prod_{j=1}^{n+m} X_j^{h_{ij}}. \tag{3.3}$$

In the case of a GMA system, each equation may contain one, two, or more than two terms. For instance, at a branch point where two pathways converge, two terms in the form of products of power-law functions describe the production of the metabolite. Similarly, at a branch point where three pathways diverge, three products of power-law functions describe the degradation of the metabolite. No matter how many terms are included in a GMA equation, the rules for constructing the power-law terms are always the same. Each term contains a rate constant γ_{ik}, which may be positive or negative. The index i refers to the equation in question, and the index k enumerates the multiple processes of production or degradation. Each term also contains all dependent and independent variables that *directly* affect the given process. Each variable X_j is raised to an individual kinetic order f_{ijk}. The first index refers to the equation in question, the second index to the dependent or independent variable that exerts the direct effect, and the third index to the particular process. Each GMA equation (for $i = 1, 2, \ldots, n$) thus reads

$$\dot{X}_i = \gamma_{i1} \prod_{j=1}^{n+m} X_j^{f_{ij1}} + \gamma_{i2} \prod_{j=1}^{n+m} X_j^{f_{ij2}} + \cdots + \gamma_{ik} \prod_{j=1}^{n+m} X_j^{f_{ijk}} + \cdots. \quad (3.4)$$

The following examples illustrate the procedure of constructing S-systems and GMA systems.

Examples

1. S-system representation of a linear pathway. (See Exercise 5 for the GMA representation.) Consider the simple linear pathway of Fig. 3.1 that is fed by a constant input. There are three dependent variables, representing the dependent metabolites, and one independent variable, representing the input. For each metabolite, only the precursor directly affects its synthesis, while its degradation depends on its current concentration but is not directly affected by any other constituent of the system. For instance, the production of C is a (direct) function only of B. Of course, changes in input eventually filter through to C, but these effects are indirect and will be taken into account by the dynamics of the model. Only direct influences are to be included in the model equations.

We code $X_1 = A$, $X_2 = B$, $X_3 = C$, $X_4 =$ input, and thus specify three terms V_i^+ and three terms V_i^- for the three dependent variables. Each term V_i^+ is associated with a rate constant α_i, and each term V_i^- is associated with a rate constant β_i.

Next, we list for each term the dependent and independent variables that have a direct effect on it (Table 3.1). Each of these variables is associated with one and only one kinetic order. For α-terms, these kinetic orders are called g, and for β-terms, they are called h. Each g and h has two indices, one identifying the equation and the other one identifying the effector variable, whether it is dependent or independent. Note that these pairs of indices can be read off directly from columns 1 and 3 of Table 3.1. The resulting system equations are

$$\dot{X}_1 = V_1^+ - V_1^- = \alpha_1 X_4^{g_{14}} - \beta_1 X_1^{h_{11}},$$
$$\dot{X}_2 = V_2^+ - V_2^- = \alpha_2 X_1^{g_{21}} - \beta_2 X_2^{h_{22}}, \quad (3.5)$$
$$\dot{X}_3 = V_3^+ - V_3^- = \alpha_3 X_2^{g_{32}} - \beta_3 X_3^{h_{33}},$$
$$X_4 = \text{constant}.$$

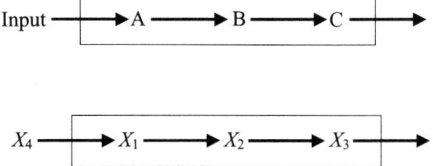

Figure 3.1. A simple linear pathway. Top panel: Map with *ad hoc* symbols, which represent some metabolites. Bottom panel: Indexed notation in which the dependent variables are listed first and the independent variable last.

FROM MAPS TO EQUATIONS

Table 3.1. Rate Constants and Variables in Fluxes of the Pathway in Fig. 3.1

Flux	Rate Constant	Variables to Be Included	Kinetic Orders
V_1^+	α_1	X_4	g_{14}
V_1^-	β_1	X_1	h_{11}
V_2^+	α_2	X_1	g_{21}
V_2^-	β_2	X_2	h_{22}
V_3^+	α_3	X_2	g_{32}
V_3^-	β_3	X_3	h_{33}

Note that even though we may not know what the values of the parameters are, we can still formulate the system equations in an unambiguous fashion.

Precursor–product constraints. Studying the model equations in comparison with the map of the pathway, we observe that the degradation of X_1 constitutes the same process as the synthesis of X_2, whereas the corresponding terms V_1^- and V_2^+ have a different appearance. The same is true for the conversion of X_2 into X_3. Formally, our mathematical representation is still correct, but each *precursor–product relationship*, i.e., equivalence of two processes like V_1^- and V_2^+, imposes *constraints* on the choice of values for the involved parameters. Specifically, we must constrain the parameter values of the system to satisfy the conditions $\alpha_2 = \beta_1$, $g_{21} = h_{11}$, $\alpha_3 = \beta_2$, and $g_{32} = h_{22}$, because that is the only general way to assure that the two pairs (V_1^-, V_2^+) and (V_2^-, V_3^+) are equivalent. Violating these constraints would mean that the amount of material leaving the pool X_1 or X_2 was different from the amount of material reaching the pools X_2 and X_3, respectively. The precursor–product constraints ensure *conservation of flux* and provide an important piece of information that needs to be included in the system representation, since it reduces the number of parameters that we need to estimate for a numerical implementation of the model. We will return to this issue many times – for instance, when we estimate parameters in Chapter 5.

To obtain a feel for this relatively simple model of a linear pathway, it is useful to implement it numerically in PLAS. The file *Linear1.plc* contains the following system, which is a model for the linear pathway in Fig. 3.1:

$$\dot{X}_1 = 0.5 X_4 - 0.5 X_1^{0.5}, \quad X_1(0) = 1,$$
$$\dot{X}_2 = 0.5 X_1^{0.5} - 4 X_2, \quad X_2(0) = 1, \quad (3.6)$$
$$\dot{X}_3 = 4 X_2 - 2 X_3^{0.75}, \quad X_3(0) = 1,$$
$$X_4 = 0.5.$$

Note that in addition to the rate constants and kinetic orders, we need to specify initial conditions of the dependent and independent variables, i.e., the concentrations of X_1, X_2, X_3, and X_4 at the beginning of the experiment. We also have to set the initial

Figure 3.2. Dynamical solution to system (3.2), describing a linear pathway. X_1 decreases comparatively slowly, X_2 decreases faster, and X_3 overshoots before approaching the same level as X_2.

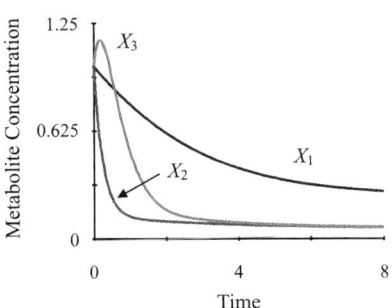

time, here $t0 = 0$, a final time, here $tf = 8$, and the report interval, here $hr = 0.05$, which specifies at which time points PLAS will actually display the solution.

Without knowing much about the system, we start with the initial conditions 1, 1, and 1 for X_1, X_2, and X_3. X_4 is an independent input, whose value we supposedly know; for the given example, we set it at 0.5. There is no particular reason to choose these values, and you should try other possibilities (see Exercises 7 and 8).

A plot of the solution is shown in Fig. 3.2. X_1 and X_2 decrease, at different speeds, while X_3 initially overshoots and then decreases. Apparently the solution soon approaches a situation where none of the variables changes anymore: The system reaches a *steady state*. Steady states are of great importance, and we will discuss them throughout the book and, in particular, in Chapter 6. Many natural systems are at a steady state or close to it.

It is instructive to study how the system responds to changes in input. For this purpose, initialize again at $X_1 = X_2 = X_3 = 1$, but set X_4 at a different value, and investigate the dynamic responses. For instance, if $X_4 = 1.1$, then X_1 increases, and X_2 and X_3 approach different values (see Exercise 6).

2. Linear pathway with end product inhibition. The second example represents a pathway in which the end product inhibits the synthesis of the first intermediate of the pathway (Fig. 3.3). The only difference from the previous case is the direct effect of X_3 on V_1^+, which requires inclusion of X_3 in the first power-law term of the first equation. Thus, the system reads

$$\begin{aligned}
\dot{X}_1 &= V_1^+ - V_1^- = \alpha_1 X_3^{g_{13}} X_4^{g_{14}} - \beta_1 X_1^{h_{11}}, \\
\dot{X}_2 &= V_2^+ - V_2^- = \alpha_2 X_1^{g_{21}} - \beta_2 X_2^{h_{22}}, \\
\dot{X}_3 &= V_3^+ - V_3^- = \alpha_3 X_2^{g_{32}} - \beta_3 X_3^{h_{33}}, \\
X_4 &= \text{constant}.
\end{aligned} \qquad (3.7)$$

We do not know what the precise value of g_{13} may be, but since it represents an inhibitory effect, its value is negative (see Chapters 2, 4, and 5). The inclusion of

Figure 3.3. Linear pathway with end product inhibition.

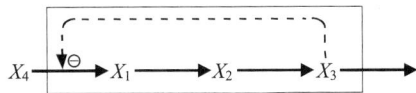

FROM MAPS TO EQUATIONS

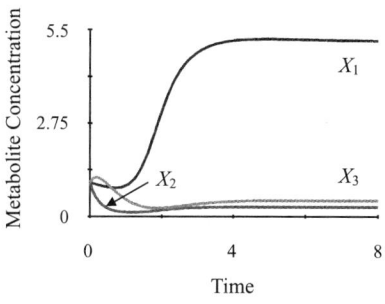

Figure 3.4. Dynamical to solution system (3.6), describing a linear pathway with feedback. A comparison with Fig. 3.2 demonstrates the strong effect of the inhibition by X_3.

feedback inhibition by the end product illustrates how easy it is to amend an S-system model. While many traditional approaches would require precise knowledge of the mechanism of production of the first intermediate in the pathway, we merely had to add a term in a uniquely defined way. It is noted, however, that the inclusion of an additional modulation in a real-world application may require a change in the value of corresponding rate constant, here α_1 (see section *Controlled Comparisons* in Chapter 4).

The inclusion of end product inhibition is readily implemented in PLAS. According to the first system equation, we simply add the term X3^g13 and a definition for g_{13}. As a first experiment, set $g_{13} = 0$ (which corresponds to no inhibition) and run the system under the same conditions as before. The result should be the same, since the term $X_3^{g_{13}}$ with $g_{13} = 0$ has no effect on V_1^+. Now change g_{13} to various negative values, such as -0.5, -1, and -2, and study the effects of increased inhibition. For simplicity, we leave the rate constant unchanged. The system for $g_{13} = -2$ thus reads

$$\begin{aligned}
\dot{X}_1 &= 0.5 X_3^{-2} X_4 - 0.5 X_1^{0.5}, & X_1(0) &= 1, \\
\dot{X}_2 &= 0.5 X_1^{0.5} - 4 X_2, & X_2(0) &= 1, \\
\dot{X}_3 &= 4 X_2 - 2 X_3^{0.75}, & X_3(0) &= 1, \\
X_4 &= 0.5.
\end{aligned} \quad (3.8)$$

It is implemented in PLAS as *Linear2.plc*. As Fig. 3.4 demonstrates, the effect of this inhibition is dramatic (see Exercise 9). It would have been difficult to predict this result intuitively, that is, without a mathematical model. Thorough analyses of this type are executed in Chapter 4.

3. Pathways with branch points. Figure 3.5 illustrates a branched pathway that is controlled by four inhibitory mechanisms, two that affect the synthesis of the

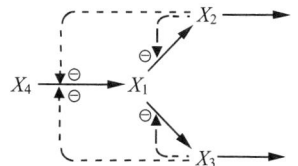

Figure 3.5. Branched pathway with several feedback inhibitions.

Table 3.2. Rate Constants and Variables in Fluxes of the Pathway in Fig. 3.3

Flux	Rate Constant	Variables to Be Included	Kinetic Orders
V_1^+	α_1	X_2, X_3, X_4	g_{12}, g_{13}, g_{14}
V_1^-	β_1	X_1, X_2, X_3	h_{11}, h_{12}, h_{13}
V_2^+	α_2	X_1, X_2	g_{21}, g_{22}
V_2^-	β_2	X_2	h_{22}
V_3^+	α_3	X_1, X_3	g_{31}, g_{33}
V_3^-	β_3	X_3	h_{33}

first intermediate and two that allow the organism to channel more or less material into either one of the two branches. The system contains one independent and four dependent variables, and the fluxes contain variables representing substrates and inhibitors, as detailed in Table 3.2.

The S-system equations are readily constructed from Table 3.2 by inclusion of single-indexed rate constants and appropriate double-indexed kinetic orders. The resulting system is

$$\dot{X}_1 = \alpha_1 X_2^{g_{12}} X_3^{g_{13}} X_4^{g_{14}} - \beta_1 X_1^{h_{11}} X_2^{h_{12}} X_3^{h_{13}},$$
$$\dot{X}_2 = \alpha_2 X_1^{g_{21}} X_2^{g_{2}} - \beta_2 X_2^{h_{22}}, \qquad (3.9)$$
$$\dot{X}_3 = \alpha_3 X_1^{g_{31}} X_3^{g_{33}} - \beta_3 X_3^{h_{33}},$$
$$X_4 = \text{constant}.$$

Even though this model description is entirely symbolic, we have some knowledge about its parameter values. Of course, all rate constants are positive, but we also know that kinetic orders representing inhibitions are negative. These kinetic orders are $g_{12}, g_{13}, h_{12}, h_{13}, g_{22}$, and g_{33}. All other kinetic orders signify the flow of material and are positive.

Branch point constraints. While this model set-up is formally correct, we must be aware that the numerical values of some of the parameters are constrained by conservation of flux between X_1 and X_2 at the branch point. This constraint is similar to the precursor–product constraint discussed in the previous example, but it is mathematically a little bit more complicated. We develop it in two steps. First, we formulate the corresponding GMA model, and subsequently, we derive the constraints.

The GMA model differs from the S-system model only at the branch point – specifically, in the term V_1^-. Whereas the S-system model aggregates the two fluxes toward X_2 and X_3 in V_1^-, the GMA model keeps them separate. In mathematical terms, the first equation of the GMA model reads

$$\dot{X}_1 = \gamma_{11} X_2^{f_{121}} X_3^{f_{131}} X_4^{f_{141}} - \gamma_{12} X_1^{f_{112}} X_2^{f_{122}} - \gamma_{13} X_1^{f_{113}} X_3^{f_{133}}. \qquad (3.10)$$

It is easy to see that the first term is the same as in the S-system model. Thus,

$$\dot{X}_1 = \alpha_1 X_2^{g_{12}} X_3^{g_{13}} X_4^{g_{14}} - \gamma_{12} X_1^{f_{112}} X_2^{f_{122}} - \gamma_{13} X_1^{f_{113}} X_3^{f_{133}}. \tag{3.11}$$

The other equations are exactly the same as in the S-system model, except possibly for the names of the parameters.

One advantage of the GMA representation is that the branch point constraints are nothing but precursor–product constraints. These are

$$\begin{aligned} \gamma_{12} X_1^{f_{112}} X_2^{f_{122}} &= \alpha_2 X_1^{g_{21}} X_2^{g_{22}}, \\ \gamma_{13} X_1^{f_{113}} X_3^{f_{133}} &= \alpha_3 X_1^{g_{31}} X_2^{g_{33}}, \end{aligned} \tag{3.12}$$

and require $\gamma_{12} = \alpha_2$, $\gamma_{13} = \alpha_3$, $f_{112} = g_{21}$, $f_{122} = g_{22}$, $f_{113} = g_{31}$, $f_{133} = g_{33}$. Thus, all information about the aggregated term is already in the system.

In the S-system model, the flux out of X_1 must equal the flux into X_2 plus the flux into X_3. The constraint is therefore

$$\beta_1 X_1^{h_{11}} X_2^{h_{12}} X_3^{h_{13}} = \alpha_2 X_1^{g_{21}} X_2^{g_{22}} + \alpha_3 X_1^{g_{31}} X_3^{g_{33}}. \tag{3.13}$$

The functions on the left-hand side and on the right-hand side of the equation have different structures, and strict equality can be assured only at one point, such as the steady state. If we deviate from the steady-state point, we do not strictly preserve the stoichiometry of fluxes. However, several analyses have shown that the introduced error is very small if the deviation from the steady state is not too large (Voit and Savageau, 1987; Sorribas and Savageau, 1989a–c; Curto et al., 1998a). In fact, comparisons with Michaelis–Menten rate laws have demonstrated that the S-system represents changes in pools more accurately than the GMA representation (Voit and Savageau, 1987). This is due to the cancellation of slight inaccuracies that stem from the approximate character of the power-law representation in the GMA and S-system models on one hand and the slight deviation from stoichiometry on the other. At the operating point, typically the steady state, both models, GMA and S, are the same and they are exact. Close to this point, they differ very slightly, and only if the deviations are substantial are the differences significant. In such cases, however, one must wonder whether the power-law representation is still valid.

Whichever model may be more accurate, it is clear that some analytical evaluations are only possible in the S-system form. Such analyses include the algebraic determination of steady states (e.g. Savageau, 1969b) and biotechnological optimization (e.g., Voit, 1992a; Torres et al., 1996, 1997). It is furthermore noted that all information given in the GMA model is also explicitly present in the S-system model (Savageau et al., 1987ab). The issue of flux aggregation has been the subject of heated discussions in the literature since the mid-1980s, because it constitutes one of the few real differences between S-system modeling and GMA modeling, the latter of which is closer to the alternative approach of *metabolic control analysis* (e.g., Savageau et al., 1987ab; Cornish-Bowden and Cárdenas, 1990; Fell, 1997).

The branch point constraint (3.13) translates into the following constraints on the S-system parameters:

$$h_{11} = \frac{g_{21} V_2^+ + g_{31} V_3^+}{V_2^+ + V_3^+},$$

$$h_{12} = \frac{g_{22} V_2^+}{V_2^+ + V_3^+},$$

$$h_{13} = \frac{g_{33} V_3^+}{V_2^+ + V_3^+},$$

$$\beta_1 = \frac{V_2^+ + V_3^+}{X_1^{h_{11}} X_2^{h_{12}} X_3^{h_{13}}},$$

(3.14)

In these constraints, the fluxes and variables are taken at the operating point, which is usually the steady state. The derivation of these constraints is given in a later section of this chapter.

Note that the kinetic orders in the term V_1^- are averaged kinetic orders of the terms V_2^+ and V_3^+ that are weighted with the magnitudes of the fluxes in steady state. As a consequence, a kinetic order like h_{11} can never be smaller than both kinetic orders g_{21} and g_{31}, and it can never be larger than both. If the two fluxes V_2^+ and V_3^+ are equal at steady state, the constraints on the kinetic orders simplify to

$$h_{11} = \frac{g_{21} + g_{31}}{2},$$

$$h_{12} = \frac{g_{22}}{2},$$

$$h_{13} = \frac{g_{33}}{2}.$$

(3.15)

At the other extreme, if one of the fluxes is zero, the constraints reduce to precursor–product relationships.

As an illustration of the differences and similarities between GMA and S-systems, the branched pathway of Fig. 3.5 is implemented in PLAS as a GMA system (file *BranchG.plc*) in the following numerical form:

$$\dot{X}_1 = 0.8 X_2^{-1} X_3^{-1} X_4^{0.5} - 3 X_1^{0.5} X_2^{-0.1} - 2 X_1^{0.75} X_3^{-0.2}, \quad X_1(0) = 0.5,$$

$$\dot{X}_2 = 3 X_1^{0.5} X_2^{-0.1} - 1.5 X_2^{0.5}, \quad X_2(0) = 0.5,$$

$$\dot{X}_3 = 2 X_1^{0.75} X_3^{-0.2} - 5 X_3^{0.5}, \quad X_3(0) = 1,$$

$$X_4 = 0.25.$$

(3.16)

Again, the numerical values were chosen as reasonable (see Chapter 5), but without a particular rationale. The dynamical solution is plotted in Fig. 3.6. It exhibits various overshoots and undershoots before it approaches a steady state. This steady state cannot be calculated analytically for the GMA form, but can be obtained numerically by setting the final time for solution, *tf*, to a large value, such as 50. PLAS also allows us to compute the fluxes V_1^-, V_2^+, and V_3^+ as *transformations* (see *BranchG.plc*)

FROM MAPS TO EQUATIONS

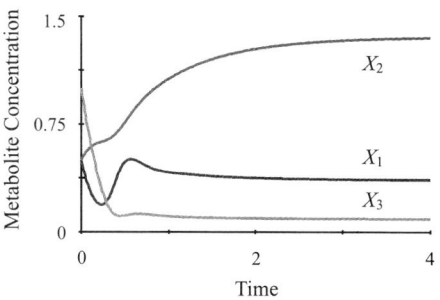

Figure 3.6. Dynamics of the branched pathway with feedback, modeled as a GMA system [Eq. (3.16)].

that read

$$V_1 = 3X1^{0.5}X2^{(-0.1)} + 2X1^{0.75}X3^{(-0.2)}$$
$$V_2 = 3X1^{0.5}X2^{(-0.1)} \quad (3.17)$$
$$V_3 = 2X1^{0.75}X3^{(-0.2)}$$

The numerical steady-state solution is

$X_{1S} = 0.3607566,$
$X_{2S} = 1.357465,$
$X_{3S} = 0.0905943,$
$V_1^- = 3.252599,$
$V_2^+ = 1.747654,$
$V_3^+ = 1.504945.$

As was to be expected, the degradative flux V_1^- is equal to the sum of V_2^+ and V_3^+. Using the formulas in Eq. (3.14), the parameters of the aggregated S-system term are computed as

$h_{11} = 0.615672498,$
$h_{12} = -0.053731001,$
$h_{13} = -0.092537998,$
$\beta_1 = 4.959813621.$

Note that the *aggregated* kinetic order $h_{11} = 0.615672498$ lies between the corresponding parameter values, $g_{21} = 0.5$ and $g_{31} = 0.75$, of V_2^+ and V_3^+. The corresponding S-system model is thus

$$\dot{X}_1 = 0.8 X_2^{-1} X_3^{-1} X_4^{0.5} - 4.959813621 X_1^{0.615672498}$$
$$\times X_2^{-0.053731001} X_3^{-0.092537998}, \quad X_1(0) = 0.5,$$
$$\dot{X}_2 = 3 X_1^{0.5} X_2^{-0.1} - 1.5 X_2^{0.5}, \quad X_2(0) = 0.5, \quad (3.18)$$
$$\dot{X}_3 = 2 X_1^{0.75} X_3^{-0.2} - 5 X_3^{0.5}, \quad X_3(0) = 1,$$
$$X_4 = 0.25.$$

It is implemented as *BranchS.plc*. Even though the system is initialized relatively far away from the steady state, the dynamical solution, shown in Fig. 3.7, is essentially indistinguishable from the GMA solution.

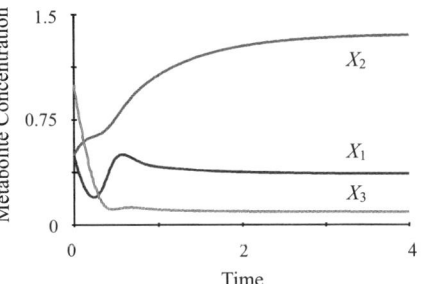

Figure 3.7. Dynamics of the branched pathway with feedback, modeled as an S-system model [Eq. (3.18)].

4. Cascades. Beginning in the 1960s and 1970s, numerous researchers have investigated the role of cascaded systems as amplifiers for biochemical signals (for references, see Savageau, 1976). Some input leads to increased production of a metabolite at the first level, and this metabolite serves as an activator of a process at a second level. Savageau (1976: Chapter 13) reviewed a number of cascaded mechanisms and pointed out their significance for amplification, speed of response, control, and efficiency. Cascaded mechanisms are found in diverse areas of biochemistry and physiology, including hormonal control, gene regulation, immunology, blood clotting, and visual excitation. Purely biochemical examples are the phosphorylase activation cascade in muscle and the bicyclic glutamine synthase cascade (cf. Rubin and Rosen, 1975; Chock et al., 1980; Fell, 1997: pp. 238–241).

A generic example of a cascade with feedback is given in Fig. 3.8. Three precursors ($X_4 - X_6$) are converted into metabolites ($X_1 - X_3$) at three different stages, and the synthesized metabolites early in the cascade affect synthesis at the next step of the cascade. In addition, the first process in the cascade is triggered by a stimulating input, represented by X_7 and g_{17}. The amplification process is slowed down when the end product, X_3, is available in a sufficient quantity. This feedback regulation is represented by the kinetic order g_{13}.

The map in Fig. 3.8 tells us immediately which variables need to be included in each power-law term of the three equations for X_1, X_2, and X_3. Again, flow of material requires the corresponding kinetic order to be positive, while inhibition implies negative powers. The describing equations read

$$\dot{X}_1 = \alpha_1 X_3^{g_{13}} X_4^{g_{14}} X_7^{g_{17}} - \beta_1 X_1^{h_{11}},$$
$$\dot{X}_2 = \alpha_2 X_1^{g_{21}} X_5^{g_{25}} - \beta_2 X_2^{h_{22}}, \qquad (3.19)$$
$$\dot{X}_3 = \alpha_3 X_2^{g_{32}} X_6^{g_{36}} - \beta_3 X_3^{h_{33}},$$
$$X_4, X_5, X_6, X_7 = \text{constant}.$$

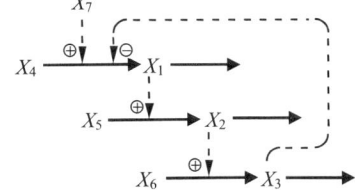

Figure 3.8. Cascade with three steps and feedback. The dependent variables X_1 through X_3 are produced from precursors, X_4 through X_6, which are considered independent. The input X_7 triggers the amplification process, which is slowed down once enough end product X_3 is available.

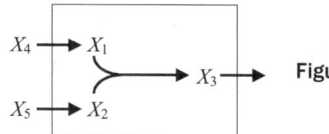

Figure 3.9. Bimolecular reaction with inputs.

(See Exercise 10). Detailed analyses of S-system models of cascaded systems can be found in (Savageau 1976: Chapter 13), Irvine and Savageau (1985ab), and Savageau and Sands (1991).

5. Bimolecular reactions. In contrast to two converging reactions that represent different pathways, a bimolecular reaction requires the presence of both precursors. This situation is reflected in a map by a double-tailed arrow (cf. Figs. 1.11 and 3.9). For instance, if X_3 is produced from X_1 and X_2 via two separate processes, the degradation term of X_1 will contain X_1 as the only variable and the degradation term of X_2 will contain X_2 as the only variable. By contrast, if X_3 is synthesized from X_1 and X_2 in a bimolecular reaction, the degradation of X_1 also depends on X_2, and the degradation of X_2 also depends on X_1. Without X_1, X_3 cannot be produced from X_2, and without X_2, X_3 cannot be produced from X_1. The equations for this scenario read

$$\begin{aligned}
\dot{X}_1 &= \alpha_1 X_4^{g_{14}} - \beta_1 X_1^{h_{11}} X_2^{h_{12}}, \\
\dot{X}_2 &= \alpha_2 X_5^{g_{25}} - \beta_2 X_1^{h_{21}} X_2^{h_{22}}, \\
\dot{X}_3 &= \alpha_3 X_1^{g_{31}} X_2^{g_{32}} - \beta_3 X_3^{h_{33}}, \\
X_4, X_5 &= \text{constant}.
\end{aligned} \qquad (3.20)$$

Obviously, some of the parameters in this model are constrained, as the production of X_3 must equal the degradation of X_1 and X_2, at least at the steady state. The derivation of the constraints is given in the next section. The constraints are

$$g_{31} = \frac{h_{11} V_1^- + h_{21} V_2^-}{V_1^- + V_2^-},$$

$$g_{32} = \frac{h_{12} V_1^- + h_{22} V_2^-}{V_1^- + V_2^-},$$

$$\alpha_3 = \frac{V_1^- + V_2^-}{X_1^{g_{31}} X_2^{g_{32}}}.$$

As in the case of a modulated branch point, the kinetic orders of the aggregated flux are equal to the kinetic orders of the individual fluxes, weighted by the magnitudes of the fluxes at steady state.

GENERAL DERIVATION OF CONSTRAINTS

We have encountered constraints between parameters in several of the above examples. While these may have appeared to be different for each case, they are always derived in the same mathematical fashion. The principles for this derivation are based on the fundamental concept behind the power-law representation, which is the Taylor approximation in logarithmic coordinates (see *Complements* to Chapter 2, and Appendix).

In order to keep this derivation simple, let's consider the case of two variables and reduce indices to a minimum. Specifically, suppose that a constraint requires the term $\alpha X^p Z^q$ to be equivalent to a function $V(X, Z)$ at an operating point, such as a steady state. The theory behind the power-law formalism then asserts that p and q are computed as partial derivatives of V with respect to X or Z, respectively, and these partial derivatives are to be multiplied by X or Z and divided by V (see Appendix). This is formulated mathematically as

$$p = \frac{\partial V}{\partial X}\frac{X}{V} \quad \text{and} \quad q = \frac{\partial V}{\partial Z}\frac{Z}{V}, \tag{3.21}$$

and these expressions are to be evaluated at the operating point. If these equations are satisfied, the function V and the power-law representation have the same slopes. The rate constant α is computed as

$$\alpha = \frac{V(X, Z)}{X^p Z^q}, \tag{3.22}$$

which is also evaluated at the operating point. The formula for α is easy to understand, because it is a simple reformulation of the requirement that $\alpha X^p Z^q$ and $V(X, Z)$ be equal at the operating point. All constraints in the previous section follow these principles.

Nothing is special about the fact that V in this case depends on only two variables. The same concepts apply for functions of one variable and functions of n variables. Expressed in general terms, if the constraint equation is

$$\alpha_i X_1^{g_{i1}} X_2^{g_{i2}} \cdots X_n^{g_{in}} = V_i(X_1, X_2, \ldots, X_n), \tag{3.23}$$

one obtains any kinetic order g_{ij} and the rate constant α_i as

$$g_{ij} = \frac{\partial V_i}{\partial X_j}\frac{X_j}{V_i} \tag{3.24}$$

and

$$\alpha_i = \frac{V_i}{X_1^{g_{i1}} X_2^{g_{i2}} \cdots X_n^{g_{in}}}. \tag{3.25}$$

These expressions are again evaluated at an operating point, such as a steady state.

Constraints for Bimolecular Reactions

For the case of a bimolecular reaction, we formulated as the constraint that *the production of X_3 must equal the degradation of X_1 and X_2 at the steady state* (see above). Expressed in terms of the S-system model, this constraint reads

$$\begin{aligned}\alpha_3 X_1^{g_{31}} X_2^{g_{32}} &= V(X_1, X_2) \\ &= \beta_1 X_1^{h_{11}} X_2^{h_{12}} + \beta_2 X_1^{h_{21}} X_2^{h_{22}}.\end{aligned} \tag{3.26}$$

FROM MAPS TO EQUATIONS

According to the general rule for kinetic orders in constraint equations, we compute partial derivatives of the right-hand side with respect to X_1 and X_2. Specifically, we compute

$$g_{31} = \frac{\partial V}{\partial X_1} \frac{X_1}{V}$$

$$= \left(h_{11}\beta_1 X_1^{h_{11}-1} X_2^{h_{12}} + h_{21}\beta_2 X_1^{h_{21}-1} X_2^{h_{22}}\right) \frac{X_1}{V}$$

$$= \frac{h_{11}\beta_1 X_1^{h_{11}} X_2^{h_{12}} + h_{21}\beta_2 X_1^{h_{21}} X_2^{h_{22}}}{\beta_1 X_1^{h_{11}} X_2^{h_{12}} + \beta_2 X_1^{h_{21}} X_2^{h_{22}}}, \qquad (3.27)$$

$$g_{32} = \frac{\partial V}{\partial X_2} \frac{X_2}{V}$$

$$= \left(h_{12}\beta_1 X_1^{h_{11}} X_2^{h_{12}-1} + h_{22}\beta_2 X_1^{h_{21}} X_2^{h_{22}-1}\right) \frac{X_2}{V}$$

$$= \frac{h_{12}\beta_1 X_1^{h_{11}} X_2^{h_{12}} + h_{22}\beta_2 X_1^{h_{21}} X_2^{h_{22}}}{\beta_1 X_1^{h_{11}} X_2^{h_{12}} + \beta_2 X_1^{h_{21}} X_2^{h_{22}}}, \qquad (3.28)$$

and

$$\alpha_3 = \frac{\beta_1 X_1^{h_{11}} X_2^{h_{12}} + \beta_2 X_1^{h_{21}} X_2^{h_{22}}}{X_1^{g_{31}} X_2^{g_{32}}}. \qquad (3.29)$$

These are the constraints given in the previous section, expressed not in terms of fluxes V but, equivalently, in terms of S-system parameters (convince yourself of this).

Constraints for Branch Points

For the modulated branch point, the original constraint is

$$\beta_1 X_1^{h_{11}} X_2^{h_{12}} X_3^{h_{13}} = \alpha_2 X_1^{g_{21}} X_2^{g_{22}} + \alpha_3 X_1^{g_{31}} X_3^{g_{33}} \qquad (3.30)$$

[cf. Eq. (3.13)]. In this case, three variables are involved, but the principle remains the same. For instance, h_{12} is obtained by partial differentiation of the right-hand side with respect to X_2 and subsequent multiplication by X_2/V, where V is the sum on the right-hand side:

$$h_{12} = \frac{\partial V}{\partial X_2} \frac{X_2}{V}$$

$$= g_{22}\alpha_2 X_1^{g_{21}} X_2^{g_{22}-1} \frac{X_2}{V}$$

$$= \frac{g_{22}\alpha_2 X_1^{g_{21}} X_2^{g_{22}}}{\alpha_2 X_1^{g_{21}} X_2^{g_{22}} + \alpha_3 X_1^{g_{31}} X_3^{g_{33}}}, \qquad (3.31)$$

which is equivalent to the constraint given in the previous section. The derivation of h_{11}, h_{13}, and β_1 is left as an exercise (Exercise 11).

Stoichiometric Constraints

In splitting reactions, the degradation of one substrate molecule results in two molecules of product (cf. Chapter 1). An example is the conversion of fructose diphosphate, X_1, into two molecules of phosphoenol pyruvate, X_2, during glycolysis. Ignoring modulating effects, the degradation of X_1 is given as

$$V_1^- = \beta_1 X_1^{h_{11}}, \tag{3.32}$$

and the synthesis of X_2 is

$$V_2^+ = \alpha_2 X_1^{g_{21}}. \tag{3.33}$$

Simple precursor–product relationships would mandate $g_{21} = h_{11}$ and $\alpha_2 = \beta_1$. However, because of stoichiometry, the constraints in this case are

$$g_{21} = h_{11} \tag{3.34}$$

and

$$\alpha_2 = 2\beta_1. \tag{3.35}$$

The kinetic order of the process is unaffected, while the rate constants reflect the doubling in the number of molecules.

Conserved Masses

All constraints so far addressed the equivalence of fluxes. There are also instances where masses are conserved. An example is the pooling of closely related metabolites, such as ATP, ADP, and AMP, in one variable. Such an aggregation has the advantage that only one variable needs to be considered instead of three, which reduces complexity in general and the quantification of conversion processes in particular. Another prominent example is an enzyme that can be present in a bound or an unbound form. If the enzyme concentrations in the two forms do not change during the experiment, the constraint is no issue, since it deals with independent variables whose values are simply set in such a way that the bound and unbound concentrations sum to the total enzyme concentration. By contrast, if the relative amounts of the two forms change appreciably, the bound and unbound fractions need to be represented by constrained dependent variables.

Suppose the total enzyme concentration of the system is constant during the experiment and represented by the independent variable X_6. Suppose further that the enzyme can be present in a bound form, coded as X_4, or a free form, coded as X_5, and that the dynamics of the bound form is driven by the system. Consequently, X_4 is a dependent variable, while X_5 is constrained as $X_5 = X_6 - X_4$. This constraint allows us to eliminate X_5. Direct substitution of the difference would destroy the homogeneous form of the equation, and it has proven more convenient, and sufficiently accurate, to use a substitution based upon power-law approximation of the difference (Savageau, 1969a, 1979b). This is accomplished in the same manner as for other constraints, namely by partial differentiation. Specifically, we compute

$$X_5 = X_6 - X_4 \approx \gamma_5 X_4^{g_{54}} X_6^{g_{56}}, \tag{3.36}$$

$$g_{54} = \frac{\partial(X_6 - X_4)}{\partial X_4} \frac{-X_4}{X_6 - X_4} = -\frac{X_4}{X_5}, \tag{3.37}$$

$$g_{56} = \frac{\partial(X_6 - X_4)}{\partial X_6} \frac{X_6}{X_6 - X_4} = \frac{X_6}{X_5}, \tag{3.38}$$

$$\gamma_5 = (X_6 - X_4) X_4^{-g_{54}} X_6^{-g_{56}} = X_4^{-g_{54}} X_5 X_6^{-g_{56}}, \tag{3.39}$$

and these quantities are again evaluated at an operating point of our choice, such as a steady state.

Once the constraint is established in this manner, X_5 is eliminated from the system by substitution of X_5 with the power-law term $\gamma_5 X_4^{g_{54}} X_6^{g_{56}}$. Detailed analyses, using these types of constraints, can be found in Sorribas and Savageau (1989a) and in Torres (1994bc). As an alternative to this strategy for handling conserved masses, one may use so-called *buffer boxes* (Voit and Ferreira, 1998), which are discussed in the *Complements* to Chapter 4.

REVERSIBLE PATHWAYS

So far, we have only considered irreversible or essentially irreversible pathways, in which the reverse reaction is negligible (for further discussion, see Chapter 1). In these cases, the recipes for setting up S-system equations were sufficient and unique. The situation is slightly more complicated for reversible reactions, because there are two options for aggregating fluxes. As an illustration, consider the pool X_2 of the pathway in Fig. 3.10. There is influx of material from X_1 and efflux toward X_3, just as in the corresponding irreversible pathway. However, here we have also the reverse reactions, which constitute an efflux toward X_1 and an additional influx from X_3. In the general notation of mass balance equations, we can describe the change in X_2 as

$$\dot{X}_2 = (v_{12} - v_{21}) - (v_{23} - v_{32}) \tag{3.40}$$

or as

$$\dot{X}_2 = (v_{12} + v_{32}) - (v_{21} + v_{23}). \tag{3.41}$$

Obviously, these two representations are mathematically equivalent, but semantically they suggest a different focus. In the first case, we describe the change in X_2 as the difference between net fluxes: $v_{12} - v_{21}$ describes the net flux between X_1 and X_2, and $v_{23} - v_{32}$ describes the net flux between X_2 and X_3. By contrast, the second representation focuses on the difference between the total influx and the total efflux. This semantic difference suggests two different options of aggregation of fluxes. In the first case, $V_2^+ = v_{12} - v_{21}$ is a function of X_1 and X_2, and $V_2^- = v_{23} - v_{32}$

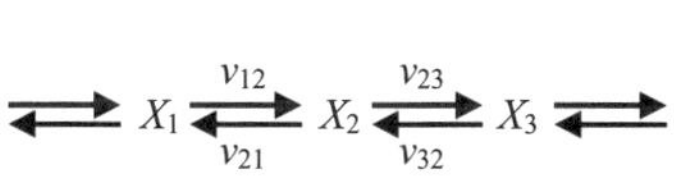

Figure 3.10. Map of a generic reversible pathway.

is a function of X_2 and X_3. In the second case, both fluxes $V_2^+ = v_{12} + v_{32}$ and $V_2^- = v_{21} + v_{23}$ are functions of all three variables. One could surmise that the second option should be superior because both fluxes are functions of all three variables, and thus contain more parameters, which in turn would offer greater flexibility. While there are indeed more parameters, several of them are constrained, and the total number of independent parameters is exactly the same in both cases (Sorribas and Savageau, 1989c). This criterion being indiscriminate, there is no way to find out by casual inspection which option is better.

Sorribas and Savageau (1989c) concluded from a careful analysis that, at least for reactions close to thermodynamic equilibrium, the second option [Eq. (3.41)] is clearly superior. They call this mode of aggregation of fluxes the *reversible strategy*. The superiority of this strategy was established on the basis of several criteria, which included accuracy in describing fluxes, accuracy in modeling transient responses, and overall robustness. Furthermore, only the reversible strategy was found to be able to capture the principal feature of flux reversal in amphibolic pathways.

If the reversible reaction occurs far from thermodynamic equilibrium, the superiority of the reversible strategy becomes insignificant, and the *irreversible strategy* of Eq. (3.40) may be the choice of convenience. According to this strategy, fluxes are aggregated into net fluxes, and this philosophy is more easily reconciled with traditional rate laws. We shall see in Chapter 5 how this fact can be utilized in the estimation of parameter values.

As an illustration of the reversible strategy, consider the amphibolic pathway in Fig. 3.11, in which the independent concentrations X_3 and X_4 are held constant by the environment of the experimental set-up. Thus, there are two dependent and two independent variables, and the mass balance and the reversible S-system representations are

$$\dot{X}_1 = (v_{41} + v_{21}) - (v_{14} + v_{12}) = V_1^+ - V_1^-,$$
$$\dot{X}_2 = (v_{12} + v_{32}) - (v_{21} + v_{23}) = V_2^+ - V_2^-, \quad (3.42)$$
$$X_3, X_4 = \text{constant}$$

and

$$\dot{X}_1 = \alpha_1 X_1^{g_{11}} X_2^{g_{12}} X_4^{g_{14}} - \beta_1 X_1^{h_{11}} X_2^{h_{12}} X_4^{h_{14}},$$
$$\dot{X}_2 = \alpha_2 X_1^{g_{21}} X_2^{g_{22}} X_3^{g_{23}} - \beta_2 X_1^{h_{21}} X_2^{h_{22}} X_3^{h_{23}}, \quad (3.43)$$
$$X_3, X_4 = \text{constant},$$

The corresponding irreversible representations are

$$\dot{X}_1 = (v_{41} - v_{14}) - (v_{12} - v_{21}) = V_1^+ - V_1^-,$$
$$\dot{X}_2 = (v_{12} - v_{21}) - (v_{23} - v_{32}) = V_2^+ - V_2^-, \quad (3.44)$$
$$X_3, X_4 = \text{constant},$$

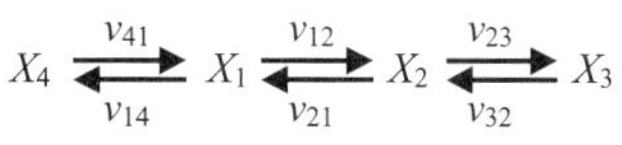

Figure 3.11. Clamped reversible pathway. $X_4 \underset{v_{14}}{\overset{v_{41}}{\rightleftarrows}} X_1 \underset{v_{21}}{\overset{v_{12}}{\rightleftarrows}} X_2 \underset{v_{32}}{\overset{v_{23}}{\rightleftarrows}} X_3$

and

$$\dot{X}_1 = \alpha_1 X_1^{g_{11}} X_4^{g_{14}} - \beta_1 X_1^{h_{11}} X_2^{h_{12}},$$
$$\dot{X}_2 = \alpha_2 X_1^{g_{21}} X_2^{g_{22}} - \beta_2 X_2^{h_{22}} X_3^{h_{23}}, \quad (3.45)$$
$$X_3, X_4 = \text{constant}.$$

Note that the parameters in the reversible and the irreversible representations have a different meaning and, in general, different values (see Exercise 12).

CLOSED SYSTEMS, OPEN SYSTEMS

We call a system closed if no material enters or leaves it. There may be signals affecting the system from the outside, but all material is conserved. In contrast, an open system receives material, for instance, in the form of an exogenous supply of substrate, or it loses material to the outside. If material is supplied from the outside, while no material leaves the system, the system grows; that is, the totality of all concentrations increases. If no material is supplied, but some material leaks out, the totality of all concentrations decreases. An open system remains constant in total size if input and output are in balance.

Whether a system is open or closed can be read directly off the map: If all flow arrows originate and terminate at a dependent variable, the system is closed. If any flow arrow originates at an independent variable, or if no variable is associated with the head or tail of a flow arrow, the system is open. To illustrate this difference, compare two systems that superficially are equivalent (Fig. 3.12). The only difference between the two is that X_5 in the first map (top panel) is an independent variable, whereas X_5 in the second map (bottom panel) is a dependent variable. Thus, the system depicted in the top panel is open, whereas the system in the bottom panel is closed. This difference has significant implications for the formulation of equations and for dynamics of the system. In the first case (top), X_5 is supplied exogenously at a constant rate, and since no material leaves the system, the totality of all concentrations will grow continuously. By contrast, the initial amount of X_5 in the closed system (bottom) is being used up and there is no mechanism to replenish it. Consequently, X_5 will disappear, and all the material of the system will eventually cycle among the pools X_2, X_3, and X_4. The overall amount of material remains the same.

The difference in the two systems is also reflected in the system equations. In the top case, X_5 is independent and represented by a statement of the type

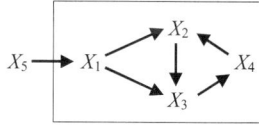

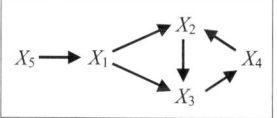

Figure 3.12. In the top panel, X_5 is an independent variable outside the system. In the bottom panel, X_5 is a constituent within the system and is modeled as a dependent variable.

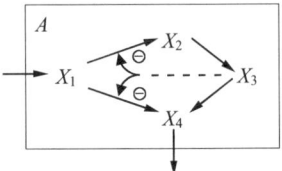

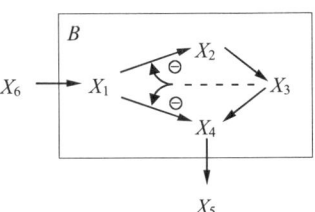

Figure 3.13. The two maps A (top) and B (bottom) differ only by the fact that two variables outside the system are made explicit.

"$X_5 = $ constant." In the bottom case, X_5 is represented by a differential equation (Exercise 13).

The variables of a system are not necessarily affected by the question of whether a system is open or closed. For instance, the maps in Fig. 3.13 depict two similar systems. In map A, substrate is supplied at a constant rate from outside for the production of X_1, and the degradation product of X_4 leaves the open system. In map B, the exogenous supply of substrate is represented by the independent variable X_6, and the degradation product of X_4 is collected in the form of the additional dependent variable X_5. X_5 is a dependent variable, because it changes over time in response to changes in X_4, whereas X_6 is an independent variable, because otherwise the production of X_1 would be time-dependent, and maps A and B would represent different situations.

The S-system descriptions for the two pathways are the same except for the production term of the first equation and the additional equation for X_5 in the closed-system description (see Exercise 14). The production terms of the first equations for the two systems have a different appearance but actually are equivalent, since the independent variable X_6 is constant and can be merged with the rate constant. The additional equation for X_5 shows how much of X_4 is degraded over a period of time. This information may be useful for bookkeeping but it is irrelevant for the dynamics of the remaining dependent variables. Mathematically, this is evident from the fact that X_5 does not appear in any of the system equations, and that therefore the remaining variables are independent of X_5. As far as the remaining variables are concerned, the equation for X_5 can be eliminated without loss of information.

EXERCISES

1. Is an S-system model always a special case of a GMA model with the same number of variables? Is the opposite true? Explain your answers.
2. What would it mean (biochemically and mathematically) if both α_i and β_i in the ith equation of an S-system were zero?

3. What does it mean (biochemically and mathematically) if a power-law term in an S-system or GMA system consists exclusively of the rate constant and does not contain any variables?

4. What would it mean (biochemically and mathematically) if both power-law terms in an S-system consisted exclusively of the rate constants and these had the same value?

5. Example 1 developed an S-system representation of a linear pathway. Develop the corresponding GMA representation. Discuss your answer in light of Exercise 1.

6. Explore the linear pathway in Example 1. In the PLAS program *Linear1.plc* change the independent concentration X_4 to different values, such as 0.001, 0.1, 1, and 1.1. Solve the system by clicking the solution button . Describe and interpret the result. Keep a detailed logbook of results.

7. Explore the linear pathway in Example 1 further. In the PLAS program *Linear1.plc* change the initial values X_1, X_2, and X_3 to other values, such as 0.01, 0.5, and 1, respectively. Solve the system by clicking the solution button . Describe and interpret the result. Try other sets of initial values. Keep a detailed logbook of results.

8. Explore the linear pathway in Example 1 further. In the PLAS program *Linear1.plc* change the initial values X_1, X_2, and X_3 to 0.25, 0.0625, and 0.0625, respectively. Solve the system by clicking the solution button . Describe and interpret the result in your logbook.

9. Compare Figs. 3.2 and 3.4 in words. Change g_{13} to -1, solve the system, and compare again.

10. Which kinetic orders in the model (3.19) of a cascaded pathway are positive and which are negative?

11. Compute the parameters h_{11}, h_{13}, and β_1 from the constraint equation (3.30).

12. Interpret the kinetic orders in the reversible and the irreversible strategies for modeling reversible pathways.

13. Formulate S-system equations for the open and closed systems in Fig. 3.12. Which constraints must be considered?

14. Formulate S-system equations for the two systems in Fig. 3.13. Which constraints must be considered?

15. Develop constraint relationships for a pathway with a branch point of three converging reactions.

16. Develop constraint relationships for a pathway with a branch point of three diverging reactions.

17. Does an S-system model, including parameter values and initial conditions, uniquely determine one particular pathway, or is it imaginable that two different pathways have the same S-system representation? Provide a proof of uniqueness or a counterexample.

18. Set up equations for the following pathway:

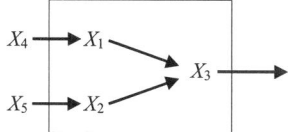

19. Set up equations for the following pathway:

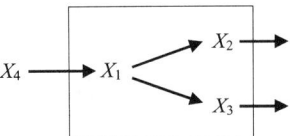

20. Sorribas and Savageau (1989c) analyzed reversible and irreversible models for the clamped pathway shown below

$$X_4 \underset{v_{14}}{\overset{v_{41}}{\rightleftarrows}} X_1 \underset{v_{21}}{\overset{v_{12}}{\rightleftarrows}} X_2 \underset{v_{32}}{\overset{v_{23}}{\rightleftarrows}} X_3$$

For comparison they assumed that the forward reactions v_{ij} (v_{41}, v_{12}, and v_{23}) had the mathematical form

$$v_{i-1,i} = \frac{V_i X_{i-1}/K_i}{1 + X_{i-1}/K_i + X_i/K_{-i}}, \quad i = 1, 2, 3; i - 1 = 4 \text{ if } i = 1.$$

The backward reactions (v_{14}, v_{21}, and v_{32}) were assumed to have the form

$$v_{i,i-1} = \frac{V_{-i} X_i/K_{-i}}{1 + X_{i-1}/K_i + X_i/K_{-i}}, \quad i = 1, 2, 3; i - 1 = 4 \text{ if } i = 1.$$

The parameters used are listed in the following table:

Reaction (i)	V_i	K_i	V_{-i}	K_{-i}
1	60	1	480	8
2	50	1	300	6
3	70	0.5	700	5

Compute parameter values for reversible and irreversible S-system models. Select as the concentrations of $X_1, \ldots, X_4$ at the operating points 8, 6, 5, and 10 μM, respectively.

REFERENCES

[52], [60], [68], [87], [165–6], [297], [310–11], [319], [322], [349], [352–4], [394–6], [417–18], [422–3], [439–40], [455], [466].

CHAPTER FOUR

Computer Simulation

In Chapter 3 we discussed how a biochemical map is represented mathematically as a set of equations in S-system form or GMA form. We also implemented some of these systems in PLAS to gain a first intuitive impression of how these types of equations relate to dynamic responses. It is now time to begin analyzing S-system and GMA equations more systematically. There are essentially two ways to do this. One can study the mathematical properties of the equations with paper and pencil, and that will be explained in Chapters 6 and 7. Alternatively, or in addition, one can gain insight into the system by *simulation*, that is, by evaluating the equations with the computer under different parameter settings. Ideally, one will do both, thus obtaining a comprehensive understanding of the pathway. In practice, we can neither perform a complete mathematical analysis, because the equations are usually quite complicated, nor do enough simulations to cover every conceivable situation of interest, because there would be too many. For example, over sixty million simulations would be required to obtain all responses of a pathway with ten parameters, if one were to explore just six different values for each of the parameters. The strategy for exploring a metabolic pathway is in most cases a combination of some mathematical analysis and some computer simulation. Many features of S-systems can be evaluated with strictly analytical means, but when the analysis becomes too complicated or impossible, simulations provide a welcome tool for exploring complex systems. And with an interactive and user-friendly program like PLAS, simulations can be a lot of fun, and they can even become borderline addictive. As Yates (1978) said, "we hunger for some dose of complexity in many of our endeavors," and as a skier, after developing some expertise, gets tired of slopes for novices, the biochemical systems modeler will soon enjoy the challenges of unexpected and seemingly unexplainable simulation results. It is very difficult to make PLAS crash, so there is no need to fear that something bad will happen when new scenarios are explored.

While there are some guidelines and good simulation practices, the selection of conditions to be simulated is a type of expertise that can only be learned by doing – a lot of doing. In the analyses and examples of this chapter, we will demonstrate a few typical simulations, but these are far from sufficient for really understanding the analyzed systems. The reader is strongly encouraged not only to follow the examples in the text and to do the recommended exercises, but to ask further questions and to execute them with simulations. Except for the investment of time, not much is lost by doing too many simulations, even if they are, in hindsight, trivial or nonsensical, but a lot may be gained. It is often the unusual situations that shed light on some of the most interesting system properties. How many simulations are necessary or desirable? As obvious as the question is, there is no good answer. A rule of thumb may be the following. Before every simulation, predict how the system will respond. If you come to a point where the predictions are consistently correct, you may have done enough simulations.

In contrast to many mathematical analyses, which can be performed in symbolic form – i.e., without the numerical specification of all parameter values – computer simulations almost always require that we specify values for all parameters (for limited exceptions see Weinberger, 1991; Sorribas, 1996). So far, we have not discussed parameter values much, and we postpone the description of methods for obtaining parameter values from experimental measurements until Chapter 5. For now, we just assume that we know – or have previously determined – all required parameter values.

The later parts of this chapter refer to some steady-state characteristics of biochemical systems, like stability and sensitivities, that we have not yet encountered and that are the subject of Chapters 6 and 7. While these system features are firmly rooted in precise mathematical definitions, it is sufficient for now to provide intuitive working definitions. We have done a lot of theory in the preceding chapters, and it is time to begin experiencing the wonderworld of non-linear systems, before we present a rigorous algebraic treatment.

The organization of the chapter loosely follows the three steps that constitute a typical computer simulation of a biochemical pathway:

1. Enter all model specifications in the computer.
2. Perform various simulations and analyses.
3. Interpret the results.

We discuss these steps primarily with the example of Chapter 1 that describes some features of glucose-6-phosphate metabolism (cf. Fig. 1.19; see also Chapter 11 for algebraic analyses), but also with some new examples.

Suppose the glucose-6-phosphate pathway of interest is represented by the S-system

$$\begin{aligned}
\dot{X}_1 &= \alpha_1 X_4^{g_{14}} X_6^{g_{16}} - \beta_1 X_1^{h_{11}} X_2^{h_{12}} X_7^{h_{17}}, \\
\dot{X}_2 &= \alpha_2 X_1^{g_{21}} X_2^{g_{22}} X_5^{g_{25}} X_7^{g_{27}} X_{10}^{g_{2,10}} - \beta_2 X_2^{h_{22}} X_3^{h_{23}} X_8^{h_{28}}, \\
\dot{X}_3 &= \alpha_3 X_2^{g_{32}} X_3^{g_{33}} X_8^{g_{38}} - \beta_3 X_3^{h_{33}} X_9^{h_{39}},
\end{aligned} \qquad (4.1)$$

where X_1, X_2, and X_3 are dependent variables, and X_4 through X_{10} are independent.

COMPUTER SIMULATION

For illustrative purposes, we select the following parameter values:

$$
\begin{aligned}
&\alpha_1 = 1.75817 \times 10^{-3}, &&g_{14} = 0.63, &&h_{11} = 2, \\
&\alpha_2 = 6.04276 \times 10^{-2}, &&g_{16} = 1, &&h_{12} = -1.05, \\
&\alpha_3 = 1.93417 \times 10^{-4}, &&g_{21} = 0.75, &&h_{17} = 1, \\
&\beta_1 = 1.44890, &&g_{22} = -0.45625, &&h_{22} = 4.65, \\
&\beta_2 = 1.93417 \times 10^{-4}, &&g_{25} = 0.5125, &&h_{23} = -4.29, \\
&\beta_3 = 3.46570 \times 10^{-2}, &&g_{27} = 0.375, &&h_{28} = 1, \\
& &&g_{2,10} = 0.625, &&h_{33} = 0.3, \\
& &&g_{32} = 4.65, &&h_{39} = 1, \\
& &&g_{33} = -4.29, && \\
& &&g_{38} = 1. &&
\end{aligned}
\quad (4.2)
$$

Concentration values for the independent variables are assumed to be $X_4 = 10$ mM, $X_5 = 5$ mM, and $X_6 = X_7 = X_8 = X_9 = X_{10} = 1$ mM.

Our first task is to implement the system in PLAS. While the enclosed disk contains a file with the system (*G6P.plc*), it is a good learning experience to create the file from scratch. We start with the template file *Template.plc*. Opening this file, the screen shows with an example what pieces of information are needed (Fig. 4.1).

Specification of S-System Equations

PLAS lets us formulate the system in two extreme ways and in any combination of the two. Either we formulate the equations symbolically, using parameter names like g_{32} and α_2 plus definition statements for all parameters, or we may immediately substitute numerical values for some or all the parameters. The latter requires less typing, but the former has some advantage of transparency, comparison with input settings, and later changes. While irrelevant in this particular case, it also facilitates the inclusion of constraints. For instance, if precursor–product relationships between X_1 and X_2 are to be satisfied, PLAS statements like $\alpha_2 = \beta_1$ and $g_{21} = h_{11}$ ensure that we don't violate these constraints.

The user of PLAS may use common names, abbreviated names, or symbols for the constituents of the model. As discussed in Chapter 1, each choice offers advantages and disadvantages. In the following, we use symbolic names with indices, but each symbol or the entire set could be replaced with other variable names. Whatever alternative is chosen, PLAS internally translates the variables into X's with indices, according to their order of appearance in the file.

With few exceptions, PLAS does not worry much about the order in which commands are presented. Again, it is useful in the beginning to follow the same routine. The standard order begins with a definition of the differential equations. For our example, the equations may read as shown in Fig. 4.2. The symbols chosen here correspond directly to the original equations, except that the Greek letters are replaced with Latin letters and subscripts and superscripts are typed in regular font. Instead of *a* and *b*, we could spell out *alpha* and *beta*. Save this file as *Phospho1.plc*.

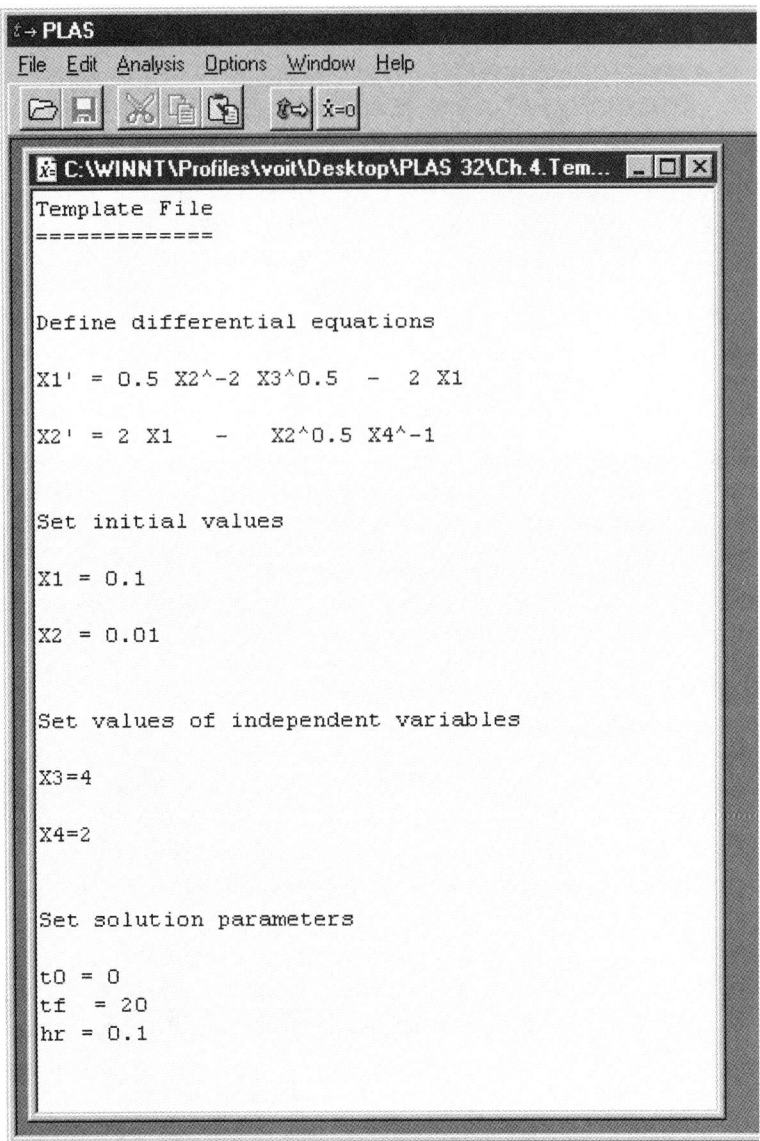

Figure 4.1 (see Color Plate I). Template file (*Template.plc*) shows with a simple example how equations, initial values, and other settings are implemented in PLAS.

As mentioned earlier, PLAS does not force us to use any of these symbols. For instance, we could also use mnemonic names like G1P (glucose-1-phosphate; X_1), G6P (glucose-6-phosphate; X_2), F6P (fructose-6-phosphate; X_3), PP (phosphate; X_4), GLC (glucose; X_5), PPA (phosphorylase a; X_6), PGM (phosphoglucomutase; X_7), PGI (phosphoglucose-isomerase; X_8), PFK (phosphofructokinase; X_9), GK (glucokinase; X_{10}), and GLYC (glycogen; X_{11}). In fact, we could use shorter or longer names. Furthermore, we could follow the alphabet for the kinetic orders. Such an

COMPUTER SIMULATION

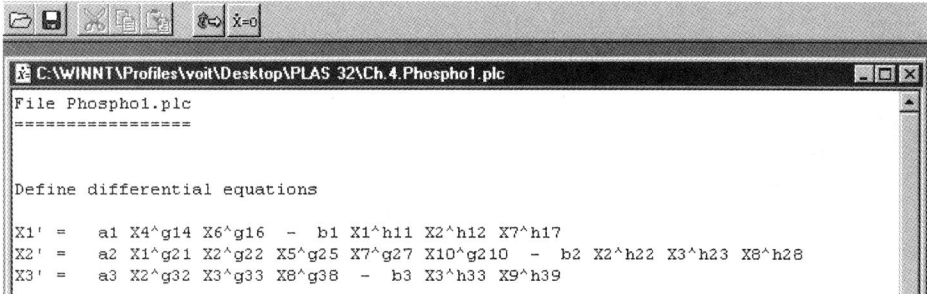

Figure 4.2 (see Color Plate II). Equations (4.1) as defined and implemented in PLAS.

implementation in PLAS would read

$$G1P' = a1\ PP\hat{\ }a\ PPA\hat{\ }b - b1\ G1P\hat{\ }c\ G6P\hat{\ }d\ PGM\hat{\ }e$$
$$G6P' = a2\ G1P\hat{\ }f\ G6P\hat{\ }g\ GLC\hat{\ }h\ PGM\hat{\ }i\ GK\hat{\ }j$$
$$- b2\ G6P\hat{\ }k\ F6P\hat{\ }l\ PGI\hat{\ }m \qquad (4.3)$$
$$F6P' = a3\ G6P\hat{\ }n\ F6P\hat{\ }o\ PGI\hat{\ }p - b3\ F6P\hat{\ }q\ PFK\hat{\ }r$$

Open *Template.plc* and enter these equations. Note that with only three equations, we are already almost running out of single-letter names for the kinetic orders. Nonetheless, as pointed out in Chapter 1, for some purposes such a representation may be advantageous, and PLAS does allow us to use it. For later comparisons, save this file as *Phospho2.plc*.

After entering the equations, we set numerical values for all parameters that were not assigned in the differential equations themselves. The values may be specified in decimal notation, in terms of powers of ten, or even as algebraic expressions. For instance, the value for α_1 may be entered as

$$a1 = 1.75817*10\hat{\ }\ 3$$

as

$$a1 = 0.00175817$$

or as

$$a1 = 1.75817/1000$$

Enter all kinetic orders and rate constants in *Phospho1.plc* and *Phospho2.plc*, in each case using the correct symbols and the parameter values given above. Note that the double indices of the kinetic orders in PLAS are not separated by commas or spaces. For instance, $g_{1,4}$ is entered as *g14*, and $g_{2,10}$ as *g210*. In cases of possible confusion, such as $g_{2,10}$ and $g_{21,0}$, different symbols like *g2_10* and *g21_0* must be introduced.

Once all parameters are defined, save the two files (using the *File* menu) under their original names. In order to avoid some typing, the entire block of definitions can be copied from *Phospho1.plc* to *Phospho2.plc*; then the only task left is to substitute the correct parameter names.

Next is the assignment of values for the independent variables. Obviously, these need to match the symbols in the differential equations. Units are omitted. Choosing the first representation (Fig. 4.2), the assignments take the form

X4 = 10
X5 = 5
X6 = 1
X7 = 1
X8 = 1
X9 = 1
X10 = 1

If the differential equations are given in the form of Eq. (4.3), the independent variables are specified accordingly. For instance, the first two assignments read

PP = 10
GLC = 5

The section on independent variables is followed by the specification of initial values for all dependent variables. They are entered in the same format as settings for independent variables. Corresponding to the differential equations in Eq. 4.1, the specification may read

X1 = 0.1
X2 = 0.4
X3 = 0.2

and the analogous settings are necessary if we work with the representation (4.3). Enter these specifications in *Phospho1.plc* and *Phospho2.plc*, and save the files under their original names.

At this point, we may also add auxiliary variables that are derived from dependent and independent variables, such as the total mass of the system. For instance, we could add in *Phospho1.plc*

$$\text{Sum} = X1 + X2 + X3 + X4 + X5 + X6 + X7 + X8 + X9 + X10 \tag{4.4}$$

or we could be interested in the sum of the dependent variables

$$\text{Sum} = X1 + X2 + X3 \tag{4.5}$$

Enter the latter sum (4.5) in *Phospho1.plc*.

Definitions of such additional variables do not have to adhere to the format of products of power-law functions. For instance, we could define a variable for the natural logarithm of X_1 as

$$XY1 = \ln[X1] \tag{4.6}$$

We just have to make sure that the variables and parameters on the right-hand side of the definition are defined and that we use brackets (instead of parentheses) in algebraic functions.

COMPUTER SIMULATION

Finally, we set the so-called *solution* or *run-time parameters*. These specify the start and finish times of the experiment, as well as the *report interval*, which determines for which time points PLAS should save and report the solution. Typical specifications are

$$t0 = 0$$
$$tf = 10 \qquad (4.7)$$
$$hr = 0.1$$

which means "Start at time 0, solve until time 10, and report the solution at time points 0, 0.1, 0.2, ..., 9.9, 10."

Because the rate constants are computed from steady-state fluxes that were originally expressed in time units of minutes, *tf* reflects a final time of 10 minutes and *hr* represents a time interval of 10% of a minute, i.e., 6 seconds. If the results are needed for every second, *hr* may be decreased, or one may redefine the time units by dividing all rate constants by 60, and setting *hr* to 1 and *tf* to 600. Note that the names *t0*, *tf*, and *hr* are mandatory and cannot be replaced with other names. By the same token, *these symbols must not be assigned for other purposes*.

Enter all parameter settings in *Phospho1.plc* and *Phospho2.plc*; your program should resemble that in Fig. 4.3. Once that has happened, it is possible to check the *internal representation* of the system, which PLAS uses for all further analyses. It is called up with the steady-state button, [x=0], and shows under the tab *General Information* the equations and other settings in a standardized form. Even in *Phospho2.plc*, the internal representation uses X's with indices, and this is the notation PLAS uses for all computational purposes. Nonetheless, the output results will be presented in terms of the user-chosen variable names.

It is noted that GMA systems are entered in exactly the same fashion. Of course, their differential equations may contain any number of power-law terms, but the implementation in PLAS follows exactly the same pattern (see Exercise 6).

Simulations and Computer-Aided Analyses

The S-system or GMA model is now entirely specified, and we are ready for some exploratory simulations. As said in the introduction to this chapter, we shall usually not be able to do all-encompassing simulations that cover every possible situation. The art of simulation consists in identifying relevant situations that reflect as accurately as possible the typical responses of the system. Again, this art is best learned by doing.

The typical computational analysis has two parts that are different in emphasis and complement each other. One part addresses the *dynamics* of the system and focuses on the transient responses that the system exhibits as a reaction to a perturbation or to new environmental conditions. The generic question asked here is: *How do influences from outside affect the short-term and long-term responses of the system?* Specific questions could be: *How long does it take the system to recover from a sudden 20% drop in glucose-6-phosphate?* Or: *What happens if the glucose-6-phosphate pathway described above receives an exogenous bolus of glucose-1-phosphate?*

```
File Phospho1.plc           //Example in Chapter 4
==================

Differential equations:

X1'  =  a1 X4^g14 X6^g16   -  b1 X1^h11 X2^h12 X7^h17
X2'  =  a2 X1^g21 X2^g22 X5^g25 X7^g27 X10^g210 - b2 X2^h22 X3^h23 X8^h28
X3'  =  a3 X2^g32 X3^g33 X8^g38   -  b3 X3^h33 X9^h39
a1   =  1.75817*10^-3
a2   =  6.04276*10^-2
a3   =  1.93417*10^-4
b1   =  1.44890
b2   =  1.93417*10^-4
b3   =  3.46570*10^-2
g14  =  0.63
g16  =  1
g21  =  0.75
g22  =  -0.45625
g25  =  0.5125
g27  =  0.375
g210 =  0.625
g32  =  4.65
g33  =  -4.29
g38  =  1
h11  =  2
h12  =  -1.05
h17  =  1
h22  =  4.65
h23  =  -4.29
h28  =  1
h33  =  0.3
h39  =  1

Initial values:
X1  =  0.1
X2  =  0.4
X3  =  0.2

Values of independent variables:
X4 = 10
X5 = 5
X6 = 1
X7 = 1
X8 = 1
X9 = 1
X10 = 1

Transformations:
Sum = X1 + X2 + X3

Solution parameters:
t0 = 0        // initial time
tf = 10       // final time
hr = .1       // report interval
```

Figure 4.3. PLAS listing of *Phospho1.plc*. All lines that don't follow the syntax of equations and definitions of initial values, independent variables, or parameters are interpreted in PLAS as comments. Comments can also be entered within lines, where they are identified with two slashes (//).

In contrast to these questions concerning response times and features of the transient responses, the second part of the analysis is concerned with the *steady state*. A precise definition will be given in a later chapter, but a simple illustrative example may suffice to develop an intuitive feel. The example is that of a bucket with a hole at the bottom. Without the addition of more water, the water that is currently in the bucket runs out. Adding a little bit of water does not offset the leaking, and the water level still sinks. However, if one replenishes just as much water as is leaking out, the net amount of water in the bucket does not change. Even though there is flow through the bucket, the amount of water is constant, and the system is said to be in a steady state. As a loose definition, one may state that a steady state is characterized by a set of concentrations (or, in general, values of the dependent variables) that does not change over time.

Typical questions about steady states are: *How does the steady state change if one of the independent variable concentrations is altered?* Or: *How does the steady state change if one of the rate constants or kinetic orders is changed, for instance, as a consequence of a mutated enzyme?* Other questions about the steady state that are important, but not always asked, address the stability of the system.

In almost all relevant cases, questions about the dynamics of a pathway can only be explored with computer simulations, since mathematical solutions don't exist, are unknown, or are too complicated to provide satisfactory answers. In contrast, many steady-state issues can be analyzed with paper and pencil (see Chapters 6 and 7), but they can also be treated numerically with PLAS. Let's explore PLAS with a numerical analysis of the glucose-6-phosphate pathway and with some other examples.

DYNAMICS

Transient Responses

The dynamics of a pathway is determined by solutions of the S-system or GMA system equations under different conditions. These solutions quantify how the dependent variables change over a period of time. In other words, when we solve the system, the result is a set of values of all variables (concentrations) over some time interval. Solutions are obtained in PLAS by clicking the solution button ▣ (see Fig. 4.4).

Load one of the two files, *Phospho1.plc* or *Phospho2.plc*, which we established above, or load the file *G6P.plc*, which is provided on the CD. All three should be essentially the same, except for differences in variable and parameter names. Click the solution button ▣ to initiate the solution.

The solution appears in the form of a graph showing the three variables X_1, X_2, and X_3, as well as the sum of the three, plotted versus time. The output is not very dramatic (cf. Fig. 4.5). At the beginning of the experiment (time $t = 0$) the variables exhibit the initial values that we specified. Subsequently, the variables change with time according to the model equations: X_1 and X_3 decrease monotonically, whereas X_2 increases monotonically. The sum decreases slightly.

During this simulated ten-minute experiment, the system seems to approach some steady state, without actually reaching it: The curves are still changing at the final

106 COMPUTATIONAL ANALYSIS OF BIOCHEMICAL SYSTEMS

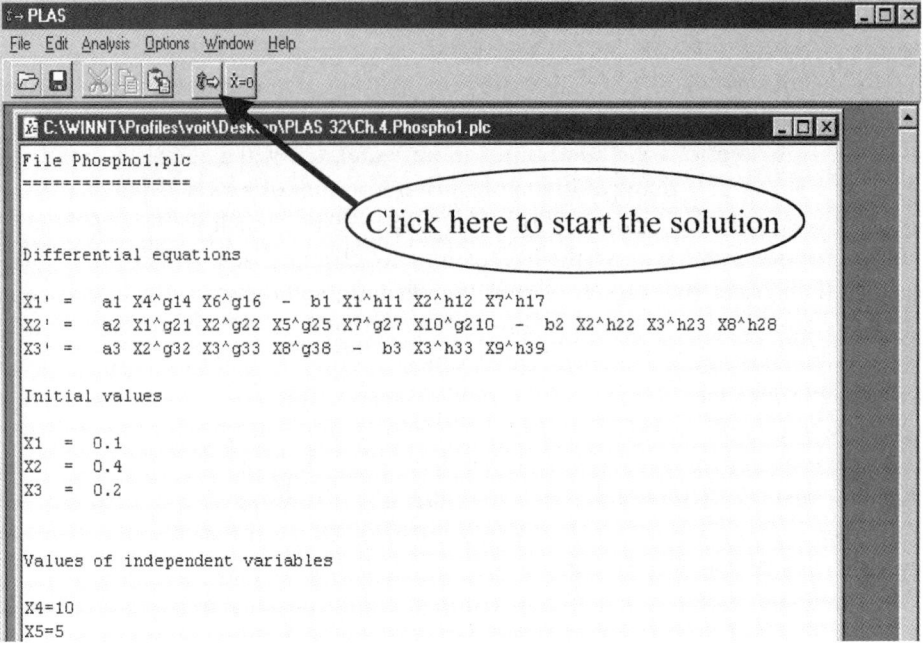

Figure 4.4 (see Color Plate III). Initiate the solution in PLAS by clicking the button 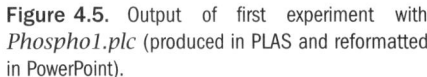.

time. How long will it take until they reach a constant value? To be mathematically precise, it takes infinitely long to reach the true steady-state point, but after some time the changes in concentrations are so small that they are no longer visible on the graph and eventually become smaller than the precision of the computer. How long does that take? We can find out by solving the model again, this time over a longer time period. For instance, we can try a 20-min experiment by resetting *tf* to 20. Two observations can be made from the output: (i) All three variables and their sum seem to be nearly constant now; (ii) X_3 actually drops first, but then starts to increase again.

More details are hard to read off the graph, but it is possible to obtain numerical results by selecting the *Table* option under the *Results* menu while the graph is being displayed. The table of numerical results shows that the variables still have not settled down completely, but that the changes from one time point to the next are getting rather small. Also, X_3 starts to rise again at about $t = 6.8$. Solve the system again,

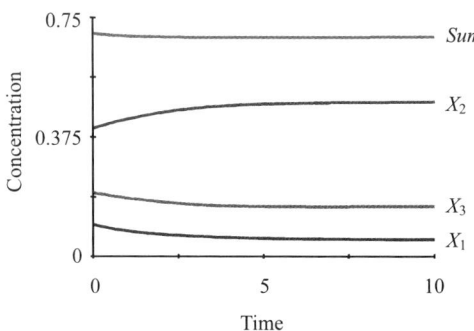

Figure 4.5. Output of first experiment with *Phospho1.plc* (produced in PLAS and reformatted in PowerPoint).

this time setting $tf = 1000$ and $hr = 1$. The last rows of the table of results indicate that the system has reached a steady state at $X_{1S} = 0.05$, $X_{2S} = 0.5$, $X_{3S} = 0.16$ within the accuracy of the display. We use the additional subscript S to indicate that these values are steady-state values.

As another example, we study what happens if X_3 is initiated at ten times its steady state. Thus, we set the initial values as $X_1 = 0.05$, $X_2 = 0.5$, and $X_3 = 1.6$ and solve up to $tf = 10$. The results graph does not show X_3 anymore. This is easily remedied by adjusting the scale of the Y-axis. This is accomplished by double-clicking somewhere on the graph and specifying an appropriate maximum value for Y in the edit control. The adjusted graph indicates that X_3 is decreasing, while X_2 is clearly increasing. The trend in X_1 seems to be a slight increase. However, it is clear that not enough time has passed to see the entire transient dynamics. Try $tf = 100$. The results now exhibit an almost bi-phasic response. Initially, X_3 drops rapidly and X_2 rises concomitantly. After about 30 or 35 min, the graph of X_3 bends almost abruptly, X_2 begins to decrease, and X_1 shows a maximum at about the same time. Solving the system with $tf = 1000$ shows the entire transient response. Exercise 1 explores this type of simulation some more.

Bolus Experiments

With this baseline information about the system we can explore its responses under new conditions. For example, we can simulate the following scenario: The system is at its steady state, and at time $t = 2$ we inject a single bolus of 0.1 mM glucose-1-phosphate. Biochemical intuition tells us that the system should eventually return to its steady state, but the transient response is not as easy to predict. Here is how we simulate the situation:

1. Initialize the system at the steady state 0.05, 0.5, and 0.16, and retain the original values of all independent variable concentrations, namely 10, 5, 1, 1, 1, 1, respectively. Set $t0 = 0$, $tf = 30$, and $hr = 0.01$ for better resolution.
2. Add to the first differential equation a bolus term, so that it reads

    X1' = a1 X4^g14 X6^g16 − b1 X1^h11 X2^h12 X7^h17 + bolus

3. Define the value of *bolus* by adding the statement

    bolus = 0

    to the section of the program where parameter and initial values are defined.
4. Enter two new statements, just before the solution parameters *tf*, *t0*, and *hr* are set. They read

    @ 2 bolus = 100
    @ 2.01 bolus = 0

    The first statement specifies that the variable bolus is set to 100 at time 2, and the second statement resets the value to 0 at time 2.01. Since a bolus of 100 is added for a period of 0.01 min, this construction roughly corresponds to instantaneously adding 1 mM of glucose-1-phosphate.
5. Solve the system as before.

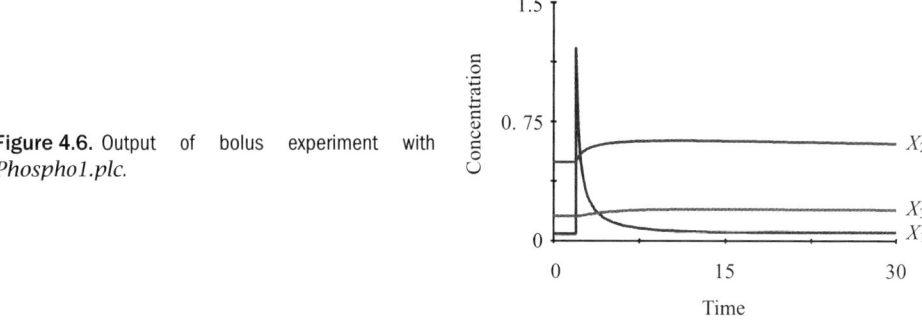

Figure 4.6. Output of bolus experiment with Phospho1.plc.

The plot (Fig. 4.6) shows the three variables over the entire time period of 30 min. At time $t = 2$, we see the exogenously induced, instantaneous increase in X_1. After reaching the peak concentration essentially instantaneously, X_1 drops back in a quasi-exponential fashion. Actually, the decrease is more complicated. It is given by the degradation term (β-term) of the equation for X_1, and this term depends on X_1 and X_2. X_2 reacts to the increased concentration of glucose-1-phosphate with an immediate increase and a subsequent, very slow decrease toward its original steady-state value. X_3 follows X_2 with a slight delay, but its response is much weaker. To see the changes in X_3 more clearly, the Y-scale may be adjusted again by double-clicking the graph and resetting the maximum Y-scale value in the edit control to, e.g., 0.25.

The three variables do not reach their steady-state values within the experimental time period of 30 min. Even after 300 min, the system has not fully recovered from the bolus, even though the concentrations are clearly on the path of returning to their former steady-state values.

Other problems of a related nature are addressed in Exercises 1 and 2.

Persistent Changes

Another type of experiment that can be simulated in a similar manner is a persistent change in one or several of the independent variables. In the solution step, one simply enters a different value for the appropriate variable. Suppose we want to study the effects of reducing the constant input of glucose by 50%. We start the numerical solution, specify timing parameters, e.g., $t0 = 0, tf = 300$, and $hr = 1$, and initialize the dependent variables with the formerly obtained steady-state concentrations and the independent variables with the previous concentrations. Right before the specification of the solution parameters *t0*, *tf*, and *hr*, we add the statement

@ 30 X5 = 2.5

which signifies that, beginning at time 30 (min), the glucose concentration is permanently lowered to half the original value. The solution graph indicates the instantaneous response in all three dependent variables, X_1, X_2, and X_3. They decrease noticeably, and it appears that the reduction in X_5 affects not only the transient but also the steady state of the system. The result is shown in Fig. 4.7. A second solution with *tf* = 1000 suggests that the steady state is now about 0.03, 0.19, 0.06.

COMPUTER SIMULATION

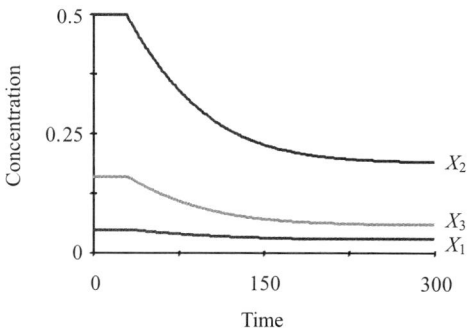

Figure 4.7. Results of a persistent decrease in X_5 to 2.5 at time $t = 30$.

A similar experiment is the simulation of a reduced activity of phosphofructokinase, which catalyzes the degradation of fructose-6-phosphate. Suppose its amount is reduced by 25%. We solve the system again, beginning with the same steady-state specifications as before, but this time we change the value of X_9 to 0.75 after 10 min. Solving for an experimental time period of 2 hours, we see constant values of all three variables up to $t = 10$, followed by noticeable increases in all three concentrations. As expected, the reduced removal of fructose-6-phosphate leads to an accumulation of all three internal variables.

Other problems of a similar nature are given in Exercises 3 and 4.

Modifications in System Characteristics

A different type of analysis involves changes in one or several of the parameter values. Such a change is conceptually different from the changes introduced above. In the previous scenarios, the structure of the model remained the same, while some of the current settings were altered. We could imagine exposing the system to a different environment which, for instance, happened to receive a temporary, exogenous supply of glucose-1-phosphate. Bolus changes in dependent variables were shown to be transient, and the system was able to absorb them. Modifications of the system structure are conceptually different. They are comparable, for instance, with a mutation that affects the affinity of an enzyme, and might be implemented with a change in the appropriate kinetic order. Of course, the situation is not all black and white. A change in an independent variable could be considered a temporary perturbation of the system, and as such would fall into the previous class of simulations. It could also be seen as a structural change in an enzyme, and as such would fall into the present type of simulation.

The parameter values are easily changed in PLAS. For instance, to alter the production characteristics of X_2, we may alter the values of α_2 and g_{22}. As a numerical example, specify

$\alpha_2 = 0.1,$
$g_{22} = -0.05.$

To see the effect, we solve the original system and save the resulting graph under a new name. Then, we solve the altered system and compare the results by *tiling* the windows (*Tile* command under the *Window* menu).

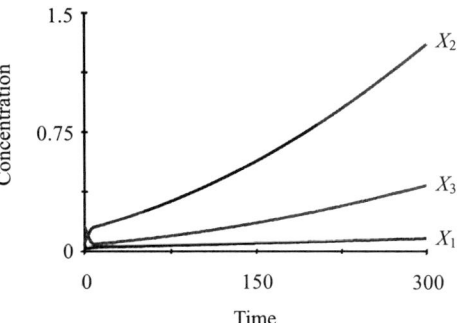

Figure 4.8. Changes in α_2 and g_{22} have a dramatic effect.

As a specific example, load the original pathway *G6P.plc* or *Phospho1.plc* and initiate with a 90% reduced concentration of glucose-6-phosphate: $X_2 = 0.05$. Solve the system with *tf* = 300, and save the graph as *G6Pa.res*. Select the *Tile* command under the *Window* menu. Now change α_2 and g_{22}, and solve and tile again. The difference between the two systems is quite dramatic (Fig. 4.8). Of course, the numerical results can also be compared, upon calling up the *Table* option under the *Results* menu, as before. See Exercise 4 for further analyses. If desired, the change in parameters could be implemented to begin at a later time point. As before, the @ statement would be used for this purpose.

Phase-Plane Plots

When the results of a simulation are oscillatory in nature, it is often useful to study the so-called *phase-plane plot*. In this representation, two variables, such as X_1 and X_2, are not plotted as functions of time, but one variable is plotted as a function of the other. A plot of this type expresses how one variable changes in response to changes in the other variable. The resulting curve is sometimes called a *trajectory*. Trajectories can be straight or curved lines, and they may also form distorted ellipses, spirals, or more complicated patterns, especially if the processes in question are oscillating.

One of the simplest examples is a harmonic oscillation, which corresponds to sine and cosine waves. One can show that X_1 follows a sine wave and X_2 follows the corresponding cosine, if they are given by a simple S-system like

$$\begin{aligned} \dot{X}_1 &= 3 - X_2, & X_1(0) &= 3, \\ \dot{X}_2 &= X_1 - 3, & X_2(0) &= 4. \end{aligned} \tag{4.8}$$

In a nutshell, the cosine is the derivative of the sine, and the sine is the negative derivative of the cosine, and these relationships can be reformulated in the above fashion. In this representation, α_1 and β_2 are the same (here, they take the value 3) and determine the midpoint of the amplitude, while the initial conditions determine phase and magnitude of the oscillation. If all rate constants are multiplied by a number greater than 1, the frequency increases, and if the multiplier is less than 1, the oscillation is slower. Note that the center of the oscillation is shifted to a value of 3 in order to prevent the dependent variables from becoming negative.

COMPUTER SIMULATION

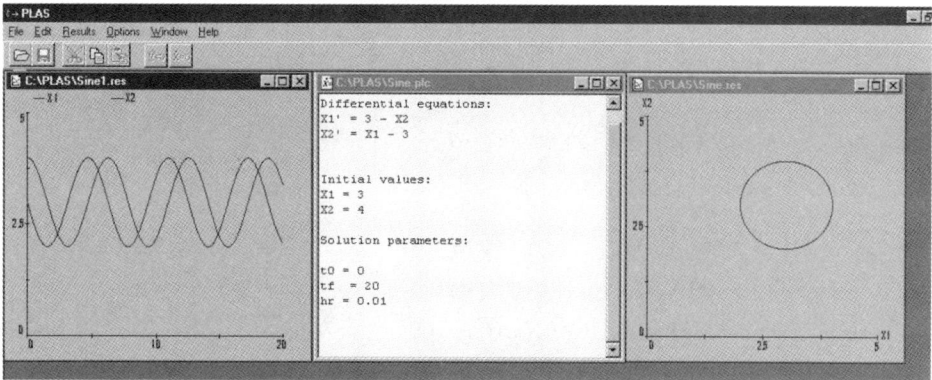

Figure 4.9 (see Color Plate IV). Analysis of a simple harmonic oscillation. The program and result windows are tiled. The left panel contains a typical time plot, while the right panel shows a phase-plane plot.

Define the system in PLAS, or call up the file *Sine.plc* and solve it from $t0 = 0$ to $tf = 20$ with a report interval $hr = 0.01$. The results are obtained very quickly, exhibiting the sine and cosine oscillations. The (phase-plane) trajectory is called up by double-clicking the graph and specifying the *Phase 2D* representation (Fig 4.9). As to be expected from the theory of harmonic oscillations, the trajectory associated with the sine wave is a circle (which may appear as an ellipse because of differences in scale between the X-axis and the Y-axis). A trajectory that is closed, as in this case, is sometimes called an *orbit*. The amplitude of the oscillation is reflected in the diameter of the circle, and the initial value determines where on the circle the oscillation starts. Demonstration of these facts is left as Exercise 5.

The speed of the oscillation is not directly visible. This is typical for phase-plane plots, which, by design, exclude time. Implicitly, time is part of the process, and one could place tick marks on the graph, showing at what time the curve runs through a particular point. However, without such marks, time is not visible. PLAS allows output graphs in which the solution is depicted only at the specified report intervals. The density of points in such a plot indicates the speed with which the system moves along a trajectory. This option is realized in PLAS by calling up *Curve options* under the *Results* menu and changing the line style of the variables by toggling to equally spaced points.

It is worth remembering that a sine wave corresponds to a circle in the phase plane. Oscillations that decrease in amplitude spiral inward, and growing oscillations spiral outward. The following examples demonstrate some oscillatory patterns in a simple branched pathway.

Example: Branched Pathway with Feedback

Consider the branched pathway in Fig. 4.10. It contains four dependent variables and one independent variable. One of the end products, X_3, inhibits the synthesis of X_1. In setting up the equations, we have to make sure that precursor–product relationships and branch point constraints are satisfied (cf. Chapter 3). A numerical

Figure 4.10. Branched pathway with feedback.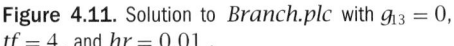

model that satisfies all requirements is

$$\dot{X}_1 = 10 X_3^{g_{13}} X_5 - 5 X_1^{0.5}, \qquad X_1(0) = 1.1,$$
$$\dot{X}_2 = 5 X_1^{0.5} - 10 X_2^{0.5}, \qquad X_2(0) = 0.5,$$
$$\dot{X}_3 = 2 X_2^{0.5} - 1.25 X_3^{0.5}, \qquad X_3(0) = 0.9, \qquad (4.9)$$
$$\dot{X}_4 = 8 X_2^{0.5} - 5 X_4^{0.5}, \qquad X_4(0) = 0.75,$$
$$X_5 = 0.5.$$

Note that the inhibitory effect of X_3 on the synthesis of X_1 was retained as a symbol, g_{13}. This was done because the primary purpose of this analysis is an exploration of the influence of the strength of this inhibition. In particular, by setting $g_{13} = 0$, we obtain a system without feedback. The system is implemented in the PLAS file *Branch.plc*. Exercise 6 asks you to formulate the pathway as a GMA system.

As a baseline analysis, we set $g_{13} = 0$ and, not knowing what the dynamic response might be, solve the system with the standard settings $t0 = 0$, $tf = 10$, $hr = 0.1$. The solution is obtained almost instantaneously. We see that the four dependent variables act quite differently (Fig. 4.11). X_1 approaches its apparent steady-state value of 1.0 in a monotonic fashion. X_2 drops rapidly and subsequently approaches a value of 0.25. X_3 initially increases, reaches a temporary maximum, and monotonically decreases to a value of 0.64. X_4 shows a similar, but not identical, pattern. It initially increases to a maximum, but its decrease leads to an undershoot, before it approaches the final value of 0.64. Confirm these results with the *Table* option under the *Results* menu. For later comparison, save the results as *Branch0.res* and tile the display of windows.

Now we'll start our exploration of the effect of inhibition. The S-system structure guarantees that the strength of inhibition is directly and uniquely represented by the value of g_{13}, so let's try different values for this parameter. We know that the value must be negative, since otherwise the effect would be activating or augmenting,

Figure 4.11. Solution to *Branch.plc* with $g_{13} = 0$, $tf = 4$, and $hr = 0.01$.

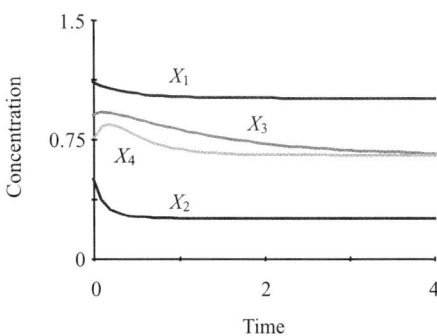

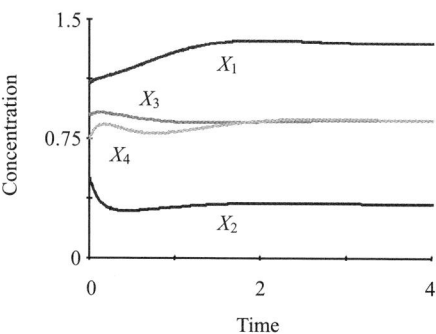

Figure 4.12. *Branch.plc* with $g_{13} = -1$.

rather than inhibiting, the synthesis of X_1. Without specific prior knowledge, we set $g_{13} = -1$ and solve.

The effect is striking (Fig. 4.12). X_1 no longer decreases monotonically toward 1, but increases to a temporary maximum, before approaching a value of about 1.35. In other words, the transient response as well as the steady-state value has changed. The other three variables also exhibit responses that are qualitatively different from the baseline experiment. For instance, X_4 increases to a first temporary maximum, then decreases, then increases to a slightly higher maximum than before, and finally approaches a new steady state. Again, the dynamic features and steady-state concentrations are affected. Save the results as *Branch1.res*.

Why stop here? In a series of experiments, set $g_{13} = -2, -4, -8,$ and -16, solve, save the results under new names, and tile the window display. The comparison nicely demonstrates the effects of the feedback inhibition exerted by X_3 (cf. Figs. 4.11, 4.12, and 4.13). It leads to increasingly stronger oscillations, and eventually the system seems to be out of control. In fact, if g_{13} is set to -64, the first oscillation leads to X_1 and X_2 crashing, and the solution is prematurely terminated. A value of -64 is not very realistic for a biochemical system, but a mathematical analysis might include such a case, because it might deepen our understanding of what these types of systems are able to do. In actual analyses of biochemical pathways, inhibition parameters are usually close to zero and seldom fall below -2 (e.g., cf. Shiraishi and Savageau, 1992a–d, 1993; Torres, 1994ad; Curto et al., 1995, 1998a; Ni and Savageau, 1996ab).

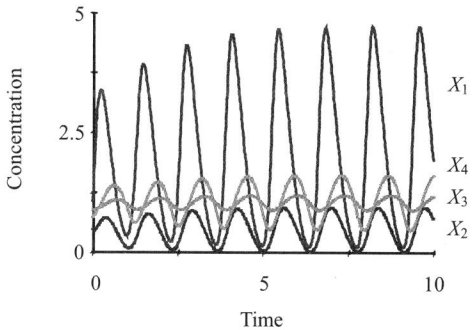

Figure 4.13. *Branch.plc* with $g_{13} = -16$.

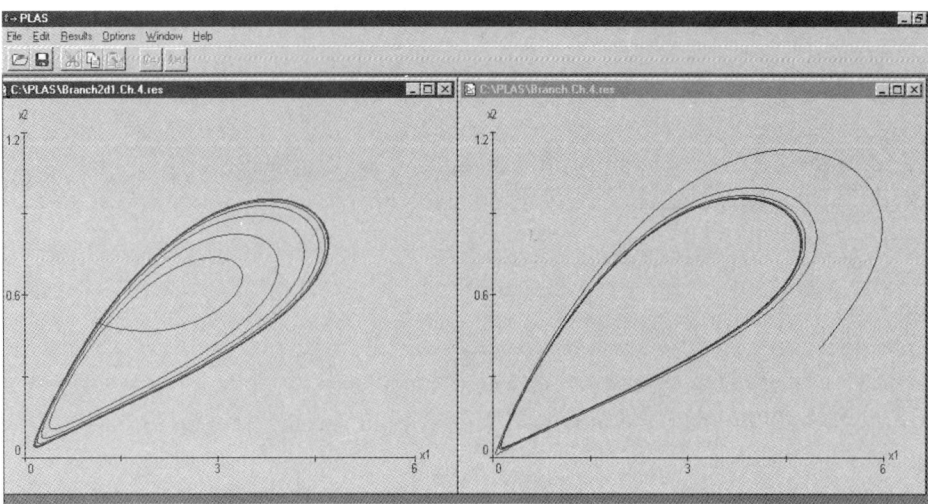

Figure 4.14 (see Color Plate V). Tiled PLAS output of *Branch16.plc*, showing how a stable limit cycle is approached from either the inside or the outside.

Since the behavior of the pathway exhibits some oscillations, it might be useful to study the phase planes of the system. For instance, we could pick X_1 and X_2, or X_3 and X_4, as pairs that are plotted against each other. It is again useful to begin with the baseline system ($g_{13} = 0$). Load the graph of *Branch0.res*, if it isn't still on the screen, and specify the *Phase-2D* option of the *Results* menu. The result is not very exciting: the trajectory begins at the initial value (1.1, 0.5) and directly approaches the steady-state point (1, 0.25).

The situation is different for the damped oscillation in the case $g_{13} = -8$ (*Branch8.res*). The phase-plane plot in this case shows a pretty spiral. In *Branch16.res*, the oscillation initially grows, but then seems to settle into a regular pattern without growth or decay. The corresponding phase-plane plot reflects this dynamic response. The trajectory spirals outwards but never goes outside some imaginary ellipse. The sustained oscillation corresponds to a so-called *limit cycle*. If the system starts inside such a limit cycle, it spirals out, approaching a sustained oscillation in the time plot and some quasi-elliptical orbit in the phase-plane plot. Otherwise, if the system starts outside the limit cycle, it spirals inwards toward the same limit cycle (see Fig. 4.14). We can test this with the pathway in *Branch16.plc* by initializing it with $X_1 = 5$, $X_2 = 0.4$, and X_3, X_4, and X_5 having values 0.9, 0.75, and 0.5, respectively, as before. The system is now outside the limit cycle, but approaches the same oscillation as with the original initial values. Show this in the time plot.

Controlled comparisons. This example provides a good opportunity to mention the *method of controlled mathematical comparisons* (Savageau, 1985b; Irvine and Savageau, 1985ab; Hlavacek and Savageau, 1995, 1996, 1997). By increasing g_{13}, the entire synthesis term of X_1 is affected, and this changes the entire system, including the steady state. This may not be quite fair if we want to compare, for instance, how a system with feedback and the corresponding system without feedback can tolerate perturbations in input. It was therefore proposed to make the two systems *externally* *equivalent*, which means as similar as possible to the outside observer except for the one feature of interest – in this case, the feedback inhibition. In other words, under normal steady-state conditions, the observer would not notice a difference between the two externally equivalent systems. In our case, this is accomplished by adjusting the rate constant α_1 in such a fashion that the synthesis term of X_1 at steady state is the same as in a system with $g_{13} = 0$. The synthesis term of the original pathway at steady state has the value

$$\alpha_1 X_3^{g_{13}} X_5 = 10 \times 0.64^0 \times 0.5 = 5. \tag{4.10}$$

If we now change g_{13}, for instance to -2, we should adjust α_1 so that this flux again has a value of 5. This implies

$$\alpha_{1adj} X_3^{g_{13}} X_5 = \alpha_{1adj} \times 0.64^{-2} \times 0.5 = 1.220703 \alpha_{1adj} = 5, \tag{4.11}$$

which leads to an adjusted rate constant $\alpha_{1adj} = 5/1.220703 = 4.096$. Indeed, without this change, the system *Branch2.plc* has the steady state $X_{1S} = 1.429078$, $X_{2S} = 0.3572696$, $X_{3S} = 0.9146101$, $X_{4S} = 0.9146101$, but with the adjusted α_{1adj} of 4.096, the steady state of the system with $g_{13} = 0$ ($X_{1S} = 1.0$, $X_{2S} = 0.25$, $X_{3S} = 0.64$, $X_{4S} = 0.64$) is regained.

Controlled comparisons have become an important tool for investigating the designs of natural systems. By controlling for all other factors, the differences in results can be attributed to the one structural difference of interest. Further comments about this method can be found in the literature cited above and in the Epilogue. See also part 17.10 of Exercise 17.

Example: A Simple Pathway with Surprise

As another example, consider the map in Fig. 4.15. The pathway is purely didactic, but it does have a structure that could represent some metabolic phenomenon. The system consists of a linear pathway with two dependent variables and one independent variable that feeds the system. The synthesis of the precursor is inhibited by the product, and the conversion of substrate to product is activated by the product. An S-system model of the pathway is set up in the straightforward fashion discussed in the previous chapter, i.e., by including in each term those and only those variables that directly contribute to the corresponding process, allowing for precursor–product relationships, and selecting appropriate parameter values. The resulting S-system representation is

$$\begin{aligned}
\dot{X}_1 &= \alpha_1 X_2^{g_{12}} X_3^{g_{13}} - \beta_1 X_1^{b_{11}} X_2^{b_{12}}, \\
\dot{X}_2 &= \beta_1 X_1^{b_{11}} X_2^{b_{12}} - \beta_2 X_2^{b_{22}}, \\
X_3 &= \text{contant},
\end{aligned} \tag{4.12}$$

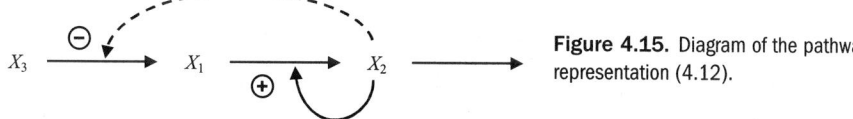

Figure 4.15. Diagram of the pathway with the representation (4.12).

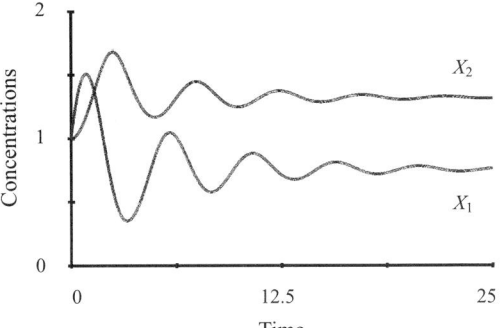

Figure 4.16. First exploration of the surprise pathway [Eq. (4.13)].

and with typical parameter values (cf. Chapter 5) it reads

$$\dot{X}_1 = X_2^{-2} X_3 - X_1^{0.5} X_2, \quad X_1(0) = 1,$$
$$\dot{X}_2 = X_1^{0.5} X_2 - X_2^{0.5}, \quad X_2(0) = 1, \quad (4.13)$$
$$X_3 = 2.$$

The system is implemented with numerical specifications in the PLAS file *Surprise.plc*. As a first exploration, we solve it for a time interval of 0 to 25 with $hr = 0.1$. After some damped oscillations, the system approaches a steady state at about (0.76, 1.32) within about 25 time units (Fig. 4.16). Changing X_3 up or down a bit doesn't change the pattern. The steady-state values are somewhat different, but the overall transient response is unaffected.

Since the system is oscillating, it is useful to look at the phase-plane plot. As expected from the time course plot, the phase plane shows an inward spiral toward the steady state (Fig. 4.17).

If we change the initial values of X_1 or X_2 within reason, the responses are essentially the same. The amplitudes and shapes of the oscillations may change a bit, but the overall appearance of the response is preserved. It is also instructive to change the input. Again, the results are comparable, even though the system now reaches different steady-state values. As an example, Fig. 4.18 shows the responses to setting $X_3 = 1.5$ and also starting X_1 and X_2 at 1.5. Check other scenarios in PLAS. Before execution, predict what may happen; after execution of the simulation, try to explain the results.

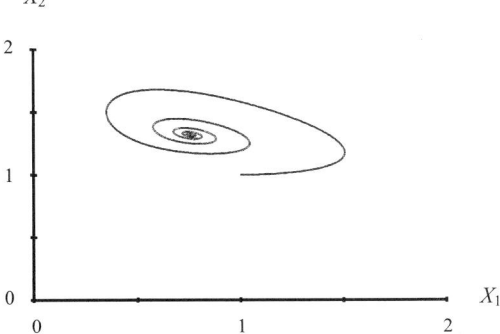

Figure 4.17. The phase-plane plot of the surprise pathway exhibits an inwardly spiraling trajectory.

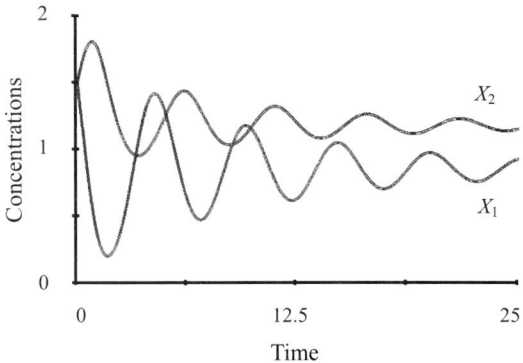

Figure 4.18. The same system (4.13) with $X_1 = X_2 = X_3 = 1.5$ at $t = 0$.

Summarizing the exploration of different combinations of initial values and the independent variable, we may be tempted to conclude that we understand the pathway. Just as a last confirmation, we decide to decrease the environmental input to half its normal value by setting $X_1 = X_2 = 1$, $X_3 = 0.75$. The result is cause for concern. All of a sudden, the system is spiraling outward (Fig. 4.19). We might have expected this behavior for very large inputs, but for *small* inputs?

Since the oscillations are growing, we need to find out when (or whether) they will lead to one or both of the variables becoming zero. The easiest way of accomplishing this is simply to run the system again, but up to a larger final time, such as $tf = 100$. The results are surprising. The oscillations stabilize at about $t = 50$ to reach a constant amplitude and frequency (Fig. 4.20). Predict what the phase-plane plots will look like for time going from 0 to 25 and for time going from 0 to 100. Check your predictions in PLAS.

What happens if we start the system with initial values "outside" this oscillation? We initialize the system with $X_1 = 2$, $X_2 = 1.5$, and $X_3 = 0.75$. The result is shown in Fig. 4.21. The amplitudes decrease and apparently approach the same oscillatory pattern as before. Again, predict what the phase-plane plot will look like and check the prediction in PLAS. Use the *Merge from File* option from the *Results* menu in PLAS to superimpose the phase-plane plots.

What have we learned from the system so far? For strong inputs ($X_3 = 2$ to 3), the system is stable. Upon temporary or (mild) persistent perturbations, it returns to the original or a similar steady state. The temporal responses are qualitatively

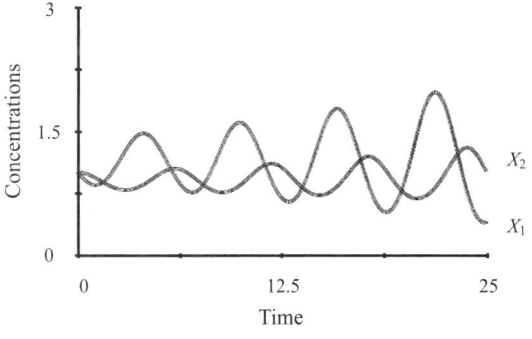

Figure 4.19. Response of the pathway (4.13) with $X_1 = X_2 = 1$, $X_3 = 0.75$. Where are the oscillations heading?

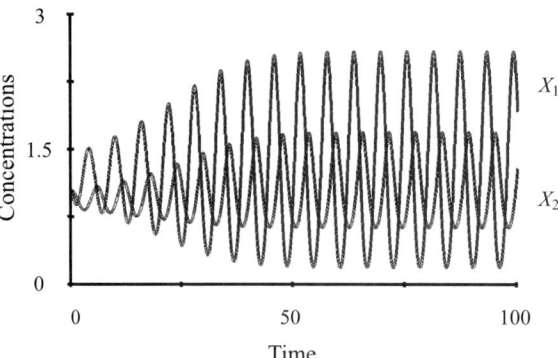

Figure 4.20. Response of the pathway (4.13) with $X_1 = X_2 = 1$, $X_3 = 0.75$ over a longer time period. The oscillations stabilize at about $t = 50$.

unchanged. However, for weak inputs (e.g., $X_3 = 0.75$), the steady state is unstable. Upon perturbation, it moves toward a stable limit cycle. Is there some cutoff value or some threshold for X_3? The answer is yes. In this particular case, the threshold is $X_3 = 1$. For this value, the oscillations neither grow nor decrease. Their phase-plane plots are cycles, but not limit cycles. On whatever trajectory the system starts, it will stay on this trajectory. Confirm this in PLAS; see also Exercise 17.

The main message from this example is that apparently very simple pathways may exhibit a surprising variety of qualitatively different behaviors. Imagine what complicated pathways with dozens of variables and parameters can do! In some cases, complex dynamic features can be discovered and characterized with algebraic means, but in most realistic cases, a complete mathematical analysis is impossible. Simulations are then the only method for exploring the system dynamics, and this example indicates how treacherous that method can be. For systems with dozens of variables and complicated dynamics, many simulations are needed, and extreme caution is still in order when we want to make general statements about the repertoire of possible responses.

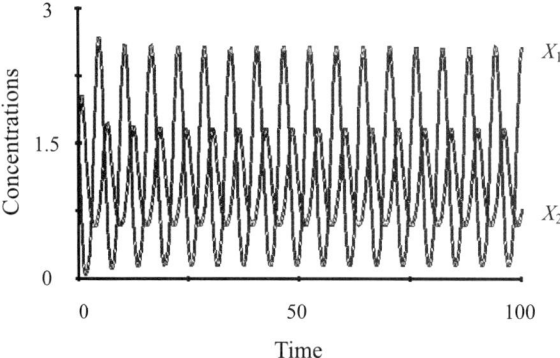

Figure 4.21. Response of pathway (4.13) with $X_1 = 2, X_2 = 1.5$, $X_3 = 0.75$. After one oscillation with higher amplitude, the oscillations approach the same pattern as before (Fig. 4.20).

In this particular example, the system was constructed to possess a stable limit cycle, which corresponds in the phase plane to a distorted circle to which the system converges from the inside or the outside in a spiraling fashion. The algebraic analysis of limit cycles in S-system models is a very interesting topic, but it is beyond our scope here. The interested reader is referred to Lewis (1991) and Voit (1993); for limit cycles in other biochemical models, see also Heinrich et al. (1977) and Heinrich and Schuster (1996). Exercises 7–11 explore this system further, and the project in Exercise 17 introduces another system with surprises.

STEADY-STATE ANALYSIS

Steady-State Concentrations

We have used the dynamical solution option in PLAS to find out which concentration values the dependent variables approach when the system initially is not in a steady state or when it is perturbed by exogenous causes. While this method yields answers, they are not always precise, and in some cases we had to rerun the solution over longer time intervals, before the system finally settled down. The theory behind S-systems allows us to compute steady states in relevant cases with paper and pencil (cf. Chapter 6), and this algebraic method has also been implemented in PLAS. It is called up with the *Steady-State* button and returns steady-state values of all dependent variables (Fig. 4.22).

As a demonstration, load the example of the pathway in Eq. (4.13), *Surprise.plc*, and solve it again to obtain approximate steady-state values. Now click the *Steady-State* button and check the steady-state values of the dependent variables. In the example, the steady-state values are 0.7578583 and 1.319508. Use these values as initial values, and confirm with a numerical solution in PLAS that they do not change over time.

The steady-state window displays additional information (see Fig. 4.22). First, there are flux values. These quantify how much material is flowing through each metabolic pool X_i at steady state. In our former, trivial example of a bucket with a hole, it corresponds to the amount of water leaking out or, equivalently, to the amount of water that needs to be replenished to keep the water level steady.

For the example of the pathway with surprise, both flux values are 1.148698. This value indicates that every minute about 1.15 units of precursor X_3 are converted into X_1 [which is expressed by the first production term, $X_2^{-2} X_3$, in Eq. (4.13); see also the PLAS file in Fig. 4.22], and that the same amount of X_1 is converted into X_2 (which is expressed by the first degradation term, $X_1^{0.5} X_2$). These two terms have the same numerical value, since the system is in steady state.

As another example, load the model of a branched pathway with $g_{13} = 0$, *Branch.plc*. The first flux value is about 5, which indicates that every minute 5 units of precursor are converted into X_1, and that the same amount of X_1 is converted into X_2. Again, these two terms have the same numerical value, since the system is in steady state. Also note that the sum of fluxes leaving the branch point is equal to the flux entering the branch point. Confirm this in PLAS.

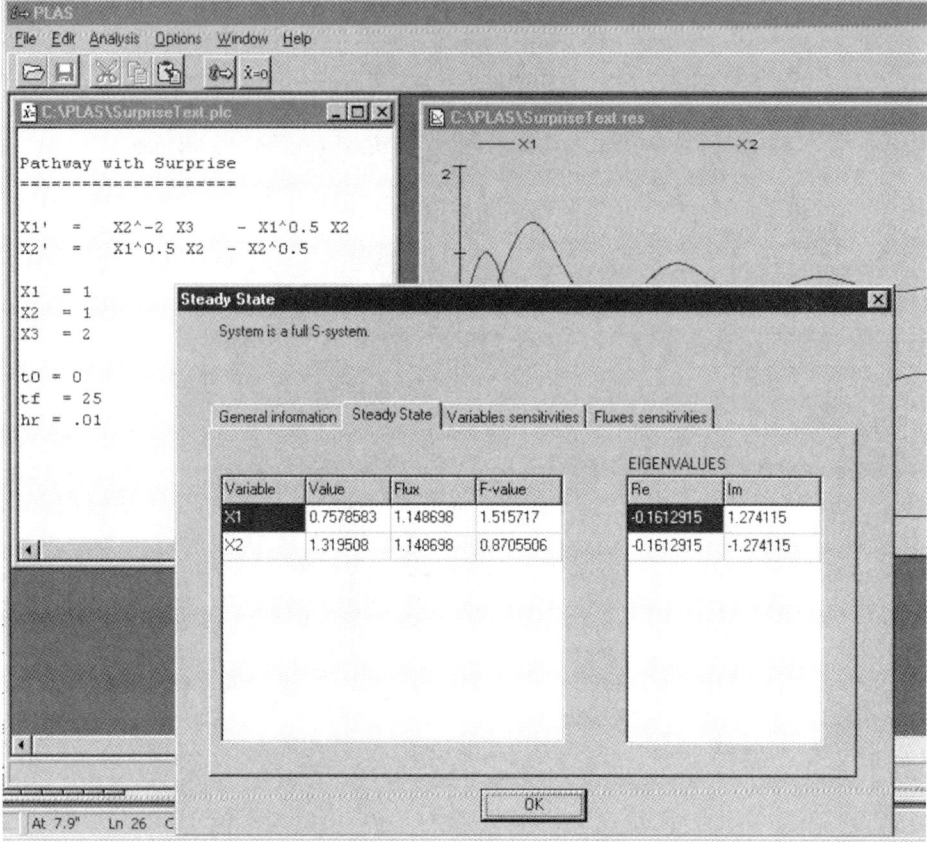

Figure 4.22 (see Color Plate VI). Steady-state solution of the pathway with surprise [Eq. (4.13)], displayed in PLAS.

As a third example, load the model *G6P.plc* of the glucose-6-phosphate pathway and multiply all rate constants by 2. In biochemical terms, this might correspond to some general increase in metabolic rate, as it may be provoked by a rise in temperature. The result confirms our expectation: The steady-state values are unchanged, but the fluxes have doubled. (See also Exercises 12 and 16.)

Logarithmic Gains

We have seen that changes in one of the independent variables can cause the system to assume a different steady state. Instead of solving the system as we did before, we can study the effect with the *Steady State* command in PLAS.

As an example, load the original system *G6P.plc* again, solve it numerically with an altered initial value for glucose, $X_5 = 2.5$, and then compute the steady state. The dynamical solution indicates that all dependent variable concentrations are affected. The steady-state computation confirms this: under this condition of reduced glucose input the entire steady state is lowered to 0.030056, 0.18964, 0.059922. Of course, changes in an independent variable have no effect on the other independent variables.

COMPUTER SIMULATION

Experiments of this type are very important, because they yield insight into the functioning of the model and because they are relatively easy to execute mathematically as well as experimentally in the lab. We shall see in Chapter 5 that the result of this type of experiment can even be crucial in the identification of parameter values of a model. Responses to small variations in independent variables can be computed algebraically in the S-system methodology. They are characterized by so-called *logarithmic gains*, which we will discuss in detail in Chapter 7. In a nutshell, a logarithmic gain is a measure for the relative change in a dependent variable that is triggered by a 1% change in an independent variable. Logarithmic gain computations are implemented in PLAS and called up with the *Variable Sensitivities* tab under the *Steady-State* button. Before we can see the gains, though, we have to declare which quantities are actually independent variables and not just some parameter. For the independent variable that X_5, this is accomplished by adding the declaration

&& X5

Continuing with the glucose-6-phosphate example, declare X_4 through X_{10} as independent variables and click the *Steady-State* button. The logarithmic gains of the three dependent variables with respect to the independent variables are quite different in amplitude. The first seven numbers in the first column show the logarithmic gains of X_1 with respect to X_4 through X_{10} (Fig. 4.23). For example, the second gain shown is about 0.734. This number reflects the approximate percentage change in X_1 caused by a 1% change in X_5: If X_5 is increased, the steady-state concentration of X_1 will also increase, but only by about three-quarters of the increase in X_5.

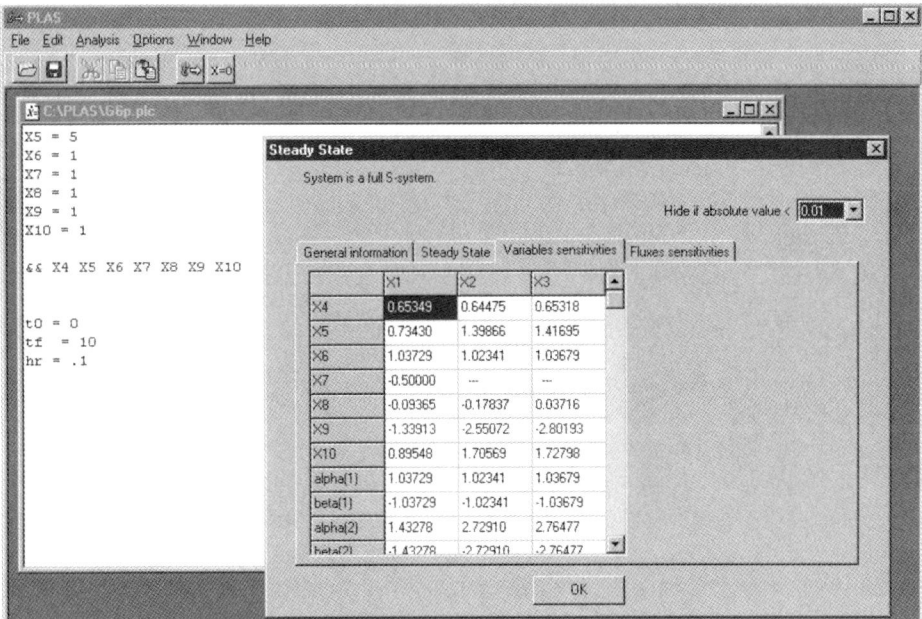

Figure 4.23 (see Color Plate VII). PLAS displays logarithmic gains and sensitivities.

The second column contains the gains with respect to X_2. For instance, the second entry, 1.399, quantifies the change in the steady-state concentration in X_2 that is caused by a 1% increase in X_5. A comparison of the gains of X_1 and X_2 indicates that glucose-6-phosphate responds about twice as strongly as glucose-1-phosphate to changes in exogenous glucose.

Another conclusion from the same output is the difference in responses of glucose-1-phosphate, glucose-6-phosphate, and fructose-6-phosphate to a change in the amount of the enzyme phosphoglucomutase, X_7. The corresponding logarithmic gains are approximately -0.5, 10^{-16}, and 10^{-16}. The latter two are not shown because of the specification *Hide if absolute value is* < 0.01 on the top right of the window. This specification can be changed with the associated drop-down combo box and will be used as default the next time sensitivities are displayed. The logarithmic gains indicate that the steady-state concentration of glucose-1-phosphate decreases by about half a percent for each percent increase in enzyme concentration, while glucose-6-phosphate and fructose-6-phosphate are essentially unaffected by changes in this enzyme. A dynamic solution of the system with all the original settings, but X_7 increased by 10% from 1 to 1.1, confirms this finding: The steady-state concentration of glucose-1-phosphate decreases from 0.05 to 0.0477, which corresponds to a drop by about 5%, whereas glucose-6-phosphate and fructose-6-phosphate are essentially unchanged.

We must be aware that logarithmic gains (in the interpretation given above) are strictly valid only for infinitesimally small variations in independent variables. However, experience has shown that they provide rather accurate predictions of responses even if an independent variable is changed by an appreciable percentage. Chapter 7 discusses these issues in greater detail. Exercise 13 asks you to perform simulations to test the quality of these extrapolations.

Parameter Sensitivities

Similar to logarithmic gains, *parameter sensitivities* quantify how a system responds to changes in parameter values. As before, these changes only affect the dependent variables.

An example is the change in a dependent variable that is evoked by changing the strength of end product inhibition, which is represented by some kinetic order. Continuing the analysis of the glucose-6-phosphate pathway, we ask what are the effects of changes in the activity of phosphofructokinase. This activity is characterized by the kinetic order h_{33}.

Clicking the [x=0] button and selecting the tab *Variable Sensitivities* displays the sensitivities of the dependent variables. For example, the sensitivities of X_1, X_2, and X_3 with respect to h_{33} are 0.736, 1.402, and 1.540, respectively. This implies that a 1% increase in enzyme activity results in increases of about $\frac{3}{4}$% in X_1, and about $1\frac{1}{2}$% in X_2 and X_3.

Sensitivities with respect to rate constants are computed in a similar manner. For instance, X_3 responds to a 1% increase in α_2 with an increase of about 2.8%. In Fig. 4.23, the more precise value 2.76477 is listed in the column denoted $X3$ and

COMPUTER SIMULATION

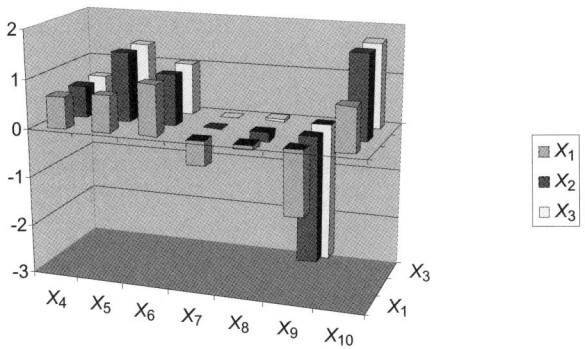

Figure 4.24 (see Color Plate VIII). Pseudo-three-dimensional representation of gains of the glucose-phosphate pathway, using Excel®.

row denoted *alpha(2)*. It is noted that these computations do not take constraints into account. For instance, the sensitivity with respect to α_2 is computed without consideration of β_1, even if the synthesis of X_2 is constrained by the degradation of X_1. Chapter 7 discusses these issues in more detail; see also Exercise 4.

Logarithmic gains and sensitivities provide an impression of how *robust* a system is. When all gains and sensitivities are low in magnitude, perturbations tend to have little effect. By contrast, if some of the gains or sensitivities are very high, even small changes in environmental conditions or the system structure may lead to significant changes. An appealing way for representing all gains or all sensitivities at once is a pseudo-three-dimensional representation in which columns on a grid of independent and dependent variables show the magnitudes of all gains. As an example, the logarithmic gains of the glucose-6-phosphate pathway are shown in Fig. 4.24. The graph was produced by copying the gain data (per selection from the gain/sensitivity table in PLAS and the standard *Ctrl-C* command) and creating a bar graph in Excel®.

Stability

Implement in PLAS the simple S-system

$$\dot{X}_1 = X_1 - X_1^2, \qquad X_1(0) = 0.05, \tag{4.14}$$

which could describe a logistic growth function with a carrying capacity of 1. (Execute the implementation from scratch or from the *Template.plc* file. If you have problems, check the file *Logistic.plc*.) The numerical solution of this differential equation starts at the small initial value $X_1(0) = 0.05$ and gradually approaches 1 in a sigmoid fashion. If we initialize the function at a different value, such as 0.01, the graph rises later, but has the same shape. It is easy to check with the steady-state option that the non-zero steady-state value of the equation is 1. What happens if we begin above 1? Numerical solution with PLAS demonstrates that X_1 decreases and again approaches 1. No matter what the initial value, as long as it is positive, the solution approaches 1.

Now exchange the two terms on the right-hand side, and consider the S-system

$$\dot{X}_1 = X_1^2 - X_1. \tag{4.15}$$

We easily confirm with the steady-state computation in PLAS or with a dynamical solution initialized at 1 that the non-zero steady state is again 1. To explore the dynamics, let's begin with an initial value slightly above 1. Remember that in the previous case the solution decreased toward 1. The solution with the new system exhibits a rather different behavior: it grows and grows, without bound, even if we increase *tf* to a large number. Initiating below 1, e.g., $X_1(0) = 0.9$, results in the system crashing into 0. Only the solution beginning at exactly 1 remains the same as before; all other solutions diverge away from 1.

The situation encountered here is that of *instability*. If the system is exactly at the steady state, it remains there, but any minute disturbance causes the system to diverge, in our case either toward zero or to infinity.

The analysis of stability or instability is in general not a trivial matter. For S-systems, however, it can most often be executed with paper and pencil (at least in principle; see Chapter 6) or with numerical means. The indicators for stability or instability are the so-called *eigenvalues* of the system. If the system contains n dependent variables, it possesses n eigenvalues. Independent variables do not count in this context. Eigenvalues are complex numbers that consist of two parts, the *real part* and the *imaginary* part, and are written, e.g., as $5.003 + 0.6i$. Here, 5.003 is the real part and 0.6 is the imaginary part, which is identified with the imaginary "unit" i. The imaginary component 0.6 is of a different nature than the real part 5.003. Like latitude and longitude on a globe, real and imaginary parts represent different directions in the so-called *complex plane*. In particular, it is *not* valid to add the two components together to something like $5.603i$.

The good news in all the confusion is that the only aspect of interest in the context of stability is whether the real parts of all eigenvalues of the system are positive, negative, or zero. The simple rule says: *If all real parts are negative, the steady state is (locally) stable. If even one real part is positive, the steady state is unstable.* The parenthetic term *locally* again refers to the fact that stability is a feature that, in a strict mathematical sense, applies only to infinitesimally small perturbations. In other words, a locally stable system is mathematically guaranteed to tolerate a perturbation only if this perturbation is very, very small. In practice, the perturbations can often be quite large, and a stable system will still be able to tolerate it.

If one or more real parts are zero and at least one real part is positive, the system is unstable. If one, several, or all real parts are zero and any remaining real parts are negative, the decision about stability requires further analysis (cf. Lewis, 1991). These cases are rather rare in practice and can often be addressed with numerical simulations in PLAS. An example is the pathway with surprise for $X_3 = 1$.

The imaginary part does not contribute to the question of stability, but contains some information as well: Non-zero imaginary parts indicate that the system is likely to exhibit oscillations (for a detailed analysis, see Savageau, 1976; Lewis, 1991).

PLAS displays all eigenvalues along with the steady-state values upon clicking the *Steady-State* button. For the pathway with surprise (Eq. 4.13), the two eigenvalues have negative real parts (-0.1612915), indicating local stability; the imaginary parts are ± 1.274115, indicating the potential for oscillations (see Fig. 4.22).

All eigenvalues of the *G6P.Plc* example are negative, indicating stability of the system. In the case of the branched pathway with feedback, stability depends on the inhibition parameter g_{13}. For reasonable values, e.g., $g_{13} = -0.5$, the system is stable. However, extremely strong inhibition or a switch to a positive value for g_{13} can destabilize the system (Exercises 14 and 15).

The system of sine and cosine oscillations in Eq. (4.7) exhibits two eigenvalues with zero real parts. The stability of the steady state, which is at the center of the phase-plane plot, requires more sophisticated analysis. Dynamical simulations show that if the system is perturbed, e.g., by an exogenous small change in either X_1 or X_2, the oscillations change in amplitude and do not return to the previous pattern (see Exercise 5). Such an oscillation would not be very healthy for a biological system. Imagine an EKG whose oscillation characteristic is permanently changed with every perturbation! Biological systems should return to their "target" oscillations. Such stable oscillations are typically realized as stable limit cycles, as we encountered them above.

EXERCISES

Many of the following exercises begin with "Explore...," "Study...," or "Analyze...." No instructions are given as to how many simulations to execute. This is intentional, because it reflects reality. Sometimes one or two simulations may be all that is necessary to gain full insight. In other situations, even very many simulations do not assure us that we truly understand the system under study. The necessary number of simulations also depends on the level of expertise and experience with these types of systems. To hone your intuition, it is strongly recommended that you *always* predict the system responses before executing the simulation.

1. Explore how the glucose-6-phosphate pathway reacts to other initial conditions. Perform enough simulations to obtain a comprehensive understanding of this aspect of the pathway. Tabulate the results and describe them in words.

2. Study the effects of different bolus injections of glucose-6-phosphate or fructose-6-phosphate into the glucose-6-phosphate pathway. Describe the effects in words.

3. What is the steady state of the glucose-6-phosphate pathway when the activity of phosphofructokinase is reduced by 10%, 25%, 50%?

4. Study the effects of alterations in other enzyme activities of the glucose-6-phosphate pathway. Which enzyme affects the pathway the most? Support your answers with simulation results and steady-state analyses.

5. 5.1. Show that for harmonic oscillations like (4.8) the amplitude is reflected in the diameter of the circle, and the initial value determines the location on the circle where the oscillation starts.

   5.2. Show that the frequency of the oscillation does not affect the overall appearance of the phase-plane plot. Which details of the plot are affected?

5.3. The system (4.8) exhibits two eigenvalues with zero real parts. Perturb the system with exogenous small changes in either X_1 or X_2. Interpret the results. Try to introduce system alterations that stabilize the system (anything goes!).

5.4. Replace the rate constants α_1 and β_2 with an additional variable X_3 that is defined as the decreasing function $\dot{X}_3 = -\beta_3 X_3$. Execute simulations with different values for β_3 and different initial values. Interpret the results. Before you compute the solution in PLAS, make predictions about both the time course and the phase plane.

6. Implement the branched pathway with inhibition (Fig. 4.10) as a GMA system. Identify all constraints that need to be imposed. Compare the numerical results with those of S-system analyses.

7. For the pathway with surprise (4.13), show the effects of X_3 by means of phase-plane plots. Predict the appearance of each plot before you execute the simulation.

8. For the pathway with surprise (4.13), study responses for $X_3 = 1$ and for values very close to 1. Study time courses and phase-plane plots. Superimpose trajectories for $X_3 = 1$, using different sets of initial values for X_1 and X_2. Predict the responses before execution of the simulation, and explain them afterwards.

9. In the pathway with surprise (4.13), study the effects of changes in some of the system parameters. For instance, analyze the role of X_2 activating its own synthesis. Characterize the influence of the feedback inhibition on the production of X_1.

10. What happens in the pathway with surprise (4.13) when all four rate constants are changed to the same degree or to different degrees? For instance, set all or some rate constants to 1 or to 20.

11. The pathway with surprise represents a stable limit cycle oscillation. Solve the system backward in time (setting $tf < t0$ and *hr* negative), and interpret the results. Hint: Use insight from Exercise 10.

12. To review the concept of steady-state concentrations and fluxes, load any of the systems (e.g., the branched pathway) and multiply just one of the rate constants by ten. Predict the results intuitively, run the simulation, compare prediction and actual outcome, and interpret what this experiment indicates. Now multiply all the rate constants by ten. Predict the responses, and check them in PLAS.

13. In the section on logarithmic gains, we showed that a 1% increase in X_5 would evoke a 0.734% increase in X_1. By extrapolation, a 5% increase should lead to a 3.67% increase, and a 30% decrease should yield approximately a 22% decrease. Perform simulations to test the quality of these extrapolations. First, compute the expected steady-state values based on logarithmic gains. Second, determine the steady-state values of X_1 in *G6P.plc* models, in which X_5 is altered appropriately. Third, compare and interpret the results. Do the same with 50% or 100% increases or decreases.

14. Stability of the branched pathway with feedback depends on the inhibition parameter g_{13}. Find cases for which the system is unstable. Determine the threshold for stability. Confirm your findings with simulations.

15. Change the inhibition parameter g_{13} in *G6P.plc* to a positive quantity, such as $+2$. Compute steady-state and dynamic solutions. Interpret the results in terms of biochemical and mathematical terminology.

16. 16.1. Show that, for any S-system, increasing all rate constants by the same amount corresponds to changing the time scale of the experiment. Hint: Introduce a new variable for time that is twice as fast as the original variable, and express the differential equations in terms of this variable.

16.2. For different S-system models, multiply all rate constants by -1. Observe the transients, analyze steady-state characteristics, and try to formulate general insights.

16.3. Test your deductions with the branched pathway.

17. (Project) Analyze the didactic example represented by the map

17.1. Describe the pathway in words.

17.2. Set up an S-system model of the pathway. Consider X_3 an independent variable. What would a GMA model of the pathway look like?

17.3. Specify all rate constants as 10. Since X_3 is transported into the system, set its kinetic order to 1 (to see why, read Chapter 5). Set the kinetic order associated with the degradation of X_1 to 0.5. Suppose the feedforward inhibition of X_1 has a strength of -1 and the inhibition parameters g_{12} and h_{22} are -0.4.

17.4. To get started, initialize the dependent variables at 1 and assume a value of 2 for the independent variable.

17.5. Perform simulations with different initial values; record steady states and check for stability.

17.6. Change the input concentration to different values, and record the results.

17.7. Multiply all equations by $+2$. Execute dynamic and steady-state analyses, and interpret the results.

17.8. Multiply all equations by -2. Execute dynamic and steady-state analyses, and interpret the results.

17.9. Analyze the system for different values of the inhibition parameter h_{22}.

17.10. Perform controlled comparisons of systems with different values of h_{22}. Implement these in PLAS in such a way that the rate constant is automatically adjusted. Compare with the results in part 17.9.

COMPLEMENTS

Model Refinements

We would like to think that once we have done our best to design a model, the model analysis is completed. Reality teaches us otherwise. Modeling is an iterative process that switches back and forth between experimentation and mathematical analysis. Usually, experimental findings suggest a hypothesis that is formulated as a question that can be answered by the model. For instance, one might wonder which components affect the system most strongly. Once a mathematical model is developed, it is tested against observations and used to make predictions. Deviations between experimental and model results often imply that assumptions about the data or the model are wrong, and this uncertainty often suggests clarifying experiments and model refinements. Thus, modeling and experimentation go hand in hand in the elucidation of a complex phenomenon.

Amending a mathematical model does not necessarily imply that the original model is faulty, but may reflect that new aspects have gained in importance or that the focus of the analysis has changed. For example, it may become appropriate to replace a formerly independent variable with a dependent variable or vice versa. In the former case, a variable that was considered constant in fact shows some slight dynamics that in retrospect appears important. In the latter case, we may perform a new type of clamping experiment that allows us to hold a variable constant.

Changes of these types are comparatively easy to implement in S-system and GMA models. For instance, the replacement of a dependent variable with an independent variable merely requires the elimination of its differential equation and the declaration that this variable now is an independent variable with a fixed value.

The replacement of a formerly independent variable with a dependent variable is slightly more complicated in that it requires us to define a new system equation. However, this definition proceeds according to exactly the same rules and recipes that guide the development of a model from scratch. It is even possible to replace an independent variable with a time-dependent process that is yet independent of the system. For example, we may want to test the response of a system to a pulse input or to a bolus that decreases exponentially.

The variables of the model are not necessarily of the same nature. For instance, nothing prevents us from including temperature as an independent or dependent variable. Other useful quantities that may be modeled as dependent variables are measures of the productivity of a system or the total accumulation of metabolites.

It may even be useful to replace a dependent variable with an entire system of variables. Such a situation may emerge when experiments have yielded new insights into the original pathway. For instance, suppose the original model contained one combined pool of the adenylates ATP, ADP, and AMP. To allow for conversions between these metabolites in an amended model, the original S-system equation is replaced with three equations, whose parameters are computed from the new experimental information. The original variable is substituted, wherever it appears, with one, two, or all of the new variables, thus reflecting which component of the formerly combined pool actually exerts an effect. Flux terms not affected by the former variable remain unaltered. A detailed example is given in Chapter 10.

The opposite step may be relevant in situations where the model becomes overly complicated. For instance, it may turn out that the dependent variables of a model show responses at very different time scales, with some rapidly reaching a steady state while others are just at the beginning of their transient response. In such cases, it may be feasible to split the model into a fast and a slow subsystem (cf. Savageau, 1976). For analyses of the fast subsystem, one considers only the variables with fast dynamics as dependent variables; for analyses of the slow system, only the slowly

COMPUTER SIMULATION

responding variables. This separation of time scales obviously reduces complexity and simplifies algebraic and numerical analyses. It is similar to modal analysis (for metabolic examples, see Palsson and Lightfoot, 1984; Palsson et al., 1984, 1985) and, if justifiable biologically and mathematically, can greatly simplify the analysis and may ameliorate numerical problems of stiffness (Agresar and Savageau, 1998).

In some cases, the original model may, intentionally or unintentionally, have ignored secondary reactions or salvage pathways. Later inclusion of these requires us to revisit the affected flux terms while all other parts of the system remain unchanged. An example is a series of analyses of the tricarboxylic acid cycle in *Dictyostelium* (Shiraishi and Savageau, 1992a–d, 1993; see also Chapter 9).

The switching between models at different levels of organization is in general a dramatic process. For instance, the switch from a Michaelis–Menten model of a metabolic pathway to a physiological model at the level of an organ is typically accompanied by such massive changes in conceptual frameworks and terminologies that it is very difficult to integrate information from both levels. In contrast, the two models, formulated as S-systems, would have exactly the same mathematical structure, and subsystems from the lower level could be incorporated in the higher-level model with relative ease. This compatibility of models at different levels of organization has been called the *telescopic property* of S-systems (Savageau, 1979a, 1985b). With the exception of linear systems, other power-law approaches, Lotka–Volterra systems (e.g., Peschel and Mende, 1986; Hernández-Bermejo and Fairén, 1997), and fractals, almost no other modeling approaches are known to possess this feature.

Condensation of Pools

Consider a linear pathway of five metabolites with constant input, as shown in the top panel of Fig. 4C.1. In some cases, not much is known about the intermediate metabolites X_2, X_3, and X_4, and in other cases, they may not be of particular interest. An obvious question is therefore: Can they be eliminated from the system without changing its remaining features, as indicated in the middle and bottom panels of Fig. 4C.1? As expected, the answer is: That depends.... It depends on the system and on the part to be condensed, and it also depends on the system features of interest. As a rule of thumb, we might say that if nothing leaves the condensed part of the system and if nothing is added, with the exception of the output from X_1 and the

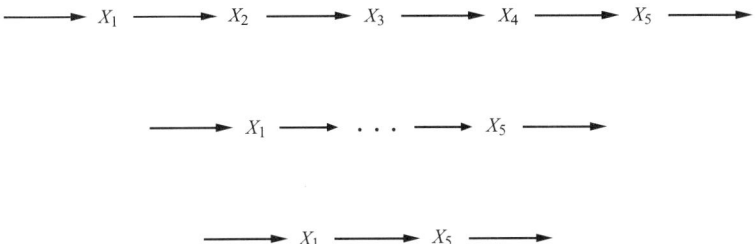

Figure 4C.1. Condensation of a linear pathway.

input to X_5, the steady state is usually preserved, but the dynamics are affected. Both aspects are easy to see intuitively. At steady state, the same amount of material enters and leaves the condensed part, and this amount is quantified by the steady-state flux between the remaining variables of the condensed system (in the linear pathway of Fig. 4C.1, the arrow between X_1 and X_5). At the same time, it is not surprising that the dynamics will be affected. Imagine a perturbation in input. In the full system, this perturbation filters through all pools in a sequential fashion, which takes more time than in the condensed system.

Since intuition is sometimes deceiving, we set up a full system and its condensed analogue and compare results of some representative simulations. The full system, with typical parameter values (see Chapter 5), may read

$$\begin{aligned} \dot{X}_1 &= 2.2 - 2X_1^{0.5}, & X_1(0) &= 0.1, \\ \dot{X}_2 &= 2X_1^{0.5} - 2.1X_2^{0.5}, & X_2(0) &= 2, \\ \dot{X}_3 &= 2.1X_2^{0.5} - 2.2X_3^{0.75}, & X_3(0) &= 1, \\ \dot{X}_4 &= 2.2X_3^{0.75} - 2.4X_4, & X_4(0) &= 2, \\ \dot{X}_5 &= 2.4X_4 - 2.5X_5^{0.5}, & X_5(0) &= 0.1. \end{aligned} \quad (4\text{C.1})$$

Note that precursor–product relationships are reflected in the same kinetic orders and rate constants of each subsequent pair of metabolites. The corresponding condensed system also must satisfy precursor–product relationships. It is convenient to implement the two systems simultaneously in one PLAS program, which is easily accomplished when we rename the variables in the second system as C_1 and C_5:

$$\begin{aligned} \dot{C}_1 &= 2.2 - 2C_1^{0.5}, & C_1(0) &= 0.1, \\ \dot{C}_5 &= 2C_1^{0.5} - 2.5C_5^{0.5}, & C_5(0) &= 0.1, \end{aligned} \quad (4\text{C.2})$$

which, again, satisfies precursor–product relationships. Execute the implementation (or load the provided PLAS file *Condense.plc*).

Without even solving the system dynamically, we can compute the steady state by clicking the [x=0] button. Our expectation is confirmed: The steady state of the full system is (1.2, 1.1, 1, 0.92, 0.77), and the condensed system has the same values for the first and last variables. The fluxes are also equivalent.

Now we analyze the dynamics by solving the two systems up to *tf* = 20. The graph at first looks confusing, with some variables increasing monotonically and others exhibiting undershoots and overshoots. For the comparison, we identify C_1; but where is X_1? The *Table* representation under the *Results* menu shows that the two variables are exactly the same, and since X_1 is plotted first, it is subsequently covered up exactly and completely by the graph of C_1. While X_1 and C_1 are the same, X_5 and C_5 behave quite differently. C_5 exhibits a slight S-shape, but climbs monotonically to its steady-state value of 25. By contrast, X_5 of the full system starts off with a rapid increase, falls back to a temporary minimum, and then begins a monotonic rise to the steady state. Clearly, the dynamics of the two systems are different.

Another illustration of the differences in dynamics starts the system at steady state and adds a bolus of input. Insert in the equations of X_1 and C_1 the term +bolus,

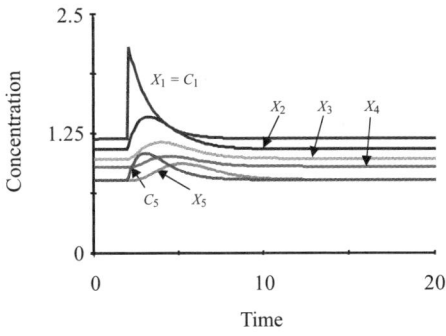

Figure 4C.2. Response to a linear pathway with variables X_1 through X_5 and the corresponding condensed pathway with variables C_1 and C_5. The responses of X_1 and C_1 are identical.

and add the statements

@ 2 bolus = 10

@ 2.1 bolus = 0

Solving the system confirms that the system starts at the steady state. At time 2, X_1 and C_1 jump by about one unit, and they subsequently drop back to the steady state in a monotonic fashion (see Fig. 4C.2). X_2, X_3, X_4, and X_5 react with successively longer delays and weaker responses, reflecting their positions in the pathway. A comparison of X_5 and C_5 again shows the effect of condensing parts of the pathway. There is essentially no delay in the response of C_5, and its maximum value (about 1.06) is noticeably higher than the maximum of X_5 (about 0.95). In fact, the response of C_5 resembles more that of X_2 than that of X_5.

There is no easy way of compensating for the difference. An entire area of systems analysis deals with *delay equations* that would be needed here (for biological examples, see Murray, 1990; Yeargers et al., 1996). However, we may focus on the positive side of this phenomenon: Turning the argument around, one easily sees that the *inclusion* of more intermediates in a pathway leads naturally to time delays. It is thus simple to account for delayed reactions, should they be observed.

The analysis of the linear pathway was rather straightforward. In general, it may be very difficult to predict how the condensation of parts of a nonlinear model may affect the dynamics, and many simulations may be necessary to develop a sufficient understanding of the possible consequences of condensation. By the same token, the expansion of a model component into a pathway or into an entire submodel must be expected to lead to new dynamic features, even if the steady-state values of the original system remain the same. Exercise C.1 discusses another example.

Productivity

Sometimes the total productivity of a system or process over a period of time is of interest. To assess the productivity of the entire system, one defines dependent variables that collect the material that otherwise would leave the system. For instance, suppose we want to know how much material is produced by the branched pathway

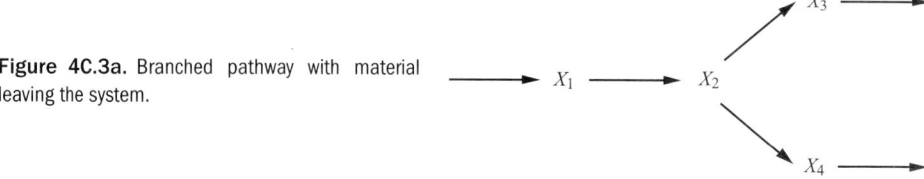

Figure 4C.3a. Branched pathway with material leaving the system.

in Fig. 4C.3a, which may be modeled by the S-system

$$\begin{aligned}
\dot{X}_1 &= 2 - 3.5 X_1^{0.5}, \\
\dot{X}_2 &= 3.5 X_1^{0.5} - 1.5 X_2, \\
\dot{X}_3 &= 0.5 X_2 - 0.2 X_3^{0.75}, \\
\dot{X}_4 &= X_2 - 0.4 X_4.
\end{aligned} \qquad (4C.3)$$

To assess the productivity of the system, we define two new dependent variables, X_5 and X_6, as collecting pools for material leaving X_3 and X_4 (Fig. 4C.3b). The system equations for the new variables are easily determined from the precursor–product relationships. They read

$$\begin{aligned}
\dot{X}_5 &= 0.2 X_3^{0.75}, \quad X_5(0) = 0, \\
\dot{X}_6 &= 0.4 X_4, \quad X_6(0) = 0.
\end{aligned} \qquad (4C.4)$$

A reasonable measure of productivity may be obtained in the following fashion. Compute the steady state of the original system with the steady-state option in PLAS. (Question: Why shouldn't you use the six-variable system? If you don't know the answer, try it and explain the result.) Initiate the six-variable model with X_1 through X_4 at the steady state and X_5 and $X_6 = 0$, and solve for a time period of one unit.

Numerical solution of the system provides in $X_5 + X_6$ and in $\dot{X}_5 + \dot{X}_6 = 0.2 X_3^{0.75} + 0.4 X_4$ two different measures of the total productivity of the pathway. The first reflects an accumulation of material over the time period examined, while the latter presents the total outflux per unit time. Note that the pools X_5 and X_6 continue to grow, because the pathway is continuously fed by the constant input $\alpha_1 = 2$. Note also that the numerical sizes of these pools depend on the initial values of the system. Because X_5 and X_6 are dependent variables, they depend on the entire dynamics of the system under the specific simulated conditions.

For computational purposes, X_5 and X_6 may be initialized at 1 instead of 0; then, after the solutions are obtained, 1 is subtracted from all values of X_5 and X_6. This is mathematically legitimate and speeds up the numerical solution, since the

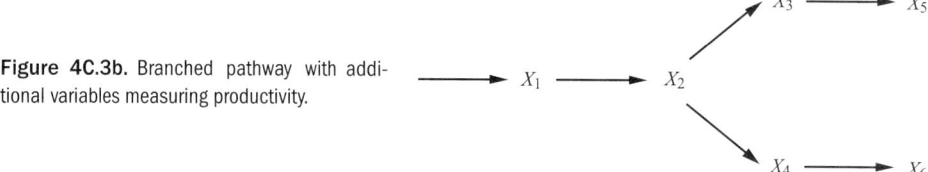

Figure 4C.3b. Branched pathway with additional variables measuring productivity.

Taylor algorithm in PLAS uses logarithmic transformations of variables, which are somewhat problematic for initial values of 0.

The characterization of productivity is slightly more complicated when we are interested in the productivity of one particular, possibly internal, process. Suppose the question is how much X_2 is being produced over a period of time. This accumulated production is described by the integral of the production term of X_2. To be specific, suppose we are interested in the total production between the time points $t = 0$ and $t = 1$, and call this production $P(0, 1)$. $P(0, 1)$ is given as

$$P(0, 1) = \int_0^1 (\text{production term of } X_2) \, dt$$
$$= \int_0^1 \alpha_2 X_1^{g_{21}} X_2^{g_{22}} \cdots X_n^{g_{2n}} \, dt$$
$$= \int_0^1 3.5 X_1^{0.5} \, dt. \qquad (4C.5)$$

Differentiating and renaming $X_7 = P(0, 1)$, we have the total production in the form of the S-system equation

$$\dot{X}_7 = \dot{P}$$
$$= \alpha_2 X_1^{g_{21}} X_2^{g_{22}} \cdots X_n^{g_{2n}}$$
$$= 3.5 X_1^{0.5}, \qquad (4C.6)$$

which we solve from 0 to 1 with the initial amount of X_2 as the initial value. This result is intuitively consistent with the previous case: Imagine that we attach an outgoing arrow to X_1, leading to a new dependent variable X_8. The S-system equation of X_8 would monitor the accumulation of material leaving X_1, and if we specified $\alpha_8 = \beta_1$ and $g_{81} = h_{11}$, it would mean that all material leaving X_1 would be counted twice, once moving into X_2 and once moving into X_8. Since no material is leaving X_8, X_8 provides a record of all efflux out of X_1 and thus of the total production of X_2. Exercise C.2 explores another example.

Time-Dependent Inputs

In all examples so far, the input to a system was essentially constant, even though we considered bolus inputs at predefined time points. A different situation emerges if the input is explicitly time-dependent. For instance, consider a pathway within a target organ that receives input from the cardiovascular system. Imagine a drug that is given intravenously at time point t_0. Because of clearance by liver and kidneys and other degrading influences, the concentration of the drug in the blood decreases in some fashion. It would clearly be unreasonable to include liver and kidney metabolism, but at the same time we must take account of the fact that the pathway is fed by a decreasing input.

The situation can be addressed nicely with canonical models through the introduction of *input modules*. Such modules are simple submodels whose only purpose

Figure 4C.4. Branched pathway with repeated, exponentially decaying input I.

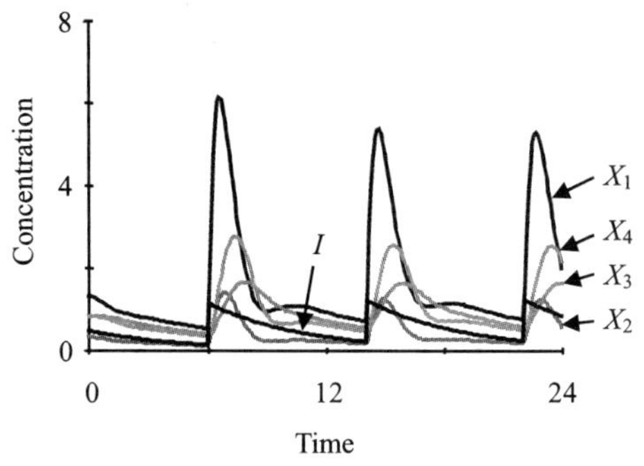

it is to provide the appropriate input. For instance, if the disappearance of drug from the blood is a first-order process, which corresponds to an exponential function of time, we could define the time-dependent input I as

$$\dot{I} = -kI, \quad I(0) = I_0, \tag{4C.7}$$

where k is the rate constant of decay and I_0 the initial concentration available to the pathway. For a drug regimen of three applications a day, this input module could be expanded to, for instance,

$$\dot{I} = -kI + \text{bolus}, \quad I(0) = I_0,$$

$bolus = 0$
@ 6 $bolus = 50$
@ 6.1 $bolus = 0$
@ 14 $bolus = 50$ (4C.8)
@ 14.1 $bolus = 0$
@ 22 $bolus = 50$
@ 22.1 $bolus = 0$

where the bolus statements reflect drug applications of 5 units at 6 a.m., 2 p.m., and 10 p.m.

As an example, load the branched pathway *Branch.plc* with $g_{13} = -1$, and initialize it at the steady state ($X_1 = 1.34652$, $X_2 = 0.33633$, $X_3 = 0.861774$, $X_4 = 0.861774$). Enter the above input equation, and set $I(0) = 0.5$, which could reflect a remainder drug concentration from the previous day. Implement the above bolus statements, and solve. In case your implementation does not work, the model is also provided as *Drug.plc*.

The graphical output nicely shows the exponentially decreasing drug concentration and the responses of all variables in the pathway (Fig 4C.4). Exercises C.3–C.5 explore input modules some more.

Buffer Boxes

An interesting variation on the use of modules is the inclusion of *buffer boxes* in a pathway model. A buffer box is designed to keep system variables from leaving certain

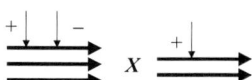

Figure 4C.5. Graphical representation of the change in X in a simple modulated pathway.

nominal ranges by absorbing excess material produced in a pathway and providing material if it is in demand. A buffer box often consists of just one equation. Details on the construction and functioning of buffer boxes can be found in Voit and Ferreira (1998).

As we discussed in Chapter 3, one task of modeling is to keep the model as manageable as possible. This is often achieved by considering variables as constant if they change very slowly, i.e., insignificantly during the mathematical experiment, or if they change so rapidly that they are essentially always at their nominal values. For instance, transitions between different states of an oligomeric enzyme are very rapid, and the different states are not modeled separately. Of course, considering variables as constants is a compromise that may lead to undesired side effects. For instance, it happens in simplified models that marginally interesting variables, such as NADH, accumulate or disappear over time. This situation is not very realistic, since any organism would be buffered against such deviations.

We can allow for such buffering action with a *buffer box* module. Suppose the change in X is represented by the simple map in Fig. 4C.5. X is synthesized by three primary processes that are subject to one activating and one inhibiting process, and it is degraded through two primary processes that are modulated by some other system constituent.

In a real organism, there are additional pathways affecting X, but these are not captured in the model, in order to keep it manageable; they are indicated as lighter, arched arrows in Fig. 4C.6. The justification for not including these additional pathways is that they are normally negligible and become important only if X is in high demand or excess. In these extreme situations, they buffer the concentration of X by returning it to its nominal range, thereby preventing harm to the organism.

A buffer box B is constructed by aggregating all secondary processes of synthesis and of degradation, respectively, and by allowing for the fact that X can be synthesized from buffer material and that excess material from pool X goes back into the buffer pool, as shown in Fig. 4C.7.

The buffer box B is implemented in the same way that canonical models are set up: B is defined as a differential equation and included in the equation of X. Several variations are possible, depending on whether B is included in the synthesis or degradation term of X or in both. Also, we have some flexibility in assigning parameter values. An implementation that seems to work well under typical conditions is

$$\begin{aligned}\dot{X} &= V^+ - B^p V^-, & X(0) &= X_N, \\ \dot{B} &= C_B(X - X_N), & B(0) &= 1,\end{aligned} \quad (4C.9)$$

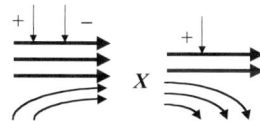

Figure 4C.6. Inclusion of secondary reactions in the simple pathway of Fig. 4C.5.

Figure 4C.7. Secondary reactions are lumped, and X is buffered by buffer box B.

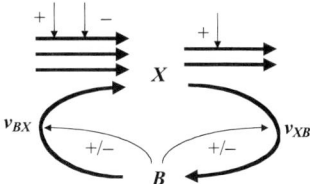

where C_B is a positive *buffer constant* and the nominal value of B is 1; i.e., $B(X = X_N) = 1$. For a concentration below the nominal value ($X < X_N$), the derivative of B is negative, and B decreases. This is the signal to increase the flow from the buffer box into the pool X and/or to decrease the flow out of pool X into B. The opposite is true for $X > X_N$. Variations on this implementation are found in Voit and Ferreira (1998).

As an example, we analyze a generic linear process in which a toxic by-product (variable *Toxic*; X_3) is generated (see Fig. 4C.8). *In vivo*, the organism must attempt to keep the concentrations of such by-products below the level of toxicity, which can be expected to involve carefully regulated detoxification mechanisms. In the model, these mechanisms are lumped into a single buffer box, *ToxBuff*; X_4.

As an illustration, the pathway is exposed to (pseudo-)randomly fluctuating input, *Stim* (X_5), which is realized as a *frequency-modulated Rössler* oscillator, whose dynamic is shown in Fig. 4C.9 (for the mathematical definition, see Exercises C.7 and C.8 and Voit, 1993). A comparison between the unbuffered and the buffered system demonstrates the efficacy of the buffer box (Fig. 4C.9).

The map of the pathway is easily translated into an S-system model. It may read, with typical parameter values,

$$\dot{X}_1 = X_5 X_2^{-1} - X_1$$
$$\dot{X}_2 = 0.8 X_1 - X_2$$
$$\dot{X}_3 = 0.2 X_1 - X_3 X_4 \quad (4C.10)$$
$$\dot{X}_4 = C_B(X_3 - 0.01)$$

$$X_1(0) = 1$$
$$X_2(0) = 1$$
$$\text{Toxic } X_3(0) = 0.01$$
$$\text{ToxBuff } X_4(0) = 1$$

The input variable *Stim* (X_5) corresponds to the variable X_{10} in Exercise C.7.

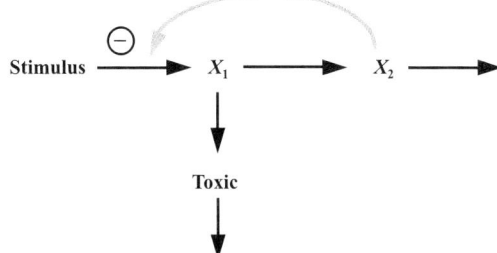

Figure 4C.8. Generic linear pathway with toxic by-product.

COMPUTER SIMULATION

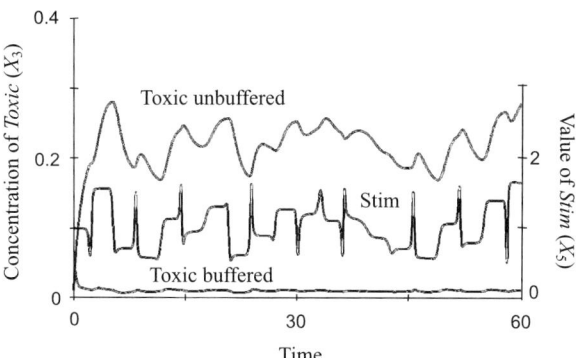

Figure 4C.9. Time courses of by-product *Toxic* (X_3) in buffered and unbuffered linear pathways. The pathway reacts to an input, *Stim* (X_5), which fluctuates according to a pseudo-random oscillator. Buffering parameters are $p = 1$ and $C_B = 1000$ (from Voit and Ferreira, 1998).

As Fig. 4C.9 clearly demonstrates, the buffer box is very efficient. Without buffering, the toxic by-product fluctuates considerably at a rather high level, while the buffer reduces both the absolute level of toxic by-product and the amplitude of its fluctuations. See also Exercise C.8.

Exact Analogues of Other Models

Introducing input modules, we used an exponential function for the time-dependent decay of the drug and formulated this function as a differential equation in order to make it conform with the overall structure of a canonical model. This was legitimate, since the exponential function and the differential equation we used are mathematically equivalent (see Appendix).

The same technique may be used to recast other algebraic functions of time as differential equations in canonical form. In fact, it has been shown that virtually any differentiable function or differential equation can be put in this form, and the two formulations are entirely equivalent (e.g., Savageau and Voit, 1987; Voit and Rust, 1992; see also the Epilogue).

As an example, consider the linear pathway in Fig. 4C.10, and suppose that the conversion of X_1 into X_2 has the sigmoid characteristics of a Hill function. Modeling the process with a single power-law term would capture the essence of the process close to the operating point, and in most cases that would be sufficient. However, suppose, the entire sigmoid characteristic is relevant. A fitting model, with the typical parameters of a Hill function, could read

$$\dot{X}_1 = 0.5 - \frac{V_{max} X_1^n}{K_M^n + X_1^n}, \qquad X_1(0) = 0.5,$$
$$\dot{X}_2 = \frac{V_{max} X_1^n}{K_M^n + X_1^n} - 0.5 X_2, \qquad X_2(0) = 0.5, \qquad (4C.11)$$
$$\dot{X}_3 = 0.5 X_2 - 1.2 X_3^{0.75}, \qquad X_3(0) = 0.5,$$

$$\longrightarrow X_1 \xrightarrow{\text{Hill} \atop \text{Process}} X_2 \longrightarrow X_3 \longrightarrow$$

Figure 4C.10. Linear pathway in which the conversion of X_1 into X_2 is described by a Hill function.

Obviously, this structure is not consistent with that of a canonical model, but it can be put in that form by means of recasting. For this purpose, we define a new *auxiliary* variable, say X_{1A}, as the denominator in the Hill function:

$$X_{1A} = K_M^n + X_1^n, \tag{4C.12}$$

so that the first equation becomes

$$\dot{X}_1 = 0.5 - V_{max} X_1^n X_{1A}^{-1}. \tag{4C.13}$$

Note that this equation is now in canonical form.

The derivative of the auxiliary variable is computed according to the chain rule as

$$\frac{dX_{1A}}{dt} = \frac{dX_{1A}}{dX_1} \frac{dX_1}{dt}, \tag{4C.14}$$

which yields a differential equation in canonical form that is added to the above system:

$$\begin{aligned}\frac{dX_{1A}}{dt} &= nX_1^{n-1}(0.5 - V_{max} X_1^n X_{1A}^{-1}) \\ &= 0.5n X_1^{n-1} - nV_{max} X_1^{2n-1} X_{1A}^{-1}.\end{aligned} \tag{4C.15}$$

The initial value of the new variable is determined by its definition as

$$X_{1A}(0) = K_M^n + X_1^n(0). \tag{4C.16}$$

Exercise C.9 asks you to implement this system and to compare it with one that approximates the Hill function with a power-law function.

While it may appear to be serendipitous that the Hill function is so easily cast in the power-law form, such exact transformation is possible for essentially any ordinary differential equation or differentiable function (Savageau and Voit, 1987; Voit, 1988b, 1990c; see also the Epilogue).

Monte Carlo Simulation

Throughout this chapter, we have used models to simulate scenarios that have some observed or imaginable analogue in the real world. For instance, we simulated the effects of increasing or decreasing the strength of inhibition in a simple pathway.

In a generic situation, we start with parameter values that somehow represent the most likely situation. This situation is consistent with data under "normal" conditions and serves as a baseline. In the second phase, simulations of other situations are executed as deviations from this baseline and compared with it. Prominent simulations in this phase are explorations of worst-case or best-case scenarios, but other scenarios can be investigated that elucidate the system under different sets of conditions. Simulations of this type will accompany us throughout the case studies and, in general, whenever the system is too complicated to permit algebraic analyses. In a metabolic context, a typical simulation may represent a disease state that is characterized by the decreased activity of a key enzyme.

In some areas of science, such as human and ecological risk assessment, *Monte Carlo simulations* have become very popular. Monte Carlo simulations were developed by two mathematicians, Stansilav Ulam and John von Neumann, in connection with the design and testing of the first hydrogen bomb (see Rugen and Callahan, 1996). The simulation method was named after the gamblers' paradise on the French Riviera, because Monte Carlo simulations make heavy use of probabilities, just like gambling.

The key idea is the following. Suppose a model contains many parameters. So far, we have considered these parameters as uniquely determined quantities. In reality, that is often an invalid assumption, as parameters are subject to natural variability and to uncertainties that are due to insufficient knowledge about the phenomenon under investigation. The only reliable information may be that the value of each parameter falls within a certain range and, possibly, that its potential values follow some statistical distribution.

As an example, suppose human body weight is one of the parameters in a model. Clearly, different individuals have different weights, and furthermore, we can only determine the weight of a given person with some measurement error. Studying the population of interest or some relevant sample from it, we may find that body weights are roughly distributed with a lognormal distribution. Executing a simulation with the model, we have to decide which value of this lognormal distribution to pick. For the baseline mode, we may select the mean or median, but it will also be interesting to explore what happens to the model responses if the body weight deviates significantly from the chosen average. A Monte Carlo simulation assesses this situation by running the system very often, sometimes many thousand times. For every run, the value of the body weight parameter is chosen in such a fashion that the parameter values of all simulations taken together form the observed lognormal distribution. A thousand choices of parameter values result in a thousand system responses. These may be some steady-state characteristics or some dynamical feature. The thousand system responses in themselves form a statistical (output) distribution. From a statistical viewpoint, the dynamical model transforms the (input) distribution of parameter values into an (output) distribution of responses (cf. Voit, 1996a).

The Monte Carlo rationale is readily extended to the simultaneous selection of values for several parameters. For instance, imagine a metabolic system in which each enzyme activity is distributed with its own distribution. The shapes and statistical features of these distributions may be the same, similar, or totally different. For each run of a Monte Carlo simulation, one randomly selects one value for each parameter, according to its statistical distribution, and obtains one or several output values of interest.

For simplicity of argument, suppose the steady-state value of the variable X_2 is of particular importance. If there are many enzyme activities and each has its own distribution, an exhaustive exploration of all possible combinations becomes infeasible, yet a good sampling design will provide a representative answer. Let's assume we have executed 1,000 runs of this sort, and thus obtained an output distribution that is composed of 1,000 values of X_2. The results can be interpreted in two ways. First, we can simply study the features of the output distribution of X_2-values, find its mean, variance, and 99th and 99.9th percentiles, characterize its shape, and consider what all these features may mean. Secondly, the results can be listed in a table with columns for every enzyme activity and an additional column for X_2. The first row of entries in the table consists of the values of each enzyme activity and of X_2 in the first run, the second row contains the values for the second run, and so forth.

It is possible to evaluate this table with methods of rank correlation analysis, and the results establish which of the enzyme activities has the strongest effect on X_2. Intuitively, this type of analysis compares scenarios in which all parameter values are roughly the same except for one. If differences in this one parameter result in greatly differing results, the parameter has a strong effect, or, in another terminology, the system is *sensitive* to this parameter. By contrast, if the system responses are about the same, no matter what the value of the parameter, the system is *insensitive* to this parameter.

It is noted that this concept of a sensitivity is slightly different from the parameter sensitivities discussed before and described in detail in Chapter 7, even though both reflect the effect of a parameter on the system. Parameter sensitivities discussed before are computed by differentiation. They describe the precise effect of changes in the parameter value close to the baseline operating point. They are *local properties*. By contrast, sensitivities obtained from Monte Carlo simulations address large changes in parameter values and are *global properties*.

For a statistical assessment of Monte Carlo-based sensitivities, one selects one parameter at a time and computes *Spearman's rank correlation coefficient* or a partial correlation coefficient between the values of this parameter and the output values for all runs of the simulation (e.g., Shott, 1990: p. 248; Norman and Streiner, 1994: p. 176). A comparison of the rank correlation coefficients gives some indication of the importance of the parameter. Detailed descriptions of methods for analyzing Monte Carlo simulations are found, for instance, in a special issue of the journal *Human and Ecological Risk Assessment* [2(4), 1996], in EPA's 1992 *Guidelines for Exposure Assessment*, and in the documentation to simulation software like *Crystal Ball*® (Sargent and Wainwright, 1996) and *@Risk*® (Palisade, 1997). For new twists on Monte Carlo modeling, the interested reader may also consult Voit et al. (1993, 1995) and Bangerter (1998).

EXERCISES

C.1. Implement the following pathway in PLAS. As default kinetic orders, you may use 0.5, except for the inhibition parameter, which you may specify as -1. Study the dynamics under different conditions. Once you essentially understand the system, eliminate X_4, X_5, and X_6 from the pathway. Perform the same simulations as before, and compare. Always make predictions of responses before you execute the simulations.

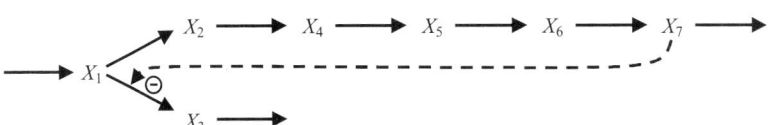

COMPUTER SIMULATION

C.2. Study the productivity of the following branched pathway under two scenarios. In the first scenario, consider X_1 as an independent variable with value 1, and in the second consider it as a dependent variable with initial value 1. Before executing the simulations, predict the responses.

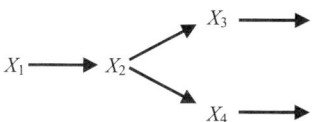

C.3. The model *Drug.plc* describes a treatment schedule with three doses per day. Use different decay constants k, and interpret the results.

C.4. The model *Drug.plc* describes a treatment schedule with three doses per day. Compare this regimen with one that uses fewer, higher doses and with one that uses more, smaller doses. Discuss pharmaceutical advantages and disadvantages of the different regimens.

C.5. Imagine a pathway that is exposed to circadian variations in input. Model these as a sine function input, and study the results.

C.6. (Project) Panetta (1998) proposed a model for the growth of heterogeneous tumors and their treatment with some chemotherapeutic drug. Its dependent variables, X_1 and X_2, describe the sensitive and the resistant cell masses, respectively. The model takes the form

$$\dot{X}_1 = [r_1 - d_1(t)]X_1,$$
$$\dot{X}_2 = b_1 d_1(t) X_1 + [r_2 - d_2(t)]X_2,$$

where r_1 and r_2 are growth rates and b_1 is the rate at which resistance to the drug is induced. $d_1(t)$ and $d_2(t)$ are functions of time, which describe the cell losses. Analyze the model for different types of periodic functions $d_1(t)$ and $d_2(t)$.

C.6.1. Set $d_2(t)$ equal to zero. What does that mean biologically?

C.6.1.1. Model $d_1(t)$ as a sine wave oscillating between values close to 0 and some maximum value D_1.

C.6.1.2. Model $d_1(t)$ as a square wave function, which switches between being a constant for some time and being 0 for some time.

C.6.1.3. Model $d_1(t)$ as a function that decreases exponentially for a while and then is restored to the original value.

C.6.2. Model both $d_1(t)$ and $d_2(t)$ as periodic functions, as indicated in parts C6.1.1–C6.1.3.

C.6.3. Write a report about your findings.

C.7. (Project) The following system was constructed by coupling two chaotic oscillators of Rössler type (see Voit, 1993). The original oscillators are fairly, though not totally, predictable, but the coupling yields very complicated dynamics. Implement the system in PLAS and study its dynamics. Analyze responses to slight changes in the initial value of X_1. Use X_{10} as input variable for a linear pathway and for a branched pathway. Record the results and interpret them.

$$\dot{X}_1 = X_2 - X_3^{25} X_4^{25}, \quad X_1(0) = 25,$$
$$\dot{X}_2 = 0.5 X_2 - X_1, \quad X_2(0) = 47,$$

$$\dot{X}_3 = 58.6 X_3^{-24} X_4^{-25} - 1.984 X_1 X_3^{-24} X_4^{-25}, \qquad X_3(0) = 1,$$
$$\dot{X}_4 = 0.04 X_1 X_4 - 1.184 X_4, \qquad X_4(0) = 50^{0.04},$$
$$\dot{X}_5 = \frac{1}{13}\left(X_2 - X_3^{25} X_4^{25}\right), \qquad X_5(0) = 0.6,$$
$$\dot{X}_6 = X_5^8 X_7 - X_5^8 X_8^{25} X_9^{25}, \qquad X_6(0) = 25,$$
$$\dot{X}_7 = 0.5 X_5^8 X_7 - X_5^8 X_6, \qquad X_7(0) = 47,$$
$$\dot{X}_8 = 58.6 X_5^8 X_8^{-24} X_9^{-25} - 1.984 X_5^8 X_6 X_8^{-24} X_9^{-25}, \qquad X_8(0) = 1,$$
$$\dot{X}_9 = 0.04 X_5^8 X_6 X_9 - 1.18 X_5^8 X_9, \qquad X_9(0) = 50^{0.04},$$
$$\dot{X}_{10} = 0.01 X_5^8 X_7 - X_5^8 X_8^{25} X_9^{25}, \qquad X_{10}(0) = 1.$$

C.8. (Project) Using the Rössler system in Exercise C.7, implement the example of a linear pathway with toxic by-product. Reproduce Fig. 4C.9, and investigate the effect of the buffering parameters p and C_B.

C.9. (Project)

C.9.1. Implement the linear system in Fig. 4C.10 in two ways. First, use the recast form of the Hill rate law as shown in the text. Second, approximate the Hill function with a power-law function with the kinetic order $h = n/2$ and the rate constant $\beta = 0.5 V_{max} K_M^{-n/2}$. For the first analysis, use $n = 2$, $V_{max} = 1$, $K_M = 2$. Determine the steady state. In a second run, initialize the system at the steady state and perturb it between $t = 2$ and $t = 2.1$ with a bolus of 10. Note that the bolus must be entered in the equation of X_{1A} according to the definition of X_{1A}. Compare the results with respect to steady states and dynamic responses.

C.9.2. For further analyses, change the parameters n, V_{max}, and K_M, while making sure that the recast and the approximated system have the same steady state. Determine under what conditions the two systems differ significantly, and try to assess the approximation error.

REFERENCES

[3], [15], [66], [68], [83], [144], [146], [149], [155–7], [164–6], [217], [252], [258–9], [262], [272–5], [277], [279], [298], [307], [319], [321], [330], [352], [374–8], [380], [389], [416], [419], [435], [438], [444], [453–5], [459], [482], [517], [519].

CHAPTER FIVE

Parameter Estimation

Steady states of S-system models and their analyses can, in principle, be executed entirely in symbolic form and with algebraic means, that is, with paper and pencil and without the specification of numerical values for the kinetic order and rate constant parameters. Symbolic evaluations have the advantage of mathematical rigor and of a degree of generality that is seldom matched by numerical simulations. Nonetheless, symbolic steady-state analyses are usually limited to moderately small systems. For instance, we shall see in Chapter 6 that the algebraic evaluation of stability becomes quite cumbersome if the model contains more than four or five variables, whereas it is easily done in PLAS for a numerically specified model. The situation is even more tilted toward numerical analyses when we are interested in dynamical simulations, which are based on solutions of the dependent variables as functions of time. If we had such solutions in an explicit, symbolic form, we could analyze almost every aspect of the system with paper and pencil. However, such solutions are very rarely available, and we must instead obtain approximate, numerical solutions. For numerical analyses of either steady states or dynamic responses, all parameters have to be assigned numerical values. This chapter discusses some standard techniques of deriving such values from experimental data.

There is no unique recipe for the estimation of parameters. In fact, the estimation problem is in general so complicated that in spite of a number of methods and computer programs, the estimation of parameter values for a given system is a major task. One could even claim that parameter estimation is currently the limiting step in biomathematical modeling. For biochemical systems, this situation is improving because biochemists, aware of the data needs for our types of analyses, have begun to produce and analyze data with systems approaches in mind. As a consequence, there are more approaches for biochemical systems than for many other biological phenomena.

Parameter values can be estimated from steady-state data or from dynamical data. The nature of suitable data for these two types of estimation is rather different, and so are the methods of analysis. Estimations of parameter values from steady-state data are generally based on experiments that measure how a biochemical system responds to small perturbations around the steady state. Two approaches can be taken. Either one tries to measure directly how the variable X_j affects the influx into or efflux out of the pool X_i (this effect, by definition, is represented by the kinetic order parameters g_{ij} and h_{ij}, respectively), or one measures logarithmic gains, which describe the influence of an independent variable on a dependent variable. Because gains are closely related to kinetic orders, they provide an indirect measure of relevant system properties and even the topological structure of the model (e.g., Savageau, 1976; Sorribas and Cascante, 1994).

Estimation from dynamic data is based on quite a different type of experimentation. In this case, measurements are needed for all dependent variables at sequential points in time. These temporal data may stem from transient responses after a perturbation from steady state (e.g., Sorribas et al., 1993), but they are more often found in the analysis of systems that do not exhibit a steady state, but exhibit growth, decay, or some other long-term dynamics.

In addition to direct estimations from experimental data, parameter values of S-system or GMA models can be computed from rate laws of fluxes that are available in some other mathematical form, such as a Michaelis–Menten rate law. This approach is very valuable in that the literature seldom presents raw data and many kinetic results are instead presented in the terminology of traditional enzyme kinetics. Estimation of parameters from rate laws is a purely mathematical exercise in (partial) differentiation, in which the S-system terms V_i^+ and V_i^- or the corresponding GMA terms are computed as power-law approximations of the rate laws and their parameter values are identified as shown in the Appendix. Of course, this approach presupposes that the assumed rate laws are true reflections of the reaction *in vivo*. Often the use of Michaelis–Menten functions and similar rate laws is so commonplace that their validity is not questioned. Supporting this practice is some evidence that processes *in vivo* actually tend to be similar to those *in vitro*. However, there is also a growing body of studies suggesting that information obtained *in vitro* is not necessarily adequate for analyses *in vivo*. Questions of relevance and validity are discussed later in this chapter.

In many cases, a mixture of two or three of these methods, augmented by some assumptions or educated guesses, is required to estimate all parameters of a large model. However, with a growing awareness of the importance of integrated analyses, the number of suitable data sets has already begun to increase, and we may have some "pure" sets of parameter values of relevant systems in a not-too-far future.

Much of the information necessary to estimate kinetic orders depends not only on steady-state measurements, but on some kind of dynamical experiments. These experiments may be simple perturbations, as we simulated them in Chapter 4, or actual time series of measurements on transients that follow some deviation from the steady state.

Whatever method is used, parameter estimation is obviously tied directly to experimental data. As mentioned in Chapter 1, information about pathways and their constituents used to be obtained exclusively from book and journal articles. This is rapidly changing with the emergence of databases that allow easy access to a vast amount of information about metabolites, enzymes, pathways, genes, and diseases (see Chapter 1 and the special section of the references dealing with web sites). Also, it should be mentioned again that the new DNA chip and microarray technologies are in the process of creating avenues of analysis unthinkable just a few years ago. For example, it will be possible to study time series of gene expression patterns, which can subsequently be mapped onto metabolic networks and show which pathways are activated or suppressed in response to some stimulus. Thus exhibiting the temporal coordination of biochemical responses, this information, taken together with flux and steady-state measurements, will provide extraordinary insights in the integrated functioning of organisms *in vivo*.

What Kinds of Estimation Results May We Expect?

Before we go into the thicket of parameter estimation, it might be useful to limit the scope of what we hope to achieve. For biological systems, it seems very unlikely that a kinetic order will ever be as high as 1,000 or even 100. Kinetic orders are closely related to stability, gains, and sensitivities (see Chapter 6), and a kinetic order of 1,000 would make a system extremely sensitive and unreliable. For example, imagine a variable X_j has a nominal value of 1 and a kinetic order of $g_{ij} = 100$ in the synthesis term of X_i. Suppose some external influence increases X_j by 5%. The effect of this relatively small increase on the synthesis of X is enormously magnified by the kinetic order from $1^{100} = 1$ to $1.05^{100} = 131.5$. That is, a 5% perturbation leads to a more than 100-fold change.

A rule of thumb is that most kinetic orders are between 0 and 1 for the flow of material and for activations, and that inhibitory effects are often represented with kinetic orders between 0 and -0.5. In fact, an analysis of over 1,000 kinetic orders in recently modeled biochemical systems (Shiraishi and Savageau, 1992a–d, 1993; Torres, 1994bc; Curto et al., 1995; Ni and Savageau, 1996ab; Curto et al., 1998a) shows that an unmodulated flow of material is often modeled with a kinetic order of 1 or 0.5, and that the kinetic orders in modulated flows are usually smaller. A kinetic order of 1 corresponds to a first-order kinetic and is also used for transport processes, and a kinetic order of 0.5 corresponds to a hyperbolic rate law, such as a Michaelis–Menten process, that operates close to the Michaelis constant K_M. Other kinetic orders between 0 and 1 can be interpreted as Michaelis–Menten-like processes operating at values less than K_M (kinetic order between 0.5 and 1) or greater than K_M (kinetic order between 0 and 0.5). Inhibitory effects in the above set of models were often modeled with a negative kinetic order of small magnitude, such as -0.1, and the value rarely fell below -1. While there is no proof, it also appears that kinetic orders are smaller in magnitude if the model contains more variables.

Thus, if a variable has a noticeable effect on a particular flux, then an appropriate kinetic order is likely to be around 0.75 or 1, and if the effect is small, a kinetic order like 0.1 maybe a good first guess. Inhibitions may be modeled with negative numbers around -0.2, with the magnitude reflecting the strength of inhibition. While not mathematically solidified, this heuristic information provides a good starting point for selecting parameters. This is especially valuable in three cases: (i) If information about the system is scarce; (ii) If we are interested in an *order-of-magnitude model* which does not attempt to model quantitative features of a phenomenon but tries to elucidate the overall dynamic repertoire of a model structure; or (iii) If we need initial estimates that are to be used in a computational estimation procedure.

A word of caution is in order. Although the above rules were extracted from representative studies that were executed by different experts in the field, there may be situations where kinetic orders are much larger in magnitude. For instance, Savageau (1995a) showed that Michaelis–Menten rate laws may lead to kinetic orders much higher than 1 if the reactions are catalyzed by enzymes that are located on surfaces or in channels.

Rate constants are not quite as sensitive, and their magnitudes depend on the unit of time. If the dynamics of an experiment is expressed in hours instead of seconds, all rate constants are multiplied by a factor of 3,600.

For quantitative systems analyses, a rule-of-thumb assignment of parameter values is not sufficient, but provides a good starting point or a quality check for results obtained with any of the estimation methods discussed in this chapter.

ESTIMATION OF KINETIC ORDERS FROM STEADY-STATE DATA

Direct Experimental Measurements of Kinetic Orders

Consider a biochemical system in steady state, and suppose we have the experimental means of varying one particular process while holding everything else constant at the nominal steady state. In the first experiment, one determines the rate of this process, which serves as the baseline measurement. Then, in a series of experiments, one systematically fixes one of the variables at slightly altered values and again measures the rate of the process. The results of these experiments are plotted as *relative concentrations* against the *relative process rate*, i.e., as the experimentally altered concentration divided by the steady-state concentration against the flux that corresponds to the altered concentration divided by the steady-state flux. While these quantities could be plotted in a regular Cartesian coordinate system, experience has shown that it is more convenient to plot them in logarithmic coordinates, where such plots often become almost linear. An example is shown in Fig. 5.1.

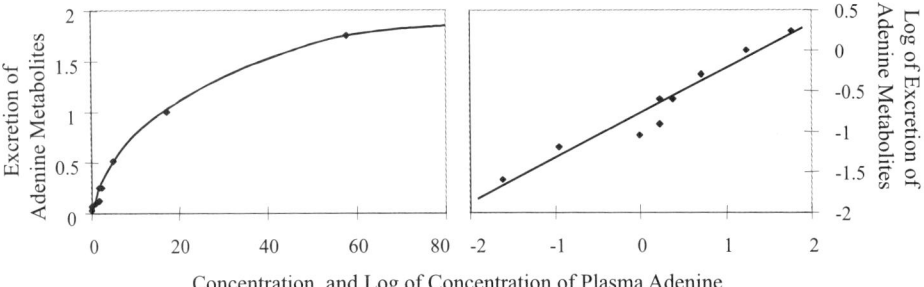

Figure 5.1. Adenine excretion as a function of plasma adenine concentration in Cartesian and in logarithmic coordinates (data from de Verdier et al., 1977; redrawn from Curto et al., 1998a). The Cartesian representation shows saturation in excretion, while the linearity of the logarithmic representation facilitates the determination of the corresponding kinetic order. Note that linearity in logarithmic coordinates holds for almost four orders of magnitude in adenine concentration.

PARAMETER ESTIMATION

What can we learn from this quasi-straight-line relationship between the logarithms of concentrations and fluxes? Suppose, we are analyzing a flux V_4^+, which depends on X_1, X_2, and X_5, in response to changes in X_2. In S-system formulation, this flux takes the symbolic form

$$V_4^+ = \alpha_4 X_1^{g_{41}} X_2^{g_{42}} X_5^{g_{45}}. \tag{5.1}$$

Even though V_4^+ is a function of three variables, our experimental set-up simplifies the situation in that we keep X_1 and X_5 constant at their nominal steady-state values. Thus, if we divide the experimentally altered V_4^+ by the steady-state flux V_{4S}^+, the rate constant α_4 and the terms $X_1^{g_{41}}$ and $X_5^{g_{45}}$ cancel out:

$$V_4^+ / V_{4S}^+ = X_2^{g_{42}} / X_{2S}^{g_{42}} = (X_2/X_{2S})^{g_{42}}. \tag{5.2}$$

When we take logarithms on both sides, the equation reduces to a linear equation:

$$\ln V_4^+ - \ln V_{4S}^+ = g_{42}(\ln X_2 - \ln X_{2S}), \tag{5.3}$$

or

$$g_{42} = \frac{\ln V_4^+ - \ln V_{4s}^+}{\ln X_2 - \ln X_{2s}}. \tag{5.4}$$

Recalling the formula for the slope m of a straight line that is given by two points (x_1, y_1) and (x_2, y_2), namely

$$m = \frac{y_2 - y_1}{x_2 - x_1}, \tag{5.5}$$

we see that the kinetic order g_{42} of the process V_4^+ with respect to X_2 is the slope in the plot of V_4^+ / V_{4S}^+ versus X_2/X_{2S}, when these are represented in logarithmic coordinates.

An alternative way of explaining this relationship is actual computation of the slope of the response function, which is achieved by partial differentiation of the relative flux with respect to a relative concentration. As discussed in the Appendix, this differentiation directly yields the kinetic order in question:

$$\frac{\partial V_4^+}{\partial X_2} \cdot \frac{X_2}{V_4^+} = g_{42}\left[\alpha_4 X_1^{g_{41}} X_2^{g_{42}-1} X_5^{g_{45}}\right] \frac{X_2}{\alpha_4 X_1^{g_{41}} X_2^{g_{42}} X_5^{g_{45}}}$$
$$= g_{42}. \tag{5.6}$$

In fact, this relationship is equivalent to the definition of the exponents g_{ij} in the power-law representation of the process V_i^+ (see Appendix), inasmuch as we recall that

$$\frac{\partial V_i^+}{\partial X_j} \cdot \frac{X_j}{V_i^+} = \frac{\partial \ln V_i^+}{\partial \ln X_j}. \tag{5.7}$$

Derivatives, whether in Cartesian or in logarithmic coordinates, refer to infinitesimal changes. Since kinetic orders are mathematically equivalent to derivatives, all direct experimental measurements of slopes, which can only be based on finite differences, constitute approximations whose precision is not known *a priori*. As Fell (1992) observed, one possibility to alleviate this problem is not to study just one alteration in enzyme activity but to make a number of balanced increases and decreases of enzyme activities and then interpolate the results to an infinitesimally small change.

Under ideal circumstances, the data include enough concentrations and corresponding fluxes to allow a statistical regression analysis. In reality, there are often only three or four points, and the variation in these points may even indicate a nonlinear trend rather than a straight-line relationship with random scatter. It would be desirable to add more experimental data points, but if these are not available, we might use the few data at hand, keeping in mind that they are subject to experimental error of an extent that is hard to identify. The theory here is further advanced than the experimental data collection, and in actual estimations one can only hope that future experiments will confirm and solidify scanty data.

The use of linear regression techniques requires that we take logarithms of metabolite concentrations and fluxes. This step may cause concern among statisticians, since the logarithmic transformation affects the error structure of the estimates. However, in biochemical systems relative quantities may be more relevant than absolute quantities, and that suggests that error estimates should also be based on relative quantities. For instance, we probably would not object to a statement of an estimated pH of 6.4 ± 0.1, which, of course, refers to a logarithmic quantity, and thus to an error with logarithmic structure.

In summary, the kinetic order g_{ij} is estimated as the slope of a function that describes the relative flux V_i^+/V_{iS}^+ as a function of relative concentrations X_j/X_{jS}, under experimental conditions where all other variables are held constant.

It may seem impossible to design experiments in which only one substrate is varied whereas all other variables are held constant. However, it can be done with reasonable accuracy in some cases. For instance, if we are interested in the first enzyme of a pathway, the concentration of the substrate can be kept constant while the product concentration is varied, and it becomes possible to determine the kinetic order associated with the product. A second scenario of practical relevance is the case where an enzyme is essentially insensitive to its product. In this case, variation of the substrate allows the estimation of the corresponding kinetic order.

Fell (1992) reviewed experimental means for estimating kinetic orders that are based on changing enzyme activities while only minimally affecting other system properties. [In Fell's paper, kinetic orders are referred to as *elasticities*, as is commonplace in Metabolic Control Analysis. Except for theoretical subtleties, elasticities and kinetic orders are essentially equivalent (Savageau et al., 1987ab; Curto et al., 1995).] Of particular importance are two direct approaches. One consists of various schemes of titration with purified enzyme or specific inhibitors, and the other involves genetic manipulation of expressed enzyme activity. Successful titration requires that the enzyme actually have access to its substrate or that the inhibitor reach the target enzyme. So far, this has been achieved experimentally in homogenates and permeabilized cells. For example, Groen et al. (1982b) used inhibitor titration to study oxidative phosphorylation in mitochondria in rat liver cells; other investigations of this nature are reviewed in Fell (1992, 1997). One illustration of this method is presented below as Example 2. A problem with the inhibitor titration method is that it requires extrapolation of the relationship between inhibitor and flux toward an inhibitor concentration of zero. This extrapolation can be quite sensitive (Small, 1988; Smith et al., 1990; Fell, 1992).

The second approach is alteration of expressed enzyme activity by classical methods of genetics. These include the breeding of homozygotes and heterozygotes for alleles that affect the enzyme and the insertion of additional gene copies on a plasmid. While the genetic methodologies themselves are well worked out, they are of limited use for the estimation of kinetic orders, since the fine tuning of enzymatic activity is difficult to accomplish. Nonetheless, this approach has been used to obtain order-of-magnitude values of kinetic orders (e.g., Flint et al., 1981; Dykhuizen et al., 1987). In particular, if this method yields very low responses, one may conclude that the corresponding kinetic orders are close to zero (e.g., Middleton and Kacser, 1983). For some systems, fine tuning can be achieved if gene expression can be controlled externally. For instance, Walsh and Koshland (1985) and Walsh et al. (1987) were able to control the gene for citrate synthase, which was inserted on a plasmid along with the *lac* repressor, by varying the inducer isopropylthiogalactoside.

Stitt and collaborators (Kruckeberg et al., 1989; Neuhaus and Stitt, 1990) studied kinetic orders of the photosynthetic production of sucrose and starch with mutants that showed substantially lowered activity. Instead of interpolating to infinitesimal changes, they computed the kinetic orders from ratios of responses. Specifically, if X_k^1 and X_k^2 are the enzyme levels under conditions 1 and 2 and V_i^1 and V_i^2 are the corresponding fluxes, Stitt and his coworkers proposed that the associated kinetic order take the form

$$g_{ik} = \frac{X_k^2}{X_k^1}\left(1 - \frac{V_i^2}{V_i^1}\right) \bigg/ \frac{V_i^2}{V_i^1}\left(1 - \frac{X_k^2}{X_k^1}\right). \tag{5.8}$$

The accuracy of this estimation is difficult to assess in general (see Exercise 6). Two comparisons with more direct methods are presented in the examples below.

Examples

Example 1: Adenine excretion. Curto et al. (1997, 1998ab; see also the Case Study in Chapter 10) analyzed in detail the kinetic and dynamic properties of purine metabolism. Their estimation of the kinetic order of adenine excretion provides a good example for the techniques discussed above. Data, obtained from the literature (de Verdier et al., 1977), are given in Table 5.1. When plotted in Cartesian coordinates, the relationship between excretion and plasma adenine concentration is saturating, but it becomes essentially linear when plotted in logarithmic coordinates (see Fig. 5.2). The logarithmic plot corresponds to relative changes in excretion as a function of relative deviations in plasma concentration and permits direct estimation of the corresponding kinetic order. In this particular case, the kinetic order can be measured as the slope of the eyeballed best-fitting straight line, or it can be obtained more rigorously with linear regression. The result is a kinetic order of 0.55. (See also Exercise 1.)

Table 5.1. Measurements of Plasma Adenine Concentration and Adenine Excretion

Adenine Concentration in Plasma (μmol)	Adenine Excretion Per Unit Body Weight (μmol min^{-1} K$_0^{\delta-1}$)
0.024	0.025
0.112	0.064
1.00	0.089
1.73	0.122
1.73	0.250
2.40	0.250
5.18	0.511
17.3	1.00
57.8	1.75

Source: de Verdier et al. (1977).

Example 2: Pyruvate kinase in rat liver. Interested in the control of gluconeogenesis, Groen at al. (1986) studied the kinetic properties of pyruvate kinase and pyruvate carboxylase in rat liver cells. It had been documented in the literature that gluconeogenesis is sensitive to these enzymes and that pyruvate carboxylase essentially controls the pathway if the glycolytic enzyme pyruvate kinase is inactive.

In one experiment, Groen and coworkers measured the activity of pyruvate kinase (PK), which catalyzes the conversion of phosphoenylpyruvate (PEP) to pyruvate (Pyr):

$$\text{PEP} + \text{ADP} \rightarrow \text{Pyr} + \text{ATP}.$$

For the determination of the desired kinetic order, parenchymal liver cells of 20–24-h-starved Wistar rats were perfused with different amounts of the gluconeogenic

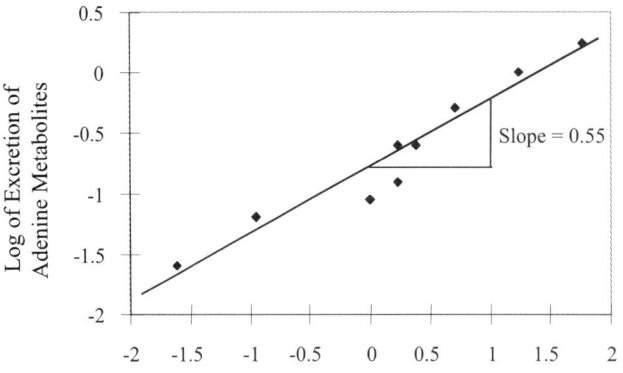

Figure 5.2. The logarithm of adenine excretion is essentially a linear function of the logarithm of plasma adenine concentration (data excerpted from de Verdier et al., 1977, and redrawn from Curto et al., 1998a). The kinetic order of adenine in the rate law for adenine excretion is obtained from the straight-line relationship graphically or through linear regression.

PARAMETER ESTIMATION

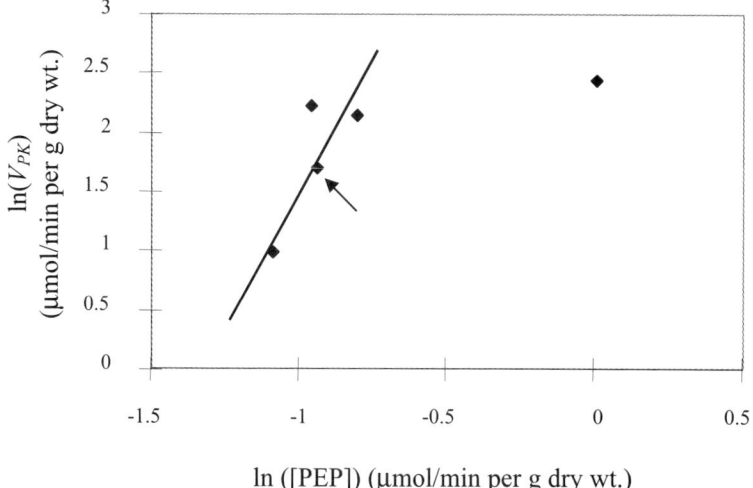

Figure 5.3. Dependence of pyruvate kinase flux V_{PK} on concentration of phosphoenolphosphate (PEP) in perfused rat liver. The point of interest, for which the ratio of lactate to pyruvate is 10, is marked by an arrow. (Logarithmic representation of data from Groen et al., 1986.)

substrate lactate in the presence of 0.5 mM pyruvate and 0.1 mM oleate. In each steady state of glucose formation, a sample of the cell suspension was taken and fractionated with the digitonin technique, as described in Groen et al. (1983). The concentration of PEP in the cytosolic fraction was measured. The values of the fluxes V_{PK}, which describe the step catalyzed by PK at different steady states of glucose formation, were derived from measurements of the glucose flux in the presence or absence of 0.1 μM glucagon. This procedure appeared to be valid, because under the experimental conditions the concentrations of the other substrates and products of PK remained constant.

From these measurements the authors obtained a collection of values of V_{PK} at different steady states that corresponded to different cytosolic concentrations of PEP (Fig. 5.3). Over the entire range of conditions, the graph of these data is nonlinear, showing that the kinetic order is affected by the experimental conditions. The domain of interest is characterized by a lactate/pyruvate ratio of about 10, which corresponds to a PEP concentration of about 0.4. In this domain, the logarithmic representation of V_{PK} versus PEP, which corresponds to relative fluxes and relative concentrations, has a slope of about 4.6, and this slope value identifies the desired kinetic order.

Computation of the kinetic order with the formula of Stitt and collaborators [Eq. (5.8); Kruckeberg et al. (1989); Neuhaus and Stitt (1990)], using the first and second, the first and third, or the first and fourth concentrations of PEP and the corresponding fluxes, produces kinetic orders of 3.61, 5.47, and 2.63, respectively (see also Exercise 6).

Example 3: Pyruvate carboxylase in rat liver. In the same study and with the same experimental technique, Groen et al. (1986) determined the kinetic order of pyruvate carboxylase (PC) with respect to its product mitochondrial oxalacetate (OAA_m). PC

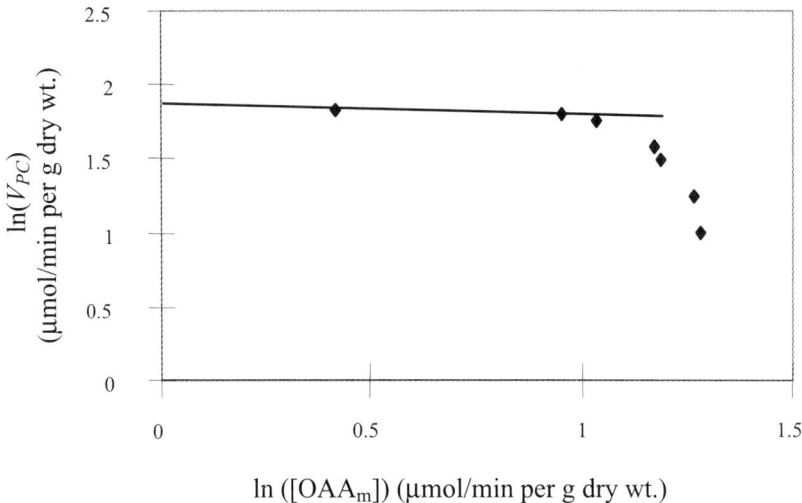

Figure 5.4. Dependence of pyruvate carboxylase flux V_{PC} on concentration of mitochondrial oxalacetate (OAA_m) in rat liver cells. (Logarithmic representation of data from Groen et al., 1986.)

catalyzes the reaction

$$Pyr + ATP \to OAA_m + ADP + P_i.$$

Theoretical considerations (see also next section) had suggested an upper limit of -0.04 for the kinetic order, and one purpose of this experiment was to test the theoretical prediction.

The experiment consisted in manipulating the steady-state rate of gluconeogenesis by adding different amounts of mercaptopicolinic acid, which is a non-competitive inhibitor of phosphoenolpyruvate carboxykinase (PEPCK). The cells were incubated with saturating concentrations of lactate and pyruvate, 1 mM oleate, and 1 μM glucagon. The concentration of OAA_m was again determined after digitonin fractionation of the cells. In these experiments, the flux through pyruvate carboxylase at different concentrations of OAA_m was indicated by the rate of glucose formation. The kinetic order was derived from data corresponding to different experimental concentrations of OAA_m (cf. Fig. 5.4). For a low concentration of OAA_m of about 1.5 ($\ln[OAA_m] = 0.4$), the kinetic order of V_{PC} is -0.05, which is indeed close to the theoretical upper limit of -0.04. It should be mentioned that in cases like this, experimental error can substantially influence the numerical value of the estimated kinetic order. Computation of the kinetic order with the formula of Stitt and collaborators [Eq. (5.8); Kruckeberg et al. (1989); Neuhaus and Stitt (1990)], using the two lower OAA_m concentrations and the corresponding fluxes, produces a value of -0.074, which is not too far off in absolute terms.

Example 4: Citrulline pathway. Wanders et al. (1984) investigated metabolic effects of a high protein diet in male Wistar rats. The diet contained 80% casein, which is more than three times the percentage in a normal diet (about 15%). High dietary protein contents increase intra-mitochondrial N-acetylglutamate, which is an essential

PARAMETER ESTIMATION

Table 5.2. Concentration of Intra-mitochondrial Carbamoyl Phosphate and Critulline Production under Varied Conditions of Malonate Inhibition

Malonate (mM)	[Carbamoyl Phosphate] (nmol/mg)	Citrullin Production (nmol/mg min)
0	1.49	24.8
1	1.42	24.0
2	1.25	20.2
3	0.923	17.1
4	0.923	14.5
5	0.710	12.4

Source: Wanders et al. (1984: Fig. 5a).

activator of carbamoyl phosphate synthetase. Carbamoyl phosphate synthetase and ornithine transcarbomylase catalyze the two consecutive steps of a pathway in which citrulline is synthesized from NH_4^+, HCO_3^-, and ATP. Wanders and coworkers were especially interested in characterizing the kinetic orders (*elasticities*, in their nomenclature) of ornithine transcarbamoylase and carbamoyl phosphate synthetase with respect to carbamoyl phosphate (CP).

In order to determine the kinetic order of ornithine transcarbamoyl phosphate, Wanders et al. (1984) varied the flux through the citrulline-synthesizing pathway by inhibiting the flux through carbamoyl phosphate synthetase. This was accomplished by decreasing the rate of ATP supply with malonate, which inhibits succinate dehydrogenase. The result was an almost linear relationship between intra-mitochondrial CP and the flux through ornithine transcarbamoylase, which was measured as citrulline production (see Table 5.2 and Fig. 5.5). The corresponding kinetic order of

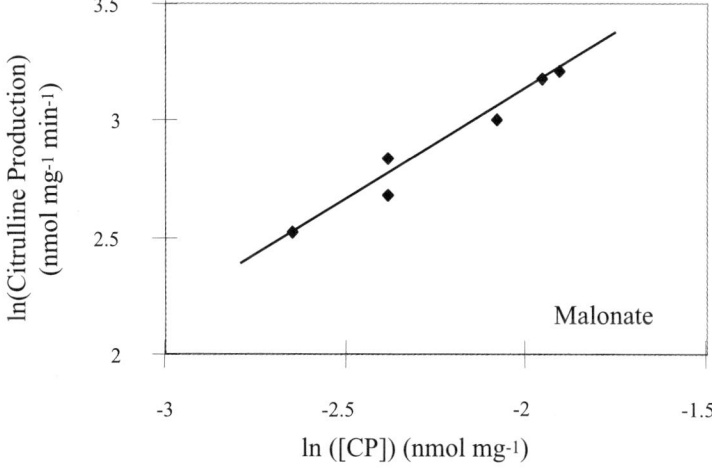

Figure 5.5. Citrulline production as a function of intra-mitochondrial carbamoyl phosphate concentration under different malonate regimens (See text for details. Data, redrawn in logarithmic coordinates, from Wanders et al., 1984.)

Table 5.3. Concentration of Intra-mitochondrial Carbamoyl Phosphate and Citrulline Production under Varied Conditions of Norvaline Inhibition

Norvaline (mM)	[Carbamoyl phosphate] (nmol/mg)	Citrulline Production (nmol/mg min)
0	0.14	24.9
1	0.63	24.0
2	1.1	23.1
3	1.58	22.1

Source: Wanders et al. (1984: Fig. 5b).

ornithine transcarbamoylase is given as the slope of this relationship. It is computed as

$$\frac{\partial \ln v_{cit}}{\partial \ln[CP]} = 0.92. \tag{5.9}$$

The kinetic order of carbamoyl phosphate synthetase with respect to CP was determined in a similar fashion, but in this case, norvaline was used to inhibit the flux through the pathway. Norvaline inhibits only ornithine transcarbamoylase in the pathway of citrulline synthesis. It inhibits competitively with respect to ornithine and non-competitively with respect to carbamoylphosphate. The data are given in Table 5.3.

Figure 5.6 shows the relationship between the flux through CP synthetase, measured as citrulline production, and intra-mitochondrial CP. The slope of the regression line, -0.01, is equal to the corresponding kinetic order. Wanders et al. (1984) approximated the kinetic order from the data in Cartesian coordinates as a ratio of

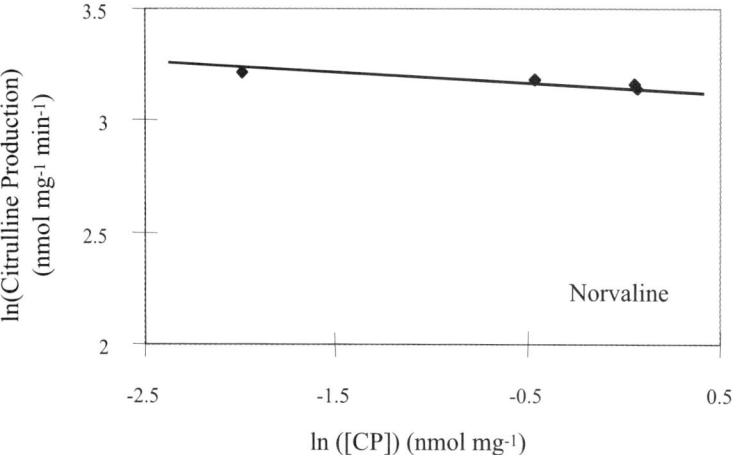

Figure 5.6. Citrulline production as a function of intra-mitochondrial carbamoyl phosphate concentration [CP] under different norvaline regimens. (See text for details. Data, redrawn in logarithmic coordinates, from Wanders et al., 1984.)

PARAMETER ESTIMATION

relative changes between the highest and lowest values:

$$\text{kinetic order} \approx \frac{\Delta(\text{citrulline production})}{\text{citrulline production}} \bigg/ \frac{\Delta[CP_m]}{[CP_m]}$$

$$= \frac{24.8 - 22.1}{24.8} \bigg/ \frac{0.15 - 1.58}{0.15} = -0.0114. \qquad (5.10)$$

If the data lie close to the regression line, this method is sufficiently accurate. The estimated kinetic order of -0.01 indicates that the enzyme is rather insensitive to changes in the concentration of its product. It is noted that under the given experimental conditions the kinetic orders of carbamoyl phosphate synthetase and ornithine transcarbamoylase with respect to CP can be determined with the enzymes left active in the pathway. This is valid when an enzyme catalyzes the first step in the pathway and one varies the concentration of product, or when the enzyme catalyzes the last step and one varies the substrate.

Experimental Measurements of Logarithmic Gains

Kacser and Burns (1979) proposed a method for determining kinetic orders that is more often applicable. This method is based on small perturbations that are experimentally introduced anywhere in the system and result in modulations of concentrations and fluxes in the interesting portion of the pathway. The rationale for using *small* perturbations is that the describing model equations then are quasi-linear, whereas for larger changes in conditions the nonlinear dynamics becomes significant and greatly complicates the analysis.

Suppose we are interested in the kinetic properties of an enzymatic process V that catalyzes some intermediate step in a pathway between X_3 and X_4, as illustrated in Fig. 5.7. The mathematical form of V is not known, but because of modulations and potential reversibility, V may be a function that involves both X_1 and X_2. Formulated as a power-law function, V includes the kinetic orders g_{21} and g_{22}, whose estimation involves two perturbation experiments of the same type. In experiment A, one introduces some persistent change in the concentration of the independent variable X_3. This change in input affects small deviations in the constituents of interest and leads to a new steady state. We denote the new steady-state values and deviations as $(X_1^A, \delta X_1^A)$, $(X_2^A, \delta X_2^A)$, and $(V^A, \delta V^A)$, respectively. Since V is a function of X_1 and X_2, we can formulate its change mathematically in the form of a total differential

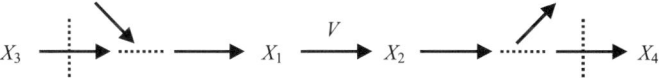

Figure 5.7. Generic pathway used by Kacser and Burns (1979) to illustrate their method of parameter estimation. The enzymatic process V converts X_1 into X_2. X_3 and X_4 are independent variables that can be modified experimentally. The arrows signify reactions that may be modulated and/or reversible. It is assumed that changes in X_4 affect the steady state of the system. See text for further details.

(see Appendix):

$$dV = \frac{\partial V}{\partial X_1} dX_1 + \frac{\partial V}{\partial X_2} dX_2. \tag{5.11}$$

According to their definitions, the kinetic orders g_{12} and g_{21} can be written as

$$g_{21} = \frac{\partial V}{\partial X_1} \cdot \frac{X_1}{V} \quad \text{and} \quad g_{22} = \frac{\partial V}{\partial X_2} \cdot \frac{X_2}{V} \tag{5.12}$$

at steady state. Turning these equations around, we obtain expressions that appear on the right-hand side of Eq. (5.11):

$$\frac{\partial V}{\partial X_1} = g_{21} \frac{V}{X_1} \quad \text{and} \quad \frac{\partial V}{\partial X_2} = g_{22} \frac{V}{X_2}. \tag{5.13}$$

We substitute these expressions in (5.11) and divide the equation by V. The result is

$$\frac{dV}{V} = g_{21} \frac{dX_1}{X_1} + g_{22} \frac{dX_2}{X_2}. \tag{5.14}$$

The differential quantities dV, dX_1, and dX_2 are strictly speaking infinitesimal, but they are rather well approximated by the corresponding quantities obtained from the small perturbation in experiment A. Thus, substitution transforms Eq. (5.14) into an equation that contains, except for the desired kinetic orders, only measured quantities:

$$\frac{\delta V^A}{V^A} = g_{21} \frac{\delta X_1^A}{X_1^A} + g_{22} \frac{\delta X_2^A}{X_2^A}. \tag{5.15}$$

Experiment B is executed in the same fashion, except that X_4 is changed. The result is

$$\frac{\delta V^B}{V^B} = g_{21} \frac{\delta X_1^B}{X_1^B} + g_{22} \frac{\delta X_2^B}{X_2^B}. \tag{5.16}$$

Again, this equation contains exclusively the desired kinetic orders and measured quantities. Eqs. (5.15) and (5.16) constitute a linear system with two equations and two unknowns, the kinetic orders g_{21} and g_{22}. This system can be solved by substitution or with matrices (see Chapter 6 and Appendix) to yield numerical values for the desired kinetic orders (see Example 5). Several authors (Small, 1988; Giersch, 1994; Fell, 1997: p. 171) have suggested locating one perturbation upstream and one downstream of the process in question.

The deviations in Eqs. (5.15) and (5.16), divided by the quantities themselves, correspond (in the limit) to logarithmic derivatives, and Eq. (5.15) can be written as

$$\delta \ln V^A \approx g_{21} \delta \ln X_1^A + g_{22} \delta \ln X_2^A, \tag{5.17}$$

where the deviations in logarithmic fluxes are defined, e.g., as $\delta \ln V^A = \ln V^A - \ln V$ (cf. Fell, 1997: p. 172). By executing several – and not just two – perturbations and plotting $\ln V$ as a function of $\ln X_1$ or $\ln X_2$, respectively, the kinetic orders can be obtained as slopes of a regression line. Fell (1997: p. 172) re-analyzed data of Giersch (1995) in this fashion.

PARAMETER ESTIMATION

As an alternative to changing external independent variables, one could alter two enzymes that affect the system, if these alterations evoke small but measurable deviations of metabolites and fluxes from their steady-state values. Obviously, combinations of changes in metabolites and enzymes are possible. Extensions and variations of the method are discussed in detail in the literature (e.g., Delgado and Liao, 1992ab; Fell, 1997: pp. 176 ff.) and, to a limited degree, in the *Complements*.

Example 5: Kinetic order estimation by double modulation. This example has a somewhat different character than most of the others. Its prime purpose is not an attempt to show how parameter values are estimated from data but a demonstration of how one can assess the quality of a method itself.

In a true experimental situation, we may apply the method of Kacser and Burns (1979) – or any other method – and obtain parameter estimates. However, we hardly have the means of evaluating the quality of the estimates, because we just don't know what the true values of the kinetic orders are. This situation is quite common, and there is no general remedy.

One strategy for assessing the performance of a method is simulation with a system that is completely known to us. This strategy has the slight disadvantage that the analyzed system is "cleaner" than a real-world system, which may lead to results that are better than can be expected in a wet lab experiment. On the other hand, the strategy has the great advantage that we can perform any manipulation we please and obtain results that are free of experimental uncertainties. In this sense, the results yield insight into the performance of a method under ideal conditions. Furthermore, simulations are very fast and cheap in comparison with biochemical experiments, and if necessary we can easily perform hundreds of simulations to explore the quality of a procedure under different conditions.

In order to evaluate the accuracy of the method of Kacser and Burns (1979), we perform a simulation with a simple pathway that we define. First, we compute the nominal steady state; then we implement two perturbations and record their effects on the steady-state characteristics of the system. The results are entered in Eqs. (5.15) and (5.16). Solving the equations yields estimates for the kinetic orders, which can be compared for accuracy with the true kinetic orders.

As a model for the simulation, consider the artificial system in Fig. 5.8. It consists of a linear pathway with feedback inhibition and two independent variables that represent the substrate of the pathway and an external inhibitor, respectively. A

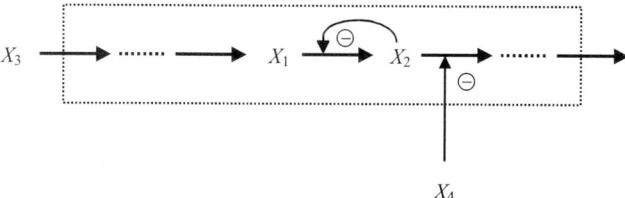

Figure 5.8. Artificial pathway for assessing performance of an estimation method suggested by Kacser and Burns (1979).

numerical realization is

$$\dot{X}_1 = 10 X_3^{0.75} - 5 X_1^{0.5} X_2^{-0.3},$$
$$\dot{X}_2 = 5 X_1^{0.5} X_2^{-0.3} - 10 X_2^{0.5} X_4^{-0.75}, \quad (5.18)$$
$$X_3 = 1,$$
$$X_4 = 1.$$

The system is readily implemented in PLAS (*Kacser.plc*), and the steady-state concentrations and flux are obtained as $X_{1S} = 4$, $X_{2S} = 1$, and $V = 10$. Suppose that for experiment A the independent variable X_3 is increased by 1% to 1.01. The resulting steady-state characteristics are again obtained in PLAS as $X_1^A = 4.15207$, $X_2^A = 1.03802$, and $V^A = 10.0749$, which directly implies $\delta X_1^A = 0.15207$, $\delta X_2^A = 0.03802$, and $\delta V^A = 0.0749$. In experiment B, the independent variable X_4 is increased by 2% to 1.02. The resulting steady-state concentrations are again obtained in PLAS as $X_1^B = 4.18225$, $X_2^B = 1.07709$; the flux turns out to be unchanged: $V^B = 10$. The differences are thus $\delta X_1^B = 0.18225$, $\delta X_2^B = 0.07709$, and $\delta V^B = 0$. Check these results in PLAS.

The numerical results are entered in the linear system for estimating the kinetic orders $g_{21} = h_{11}$ and $g_{22} = h_{12}$. This system reads

$$\frac{0.0749}{10.0749} = h_{11} \cdot \frac{0.15207}{4.15207} + h_{12} \cdot \frac{0.03802}{1.03802},$$
$$\frac{0}{10} = h_{11} \cdot \frac{0.18225}{4.18225} + h_{12} \cdot \frac{0.07709}{1.07709}. \quad (5.19)$$

Solving the second equation for h_{11}, we obtain

$$h_{11} = -0.608851634 h_{12}. \quad (5.20)$$

Substitution of this expression in the first equation and then in the second equation yields the solution

$$h_{11} = 0.51899539,$$
$$h_{12} = -0.315991192. \quad (5.21)$$

These estimated parameters are not quite equal to the true kinetic orders, 0.5 and -0.3, but they are within an error of about 5%. It is up to the analyst to judge whether this accuracy is sufficient. One possibility for such an assessment is a simulation analysis of the model with the estimated, instead of the true, parameter values (see Exercise 3).

ESTIMATION OF KINETIC ORDERS FROM TRADITIONAL RATE LAWS

Any information about a process can potentially be used for the estimation of parameter values. Sometimes, precise values can be extracted, and in other situations, experimental information may put bounds on an estimate by suggesting that a parameter is not likely to be higher or lower than certain threshold values. The work of Savageau and Sands (1991) illustrates this type of estimation with an analysis of coupled gene circuits in which enough information was collected from the literature to estimate all parameters of the model.

The simplest example of a circumstantially obtained parameter estimate is probably the observation that a process V_i^+ is proportional to some variable X_j, which immediately implies a kinetic order $g_{ij} = 1$. In a similar vein, transport processes are considered to be proportional to the concentration of the transported material, and their associated kinetic orders are set to 1. The same is true for chemical reactions of first order and for exponential growth or decay processes.

If there is reason to believe that a process is well represented by an enzymatic rate law or some other algebraic function, this rate law or function can be used to estimate kinetic orders for a GMA or S-system model. The rationale is that the GMA and S-system models can be considered power-law representations of the known rate law at an operating point of our choosing, which often corresponds to the nominal steady state.

If a rate law V is a function of just one variable X, the kinetic order with respect to this variable is obtained through differentiation of $\ln V$ with respect to $\ln X$ (cf. Appendix):

$$g = \left.\frac{d \ln V}{d \ln X}\right|_P = \left.\frac{dV}{dX}\right|_P \cdot \frac{X}{V}. \tag{5.22}$$

The expression on the right-hand side is evaluated at the chosen operating point P and thus yields a *number*, the kinetic order g.

As an example, let's compute a kinetic order for a power-law term that corresponds to the Michaelis–Menten rate law

$$V(X) = \frac{V_{max} X}{K_M + X} \tag{5.23}$$

at the operating point P. The kinetic order of g is obtained by differentiation, evaluation at P, and subsequent multiplication by X/V, as shown above:

$$g = \left.\frac{d \frac{V_{max} X}{K_M + X}}{dX}\right|_P \cdot \frac{P}{\frac{V_{max} P}{K_M + P}}. \tag{5.24}$$

We compute the derivative and evaluate the expression at the operating point P. The result is

$$\begin{aligned}g &= \frac{V_{max}(K_M + P) - V_{max} P}{(K_M + P)^2} \cdot \frac{P(K_M + P)}{V_{max} P} \\ &= \frac{K_M}{K_M + P}.\end{aligned} \tag{5.25}$$

There is discussion in the literature about how K_M and the steady-state substrate concentration (let's just call it P) relate to each other *in vivo*. Lowry and Passonneau (1964) executed *in vitro* measurements of glycolytic enzymes from brain tissue and found the ratio P/K_M to range between 0.06 and 1.6 in eight out of eleven cases, but they also observed values of 18 and 65, as well as one value given as "< 0.1."

Some authors are convinced that K_M is close to the substrate concentration *in vivo*. Hochachka and Somero (1984: p. 59) state very clearly: "... it is widely observed (in studies of different tissues and species) that for enzymes displaying saturation kinetics, *in vivo* substrate concentrations are either similar to the value of K_M or actually lower." This arrangement is advantageous, because it provides two means of regulation: If the substrate varies, the change in reaction velocity is relatively large; similarly, small changes in K_M have a greater effect if the substrate concentration is not too high. Using thermodynamic arguments *in vitro*, Cornish-Bowden (1976) came to the same conclusion that K_M and P should be similar, at least if the enzyme concentration is low. He added that inhibition and buffering against adverse conditions *in vivo* should probably lead to a K_M (slightly) exceeding P, which again is in line with Hochachka and Somero's arguments.

For enzymes where the maintenance of a steady-state flux of products is the prime objective, as well as for extracellular hydrolytic enzymes, such as pepsin and chymotrypsin, Cornish-Bowden (1976) argued that the K_M should be much lower than P. By contrast, Fersht (1974) concluded from thermodynamic and kinetic arguments that K_M should be about an order of magnitude higher than the substrate concentration. According to Crowley (1975), natural selection should lead to a K_M close to P if the substrate concentration doesn't change much on an evolutionary time scale, but to a much higher K_M if the reaction rate doesn't change.

If we choose the operating point as $P = K_M$, the corresponding kinetic order is

$$g = \tfrac{1}{2}. \tag{5.26}$$

If K_M is greater than P, the value of g increases. In the extreme if P becomes negligibly small, such that $K_M + P \approx K_M$, the value of g approaches 1. At the other extreme, if P is very large in comparison with K_M, the kinetic order g is close to 0, indicating that the power-law approximation in the saturation phase is essentially a horizontal line. Recall what a kinetic order $g = 0$ means: The substrate concentration, raised to the power 0, is 1 and thus has no effect. This is consistent with the biochemical interpretation of saturation. (See Fig. 5.9 and Exercise 4.)

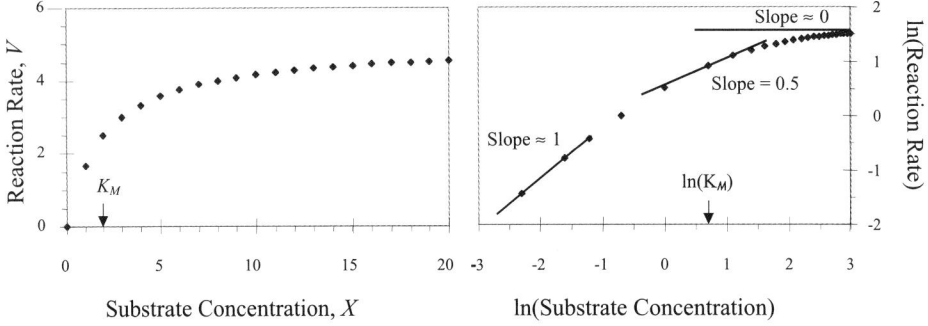

Figure 5.9. Kinetic orders in power-law terms that correspond to Michaelis–Menten rate laws. Left panel: Michaelis–Menten rate law with $K_M = 2$ and $V_{max} = 5$. Right panel: The same rate law in logarithmic coordinates. With increasing substrate concentration, the kinetic order decreases from 1 to 0. At $\ln K_M$, the kinetic order is 0.5.

It is noted that the maximal velocity V_{max} has no effect on the kinetic order. As for most other rate laws, it drops out during differentiation and scaling with X/V. This is important because using a maximal velocity that had been determined *in vitro* for an analysis of a system *in vivo* is often much more questionable than the use of a Michaelis constant (cf. Fell, 1997: p. 325; Shiraishi and Savageau, 1993: p. 16927). (See Exercises 4–8 for other examples.)

If V depends on more than one variable, the kinetic order is computed through partial, rather than ordinary, differentiation. For instance, if one of the variables is X_j, then the kinetic order g_j with respect to X_j is

$$g_j = \frac{\partial \ln V}{\partial \ln X_j} = \frac{\partial V}{\partial X_j} \cdot \frac{X_j}{V}, \tag{5.27}$$

which, again, is computed at the chosen operating point. This method is heavily employed in published studies and in some of the case studies in Chapters 8–11.

As an example, consider a rate law describing competitive inhibition, in which the substrate $S = X_1$ and the inhibitor $I = X_2$ are the only variables affecting the rate. In mathematical symbols, this rate law may take the form

$$V(X_1, X_2) = \frac{V_{max} X_1}{X_1 + K_M (1 + X_2/K_I)}. \tag{5.28}$$

The exponents g_1 and g_2 are computed via partial differentiation:

$$g_1 = \left.\frac{\partial V(X_1, X_2)}{\partial X_1}\right|_{(P_1, P_2)} \cdot \frac{P_1}{V(P_1, P_2)}, \tag{5.29}$$

$$g_2 = \left.\frac{\partial V(X_1, X_2)}{\partial X_2}\right|_{(P_1, P_2)} \cdot \frac{P_2}{V(P_1, P_2)}, \tag{5.30}$$

where the derivatives are evaluated at the operating point $P = (P_1, P_2)$. Substituting V, we obtain

$$g_1 = \left.\frac{\partial \frac{V_{max} X_1}{K_M(1+X_2/K_I)+X_1}}{\partial X_1}\right|_{(P_1, P_2)} \cdot \frac{P_1}{\frac{V_{max} P_1}{K_M(1+P_2/K_I)+P_1}}, \tag{5.31}$$

$$g_2 = \left.\frac{\partial \frac{V_{max} X_1}{K_M(1+X_2/K_I)+X_1}}{\partial X_2}\right|_{(P_1, P_2)} \cdot \frac{P_2}{\frac{V_{max} P_1}{K_M(1+P_2/K_I)+P_1}}. \tag{5.32}$$

Computation of the derivatives and selection of the operating point $P = (P_1, P_2) = (K_M, K_I)$ yields

$$g_1 = \frac{K_M(1 + P_2/K_I)}{K_M(1 + P_2/K_I) + P_1} = \frac{2}{3}, \tag{5.33}$$

$$g_2 = \frac{-V_{max} P_1 K_M/K_I}{[P_1 + K_M(1 + P_2/K_I)]^2} \cdot \frac{P_2}{\frac{V_{max} P_1}{K_M(1+P_2/K_I)+P_1}} = \frac{-P_2 K_M/K_I}{P_1 + K_M(1 + P_2/K_I)} = -\frac{1}{3}. \tag{5.34}$$

As was to be expected, the inhibitor carries a negative kinetic order (see Exercise 4 for evaluation at different operating points).

Because of its practical importance, consider as another example a reversible reaction between X_i and X_j whose forward and reverse rates, v_F and v_R, result in a net rate v_{net} given by the rate law

$$v_{net} = \frac{(V_{max(F)}/K_{Mi})(X_i - X_j/K_{eq})}{1 + X_i/K_{Mi} + X_j/K_{Mj}} \tag{5.35}$$

(Haldane, 1930), where $V_{max(F)}$ is the maximum velocity of the forward reaction, K_{Mi} and K_{Mj} are the Michaelis constants with respect to X_i and X_j, and K_{eq} is the equilibrium constant. If the mass action ratio is denoted as Γ and if $V_{max(R)}$ is the maximum velocity of the reverse reaction, the kinetic orders for the two reactions are given as

$$g_{ji} = \frac{1}{1 - \Gamma/K_{eq}} - \frac{v_F}{V_{max(F)}}, \tag{5.36}$$

$$g_{jj} = -\frac{\Gamma/K_{eq}}{1 - \Gamma/K_{eq}} - \frac{v_R}{V_{max(R)}} \tag{5.37}$$

(Groen et al., 1982a; see also Exercise 10). g_{ji} and g_{jj} can also be expressed in terms of net flux J when we employ the equalities

$$\frac{v_F}{v_R} = \frac{\Gamma}{K_{eq}} \tag{5.38}$$

(Rolleston, 1972) and

$$v_F = \frac{J}{1 - \Gamma/K_{eq}}. \tag{5.39}$$

The result is

$$g_{ji} = \frac{1}{1 - \Gamma/K_{eq}}\left(1 - \frac{J}{V_{max(F)}}\right), \tag{5.40}$$

$$g_{jj} = -\frac{\Gamma/K_{eq}}{1 - \Gamma/K_{eq}}\left(1 + \frac{J}{V_{max(R)}}\right). \tag{5.41}$$

A detailed example for these computations is presented later in this chapter (see Example on glucose-6-phosphate in *Aspergillus niger*). If the reaction occurs far from equilibrium or if the equilibrium constant is very large, the ratio Γ/K_{eq} is close to zero, and the expressions for g_{jj} and g_{ji} simplify to

$$g_{ji} \approx 1 - \frac{v_F}{V_{max(F)}}, \tag{5.42}$$

$$g_{jj} \approx -\frac{v_R}{V_{max(R)}}. \tag{5.43}$$

PARAMETER ESTIMATION

Groen et al. (1982a) also showed that similar equations apply to multi-substrate rate laws that have numerators similar to that in Eq. (5.35).

As Groen et al. (1986) point out, if Γ/K_{eq} is less than about 0.01, the kinetic order depends almost solely on v/V_{max}, that is, on the extent to which the enzyme is saturated with its substrate or product. On the other hand, if $\Gamma/K_{eq} \approx 0.5$, the enzyme operates close to equilibrium, and the kinetic order essentially depends just on the value of the ratio Γ/K_{eq}.

If Γ/K_{eq} is close to 1, the estimation of g_{ji} and g_{jj} with these methods can be problematic. For instance, if a reaction occurs close to the equilibrium, the net flux J is negligible. For the sake of argument, suppose $\Gamma/K_{eq} = 0.95$. According to the above formulas, the resulting kinetic orders are approximately

$$g_{ji} = \frac{1}{1 - \Gamma/K_{eq}} = 20, \quad (5.44)$$

$$g_{jj} = -\frac{\Gamma/K_{eq}}{1 - \Gamma/K_{eq}} = -19, \quad (5.45)$$

and that appears to be unrealistic. The reason for this problem is the use of the irreversible strategy, which is not valid near thermodynamic equilibrium (cf. Chapter 3). To correct the problem, we have to set up our model by aggregating fluxes according to the reversible strategy. This strategy may be slightly more complicated, but given that the forward and reverse fluxes are known, it is straightforward and does not change the number of parameters to be estimated.

An alternative strategy is available if the equilibrium between the two metabolites in question is achieved very rapidly. For instance, in some metabolic systems, phosphoglucoseisomerase very rapidly converts glucose-6-P into fructose-6-P, and vice versa. As a consequence, the concentrations of the two metabolites are almost always in an equilibrium that is given by the equation

$$[\text{glucose-6-P}] = K_{eq}[\text{fructose-6-P}]. \quad (5.46)$$

This equilibrium relation can be used as a constraint to eliminate one of the two metabolites. The key requirement for the validity of this procedure is that the speed of the reaction in question is fast in comparison with other reactions in the system.

Fell (1997, p. 117) notes that the kinetic order of an allosteric enzyme may be between 1 and the Hill coefficient h (see also Exercise 5). Sauro (as mentioned in Fell, 1997), suggested that the kinetic order of an irreversible enzyme in process V_i^+ is

$$g_{ij} = (1 - \bar{Y})h \quad (5.47)$$

if the rate of reaction is proportional to its fractional saturation $\bar{Y}$ with substrate X_i. For low fractional saturation, the kinetic order is thus approximately the same as the Hill coefficient, and, like the kinetic order of a Michaelis–Menten process, it decreases to zero for high saturation.

Instead of computing kinetic orders in such a fashion that the original function and the power-law approximation are identical at one chosen operating point, it is possible to spread out the approximation error over some predefined range. Intuitively, this is similar to fitting the power-law function to data (which happen to lie densely and exactly on the known rate law) by means of least-squares regression. Hernández-Bermejo et al. (1999) recently described details of this approach. Voit (1992c) used the same type of functional fitting to estimate parameters for S-distributions.

ESTIMATION OF RATE CONSTANTS

Once kinetic order parameters have been estimated from experimental data or from a known rate law, the rate constant is computed in a second step. Since the original rate law and the S-system flux term should produce the same value at the chosen operating point, the two are simply equated, and with the kinetic orders determined before, the rate constant is the only unknown and easily determined. Suppose we are dealing again with a reversible Michaelis–Menten process, as described in Eq. (5.35), and the kinetic orders of the S-system flux

$$V_j^+ = \alpha_j X_i^{g_{ji}} X_j^{g_{jj}} \tag{5.48}$$

have been determined at an operating point P. Then, at P, we require $V_j^+ = v_{net}$, and α_j is computed from a simple rearrangement of this equality as

$$\alpha_j = v_{net} X_i^{-g_{ji}} X_j^{-g_{jj}}. \tag{5.49}$$

This procedure holds in general, for any number of variables and for influxes as well as effluxes. Suppose, for instance, that the degradation rate v_{deg} of variable X_4 includes the dependent or independent variables X_1, X_2, X_4, and X_6. The corresponding power-law term is

$$v_{deg} = \beta_4 X_1^{h_{41}} X_2^{h_{42}} X_4^{h_{44}} X_6^{h_{46}}. \tag{5.50}$$

Upon computation of the kinetic orders h_{41}, h_{42}, h_{44}, and h_{46}, the rate constant is obtained as

$$\beta_4 = v_{deg} X_{1P}^{-h_{41}} X_{2P}^{-h_{42}} X_{4P}^{-h_{44}} X_{6P}^{-h_{46}}, \tag{5.51}$$

where the subscript P indicates evaluation at the operating point.

As a specific example, consider again a rate law with inhibition [Eq. (5.28)]. At the operating point $(P_1, P_2) = (K_M, K_I)$ the kinetic orders g_1 and g_2 were computed as $\frac{2}{3}$ and $-\frac{1}{3}$, respectively [see Eqs. (5.33) and (5.34)]. The corresponding rate constant α is computed by equating the original rate law and the power-law representation at the operating point and solving for the rate constant:

$$V(X_1, X_2) = \frac{V_{max} X_1}{X_1 + K_M(1 + X_2/K_I)} = \alpha X_1^{g_1} X_2^{g_2} \quad \text{at } (P_1, P_2). \tag{5.52}$$

PARAMETER ESTIMATION

Thus, for the operating point (K_M, K_I) the rate constant α is computed as

$$\alpha = \frac{V_{max} P_1}{P_1 + K_M(1 + P_2/K_I)} P_1^{-g_1} P_2^{-g_2}$$

$$= \frac{V_{max} K_M}{K_M + K_M(1 + K_I/K_I)} K_M^{-2/3} K_I^{1/3}$$

$$= \tfrac{1}{3} V_{max} K_M^{-2/3} K_I^{1/3}. \tag{5.53}$$

Example 6: Isocitrate degradation in *Dictyostelium discoideum*. Shiraishi and Savageau (1992a–d, 1993) analyzed in detail the tricarboxylic acid cycle in *D. discoideum*. One step of this pathway consists of the conversion of isocitrate (ISOC) into α-ketoglutarate, which is catalyzed by isocitrate dehydrogenase (ICDH) and involves the cofactors NAD and NADH. As a suitable traditional rate law for this reaction, the authors considered the following function:

$$V = \frac{\text{ICDH} \cdot \text{ISOC} \cdot \text{NAD}}{0.34\,\text{ISOC} + 0.13\,\text{NAD} \cdot (1 + \text{NADH}/0.02) + \text{ISOC} \cdot \text{NAD}}. \tag{5.54}$$

The corresponding power-law term for an S-system or GMA model is derived from this rate function by partial differentiation and evaluation at the nominal steady state, which is characterized by the following concentrations and activities:

$\text{ISOC}_S = 0.01\,\text{mM},$

$\text{NAD}_S = 0.072\,\text{mM},$

$\text{NADH}_S = 0.180\,\text{mM},$

$\text{ICDH}_S = 271\,\text{mM min}^{-1},$

where the subscript S reminds us that the quantity is measured at the steady state. With these specifications, the steady-state flux V_S is 2 mM/min. To establish the S-system rate law, we differentiate V with respect to the four constituents, and it is immaterial for this purpose whether these are modeled as dependent or independent variables.

The first variable is ICDH. It appears in V only once as a factor. Thus, the derivative of V with respect to ICDH consists of everything in V except for ICDH itself:

$$\frac{\partial V}{\partial \text{ICDH}} = \frac{\text{ISOC} \cdot \text{NAD}}{0.34\,\text{ISOC} + 0.13\,\text{NAD} \cdot (1 + \text{NADH}/0.02) + \text{ISOC} \cdot \text{NAD}}. \tag{5.55}$$

To compute the kinetic order of ICDH in V, the derivative is evaluated at the steady state and multiplied by ICDH/V. In this particular instance, we do not have to execute the entire numerical computation, since the derivative multiplied by ICDH equals V. Thus, by dividing the product by V, the result is simply

$$g_{\text{ICDH}} = \left.\frac{\partial V}{\partial \text{ICDH}}\right|_{\text{steady state}} \cdot \frac{\text{ICDH}_S}{V_S} = 1. \tag{5.56}$$

In other words, the process is proportional to the activity of the catalyzing enzyme.

Determination of the kinetic order associated with ISOC requires differentiation with respect to ISOC. We compute

$$\left.\frac{\partial V}{\partial \text{ISOC}}\right|_{\text{steady state}}$$

$$= \{\text{ICDH}_S \cdot \text{NAD}_S \cdot [0.34\,\text{ISOC}_S + 0.13\,\text{NAD}_S \cdot (1+\text{NADH}_S/0.02) + \text{ISOC}_S \cdot \text{NAD}_S]$$
$$- [0.34 + \text{NAD}_S] \cdot \text{ICDH}_S \cdot \text{ISOC} \cdot \text{NAD}\}$$
$$\times [0.34\,\text{ISOC}_S + 0.13\,\text{NAD}_S \cdot (1+\text{NADH}_S/0.02) + \text{ISOC}_S \cdot \text{NAD}_S]^{-2}$$

$$= \frac{271 \times 0.072[0.0034 + 0.13 \times 0.072(1+0.18/0.02) + 0.01 \times 0.072] - [0.34 + 0.072] \times 2.71 \times 0.072}{[0.0034 + 0.13 \times 0.072(1+0.18/0.02) + 0.01 \times 0.072]^2}$$

$$= \frac{1.91 - 0.08}{0.09772^2}$$

$$= 191.64. \tag{5.57}$$

This partial derivative is multiplied by ISOC_S and divided by V_S to yield the appropriate kinetic order:

$$g_{\text{ISOC}} = \left.\frac{\partial V}{\partial \text{ISOC}}\right|_{\text{steady state}} \cdot \frac{\text{ISOC}_S}{V_S}$$

$$= 191.64 \times \frac{0.01}{2}$$

$$= 0.958. \tag{5.58}$$

Computation of the kinetic order associated with NAD requires differentiation with respect to NAD. We obtain

$$\left.\frac{\partial V}{\partial \text{NAD}}\right|_{\text{steady state}}$$

$$= \{\text{ICDH}_S \cdot \text{ISOC}_S \cdot [0.34\,\text{ISOC}_S + 0.13\,\text{NAD}_S \cdot (1+\text{NADH}_S/0.02) + \text{ISOC}_S \cdot \text{NAD}_S]$$
$$- [0.13(1+\text{NADH}_S/0.02) + \text{ISOC}_S] \cdot \text{ICDH}_S \cdot \text{ISOC} \cdot \text{NAD}\}$$
$$\times [0.34\,\text{ISOC}_S + 0.13\,\text{NAD}_S \cdot (1+\text{NADH}_S/0.02) + \text{ISOC}_S \cdot \text{NAD}_S]^{-2}$$

$$= \frac{2.71[0.0034 + 0.13 \times 0.072(1+0.18/0.02) + 0.01 \times 0.072] - [0.13 \times (1+0.18/0.02) + 0.01] \times 2.71 \times 0.072}{[0.0034 + 0.13 \times 0.072(1+0.18/0.02) + 0.01 \times 0.072]^2}$$

$$= \frac{0.2648 - 0.2556}{0.09772^2}$$

$$= 0.964. \tag{5.59}$$

This partial derivative is multiplied by NAD_S and divided by V_S to yield the appropriate kinetic order:

$$g_{\text{NAD}} = \left.\frac{\partial V}{\partial \text{NAD}}\right|_{\text{steady state}} \cdot \frac{\text{NAD}_S}{V_S}$$

$$= 0.964 \times \frac{0.072}{2}$$

$$= 0.0348. \tag{5.60}$$

The kinetic order associated with NADH is computed in an analogous fashion:

$$\left.\frac{\partial V}{\partial \text{NADH}}\right|_{\text{steady state}}$$
$$= (-0.13/0.02) \cdot \text{ICDH}_S \cdot \text{ISOC}_S \cdot \text{NAD}_S$$
$$\times [0.34\,\text{ISOC}_S + 0.13\,\text{NAD}_S \cdot (1 + \text{NADH}_S/0.02) + \text{ISOC}_S \cdot \text{NAD}_S]^{-2}$$
$$= \frac{-0.0914}{0.09772^2}$$
$$= -9.572. \tag{5.61}$$

This partial derivative is multiplied by NADH_S and divided by V_S to yield the appropriate kinetic order:

$$g_{\text{NADH}} = \left.\frac{\partial V}{\partial \text{NADH}}\right|_{\text{steady state}} \cdot \frac{\text{NADH}_S}{V_S}$$
$$= -9.572 \times \frac{0.18}{2}$$
$$= -0.862. \tag{5.62}$$

Finally, the rate constant for this process is obtained by equating the power-law representation with the original rate law at steady state. The result is

$$\alpha = V_S\,\text{ICDH}_S^{-1}\,\text{ISOC}_S^{-0.958}\,\text{NAD}_S^{-0.0348}\,\text{NADH}_S^{+0.862}$$
$$= 2 \times 0.00369 \times 82.414 \times 1.0959 \times 0.228$$
$$= 0.152. \tag{5.63}$$

Thus, the power-law rate law for the degradation of isocitrate reads

$$V_{\text{ISOC}}^{-} = 0.152\,\text{ICDH} \cdot \text{ISOC}^{0.958} \cdot \text{NAD}^{0.0348} \cdot \text{NADH}^{-0.862}. \tag{5.64}$$

It is noted that symbolic algebra software like Mathematica® can perform these types of computations very efficiently and without errors.

Example 7: Synthesis and degradation of glucose-6-phosphate in *Aspergillus niger*. Torres (1994bc) proposed a model of carbohydrate metabolism during citric acid accumulation in the mold *Aspergillus niger*. The first state variable in this model represents glucose-6-phosphate (see Fig. 5.10). It is made available to the mold by a hexokinase–substrate transport step that uses glucose as the substrate. In Torres's model, glucose and hexokinase are represented as independent variables. Glucose-6-phosphate is utilized via two pathways. One is catalyzed by glucose-6-phosphate dehydrogenase. The other constitutes a reversible reaction between glucose-6-phosphate and fructose-6-phosphate, which is catalyzed by phosphoglucoseisomerase. However, under the given conditions, the process is not amphibolic but always shows a net flow toward fructose-6-phosphate.

In Torres's model, the synthesis of glucose-6-phosphate depends on the substrate glucose (X_9), ATP (X_8), and the hexokinase–substrate transport step (X_{10}), which are

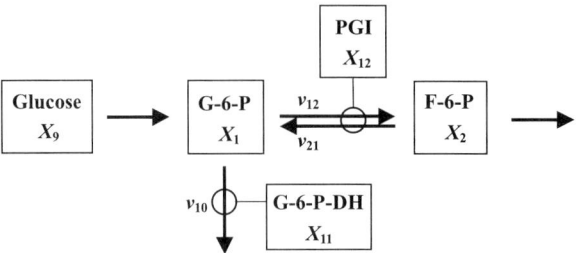

Figure 5.10. Dynamics of glucose-6-phosphate in Torres's (1994bc) model of carbohydrate metabolism in *Aspergillus niger*. See text for details.

represented by independent variables. The synthesis term thus has the symbolic form

$$\alpha_1 X_8^{g_{18}} X_9^{g_{19}} X_{10}^{g_{1,10}}. \tag{5.65}$$

Assuming Michaelis–Menten kinetics with respect to both ATP and glucose and substituting nominal steady-state concentrations of 0.5 and 290 mM, respectively, the kinetic orders g_{18} and g_{19} take the form

$$g_{18} = \frac{K_{M,\,\mathrm{ATP}}}{K_{M,\,\mathrm{ATP}} + X_{8S}} = \frac{0.15}{0.15 + 0.5} = 0.23, \tag{5.66}$$

$$g_{19} = \frac{K_{M,\,\mathrm{substrate}}}{K_{M,\,\mathrm{substrate}} + X_{9S}} = \frac{0.27}{0.27 + 290} = 0.0009. \tag{5.67}$$

The dependence of the flux on the enzyme concentration is assumed to be linear, so that the kinetic order with respect to X_{10} is unity:

$$g_{1,10} = 1. \tag{5.68}$$

The rate constant α_1 is computed from the measured flux $V_1^+ = 6.3 \times 10^{-3}$ μmol/min (mg protein) as

$$\begin{aligned}\alpha_1 &= 6.3 \times 10^{-3} X_{8S}^{-g_{18}} X_{9S}^{-g_{19}} X_{10S}^{-g_{1,10}} \\ &= 0.0063\, 0.5^{-0.23} 290^{-0.0009} 0.016^{-1} \\ &= 0.459. \end{aligned} \tag{5.69}$$

The degradation V_1^- of glucose-6-phosphate occurs via two pathways, which are catalyzed by glucose-6-phosphate dehydrogenase (X_{11}) and phosphoglucoseisomerase (X_{12}). The former is essentially irreversible, whereas the latter is reversible. In general terms, the degradation can thus be formulated as $V_1^- = v_{10} + v_{12} - v_{21}$. Because the phosphoglucoseisomerase reaction is reversible, V_1^- depends on the concentrations of both, the substrate and the product fructose-6-phosphate (X_2). In order to estimate parameter values with respect to X_2 from published rate laws, Torres defined the net flux between X_1 and X_2 as $v_{12}^n = v_{12} - v_{21}$ and used the irreversible strategy for modeling this reaction. This was considered justifiable because the process does not change direction under the given conditions. The degradation term in power-law form thus reads

$$V_1^- = \beta_1 X_1^{h_{11}} X_2^{h_{12}} X_{11}^{h_{1,11}} X_{12}^{h_{1,12}}. \tag{5.70}$$

PARAMETER ESTIMATION

Glucose-6-phosphate affects its own degradation as the substrate for two reactions. The representing kinetic order, h_{11}, therefore is computed by partial differentiation of v_{12}^n and v_{10}:

$$h_{11} = \frac{\partial V_1^-}{\partial X_1} \cdot \frac{X_1}{V_1^-} = \frac{\partial v_{12}^n}{\partial X_1} \cdot \frac{X_1}{V_1^-} + \frac{\partial v_{10}}{\partial X_1} \cdot \frac{X_1}{V_1^-}. \tag{5.71}$$

When we multiply and divide the first term on the right-hand side by v_{12}^n and the second term by v_{10}, we see that h_{11} is a weighted sum of the individual kinetic orders that characterize the two branches of the degradation of glucose-6-phosphate:

$$h_{11} = \frac{\partial v_{12}^n}{\partial X_1} \cdot \frac{X_1}{v_{12}^n} \cdot \frac{v_{12}^n}{V_1^-} + \frac{\partial v_{10}}{\partial X_1} \cdot \frac{X_1}{v_{10}} \cdot \frac{v_{10}}{V_1^-} = h_{11}^1 \cdot \frac{v_{12}^n}{V_1^-} + h_{11}^2 \cdot \frac{v_{10}}{V_1^-}, \tag{5.72}$$

where

$$h_{11}^1 = \frac{\partial v_{12}^n}{\partial X_1} \cdot \frac{X_1}{v_{12}^n} \tag{5.73}$$

and

$$h_{11}^2 = \frac{\partial v_{10}}{\partial X_1} \cdot \frac{X_1}{v_{10}} \tag{5.74}$$

are defined in the usual way and computed from traditional rate laws. Specifically, we obtain

$$h_{11}^1 = \frac{1}{1 - \Gamma/K_{eq}} \left(1 - \frac{v_{12}^n}{V_{max(F)}}\right) \tag{5.75}$$

and

$$h_{11}^2 = \frac{K_{M,G6P}}{K_{M,G6P} + X_{9S}}. \tag{5.76}$$

Substitution of the numerical values for the steady-state net flux $v_{12}^n = 0.00504$, the mass action ratio $\Gamma = 0.25$, the equilibrium constant $K_{eq} = 0.4$, the Michaelis constant $K_{M,G6P} = 0.045$, and the maximum velocities of substrate consumption, $V_{max(F)} = 15.3$, yields

$$h_{11}^1 = \frac{1}{1 - 0.25/0.4} \left(1 - \frac{0.00504}{15.3}\right) = 2.66, \tag{5.77}$$

$$h_{11}^2 = \frac{0.045}{0.045 + 0.1} = 0.31, \tag{5.78}$$

$$h_{11} = 2.66 \frac{v_{12}^n}{V_1^-} + 0.31 \frac{v_{10}}{V_1^-}$$

$$= 2.66 \times \frac{0.00504}{0.00504 + 0.00126} + 0.31 \times \frac{0.00126}{0.00504 + 0.00126} = 2.19. \tag{5.79}$$

The kinetic order with respect to the product fructose-6-phosphate, h_{12}, is computed in the same fashion. One obtains

$$h_{12} = \frac{\partial v_{12}^n}{\partial X_2} \cdot \frac{X_2}{v_{12}^n} \cdot \frac{v_{12}^n}{V_1^-} + \frac{\partial v_{10}}{\partial X_2} \cdot \frac{X_2}{v_{10}} \cdot \frac{v_{10}}{V_1^-} = h_{12}^1 \cdot \frac{v_{12}^n}{V_1^-} + h_{12}^2 \cdot \frac{v_{10}}{V_1^-}, \qquad (5.80)$$

where

$$h_{12}^1 = \frac{\partial v_{12}^n}{\partial X_2} \cdot \frac{X_2}{v_{12}^n} \qquad (5.81)$$

and

$$h_{12}^2 = \frac{\partial v_{10}}{\partial X_2} \cdot \frac{X_2}{v_{10}}. \qquad (5.82)$$

Because the branch v_{10} is independent of X_2, the corresponding kinetic order h_{12}^2 is zero. h_{12}^1 describes the contribution of fructose-6-phosphate to the reversible reaction catalyzed by phosphoglucoseisomerase. It takes the form

$$h_{12}^1 = -\frac{\Gamma/K_{eq}}{1 - \Gamma/K_{eq}} \left(1 + \frac{v_{12}^n}{V_{max(R)}}\right). \qquad (5.83)$$

Substitution of numerical values yields

$$h_{12}^1 = -1.66, \qquad (5.84)$$

$$h_{12} = -1.66 \frac{v_{12}^n}{V_1^-} = -1.33. \qquad (5.85)$$

Glucose-6-phosphate dehydrogenase, X_{11}, only catalyzes the main branch of the degradation of glucose-6-phosphate which leads to the synthesis of fructose-6-phosphate. Therefore, the partial derivative of the other branch, v_{12}^n, with respect to X_{11} is zero, and the required kinetic order is simply given as

$$h_{1,11} = h_{1,11}^1 \cdot \frac{v_{10}}{V_1^{-1}} = \frac{\partial v_{10}}{\partial X_{11}} \cdot \frac{X_{11}}{v_{10}} \cdot \frac{v_{10}}{V_1^{-1}}. \qquad (5.86)$$

Assuming direct proportionality between velocity and enzyme activity, the unweighted kinetic order of h_{11}^1 equals 1, and considering appropriate flux weighting with the term v_{10}/V_1^-, the overall kinetic order is

$$h_{1,11} = 1 \times 0.2 = 0.2. \qquad (5.87)$$

Finally, the main branch of the degradation of glucose-6-phosphate is catalyzed by phosphoglucoseisomerase, X_{12}. Because this enzyme affects only v_{12}^n, the partial derivative of v_{10} with respect to X_{12} is zero. Again assuming proportionality between velocity and enzyme activity, the unweighted kinetic order is zero, and accounting for the appropriate contribution of this branch to the total flux, the desired kinetic order is

$$h_{1,12} = \frac{\partial v_{12}^n}{\partial X_{12}} \cdot \frac{X_{12}}{v_{12}^n} \cdot \frac{v_{12}^n}{V_1^{-1}} = 0.8. \qquad (5.88)$$

The rate constant β_1 is obtained from equating the measured overall flux V_1^- and the corresponding power-law term in which the numerical values of all kinetic orders have been substituted. Simple rearrangement of this equation yields

$$\beta_1 = V_1^- X_1^{-b_{11}} X_2^{-b_{12}} X_{11}^{-b_{1,11}} X_{12}^{-b_{1,12}} = 1.833 \times 10^{-3}. \tag{5.89}$$

The equation for the dynamics of glucose-6-phosphate, in numerical form, thus reads

$$\dot{X}_1 = 0.459 X_8^{0.23} X_9^{0.0009} X_{10} - 1.833 \times 10^{-3} X_1^{2.19} X_2^{-1.33} X_{11}^{0.2} X_{12}^{0.8}. \tag{5.90}$$

A WORD OF CAUTION

At this point in time, most information about reactions and modulations stems from experiments *in vitro*. This situation poses difficult questions about the validity of our estimates for systems *in vivo*. For instance, do kinetic data obtained with purified enzymes adequately reflect the kinetic behavior of integrated enzyme systems *in vivo*? Albe et al. (1989) analyzed questions of *in vitro–in vivo* extrapolation and came to the conclusion that it is generally legitimate to use data obtained *in vitro*. Caution is required nevertheless. In some instances, the concentrations of an enzyme and its substrate are comparable *in vivo*. As a consequence, the nominal substrate concentration may be much higher than the concentration of unbound substrate, and this may result in grossly overestimated rates *in vivo* (cf. Shiraishi and Savageau, 1993). Mavrovouniotis and coworkers (Mavrovouniotis, 1988; Mavrovouniotis et al., 1990a; Bish and Mavrovouniotis, 1998) pointed out that the maximum reaction rate proposed in the Michaelis–Menten formalism is often an unrealistic consequence of the assumptions underlying the proposed mechanisms of the reactions and suggested taking into account chemical as well as physical steps. In their illustrations, quantification of kinetic and thermodynamic aspects led to upper bounds for maximal rates that sometimes differed significantly from the default V_{max}.

It is becoming increasingly evident that cells are not a homogeneous "soup" in which enzymatic reactions occur like molecular collisions in an ideal gas, but that there is *molecular trafficking* (cf. Weng et al., 1999). Often, several enzymes of the same pathway form multi-enzyme complexes, and reactants are passed from enzyme to enzyme along a *scaffold*, without reentering the bulk solution (cf. Welch, 1977; Ovadi, 1991; Faux and Scott, 1996; Dong et al., 1997; Whitmarsh et al., 1998; Schaeffer et al., 1998). The fact that reactions are tied to surfaces or confined to channels or scaffolds leads to constraints that can have unexpected consequences in that kinetic orders are much higher than would be found in homogeneous media (cf. Savageau, 1991b). For example, a bimolecular reaction in a homogeneous solution is characterized by a kinetic order of 2, whereas the same reaction on a two-dimensional grid has a kinetic order of 2.46; on a one-dimensional grid this number increases to 3 (Kopelman, 1986; Shiraishi and Savageau, 1992a; Savageau, 1995a). Cascante et al. (1994) compared non-interactive systems with interactive systems and systems of multi-enzymatic complexes and found substantial differences in kinetic properties. For instance, transition times differed by three orders of magnitude. In a comprehensive analysis of the tricarboxylic acid cycle in *Dictyostelium*, Wright et al. (1992) had to make *ad hoc* adjustments in maximal rates when kinetic properties obtained *in vitro* were integrated into an encompassing model and compared with cellular fluxes *in vivo*. Such adjustments suggest that *in vitro* data have to be evaluated with care (Shiraishi and Savageau, 1992a–d, 1993; Fell, 1997: p. 325).

ESTIMATION FROM DYNAMIC DATA

As mentioned in the introduction, estimation from dynamic data is quite different from steady-state estimation. Measurements are needed for all dependent variables at several points in time. These temporal data could stem from transient responses after a perturbation from steady state (e.g., Sorribas et al., 1993), but they are more often found in the analysis of systems that have not reached a steady state, but exhibit growth, decay, or some other long-term dynamics.

In principle, the estimation of parameter values from dynamic data seems to be a straightforward process. The dynamic data, given as measurements of all variables at several time points, are entered in a nonlinear regression program, such as BMDP AR® or Mathematica®, along with the system equations and initial guesses for all parameter values. The numerical algorithm then uses some strategy of improving the current parameter values, thereby diminishing the residual difference between data and model results at the given time points. Upon convergence, the differences are minimal, and the parameter values that led to this best fit are considered optimal. Unfortunately, this procedure only works in special cases. It has a good chance of success if the system contains only a few equations, if the error in the data is small, and if the initial guesses for the parameter values are not too far off (e.g., Kohn et al., 1979; Menten et al., 1981). In situations like our Case Studies (Chapters 8–11), which often deal with ten or more simultaneous differential equations, the currently available algorithms are usually unable to yield a solution. Nonetheless, because some dynamical data exist and algorithms are likely to improve with time, we should at least sketch their use.

The typical problems that one regularly encounters with direct estimation from dynamical data fall into two major categories. The first one is characterized by problems of size and complexity of the model, which translate into problems of computer capacity. For instance, suppose the system contains 20 parameters, which in reality is not very much. If the algorithm were to evaluate 10 values for each parameter in a systematic, all-encompassing fashion, it would have to consider 10^{20} combinations of parameter values. The algorithm would have to solve the set of differential equations numerically for each combination. If we suppose that each solution would take just one second, the algorithm would run for 3 trillion years. [Check that $10^{20}/(365 \times 24 \times 3{,}600) = 3.17 \times 10^{12}$.] Many of these trial solutions would not be satisfactory, and modern algorithms are designed to avoid them. Still, experience shows that the current algorithms often get lost in a never-ending trial-and-error chase of irrelevant combinations of parameter values.

The second category of problems includes issues of mathematical redundancy in our models. These derive from the fact that different sets of parameter values can produce responses that fit the experimental data about equally well (e.g., Sands and

Voit, 1996). As a very simple example of redundancy suppose that the two parameters p_1 and p_2 always appear in the form of the sum $p_1 + p_2$. It is quite obvious that any error in p_1 can be adjusted for by a corresponding error in p_2. Now suppose the algorithm is currently using values for p_1 and p_2 that are optimal. The algorithm has no easy way of knowing that this solution is truly optimal and attempts to improve the estimates of p_1 or p_2 by increasing one and concomitantly decreasing the other parameter, without actually improving the overall fit. Modern algorithms are equipped with rules to terminate such futile search, but then again, experience shows that algorithms often keep wandering about without making headway if the redundancy is not quite as simple.

After this sobering introduction, let's discuss some successful examples and see how the situation can be ameliorated in other cases.

1. Direct Estimation from Dynamic Data

Regression. In small systems with only a limited number of unknown parameters, the direct estimation from dynamical data is a good strategy. The equations are entered in an algorithm, and the unknown parameters, which may include initial values of the dependent variables, are specified. One also needs to enter the observed data with which the model solutions are to be compared. The only other input of interest is the specification of *initial guesses*. These are user-supplied values for all parameters, with which the numerical algorithm begins the estimation process. While most algorithms supply default values, well-chosen initial guesses are often crucial to the success of the estimation procedure.

The principle behind the estimation procedure is typically the following. Using the initial guesses, the algorithm solves the differential equations and compares the model results with observed results. It then selects other values for the unknown parameters, solves the equations again, compares computed and observed results, and assesses whether this solution is better than the previous solution. In contrast to linear regression, where optimal combinations of parameter values can be obtained with methods of linear algebra, nonlinear regression requires a cycle of solving and improving that is repeated very many times.

The key components of nonlinear regression are thus two: computation of "predicted" results with the model and selection of new parameter values to be tried out. In the case of differential equations, the former component is usually the one consuming most of the computation time, since for every iteration the algorithm has to compute numerically the solutions of the differential equation model. The selection of parameter values for the next cycle is often based on a gradient method that determines which changes in parameter values have the best chance of improving the estimates most. Many algorithms of this type originate in the work of Gauss and Newton, but many mathematicians in recent years have improved on the basic concepts.

Theoretically, obtaining the optimal estimates should simply be a matter of pushing the button. In reality, that is rarely the case. Almost all nonlinear estimations are treacherous. The results often depend on the initial guesses and on internal settings of the algorithm, and it is strongly recommended to run several estimations and to compare the results (cf. Berg et al., 1996). Nonlinear estimation is a large and very complicated branch of numerical mathematics, and there is no way for us even to cover the basics. It should be mentioned for completeness that Johnson (1988, 1991) developed methods of parameter estimation in simple stochastic S-systems.

Genetic algorithms. These algorithms provide a nonlinear estimation method that combines features of an exhaustive exploration of all possible parameter combinations with a gradient-based search that is the basis of most nonlinear regression algorithms. A genetic algorithm often converges much faster than an exhaustive search and is less likely to get stuck at a local minimum, which is a point where the error is relatively better than at all neighboring points. The key concept behind genetic algorithms is the coding of parameters as part of a *gene*, which is typically realized in the form of binary strings of zeros and ones. Each of the genes completely characterizes the behavior of the model for one particular set of parameter values. During the estimation procedure, several genes are produced and evaluated against an objective function that reflects the quality of their ability to model the data. The *fittest* genes, which correspond to the lowest residual errors among all genes, are used to produce *offspring*. This offspring consists of new genes that derive from recombining appropriate portions from pairs of old genes. The new genes again represent fully parameterized systems, which are evaluated for fitness and selection for the next generation. Good introductions to genetic algorithms were presented by Goldberg (1989), Ribeiro Filho et al. (1994), Srinivas and Patnaik (1994), and Yao and Sethares (1994).

S-system models are particularly suited for analyses with genetic algorithms because of their regular structure. This structure suggests straightforward coding rules and allows for fast numerical solution of the differential equations, which is unavoidable in any estimation approach. Although the combination of S-systems and genetic algorithms appears to be very appealing, only two limited estimation procedures have been published so far (Zhang et al., 1996; Tominaga and Okamoto, 1998). In Zhang's analysis, yield data on oil palms in Malaysia were used to parameterize a growth model with two variables. The limited experience from this and some other examples suggests that genetic algorithms may be successful for fast preliminary estimations, but that they are not always able to produce highly accurate solutions (Lorelei H. Schwacke, pers. comm.).

2. Replacing Derivatives with Slopes

The procedural part of the direct estimation method that requires the most computational effort usually is the numerical solution of the differential equations. By comparison, the evaluation of the error function and the determination of new tentative parameter values are fast. One strategy for improving efficiency is therefore to avoid the need for solving the differential equations. If the data show only a modest degree of scatter, this can be accomplished by replacing the derivatives $\dot{X}_i$ with slopes of the response curves. Studies that have successfully used this method include

PARAMETER ESTIMATION

Savageau and Voit (1982), Voit and Savageau (1982ab), Torsella and Bin Razali (1991), Voit (1992c), and Voit and Yu (1994). A slight variation of the method, in which time data and fluxes were used, can be found in Voit and Sands (1996ab). If the data are noisy, it may not be easy to obtain accurate measurements of slopes. In such a case it may still be advantageous to use this method, even with rough estimates, and in a second step to use the results as good initial guesses in direct nonlinear regression with the differential equations themselves. If the data are noisy, one is likely to obtain solutions that are sensitive to initial guesses and to slight changes in data.

If the data points follow clear, monotonic trends, the determination of slopes can even be done by hand. In the first step, the data points are fitted with a curved line, either freehand or with some smoothing algorithm of the kind implemented in many graphics programs, such as PowerPoint. In the next step the slopes of the freehand curves are measured and tabulated at all time points where values of the variables are available (see Fig. 5.11). If there are three dependent variables, measured at 12 time points, the augmented data set consists of 12 rows, each containing the time point, the concentrations of all three variables, and the slopes of the three variables at these time points. The measured slopes are surrogates for the true slopes, and these are nothing else but the derivatives $\dot{X}_i$ at the given time point. Thus, we substitute them, for each time point, on the left-hand sides of the differential equations, and substitute the measured values of X_1, X_2, and X_3 on the right-hand sides. The resulting system has 3×12 equations.

To illustrate the method, we deviate a little bit from the hand-fitting scenario and study instead a fictitious situation with a model that we know in its entirety. As discussed earlier in this chapter – in the context of the method of Kacser and Burns (1979) – this strategy allows us to evaluate the quality of the estimation method better than the analysis of some actual data would. Specifically, we have the luxury of choosing parameter values of our liking and evaluating the quality of the results

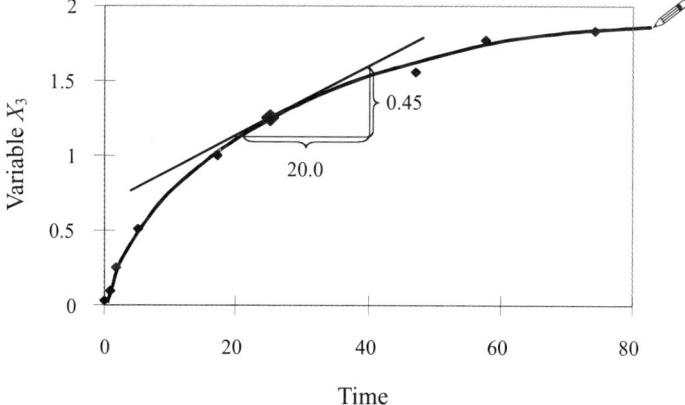

Figure 5.11. Artificial data plot illustrating parameter estimation from measured slopes. First, the data (♦) are fitted by hand with a smooth curve. Then, slopes are measured at each data point and substituted for derivatives. For instance, at $t = 25$ (enlarged symbol), $\dot{X}_3 \approx 0.45/20.0 = 0.0225$.

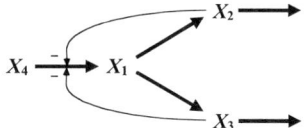

Figure 5.12. Branched pathway with constant input and feedback inhibition.

by comparison with the true solution. Furthermore, we have the option to work with data that are entirely error-free, which allows us to assess the quality of an estimation method under the best possible conditions. Finally, a completely specified model permits us to compute, rather than measure, slopes at as many points as we desire.

The method of substituting slopes for derivatives is particularly useful if the model happens to have only one product of power-law functions in each equation – in other words, if in each equation one of the two rate constants is zero. This case is special, because it allows us to take logarithms on both sides, which transforms the problem into one of simple linear regression. An example of this situation can be found in Torsella and Bin Razali (1991).

Example 1: Branched pathway. As a didactic example, consider a branched pathway with constant input X_4 and feedback inhibition, as shown in Fig. 5.12, which we model with the S-system

$$\dot{X}_1 = \alpha_1 X_2^{g_{12}} X_3^{g_{13}} X_4^{g_{14}} - \beta_1 X_1^{h_{11}},$$
$$\dot{X}_2 = \alpha_2 X_1^{g_{21}} - \beta_2 X_2^{h_{22}}, \quad (5.91)$$
$$\dot{X}_3 = \alpha_3 X_1^{g_{31}} - \beta_3 X_3^{h_{33}},$$
$$X_4 = \text{constant}.$$

In a real-life situation, we would of course not know the parameter values of the model, but we would still have some qualitative knowledge that is useful in nonlinear estimation problems. For example, all rate constants are positive by definition. Also, we know that the parameters g_{12} and g_{13} are negative, since they represent inhibitory effects, while all other kinetic orders are positive. If the processes followed first-order kinetics, the corresponding kinetic orders would be 1. If the processes resembled Michaelis–Menten kinetics, the appropriate kinetic orders would be between 0 and 1. Thus, for the initial guesses required by most nonlinear regression algorithms, values between 0 and 1 seem reasonable. Finally, there are precursor–product relationships at the branch point, as discussed in Chapter 3 (see also *Complements*).

For the numerical model, we specify, without a particular rationale, parameter values as we might expect them in an actual analysis:

$$\dot{X}_1 = 1.6 X_2^{-0.2} X_3^{-0.8} X_4 - 3.4 X_1^{0.5}, \quad X_1(0) = 0.2,$$
$$\dot{X}_2 = 2 X_1^{0.5} - 1.3 X_2^{0.75}, \quad X_2(0) = 0.5, \quad (5.92)$$
$$\dot{X}_3 = 1.4 X_1^{0.5} - 3.2 X_3^{0.9}, \quad X_3(0) = 0.1,$$
$$X_4 = 0.75.$$

PARAMETER ESTIMATION

We implement the model in PLAS (it is included as *Estimate.plc*), solve it numerically, and have the results of this "true solution" reported at time points 0, 0.1, 0.2, ..., 2.0. We also define additional variables S_i, which represent the slopes of X_i. Since $S_i = \dot{X}_i$, this is readily implemented in PLAS by entering additional definitions as

$$S_1 = 1.6 X_2^{-0.2} X_3^{-0.8} X_4 - 3.4 X_1^{0.5},$$
$$S_2 = 2 X_1^{0.5} - 1.3 X_2^{0.75}, \quad (5.93)$$
$$S_3 = 1.4 X_1^{0.5} - 3.2 X_3^{0.9}.$$

The solution, in tabular form, is given in Table 5.4; Fig. 5.13 exhibits the time courses of the three dependent variables.

From these data, the estimation equations are set up as described above. There is one equation per time point, each containing on the left side the measured (in our fictitious case, computed) slope, and on the right side the difference of power-law terms in which the numerical values from Table 5.4 are substituted for the dependent variables. The values of the independent variables are known and also substituted. The rate constants and kinetic orders are kept in their symbolic form and constitute the parameters to be estimated.

Table 5.4. "True" Solution to the Branched Pathway Model in Eqs. (5.92) and (5.93) and Figs. 5.12 and 5.13

Time	X_1	X_2	X_3	Slope of X_1	Slope of X_2	Slope of X_3
0	0.2	0.5	0.1	7.17683	0.1214426	0.2232429
0.1	0.7115766	0.5580099	0.1486528	3.327972	0.8477854	0.6053898
0.2	0.9304035	0.6519015	0.2086945	1.299082	0.9860005	0.5692981
0.3	1.005835	0.749898	0.2599952	0.3243903	0.9582272	0.4521181
0.4	1.011434	0.8415626	0.298953	−0.1558976	0.8691595	0.3285558
0.5	0.982955	0.9230369	0.3262332	−0.3833509	0.7586669	0.2203331
0.6	0.9393477	0.9932005	0.3437218	−0.4716457	0.6450341	0.1330063
0.7	0.8912408	1.052251	0.3535281	−0.4806965	0.5374916	0.06642414
0.8	0.84468	1.101063	0.3576244	−0.4451761	0.4407883	0.01835431
0.9	0.8029672	1.140846	0.3577189	−0.3865597	0.3571302	−0.0141197
1	0.7676711	1.172945	0.3552194	−0.3186864	0.2871209	−0.03402128
1.1	0.7392356	1.19871	0.3512375	−0.250533	0.2302874	−0.04422812
1.2	0.7173835	1.219403	0.3466097	−0.1877228	0.1854407	−0.04734466
1.3	0.7014044	1.236144	0.3419292	−0.1334421	0.1509622	−0.04562834
1.4	0.6903633	1.24988	0.3375824	−0.08907361	0.1250422	−0.04094767
1.5	0.6832556	1.261377	0.3337892	−0.05470001	0.1058763	−0.03476685
1.6	0.6791159	1.271225	0.3306428	−0.02952989	0.09180902	−0.02816058
1.7	0.6770861	1.279861	0.3281468	−0.01225662	0.08142098	−0.02185327
1.8	0.6764522	1.287594	0.3262476	−0.001351067	0.07356773	−0.01627632
1.9	0.6766549	1.29463	0.3248603	0.004719958	0.06737813	−0.0116338
2	0.6772817	1.301104	0.3238882	0.007349104	0.06222652	−0.007967323

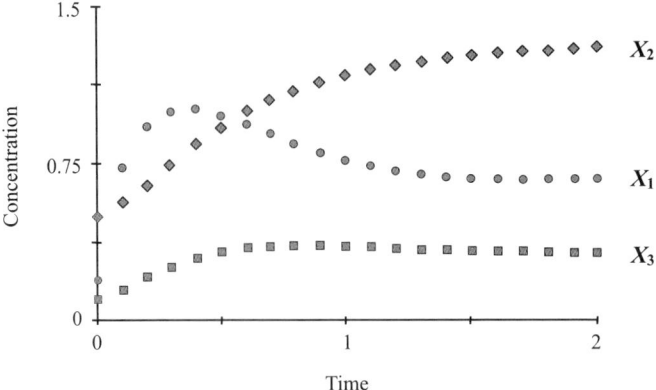

Figure 5.13. Responses of the three dependent variables in the model (5.92).

Under the assumption that X_4 is transported into the system, the kinetic order g_{14} is 1. The first four equations thus have the form

$$\begin{aligned}
\text{Slope of } X_1(t=0) &= \alpha_1 X_2(t=0)^{g_{12}} X_3(t=0)^{g_{13}} X_4 & - \beta_1 X_1(t=0)^{h_{11}}, \\
\text{Slope of } X_2(t=0) &= \alpha_2 X_1(t=0)^{g_{21}} & - \beta_2 X_2(t=0)^{h_{22}}, \\
\text{Slope of } X_3(t=0) &= \alpha_3 X_1(t=0)^{g_{31}} & - \beta_3 X_3(t=0)^{h_{33}}, \\
\text{Slope of } X_1(t=0.1) &= \alpha_1 X_2(t=0.1)^{g_{12}} X_3(t=0.1)^{g_{13}} X_4 & - \beta_1 X_1(t=0.1)^{h_{11}},
\end{aligned}$$

(5.94)

and the last equation reads

$$\text{Slope of } X_3(t=2) = \alpha_3 X_1(t=2)^{g_{31}} - \beta_3 X_3(t=2)^{h_{33}}. \tag{5.95}$$

The corresponding equations in numerical form, which are used internally in the estimation algorithm, read

$$\begin{aligned}
7.17683 &= \alpha_1 0.5^{g_{12}} 0.1^{g_{13}} 0.75 - \beta_1 0.2^{h_{11}}, \\
0.1214426 &= \alpha_2 0.2^{g_{21}} - \beta_2 0.5^{h_{22}}, \\
0.2332429 &= \alpha_3 0.2^{g_{31}} - \beta_3 0.1^{h_{33}}, \\
3.327972 &= \alpha_1 0.5580099^{g_{12}} 0.1486528^{g_{13}} 0.75 - \beta_1 0.7115766^{h_{11}}, \\
&\vdots \\
-0.007967323 &= \alpha_3 0.6772817^{g_{31}} - \beta_3 0.3238882^{h_{33}}.
\end{aligned}$$

(5.96)

In our example, we have 21 equations of this type, and these form the basis for the nonlinear regression. Since we do not have to solve differential equations any more, applicable programs are more widely available; examples include the statistics packages BMDP®, SAS®, and Statistica®. For our example, we use the "quasi-Newton" procedure in Statistica.

Statistica actually allows us to enter only one function to be minimized, and not a system. As a straightforward remediation of this problem, one may define this function as the sum of squared differences between slopes and right-hand sides,

$$F = \left[\text{Slope}_1 - \left(\alpha_1 X_2^{g_{12}} X_3^{g_{13}} X_4 - \beta_1 X_1^{h_{11}}\right)\right]^2$$
$$+ \left[\text{Slope}_2 - \left(\alpha_2 X_1^{g_{21}} - \beta_2 X_2^{h_{22}}\right)\right]^2$$
$$+ \left[\text{Slope}_3 - \left(\alpha_3 X_1^{g_{31}} - \beta_3 X_3^{h_{33}}\right)\right]^2, \quad (5.97)$$

and define its "observed" values as all zeros (Statistica actually does not allow this either, giving the message that "F has no variance." The problem can be circumvented by replacing two of the zero values of F with very small positive or negative numbers, which do not affect the error much.)

Entering this equation and using initial guesses close to the true values, the algorithm quickly converges to the true solution, which indicates that the procedure is working. Next, we start the algorithm with other initial guesses that appear reasonable. As discussed earlier, kinetic orders of material fluxes are typically between 0 and 1, and kinetic orders representing inhibition are often slightly negative. Not much is known *a priori* about rate constants. Pretending for this simulation that we have no further information about the system, we set all kinetic orders to 0.5, except for the inhibition parameters, which we specify as -0.3, and define all rate constants as 1. The algorithm converges to a solution that is not perfectly correct, but is very close to the true set of parameter values.

Finally, we initialize the procedure with Statistica's default value of 0.1 for all parameters. In this case, the estimates for the first two equations are fine, but the estimates for the third equation are

$$\alpha_3 = 0.34, \quad g_{31} = 2.184, \quad \beta_3 = 5.811, \quad h_{22} = 55.126; \quad (5.98)$$

the remaining error is about 0.0687, which is not very high. In fact, if we were dealing with actual data, which are subject to natural variation and noise, we might be very happy with an error of this low magnitude. However, knowing what the true values are, and especially considering the value for h_{22}, which is supposed to represent a kinetic order, the estimates are inadequate. The algorithm has gotten stuck at a local minimum, and small changes in parameter values would only increase the error. Local minima can cause very significant problems in nonlinear estimations. It is difficult to avoid local minima, and because the final error does not always indicate whether the solution really is optimal, it is sometimes difficult to evaluate the estimation results. The only option for assessing a solution is often to repeat the estimation procedure several times with different sets of initial values. If they all yield essentially the same estimates, the degree of confidence in the truth of the estimates is greatly increased. If we had used the default values in our first estimation, we should now start with other initial guesses, until a more reasonable solution is obtained.

Example 2: A cascade. As a second example, consider the model in Fig. 5.14, which represents a cascaded mechanism, such as is found, for instance, in gene regulation (e.g., Savageau, 1976; Savageau and Sands, 1991) and immunology (e.g., Irvine and

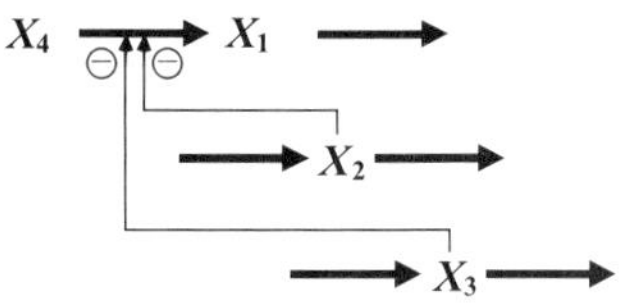

Figure 5.14. Cascaded pathway.

Savageau, 1985ab). As a numerical model, we specify

$$\begin{aligned}
\dot{X}_1 &= 10 X_2^{-0.1} X_3^{-0.05} X_4 - 5 X_1^{0.5}, & X_1(0) &= 0.2, \\
\dot{X}_2 &= 2 X_1^{0.5} - 1.44 X_2^{0.5}, & X_2(0) &= 0.5, \\
\dot{X}_3 &= 3 X_2^{0.5} - 7.2 X_3^{0.5}, & X_3(0) &= 0.1, \\
X_4 &= 0.75.
\end{aligned} \quad (5.99)$$

We implement the model in PLAS (file *Cascade.plc*) and solve it numerically at time points 0, 0.2, 0.4,..., 8.0. We also define additional variables S_i, which represent the slopes of the dependent variables, $\dot{X}_i$,

$$\begin{aligned}
S_1 &= 10 X_2^{-0.1} X_3^{-0.05} X_4 - 5 X_1^{0.5}, \\
S_2 &= 2 X_1^{0.5} - 1.44 X_2^{0.5}, \\
S_3 &= 3 X_2^{0.5} - 7.2 X_3^{0.5},
\end{aligned} \quad (5.100)$$

and implement them in PLAS, as before.

The solution, in tabular form, is given in Table 5.5; Fig. 5.15 exhibits the time courses of the three dependent variables.

From these data, the estimation equations are set up as before; in this case, there are 41 triplets of equations. The first four equations have the form

$$\begin{aligned}
\text{Slope of } X_1(t=0) &= \alpha_1 X_2(t=0)^{g_{12}} X_3(t=0)^{g_{13}} X_4 &&- \beta_1 X_1(t=0)^{h_{11}}, \\
\text{Slope of } X_2(t=0) &= \alpha_2 X_1(t=0)^{g_{21}} &&- \beta_2 X_2(t=0)^{h_{22}}, \\
\text{Slope of } X_3(t=0) &= \alpha_3 X_2(t=0)^{g_{32}} &&- \beta_3 X_3(t=0)^{h_{33}}, \\
\text{Slope of } X_1(t=0.2) &= \alpha_1 X_2(t=0.2)^{g_{12}} X_3(t=0.2)^{g_{13}} X_4 &&- \beta_1 X_1(t=0.2)^{h_{11}},
\end{aligned}$$
$$(5.101)$$

and the last equation reads

$$\text{Slope of } X_3(t=8) = \alpha_3 X_2(t=8)^{g_{32}} - \beta_3 X_3(t=8)^{h_{33}}. \quad (5.102)$$

Internally, Statistica uses the corresponding equations in numerical form for the estimation procedure. They read

$$\begin{aligned}
6.783054 &= \alpha_1\, 0.5^{g_{12}}\, 0.1^{g_{13}}\, 0.75 - \beta_1\, 0.2^{h_{11}}, \\
-0.1238066 &= \alpha_2\, 0.2^{g_{21}} - \beta_2\, 0.5^{h_{22}}, \\
-0.1555196 &= \alpha_3\, 0.5^{g_{32}} - \beta_3\, 0.1^{h_{33}}, \\
3.452052 &= \alpha_1\, 0.6226305^{g_{12}}\, 0.09867241^{g_{13}}\, 0.75^{g_{14}} - \beta_1\, 1.156611^{h_{11}}, \\
&\vdots \\
0.003464317 &= \alpha_3\, 3.507678^{g_{32}} - \beta_3\, 0.6082212^{h_{33}}.
\end{aligned}$$
$$(5.103)$$

PARAMETER ESTIMATION

Table 5.5. Numerical Values of the Dependent Variables X_1, X_2, and X_3 and the Corresponding Slopes at Time Points $t = 0, 0.2, 0.4, \ldots, 8$ in the Cascaded Pathway of Fig. 5.14 and Eqs. (5.99) and (5.100)

Time	X_1	X_2	X_3	Slope of X_1	Slope of X_2	Slope of X_3
0	0.2	0.5	0.1	6.783054	−0.1238066	−0.1555196
0.2	1.156611	0.6226305	0.09867241	3.452052	1.014657	0.1055323
0.4	1.682751	0.8560131	0.1294432	1.951368	1.262116	0.1851974
0.6	1.978119	1.11365	0.1690374	1.077092	1.293286	0.2056694
0.8	2.136435	1.367789	0.2104076	0.5496218	1.239194	0.2059199
1	2.211623	1.606964	0.2508141	0.2289	1.148873	0.1971259
1.2	2.236473	1.826448	0.2889706	0.036348	1.044864	0.1839538
1.4	2.231527	2.024772	0.3242677	−0.07516447	0.9386186	0.1688299
1.6	2.209816	2.202156	0.3564668	−0.1352009	0.8361784	0.1531503
1.8	2.179593	2.359708	0.3855471	−0.1628138	0.7406558	0.1377472
2	2.146012	2.498977	0.4116174	−0.170449	0.6534808	0.123112
2.2	2.112196	2.621687	0.4348616	−0.1662516	0.5750849	0.109517
2.4	2.079943	2.729587	0.4555025	−0.1555158	0.505311	0.09709176
2.6	2.050197	2.824354	0.4737787	−0.1416228	0.4436645	0.0858703
2.8	2.023362	2.90755	0.4899292	−0.126676	0.3894769	0.07582605
3	1.999512	2.980593	0.5041834	−0.1119297	0.3420091	0.06689506
3.2	1.978529	3.044752	0.5167558	−0.098084	0.3005166	0.05899262
3.4	1.960195	3.10115	0.5278427	−0.08548174	0.2642871	0.0520242
3.6	1.944245	3.150774	0.5376213	−0.0742454	0.2326634	0.04589355
3.8	1.930405	3.194483	0.5462498	−0.06436556	0.205052	0.04050722
4	1.918413	3.233026	0.5538681	−0.05576177	0.1809267	0.03577861
4.2	1.908023	3.267055	0.5605997	−0.04831616	0.159825	0.03162757
4.4	1.899018	3.297131	0.5665529	−0.04189813	0.1413447	0.02798228
4.6	1.891204	3.323744	0.5718221	−0.03637816	0.1251374	0.02477907
4.8	1.884415	3.347317	0.5764903	−0.0316349	0.1109027	0.02196157
5	1.878506	3.368219	0.5806295	−0.02755858	0.09838194	0.019481
5.2	1.873354	3.386769	0.5843026	−0.02405235	0.08735263	0.01729562
5.4	1.868853	3.403247	0.5875649	−0.02103234	0.0776234	0.01536707
5.6	1.864913	3.417894	0.5904644	−0.01842644	0.06902948	0.0136632
5.8	1.861458	3.430925	0.5930435	−0.01617323	0.0614288	0.01215566
6	1.858424	3.442524	0.5953386	−0.0142205	0.05469865	0.01082222
6.2	1.855753	3.452855	0.5973825	−0.01252443	0.0487329	0.009640215
6.4	1.853399	3.462062	0.5992035	−0.01104783	0.04343944	0.008591681
6.6	1.851321	3.47027	0.6008267	−0.009759293	0.03873825	0.007661516
6.8	1.849485	3.477592	0.6022746	−0.00863267	0.03455969	0.006833587
7	1.847859	3.484125	0.6035662	−0.007645349	0.03084288	0.006097474
7.2	1.846419	3.489956	0.604719	−0.006778416	0.02753459	0.005442524
7.4	1.845141	3.495163	0.6057482	−0.006015913	0.02458814	0.004858807
7.6	1.844006	3.499813	0.6066671	−0.005343823	0.02196253	0.00433978
7.8	1.842998	3.503967	0.6074879	−0.004750694	0.01962172	0.003877032
8	1.842101	3.507678	0.6082212	−0.004226489	0.01753392	0.003464317

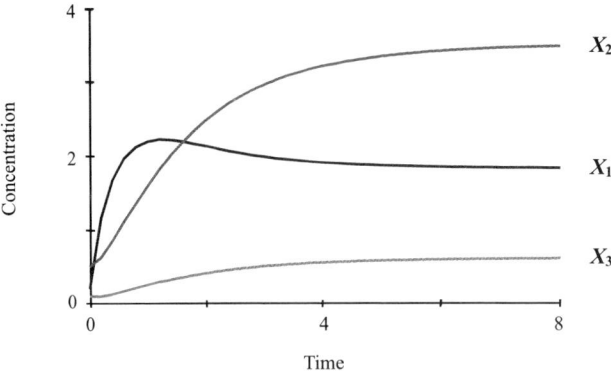

Figure 5.15. Dynamic responses of the dependent variables X_1, X_2, and X_3 in the cascaded pathway of Fig. 5.14 and Eq. (5.99).

As before, we define an estimation function of the type (5.97) that is based on these equations. The only other required input is the specification of initial guesses. Starting the algorithm with the true solution leads to immediate, successful termination with an error of zero. While rather obvious, this test provides a good check for typos in the equations and other unintended discrepancies.

Using initial guesses close to the true values, the algorithm fairly quickly converges to the following solution:

$\alpha_1 = 9.237,\quad g_{12} = -0.128441,\quad g_{13} = -0.003029,\quad \beta_1 = 5.496,\quad h_{11} = 0.4779,$
$\alpha_2 = 2.00,\quad g_{21} = 0.5,\qquad\qquad\qquad\qquad\qquad\qquad\beta_2 = 1.44,\quad h_{22} = 0.500,$
$\alpha_3 = 3.072\quad g_{32} = 0.0648,\qquad\qquad\qquad\qquad\qquad\beta_3 = 3.236,\quad h_{22} = 0.0397.$

There is good news and bad news. The good news is that the algorithm converged to a solution with an error of 9×10^{-6}. This is as close to a perfect fit as we could ever hope for in a real-world parameter estimation. The bad news is that some parameter values are rather different from the true values. Indeed, implementing this system in PLAS quickly demonstrates that the solution is dynamically not really adequate (Fig. 5.16).

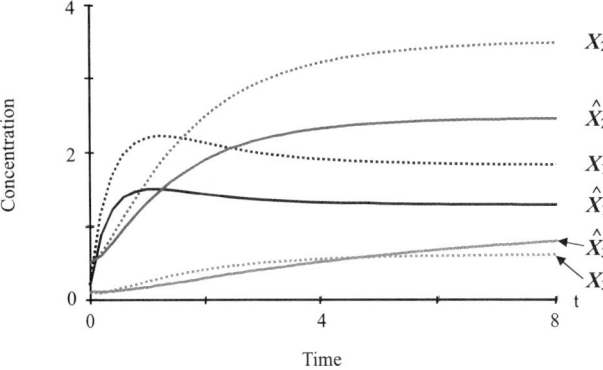

Figure 5.16. The dynamic responses of X_1, X_2, and X_3 (identified by $\wedge$) are clearly different from the "true" solution (dotted), even though the error is only 9×10^{-6} (see text for discussion).

PARAMETER ESTIMATION

Simulating the situation that we don't know anything about the system, we use the default initial value of 0.1 for all parameters to be estimated. The error upon convergence is 0.019, which may not look so bad, but the solution is unacceptable: Some of the rate constants are negative, and the kinetic order h_{22} has an estimated value of 423.074!

Next we take account of the fact that some parameters cannot be negative and set, as initial guesses, all rate constants to 1 and all kinetic orders to 0.5, except for the inhibition parameters g_{12} and g_{13}, which we specify as -0.3. The procedure converges, this time with an error of 0.011435 and the following estimates:

$\alpha_1 = 4.0539,\quad g_{12} = 0.4681,\quad g_{13} = -2.6828,\quad \beta_1 = 0.00012,\quad h_{11} = 9.9350,$
$\alpha_2 = 4.9534,\quad g_{21} = 0.2030,\quad\quad\quad\quad\quad\quad\quad\quad\quad\quad \beta_2 = 4.3462,\quad h_{22} = 0.1922,$
$\alpha_3 = 6.7284,\quad g_{31} = 0.2316,\quad\quad\quad\quad\quad\quad\quad\quad\quad\quad \beta_3 = 10.0727,\quad h_{22} = 0.2305.$

Again, the estimates are clearly different from the true values. For instance, g_{12} is positive, which implies that X_2 is activating rather than inhibiting the synthesis of X_1. Nevertheless, the fit (Fig. 5.17) is actually not bad. In particular, if we imagine a realistic situation in which the data are subject to natural variations and uncertainties, the fit could be deemed satisfactory.

As another test, we start the solution with all rate constants at 5 and all kinetic orders at 0.1, except for inhibitions, which we specify again as -0.3. The result is qualitatively similar: Some of the estimates are clearly wrong, and in this case, the data fit is essentially indistinguishable from the original. The only real difference we are able to discern (and this only because we have complete knowledge of the true system), is that the flux through X_1 is about 8.5 instead of 6.8 in the original system.

How is it possible that wrong parameter values give us the right answer, and what can we do to avoid this phenomenon? The first question is again related to redundancies between parameters and also, to a lesser degree, to the replacement of derivatives with measured slopes at a limited number of points. These more theoretical questions are discussed in some detail in the literature (Berg et al., 1996; Voit and Sands, 1996a; Sands and Voit, 1996).

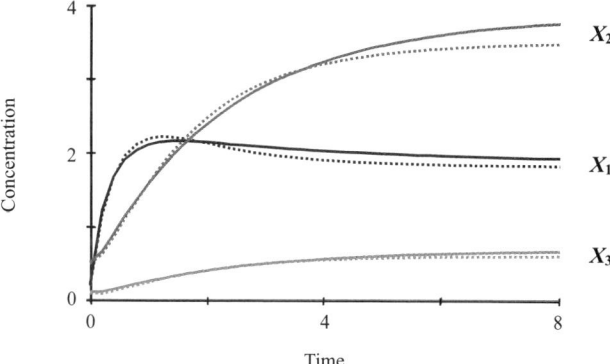

Figure 5.17. The estimation procedure may yield parameter values that are different from the true values but nevertheless correspond to an acceptable data fit (true solution dotted, estimated solution in solid lines; see text for details).

The second question is of more practical interest. A partial answer is to avoid redundancies, which is sometimes possible by constraining parameters. As we did before, we may limit inhibition parameters to negative values and impose similar constraints on other parameters. We implement such a constraint below. Other types of constraints may include information derived from a known steady state or from less obvious precursor–product relationships. Such constraints often consist of rather complicated sets of algebraic equations, but the effort of implementing them may be worthwhile. Two examples for sets of constraints are based on the steady-state equations and on the flux stoichiometry at a branch point. They are discussed in the *Complements*.

One strategy for forcing the estimation algorithm at least to keep parameters positive or negative, according to their role in the model, is to enter all parameters as squared parameters and subsequently to estimate their square roots. For instance, the α_1-terms in the two models above could be coded as $\delta_1^2 X_2^{-f_{12}^2} X_3^{-f_{13}^2} X_4$, and the β_3-terms as $\gamma_3^2 X_3^{k_{33}^2}$. Implementing this new system for the cascaded pathway and initializing it with all δ and γ equal to 1 and all f and k as 0.1, the algorithm converges. Once the estimates are squared, the results are close to the true values.

3. Transient Responses

Sorribas et al. (1993) analyzed the feasibility of estimating kinetic parameter values from transient response experiments. The key idea of this method is to study the immediate response of a system in steady state to a perturbation. To be specific, suppose the variable X_k is perturbed in a specified way and we can measure the immediate change in X_i. The quantity describing this finite perturbation is

$$\frac{\Delta \dot{X}_i}{\Delta X_k} = \frac{\dot{X}_{ip} - \dot{X}_{i0}}{X_{kp} - X_{k0}}, \tag{5.104}$$

where the index p refers to values after perturbation and the index 0 to the original values. Since the system had been at a steady state before the perturbation, $\dot{X}_{i0}$ is equal to zero, and the numerator reduces to $\dot{X}_{ip}$, which describes the instantaneous change in X_i after perturbation and is equivalent to the net flux V_{ip} after perturbation. The expression on the left-hand side directly relates to some of the kinetic orders in the system, as can be seen from the following argument. Because we are dealing with small perturbations, it is reasonable to assume that the finite ratio of differences is a good approximation of the corresponding derivative at steady state,

$$\frac{\Delta \dot{X}_i}{\Delta X_k} \approx \frac{\partial \dot{X}_i}{\partial X_k}. \tag{5.105}$$

Since $\dot{X}_i$ is equivalent to the difference of steady-state fluxes, we obtain

$$\frac{\partial \dot{X}_i}{\partial X_k} = \frac{\partial (V_i^+ - V_i^-)}{\partial X_k} \tag{5.106}$$

at steady state. We also know from the definition of kinetic orders that

$$g_{ik} = \frac{\partial V_i^+}{\partial X_k} \cdot \frac{X_k}{V_i^+} \tag{5.107}$$

PARAMETER ESTIMATION

and

$$h_{ik} = \frac{\partial V_i^-}{\partial X_k} \cdot \frac{X_k}{V_i^-} \tag{5.108}$$

at steady state. At the steady state, the influx V_i^+ is equivalent to the efflux V_i^-. Thus, we obtain

$$g_{ik} - h_{ik} = \frac{\partial (V_i^+ - V_i^-)}{\partial X_k} \cdot \frac{X_k}{V_i^+}, \tag{5.109}$$

and employing the previous relationships, we arrive at

$$g_{ik} - h_{ik} = \frac{\dot{X}_{i_p}}{X_{kp} - X_{k0}} \cdot \frac{X_{k0}}{V_{i0}^+} = \frac{V_{ip}}{X_{kp} - X_{k0}} \cdot \frac{X_{k0}}{V_{i0}^+}. \tag{5.110}$$

Thus, the difference of two kinetic orders can be determined from measurements of changes in X_i or the net flux V_{ip} after perturbation.

EXERCISES

1. Use the data in Table 5.6 (Lathem and Rodnan, 1962; see also Curto et al., 1998a) to estimate the kinetic order of uric acid in the excretion of uric acid. Is there

Table 5.6. Data for Exercise 1

Subjects	Uric Acid Concentration in Plasma	Uric Acid Excretion
Normal	13	0.015
	20	0.056
	23	0.040
	30	0.131
	35	0.085
	35	0.085
	37	0.085
	50	0.138
Symptomatic	33	0.077
	54	0.128
	81	0.473
	122	1.11
	170	2.62
	195	3.25
	257	5.55
	316	8.51
	339	9.47
	398	20.04
	502	47.21
	589	65.09
	676	130.5
	776	167.0

a difference in kinetic orders between normal and symptomatic subjects? Explain your answer. (For a formal decision about the difference between kinetic orders, see, e.g., Neter and Wasserman, 1974, for a statistical "comparison of two regression lines.")

2. The graph below shows the dependence of a fictitious process V_i^+ on some metabolite concentration X_j. Discuss whether and with which possible limitations a kinetic order can be obtained from this relationship. If feasible, determine one or several kinetic orders.

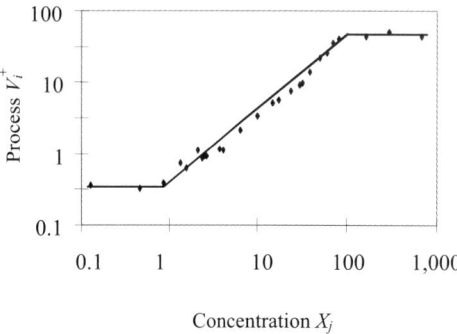

3. In the example evaluating the method of Kacser and Burns (1979), kinetic orders were estimated as $h_{11} = 0.519$ and $h_{12} = -0.316$.

   3.1 Implement the original model with these estimated kinetic orders (instead of the original parameter values), and compare steady-state and dynamical results of the two models.

   3.2 Perform the same analysis as in the example, but increase or decrease X_3 and X_4 by 5% and 10%. Discuss your results.

   3.3 Simulate five perturbations and estimate kinetic orders by means of linear regression. Discuss advantages and disadvantages of small and large perturbations in this context of regression analysis.

4. Compute kinetic orders for substrate and inhibitor for different operating points, using the traditional rate law

$$v(X_1, X_2) = \frac{V_{max} X_1}{X_1 + K_M(1 + X_2/K_I)}.$$

Describe how the kinetic orders change with increasing substrate and inhibitor concentrations. Provide rules of thumb for assigning kinetic orders if details of the process are not known.

5. Fell (1997, p. 117) states that the kinetic order of the substrate in a Hill process,

$$v = \frac{V_{max} X^n}{K_M^n + X^n},$$

may be greater than 1, but that it is always less than the Hill coefficient n. Prove these statements in two ways: (i) using the definition of kinetic orders, and (ii) using Eq. (5.47) for irreversible processes whose rates are proportional to fractional saturation with substrate.

6. 6.1 Compare mathematically Stitt's formula (5.8) with the definition of kinetic orders. Under what conditions do the two yield the same or similar results?

    6.2 Confirm the results from Stitt's formula in Examples 2 and 3.

7. Bohnensack and Halangk (1986) analyzed mitochondrial respiration and motility in ejaculated bull spermatozoa. They suggested the following rate laws for respiration (v_{resp}) and motility (v_{mot}):

$$v_{resp} = \frac{V_{resp}}{1 + (X + K_{resp})^n},$$

$$v_{mot} = \frac{V_{mot}}{1 + K_{mot}/X}.$$

In these equations, X is the cellular ATP/ADP ratio, which is generally accepted to be the most important signal in the coordination of ATP turnover. V_{resp} and V_{mot} are the maximum velocities, K_{resp} and K_{mot} are the ATP/ADP ratios for half-maximum rates, and n is an apparent cooperativity coefficient. Determine the kinetic orders for this system. Use as parameter values $V_{resp} = 9.3$ (nmol O_2) min^{-1} (μl cells)$^{-1}$, $K_{resp} = 4.7$ ATP/ADP, $n = 2.9$, $V_{mot} = 17\%$ motility/min, $K_{mot} = 1.7$ ATP/ADP, $m = 0.14$ (nmol O_2) (μlcells)$^{-1}$ (percent motile cells)$^{-1}$. Compute the kinetic orders for different values of ATP/ADP. Sketch plots of the kinetic orders as functions of ATP/ADP.

8. 8.1 Kohn and Lemieux (1991) modeled the hexokinase and phosphohexose isomerase steps of glycolysis in ascites cells with the following rate laws:

$$v_{hex} = \frac{1}{\left(1 + \frac{50}{\text{Glc}} + \frac{440}{\text{MgATP}}\right)\left(1 + \frac{\text{G6P}}{60(1+\text{Pi}/1900)}\right)},$$

$$v_{phi} = \frac{1 - \frac{\text{F6P}/\text{G6P}}{70/200}}{1 + \frac{200}{\text{G6P}}\left(1 + \frac{\text{F6P}}{70}\right)}.$$

In the absence or presence of exogenous glucose, respectively, the following nominal concentrations were given:

Metabolite	Concentration (μM)	
	0 mM Glucose	12.5 mM Glucose
Glc	1	2000
G6P	51	592
F6P	44	215
MgATP	2901	2226
Pi	2000	2683

Develop power-law representations for v_{hex} and v_{phi}. Is it relevant whether any of the quantities are modeled as dependent or independent variables? Kohn and Lemieux (1991) list several more rate laws that are part of glycolysis. Use them as additional exercises.

    8.2 Galazzo and Bailey (1990) modeled the glycogen synthetase reaction

UDPG + ATP → glycogen + ATP

with the rate law

$$v_{gly} = \frac{V_{max}^{gly} G6P^n}{K_{gly}^n + G6P^n} \bigg/ \left[\frac{K_{M0}}{UDPG} \left(1 + \frac{K_M}{GP6}\right) + 1 \right].$$

This formulation implies that glucose-6-phosphate (G6P) is an effector of the UDPG binding constant and of the maximal velocity of the reaction. V_{max}^{gly} and n were estimated from ^{31}P-RMN measurements of ATP + ADP, UDPG, and G6P. The remaining parameter values were taken from the literature. The numerical specifications are: $V_{max}^{gly} = 14.31$, $K_{gly} = 2$, $K_M = 1.1$, $K_{M0} = 1$, $n = 8.26$. At steady state, the rate and concentration were measured as $v_{gly} = 0.014$, G6P = 1.011, respectively.

Formulate the process in S-system terminology. Compute the kinetic orders and the rate constant. Are the results the same for the corresponding GMA model?

9. 9.1 In Shiraishi and Savageau's (1992a–d, 1993) model of the tricarboxylic acid cycle in *Dictyostelium discoideum*, citrate is produced from a reaction of acetyl-CoA (ACO) and oxalacetate (OAA). The process is catalyzed by citrate synthetase (CS) and affected by CoA and has the following traditional form:

$$v_1 = \frac{CS \cdot ACO \cdot OAA}{0.007 \cdot ACO + 0.01 \cdot OAA(1 + CoA/0.11) + ACO \cdot OAA}.$$

Develop the corresponding power-law representation. Use as steady-state values of ACO, OAA, CoA, and CS, respectively: 0.0593 mM, 0.0025 mM, 0.1 mM, and 8.24 mM/min.

9.2 In the same model, the degradation of succinate (SUC) occurs via two pathways, one (v_2) leading to glutamate (GLU) and the other (v_3) to fumarate (FUM). The authors considered the conversion into GLU as proportional to substrate (SUC) and enzyme activity (ENZ). The conversion to FUM was assumed to follow the following rate law:

$$v_3 = \frac{SDH \cdot (SUC - FUM/10)}{0.1 + SUC + FUM/10}.$$

Develop the power-law term for the degradation of SUC. According to the authors, the following concentrations and activities provide appropriate steady-state values for SUC, GLU, FUM, ENZ, and succinate dehydrogenase (SDH), respectively: 0.801 mM, 6.03 mM, 0.04 mM, 1 mM/min, 3.15 mM/min.

10. Derive Eqs. (5.36), (5.37), (5.40), and (5.41) from the definition of kinetic orders. Assume that the forward and reverse reactions are defined as

$$v_F = \frac{V_{max(F)} \cdot X_i / K_{Mi}}{1 + X_i / K_{Mi} + X_j / K_{Mj}} \quad \text{and} \quad v_R = \frac{V_{max(R)} \cdot X_j / K_{Mj}}{1 + X_i / K_{Mi} + X_j / K_{Mj}}$$

(e.g., Sorribas and Savageau, 1989c).

11. Suppose a traditional rate law depending on one variable X is given as the ratio of some numerator N and some denominator D:

$$V = \frac{N}{D}.$$

Show that the kinetic order g of a power function αX^g that approximates V can be

PARAMETER ESTIMATION

computed as

$$\frac{N'}{N} - \frac{D'}{D},$$

where N' and D' denote derivatives with respect to X, and all quantities are evaluated at the steady-state operating point. Prove or disprove the analogous assertion for a rate law with several variables.

12. (Requires linear or nonlinear regression.) Torsella and Bin Razali (1991) described the growth of trees as a function of current tree size, coded as X_1 and measured as mean diameter at breast height (DBH), and the number of trees per unit area, X_2, which was considered a substitute for spacing and competition between trees. Their model had the form

$$\dot{X}_1 = \alpha_1 X_1^{g_{11}} X_2^{g_{12}},$$
$$\dot{X}_2 = -\beta_2 X_1^{h_{21}} X_2^{h_{22}},$$

indicating that both variables affect each other and that size increases, whereas the number of trees decreases due to competition and other causes. Use the (simplified) data in Table 5.7 to estimate parameter values with linear or nonlinear regression.

Table 5.7. Data for Exercise 12

Time (years)	Size (DBH)	Number
0	0.36	1000
1	1.5	998
2	2.2	984
3	2.8	966
4	3.5	950
5	4	930
6	4.3	902
7	4.7	872
8	5.2	846
9	5.5	816
10	6.0	777
11	6.2	744
12	6.9	710
13	6.8	669
14	7.3	630
15	7.5	590
16	7.9	550
17	8.1	511
18	8.3	471
19	8.5	432
20	8.8	393
21	9.1	355
22	9.2	318
23	9.4	280
24	9.6	248
25	9.7	215

COMPLEMENTS

Mathematical Explanation of Kacser and Burns's Method for Determining Kinetic Orders

In the text, we used Kacser and Burns's (1979) method without explaining why it is legitimate. The proof is a consequence of the properties of logarithmic gains ("log gains"), as we encountered them briefly in Chapter 4 and as they are discussed in greater detail in Chapter 7. Since background material of Chapter 7 is required to understand the following argument fully, the reader is advised to return to this section after having worked through Chapter 7.

Mathematically, a log gain describes the relative change in a dependent variable or flux that is the consequence of a relative change in an independent variable. A log gain is a quantity obtained through differentiation, and thus refers to infinitesimally small changes. The experimental analogue to a log gain associated with a quantity Z is a small, but not infinitesimal, relative change of the type $\delta Z/Z$ that is caused by a small, but not infinitesimal, relative change in an independent variable. Thus, a logarithmic flux gain $L(V, X_j)$ is approximated by $\delta V^A/V^A$, and a logarithmic concentration gain $L(X_i, X_j)$ is approximated by $\delta X_i^A/X_i^A$. Substitution of these approximate quantities produces directly equations like (5.15) and (5.16).

In the case of two independent variables in Kacser's generic pathway (see Fig. 5.10), the S-system equations for X_1 and X_2 read

$$\dot{X}_1 = \alpha_1 X_3^{g_{13}} - \beta_1 X_1^{h_{11}} X_2^{h_{12}} X_3^{h_{13}} X_4^{h_{14}},$$
$$\dot{X}_2 = \alpha_2 X_1^{g_{21}} X_2^{g_{22}} X_3^{g_{23}} X_4^{g_{24}} - \beta_2 X_2^{h_{22}} X_4^{h_{24}}, \tag{5C.1}$$

X_3, X_4 independent.

Because of precursor–product relationships, the fluxes V_1^- and V_2^+ are equal, which implies $\alpha_2 = \beta_1$, $g_{21} = h_{11}$, $g_{22} = h_{12}$, $g_{23} = h_{13}$, and $g_{24} = h_{14}$. Since nothing is known about the mechanistic details of the pathway, the formulation of the flux $V = V_1^- = V_2^+$ between X_1 and X_2 includes possible effects of X_2, X_3, and X_4, which could be related to modulation or reversibility.

The log gains of the flux V_2^+ between X_1 and X_2 with respect to changes in the independent variables X_3 and X_4 are formally given as

$$L(V_2^+, X_3) = g_{23} + g_{21} L(X_1, X_3) + g_{22} L(X_2, X_3),$$
$$L(V_2^+, X_4) = g_{24} + g_{21} L(X_1, X_4) + g_{22} L(X_2, X_4) \tag{5C.2}$$

(cf. Chapter 7). If it is known that X_3 and X_4 do not affect V_2^+ directly, but only indirectly through precursors or other modulators, the kinetic orders g_{23} and g_{24} are zero. In this case, the system has two equations in the two unknowns, g_{21} and g_{22}, and can be solved by algebraic means (see Chapter 6 and Appendix).

If the number of unknown kinetic order parameters is lower than the number of independent log gain equations, the system is typically *overdetermined*. In such a case, there may not be a precise algebraic solution, but we can estimate the kinetic orders through linear regression. If the number of unknown kinetic orders is larger than the number of independent log gain equations, then each log gain equation constitutes a constraint that the system has to satisfy. In this case, we are not able to estimate all kinetic orders, but we have a set of p independent linear equations with q unknowns ($p < q$). With methods of linear algebra, this set can be partially solved, and this solution eliminates p degrees of freedom from the system. For instance, suppose the system has six unknown kinetic order parameters, and we are able to perform four independent log gain measurements. To identify the entire system, we must determine two kinetic orders of the system with other methods. Knowing their values, we can deduce the remaining four kinetic orders from the log gain equations.

Two Examples of the Use of Constraints in Parameter Estimation

1. Constraints based on flux stoichiometry. In the example of a branched pathway, the processes are not independent. As discussed in Chapter 3, the degradation of X_1 must equal the sum of the two production terms of X_2 and X_3 at the steady state:

$$\beta_1 X_1^{h_{11}} = \alpha_2 X_1^{g_{21}} + \alpha_3 X_1^{g_{31}}. \tag{5C.3}$$

The branch point constraint is derived from the definition of kinetic orders as logarithmic derivatives (see Chapter 3 and Appendix) and translates into the following constraints on the parameters:

$$\begin{aligned} h_{11} &= \frac{\partial \left(\alpha_2 X_1^{g_{21}} + \alpha_3 X_1^{g_{31}}\right)}{\partial X_1} \cdot \frac{X_1}{\alpha_2 X_1^{g_{21}} + \alpha_3 X_1^{g_{31}}} \\ &= \frac{\alpha_2 g_{21} X_1^{g_{21}} + \alpha_3 g_{31} X_1^{g_{31}}}{\alpha_2 X_1^{g_{21}} + \alpha_3 X_1^{g_{31}}}, \end{aligned} \tag{5C.4}$$

which is computed at the steady state. Furthermore, the rate constant β_1 is given as

$$\beta_1 = \alpha_2 X_1^{g_{21} - h_{11}} + \alpha_3 X_1^{g_{31} - h_{11}}, \tag{5C.5}$$

which also is evaluated at the steady state and involves the constraint on h_{11}.

If steady-state values are known, the equations become relationships between parameter values, and h_{11} and β_1 can be replaced by the right-hand side, which eliminates the need to estimate these two parameters. Many estimation algorithms allow constraint equations, and if such an option is not explicitly given, h_{11} and β_1 can be replaced manually in the model equations. The computational savings through constraints can be enormous.

2. Constraints based on steady-state equations. Chapter 6 deals with the algebraic characterization of steady states in S-system models. In contrast to most other nonlinear models, S-systems normally have linear steady-state equations if represented in logarithmic coordinates. These transformed equations can be used as constraints on the rate constants.

For simplicity, let's consider an S-system without independent variables. If a steady state exists in which no concentration is zero, this steady state satisfies the equations

$$\alpha_i \prod_{j=1}^{n} X_j^{g_{ij}} = \beta_i \prod_{j=1}^{n} X_j^{h_{ij}}. \qquad (5C.6)$$

for all $i = 1, 2, \ldots, n$. By dividing both sides by one of the products of power-law functions, one of the rate constants can be separated. For instance,

$$\alpha_i = \beta_i \frac{\prod_{j=1}^{n} X_j^{h_{ij}}}{\prod_{j=1}^{n} X_j^{g_{ij}}} = \beta_i \prod_{j=1}^{n} X_j^{h_{ij}-g_{ij}} \qquad (5C.7)$$

Thus, if the steady-state concentrations are known, the equation constitutes a constraint between parameters. This constraint can be exploited for estimation purposes if all α_i are replaced with the corresponding terms on the right-hand sides. This effectively reduces the number of parameters to be estimated by n.

REFERENCES

[4], [27], [30–31], [33], [46], [58], [63], [66–9], [71–2], [74–5], [78], [85–7], [90], [92], [100–1], [117–18], [120], [126–9], [132], [150], [158], [165–6], [172–3], [179], [200–3], [206], [211], [225], [233], [235–6], [242], [248], [255–6], [258–9], [270], [290], [293], [306], [308], [319], [334], [342], [349–51], [353–4], [360], [374–8], [385–6], [390], [393], [398], [401], [403–4], [414], [417–18], [426], [439], [442], [460–3], [473], [479–81], [483], [485], [490], [502], [515], [522].

CHAPTER SIX

Analytical Steady-State Evaluation

The full dynamic behavior of a model system can only be analyzed when the describing differential equations are solved and evaluated symbolically or numerically. In contrast, many important aspects of the system at or close to a steady state can be analyzed with much simpler means that only require elementary knowledge of matrices and differentiation. These aspects are of great importance, since most biochemical systems in nature operate close to a steady state, in which all inputs and outputs are in balance. Even in a disease state, a metabolic system typically is in a steady state, albeit some of the steady-state concentrations may be different from normal.

This chapter discusses algebraic steady-state evaluations and provides the mathematical background for the corresponding computations with PLAS that were discussed in Chapter 4. This mathematical background is important for readers who are interested in really understanding the concepts of steady states, gains, and sensitivities, and who may be interested in theoretical questions about S-systems. Readers who want to restrict their use of S-systems to applications and their numerical evaluations with PLAS may skip this chapter and the following or return to them at a later time.

Several sections of this chapter make use of vectors, matrices, and determinants. Readers not familiar with these concepts are advised to study the Appendix before proceeding.

Steady-State Equations

Consider a simple pathway with two dependent variables in which the production of the first variable is inhibited by the second (Fig. 6.1). This linear pathway with feedback is represented by the S-system

$$\begin{aligned}\dot{X}_1 &= \alpha_1 X_2^{g_{12}} - \beta_1 X_1^{h_{11}}, \\ \dot{X}_2 &= \beta_1 X_1^{h_{11}} - \beta_2 X_2^{h_{22}},\end{aligned} \quad (6.1)$$

where all parameters except for the inhibition parameter g_{12} are positive. Note that precursor–product relationships are made explicit by replacing α_2 with β_1 and g_{21} with h_{11}.

Figure 6.1. Linear pathway with feedback.

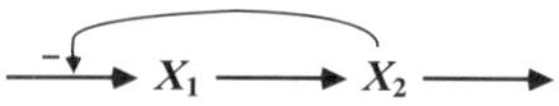

Suppose we have determined the parameter values of the pathway, and the numerical model reads

$$\dot{X}_1 = 0.4 X_2^{-2} - 2 X_1,$$
$$\dot{X}_2 = 2 X_1 - X_2^{0.5}. \quad (6.2)$$

We are interested in concentrations of X_1 and X_2 for which the system is in a steady state, that is, for which the influx into X_1 is equal to the efflux of X_1 and the influx into X_2 is equal to the efflux of X_2. Under these conditions, the concentrations of X_1 and X_2 do not change with time. Hence, $\dot{X}_1$ and $\dot{X}_2$ are equal to zero, since the dot, by definition, indicates the time rate of change. To be sure, stating $\dot{X}_1 = 0$ and $\dot{X}_2 = 0$ does *not* mean that nothing is happening in the pathway. Material is flowing through the system, but all fluxes are perfectly in balance, and there are no temporal *net changes* in any of the concentrations.

The equations characterizing the steady-state situation are thus obtained by requiring $\dot{X}_1 = 0$ and $\dot{X}_2 = 0$ in the S-system (6.2), and this yields

$$0 = 0.4 X_2^{-2} - 2 X_1,$$
$$0 = 2 X_1 - X_2^{0.5}. \quad (6.3)$$

The equations in (6.3) can be simplified mathematically: We first add the β-terms to both sides:

$$2 X_1 = 0.4 X_2^{-2},$$
$$X_2^{0.5} = 2 X_1, \quad (6.4)$$

and then take the logarithm on both sides of the equations. Next we define new variables $y_1 = \ln X_1$ and $y_2 = \ln X_2$, and the result is a set of two linear equations:

$$\ln 2 + y_1 = \ln 0.4 - 2 y_2,$$
$$0.5 y_2 = \ln 2 + y_1, \quad (6.5)$$

which simplifies to

$$\ln 5 = -y_1 - 2 y_2,$$
$$-\ln 2 = y_1 - 0.5 y_2. \quad (6.6)$$

Note that the result of these procedures is always a linear system, even if the S-system contains many equations, as long as every differential equation contains two terms with rate constants α and β that are not zero. Note also that this method does not work for GMA systems, unless they happen to have the form of an S-system (Exercises 5 and 6).

If we can solve this system, that is, if we can determine y_1 and y_2 such that Eq. (6.6) is satisfied, then we know the concentrations X_1 and X_2 for which the pathway is

ANALYTICAL STEADY-STATE EVALUATION

in a steady state, because we easily retrieve X_1 and X_2 through application of the exponential function: $X_1 = \exp(y_1)$, $X_2 = \exp(y_2)$.

The computation of a steady state thus reduces to solving sets of linear equations. If the sets consist of only a small number of equations, we can solve by substitution. For instance, in our example we can first separate y_1 and y_2 in the two equations of (6.6),

$$
\begin{aligned}
y_1 &= -\ln 5 - 2y_2, \\
y_1 &= -\ln 2 + 0.5 y_2,
\end{aligned}
\tag{6.7}
$$

and then equate the results, which leads to a linear equation with y_2 as the only unknown variable:

$$
\begin{aligned}
-\ln 5 - 2y_2 &= -\ln 2 + 0.5 y_2, \\
y_2 &= \frac{\ln 2 - \ln 5}{2.5} \\
&= -0.36652.
\end{aligned}
\tag{6.8}
$$

Using either one of the equations in (6.7), we determine the value of y_1 as

$$y_1 = -\ln 5 - 2(-0.36652) = -0.87641$$

or as

$$y_1 = -\ln 2 - 0.5 \times 0.36652 = -0.87641. \tag{6.9}$$

Substitution of y_1 and y_2 in Eq. (6.6) confirms that y_1 and y_2 satisfy the linear system.

Of course, the variables of real interest are X_1 and X_2. They are obtained by exponentiation:

$$
\begin{aligned}
X_1 &= \exp(y_1) = 0.41628, \\
X_2 &= \exp(y_2) = 0.69314.
\end{aligned}
\tag{6.10}
$$

Substitution of X_1 and X_2 in the S-system (6.2) confirms that the steady-state equations (6.3) are satisfied, which means that X_1 and X_2 really represent steady-state concentrations:

$$
\begin{aligned}
0 &= 0.4 \times 0.69314^{-2} - 2 \times 0.41628, \\
0 &= 2 \times 0.41628 - 0.69314^{0.5}.
\end{aligned}
\tag{6.11}
$$

It is instructive to enter the system in PLAS and to confirm that these concentrations constitute a steady state. This may be done with a numerical solution of the differential equations, as was done in Chapter 4 several times, or with the steady-state option in PLAS. Note that the steady state is largely independent of the initial values.

The algebraic method of computing steady states is not much affected by independent variables. For instance, if we slightly alter the previous example by identifying an independent input variable X_3 and an independent inhibitor X_4 for the degradation of X_2, as shown in Fig. 6.2, the S-system description is

$$
\begin{aligned}
\dot{X}_1 &= \tilde{\alpha}_1 X_2^{-2} X_3^{g_{13}} - 2X_1, \\
\dot{X}_2 &= 2X_1 - \tilde{\beta}_2 X_2^{0.5} X_4^{h_{24}},
\end{aligned}
\tag{6.12}
$$

Figure 6.2. Linear pathway with feedback and two independent variables.

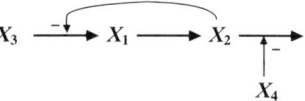

where the tildes indicate that the rate constants may have different numerical values than before. We can now argue in two ways. We may acknowledge that the constant input concentrations of X_3 and X_4 can be included mathematically in the rate constants. In other words, we redefine the rate constants as $\alpha_1 = \tilde{\alpha}_1 X_3^{g_{13}}$ and $\beta_2 = \tilde{\beta}_2 X_4^{h_{24}}$, which yields exactly the same system we dealt with previously [see Eq. (6.2)]. Or we may retain the independent variables and perform the same type of logarithmic transformation as before and obtain

$$\begin{aligned} \ln 2 + y_1 &= \ln \tilde{\alpha} - 2y_2 + g_{13} y_3, \\ \ln \tilde{\beta} + 0.5 y_2 + h_{24} &= \ln 2 + y_1. \end{aligned} \tag{6.13}$$

These equations reduce to

$$\begin{aligned} y_1 &= (\ln \tilde{\alpha} - \ln 2 + g_{13} y_3) - 2y_2, \\ y_1 &= (\ln \tilde{\beta} - \ln 2 + h_{24} y_4) + 0.5 y_2, \end{aligned} \tag{6.14}$$

which is in the same mathematical form as Eq. (6.7), when we recognize the independent variables y_3 and y_4 as constants. In either case, the inclusion of X_3 and X_4 does not alter the mechanism of computing steady-state concentrations, but the explicit inclusion of these independent variables makes their role in the dynamics of the system more transparent.

The transformation of the steady-state computation problem into one of solving a set of linear equations is exactly the same when the pathway consists of more than two or three dependent variables (Savageau, 1969b). Suppose, the system has n dependent variables and no independent variables. The describing equations are thus

$$\dot{X}_i = \alpha_i \prod_{j=1}^{n} X_j^{g_{ij}} - \beta_i \prod_{j=1}^{n} X_j^{h_{ij}}, \qquad i = 1, 2, \ldots, n. \tag{6.15}$$

We proceed in the same fashion as before. All derivatives must be zero for a steady state, which implies

$$0 = \alpha_i \prod_{j=1}^{n} X_j^{g_{ij}} - \beta_i \prod_{j=1}^{n} X_j^{h_{ij}}, \qquad i = 1, 2, \ldots, n, \tag{6.16}$$

and

$$\alpha_i \prod_{j=1}^{n} X_j^{g_{ij}} = \beta_i \prod_{j=1}^{n} X_j^{h_{ij}}, \qquad i = 1, 2, \ldots, n. \tag{6.17}$$

ANALYTICAL STEADY-STATE EVALUATION

Given that none of the rate constants α_i and β_i is zero, we can take the logarithm on both sides and obtain

$$\ln \alpha_i + \ln \left(\prod_{j=1}^{n} X_j^{g_{ij}} \right) = \ln \beta_i + \ln \left(\prod_{j=1}^{n} X_j^{h_{ij}} \right). \tag{6.18}$$

Since the logarithm of a product converts into a sum of logarithmic terms, and since for positive X and any real parameter p

$$\ln X^p = \ln[\exp(p \ln X)] = p \ln X, \tag{6.19}$$

the steady-state equations become

$$\ln \alpha_i + \sum_{j=1}^{n} g_{ij} \ln X_j = \ln \beta_i + \sum_{j=1}^{n} h_{ij} \ln X_j, \quad i = 1, 2, \ldots, n. \tag{6.20}$$

As before, we define $y_i = \ln X_i$ and move all terms including y_i to the left side and all terms without y_i to the right side. The result is

$$\sum_{j=1}^{n} g_{ij} y_j - \sum_{j=1}^{n} h_{ij} y_j = \ln \beta_i - \ln \alpha_i, \quad i = 1, 2, \ldots, n. \tag{6.21}$$

The only step left is to rename $a_{ij} = g_{ij} - h_{ij}$ and $b_i = \ln \beta_i - \ln \alpha_i = \ln(\beta_i/\alpha_i)$ for all i and all j. With that, a general S-system with n dependent variables and no independent variables has a steady state that is characterized by a set of n linear equations of the form

$$\begin{aligned}
a_{11} y_1 + a_{12} y_2 + a_{13} y_3 + \cdots + a_{1n} y_n &= b_1, \\
a_{21} y_1 + a_{22} y_2 + a_{23} y_3 + \cdots + a_{2n} y_n &= b_2, \\
a_{31} y_1 + a_{32} y_2 + a_{33} y_3 + \cdots + a_{3n} y_n &= b_3, \\
&\vdots \\
a_{n1} y_1 + a_{n2} y_2 + a_{n3} y_3 + \cdots + a_{nn} y_n &= b_n,
\end{aligned} \tag{6.22}$$

and this is true as long as no rate constant is zero.

When there are m independent variables $X_{n+1}, \ldots, X_{n+m}$, exactly the same procedures apply, and the result is

$$\begin{aligned}
a_{11} y_1 + a_{12} y_2 + \cdots + a_{1n} y_n + a_{1\,n+1} y_{n+1} + \cdots + a_{1\,n+m} y_{n+m} &= b_1, \\
a_{21} y_1 + a_{22} y_2 + \cdots + a_{2n} y_n + a_{2\,n+1} y_{n+1} + \cdots + a_{2\,n+m} y_{n+m} &= b_2, \\
a_{31} y_1 + a_{32} y_2 + \cdots + a_{3n} y_n + a_{3\,n+1} y_{n+1} + \cdots + a_{3\,n+m} y_{n+m} &= b_3, \\
&\vdots \\
a_{n1} y_1 + a_{n2} y_2 + \cdots + a_{nn} y_n + a_{n\,n+1} y_{n+1} + \cdots + a_{n\,n+m} y_{n+m} &= b_n.
\end{aligned} \tag{6.23}$$

Confirm this result.

It is important to note that the coefficients in these linear steady-state equations are directly related to the parameters of the S-system: The coefficient associated with the first variable, y_1, in the first equation is $a_{11} = g_{11} - h_{11}$; the coefficient associated with the second variable, y_2, in the first equation is $a_{12} = g_{12} - h_{12}$; the coefficient associated with the first independent variable in the first equation is $a_{1\,n+1} = g_{1\,n+1} - h_{1\,n+1}$.

In general, the coefficient in equation i that is associated with variable j is $a_{ij} = g_{ij} - h_{ij}$. Furthermore, the coefficients on the right-hand side are the logarithms of the ratios of rate constants: for the ith equation the coefficient is $b_i = \ln(\beta_i/\alpha_i)$. (For further discussions, see Exercises 9 and 10.) With this knowledge, we can directly write down the steady-state equations for an S-system, without explicitly executing the above steps of equating derivatives with zero and taking logarithms of the products of power-law functions: We simply define the quantities $y_i = \ln X_i$, $a_{ij} = g_{ij} - h_{ij}$, and $b_i = \ln(\beta_i/\alpha_i)$ and substitute them in Eq. (6.22) or (6.23). For example, the steady-state representation of the system (6.2),

$$\dot{X}_1 = 0.4 X_2^{-2} - 2X_1,$$
$$\dot{X}_2 = 2X_1 - X_2^{0.5}, \tag{6.24}$$

has the coefficients $a_{11} = 0 - 1 = -1$, $a_{12} = -2 - 0 = -2$, $a_{21} = 1 - 0 = 1$, $a_{22} = 0 - 0.5 = -0.5$, and $b_1 = \ln 2 - \ln 0.4 = \ln(2/0.4) = \ln 5$, $b_2 = \ln 1 - \ln 2 = -\ln 2$. For other examples, see Exercises 1 and 2.

Zero Steady States

The steady states computed in the previous sections consisted of solutions in which none of the variables were zero. These *non-trivial* solutions are usually the ones of interest. Nevertheless, other steady states may exist in which one, several, or all dependent variables are zero. For instance, consider the S-system

$$\dot{X}_1 = 2X_2^{0.75} - 2X_1,$$
$$\dot{X}_2 = 1.2X_1 - 1.2X_2^{0.5}. \tag{6.25}$$

Inspection of the equations or computation as discussed above reveals the steady state as $X_1 = 1$, $X_2 = 1$. However, it is not difficult to see that $X_1 = 0$, $X_2 = 0$ also is a solution, since it satisfies the steady-state conditions $\dot{X}_1 = 0$, $\dot{X}_2 = 0$.

In other cases, only some of the variables may be zero, while others are not. For instance, consider the pathway with recycling and feedback activation depicted in Fig. 6.3. The defining S-system equations with typical numerical values may read

$$\dot{X}_1 = X_3 - 0.1 X_1 X_2^{0.2}, \quad X_1(0) = 60,$$
$$\dot{X}_2 = 0.1 X_1 X_2^{0.2} - 9 X_2, \quad X_2(0) = 30, \tag{6.26}$$
$$\dot{X}_3 = 9 X_2 - X_3, \quad X_3(0) = 10.$$

Implement the system in PLAS (the model is also provided under the name *Recycle.plc*) and solve for the dynamics. X_1 increases in a sigmoid fashion to a final value of 90, X_2 quickly drops to 1, and X_3 reaches a temporary maximum before decreasing to its final value 9.

Figure 6.3. Pathway with recycling and feedback activation.

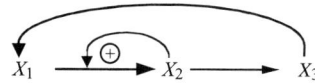

ANALYTICAL STEADY-STATE EVALUATION

The steady-state equations in linear form are

$$-y_1 - 0.2\ y_2 + y_3 = \ln 0.1,$$
$$y_1 - 0.8\ y_2 \qquad = \ln 90, \qquad (6.27)$$
$$y_2 - y_3 = \ln \tfrac{1}{9}.$$

Adding the first two equations eliminates y_1 and yields

$$-y_2 + y_3 = \ln 0.1 + \ln 90 = \ln 9. \qquad (6.28)$$

Multiplication of (6.28) by -1 reveals that this equation is the same as the third steady-state equation. Thus, there is no unique solution. When we click the steady-state button, PLAS indeed displays a warning that there are several solutions (Fig. 6.4). One solution is given, but it is *not* the one found numerically.

Solve the system again, but start with a different set of initial values; e.g., $X_1(0) = 95$, $X_2(0) = 3$, $X_3(0) = 2$. Although the transient dynamics are different, the system approaches the same values as before, and we conclude that the numerically determined final values in fact constitute a steady state and that, furthermore, this steady state is stable. However, starting with $X_1(0) = 90$, $X_2(0) = 1$, $X_3(0) = 1$, which is not very far away from the steady state, a different steady-state solution is obtained. Something is strange here! Studying the map, we see that no material is leaving or entering the system, and this implies a *conservation law* forcing the total mass to remain constant. Since we started with $60 + 30 + 10 = 100$, the final values

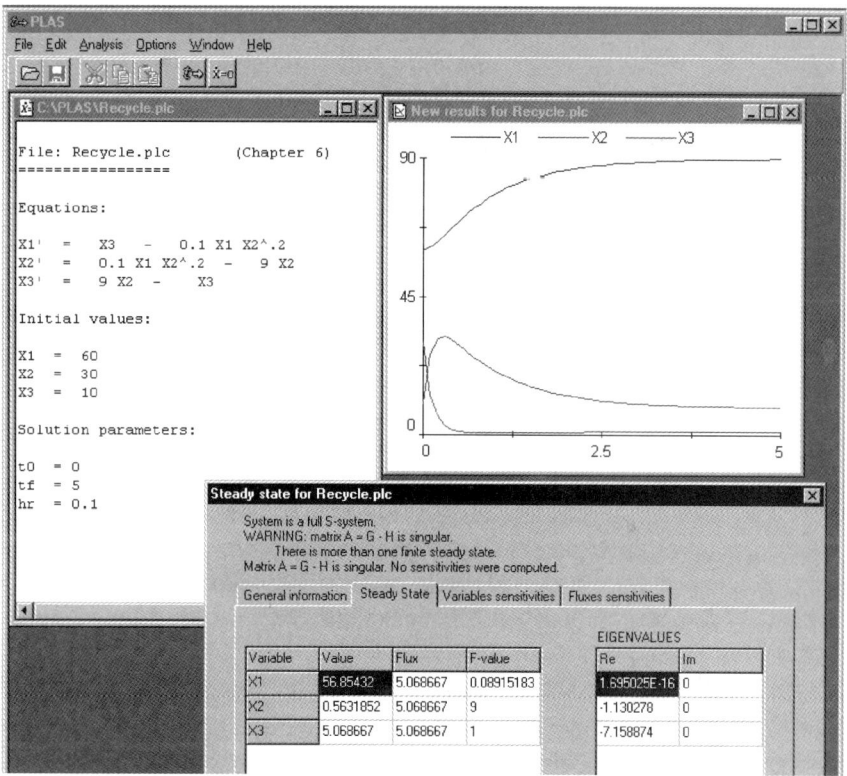

Figure 6.4 (see Color Plate IX). PLAS computes a steady state, but warns that it is just one of many.

also must add to 100, and since $X_1 = 90$ and $X_3 = 9X_2$, the solution is $X_3 = 9$, $X_2 = 1$, which is what the dynamics produced. For initial values like $X_1(0) = 90$, $X_2(0) = 1$, $X_3(0) = 1$, the final state is different, but we see that the total of the final values again equals the total of the initial values. The algebraic method for computing the steady state in PLAS does not consider such constraints and suggests that many – in fact, infinitely many – solutions exist.

What happens if we start with $X_1 = 100$, $X_2 = 0$, $X_3 = 0$? The Taylor method in PLAS does not permit zero initial values, but going through the algebra of the steady-state equations or switching to the Gear method (under *Options/Solver*) shows that the system does not change. It is at a steady state in which two variables are zero and one is not.

Matrix Notation

It doesn't take much imagination to see that the notation in (6.22) and (6.23) is somewhat redundant, because the same variables appear in each equation and we have to write a lot of plus and minus signs and indices. A condensed notation that eliminates these redundancies is based on *vectors* and *matrices*. Among non-mathematicians, vectors and matrices sometimes have the reputation of sophisticated but unnecessary mathematical complication, but in fact they are simply a convenient notation. We will use this notation for the remainder of this chapter and in the following chapter. A brief refresher on vectors, matrices, and determinants is provided in the Appendix.

In matrix notation, Eqs. (6.22) and (6.23) are written succinctly as

$$\mathbf{A} \cdot \vec{y} = \vec{b}. \tag{6.29}$$

Here the matrix $\mathbf{A}$ consists of the coefficients $a_{ij} = g_{ij} - h_{ij}$. The vector $\vec{y}$ is a row vector consisting of the n components $y_j = \ln X_j$ ($j = 1, \ldots, n$) if the system does not contain independent variables [cf. Eq. (6.22)], or of the $n + m$ components $y_j = \ln X_j$ ($j = 1, \ldots, n + m$) if the system contains m independent variables [cf. Eq. (6.23)]. $\vec{b}$ is a row vector containing the solution coefficients $b_i = \ln(\beta_i/\alpha_i)$ ($i = 1, \ldots, n$), as we discussed before.

The following sections show in detail how a steady-state solution is obtained without explicitly formulating systems of equations. In the first step, we will compute the coefficients of the matrix $\mathbf{A}$ as differences of kinetic orders ($a_{ij} = g_{ij} - h_{ij}$), and the coefficients of the vector $\vec{b}$ as logarithms of the ratios of rate constants, $b_i = \ln(\beta_i/\alpha_i)$. In the second step, the matrix equation is solved. For small systems, we will use Cramer's procedure or the method of substitution. For larger systems, we enter the matrix equation in either a numerical or a symbolic software package that executes matrix operations. The former class of software includes MathCad® and Gauss®; the latter includes Mathematica® and Maple®. In either case, the result is a vector $\vec{y}$ whose components are the logarithms of the desired steady-state values. Of course, numerical (but not symbolic) results are most easily obtained with PLAS. It is noted again, that steady states of GMA systems typically cannot be computed in this fashion.

Steady-State Solution of a Regular S-System

We have seen that the computation of a steady-state solution reduces to solving sets of linear equations if each S-system equation has an α-term and a β-term, that is, if none of the rate constants is zero. Three scenarios are possible: The linear system has no solution at all, there is exactly one solution, or there are many solutions. Without mathematical analysis, it is usually not possible to determine which scenario applies. However, it can be deduced directly from the *determinant* associated with the system matrix (see Appendix): There is a unique solution with positive metabolite concentrations $X_1, \ldots, X_n$ if and only if the determinant is different from zero: $\det \mathbf{A} \neq 0$. We call an S-system with non-zero rate constants and a non-zero determinant *regular*. In contrast, if some of the rate constants are zero or if the determinant vanishes, there may be no such solution or there may be infinitely many. We call such an S-system *irregular*. One can mathematically construct any number of irregular S-systems, but since they are of secondary importance for applied biochemical analysis, we shall only discuss them very briefly later in this chapter. The reader interested in irregular systems is referred to Savageau (1976: Chapter 6).

By executing the logarithmic transformation of variables and the components of the solution vector, we have implicitly assumed that all quantities of interest are positive. In fact, if any of the variables X_i vanishes, we are not to take the logarithm and have to study these *zero solutions* by other means, as we did above.

Regular systems without independent variables. Suppose the S-system of interest is regular and has n dependent and no independent variables. In this case, the system matrix $\mathbf{A}$ is an $n \times n$ matrix whose determinant is not zero. To compute the unique non-zero steady-state values, we logarithmically transform all variables to $y_i = \ln X_i$ $(i = 1, \ldots, n)$, define $b_i = \ln(\beta_i/\alpha_i)(i = 1, \ldots, n)$, and solve the resulting matrix equation, for instance, with Cramer's rule. According to this method, which is discussed in the Appendix, each component y_i of the solution vector $\vec{y}$ is computed by division of two determinants: $\det \mathbf{A}$, which must be non-zero, and the determinant of the matrix $\mathbf{A}_i$ that is equal to $\mathbf{A}$ except that the ith column is replaced with the vector $\vec{b}$:

$$y_i = \frac{\det \mathbf{A}_i}{\det \mathbf{A}} \qquad (i = 1, \ldots, n). \tag{6.30}$$

Recall that determinants are numbers, even though they are associated with matrices. We illustrate the procedure with two examples.

Example 1. The steady state of the simple feedback system in Fig. 6.1 [Eqs. (6.1) and (6.2)] is characterized by the system matrix $\mathbf{A}$ and solution vector $\vec{b}$:

$$\mathbf{A} = \begin{pmatrix} -1 & -2 \\ 1 & -0.5 \end{pmatrix}, \qquad \vec{b} = \begin{pmatrix} \ln 5 \\ -\ln 2 \end{pmatrix}. \tag{6.31}$$

The components of $\mathbf{A}$ are simply the differences between the appropriate g and h values. The determinant is computed directly by multiplying along the two diagonals and subtracting the results, as shown in the Appendix. The numerical value of the

Figure 6.5. Branched pathway with feedback.

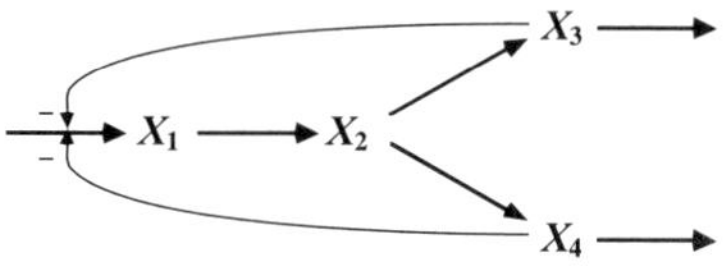

determinant is $|\mathbf{A}| = (-1)(-0.5) - (-2)(1) = 2.5$. To use Cramer's rule, we compute the two determinants $|\mathbf{A}_1|$ and $|\mathbf{A}_2|$, which result from replacing, respectively, the first and the second column with the solution vector $\vec{b}$:

$$|\mathbf{A}_1| = \begin{vmatrix} \ln 5 & -2 \\ -\ln 2 & -0.5 \end{vmatrix} = -0.5 \ln 5 - 2 \ln 2 = -2.19101,$$

$$|\mathbf{A}_2| = \begin{vmatrix} -1 & \ln 5 \\ 1 & -\ln 2 \end{vmatrix} = \ln 2 - \ln 5 = -0.91629.$$

(6.32)

Division of $|\mathbf{A}_1|$ and $|\mathbf{A}_2|$ by $|\mathbf{A}|$ generates the solution $y_1 = -0.87641$, $y_2 = -0.36652$ in logarithmic coordinates and the desired solution $X_1 = \exp(y_1) = 0.41628$, $X_2 = \exp(y_2) = 0.69314$ in Cartesian coordinates. As it should be, this is the same solution we obtained by the elementary substitution method.

Example 2. The branched feedback system in Fig. 6.5 is described by the four-variable S-system

$$\dot{X}_1 = 5 X_3^{-1} X_4^{-1} - 5 X_1^{1/2},$$
$$\dot{X}_2 = 5 X_1^{1/2} - 10 X_2^{1/2},$$
$$\dot{X}_3 = 2 X_2^{1/2} - 1.25 X_3,$$
$$\dot{X}_4 = 8 X_2^{1/2} - 5 X_4.$$

(6.33)

The steady-state matrix equation is

$$\begin{pmatrix} 0.5 & 0 & 1 & 1 \\ -0.5 & 0.5 & 0 & 0 \\ 0 & -0.5 & 1 & 0 \\ 0 & -0.5 & 0 & 1 \end{pmatrix} \cdot \vec{y} = \begin{pmatrix} 0 \\ \ln 0.5 \\ \ln 1.6 \\ \ln 1.6 \end{pmatrix}.$$

(6.34)

Confirm this equation by developing the nonlinear and the linear steady-state equations in X_i and y_i, respectively (see also Exercise 2). The determinant associated with the system matrix $\mathbf{A}$ can be computed by hand with the method of expansion (cf. Appendix). There are also many commercial software packages that make evaluations of determinants as simple as operations with a pocket calculator. The numerical value of $|\mathbf{A}|$ is

$$\det \mathbf{A} = 0.75.$$

(6.35)

(Confirm this.)

Next we compute the determinants $|\mathbf{A}_1|, \ldots, |\mathbf{A}_4|$, in which one column is replaced with the solution vector $\vec{b}$. For simplicity of notation, we call the non-zero components of this vector $p = \ln 0.5$, $q = \ln 1.6$, and $r = \ln 1.6$. Even though q

ANALYTICAL STEADY-STATE EVALUATION

and r have the same value, we use different symbols to make the computation of the determinants as lucid as possible. Each of the determinants is 4×4 and thus needs to expanded. For instance, for $|A_1|$ we obtain through expansion by the first row

$$\det A_1 = \begin{vmatrix} 0 & 0 & 1 & 1 \\ p & 0.5 & 0 & 0 \\ q & -0.5 & 1 & 0 \\ r & -0.5 & 0 & 1 \end{vmatrix} = 1 \cdot \begin{vmatrix} p & 0.5 & 0 \\ q & -0.5 & 0 \\ r & -0.5 & 1 \end{vmatrix} - 1 \cdot \begin{vmatrix} p & 0.5 & 0 \\ q & -0.5 & 1 \\ r & -0.5 & 0 \end{vmatrix}$$

$$= -0.5p - 0.5q - 0.5r - 0.5p$$
$$= -\ln 0.5 - \ln 1.6 = 0.22314. \tag{6.36}$$

Similarly, one computes

$$\det A_2 = -0.816577,$$
$$\det A_3 = -0.055786, \tag{6.37}$$
$$\det A_4 = -0.055786.$$

(Confirm these results.) Each of these determinants is divided by $|A|$ to yield the solutions y_i in logarithmic coordinates, and subsequent exponentiation results in the concentrations of interest, X_i. The result is

$$\begin{aligned} y_1 &= 0.29752, & X_1 &= 1.3465, \\ y_2 &= -1.0888, & X_2 &= 0.33663, \\ y_3 &= -0.074381, & X_3 &= 0.92832, \\ y_4 &= -0.074381, & X_4 &= 0.92832. \end{aligned} \tag{6.38}$$

These results can be checked by substitution in the steady-state equations, which are obtained by setting the differential equations in (6.33) equal to zero, or with the steady-state option in PLAS.

As an alternative to Cramer's rule, the matrix equation

$$A\vec{y} = \vec{b} \tag{6.39}$$

can be solved through computation of the inverse of A (see Appendix). In analogy with real numbers, the inverse of a matrix is again a matrix, and if the two are multiplied, the result is the *identity matrix* I. Like the number 1, which multiplied by another number doesn't change that number, I multiplied by another matrix doesn't change that matrix. The computation of the inverse matrix is mathematically legitimate if A is regular (see Appendix). That is the case here, and we can write

$$A^{-1}A\vec{y} = A^{-1}\vec{b}. \tag{6.40}$$

By definition of the inverse we know

$$A^{-1}A\vec{y} = I\vec{y} = \vec{y}, \tag{6.41}$$

and obtain

$$\vec{y} = A^{-1}\vec{b} \tag{6.42}$$

(Savageau, 1969b). This procedure explicitly expresses the solution vector $\vec{y}$ in terms of the known coefficients of the system, namely $a_{ij} = g_{ij} - h_{ij}$ and $b_i = \ln(\beta_i/\alpha_i)(i = 1, \ldots, n)$. For small matrices, the inverse can be computed by hand or with a modern pocket calculator. The computation by hand involves several steps, but nothing more complicated than the evaluation of determinants (see Appendix).

S-Systems with independent variables. The explicit inclusion of independent variables does not affect the mechanics of computing a steady state much, if the independent variables are aggregated with the solution coefficients, as we discussed in Eqs. (6.12) and (6.13). Recalling Eq. (6.13), we can write the steady-state equations for the linear pathway with feedback and two independent variables (Fig. 6.2) as

$$\ln 2 + y_1 = \ln \tilde{\alpha} - 2y_2 + g_{13}y_3,$$
$$\ln \tilde{\beta} + 0.5y_2 + h_{24}y_4 = \ln 2 + y_1. \tag{6.43}$$

The corresponding matrix equation reads

$$\begin{pmatrix} -1 & -2 \\ 1 & -0.5 \end{pmatrix} \cdot \vec{y} = \begin{pmatrix} \ln 2 - \ln \tilde{\alpha} - g_{13}y_3 \\ \ln \tilde{\beta} + h_{24}y_4 - \ln 2 \end{pmatrix}, \tag{6.44}$$

where the solution vector includes the constant values of the independent variables. Notice that the system matrix $\mathbf{A}$ is unaffected by the independent variables and that therefore $|\mathbf{A}| = 2.5$, as before. If, for simplicity of notation, we call the two components of the solution vector $b_1 = \ln 2 - \ln \tilde{\alpha} - g_{13}y_3$ and $b_2 = \ln \tilde{\beta} + h_{24}y_4 - \ln 2$, the solution according to Cramer's rule is

$$y_1 = \frac{\begin{vmatrix} b_1 & -2 \\ b_2 & -0.5 \end{vmatrix}}{|\mathbf{A}|} = \frac{2b_2 - 0.5b_1}{2.5},$$
$$y_2 = \frac{\begin{vmatrix} -1 & b_1 \\ 1 & b_2 \end{vmatrix}}{|\mathbf{A}|} = \frac{-b_1 - b_2}{2.5}, \tag{6.45}$$

which, of course, is equivalent to the result obtained from Eq. (6.32). The desired steady-state concentrations X_1 and X_2 are obtained through exponentiation:

$$\begin{aligned} X_1 &= \exp(0.2b_1 - 0.8b_2) \\ &= \exp[0.2(\ln \tilde{\alpha} + g_{13}y_3 - \ln 2) - 0.8(\ln 2 - \ln \tilde{\beta} - h_{24}y_4)] \\ &= 2^{-1}\tilde{\alpha}^{0.2}\tilde{\beta}^{0.8} X_3^{0.2g_{13}} X_4^{0.8h_{24}}, \\ X_2 &= \exp[0.4(b_1 + b_2)] \\ &= \exp[0.4(\ln \tilde{\alpha} + g_{13}y_3 - \ln \tilde{\beta} - h_{24}y_4)] \\ &= \tilde{\alpha}^{0.4}\tilde{\beta}^{-0.4} X_3^{0.4g_{13}} X_4^{-0.4h_{24}}. \end{aligned} \tag{6.46}$$

ANALYTICAL STEADY-STATE EVALUATION

As a numerical example, let's set $X_3 = 10$, $X_4 = 4$, $\tilde{\alpha} = 0.04$, $g_{13} = 1$, $\tilde{\beta} = 0.5$, and $h_{24} = 0.5$. The numerical solution for X_1 and X_2 is

$$X_1 = 0.5 \times 0.52531 \times 0.57435 \times 1.5849 \times 1.7411 = 0.41628,$$
$$X_2 = 0.275946 \times 1.3195 \times 2.5119 \times 0.75786 = 0.69314.$$
(6.47)

These are the same values we obtained in Eq. (6.10), which is consistent with the fact that in this numerical example $\alpha_1 = \tilde{\alpha}_1 X_3^{g_{13}}$ and $\beta_2 = \tilde{\beta}_2 X_4^{h_{24}}$. However, Eq. (6.46) allows us to compute steady states directly for other values of the independent variables. Execute such computations, and confirm the results in PLAS.

Aggregation of independent variables with rate constants is a feasible approach for numerical evaluations. In contrast, many symbolic analyses, such as gain computations, must be performed with an explicit distinction between constants and independent variables. Let us, therefore, study the general case of an S-system with n dependent variables, $X_1, \ldots, X_n$, and m independent variables, $X_{n+1}, \ldots, X_{n+m}$. Such a system consists of n equations, one for each dependent variable, and each product of power functions may contain dependent and independent variables. In most general terms, we have

$$\dot{X}_i = \alpha_i \prod_{j=1}^{n+m} X_j^{g_{ij}} - \beta_i \prod_{j=1}^{n+m} X_j^{h_{ij}}, \qquad i = 1, 2, \ldots, n, \tag{6.48}$$

where some of the exponents g_{ij} and h_{ij} can be zero. As before, we assume that all rate constants and variables are positive, introduce new variable names $y_j = \ln X_j$ for $j = 1, 2, \ldots, n+m$, and define coefficients $a_{ij} = g_{ij} - h_{ij}$ and $b_i = \ln(\beta_i/\alpha_i)$. The result consists of n linear equations in $n+m$ variables [cf. Eq. (6.23)]:

$$\begin{aligned}
a_{11}y_1 + a_{12}y_2 + \cdots + a_{1n}y_n &= b_1 - a_{1,n+1}y_{n+1} - \cdots - a_{1,n+m}y_{n+m}, \\
a_{21}y_1 + a_{22}y_2 + \cdots + a_{2n}y_n &= b_2 - a_{2,n+1}y_{n+1} - \cdots - a_{2,n+m}y_{n+m}, \\
a_{31}y_1 + a_{32}y_2 + \cdots + a_{3n}y_n &= b_3 - a_{3,n+1}y_{n+1} - \cdots - a_{3,n+m}y_{n+m}, \\
&\vdots \\
a_{n1}y_1 + a_{n2}y_2 + \cdots + a_{n,n}y_n &= b_n - a_{n,n+1}y_{n+1} - \cdots - a_{n,n+m}y_{n+m}.
\end{aligned}$$
(6.49)

Note that we have arranged the equations in such a way that the (unknown) dependent variables are separated from the (known) independent variables. The left-hand side is exactly the same as the expression $\mathbf{A}\vec{y}$ that we developed for systems without independent variables. To indicate that we are here talking about dependent variables, we attach the index D. The right-hand side contains the solution vector $\vec{b}$, as before, but also independent variables and coefficients associated with independent variables. We refer to these collectively as $\vec{y}_I$ and $\mathbf{A}_I$, and obtain the steady-state matrix equation

$$\mathbf{A}_D \vec{y}_D = \vec{b} - \mathbf{A}_I \vec{y}_I. \tag{6.50}$$

As in the case without independent variables, we can use the matrix inverse $\mathbf{A}_D^{-1}$ to express the solution formally as

$$\vec{y}_D = \mathbf{A}_D^{-1}\vec{b} - \mathbf{A}_D^{-1}\mathbf{A}_I\vec{y}_I \tag{6.51}$$

(Savageau, 1971a). This formulation demonstrates explicitly how the dependent variables are affected by the system characteristics, as represented by the coefficients $a_{ij} = g_{ij} - h_{ij}$ in the matrices $\mathbf{A}_D$ and $\mathbf{A}_I$, the solution coefficients $b_i = \ln(\beta_i/\alpha_i)$, and the independent variables, given in logarithmic coordinates as $\vec{y}_I$. Obviously, when there are no independent variables, this equation reduces to the corresponding equation (6.42). We shall return to this fundamental equation in Chapter 7 when we discuss how changes in parameter values affect the steady state of a pathway.

Steady-State Solutions of Irregular S-systems

If any rate constants of the S-system are zero, the S-system has no steady state with positive concentrations and at least one component dies out or grows indefinitely. This makes intuitive sense when we recall the meaning of a zero rate constant: Consider some generic S-system, and suppose that α_2 is zero and all other rate constants are positive. Then there is no process that would synthesize, increase, or augment X_2, and all the material that initially is available will be depleted if we wait long enough. Similarly, if a rate constant like β_4 is zero and all other rate constants are positive, the component X_4 cannot be degraded or depleted, and since there is influx into this pool (α_4 is positive), X_4 will grow without bounds. A rather irrelevant exception is the case that both rate constants of an equation are zero. The component characterized by this equation then never changes and can be replaced with an independent variable. Various scenarios are discussed in the Exercises.

In addition to irregular S-systems with zero rate constants, seemingly normal S-systems may not have unique steady-state solutions. We discussed as an example the linear pathway with recycling and feedback activation (Fig. 6.3). These systems are mathematically characterized by a zero system determinant: $\det \mathbf{A} = 0$. As an example for such an irregular S-system in which all rate constants are strictly positive, consider

$$\begin{aligned}\dot{X}_1 &= X_2^{0.5} - 2X_1, \\ \dot{X}_2 &= 2X_1 - X_2^{0.5},\end{aligned} \tag{6.52}$$

which could be a realization of the very simple cyclic pathway depicted in Fig. 6.6. Using the *Steady-state* menu in PLAS, we receive the information that the system

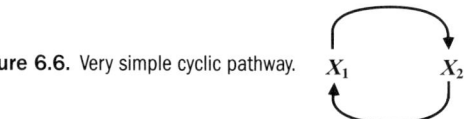

Figure 6.6. Very simple cyclic pathway.

ANALYTICAL STEADY-STATE EVALUATION

matrix is singular – which corresponds to a zero determinant. This is also easy to check by hand, since the 2×2 determinant is given as

$$|\mathbf{A}| = \begin{vmatrix} -1 & 0.5 \\ 1 & -0.5 \end{vmatrix} = (-1)(-0.5) - (1)(0.5) = 0. \tag{6.53}$$

Is there no steady state? Let's compute the numerical solution in PLAS (file *Cycle.plc*), specifying initial values $X_1 = 1$ and $X_2 = 1$ at $t = 0$. The system clearly approaches a steady state, with concentrations of about $X_1 = 0.59307$ and $X_2 = 1.4069$. We solve the system again, this time starting with the initial values $X_1 = 2$ and $X_2 = 1$ at $t = 0$. The final values for X_1 and X_2 now are 0.7500 and 2.2500. If we start at (10,1), the final values are 1.5380 and 9.4620, and if we start at (1,0.1), they are 0.41410 and 0.68590. In fact, every time we start the system at a different point, the final results are different, and algebraic analysis of the system confirms that the system has infinitely many steady states. Not every point is a steady-state point, though – otherwise the system would stay at the initial values, whatever they might be – but instead of the single steady-state point that we find in regular S-systems, there is an entire *solution space*. In this particular case, the sum of X_1 and X_2 always remains at the value of $X_1 + X_2$ at the beginning of the experiment, i.e., at $t = 0$. This conservation law reflects that the system in Fig. 6.6 is closed and that no material is added or lost. Steady-state solutions for systems of this type are described in Savageau (1976).

Cramer's rule cannot be used, because $|\mathbf{A}| = 0$. But what about substitution? The steady-state equations in terms of X_1 and X_2 are

$$\begin{aligned} 0 &= X_2^{0.5} - 2X_1, \\ 0 &= 2X_1 - X_2^{0.5}. \end{aligned} \tag{6.54}$$

Obviously, both equations are the same and correspond to

$$X_2^{0.5} = 2X_1, \quad X_2 = 4X_1^2, \quad y_2 = \ln 4 + 2y_1. \tag{6.55}$$

These results mean that for any X_1 or y_1, we can find an X_2 or y_2 so that the steady-state equations are satisfied. For instance, if we choose $X_1 = 0.75$, then X_2 is 2.25. Thus, we obtain *infinitely many* solutions, which all lie on a power-law curve when we are using X's, and all lie on a straight line when we are using y's.

Finally, there are S-systems in which all rate constants are strictly positive, but a steady-state solution can only be found if at least one concentration is equal to zero. An example is

$$\begin{aligned} \dot{X}_1 &= 2X_1 - 2X_2, \\ \dot{X}_2 &= X_1 - 2X_2, \\ \dot{X}_3 &= X_3 - X_3^2. \end{aligned} \tag{6.56}$$

For Cramer's rule to apply, we need to compute the determinant $|\mathbf{A}|$. According to the rule of Sarrus (see Appendix), we obtain

$$|\mathbf{A}| = \begin{vmatrix} 1 & -1 & 0 \\ 1 & -1 & 0 \\ 0 & 0 & 1 \end{vmatrix}$$
$$= -1 + 0 + 0 - 0 - 0 + 1 = 0. \tag{6.57}$$

That again excludes Cramer's rule from consideration. The substitution yields for the first equation $X_1 = X_2$, and for the second equation $X_1 = 2X_2$. These two constraints can only be satisfied if X_1 and X_2 are both equal to zero. The third equation in this case yields the steady-state concentrations $X_3 = 0$ or $X_3 = 1$.

Local Stability

Most biochemical systems operate at a steady state, but they are regularly exposed to influences that temporarily remove them from this state. For instance, blood serum is characterized by normal (steady state) concentrations of its constituents, but these concentrations are always subject to short-term changes, due to the intake and absorption of water and food and the action of kidneys and liver. Normal concentrations in living systems are maintained by regulatory mechanisms that make the steady states *stable*. This means that effective regulation causes the concentrations to return to their nominal steady-state levels after outside forces have temporarily changed them. As a mechanical analogue one can imagine a ball held by two springs. If it is forced out of its resting position, the springs will pull it back. The prevalent regulatory mechanism in biochemical systems is feedback inhibition: As soon as end product is available in excess, it begins to slow down production of its precursors, until the excess is annihilated. If the concentration of the end product is below the target level, inhibition is relaxed and more product is synthesized.

The mathematical definition of stability is similar to these intuitive ideas. Strictly speaking, however, typical stability analyses only address infinitesimally small deviations from steady state, and therefore determine *local* stability. In most biochemical systems, local stability implies buffering against realistic perturbations as long as they are not too large. How large they may be depends on the specifics of a system and cannot be said in general. Of course, stability with respect to large perturbations is an important feature of a system, and it would be nice if we could easily evaluate it. However, the analytical treatment of *global* stability in nonlinear systems is a very complicated problem that is largely unsolved, and in most cases we resort to simulations to determine whether a particular system can tolerate a large perturbation. Guckenheimer and Holmes (1983: p. 12) begin their treatise on the algebraic analysis of nonlinear systems by stating "We must start by admitting that almost nothing beyond general statements can be made about most nonlinear systems."

If a system does not return from an ever so slight perturbation to the nominal steady state, this steady state is either *unstable* or *marginally stable*. In the unstable case, one or several variables may go to zero, grow without bounds, or assume different steady-state values. In the marginally stable case, the steady state actually is one of an infinite number of connected steady states, and perturbations cause the system to assume another steady state. Discussing irregular S-systems with a zero determinant, we have already encountered a marginally stable system [Eq. (6.52)]. Sometimes, a system at a stable steady state is compared to a ball in a bowl. If the ball is moved out of its stable resting position, it returns to the bottom of the bowl. An unstable system is compared to two balls, one sitting on top of the other. Theoretically, the ball on top can rest in this position, but the slightest perturbation will cause it to fall down. A marginally stable system can be envisioned as a ball on a flat surface. Every time it is perturbed, it will roll a bit and then assume a new resting position.

For S-systems with unique positive steady states, the mathematical analysis of local stability can be performed with standard techniques of linear algebra. This is justified by a fundamental theorem of Hartman and Grobman (see Guckenheimer and Holmes, 1983: p. 13), which asserts that (under conditions that are satisfied here) a nonlinear system can be approximated by a linear system, and that analysis of the linear system yields valid insights into the nonlinear system.

The standard approach for linear systems evaluates the *eigenvalues* of the system, which are complex numbers that characterize the stability and oscillatory behavior of the system. We have encountered them already in Chapter 4. Eigenvalues may appear to be mathematical constructs beyond intuitive interpretation. However, one can understand them in the following way. Imagine a system is at the point P. If the system has three dependent variables, this point is in a three-dimensional space. It has three coordinates, which correspond to the concentrations of these variables. When the system moves away from this point, it moves along a *trajectory*. This trajectory is some (possibly very complicated) curve in the three-dimensional space, which is described by the differential equations that define the model. Nevertheless, very small movements along the trajectory are essentially characterized by one main direction, the so-called *tangent vector*. This tangent vector is characterized by the eigenvalues of the system. Since the vector can be decomposed into components along the axes, the eigenvalues describe the speed with which each concentration moves away from the point P. For instance, if a vector in our usual three-dimensional space has a large x component and small y and z components, the concentration X_1 will change quite drastically, while X_2 and X_3 are not much affected by the motion of the system.

In a different interpretation of the same phenomenon, the analysis of eigenvalues allows us to study the time scales on which metabolites react. In the previous example, X_1 has a faster time scale than X_2 and X_3. An analysis of this type can be found in Shiraishi and Savageau (1992c, p. 22931) and in condensed form in Case Study 2 (Chapter 9). The reader interested in this topic may also study the method of *modal analysis*, which is based on the local characterization of the movement of a dynamical system through eigenvalues and *eigenvectors* (for modal analyses in a metabolic context see Palsson and Lightfoot, 1984; Palsson et al., 1984, 1985).

The standard analytical technique for computing eigenvalues is not especially straightforward and requires quite a bit of computation. It involves several steps, including the evaluation of a determinant. For large systems, this computation becomes rather hairy, leading to a polynomial of high degree, for which no analytical solutions are known in general.

Fortunately, we don't really need complete information about the eigenvalues to determine stability. As we discussed in Chapter 4, we only need to know whether any of the eigenvalues have real parts that are positive. (Recall that eigenvalues, as complex numbers, have a real part and an imaginary part.)

Several decades ago, a clever method was developed that determines the existence of positive real parts without actually computing the eigenvalues. The method is known as the Routh–Hurwitz method (Routh, 1930: p. 226; for this method in the context of S-systems, see Savageau, 1976; Voit, 1991). Like other methods, it establishes a particular determinant and leads to a polynomial of high order. However, it evaluates the polynomial in a fashion that requires only elementary operations. Still, the computational effort is quite large, and there is no doubt that a numerical evaluation in PLAS is usually the method of choice.

An algebraic analysis of stability makes sense primarily in two instances: First, suppose the system is small, and we are interested not just in a single, numerically specified system but in a whole family of systems (e.g., Savageau and Sands, 1991). As an example, imagine a linear pathway with feedback, and suppose we are interested in determining exactly the threshold value of the feedback strength for which the system becomes unstable (see Example in Chapter 4). In this case, the Routh–Hurwitz method can be executed by using numerical values for all parameters except the feedback parameter, which one retains as a symbol (in Chapter 4, this parameter was g_{13}; see also Example 2 below). Once the analysis is completed, the results can be expressed in terms of g_{13} in order to determine the threshold of stability. The second instance of usefulness of the Routh–Hurwitz analysis is a larger system that has a very regular and convenient structure, which permits a simplified computation of the determinant and the associated polynomial. An impressive example for this scenario is the complete stability analysis of all imaginable linear pathways with feedback, which Savageau (1976) discussed in detail.

The Routh–Hurwitz analysis for S-systems begins with the computation of so-called F-factors, which are defined as relative fluxes at steady state. In steady state, each α-term is equivalent to the corresponding β-term, and therefore, F-factors may be computed from either an α-term or a β-term:

$$F_i = X_{iS}^{-1} \left(\alpha_i \prod_{j=1}^{n+m} X_{jS}^{g_{ij}} \right)$$

$$= X_{iS}^{-1} \left(\beta_i \prod_{j=1}^{n+m} X_{jS}^{h_{ij}} \right). \qquad (6.58)$$

The additional subscript S indicates that the variables are evaluated at the steady state.

ANALYTICAL STEADY-STATE EVALUATION

The F-factors are used, along with the coefficients $a_{ij} = g_{ij} - h_{ij}$ of the system matrix, to form a determinant of the following form:

$$\text{DET} = \begin{vmatrix} F_1 a_{11} - \lambda & F_1 a_{12} & F_1 a_{13} & \cdots & F_1 a_{1n} \\ F_2 a_{21} & F_2 a_{22} - \lambda & F_2 a_{23} & \cdots & F_2 a_{2n} \\ F_3 a_{31} & F_3 a_{32} & F_3 a_{33} - \lambda & \cdots & F_3 a_{3n} \\ \vdots & \vdots & \vdots & & \vdots \\ F_n a_{n1} & F_n a_{n2} & F_n a_{n3} & \cdots & F_n a_{nn} - \lambda \end{vmatrix}. \tag{6.59}$$

Here λ is treated as a variable.

The next step consists in evaluating this determinant. This is typically accomplished through expansion by rows or columns or through the rule of Sarrus (see Appendix). In most cases, the symbols F_i and a_{ij} are by now replaced with numbers, but λ remains as a symbol. Whether totally symbolic or numerically specified, the result of this step is a polynomial with λ as its variable and coefficients ϕ_i that are quantities computed from the F_i and a_{ij}. This *characteristic polynomial* takes the form

$$\lambda^n + \phi_1 \lambda^{n-1} + \phi_2 \lambda^{n-2} + \phi_3 \lambda^{n-3} + \cdots + \phi_{n-1} \lambda + \phi_n. \tag{6.60}$$

It would be nice if one could set this polynomial equal to zero and solve for the n solutions that λ can have in satisfying the equation, because these n solutions are the eigenvalues of the system. However, as soon as n exceeds three, such solutions are difficult to obtain with algebraic means.

The next step of the Routh–Hurwitz procedure avoids the computation of solutions. It asks, instead, for arranging the coefficients of the polynomial and some additional quantities Δ_{ij} in the following array:

$$\begin{array}{cccccc} 1 & \phi_2 & \phi_4 & \phi_6 & \cdots \\ \phi_1 & \phi_3 & \phi_5 & \phi_7 & \cdots \\ \Delta_{11} & \Delta_{13} & \Delta_{15} & \Delta_{17} & \cdots \\ \Delta_{21} & \Delta_{23} & \Delta_{25} & \Delta_{27} & \cdots \\ \vdots & \vdots & \vdots & & \\ \Delta_{i1} & \Delta_{i3} & \Delta_{i5} & \cdots & \\ \vdots & \vdots & & & \\ \Delta_{n-3,1} & \Delta_{n-3,3} & & & \\ \Delta_{n-2,1} & & & & \\ \Delta_{n-1,1} & & & & \end{array} \tag{6.61}$$

Each Δ_{ij} in this array is computed:

- from the two elements in the rows just above and the column just to the right and
- from the two elements in the rows just above and the first column,
- by multiplying crosswise and subtracting the results, and
- by dividing the result by the element in the row just above and in the first column.

This sounds rather complicated, but is actually readily executed. Here are some examples:

$$\Delta_{11} = (\phi_1\phi_2 - \phi_3)\phi_1^{-1},$$
$$\Delta_{13} = (\phi_1\phi_4 - \phi_5)\phi_1^{-1},$$
$$\vdots$$
$$\Delta_{21} = (\Delta_{11}\phi_3 - \phi_1\Delta_{13})\Delta_{11}^{-1}, \quad (6.62)$$
$$\Delta_{23} = (\Delta_{11}\phi_5 - \phi_1\Delta_{15})\Delta_{11}^{-1},$$
$$\vdots$$

The mathematics behind the procedure guarantees that the array becomes narrower and eventually comes to an end. Once this happens, one counts the number of sign changes from plus to minus and from minus to plus among the elements in the first column. This number equals the number of eigenvalues with positive real parts. If this number is zero, the system is stable, but if there is even one sign change, the system is unstable. The procedure fails if one or more of the ϕ_i are zero. In this case, the system may be marginally stable or unstable, but it is not stable.

An alternative to the Routh–Hurwitz criterion is the *Jury test* (Jury, 1971; see also Edelstein–Keshet, 1998), which also uses an array of quantities that are recursively computed from the coefficients of the characteristic polynomial.

Example 1: Stability of a two-variable S-system. A regular S-system with two dependent variables is stable if the following two conditions are satisfied:

$$F_1 a_{11} + F_2 a_{22} < 0,$$
$$a_{11}a_{22} - a_{12}a_{21} > 0 \quad (6.63)$$

(see Exercise 17).

Example 2: A linear pathway with feedback. Consider the simple linear pathway with feedback in Fig. 6.7. An S-system model with typical parameter values might read

$$\dot{X}_1 = 2X_3^{-1} - 2X_1^{0.5},$$
$$\dot{X}_2 = 2X_1^{0.5} - 2X_2^{0.5}, \quad (6.64)$$
$$\dot{X}_3 = 2X_2^{0.5} - 2X_3^{0.5}.$$

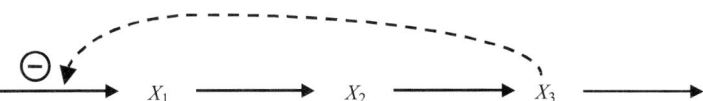

Figure 6.7. Simple linear pathway with feedback.

ANALYTICAL STEADY-STATE EVALUATION

The ingredients for the stability assessment are a_{ij}- and F-values. These are straightforwardly computed as $a_{11} = a_{22} = a_{33} = -0.5, a_{21} = a_{32} = 0.5, a_{13} = -1$, and $F_1 = F_2 = F_3 = 2$. The system determinant thus is

$$\text{DET} = \begin{vmatrix} -1-\lambda & 0 & -2 \\ 1 & -1-\lambda & 0 \\ 0 & 1 & -1-\lambda \end{vmatrix}. \tag{6.65}$$

Application of the rule of Sarrus (see Appendix) yields the characteristic polynomial

$$(-1-\lambda)^3 - 2. \tag{6.66}$$

Setting this polynomial equal to zero and multiplying out leads to

$$\lambda^3 + 3\lambda^2 + 3\lambda + 3 = 0. \tag{6.67}$$

Thus, $\phi_1 = \phi_2 = \phi_3 = 3$, $\Delta_{11} = (3 \times 3 - 3)/3 = 2$, $\Delta_{21} = (2 \times 3 - 0)/2 = 3$, and all other ϕ and Δ coefficients are zero. The Routh array is

$$\begin{matrix} 1 & 3 \\ 3 & 3 \\ 2 & \\ 3 & \end{matrix} \tag{6.68}$$

There are no sign changes, and we conclude that there are no eigenvalues with positive real parts: the system is stable. (Confirm this with PLAS.)

Changing the inhibition parameter g_{13} to -6 only affects the entry in the first row and third column of the determinant. The new entry is -12. Subsequently, one obtains $\phi_3 = 13$, $\Delta_{11} = (3 \times 3 - 13)/3 = -4/3$, and $\Delta_{21} = 13$. Now there are two sign changes, and the system is unstable. Confirm these computations and check the conclusion in PLAS.

Can we determine the threshold for stability? We see that Δ_{11} becomes negative as soon as ϕ_3 exceeds 9. Furthermore, ϕ_3 is directly related to g_{13}, namely by the equation $\phi_3 = 1 + 2(-g_{13})$. (Again, confirm this.) Thus, for inhibition parameters exceeding 4 in magnitude, the system is unstable. PLAS confirms that the real parts of two of the eigenvalues cross zero for $g_{13} = -4$.

Example 3: A pathway with feedback and cross-activation. The comparison of three very similar examples shows that it is hardly possible to predict the stability behavior of a nonlinear system without thorough mathematical analysis. Consider again the simple pathway with feedback in Fig. 6.1 [Eq. (6.2)]. PLAS reveals that this system is stable, because there is only one pair of complex eigenvalues with negative real parts. Similarly, the steady state of the branched pathway in Fig. 6.5 [Eq. (6.33)] is stable, since all eigenvalues have negative real parts. By contrast, we observed in the system of Fig. 6.6 [Eq. (6.52)] that the steady state depends on the chosen initial conditions. PLAS computes the eigenvalues as about $-2.3 + 0i$ and $0 + 0i$, where the real part of zero indicates marginal stability.

Figure 6.8. Pathway with several activations and inhibitions.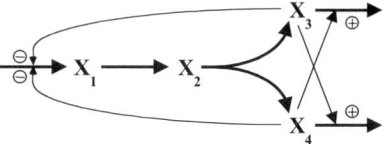

Now consider the pathway in Fig. 6.8, which is similar to those discussed before, but includes two activating processes. It may be represented as the S-system

$$\begin{aligned}
\dot{X}_1 &= X_3^{-1} X_4^{-1} - X_1^{1/2}, \\
\dot{X}_2 &= X_1^{1/2} - X_2^{1/2}, \\
\dot{X}_3 &= X_2^{1/2} - X_3 X_4^{h_{34}}, \\
\dot{X}_4 &= X_2^{1/2} - X_3^{h_{43}} X_4.
\end{aligned} \tag{6.69}$$

The cross-activation of the degradation of X_3 and X_4 is modeled with the positive parameters h_{34} and h_{43}, which we have so far left numerically unspecified.

Let us characterize the stability of the system algebraically. By choosing all rate constants as 1, the non-zero steady state of the system is always at $X_1 = X_2 = X_3 = X_4 = 1$. Furthermore, all F-factors are equal to 1, because they consist of products of steady-state values, raised to kinetic orders, that are multiplied by a rate constant. This dramatically simplifies the analysis (see Exercise 19).

The system determinant is thus

$$\text{DET} = \begin{vmatrix} -0.5 - \lambda & 0 & -1 & -1 \\ 0.5 & -0.5 - \lambda & 0 & 0 \\ 0 & 0.5 & -1 - \lambda & -h_{34} \\ 0 & 0.5 & -h_{43} & -1 - \lambda \end{vmatrix}, \tag{6.70}$$

which is evaluated, for instance, via expansion by the first column (see Appendix and Exercise 19). The result is

$$\begin{aligned}
\text{DET} = &\lambda^4 + 2.5\lambda^3 + (2.25 - h_{34}h_{43})\lambda^2 \\
&+ (1.5 - 0.5 h_{34}h_{43})\lambda + 0.75 - 0.25(h_{34}h_{43} + h_{34} + h_{43}).
\end{aligned} \tag{6.71}$$

Next, we set up the array for the Routh–Hurwitz procedure. In our case, it contains the quantities

$$\begin{array}{lll} 1 & \phi_2 & \phi_4 \\ \phi_1 & \phi_3 & \\ \Delta_{11} & \phi_4 & \\ \Delta_{21} & & \\ \phi_4 & & \end{array} \tag{6.72}$$

ANALYTICAL STEADY-STATE EVALUATION

which are given as

$$\begin{aligned}
\phi_1 &= 2.5, \\
\phi_2 &= 2.25 - h_{34}h_{43}, \\
\phi_3 &= 1.5 - 0.5h_{34}h_{43}, \\
\phi_4 &= 0.75 - 0.25(h_{34}h_{43} + h_{34} + h_{43}), \\
\Delta_{11} &= 1.65 - 0.8h_{34}h_{43}, \\
\Delta_{21} &= 1.5 - 0.5h_{34}h_{43} - 2.5\phi_4\Delta_{11}^{-1}.
\end{aligned} \qquad (6.73)$$

We need to check whether there are sign changes in the first column of the array. Since the first column begins with $+1$, we check whether – or under what conditions – any of the first column elements might be negative. More precisely, we test when these elements are exactly 0, which provides the threshold of instability.

The first element is ϕ_1, which is positive, independently of what h_{34} and h_{43} are. Δ_{11} equals 0 if $1.65 = 0.8h_{34}h_{43}$, or if $h_{34}h_{43} = 2.0625$. The threshold values thus lie on the hyperbola $h_{43} = 2.0625/h_{34}$. For the *symmetric* case $h_{34} = h_{43}$, in which both cross-activation processes are equally strong, the threshold is thus $h_{34} = h_{43} = 1.4361$. Are these the minimal values for which the system becomes unstable? Let's analyze ϕ_4, which also is an element of the first column of the Routh–Hurwitz array. Setting ϕ_4 equal to zero provides a threshold for h_{34} and h_{43} as

$$h_{34}h_{43} + h_{34} + h_{43} = 3. \qquad (6.74)$$

If the value 3 is exceeded, the system becomes unstable. An example combination for the threshold is $h_{34} = h_{43} = 1$, but combinations like 0.5 and 2 also push the system over the threshold. (Check it in PLAS.) Since $h_{34}h_{43}$ in this case is less than 2.0625, we have found a more stringent condition for stability, at least in the symmetric case $h_{34} = h_{43}$.

The only other possible threshold of stability is given by $\Delta_{21} = 0$. We have avoided it so far because it involves directly or indirectly all quantities of the array. This threshold is given by

$$1.5 - 0.5h_{34}h_{43} - 2.5[0.75 - 0.25(h_{34}h_{43} + h_{34} + h_{43})][1.65 - 0.8h_{34}h_{43}]^{-1} = 0 \qquad (6.75)$$

and is not easy to evaluate. However, we can gain some insight upon inspection if we again study the symmetric case $h_{34} = h_{43}$. If $h_{34} = h_{43} = 1$, the entire term starting with 2.5 is zero, and the condition reduces to

$$1.5 - 0.5h_{34}h_{43} = 0. \qquad (6.76)$$

For $h_{34} = h_{43} = 1$, Δ_{21} is positive, thus not implying instability. Some calculus or a graph of the constraint shows that for smaller positive values, which correspond to activation rather than inhibition, the value of Δ_{21} decreases, but never reaches 0. Even for the extreme case of $h_{34} = h_{43} = 0$, Δ_{21} is still positive with a value of 0.36363. On the other hand, if $h_{34}h_{43}$ approaches 2.0625, the term in the denominator of Eq. (6.75) approaches zero and approaches ∞. If $h_{34}h_{43}$ exceeds

2.0625, the denominator becomes negative, and eventually Δ_{21} vanishes. However, under those conditions, the constraint on ϕ_4 is already in effect, and the system is unstable anyway (see also Exercise 19).

It is instructive to run some simulations with PLAS for different values of the activation parameters h_{34} and h_{43}. For moderate activation, say $h_{34} = h_{43} = 0.5$, the system is stable with a unique positive steady state at $X_1 = X_2 = X_3 = X_4 = 1$. All eigenvalues have negative real parts; one pair of eigenvalues with non-zero imaginary parts indicates oscillatory behavior. Numerical solutions with PLAS, e.g., initialized at $X_1 = X_2 = X_3 = 1$, $X_4 = 0.5$ or 1.5, confirm the oscillatory approach to the stable steady state from below or above the steady state.

Given a slightly stronger activation ($h_{34} = h_{43} = 1$), the system is marginally stable, with a determinant of zero. Initiating the system at $X_1 = X_2 = X_3 = 1$, $X_4 = 0.5$ or 1.5, we observe some oscillations, but the variables approach different steady-state values.

For even stronger activation ($h_{34} = h_{43} = 2$), the steady state at $X_1 = X_2 = X_3 = X_4 = 1$ is unstable, and one of the system constituents dies out. The instability is indicated by one eigenvalue with a positive real part.

One could surmise that the difference in activation strengths between 1 and 2 is in fact rather significant. However, the magnitude simply determines how long it takes before one variable dies out: At exactly 1, the steady state of the system is marginally stable, for values smaller than 1 (even 0.9999) the system is locally stable, and for values greater than 1 the system is unstable. The key issue to remember is that approximate numerical information about a pathway is not always sufficient to predict one of the fundamental features of a pathway, namely its stability. Exercise 2 addresses this example.

Transition Time

The transition time τ is the average time that a molecule remains in a pathway. τ provides a measure for the *throughput*, or efficiency, of the system. The transition time is not always easy to determine, and in fact several definitions exist. For a pathway with one output V_n^-, the transition time is sometimes defined as the sum of all metabolites at steady state, divided by the steady-state efflux. In mathematical terminology we can write

$$\tau = \frac{X_1 + X_2 + \cdots + X_n}{V_n^-} \tag{6.77}$$

(cf. Easterby, 1981; Cascante et al., 1991b; Torres, 1994d). As an example, consider a linear pathway with feedback, as depicted in Fig. 6.9.

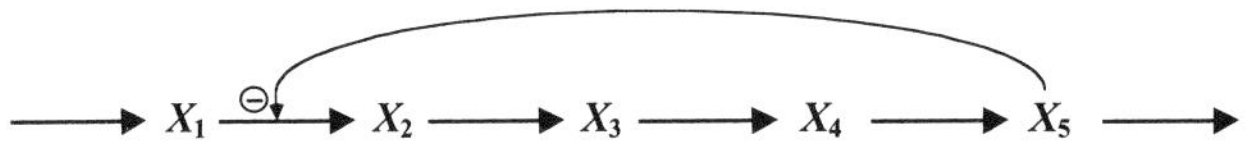

Figure 6.9. Linear pathway with feedback.

ANALYTICAL STEADY-STATE EVALUATION

The five dependent variables could be represented by the S-system

$$\dot{X}_1 = 2 - 2X_1^{0.5} X_5^{-1},$$
$$\dot{X}_2 = 2X_1^{0.5} X_5^{-1} - 4X_2^{0.5},$$
$$\dot{X}_3 = 4X_2^{0.5} - 4X_3^{0.8}, \qquad (6.78)$$
$$\dot{X}_4 = 4X_3^{0.8} - 1X_4,$$
$$\dot{X}_5 = 1X_4 - 4X_5^{0.5}.$$

Steady-state analysis, algebraically or in PLAS (see file *Transit.plc*), produces the non-trivial steady state at $(0.0625, 0.25, 0.420448, 2, 0.25)$ and an efflux $V_5^- = 4X_5^{0.5}$ of 2. Thus, the transit time of the system is

$$\tau = \frac{0.0625 + 0.25 + 0.420448 + 2 + 0.25}{2} = 1.491474. \qquad (6.79)$$

If we multiply all rate constants by 10, the steady state is unchanged, but the efflux is 10 times as high. Since the efflux appears in the denominator, the transit time is 10% of the former value. This makes sense, since all processes run at a 10 times increased speed.

EXERCISES

1. Formulate steady-state equations for the following S-systems. Execute the step-by-step derivation (setting derivatives equal to zero, taking logarithms, etc.) and also the shortcut described in Eqs. (6.22)–(6.23). Solve the linear equations. Check your answers in PLAS.

    1.1.
    $$\dot{X}_1 = 0.4 X_1 X_2^{-2} - 20 X_1 X_2,$$
    $$\dot{X}_2 = 1.5 X_1^{0.5} - 6 X_1^{0.1} X_2^{0.5}$$

    1.2.
    $$\dot{X}_1 = 2X_3^{-1} - 2X_1^{0.5} X_2^{-2},$$
    $$\dot{X}_2 = 2X_1^{0.5} X_2^{-2} - 4X_1^{-1} X_2^{0.5},$$
    $$\dot{X}_3 = 4X_1^{-1} X_2^{0.5} - X_1 X_3^{0.5}.$$

    1.3.
    $$\dot{X}_1 = 2X_2^{-1} - 2X_1^{0.5},$$
    $$\dot{X}_2 = 2X_1^{0.5} - 2X_1 X_2,$$
    $$\dot{X}_3 = 4X_2^{-1} X_3 - X_2^{-1} X_3^2.$$

    1.4.
    $$\dot{X}_1 = 2X_3^{-0.8} X_4^{-0.1} X_5 - X_1^{0.5},$$
    $$\dot{X}_2 = X_1^{0.5} X_2^{-0.2} - 4X_1^{-1} X_2^{0.5},$$
    $$\dot{X}_3 = 2.5 X_1^{-1} X_2^{0.5} - X_2^{-1} X_3^2,$$
    $$\dot{X}_4 = 1.5 X_1^{-1} X_2^{0.5} - X_2^{-1} X_4^2,$$
    $$X_5 = 2.$$

1.5.

$$\dot{X}_1 = 2X_3^{-0.8} X_4^{-0.1} X_5 - X_1^{0.5},$$
$$\dot{X}_2 = X_1^{0.5} X_2^{-0.2} - 4X_1^{-1} X_2^{0.5},$$
$$\dot{X}_3 = 2.5 X_1^{-1} X_2^{0.5} - X_2^{-1} X_3^2,$$
$$\dot{X}_4 = 1.5 X_1^{-1} X_2^{0.5} - X_2^{-1} X_4^2,$$
$$\dot{X}_5 = 2.4 - 1.2 X_5.$$

2. Confirm that the steady-state matrix equation of the system (6.33) is

$$\begin{pmatrix} -0.5 & 0 & -1 & -1 \\ 0.5 & -0.5 & 0 & 0 \\ 0 & 0.5 & -1 & 0 \\ 0 & 0.5 & 0 & -1 \end{pmatrix} \cdot \vec{y} = \begin{pmatrix} 0 \\ \ln 2 \\ \ln 0.625 \\ \ln 0.625 \end{pmatrix}.$$

Show that the equation can also be written as

$$\begin{pmatrix} 0.5 & 0 & 1 & 1 \\ -0.5 & 0.5 & 0 & 0 \\ 0 & -0.5 & 1 & 0 \\ 0 & -0.5 & 0 & 1 \end{pmatrix} \cdot \vec{y} = \begin{pmatrix} 0 \\ \ln 0.5 \\ \ln 1.6 \\ \ln 1.6 \end{pmatrix}.$$

3. Compute the following determinants:

3.1.

$$|A| = \begin{vmatrix} 1 & 2 & 3 \\ 2 & 4 & 6 \\ 4 & 8 & 12 \end{vmatrix}.$$

3.2.

$$|B| = \begin{vmatrix} 2 & 2 & 2 \\ 2 & 2 & 2 \\ 2 & 2 & 2 \end{vmatrix}.$$

3.3.

$$|C| = \begin{vmatrix} 1 & 0 & 0 & 0 \\ 0 & 1 & 0 & 0 \\ 0 & 0 & 1 & 0 \\ 0 & 0 & 0 & 1 \end{vmatrix}.$$

3.4.

$$|D| = \begin{vmatrix} 2 & .5 & 1 & 2 \\ 0 & 0 & 0 & -2 \\ 3 & 3 & 1 & 0 \\ 0 & -2 & -1 & -4 \end{vmatrix}.$$

4. Use the step-by-step derivation and also the shortcut as discussed in Eqs. (6.22)–(6.23) to demonstrate the influence of a metabolite whose kinetic order is zero or close to zero.

ANALYTICAL STEADY-STATE EVALUATION

5. Give mathematical reasons why linear steady-state equations cannot be formulated if a rate constant is zero.

6. Explain why a steady state cannot always be computed algebraically for a GMA system, even if it does have a steady state. Construct examples and check them with PLAS.

7. Is an S-system uniquely determined from its steady-state equations? In other words, given the steady-state equations, can one retrieve the S-system? Justify your answer intuitively and mathematically. If it is useful, provide examples.

8. Suppose a regular S-system has the steady-state concentrations $X_{1s}, X_{2s}, \ldots, X_{ns}$. What happens when all variables are divided by their steady-state value, i.e., when one defines new variables $Z_i = X_i/X_{is}$ for $i = 1, 2, \ldots, n$? Approach the question mathematically, run simulations with PLAS, and explain the results.

9. In the ith steady-state equation of an S-system, the coefficient associated with variable y_j is $a_{ij} = g_{ij} - h_{ij}$. What happens if some or all g_{ij} and h_{ij} are increased by the same (additive) amount? What happens if they are multiplied by the same factor? Analyze steady states and dynamics.

10. In the ith steady-state equation of an S-system, the coefficients b_i on the right-hand side are the logarithms of the ratios of rate constants: $b_i = \ln(\beta_i/\alpha_i)$. What happens if some or all α_i and β_i are increased by the same (additive) amount? What happens if they are multiplied with the same factor? Analyze steady state and dynamics.

11. 11.1. Prove or disprove: An S-system without independent variables has a steady state of all 1's if and only if all rate constants have the same value.

    11.2. Prove or disprove: An S-system without independent variables has a steady state of all 1's if and only if α_i is equal to β_i for every $i = 1, 2, \ldots, n$.

    11.3. Is the inclusion of independent variables relevant in this context?

12. If one goes "backwards in time," mathematically replacing the differentiation variable t with $-t$, all α-terms and β-terms of an S-system are interchanged.

    12.1. Give a mathematical proof of this statement.

    12.2. In PLAS, solve some S-systems of your choice forwards and backwards in time; good examples might be a single S-system equation, a linear pathway with feedback, a harmonic oscillation, or some limit cycle, as they were discussed in Chapter 4.

    12.3. Explore the effects of this change on the steady-state concentrations, stability, and log gains.

13. Construct different scenarios in which an S-system has no algebraic steady-state solution. Differentiate cases in which a numerical steady-state solution exists or does not exist.

14. Discuss all steady states of the following systems, including those for which one or more variables are zero:

    14.1.
    $$\dot{X}_1 = 2X_1 - 2X_2,$$
    $$\dot{X}_2 = X_1 - 2X_2,$$
    $$\dot{X}_3 = X_3 - X_3^2.$$

14.2.
$$\dot{X}_1 = 1.5 X_1 - 1.5 X_2,$$
$$\dot{X}_2 = 2 X_1 - 2 X_2,$$
$$\dot{X}_3 = 2.5 X_1 - 2.5 X_2.$$

14.3.
$$\dot{X}_1 = 2 X_3 - X_1,$$
$$\dot{X}_2 = X_1 - 2 X_2,$$
$$\dot{X}_3 = X_2 - 2 X_3.$$

Interpret your results in words.

15. Algebraically compute the steady-state solution for the pathway with recycling and feedback activation (Fig. 6.3) if $X_2 = 0$. What is the system determinant?

16. The Gompertz function is frequently used to describe the growth of animals and plants. It has the form $W(t) = W_f \exp[-\beta \exp(-\alpha t)]$. The function can also be formulated as an S-system with the two differential equations

$$\dot{W} = WZ,$$
$$\dot{Z} = -\alpha Z,$$

where Z is an auxiliary variable, defined as $Z = \alpha\beta \exp(-\alpha t)$ (cf. Savageau, 1980). Graph the original Gompertz function $W(t)$. Interpret its parameters. Define parameter values. Implement the differential equation version in PLAS, and solve. Compare the results. Analyze the steady state. Compute the derivatives of Z and W, and show that they satisfy the above differential equations.

17. Derive the stability conditions for a regular S-system with two dependent variables from the Routh–Hurwitz procedure.

18. Derive in symbolic form stability conditions for a linear pathway of four metabolites with feedback inhibition of the first intermediate by the final product. Once the conditions are established, check the results with numerical analyses. Compare in words the results with those obtained for a pathway with three metabolites.

19. 19.1 Example 3 of the section on local stability provided selected steps of the total stability analysis. Derive all missing steps.

    19.2 Change the model to include rate constants different from 1. Sketch the stability analysis, and discuss in detail where this generalization creates serious mathematical problems.

20. In Example 3 of the section on local stability, replace activation of degradation with inhibition of degradation. Execute algebraic and numerical stability analyses.

21. (Project) In a now classical model, Kermack and McKendrick (1927) described how an infectious disease spreads through a population. The model contains as variables the number of susceptible individuals, S; the number of infected individuals, I; and the number of *removed* individuals, R, who no longer spread the disease. One implementation of the model reads

$$\dot{S} = \gamma R - \beta I S,$$
$$\dot{I} = \beta I S - \nu I,$$
$$\dot{R} = \nu I - \gamma R.$$

ANALYTICAL STEADY-STATE EVALUATION

Draw a map of the disease dynamics, and interpret the parameters. Try to compute the steady states algebraically, first using the logarithmic transformation discussed in this chapter, and secondly without such transformation. Analyze steady states in which one or more variables are zero. Implement the system in PLAS, and perform dynamic and steady-state analyses. Interpret the results. Do the same analyses and interpretations for the special case $\gamma = 0$. Analyze other special cases algebraically and numerically. Does the pool R include individuals that die from the disease? Explain your answer. What needs to be changed in the model to include/exclude birth and death?

REFERENCES

[47], [79], [80], [110], [130], [175], [193], [229], [232–3], [273–5], [296], [311], [313], [319], [324], [349], [376], [419], [439].

CHAPTER SEVEN

Sensitivity Analysis

Stability analysis shows how a biochemical system responds to short-term perturbations that remove it from its steady state. In this chapter we want to analyze how a system responds to more permanent influences. The branch of mathematics and engineering that is concerned with such responses is *sensitivity analysis*. Sensitivity analysis has a long history, and we shall hardly scratch the surface of this interesting topic. The reader who would like to learn more about it is referred to standard treatises, such as Šiljak (1969), Frank (1978), and Rabitz et al. (1983).

One distinguishes conceptually two types of influences: (i) persistent changes in the numerical values of independent variables, and (ii) permanent changes in system parameters, i.e., in kinetic orders or rate constants. The first type of change describes a scenario in which an organism is transferred from one environment into another. Nothing changes within the genetic or metabolic structure of the organism, but resources and stimuli are different in the new environment, and the corresponding changes in independent variables may cause the system to assume a new steady state. The relative change of a system property in response to a relative change in an independent variable is called a *logarithmic gain*, a *logarithmic amplification*, or, in the jargon of the field, a *log gain*. In the analysis of actual systems, log gains are very useful for the assessment of the robustness of a model and for estimation of parameter values.

In contrast to log gains, the change in a system parameter corresponds to a structural change inside the organism, irrespective of the environmental conditions. Such a change could be caused by a mutation or a disease and, for instance, result in an altered enzyme activity. The change of a system property in response to a change in a parameter is called a *parameter sensitivity* or, for short, a *sensitivity*.

In the long run, organisms may of course adapt to altered environmental conditions, and these evolutionary adaptations may be manifold, complex, and utterly unpredictable. What we are interested in here are *immediate responses* to *small but persistent alterations*. In principle, these responses could be analyzed for systems in any situation, but we restrict our discussion to the most important case, which is a system at steady state.

TYPES OF GAINS AND SENSITIVITIES

We distinguish five types of gains and sensitivities (Savageau, 1971b; 1972):

1. Log gain of the steady-state concentration of a metabolite with respect to a change in an independent variable.
2. Log gain of a flux with respect to a change in an independent variable.
3. Log gain of the transition time with respect to a change in an independent variable.
4. Sensitivity of a steady-state concentration of a metabolite with respect to a change in a parameter.
5. Sensitivity of a flux with respect to a change in a parameter.

Besides these sensitivities and gains, other types of sensitivities have been studied to highlight specific aspects of systems. For instance, Savageau (1976) used sensitivities of log gains with respect to changes in a parameter in order to measure the effect of end product demand on the levels of intermediates in feedback-inhibited systems. Heinrich and Schuster (1996) and Fell (1997) discussed a number of related types of sensitivities in the analysis of the regulation and control of metabolic and cellular systems. Salvador (1996, 1997, 2000ab) investigated the sensitivity of systems to simultaneous changes in several parameters.

In vivo, small changes in parameter values or input variables are the rule. They occur continually and propagate throughout the system. The organism may respond to these changes by returning to the original physiological steady state or by assuming a slightly changed steady state, but usually, a small *perturbation* in a parameter value or input has no dramatic effect on the structure of the system. Turning this argument around, the analysis of sensitivities and gains provides a good test of the quality of a biochemical model. An outcome demonstrating small sensitivities – i.e., moderate responses – is no proof of the correctness of the model, but it does support its mathematical structure. On the other hand, very high sensitivities or gains in a model are often a sign of structural inadequacies. They pinpoint parts of a model lacking robustness and may even provide a tool for the identification of inconsistencies in the data.

Okamoto and Savageau (1984a), for instance, studied the isoleucyl-tRNA synthetase proofreading system. When they encountered inconsistencies, they ranked the sensitivities of the system. Re-examining the most sensitive parameters, they indeed detected a 10-fold error in published data. Shiraishi and Savageau (1992a–d, 1993) were able to pinpoint problem areas in a model of the tricarboxylic acid cycle in *Dictyostelium discoideum* (see Chapter 9). Upon identification of the trouble spots, they were able to reformulate the model in a more robust and reliable form. Curto et al. (1997, 1998ab) relied heavily on the analysis of sensitivities in refining models of purine metabolism (see Chapter 10). Torres (1994bc; see also Torres et al., 1996, 1997) evaluated sensitivities of a biotechnological system with the goal of improving yield, arguing that the most sensitive system components promised the highest probability for effective alterations.

Logarithmic Gain of a Metabolite

As an illustrative example, we use again the simple pathway in Fig. 6.2, which is redrawn here as Fig. 7.1. The analysis in Chapter 6 showed that the steady-state concentrations of metabolites depend on the concentrations of the independent variables. PLAS may refresh our insight into the roles of independent and dependent variables. We implement the system (6.12), specifying $\tilde{\alpha}_1 = 0.5, g_{13} = 0.5, \tilde{\beta}_2 = 1$, and $h_{24} = -1$, set the independent variables as $X_3 = 4$ and $X_4 = 2$, and initialize the system with $X_1 = 1$ and $X_2 = 1$ at $t = 0$. The system thus reads

$$\dot{X}_1 = 0.5 X_2^{-2} X_3^{0.5} - 2X_1,$$
$$\dot{X}_2 = 2X_1 - X_2^{0.5} X_4^{-1}, \tag{7.1}$$
$$X_3 = 4, \quad X_4 = 2$$

(file *Feedback.plc*). Numerically solving the system with the time course command, , we observe that, after some transient oscillations, the system rapidly approaches a steady state at $X_1 = 0.28717$, $X_2 = 1.3195$. Now we reset either a dependent or an independent variable and compare the effects of changes in metabolite concentrations on the system.

First we study effects of changes in a dependent variable. We initialize the system at $t = 0$ with X_2 at its steady-state value, 1.3195, but with $X_1 = 0.5$. As before, the independent variables are specified as $X_3 = 4$ and $X_4 = 2$. The system overshoots and then quickly returns to the original steady state. A similar response follows a change in X_2. This behavior makes sense, because the steady state is stable.

Now we execute a similar change in an independent variable. We initialize at $t = 0$ with the steady state values $X_1 = 0.28717$ and $X_2 = 1.3195$, but change the value of the independent variable X_3 from 4 to 6. In contrast to the previous experiment, the system assumes a new steady state at $X_1 = 0.29906$, $X_2 = 1.4310$. A similar response follows a change in X_4.

Our goal is to predict, without solving the differential equations, how strongly changes in independent variables affect the steady state of the system. By comparing the effects, we can rank the independent variables from most to least influential.

The general procedure for this type of analysis is to study the steady-state equations (Savageau, 1969b) and to express the dependent variables as functions of the independent variables. Subsequently, one uses elementary differential calculus to figure out how strongly a metabolite reacts to changes in each of the independent variables.

We perform the analysis again with the pathway in Fig. 7.1 and the corresponding equations (6.12) and (7.1). The steady-state equations of this system were given in Eq. (6.43) in symbolic form, where the variables y_i were defined as the natural logarithms of the original variables X_i. With substitution of the parameter values $\tilde{\alpha}_1, g_{13}, \tilde{\beta}_2$, and $h_{24} = -1$, the solution in terms of the independent variables y_3

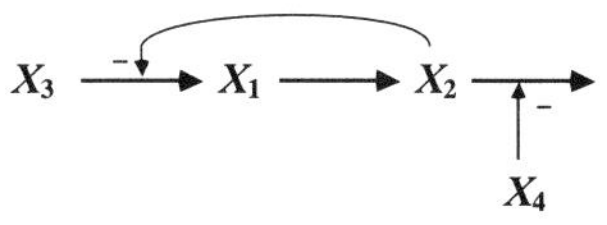

Figure 7.1. Linear pathway with feedback and two independent variables.

and y_4 is

$$\begin{aligned}
y_1 &= 0.2(\ln 0.5 + 0.5y_3 - \ln 2) - 0.8[\ln 2 - \ln 1 - (-1)y_4] \\
&= -0.83178 + 0.1y_3 - 0.8y_4, \\
y_2 &= 0.4[\ln 0.5 + 0.5y_3 - \ln 1 - (-1)y_4] \\
&= -0.27726 + 0.2y_3 + 0.4y_4.
\end{aligned} \quad (7.2)$$

Suppose we are interested in the relative (percentage) change in the first metabolite, X_1, that is caused by a relative (percentage) change in X_4. According to the *Complements* to Chapter 2, a relative change in a variable corresponds to an absolute change in its logarithm. In other words, we can characterize the system response by describing how $y_1 = \ln X_1$ responds to changes in $y_4 = \ln X_4$. Under the condition that $X_3 = 4$ is unaltered, the term $0.1y_3 = 0.1\ln 4$ in the first equation of Eq. (7.2) is constant and can be merged with the constant term -0.83178. Thus, we can rewrite this equation as

$$y_1 = -0.69315 - 0.8y_4. \quad (7.3)$$

The change in y_1 per unit change in y_4 is given as the derivative of y_1 with respect to y_4 and called a *logarithmic gain*:

$$\begin{aligned}
\text{Logarithmic gain } L_{14} &= L(X_1, X_4) \\
&= \text{change in } y_1 \text{ per unit change in } y_4 \\
&= \text{derivative of } y_1 \text{ with respect to } y_4 \\
&= \partial y_1 / \partial y_4 \\
&= -0.8.
\end{aligned} \quad (7.4)$$

Several remarks are in order:

1. We use the curly ∂ for the derivative instead of the usual d to indicate that y_1 is in fact a function of several variables (in this case, y_3 and y_4). In mathematical terminology, this type of derivative is called a *partial derivative*. It is computed just like an ordinary derivative when all independent variables, except the one of interest, are considered constant (cf. Appendix).
2. Since y_i is a linear function of y_k, the derivative is equivalent with the ratio of non-infinitesimal changes:

$$\frac{\partial y_i}{\partial y_k} = \frac{\hat{y}_i - y_i}{\hat{y}_k - y_k}, \quad (7.5)$$

   where the hat (^) denotes the quantity in the altered system.
3. If the S-system model is an *exact* description of the investigated phenomenon, log gains correctly predict changes in steady-state values for changes in independent variables of any size. Typically, though, the S-system model is an approximation of the system of interest. In this case, log gains are – strictly speaking – only exact for infinitesimally small changes in independent variables. Experience has shown that they are usually good predictors for changes of moderate magnitude.
4. The definition above states that log gains describe the change in the logarithm of a dependent variable X_i that is evoked by the change in the logarithm of an

independent variable X_k. If the relative change in the X_k is small, the log gain predicts rather accurately the relative (percentage) change in X_i. This can be seen from the following argument. Suppose the independent variable is altered to

$$\hat{X}_k = X_k + \Delta X_k \tag{7.6}$$

and the newly achieved value of the dependent variable of interest is $\hat{X}_i$, which is expressed as

$$\hat{X}_i = X_i + \Delta X_i. \tag{7.7}$$

The log gain is given as

$$L_{ik} = \frac{\hat{y}_i - y_i}{\hat{y}_k - y_k} = \frac{\ln \hat{X}_i - \ln X_i}{\ln \hat{X}_k - \ln X_k} = \frac{\ln(\hat{X}_i/X_i)}{\ln(\hat{X}_k/X_k)} \tag{7.8}$$

The logarithm in the numerator of the last ratio can be written as

$$\ln\left(\frac{\hat{X}_i}{X_i}\right) = \ln\left(\frac{X_i + \Delta X_i}{X_i}\right) = \ln\left(1 + \frac{\Delta X_i}{X_i}\right), \tag{7.9}$$

and the analogous statement is true for the denominator. For small ΔX_i, the last expression is well approximated by $\Delta X_i/X_i$, which can be seen from its Taylor expansion (see Appendix). In this case, the log gain becomes

$$L_{ik} \approx \frac{\Delta X_i/X_i}{\Delta X_k/X_k} = \frac{\text{relative change in } X_i}{\text{relative change in } X_k}. \tag{7.10}$$

Thus, for small changes in an independent variable, the log gain directly predicts the relative change in a dependent variable (see also Exercise 1).

Example. Suppose the log gain L_{14} is -0.8 and X_4 is increased by 1%. The prediction thus is that the steady-state concentration of the metabolite X_1 decreases by approximately 0.8%. The result is confirmed by PLAS, which we can check under the *Sensitivities* tab of the *Steady-State* menu, . The accuracy of predictions of larger changes in X_4 can be tested numerically with PLAS in the following way: First, compute the steady state for the baseline concentration of $X_4 = 2$. Then increase X_4 by p% and compute the steady state again. For instance, if the independent variable X_4 is increased by 5% (X_4 is set to 2.1), Eq. (7.4) predicts a $5 \times 0.8\% = 4\%$ decrease in the steady-state concentration of X_1. The concentration actually obtained from a steady-state computation in PLAS is -3.83%, which is close to the prediction. Table 7.1 shows predicted and actual increases and decreases in X_1 and X_2 that are caused by various percentage changes in X_4. For small changes in X_4, the predicted and actual changes are almost indistinguishable, and even for larger changes the prediction still gives a reasonable estimate of the actual change. It is emphasized that the small – or larger – differences between prediction and observation are not due to computational inaccuracies, but to the fact that the linear extrapolation of results for small gains is an approximation of the nonlinear responses (see remark 4 above). Using the original definition of the log gain as the ratio of changes in logarithms, the observed values are reproduced exactly. See also Exercise 2.

SENSITIVITY ANALYSIS

Table 7.1. Predicted and Observed Changes in X_1 and X_2 Caused by Changes in X_4

X_4	%Δ	X_{1Prd}	X_{1Obs}	%Δ_{Prd}	%Δ_{Obs}	X_{2Prd}	X_{2Obs}	%Δ_{Prd}	%Δ_{Obs}
2.0	0	0.28717	0.28717	0	0	1.3195	1.3195	0	0
2.02	1	0.28487	0.28490	−0.8	−0.8	1.3248	1.3248	0.4	0.4
2.1	5	0.27568	0.27618	−4.0	−3.8	1.3459	1.3455	2.0	2.0
2.2	10	0.26420	0.26609	−8.0	−7.3	1.3723	1.3708	4.0	3.9
3.0	50	0.17230	0.20762	−40.0	−27.7	1.5834	1.5518	20.0	17.6
1.98	−1	0.28947	0.28949	0.8	0.8	1.3142	1.3142	−0.4	−0.40
1.9	−5	0.29866	0.29920	4.0	4.2	1.2931	1.2927	−2.0	−2.0
1.8	−10	0.31014	0.31243	8.0	8.8	1.2667	1.2651	−4.0	−4.1
1.0	−50	0.40204	0.50000	40.0	74.1	1.0556	1.0000	−20.0	−24.2

The starting point for symbolic analyses of log gains is the steady-state matrix equation for systems with independent variables:

$$\vec{y}_D = \mathbf{A}_D^{-1}\vec{b} - \mathbf{A}_D^{-1}\mathbf{A}_I\vec{y}_I \tag{7.11}$$

[cf. Eq. (6.51)]. Let's recall what this equation says. It expresses the steady-state values of the dependent variables, collected as logarithms in the vector $\vec{y}_D$, in terms of two contributions. The first contribution, $\mathbf{A}_D^{-1}\vec{b}$, expresses how the steady-state value of a dependent variable is affected by other *dependent* variables. These effects are expressed through the inverse of the matrix $\mathbf{A}_D$, which contains all kinetic order parameters associated strictly with dependent variables, and through the vector $\vec{b}$, which is determined by all rate constants of the system. The second contribution to the steady-state solution, $\mathbf{A}_D^{-1}\mathbf{A}_I\vec{y}_I$, represents how the *independent* variables affect the steady state. The (logarithms of) independent variables constitute the vector $\vec{y}_I$. Their influences are characterized again by the inverse of the matrix $\mathbf{A}_D$, but also by the matrix $\mathbf{A}_I$, which contains all kinetic order parameters associated with the independent variables.

We are interested in the relative change in a dependent variable X_j caused by a relative change in an independent variable X_k ($k = n+1, \ldots, n+m$) or, equivalently, in the derivative of y_j with respect to y_k:

$$L(X_j, X_k) = \partial y_j / \partial y_k. \tag{7.12}$$

According to Eq. (7.11), the jth component y_j of the solution vector $\vec{y}_D$ has two parts: the jth row of $\mathbf{A}_D^{-1}$ multiplied by $\vec{b}$, and the jth row of $\mathbf{A}_D^{-1}\mathbf{A}_I$ multiplied by $\vec{y}_I$. The first part is constant and therefore does not contribute to $\partial y_j/\partial y_k$, and the only contribution of the second part is the kth element in the jth row of $\mathbf{A}_D^{-1}\mathbf{A}_I$, multiplied by -1:

$$L(X_j, X_k) = (-1) \cdot \text{element } (j, k) \text{ of } \mathbf{A}_D^{-1}\mathbf{A}_I. \tag{7.13}$$

Turning this relationship around, we see that the $m \times n$ matrix $\mathbf{A}_D^{-1}\mathbf{A}_I$ is entirely composed of logarithmic gains.

In most systems, the dependent variables are not influenced by just one independent variable but by several, and it is convenient to summarize them in the form of vectors and matrices. For instance, we can define $\vec{L}(X_j)$ as the row vector with the m components $L(X_j, X_k)$, that is, as the vector that contains the effects of all independent variables X_k ($k = n+1, \ldots, n+m$) on one particular dependent variable X_j:

$$\vec{L}(X_j) = (L(X_j, X_{n+1}), L(X_j, X_{n+2}), \ldots, L(X_j, X_{n+m})). \tag{7.14}$$

In the previous example, the vectors $\vec{L}(X_1)$ and $\vec{L}(X_2)$ are

$$\begin{aligned}\vec{L}(X_1) &= (0.1, -0.8), \\ \vec{L}(X_2) &= (0.2, 0.4).\end{aligned} \tag{7.15}$$

Furthermore, we can look at all log gains $\vec{L}(X_j)$ at the same time. There are n dependent variables and m independent variables in the system. Thus, we obtain an $n \times m$ log gain matrix $\mathbf{L}(\mathbf{X}_D, \mathbf{X}_I)$ of the form

$$\mathbf{L}(\mathbf{X}_D, \mathbf{X}_I) = \begin{pmatrix} L(X_1, X_{n+1}) & L(X_1, X_{n+2}) & \cdots & L(X_1, X_{n+m}) \\ L(X_2, X_{n+1}) & L(X_2, X_{n+2}) & \cdots & L(X_2, X_{n+m}) \\ \vdots & \vdots & & \vdots \\ L(X_n, X_{n+1}) & L(X_n, X_{n+2}) & \cdots & L(X_n, X_{n+m}) \end{pmatrix}. \tag{7.16}$$

The rows of this matrix are simply the row vectors $\vec{L}(X_j)$. For our example, the log gain matrix is

$$\mathbf{L}(\mathbf{X}_D, \mathbf{X}_I) = \begin{pmatrix} 0.1 & -0.8 \\ 0.2 & 0.4 \end{pmatrix}. \tag{7.17}$$

Using our insight from Eq. (7.13), we can express $\mathbf{L}(\mathbf{X}_D, \mathbf{X}_I)$ in terms of the coefficient matrices $\mathbf{A}_D^{-1}$ and $\mathbf{A}_I$ as

$$\mathbf{L}(\mathbf{X}_D, \mathbf{X}_I) = -\mathbf{A}_D^{-1}\mathbf{A}_I \tag{7.18}$$

(see Exercise 3). This equation is another illustration of the power and beauty of matrix notation. It relates overall properties (gains) of the system to its underlying local properties (kinetic orders g_{ij} and h_{ij} that constitute the matrices $\mathbf{A}_D^{-1}$ and $\mathbf{A}_I$).

The log gain matrix can be visualized as a pseudo-three-dimensional graph (cf. Fig. 7.2). The figure shows the gains of a model of glycolysis that is discussed in Chapter 11. Some of the gains are positive and some negative. Depending on how one slices the graph, one sees the effects of one particular independent variable on all dependent variables, or the effects on one particular dependent variable as they are exerted by the various independent variables. The figure was produced by copying (selecting and right-clicking) the sensitivities from PLAS into Excel©, creating a "3-D perspective column chart," and doing some cosmetics in PowerPoint© (see Exercise 4).

SENSITIVITY ANALYSIS

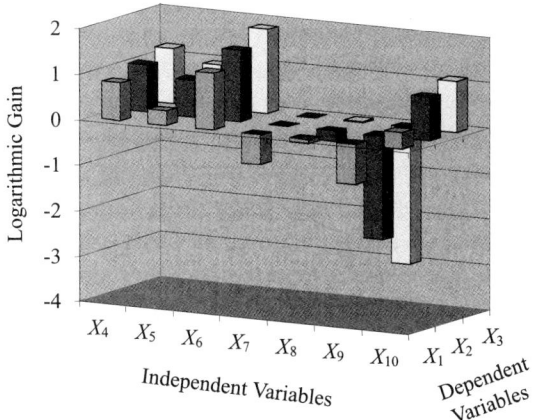

Figure 7.2 (see Color Plate X). Logarithmic gain profile of a model of glycolysis discussed in Chapter 11. Note the difference in magnitude between the effects of X_8 and X_9.

Logarithmic Gain of a Flux

The flux through the dependent variable X_i at steady state is given as the α-term of the ith equation. Because the system is at steady state, it is equivalent to the β-term of the same equation:

$$(\text{Steady-state flux through } X_i) = V_i^+ = \alpha_i \prod_{j=1}^{n+m} X_j^{g_{ij}}$$

$$= V_i^- = \beta_i \prod_{j=1}^{n+m} X_j^{h_{ij}} \quad \text{for} \quad i = 1, 2, \ldots, n, \tag{7.19}$$

where all terms are evaluated at the steady state. Taking logarithms, the first equation of (7.19) becomes a linear equation that expresses the logarithm of the flux in terms of the logarithms of the dependent and the independent variables:

$$\ln V_i^+ = \ln \alpha_i + \sum_{j=1}^{n+m} g_{ij} \ln X_j$$

$$= \ln \alpha_i + \sum_{j=1}^{n+m} g_{ij} y_j, \tag{7.20}$$

where again $y_j = \ln X_j$ ($j = 1, \ldots, m+n$). Given this equation, it is easy to revisit the question "how do changes in an independent variable affect a flux?" because computing a change in $\ln V_i^+$ with respect to a change in y_k corresponds to simple differentiation.

A change in an independent variable, e.g., in y_k, has two effects. The first one is a *direct* effect of y_k on the flux V_i^+ and the other one is an *indirect* effect that is caused by responses of the other metabolites to the change in y_k. It is convenient to split up the sum in the Eq. (7.20) in such a way that the different contributions to the total effect become explicit:

$$\ln V_i^+ = \ln \alpha_i + \sum_{j=1}^{n} g_{ij} y_j + g_{ik} y_k + \sum_{j=n+1, j\neq k}^{n+m} g_{ij} y_j. \quad (7.21)$$

When we differentiate $\ln V_i^+$ with respect to y_k, the rate constant term $\ln \alpha_i$ disappears. Furthermore, the second sum on the right-hand side disappears entirely, since the independent variables $y_{n+1}, \ldots, y_{n+m}$ are unaffected by a change in y_k. The direct affect of y_k on $\ln V_i^+$, given by $g_{ik} y_k$ in Eq. (7.21), simply contributes g_{ik} to the derivative. Finally, we need to consider the first sum. This term accounts for the dependent variables, which themselves are affected by the change in y_k. The derivative of each term $g_{ij} y_j$ with respect to y_k equals g_{ij} times the change in y_j caused by the change in y_k. This change is thus given as $g_{ij} \, \partial y_j / \partial y_k$, where the partial derivative is equal to the definition of the log gain $L_{jk} = L(X_j, X_k)$, as we discussed before. The total change in flux is therefore

Change in $\ln V_i^+$ with respect to a change in y_k
= logarithmic gain of flux V_i^+ with respect to X_k

$$= L(V_i^+, X_k) = g_{ik} + \sum_{j=1}^{n} g_{ij} L_{jk}. \quad (7.22)$$

Since an absolute change in the logarithm of a quantity corresponds to a relative change in the quantity itself, we can also state: A relative (small) change in a variable X_k effects a relative (small) change in the flux V_i^+ whose magnitude is composed of two contributions: a direct effect whose magnitude is the same as the kinetic order g_{ik}, plus the sum of indirect effects which are given as products of kinetic orders and log gains of metabolites.

For our previous example [Fig. 7.1, Eq. (7.1)], the logarithmic gain of flux V_1^+ with respect to X_3 is

$$L(V_1^+, X_3) = g_{13} + \sum_{j=1}^{2} g_{1j} L_{j3}$$
$$= 0.5 + (-2)0.2 = 0.1. \quad (7.23)$$

Thus, if X_3 is increased by 1%, then the steady-state flux into the metabolite pool X_1 increases by approximately 0.1%. The result can be checked in the steady-state dialog box in PLAS. As before, gains are defined for changes in logarithms, but direct predictions for moderate changes in the original variables and fluxes are usually acceptable. For instance, if the independent variable X_3 is increased by 10% (X_3 is set from 4.0 to 4.4), Eq. (7.23) predicts a 1% change in V_1^+ from a baseline level of 0.57435 to the value 0.58009. The steady-state flux that is actually observed in a simulation with PLAS is 0.57985, which corresponds to a change of 0.96%.

The change in the flux V_i^- caused by a relative change in an independent variable X_k is given in an analogous fashion as

$$L(V_i^-, X_k) = h_{ik} + \sum_{j=1}^{n} h_{ij} L_{jk}, \quad (7.24)$$

SENSITIVITY ANALYSIS

but we can also convince ourselves that this gain is equal to the gain of the influx:

$$L(V_i^-, X_k) = L(V_i^+, X_k). \tag{7.25}$$

For instance, without evoking any mathematical theorems, one can argue that before and after the perturbation the system is in a steady state and that therefore a change in V_i^+ must be accompanied by a concomitant change in V_i^- of the same magnitude.

In the previous example [Fig. 7.1, Eq. (7.1)], the logarithmic gain of flux V_2^- with respect to X_4 is

$$L(V_2^-, X_4) = L(V_2^+, X_4) = -0.8. \tag{7.26}$$

Most fluxes depend on more than one independent variable, and we can use vector and matrix notation to summarize all these dependences. First, we define $\vec{L}(V_i^+)$ as the row vector with the components $L(V_i^+, X_k)$:

$$\vec{L}(V_i^+) = \left(L(V_i^+, X_{n+1}), L(V_i^+, X_{n+2}), \ldots, L(V_i^+, X_{n+m})\right). \tag{7.27}$$

Second, we can further condense information when we look at all fluxes V_i^+ at the same time. In this case, we have n fluxes, one for each dependent variable, and m independent variables in the system. Thus, we obtain a $n \times m$ matrix $\mathbf{L}(\mathbf{V}^+, \mathbf{X})$ of the form

$$\mathbf{L}(\mathbf{V}^+, \mathbf{X}) = \begin{pmatrix} L(V_1^+, X_{n+1}) & L(V_1^+, X_{n+2}) & \cdots & L(V_1^+, X_{n+m}) \\ L(V_2^+, X_{n+1}) & L(V_2^+, X_{n+2}) & \cdots & L(V_2^+, X_{n+m}) \\ \vdots & \vdots & & \vdots \\ L(V_n^+, X_{n+1}) & L(V_n^+, X_{n+2}) & \cdots & L(V_n^+, X_{n+m}) \end{pmatrix}. \tag{7.28}$$

Each element $L(V_i^+, X_k)$ of this matrix is a log gain of a flux that can be written in terms of kinetic orders, g_{ij} or h_{ij}, and log gains of metabolites, L_{jk}, as described in Eq. (7.22). When we separate the contributions to the flux gain into direct influences and indirect influences, as we did in the derivation of Eqs. (7.21) and (7.22), we can represent the matrix of flux gains in form of the matrix equation

$$\mathbf{L}(\mathbf{V}^+, \mathbf{X}) = \begin{pmatrix} g_{1\,n+1} & g_{1\,n+2} & \cdots & g_{1\,n+m} \\ g_{2\,n+1} & g_{2\,n+2} & \cdots & g_{2\,n+m} \\ \vdots & \vdots & & \vdots \\ g_{n\,n+1} & g_{n\,n+2} & \cdots & g_{n\,n+m} \end{pmatrix} + \begin{pmatrix} g_{11} & g_{12} & \cdots & g_{1n} \\ g_{21} & g_{22} & \cdots & g_{2n} \\ \vdots & \vdots & & \vdots \\ g_{n1} & g_{n2} & \cdots & g_{nn} \end{pmatrix} \mathbf{L}(\mathbf{X}_D, \mathbf{X}_I). \tag{7.29}$$

In this equation, the first matrix on the right-hand side contains the kinetic orders $g_{i,n+k}$ that represent the direct effects of independent variables on the influx of the dependent variables X_i. The second matrix represents the kinetic orders g_{ij} that represent the direct effects of dependent variables on the influx into other dependent variables X_i. The third matrix, $\mathbf{L}(\mathbf{X}_D, \mathbf{X}_I)$, contains the log gains that quantify how a persistent change in an independent variable evokes a change in the steady-state concentration of a dependent metabolite. See also Exercise 5.

The flux gains of effluxes are given in a similar fashion, with g-parameters replaced by h-parameters. The result is

$$\mathbf{L}(\mathbf{V}^-, \mathbf{X}) = \begin{pmatrix} L(V_1^-, X_{n+1}) & L(V_1^-, X_{n+2}) & \cdots & L(V_1^-, X_{n+m}) \\ L(V_2^-, X_{n+1}) & L(V_2^-, X_{n+2}) & \cdots & L(V_2^-, X_{n+m}) \\ \vdots & \vdots & & \vdots \\ L(V_n^-, X_{n+1}) & L(V_n^-, X_{n+2}) & \cdots & L(V_n^-, X_{n+m}) \end{pmatrix}$$

$$= \begin{pmatrix} h_{1\,n+1} & h_{1\,n+2} & \cdots & h_{1\,n+m} \\ h_{2\,n+1} & h_{2\,n+2} & \cdots & h_{2\,n+m} \\ \vdots & \vdots & & \vdots \\ h_{n\,n+1} & h_{n\,n+2} & \cdots & h_{n\,n+m} \end{pmatrix} + \begin{pmatrix} h_{11} & h_{12} & \cdots & h_{1n} \\ h_{21} & h_{22} & \cdots & h_{2n} \\ \vdots & \vdots & & \vdots \\ h_{n1} & h_{n2} & \cdots & h_{nn} \end{pmatrix} \mathbf{L}(\mathbf{X}_D, \mathbf{X}_I).$$

(7.30)

Because of the equivalence between log gains of influxes and effluxes [cf. Eq. (7.25)], the matrices in (7.28) and (7.30) are the same:

$$\mathbf{L}(\mathbf{V}^+, \mathbf{X}) = \mathbf{L}(\mathbf{V}^-, \mathbf{X}) \tag{7.31}$$

Continuing the above example, we determine the flux gain matrix as

$$\mathbf{L}(\mathbf{V}^+, \mathbf{X}) = \begin{pmatrix} 0.1 & -0.8 \\ 0.1 & -0.8 \end{pmatrix}. \tag{7.32}$$

Note that the two rows in this matrix are the same. This is not generally the case. It is not a coincidence either, but reflects the fact that in our particular application the flux out of X_1 is identical with the flux into X_2.

Logarithmic Gain with Respect to the Transition Time

The transition time τ is the average time that a molecule remains in a pathway. As discussed in Chapter 6, τ provides a measure for the throughput, or efficiency, of the system. Following Easterby (1981), we define, for a pathway with one output V_n^-, the transition time as the sum of all metabolites at steady state, divided by the steady-state efflux. In mathematical notation we can write

$$\tau = \frac{\sigma}{V_n^-}, \tag{7.33}$$

where σ is the sum of all metabolite concentrations at the steady state:

$$\sigma = X_1 + X_2 + \cdots + X_n. \tag{7.34}$$

It is rather evident that the transition time depends on the features of the entire system. In particular, it will change if any of the independent variables are changed. We can thus study the logarithmic gain of the transition time with respect to the independent variable X_k in the usual fashion as

$$L(\tau, X_k) = \frac{1}{\tau} \cdot \frac{\partial \tau}{\partial y_k} \tag{7.35}$$

(see also Cascante et al., 1991b; Torres, 1994d; and Exercise 18).

SENSITIVITY ANALYSIS

The log gain of the transition time can be computed from log gains computed before, since the transition time is a simple function of metabolite concentrations and a flux at steady state. We obtain the following:

$$L(\tau, X_k) = \frac{\partial \ln \tau}{\partial y_k} = \frac{\partial (\ln \sigma - \ln V_n^-)}{\partial y_k} = \frac{\partial \ln \sigma}{\partial y_k} - \frac{\partial V_n^-}{\partial y_k}$$

$$= \frac{\partial \ln (X_1 + X_2 + \cdots + X_n)}{\partial y_k} - \frac{\partial V_n^-}{\partial y_k}$$

$$= \frac{\partial (X_1 + X_2 + \cdots + X_n)}{\partial X_k} \cdot \frac{X_k}{X_1 + X_2 + \cdots + X_n} - \frac{\partial V_n^-}{\partial y_k}$$

$$= \frac{\sum_{i=1}^n L_{ik} X_i}{\sigma} - \frac{\partial V_n^-}{\partial y_k}. \quad (7.36)$$

All log gains of the transition time with respect to independent variables can be expressed summarily in a vector:

$$\vec{L}(\tau, X_I) = \left(\frac{X_1}{\sigma}, \frac{X_2}{\sigma}, \ldots, \frac{X_n}{\sigma}\right) \mathbf{L}(\mathbf{X}_D, \mathbf{X}_I) - \vec{L}(V_n^-, X_I). \quad (7.37)$$

The confirmation of this result is left as Exercise 6; see also Exercise 7.

Sensitivity of a Metabolite Concentration

Mutations and diseases can permanently alter the kinetic properties and capabilities of a biochemical system. Such alterations are reflected in canonical models as numerical changes in one or some of the system parameters. As an example, the sudden availability of an enzyme with increased activity may translate into an increase in one or several of the rate constants. The effects of permanent changes in parameters are studied with methods of sensitivity analysis. In the previous sections, we studied persistent changes in the numerical values of independent variables. Here we discuss how permanent changes in parameters affect the steady state of a system. For mathematical reasons, it is convenient to distinguish (i) sensitivities of steady-state concentrations of metabolites from sensitivities of fluxes, and (ii) sensitivities with respect to changes in rate constants from those in kinetic orders. Sensitivities of a system at steady state can be analyzed with PLAS, and they can also be evaluated analytically with elementary methods of differentiation.

Before we proceed, it is important to issue a warning about the use of parameter sensitivities. Unless explicitly stated, the following sensitivity computations require that the parameter of interest be independent of the remaining parameters in the system. As soon as a parameter is functionally dependent upon other parameters, it is necessary to make all dependences explicit before parameter sensitivities can be computed. In the beginning, we will assume that all parameters of interest are independent of each other, but we shall see in a later section how to treat systems with constraints among the parameters.

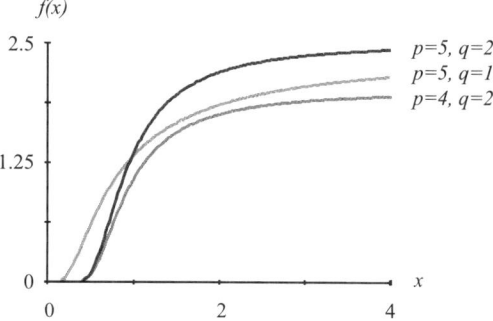

Figure 7.3. Function $f(x)$ under normal conditions ($p = 5, q = 2$) and under two sets of altered conditions ($p = 5, q = 1$ and $p = 4, q = 2$).

In order to obtain a first impression of sensitivity analysis, consider a simple function $f(x)$ with two parameters p and q. For instance,

$$f(x) = \frac{p}{1 + \exp(x^{-q})}. \tag{7.38}$$

Suppose typical parameter values are $p = 5$ and $q = 2$, and the nominal value of x is 2. In other words, we are primarily interested in the function f for values of x close to 2. The question asked in sensitivity analysis is how f is affected by alterations in p or q. Figure 7.3 shows the function under normal and two altered conditions.

Maybe the simplest means of assessing the effects of changes in either p or q is to fix x at its nominal value of 2 and to consider f as a function of either p or q. Figures 7.4 and 7.5 display the corresponding graphs. The figures demonstrate that f at the nominal value $x = 2$ depends linearly on p and nonlinearly on q. The slope of the graphs furthermore provide a measure of the degree to which f is affected: the steeper the slope, the more dramatic the change in f in response to an alteration in the parameter. In the linear case of Fig. 7.4, the slope is sufficient to make predictions for any alterations in p. Typically, however, the dependence is nonlinear, as in the case of q. In this case, the slope of the tangent, shown as a dotted line in Fig. 7.5, quantifies the effect of altering q, as long as q does not deviate too far from its original value of 2. Thus, even without producing graphs like Figs. 7.4 and 7.5, computation of the derivative of f with respect to p or q yields an immediate impression for the sensitivity of f with respect to one of these parameters. Indeed, differentiating f with

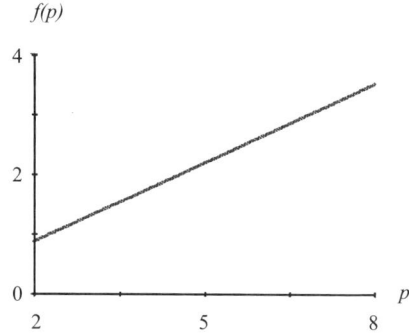

Figure 7.4. To assess the effects of altering p, the quantities x and q are held at their nominal value of 2, and f is considered a function of p over some relevant range. In this particular case, the dependence is linear. The slope of the graph indicates how strongly f reacts to an alteration in p.

SENSITIVITY ANALYSIS

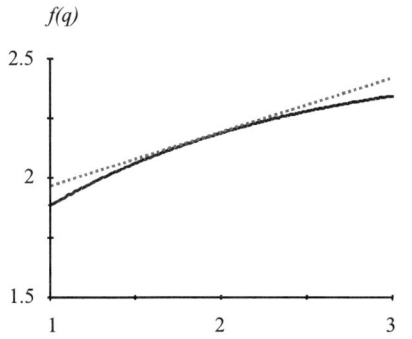

Figure 7.5. Again, x is held at its nominal value of 2. This time, p is held at 5, and f is considered a function of q over some relevant range. The dependence is nonlinear, but the slope of the tangent (dotted) at the original value ($q = 2$) indicates how strongly f reacts to a (small) alteration in q.

respect to p, holding $x = 2$ and $q = 2$, produces the slope of the graph in Fig. 7.4 as about 0.438, and differentiating f with respect to q, holding $x = 2$ and $p = 5$, yields the slope of the dotted line in Fig. 7.5 as about 0.213.

As a variation of this concept, one may compute the *relative change* of f in response to a *relative change* in p or q. The advantage of this variant is that the resulting sensitivities are scaled with respect to the absolute magnitudes of the parameters and of f, which facilitates their comparison.

The same concepts apply if the function of interest is multivariate, thus depending on variables $X_1, X_2, \ldots, X_n$. Again, the function is differentiated with respect to the parameter of interest at some value of choice. For the sensitivities in the above example, this value was chosen to be $x = 2$. In the multivariate case, the corresponding point consists of values for all variables. Typically, the function of interest is one component of the steady-state solution, such as X_{iS} or its logarithm y_{iS}. Then y_{iS} is a function of the independent variables and parameters, and this function is assessed at the steady state.

Sensitivities with respect to rate constants. The question to be studied is *how does a relative change in a rate constant affect the steady-state concentration of a metabolite?* As in the case of log gains, the analysis of the sensitivities $S(X_i, \alpha_j)$ and $S(X_i, \beta_j)$ begins with the steady-state solution of the system and requires the computation of (partial) derivatives. We define

$$S(X_i, \alpha_j) = \frac{\partial \ln X_i}{\partial \ln \alpha_j} \tag{7.39}$$

and

$$S(X_i, \beta_j) = \frac{\partial \ln X_i}{\partial \ln \beta_j}. \tag{7.40}$$

Before we discuss the computation of sensitivities in general, we illustrate the procedure with the simple pathway in Fig. 7.6. The S-system representation of the pathway is

$$\begin{aligned}\dot{X}_1 &= \alpha_1 X_0^{g_{10}} X_2^{g_{12}} - \beta_1 X_1^{h_{11}}, \\ \dot{X}_2 &= \alpha_2 X_0^{g_{20}} - \beta_2 X_2^{h_{22}},\end{aligned} \tag{7.41}$$

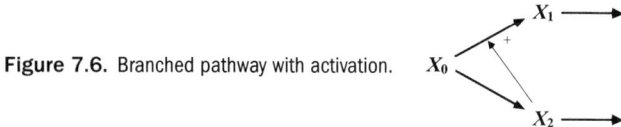

Figure 7.6. Branched pathway with activation.

and with reasonable parameter values it may read

$$\dot{X}_1 = 0.5 X_0 X_2^2 - 2 X_1^{0.5},$$
$$\dot{X}_2 = 2 X_0 - X_2^{0.5}, \quad (7.42)$$
$$X_0 = 1.$$

The steady-state solution of the system (7.41) in symbolic form is, according to Cramer's rule (see Appendix),

$$y_1 = \frac{|\mathbf{A}_1|}{|\mathbf{A}|} = \frac{\begin{vmatrix} b_1 & a_{12} \\ b_2 & a_{22} \end{vmatrix}}{\begin{vmatrix} a_{11} & a_{12} \\ a_{21} & a_{22} \end{vmatrix}} = \frac{1}{0.25} \begin{vmatrix} \ln(b_1/a_1) & 2 \\ \ln(b_2/a_2) & -0.5 \end{vmatrix}$$

$$= 4[-0.5 \ln(\beta_1/\alpha_1) - 2 \ln(\beta_2/\alpha_2)], \quad (7.43)$$

$$y_2 = \frac{|\mathbf{A}_2|}{|\mathbf{A}|} = \frac{\begin{vmatrix} a_{11} & b_1 \\ a_{21} & b_2 \end{vmatrix}}{\begin{vmatrix} a_{11} & a_{12} \\ a_{21} & a_{22} \end{vmatrix}} = \frac{1}{0.25} \begin{vmatrix} -0.5 & \ln(b_1/a_1) \\ 0 & \ln(b_2/a_2) \end{vmatrix}$$

$$= 4[-0.5 \ln(\beta_2/\alpha_2)]. \quad (7.44)$$

We could substitute numerical values for the rate constant parameters and compute the numerical values of y_1 and y_2, but for sensitivity calculations we need to retain the solution in a symbolic form that shows how y_1 and y_2 *functionally* depend on the rate constant parameters.

Suppose we would like to know how the steady-state concentration of X_1 changes if the rate of production of X_2 is altered. Thus we are interested in the effect of α_2 on X_1. The appropriate sensitivity is

$$S(X_1, \alpha_2) = \frac{\partial y_1}{\partial \ln \alpha_2}. \quad (7.45)$$

This sensitivity is readily computed when we differentiate Eq. (7.43) with respect to $\ln \alpha_2$ while keeping all other rate constants fixed. Writing $\ln(\beta_2/\alpha_2)$ as $\ln \beta_2 - \ln \alpha_2$, we see that upon differentiation only the constants 4 and 2 are left, and the result is

$$S(X_1, \alpha_2) = +8. \quad (7.46)$$

Thus, if α_2 is increased by 1%, then the steady-state concentration of the metabolite X_1 increases by approximately 8%. The result is confirmed with the computation of the steady state in PLAS. As in the case of log gains, this result is strictly true only for changes in logarithms. However, the prediction is also rather accurate for small

SENSITIVITY ANALYSIS

relative changes in the parameters themselves. For instance, if the rate of production of X_2 is decreased by 2% (α_2 is set to 1.96), the change in X_1 as predicted from Eq. (7.46) is about 16%. The steady-state concentration of X_1 that is actually observed in a simulation with PLAS is 13.612, which corresponds to a change of 15%. If α_2 is increased by 5%, the predicted change in X_1 is 40%, while the observed change is about 48%.

As with logarithmic gains, we can collect all sensitivities with respect to rate constants α_i in a matrix $S(X, \alpha)$ that has the form

$$S(X, \alpha) = \begin{pmatrix} \frac{\partial y_1}{\partial \ln \alpha_1} & \frac{\partial y_1}{\partial \ln \alpha_2} \\ \frac{\partial y_2}{\partial \ln \alpha_1} & \frac{\partial y_2}{\partial \ln \alpha_2} \end{pmatrix}. \tag{7.47}$$

For the pathway in Fig. 7.6, the matrix of rate constant sensitivities is

$$S(X, \alpha) = \begin{pmatrix} 2 & 8 \\ 0 & 2 \end{pmatrix}. \tag{7.48}$$

The analogous matrix for sensitivities with respect to β's is

$$S(X, \beta) = \begin{pmatrix} \frac{\partial y_1}{\partial \ln \beta_1} & \frac{\partial y_1}{\partial \ln \beta_2} \\ \frac{\partial y_2}{\partial \ln \beta_1} & \frac{\partial y_2}{\partial \ln \beta_2} \end{pmatrix}. \tag{7.49}$$

It can also be determined from $S(X, \alpha)$ as

$$S(X, \alpha) = -S(X, \beta). \tag{7.50}$$

In other words, the sensitivity of a variable with respect to the rate constant α_j is the negative of its sensitivity with respect to β_j.

In our numerical example

$$\begin{aligned} S(X_1, \beta_2) &= \frac{\partial y_1}{\partial \ln \beta_2} \\ &= \frac{\partial \{4[-0.5 \ln(\beta_1/\alpha_1) - 2 \ln(\beta_2/\alpha_2)]\}}{\partial \ln \beta_2} \\ &= -8 \\ &= -S(X_1, \alpha_2). \end{aligned} \tag{7.51}$$

Like logarithmic gains, rate constant sensitivities can be displayed in a pseudo-three-dimensional graph like Fig. 7.7. The default for such a display is again a profile of unconstrained sensitivities. See also Exercise 8.

It is not too difficult to show (cf. Savageau, 1972; Savageau et al., 1987ab) that the sensitivity matrices $S(X, \alpha)$ and $S(X, \beta)$ are directly related to the steady-state equation of an S-system. Specifically, $S(X, \beta)$ is equal to the inverse A_D^{-1} of the portion of the system matrix A that corresponds to the dependent variables. For the case that there are no independent variables, one therefore finds

$$\vec{y}_D = S(X, \beta) \cdot \vec{b} = -S(X, \alpha) \cdot \vec{b} \tag{7.52}$$

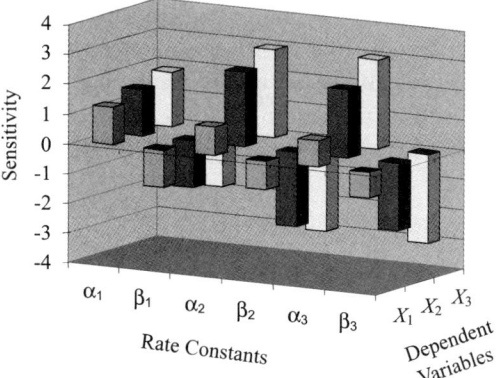

Figure 7.7 (see Color Plate XI). Unconstrained rate constant sensitivities of the model of glycolysis discussed in Chapter 11.

[cf. Eq. (6.42)], and if the system has independent variables, the steady state is characterized by the matrix equation

$$\vec{y}_D = S(X, \beta) \cdot \vec{b} + L(X_D, X_I) \cdot \vec{y}_I. \tag{7.53}$$

These relationships between the concentrations of the dependent and independent variables on one hand, and the system properties of sensitivities and gains on the other, constitute *the* fundamental representation of the steady state of an S-system that explicitly or implicitly incorporates all properties of the steady state.

In order to become more familiar with sensitivities, it is instructive to prove the relationship (7.50) mathematically. Again, we start out with the solution of the steady-state matrix equation

$$A\vec{y} = \vec{b}. \tag{7.54}$$

According to Cramer's rule, this solution has the components

$$y_i = \frac{\det A_i}{\det A}$$

$$= \frac{\begin{vmatrix} a_{11} & a_{12} & a_{13} & \cdots & b_1 & \cdots & a_{1n} \\ a_{21} & a_{22} & a_{23} & \cdots & b_2 & \cdots & a_{2n} \\ \vdots & \vdots & \vdots & & \vdots & & \vdots \\ a_{j1} & a_{j2} & a_{j3} & \cdots & b_j & \cdots & a_{jn} \\ \vdots & \vdots & \vdots & & \vdots & & \vdots \\ a_{n1} & a_{n2} & a_{n3} & \cdots & b_n & \cdots & a_{nn} \end{vmatrix}}{\begin{vmatrix} a_{11} & a_{12} & a_{13} & \cdots & a_{1i} & \cdots & a_{1n} \\ a_{21} & a_{22} & a_{23} & \cdots & a_{2i} & \cdots & a_{2n} \\ \vdots & \vdots & \vdots & & \vdots & & \vdots \\ a_{j1} & a_{j2} & a_{j3} & \cdots & a_{ji} & \cdots & a_{jn} \\ \vdots & \vdots & \vdots & & \vdots & & \vdots \\ a_{n1} & a_{n2} & a_{n3} & \cdots & a_{ni} & \cdots & a_{nn} \end{vmatrix}} \tag{7.55}$$

SENSITIVITY ANALYSIS

(cf. Appendix). The rate constants α_j and β_j appear on the right-hand side in the form of the parameter b_j, which by definition equals $b_j = \ln(\beta_j/\alpha_j) = \ln\beta_j - \ln\alpha_j$. When we replace b_j with $\ln\beta_j - \ln\alpha_j$, Eq. (7.55) becomes

$$y_i = \frac{\begin{vmatrix} a_{11} & a_{12} & a_{13} & \cdots & b_1 & \cdots & a_{1n} \\ a_{21} & a_{22} & a_{23} & \cdots & b_2 & \cdots & a_{2n} \\ \vdots & \vdots & \vdots & & \vdots & & \vdots \\ a_{j1} & a_{j2} & a_{j3} & \cdots & \ln\beta_j - \ln\alpha_j & \cdots & a_{jn} \\ \vdots & \vdots & \vdots & & \vdots & & \vdots \\ a_{n1} & a_{n2} & a_{n3} & \cdots & b_n & \cdots & a_{nn} \end{vmatrix}}{|\mathbf{A}|}. \tag{7.56}$$

Without actually executing the operation, let's imagine we expand the determinant $|\mathbf{A}_i|$ in the numerator. The result will be pretty complicated, but it is rather easy to intuit that the expression $\ln\beta_j - \ln\alpha_j$ always remains intact and that each term in the end will either contain the entire expression $\ln\beta_j - \ln\alpha_j$ as a multiplicative factor or not at all. Therefore, we can write symbolically

$$y_i = C_1 + (\ln\beta_j - \ln\alpha_j) \cdot C_2, \tag{7.57}$$

where C_1 and C_2 are sums of products of the coefficients a_{ij} and of solution coefficients $b_i \neq b_j$, which are all independent of $\ln\beta_j$ and $\ln\alpha_j$. When we now differentiate y_i with respect to $\ln\alpha_j$, the result is the sensitivity of X_i with respect to α_j,

$$S(X_i, \alpha_j) = -C_2, \tag{7.58}$$

and when we differentiate with respect to $\ln\beta_j$, the result is

$$S(X_i, \beta_j) = +C_2. \tag{7.59}$$

Thus, we have proven that the sensitivities of a variable with respect to a rate constant α_j and to a rate constant β_j are the same in magnitude but opposite in sign, as proposed in Eq. (7.50).

The proof reiterates another interesting result. It shows that sensitivities with respect to rate constants are functions of the entire system, since the terms in C_2 are composed of potentially all kinetic orders (recall that $a_{ij} = g_{ij} - h_{ij}$) and of potentially all rate constants [recall that $b_i = \ln(\beta_i/\alpha_i)$]. In other words, if we want to study the effect of changes in the rate of one particular process it is not sufficient to focus on just this process or on processes that are "close by" on the biochemical map, but the entire system must be included in the analysis.

Systems with constraints. If parameters depend on each other, their mathematical relationships must be made explicit before sensitivities can be validly computed. As an example, let's return to the simple pathway with feedback [Fig. 6.1; Eqs. (6.1) and (6.2)] that we have used as an example several times. The steady-state solution of the system (6.1) in symbolic form is

$$y_1 = \frac{|\mathbf{A}_1|}{|\mathbf{A}|} = \frac{\begin{vmatrix} b_1 & a_{12} \\ b_2 & a_{22} \end{vmatrix}}{\begin{vmatrix} a_{11} & a_{12} \\ a_{21} & a_{22} \end{vmatrix}} = \frac{1}{2.5}\begin{vmatrix} \ln(\beta_1/\alpha_1) & -2 \\ \ln(\beta_2/\alpha_2) & -0.5 \end{vmatrix}$$

$$= 0.4[-0.5\ln(\beta_1/\alpha_1) + 2\ln(\beta_2/\alpha_2)], \qquad (7.60)$$

$$y_2 = \frac{|\mathbf{A}_2|}{|\mathbf{A}|} = \frac{\begin{vmatrix} a_{11} & b_1 \\ a_{21} & b_2 \end{vmatrix}}{\begin{vmatrix} a_{11} & a_{12} \\ a_{21} & a_{22} \end{vmatrix}} = \frac{1}{2.5}\begin{vmatrix} -1 & \ln(\beta_1/\alpha_1) \\ 1 & \ln(\beta_2/\alpha_2) \end{vmatrix}$$

$$= 0.4[-\ln(\beta_2/\alpha_2) - \ln(\beta_1/\alpha_1)]. \qquad (7.61)$$

Suppose we are interested in the effect of α_1 on X_2. The appropriate sensitivity is defined as

$$S(X_2, \alpha_1) = \frac{\partial y_2}{\partial \ln \alpha_1}, \qquad (7.62)$$

which is readily computed from Eq. (7.61) when we differentiate y_2 with respect to $\ln \alpha_1$ while keeping all other rate constants fixed. This is legitimate because α_1 is independent of all other parameters. The result is

$$S(X_2, \alpha_1) = +0.4. \qquad (7.63)$$

Now let's compute the sensitivity of X_2 with respect to α_2. It is not legitimate to treat all other parameters as constant, since the precursor–product relationship between X_1 and X_2 implies that $\alpha_2 = \beta_1$ and $g_{21} = h_{11}$. We make this constraint explicit by rewriting Eq. (7.61) as

$$y_2 = 0.4[-\ln(\beta_2/\alpha_2) - \ln(\alpha_2/\alpha_1)]. \qquad (7.64)$$

Differentiation of y_2 with respect to $\ln \alpha_2$ yields

$$S(X_2, \alpha_2) = \frac{\partial y_2}{\partial \ln \alpha_2} = 0.4(1-1) = 0. \qquad (7.65)$$

Thus, a change in the rate of degradation of X_1 to X_2 does not affect the steady-state concentration of X_2. (Explain biochemically why this is so.)

PLAS does not automatically allow for constraints between rate constants, but it can be forced to do so if the constraints are made explicit. This is accomplished by using a symbol like a2 (rather than the numerical value 2) wherever the rate constant appears, namely in the β-term of the first equation and the α-term of the second equation. Also, one adds the definition line $a2 = 2$ as well as a line that identifies a2 as an independent variable:

&& a2

Calling up the sensitivities of the system, PLAS displays first the log gains (which in this case are constrained rate constant sensitivities) with respect to a2, namely

−1.000 and 0.000, and then the unconstrained parameter sensitivities, which for α_2 are given as −0.8 and +0.4 (see Exercise 10).

The PLAS procedure alludes to a close relationship between rate constants and independent variables and, thereby, between rate constant sensitivities and log gains. Although rate constants and independent variables conceptually address different aspects of a system – such as environmental effectors as opposed to intrinsic parts of the structure – they have mathematically a lot in common. Both are constant during each experiment, and we saw in Chapter 6 that it is mathematically indifferent whether we keep an independent variable explicit or whether we merge it with the rate constant. In fact, each rate constant could be called an independent variable, and rate constant sensitivities would mathematically become logarithmic gains. However, as said before, for conceptual interpretation it is useful to distinguish the two. (See also Exercise 11.)

It is noted that sensitivities with respect to constrained parameters may be computed in a rather different fashion, using implicit differentiation. This procedure is outlined in Section C2 of the *Complements* for GMA systems, which, of course, contain the S-system structure as a special case. The results are identical, and it may be a matter of taste whether the strategy here or the strategy based on implicit differentiation is simpler. Further examples for the computation of constrained rate constant sensitivities can be found in the case studies of Chapters 9 and 11.

Sensitivities with respect to kinetic orders. The question of interest is: *how does a relative change in a kinetic order affect the steady-state concentration of a metabolite?* It is again addressed with the computation of (partial) derivatives and under the assumption of independence of parameters. If dependences do exist, they must be made explicit in the steady-state equations and taken into account when the partial derivatives are computed. In analogy to rate constant sensitivities, sensitivities with respect to kinetic orders are defined as

$$S(X_k, g_{ij}) = \frac{\partial \ln X_k}{\partial \ln g_{ij}} = \frac{\partial X_k}{\partial g_{ij}} \cdot \frac{g_{ij}}{X_k} = \frac{\partial y_k}{\partial g_{ij}} \cdot g_{ij} \qquad (7.66)$$

and

$$S(X_k, h_{ij}) = \frac{\partial \ln X_k}{\partial \ln h_{ij}} = \frac{\partial X_k}{\partial h_{ij}} \cdot \frac{h_{ij}}{X_k} = \frac{\partial y_k}{\partial h_{ij}} \cdot h_{ij}. \qquad (7.67)$$

Consistent with former gains and sensitivities, these quantities are interpreted in the following way: If a kinetic order g_{ij} is changed by 1%, then the steady-state concentration of the metabolite X_k changes by approximately $S(X_k, g_{ij})$%.

As an alternative, one can define kinetic order sensitivities as relative changes in y_k (rather than X_k) that are caused by a change in a kinetic order:

$$S^*(y_k, g_{ij}) = \frac{\partial \ln y_k}{\partial \ln g_{ij}} = \frac{\partial y_k}{\partial g_{ij}} \cdot \frac{g_{ij}}{y_k} \qquad (7.68)$$

and

$$S^*(y_k, h_{ij}) = \frac{\partial \ln y_k}{\partial \ln h_{ij}} = \frac{\partial y_k}{\partial h_{ij}} \cdot \frac{h_{ij}}{y_k}. \tag{7.69}$$

These definitions have the advantage that they can be interpreted as weighted sensitivities of the linear steady-state system (cf. Savageau, 1976: Ch. 9; Sorribas and Savageau, 1989a). These sensitivities are easily obtained from those in Eqs. (7.66) and (7.67) through division by y_k (see also Exercise 12).

As in the case of log gains and rate constant sensitivities, the analysis of the sensitivities $S(X_k, g_{ij})$ and $S(X_k, h_{ij})$ begins with the steady-state solution of the system. We illustrate the procedure again with the simple feedback-inhibited pathway [Fig. 6.1; Eqs. (6.1) and (6.2)].

The steady-state solution of the system (6.1) in symbolic form is

$$y_1 = \frac{|\mathbf{A}_1|}{|\mathbf{A}|} = \frac{\begin{vmatrix} b_1 & a_{12} \\ b_2 & a_{22} \end{vmatrix}}{\begin{vmatrix} a_{11} & a_{12} \\ a_{21} & a_{22} \end{vmatrix}} = \frac{b_1(g_{22} - h_{22}) - b_2(g_{12} - h_{12})}{(g_{11} - h_{11})(g_{22} - h_{22}) - (g_{12} - h_{12})(g_{21} - h_{21})},$$

$$\tag{7.70}$$

$$y_2 = \frac{|\mathbf{A}_2|}{|\mathbf{A}|} = \frac{\begin{vmatrix} a_{11} & b_1 \\ a_{21} & b_2 \end{vmatrix}}{\begin{vmatrix} a_{11} & a_{12} \\ a_{21} & a_{22} \end{vmatrix}} = \frac{b_2(g_{11} - h_{11}) - b_1(g_{21} - h_{21})}{(g_{11} - h_{11})(g_{22} - h_{22}) - (g_{12} - h_{12})(g_{21} - h_{21})}.$$

$$\tag{7.71}$$

Suppose we are interested in the effect of g_{12} on X_2. The appropriate sensitivity is defined as

$$S(X_2, g_{12}) = \frac{\partial y_2}{\partial g_{12}} \cdot g_{12}. \tag{7.72}$$

It is helpful to rewrite Eq. (7.71) so that terms including g_{12} are separated from those independent of g_{12}, i.e.,

$$y_2 = \frac{b_2(g_{11} - h_{11}) - b_1(g_{21} - h_{21})}{g_{12}(h_{21} - g_{21}) + (g_{11} - h_{11})(g_{22} - h_{22}) + h_{12}(g_{21} - h_{21})} \tag{7.73}$$

$$= \frac{C_1}{g_{12}(h_{21} - g_{21}) + C_2}, \tag{7.74}$$

where $C_1 = -(\ln 5 - \ln 2) = 0.91629$, $C_2 = 0.5$, and $h_{21} - g_{21} = -1$ are independent of the quantity of interest, g_{12}. The derivative of y_2 with respect to g_{12} is

$$\frac{\partial y_2}{\partial g_{12}} = \frac{-C_1(h_{21} - g_{21})}{[g_{12}(h_{21} - g_{21}) + C_2]^2}, \tag{7.75}$$

SENSITIVITY ANALYSIS

which for $g_{12} = -2$ yields

$$\frac{\partial y_2}{\partial g_{12}} = \frac{-0.91619}{2.5^2} = -0.14661. \tag{7.76}$$

The sensitivity in question therefore is

$$S(X_2, g_{12}) = -0.14661 g_{12} = 0.29321. \tag{7.77}$$

Thus, if g_{12} is changed by 1%, then the steady-state concentration of the metabolite X_2 increases by approximately 0.3%. The result is confirmed by results under the *Sensitivities* tab in PLAS. Again, we can compare the predictions with actual numerical results of moderately large changes. For instance, if the feedback inhibition is increased in magnitude by 10% (g_{12} is set to -2.2), then the change in X_2 as predicted from Eq. (7.77) is about 2.93%, and the steady-state concentration of X_2 that is actually observed in a simulation with PLAS is 0.71222, which corresponds to a change of 2.75% (see also Exercise 13).

With mathematical arguments similar to those in Eqs. (7.51)–(7.59) one can show that there is a simple relationship between sensitivities with respect to changes in corresponding g and h parameters. In this case, the result is

$$\frac{S(X_i, g_{jk})}{g_{jk}} = -\frac{S(X_i, h_{jk})}{h_{jk}} \tag{7.78}$$

for all relevant indices i, j, and k.

As before, sensitivities with respect to kinetic orders may be displayed in a pseudo-three-dimensional graph (Fig. 7.8).

As mentioned before, sensitivities with respect to constrained parameters may be computed through implicit differentiation. The procedure is outlined in Section 2 of the *Complements*.

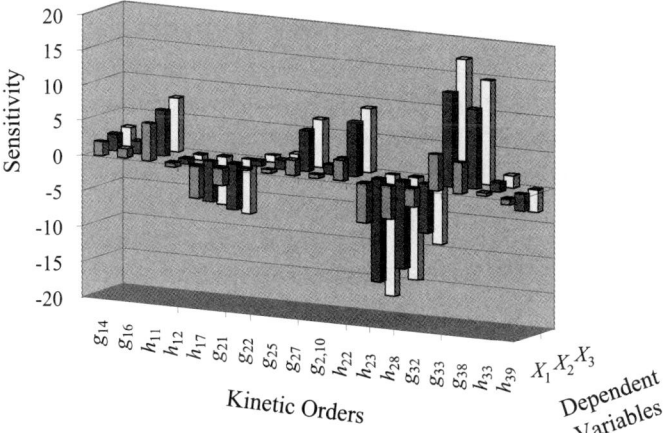

Figure 7.8 (see Color Plate XII). Unconstrained kinetic order sensitivities of the model of glycolysis discussed in Chapter 11. Unconstrained, some of the kinetic orders appear to be high, but if constraints are taken into account, they are within a reasonable range. For instance, the highest sensitivities shown are associated with h_{2i} and g_{3i} ($i = 2, 3, 8$), which in fact describe the same process and cancel each other.

Sensitivity of a Flux with Respect to a Change in a Parameter

The flux through the dependent variable X_l at steady state,

$$V_l = \alpha_l \prod_{j=1}^{n+m} X_j^{g_{ij}} = \beta_l \prod_{j=1}^{n+m} X_j^{h_{ij}}, \tag{7.79}$$

is a function of the rate constants and the kinetic orders, and we can ask how strongly it is affected by these parameters. Again, the answer is given by the appropriate sensitivities

$$S(V_l, \alpha_i) = \frac{\partial \ln V_l}{\partial \ln \alpha_i} = \frac{\partial V_l}{\partial \alpha_i} \cdot \frac{\alpha_i}{V_l}, \tag{7.80}$$

$$S(V_l, \beta_i) = \frac{\partial \ln V_l}{\partial \ln \beta_i} = \frac{\partial V_l}{\partial \beta_i} \cdot \frac{\beta_i}{V_l} \tag{7.81}$$

and

$$S(V_l, g_{ij}) = \frac{\partial \ln V_l}{\partial \ln g_{ij}} = \frac{\partial V_l}{\partial g_{ij}} \cdot \frac{g_{ij}}{V_l}, \tag{7.82}$$

$$S(V_l, h_{ij}) = \frac{\partial \ln V_l}{\partial \ln h_{ij}} = \frac{\partial V_l}{\partial h_{ij}} \cdot \frac{h_{ij}}{V_l}. \tag{7.83}$$

Before we execute the differentiation, for instance to compute the sensitivity in Eq. (7.80), it is again convenient to take logarithms in Eq. (7.79):

$$\ln V_l = \ln \alpha_l + \sum_{j=1}^{n+m} g_{lj} \ln X_j. \tag{7.84}$$

As in the computation of flux gains, the differentiation here must allow for the direct effect of changes in the rate constant on the flux and for the indirect effects caused by changes in the metabolites in response to changes in the rate constant. Inspection of Eq. (7.84) shows immediately that α_i has a direct effect only if $l = i$. Furthermore, the indirect effects derive from differentiation of each term in the sum with respect to $\ln \alpha_i$, and these derivatives are exactly the sensitivities $S(X_j, \alpha_i)$ multiplied by the kinetic orders g_{ij}. Thus, the flux sensitivity in (7.80) is related to the metabolite sensitivities as

$$S(V_l, \alpha_i) = 1 + \sum_{j=1}^{n} g_{lj} S(X_j, \alpha_i) \quad \text{if} \quad l = i, \tag{7.85}$$

or as

$$S(V_l, \alpha_i) = \sum_{j=1}^{n} g_{lj} S(X_j, \alpha_i) \quad \text{if} \quad l \neq i \tag{7.86}$$

(Exercise 14). The sums in these equations do not include the independent variables, because they are unaffected by changes in rate constants and their sensitivities therefore are zero.

The sensitivities with respect to β follow analogously with g-parameters replaced by h-parameters, but as an alternative to expressing them in terms of h_{kj}, one can also express them in terms of g_{kj}, and the result is

$$S(V_l, \beta_i) = \sum_{j=1}^{n} g_{lj} S(X_j, \beta_i) \qquad (l = i \text{ or } l \neq i) \tag{7.87}$$

(see also Exercise 15).

In analogy to previous cases, flux sensitivities with respect to constrained parameters may be computed through implicit differentiation. The procedure is outlined in Section C2 of the *Complements*.

SUMMATION AND CONNECTIVITY RELATIONSHIPS

It follows from the results in Eqs. (7.50), (7.58)–(7.59), and (7.85)–(7.87) that the sum of all sensitivities of a particular flux with respect to all rate constants is always equal to 1:

$$\sum_{i=1}^{n} [S(V_l, \alpha_i) + S(V_l, \beta_i)] = 1. \tag{7.88}$$

In words, the percentage change in a given flux resulting from a 1% change in the rate constant of a net reaction or process, summed over all such reactions or processes, is always equal to 1 (cf. Savageau et al., 1987b).

The validity of this result can be seen when we substitute the definitions of the flux sensitivities in Eq. (7.88) and symbolically distinguish the two cases that the indices of the variable and the kinetic order are the same or different. We obtain

$$\sum_{i=1, i\neq l}^{n} \left(\sum_{j=1}^{n} g_{lj} S(X_j, \alpha_i) + \sum_{j=1}^{n} g_{lj} S(X_j, \beta_i) \right) + S(V_l, \alpha_l) + S(V_l, \beta_l)$$

$$= \sum_{i=1, i\neq l}^{n} \left(\sum_{j=1}^{n} g_{lj} [S(X_j, \alpha_i) + S(X_j, \beta_i)] \right)$$

$$+ 1 + \sum_{j=1}^{n} g_{lj} [S(X_j, \alpha_l) + S(X_j, \beta_l)], \tag{7.89}$$

which reduces to 1, because we know that for all i and j

$$S(X_j, \alpha_i) = -S(X_j, \beta_i). \tag{7.90}$$

Equations like Eq. (7.88) are called *summation relationships*. They are interesting consequences of the linear structure of the steady-state equations and have been studied extensively in a branch of biochemical systems analysis called *metabolic control analysis* (MCA). Since the original papers (Kacser and Burns, 1973; Heinrich and Rapoport, 1974), dozens of articles have dealt with this type of analysis. Recent texts on the subject include Cornish-Bowden and Cárdenas (1990), Heinrich and Schuster (1996), and Fell (1997); see also issue **182**(3) of the *Journal of*

Theoretical Biology (1996). The notation in MCA is different from our notation, but several articles have been written to show the equivalence of MCA and our approach (e.g., Savageau et al., 1987ab; Savageau and Sorribas, 1989; Sorribas and Savageau, 1989a; Savageau, 1991a; Cascante et al., 1995b; Curto et al., 1995; Puigjaner et al., 1995; Sorribas et al., 1995).

Summation relationships have attracted much attention because they have the appearance of conservation laws. For instance, the summation relationship for metabolite concentrations,

$$\sum_{i=1}^{n} [S(X_l, \alpha_i) + S(X_l, \beta_i)] = 0, \tag{7.91}$$

means: The percentage change in some dependent variable X_l resulting from a 1% change in the rate constant for a given net reaction or process, summed over all such reactions in the system, is always zero (Savageau et al., 1987b; see Exercise 16).

The implication is that when one changes a net rate and measures all concentration or flux sensitivities, their sums should be 0 or 1, respectively, lest some effects had not been accounted for. While this implication is true, a seemingly similar implication is not: One might be tempted to surmise that if the sum of measured concentration or flux sensitivities actually has a value of 0 or 1, respectively, one could be assured that all effects had been accounted for. This implication is faulty, since sensitivities can be positive, negative, or zero. Consequently, an entire set of sensitivities may have been overlooked if some negative and some positive sensitivities happen to add up to zero. In *homogeneous* systems, in which the enzyme activities appear as linear functions, the sensitivities are positive, and no problems arise. Since Michaelis–Menten rate laws and many other functions actually fall into this category, the character of summation relationships as conservation laws was originally claimed as entirely general (e.g., Kacser and Burns, 1981). However, this claim is not valid, for instance, in cases of enzyme–enzyme interactions. These issues have been discussed extensively in the literature, and representative examples have been developed to illustrate when the interpretation of the summation relationships as conservation laws does or does not hold (e.g., Hofmeyer et al., 1986; Cornish-Bowden, 1989; Savageau and Sorribas, 1989; Sorribas and Savageau, 1989a; Kacser et al., 1990; Meléndez-Hevia et al., 1990; Kacser, 1991; Savageau, 1992a; see also Savageau, 1991b; Shiraishi and Savageau, 1993; Kholodenko et al., 1995, 1998). Rebutting the practice of using summation relationships to prove validity and consistency of the steady state of a model, Shiraishi and Savageau (1996) caution: "Unfortunately, obtaining zeros and ones for summation properties is neither necessary nor sufficient for a valid steady-state analysis."

In addition to summation relationships, MCA has established *connectivity relationships* that show connections between rate constant sensitivities and kinetic orders. The connectivity relationships for metabolite concentrations are

$$\sum_{i=1}^{n} [S(X_l, \alpha_i)g_{ik} + S(X_l, \beta_i)h_{ik}] = -1 \quad \text{if} \quad k = l \tag{7.92}$$

and

$$\sum_{i=1}^{n}[S(X_l, \alpha_i)g_{ik} + S(X_l, \beta_i)h_{ik}] = 0 \quad \text{if} \quad k \neq l \tag{7.93}$$

(see also Exercises 17 and 18). The connectivity relationship for steady-state fluxes states

$$\sum_{i=1}^{n}[S(V_l, \alpha_i)g_{ik} + S(V_l, \beta_i)h_{ik}] = 0, \tag{7.94}$$

where the index k refers to one of the independent variables. Like summation relationships, connectivity relationships have been studied with interest because of their interpretation as conservation laws. However, the same caveat applies as above (see also Exercises 19 and 20).

All summation and connectivity relationships are direct consequences of the S-system equations at steady state (e.g., Savageau et al., 1987b). This becomes evident when we separate the dependent and independent variables in the steady-state equations of the system by formulating the solution for each dependent variable y_i as

$$\begin{aligned} y_i &= \sum_{j=1}^{n} M_{ij} b_j + \sum_{l=n+1}^{n+m} L_{il} y_l \\ &= \sum_{j=1}^{n} M_{ij}(\ln \beta_j - \ln \alpha_j) + \sum_{l=n+1}^{n+m} L_{il} y_l. \end{aligned} \tag{7.95}$$

Using this equation to compute gains and sensitivities, one can see immediately that the coefficients M_{ij} are rate constant sensitivities (since differentiation of y_i with respect to β_j yields M_{ij}) and that the coefficients L_{il} are logarithmic gains (since differentiation of y_i with respect to y_l yields L_{il}). Furthermore, the sensitivities with respect to α and with respect to β are of the same in magnitude but opposite in sign, confirming directly the summation relationship (7.91).

An Example of the Use of Summation and Connectivity Relationships in Ranking the Importance of Processes

Bohnensack and Halangk (1986) used summation and connectivity relationships in an analysis of respiration and motility in ejaculated bull spermatozoa. Motility (measured as the percentage of cells moving per minute from the bottom of the cuvette into the light path) had been considered to be the most important process of ATP turnover in spermatozoa, but it was the authors' intent to quantify to what degree ATP turnover was actually controlled at the motility (consumption) step and to what degree at the respiration (production) step. Since motility can be measured indirectly by the uptake of oxygen, the authors performed inhibition studies of both motility and respiration in the presence of lactate, which permitted high rates of aerobic ATP formation. It is generally accepted that the inhibitory effects are linked to changes in the cellular adenine nucleotide pattern, and that the cellular ATP/ADP ratio is the most important signal in the coordination of ATP turnover. Therefore, the authors designed experiments in which respiration and motility were studied as functions of this ratio, under different inhibition regimes with vanadate and cyanide.

Their strategy was to compute and compare the sensitivities of respiration and motility with respect to the ATP/ADP ratio under different conditions, since these sensitivities quantify how the control over ATP turnover is shared. The quantification of sensitivities was to be achieved by characterizing the rates of respiration (v_{resp}) and motility (v_{mot}) with appropriate rate laws, deriving kinetic orders from these rate laws, and finally evoking summation and connectivity relationships between the flux sensitivities on one hand and the ATP/ADP ratio and the involved kinetic orders on the other.

Bohnensack and Halangk suggested the following rate laws for mitochondrial respiration (v_{resp}) and motility (v_{mot}):

$$v_{resp} = \frac{V_{resp}}{1 + (X + K_{resp})^n}, \tag{7.96}$$

$$v_{mot} = \frac{V_{mot}}{1 + K_{mot}/X}. \tag{7.97}$$

In these equations, X is the cellular ATP/ADP ratio, V_{resp} and V_{mot} are the maximum velocities, K_{resp} and K_{mot} are the ATP/ADP ratios for half-maximum rates, and n is an apparent cooperativity coefficient.

Motility was measured directly with a turbidimetric method, while the rate of respiration was taken as the difference of the measured rate of total oxygen uptake and the portion that was insensitive to inhibition with vanadate or cyanide. The parameters of the rate laws were determined with a least-squares fitting routine. Bohnensack and Halangk determined V_{mot} and K_{mot} under different regimens of inhibition. They also found that motility was proportional to respiration,

$$v_{resp} = m v_{mot}, \tag{7.98}$$

with a proportionality constant m that was interpreted as the oxygen equivalent of motility (nanomoles O_2 per microliter of cells per percent of motile cells) and depended on the type of inhibition exerted. The authors concluded that both rates, v_{resp} and v_{mot}, were proportional to ATP turnover.

Once the parameters of the rate laws were determined, the kinetic orders were computed through differentiation. Following Bohnensack and Halangk, but adapting their notation to ours (e.g., calling the ATP/ADP-ratio $X_1 = X$), the kinetic order for consumption is given as

$$h_{11} = \frac{d \ln v_{mot}}{d \ln X_1} \quad \text{at an operating point } P. \tag{7.99}$$

The kinetic order for production, g_{10}, is computed analogously. The actual execution of these computations was subject of an exercise in Chapter 5.

If V_1 denotes the flux through X_1, the desired flux sensitivities are defined as $S(V_1, \alpha_1)$ for respiration and $S(V_1, \beta_1)$ for motility. These sensitivities satisfy the

SENSITIVITY ANALYSIS

summation relationship

$$S(V_1, \alpha_1) + S(V_1, \beta_1) = 1 \qquad (7.100)$$

and the connectivity relationship

$$g_{10} S(V_1, \alpha_1) + h_{11} S(V_1, \beta_1) = 0. \qquad (7.101)$$

One can solve these two linear equations in the two unknowns $S(V_1, \alpha_1)$ and $S(V_1, \beta_1)$, and the result is

$$S(V_1, \alpha_1) = \frac{g_{10}}{g_{10} - h_{11}}, \qquad (7.102)$$

$$S(V_1, \beta_1) = \frac{-h_{11}}{g_{10} - h_{11}}. \qquad (7.103)$$

Computing these quantities for different ATP/ADP ratios, Bohnensack and Halangk demonstrated that the sensitivity of the consumption step, $S(V_1, \beta_1)$, increases in a sigmoidal fashion with increase in the ratio ATP/ADP, and that the sensitivity of the respiration step decreases concomitantly. This suggests that normally both steps share control over ATP turnover. However, the production step dominates the control at low ratios of ATP, whereas utilization assumes primary control for high ratios. Other routes of ATP utilization were not included in the analysis, and in addition to lactate, metabolites like pyruvate, acetate, and glycerol can serve as substrates for the respiration step, which implies that the control of production is shared by these contributors. In spite of these limitations, the authors concluded that "the reactions of oxidative phosphorylization exert only a small control on the ATP turnover, so that the motility is nearly independent of small fluctuations in the substrate supply. This indicates the perfect adaptation of the metabolism in spermatozoa to be an active carrier of the male genome."

Numerous analyses of a similar flavor have been executed in MCA. Many of them are discussed in Fell (1997).

EXERCISES

1. Discuss the effects of simultaneous changes in two independent variables on a dependent variable. Check the results of your conjectures in PLAS.
2. Compose a table similar to Table 7.1 for the effects of changes in X_3. What can you say about the relative importance (as measured by log gains) of X_3 and X_4?
3. Confirm Eq. (7.18) numerically by computing $\mathbf{A}_D^{-1}$ and $\mathbf{A}_I$ and comparing the result with the matrix in Eq. (7.17).
4. Create a pseudo-three-dimensional graph of the log gains in Eq. (7.17).
5. Numerically confirm Eqs. (7.29) and (7.30), using the model in Eq. (7.1).
6. Algebraically confirm Eq. (7.37).
7. Compute the log gains of the transition time for the model in Eq. (7.1).
8. Graph the profile of rate constant sensitivities for the models in Eqs. (7.1) and (7.42).

9. Using the model in Eq. (7.1), confirm Eq. (7.53) numerically.
10. Using the model in Eq. (7.1), show how constrained sensitivities can be computed in PLAS by considering them as independent variables.
11. Suppose an independent variable enters the model with some kinetic order. Discuss mathematical and conceptual differences and similarities between the log gains with respect to the independent variable and sensitivities with respect to its kinetic order.
12. For the model in Eq. (7.1), compare numerically the definitions of kinetic order sensitivities in Eqs. (7.66)–(7.69). Is it possible in general that the two sets of definitions yield different rankings of sensitivities?
13. Compose a table similar to Table 7.1 that compares predicted and observed effects of changes in kinetic orders in Eq. (7.1).
14. Using the model in Eq. (7.1), confirm Eqs. (7.85) and (7.86) numerically.
15. Graph the profiles of all flux sensitivities for the model in Eqs. (7.1).
16. Using the model in Eq. (7.1), confirm Eqs. (7.88) and (7.91) numerically.
17. Using the model in Eq. (7.1), confirm Eqs. (7.92) and (7.93) numerically.
18. Write Eqs. (7.92) and (7.93) as a matrix equation.
19. Using the model in Eq. (7.1), confirm Eq. (7.94) numerically.
20. Write Eq. (7.94) as a matrix equation.

COMPLEMENTS

C1. Trajectories

The logarithmic gain of a metabolite quantifies how the change in an independent variable evokes a change in a dependent variable. A question similar in concept is how *two dependent* variables relate to each other. The problem here is that the two variables of interest are dynamically interdependent and we therefore cannot restrict our analysis to a steady state. Nonetheless, we can study this question numerically in the *phase plane*, which graphs one dependent variable as the function of another dependent variable. This type of representation is implemented in PLAS as the Phase Plot 2D option and called up by solving the system, displaying the default solution graph, and selecting **Phase Plot 2D** under **Results** on the menu.

Mathematically one can formulate the problem as a new set of differential equations, which however may not be in canonical form. For example, to express X_1 as a function of X_2, one divides each equation of the S-system by the second equation, as long as it is not zero. The new first equation reads

$$\frac{dX_1}{dX_2} = \frac{\alpha_1 \prod_{j=1}^n X_j^{g_{1j}} - \beta_1 \prod_{j=1}^n X_j^{h_{1j}}}{\alpha_2 \prod_{j=1}^n X_j^{g_{2j}} - \beta_2 \prod_{j=1}^n X_j^{h_{2j}}} \tag{7C.1}$$

and expresses X_1 as a function of X_2, in the form of a differential equation. A function of this type is called a *trajectory*. We will not deal with trajectories in this form, and

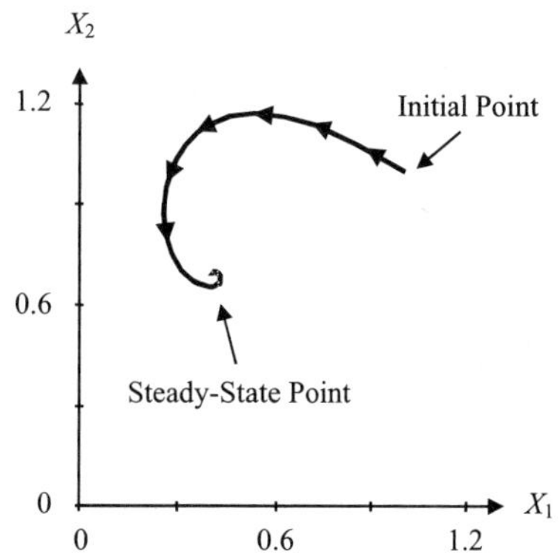

SENSITIVITY ANALYSIS

Figure 7C.1. Trajectory $X_2(X_1)$ of pathway with feedback [Eq. (7C.2)]. The arrowheads were added to the PLAS graph to indicate the flow of the system along the trajectory.

in cases where we want to demonstrate trajectorial behavior, we resort to numerical analyses with PLAS. A few examples of S-system such trajectories can be found in (Voit, 1988a, 1991, 1996a; Sands and Voit, 1996; Voit and Sands, 1996ab).

For the simple pathway with feedback [cf. Fig. 6.1, Eq. (6.2)],

$$\dot{X}_1 = 0.4 X_2^{-2} - 2 X_1,$$
$$\dot{X}_2 = 2 X_1 - X_2^{0.5},$$
(7C.2)

the trajectory $X_2(X_1)$ is illustrated in Fig. 7C.1. It starts at the point whose coordinates are the initial conditions $[X_1(t_0) = 1, X_2(t_0) = 1]$ and approaches the steady-state point (0.41628, 0.69314) after a spiral motion that corresponds to the oscillatory behavior of X_1 and X_2 as functions of time.

As an additional exercise, predict what the trajectory $X_1(X_2)$ would look like. Check your prediction in PLAS.

C2. Sensitivity Analysis in GMA Models

This section was prepared by Dr. Albert Sorribas of the University of Lleida, Spain. It shows that logarithmic gains and sensitivities can be computed for GMA models, even though it is not possible in general to solve the steady-state equations in an explicit manner. The crucial step in circumventing this problem is the differentiation of the system equations in their differential form (see also Reder, 1988; Heinrich and Schuster, 1998). The presentation in the following uses the notation of a stoichiometric matrix equation in which the "variables" are products of power-law functions. A brief comparison between this approach and the computation with S-systems can be found in Cascante et al. (1991a).

Terminology. Any GMA model of a metabolic pathway can be formulated as

$$\vec{\dot{X}} = N\vec{V},$$
(7C.3)

where $\vec{\dot{X}}$ is the vector of derivatives $\dot{X}_i$, N is a *stoichiometric matrix* (e.g., Gavalas, 1968; Heinrich et al., 1977) and $\vec{V}$ is a vector containing the fluxes v_r, each of which has the form

$$v_r = \gamma_r \prod_{j=1}^{n+m} X_j^{f_{rj}}. \tag{7C.4}$$

Instead of α_r or β_r in the S-system model, we are using the symbol γ_r for the rate constant. The symbols g_{rj} and h_{rj} are replaced generically by f_{rj}. The fluxes are functions of the dependent variables, which are collected in the vector $\vec{X}_D$. The dependent variables, in turn, depend on the independent variables, which are collected in the vector $\vec{X}_I$; on the matrices F_D and F_I of kinetic orders associated with dependent and independent variables, respectively; and on the vector of rate constants, $\vec{\Gamma}$:

$$v_r = v_r(\vec{X}_D(\vec{X}_I, \vec{\Gamma}, F_D, F_I)). \tag{7C.5}$$

As for many analyses of linear algebra, it is convenient to represent the vectors $\vec{V}$, $\vec{X}$, $\vec{X}_D$, $\vec{X}_I$, and $\vec{\Gamma}$ in the form of square matrices V, X, X_D, X_I, and Γ by putting their elements in the diagonals and setting all other matrix elements equal to zero. For instance, if there are z fluxes, the matrix V reads

$$V = \begin{pmatrix} v_1 & 0 & 0 & \cdots & 0 \\ 0 & v_2 & 0 & \cdots & 0 \\ 0 & 0 & v_3 & \cdots & 0 \\ \vdots & \vdots & \vdots & \ddots & \vdots \\ 0 & 0 & 0 & \cdots & v_z \end{pmatrix}. \tag{7C.6}$$

As usual, V^{-1}, X_D^{-1}, etc. denote the inverses of the newly defined matrices, and matrix products like VV^{-1} and $X_D^{-1}X_D$ equal the identity matrix of appropriate dimension (see Appendix). With this trick of making vectors into square matrices, the kinetic order matrices F_D and F_I can be written as logarithmic derivatives of the individual fluxes with respect to dependent or independent variables, namely,

$$\begin{aligned} F_D &= V^{-1} \frac{\partial \vec{V}}{\partial \vec{X}_D} X_D, \\ F_I &= V^{-1} \frac{\partial \vec{V}}{\partial \vec{X}_I} X_I \end{aligned} \tag{7C.7}$$

(cf. Cascante et al., 1991a).

Logarithmic Gains of Concentrations. If one changes an independent variable, the effect on the system can be computed as a total derivative. Because $\vec{V}$ is a function of different quantities, the chain rule is applied and yields

$$\begin{aligned} \frac{d\vec{X}}{d\vec{X}_I} &= N \left(\frac{\partial \vec{V}}{\partial \vec{X}_D} \frac{d\vec{X}_D}{d\vec{X}_I} + \frac{\partial \vec{V}}{\partial \vec{X}_I} \frac{d\vec{X}_I}{d\vec{X}_I} \right) \\ &= N \frac{\partial \vec{V}}{\partial \vec{X}_D} \frac{d\vec{X}_D}{d\vec{X}_I} + N \frac{\partial \vec{V}}{\partial \vec{X}_I}. \end{aligned} \tag{7C.8}$$

SENSITIVITY ANALYSIS

At the steady state,

$$\dot{\vec{X}} = N\vec{V} = 0, \tag{7C.9}$$

the derivative of $\vec{X}$ is zero, and so is the derivative of $\dot{\vec{X}}$ on the left-hand side of Eq. (7C.8). This simplifies Eq. (7C.8) to

$$N \frac{\partial \vec{V}}{\partial \vec{X}_D} \frac{d\vec{X}_D}{d\vec{X}_I} + N \frac{\partial \vec{V}}{\partial \vec{X}_I} = 0. \tag{7C.10}$$

Since VV^{-1} and $X_D X_D^{-1}$ equal the unit matrix, they can be multiplied into Eq. (7C.10) without consequence, except that this operation allows us to regroup terms. The result is

$$NVV^{-1} \frac{\partial \vec{V}}{\partial \vec{X}_D} X_D \cdot X_D^{-1} \frac{d\vec{X}_D}{d\vec{X}_I} X_I + NVV^{-1} \frac{\partial \vec{V}}{\partial \vec{X}_I} X_I = 0. \tag{7C.11}$$

Using the definitions of F_D and F_I from Eq. (7C.7) yields

$$NVF_D X_D^{-1} \frac{d\vec{X}_D}{d\vec{X}_I} X_I + NVF_I = 0. \tag{7C.12}$$

As discussed in the body of the chapter, the logarithmic gain of a dependent variable with respect to an independent variable is given in matrix notation as the relative derivative,

$$L(X_D, X_I) = X_D^{-1} \frac{d\vec{X}_D}{d\vec{X}_I} X_I. \tag{7C.13}$$

With this general definition, Eq. (7C.12) becomes

$$NVF_D L(X_D, X_I) + NVF_I = 0. \tag{7C.14}$$

This matrix equation can be solved for the logarithmic gain, yielding

$$L(X_D, X_I) = -(NVF_D)^{-1} NVF_I. \tag{7C.15}$$

Logarithmic gains of fluxes. A similar strategy leads to the logarithmic flux gains. In this case, we begin with the total derivative of the flux vector $\vec{V}$ with respect to $\vec{X}_I$, which is given as

$$\frac{d\vec{V}}{d\vec{X}_I} = \frac{\partial \vec{V}}{\partial \vec{X}_D} \frac{d\vec{X}_D}{d\vec{X}_I} + \frac{\partial \vec{V}}{\partial \vec{X}_I}. \tag{7C.16}$$

Normalization via multiplication by X_I and division by V yields

$$V^{-1} \frac{d\vec{V}}{d\vec{X}_I} X_I = V^{-1} \frac{\partial \vec{V}}{\partial \vec{X}_D} X_D \cdot X_D^{-1} \frac{d\vec{X}_D}{d\vec{X}_I} X_I + V^{-1} \frac{\partial \vec{V}}{\partial \vec{X}_I} X_I. \tag{7C.17}$$

Employing the definitions of flux and concentration gains as relative derivatives and using again the kinetic order matrices F_D and F_I from Eq. (7C.7), the flux gains are obtained as

$$L(V, X_I) = F_D \cdot L(X_D, X_I) + F_I \qquad (7C.18)$$

(see Exercise C.1).

Sensitivities of metabolites with respect to rate constants. These sensitivities describe the effects of changes in one of the rate constants on the dependent variables. The strategy of characterizing them is the same as before. In this case, we begin with the derivative of $\vec{X}$ with respect to the rate constant vector $\vec{\Gamma}$:

$$\frac{d\vec{X}}{d\vec{\Gamma}} = N \frac{\partial \vec{V}}{\partial \vec{X}_D} \frac{d\vec{X}_D}{d\vec{\Gamma}} + N \frac{\partial \vec{V}}{\partial \vec{\Gamma}} = 0. \qquad (7C.19)$$

Normalization as in the previous case yields

$$NVV^{-1} \frac{\partial \vec{V}}{\partial \vec{X}_D} X_D \cdot X_D^{-1} \frac{d\vec{X}_D}{d\vec{\Gamma}} \Gamma + NVV^{-1} \frac{\partial \vec{V}}{\partial \vec{\Gamma}} \Gamma = 0, \qquad (7C.20)$$

and using previous definitions, we obtain

$$NVF_D S(X_D, \Gamma) + NVV^{-1} \frac{\partial \vec{V}}{\partial \vec{\Gamma}} \Gamma = 0. \qquad (7C.21)$$

This equation reduces to

$$S(X_D, \Gamma) = -(NVF_D)^{-1} NV, \qquad (7C.22)$$

because

$$V^{-1} \frac{\partial \vec{V}}{\partial \vec{\Gamma}} \Gamma = I, \qquad (7C.23)$$

with **I** the identity matrix. Using Eq. (7C.15), we can relate the sensitivity to a product of the appropriate logarithmic gain and the matrix F_I of kinetic orders associated with independent variables:

$$L(X_D, X_I) = -(NVF_D)^{-1} NVF_I = S(X_D, \Gamma) F_I. \qquad (7C.24)$$

Sensitivities of fluxes with respect to rate constants. The total derivative of the flux vector with respect to $\vec{\Gamma}$ is

$$\frac{d\vec{V}}{d\vec{\Gamma}} = \frac{\partial \vec{V}}{\partial \vec{X}_D} \frac{d\vec{X}_D}{d\vec{\Gamma}} + \frac{\partial \vec{V}}{\partial \vec{\Gamma}}. \qquad (7C.25)$$

Normalization yields

$$V^{-1} \frac{d\vec{V}}{d\vec{\Gamma}} \Gamma = V^{-1} \frac{\partial \vec{V}}{\partial \vec{X}_D} X_D \cdot X_D^{-1} \frac{d\vec{X}_D}{d\vec{\Gamma}} \Gamma + V^{-1} \frac{\partial \vec{V}}{\partial \vec{\Gamma}} \Gamma, \qquad (7C.26)$$

SENSITIVITY ANALYSIS

which is directly rewritten as

$$S(V, \Gamma) = F_D \cdot S(X_D, \Gamma) + I. \tag{7C.27}$$

Sensitivities of metabolites with respect to kinetic orders. The effect of a change in a kinetic order can be formulated as

$$\frac{d\dot{X}}{df_{ij}} = N\frac{\partial \vec{V}}{\partial \vec{X}_D}\frac{d\vec{X}_D}{df_{ij}} + N\frac{\partial \vec{V}}{\partial f_{ij}} = 0. \tag{7C.28}$$

Normalizing yields

$$NVV^{-1}\frac{\partial \vec{V}}{\partial \vec{X}_D}X_D \cdot X_D^{-1}\frac{d\vec{X}_D}{df_{ij}}f_{ij} + NVV^{-1}\frac{\partial \vec{V}}{\partial f_{ij}}f_{ij} = 0, \tag{7C.29}$$

which simplifies to

$$NVF_D S(X_D, f_{ij}) + NVV^{-1}\frac{\partial \vec{V}}{\partial f_{ij}}f_{ij} = 0. \tag{7C.30}$$

Solving for the desired sensitivity leads to different, equivalent representations:

$$\begin{aligned} S(X_D, f_{ij}) &= -(NVF_D)^{-1}NVV^{-1}\frac{\partial \vec{V}}{\partial f_{ij}}f_{ij}, \\ S(X_D, f_{ij}) &= S(X_D, \Gamma) \cdot V^{-1}\frac{\partial \vec{V}}{\partial f_{ij}}f_{ij}, \\ S(X_D, f_{ij}) &= S(X_D, \Gamma)\vec{\text{Log}}(_j X_i) \cdot f_{ij}, \end{aligned} \tag{7C.31}$$

where $\vec{\text{Log}}(_j X_i)$ is a z-dimensional vector of zeros except for the jth element, which is equal to the natural logarithm of the ith variable.

Sensitivities of fluxes with respect to kinetic orders. The total derivative of the flux vector with respect to the kinetic order f_{ij} is

$$\frac{d\vec{V}}{df_{ij}} = \frac{\partial \vec{V}}{\partial X_D}\frac{dX_D}{df_{ij}} + \frac{\partial \vec{V}}{\partial f_{ij}}. \tag{7C.32}$$

Normalization results in

$$V^{-1}\frac{d\vec{V}}{df_{ij}}f_{ij} = V^{-1}\frac{\partial \vec{V}}{\partial \vec{X}_D}X_D \cdot X_D^{-1}\frac{d\vec{X}_D}{df_{ij}}f_{ij} + V^{-1}\frac{\partial \vec{V}}{\partial f_{ij}}f_{ij} \tag{7C.33}$$

and

$$S(V, f_{ij}) = F_D \cdot S(X_D, f_{ij}) + \vec{\text{Log}}(_j X_i)f_{ij} \tag{7C.34}$$

where $\vec{\text{Log}}(_j X_i)$ is again a z-dimensional vector of zeros except for the jth element, which is equal to the natural logarithm of the ith variable. Using the previous result from Eqs. (7C.27) and (7C.31) yields

$$S(V, f_{ij}) = F_D S(X_D, \Gamma)\vec{\text{Log}}(_j X_i) \cdot f_{ij}. + \vec{\text{Log}}(_j X_i) \cdot f_{ij} \tag{7C.35}$$

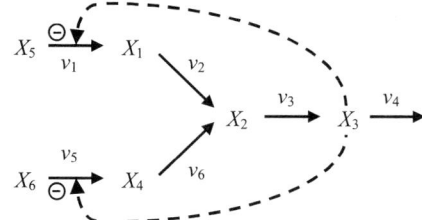

Figure 7C.2. Didactic pathway with branch point and feedback inhibition.

and

$$S(V, f_{ij}) = S(V, \Gamma) \vec{\text{Log}}(_j X_i) \cdot f_{ij}. \tag{7C.36}$$

Example. Consider the system shown in Fig. 7C.2. The differential equations in stoichiometric form are given as

$$\begin{aligned}
\dot{X}_1 &= v_1 - v_2, \\
\dot{X}_2 &= v_2 + v_6 - v_3, \\
\dot{X}_3 &= v_3 - v_4, \\
\dot{X}_4 &= v_5 - v_6.
\end{aligned} \tag{7C.37}$$

The same equations in the form of the matrix equation (7C.3) read

$$\vec{\dot{X}} = \mathbf{N}\vec{V}$$

$$= \begin{pmatrix} 1 & -1 & 0 & 0 & 0 & 0 \\ 0 & 1 & -1 & 0 & 0 & 1 \\ 0 & 0 & 1 & -1 & 0 & 0 \\ 0 & 0 & 0 & 0 & 1 & -1 \end{pmatrix} \begin{pmatrix} v_1 \\ v_2 \\ v_3 \\ v_4 \\ v_5 \\ v_6 \end{pmatrix}. \tag{7C.38}$$

The kinetic order matrices are

$$\mathbf{F}_D = \begin{pmatrix} 0 & 0 & f_{13} & 0 \\ f_{21} & 0 & 0 & 0 \\ 0 & f_{32} & 0 & 0 \\ 0 & 0 & f_{43} & 0 \\ 0 & 0 & f_{53} & 0 \\ 0 & 0 & 0 & f_{64} \end{pmatrix},$$

$$\mathbf{F}_I = \begin{pmatrix} f_{15} & 0 \\ 0 & 0 \\ 0 & 0 \\ 0 & 0 \\ 0 & f_{56} \\ 0 & 0 \end{pmatrix}. \tag{7C.39}$$

SENSITIVITY ANALYSIS

A complete characterization of gains and sensitivities, either algebraic or numerical, can be obtained by computing

$$\begin{aligned}
S(X_D, \Gamma) &= -(NVF_D)^{-1}NV, \\
S(V, \Gamma) &= F_D S(X_D, \Gamma) + I, \\
L(X_D, X_I) &= S(X_D, \Gamma) F_I, \\
L(V, X_I) &= F_D L(X_D, X_I) + F_I, \\
S(X_D, f_{ij}) &= S(X_D, \Gamma) \vec{\text{Log}}(_j X_i) \cdot f_{ij}, \\
S(V, f_{ij}) &= S(V, \Gamma) \vec{\text{Log}}(_j X_i) \cdot f_{ij}
\end{aligned} \qquad (7C.40)$$

(see Exercise C.2).

Systems with constraints. Constraint relationships between metabolites, such as the conservation of mass, complicate the computation of sensitivities. The strategy for dealing with such relationships in the present context is the determination of a *reduced set of dependent variables*, which is denoted in vector notation as $\vec{X}_d^R$. The computation of sensitivities with respect to a parameter vector $\vec{p}$ is executed via differentiation of the steady-state equation according to the chain rule and normalization, as before. The results are

$$\begin{aligned}
N \frac{dV}{d\vec{p}} &= N \frac{\partial \vec{V}}{\partial \vec{X}_D} \frac{\partial \vec{X}_D}{\partial \vec{X}_D^R} \frac{d\vec{X}_D^R}{d\vec{p}} + N \frac{\partial \vec{V}}{\partial \vec{p}} = 0, \\
NVV^{-1} \frac{\partial \vec{V}}{\partial \vec{X}_D} X_D \cdot X_D^{-1} \frac{\partial \vec{X}_D}{\partial \vec{X}_D^R} \vec{X}_D^R (\vec{X}_D^R)^{-1} \frac{d\vec{X}_D^R}{d\vec{p}} \vec{p} + NVV^{-1} \frac{\partial \vec{V}}{\partial \vec{p}} \vec{p} &= 0, \\
NVF_D S(X_D, X_D^R) S(X_D^R, \vec{p}) + NVV^{-1} \frac{\partial \vec{V}}{\partial \vec{p}} \vec{p} &= 0, \\
S(X_D^R, \vec{p}) &= -[NVF_D S(X_D, X_D^R)]^{-1} NVV^{-1} \frac{\partial \vec{V}}{\partial \vec{p}} \vec{p}, \\
S(X_D^R, \vec{p}) &= S(X_D^R, \Gamma) \cdot V^{-1} \frac{\partial \vec{V}}{\partial \vec{p}} \vec{p}.
\end{aligned} \qquad (7C.41)$$

The matrix $S(X_D, X_D^R)$ has been called the *link matrix* (Reder, 1988).

It is useful to demonstrate these computations with a specific example. Consider the scheme with conserved quantities shown in Fig. 7C.3. The set of dependent variables consists of X_1, X_2, X_4, and X_5, but because of conservation of mass, only two of these are in the reduced set. For instance, we may define this set as

$$X_D^R = \{X_1, X_2\}. \qquad (7C.42)$$

An equally valid alternative would be

$$X_D^R = \{X_1, X_5\}. \qquad (7C.43)$$

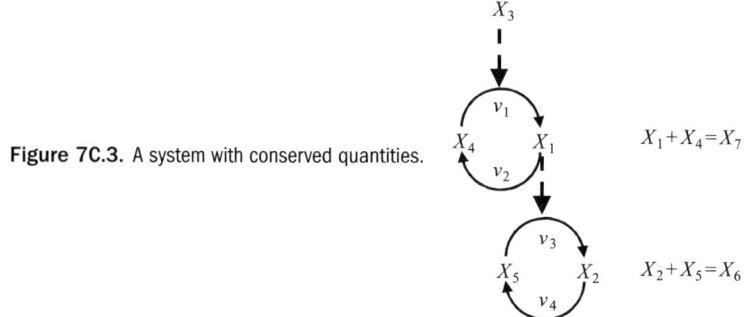

Figure 7C.3. A system with conserved quantities.

The elements of the link matrix $S(X_D, X_D^R)$ are the relative derivatives of the constraint relationships. For instance,

$$X_4 = X_7 - X_1,$$
$$\frac{dX_4}{dX_1} = -1, \qquad (7C.44)$$
$$\frac{dX_4}{dX_1}\frac{X_1}{X_4} = -\frac{X_1}{X_7 - X_1}.$$

The link matrix for $X_D^R = \{X_1, X_2\}$ is

$$S(X_D, X_D^R) = \begin{pmatrix} 1 & 0 \\ 0 & 1 \\ -\frac{X_1}{X_7 - X_1} & 0 \\ 0 & -\frac{X_2}{X_6 - X_2} \end{pmatrix}. \qquad (7C.45)$$

In addition to the link matrix, one needs a matrix $S(X_I^A, X_D^R)$ of sensitivities of an augmented set of independent variables $X_I^A = \{X_{n+1}, \ldots, X_{n+m}, X_{n+m+1}, \ldots, X_{n+m+k}\}$, where k is the number of constraints. As the definition indicates, each constraint relationship introduces one new independent variable.

With $S(X_D, X_D^R)$ and $S(X_I^A, X_D^R)$, the model can be expressed in terms of the reduced set $\vec{X}_d^R$. Specifically,

$$\begin{aligned} F_D^R &= F_D S(X_D, X_D^R), \\ F_I^A &= F_I S(X_I^A, X_D^R). \end{aligned} \qquad (7C.46)$$

With these transformations, a complete characterization in terms of the reduced set of dependent variables can be obtained. Details are given in Heinrich and Schuster (1996).

EXERCISES

C.1. Discuss the relationships between Eq. (7C.18) and Eq. (7.22) in the body of the chapter.

SENSITIVITY ANALYSIS

C.2. (Project) Select parameter values for the pathway in the first example [Eq. (7C.37)]. Compute gains and sensitivities with respect to independent variables and parameters, using Eq. (7C.40). Aggregate the fluxes v_2 and v_6 to construct the corresponding S-system model. Compute gains and sensitivities, and compare with the results from Eq. (7C.40).

REFERENCES

[33], [44–5], [47], [59–60], [66–9], [79], [87], [96], [111], [144–7], [162], [176], [178], [180], [181], [195–6], [240], [266], [283–5], [302–5], [306], [311], [314–15], [319], [333–4], [338], [350], [353–4], [374–9], [381], [391], [394], [417–19], [422–3], [434], [439], [445], [460–1].

CHAPTER EIGHT

Case Study 1 – Anaerobic Fermentation Pathway in *Saccharomyces cerevisiae*

The main emphasis of this case study is on setting up model equations, both in GMA and S-system form, and on specifying them numerically with some of the methods discussed in Chapter 5. The chapter also demonstrates various numerical analyses that are a standard component of biochemical modeling. It does not demonstrate very many algebraic aspects of steady-state analyses. These are the focus of Chapter 11.

For practical and academic reasons, the fermentation pathway in yeast (*Saccharomyces cerevisiae*) has been studied extensively. Numerous studies have investigated different experimental set-ups and conditions under which yeast cells produce ethanol, which is needed in great quantities for industrial purposes. The pathway is fairly easily accessible and therefore rather well understood in terms of metabolites and kinetic processes.

The chapter excerpts several analyses that have approached the fermentation pathway with methods of canonical modeling. Two papers by Galazzo and Bailey (1990, 1991) establish the experimental basis and also provide the kinetic equations, albeit in the traditional form of Michaelis–Menten reactions and their derivatives. From these traditional equations, Cascante, Curto, and Sorribas (Curto et al., 1995; Sorribas et al., 1995; Cascante et al., 1995ab) derived GMA and S-system models and executed most standard procedures of biochemical systems analysis. Torres et al. (1997) used the S-system model to illustrate methods of flux optimization in a biotechnological setting (see also Voit, 1992a; Torres et al., 1998). Because most information of the chapter derives from these papers, they are referenced again only in exceptional cases. Galazzo and Bailey, as well as Cascante, Curto, and Sorribas, analyzed the pathway with suspended or alginate-immobilized cells and under two different pH regimens, because Doran (1985) had shown that these conditions greatly affected ethanol production. This chapter focuses on only one of the four combinations, namely suspended cells at pH 4.5.

The structure of this chapter follows the historical development of the case. It begins with some biochemical background and the establishment of traditional rate laws. From these, GMA and S-system models are developed, and typical analyses of consistency and robustness are executed. The dynamics of the models will be

Features of the Fermentation Pathway

Overall, the fermentation pathway describes how yeast and other organisms can use glucose to produce ethanol as well as glycerol, glycogen, and trehalose. The map of the pathway is shown in Fig. 8.1. In the first step, external glucose is transported into the cell. This step should not be ignored, because it has a direct bearing on fermentation experiments, in which external – but not necessarily internal – glucose can be supplied. Also, this step is inhibited by glucose-6-phosphate and thus constitutes an important point of regulation.

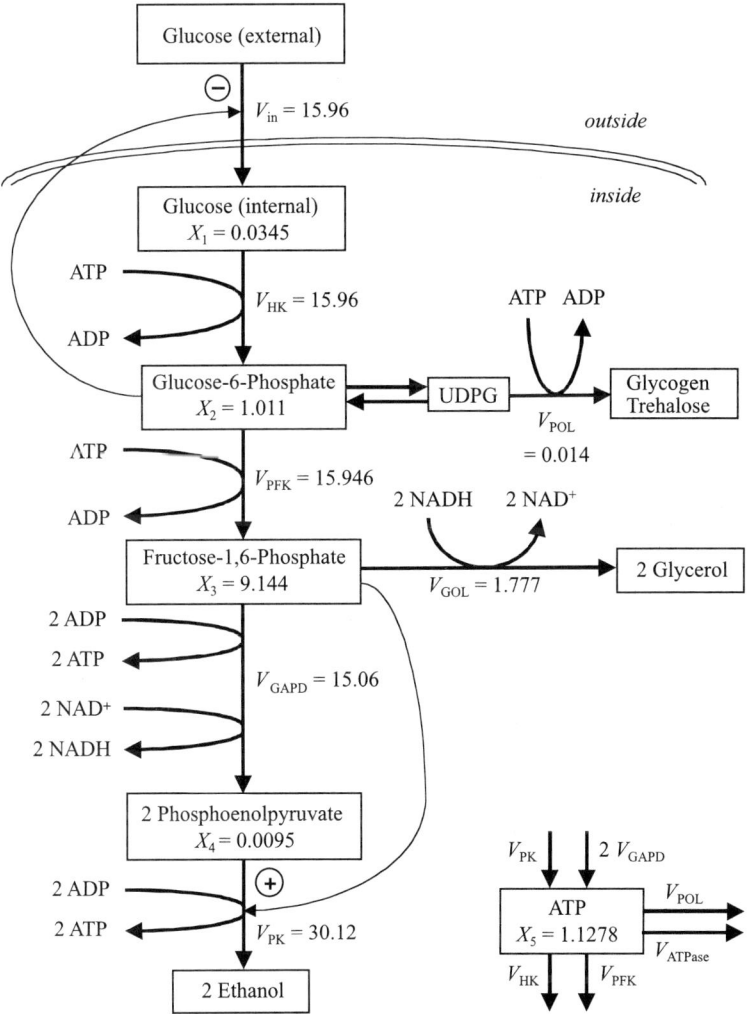

Figure 8.1. Simplified model of fermentation pathway in *Saccharomyces cerevisiae*, adapted from Galazzo and Bailey (1990) and Curto et al. (1995).

Hexokinase phosphorylates internal glucose to glucose-6-phosphate. *In vivo*, glucose-6-phosphate can flow into different pathways. Here, only two are considered. One is toward ethanol, and the other toward the polysaccharides glycogen and trehalose. The flow into the oxidative and nonoxidative pentose pathways is insignificant under the experimental conditions of interest and therefore omitted. This omission constitutes a typical modeling step. We know that some glucose-6-phosphate in reality is channeled into the pentose pathways, but make the executive decision to ignore these flows anyway. Is this good, bad, justifiable? The omission clearly has the advantage that less kinetic information is necessary to establish the model. It has the drawback that model predictions are limited to conditions under which the pentose branches in fact are negligible, at least in comparison with the ethanol branch. As is typical for such a decision, one must weigh the gained reduction in mathematical complexity with the limitation in scope. It is a step that (hopefully) can be justified, but that can also be criticized by others and makes the model and all insights from it susceptible to criticism.

Glucose-6-phosphate and fructose-6-phosphate are considered to be at equilibrium. Since they are very readily converted into each other, they are pooled in the model (see Exercise 1). In the next step, phosphofructokinase phosphorylates fructose-6-phosphate. If the enzyme fructose-1,6-diphosphatase is active, there may be a futile cycle between fructose-6-phosphate and fructose-1,6-diphosphate. However, under the experimental conditions of interest this enzyme is typically inactive, and this cycle is not included in the model – another simplification that affects the structure, complexity, and scope of the model. Fructose-1,6-diphosphate can be used to produce glycerol or phosphoenolpyruvate. Finally, pyruvate kinase catalyzes the production of ethanol. This step is activated by fructose-1,6-diphosphate. In addition to these reactions of the main pathway, several exchanges occur between ADP and ATP levels.

Definition of Variables

The next step in model design is the decision about which metabolites are to be considered as dependent or as independent variables. Clearly, internal glucose, glucose-6-phosphate, fructose-1,6-diphosphate, and phosphoenolpyruvate are dependent, and external glucose is independent. But what about ethanol, glycogen, trehalose, and glycerol? What about ADP, ATP, NAD$^+$, and NADH? It is again necessary to make executive decisions. As we discussed in Chapter 3, the inclusion of end products, such as ethanol and glycerol, does not affect the structure of the equations. If they are not explicitly included, the dynamics of the other dependent variables is unaffected. The disadvantage of inclusion is an additional equation for each variable. However, no additional data are necessary for this inclusion, because the production step of a metabolite like ethanol is already in the model as the degradation of phosphoenolpyruvate. What do we gain from the inclusion of ethanol or glycerol? The model would automatically keep track of the total accumulation of these end products, without however taking account of any degradation that might take place within the cells. Since production is of greater interest than accumulation, we formulate the first model without ethanol, polysaccharides, and glycerol as dependent variables (however, see Exercise 16).

A complicated question in many analyses of this type is what to do with secondary substrates like ADP, ATP, and NAD^+. On one hand, they actively contribute to the dynamics of the pathway and cannot be ignored. On the other hand, exchanges within the pool of adenylates are very difficult to quantify. There is no generally applicable, perfect solution to this dilemma. Nonetheless, a decision must be made, and it will depend on the focus of the analysis. In the present case, the ratio of NAD^+ to NADH appears to be fairly constant, and this ratio is therefore considered an independent variable with the value 0.042. For some rate laws of the model, knowledge of the ratio is not sufficient, since an actual concentration of either NAD^+ or NADH is needed. Experimental evidence suggests that the sum of the two equals 2 mM. Thus, there are two equations, from which the individual concentrations may be obtained:

$$\begin{aligned} &NAD^+ + NADH = 2, \\ &NADH/NAD^+ = 0.042, \\ &NAD^+ = 1.919, \\ &NADH = 0.081. \end{aligned} \quad (8.1)$$

ADP and ATP show some important dynamics, but the total pool of adenylates does not fluctuate much. Thus,

$$AMP + ADP + ATP \approx constant. \quad (8.2)$$

Experiments under conditions that are relevant in the present context suggest that the constant sum is about 3 mM.

In addition, the adenylates are catalyzed by adenylate kinase reaction, and this yields the relationship

$$K_{eq} = \frac{ADP^2}{ATP \cdot AMP} \quad (8.3)$$

with an experimentally determined $K_{eq} = 1$ (Su and Russel, 1968). Thus, if one of the adenylate concentrations is known, the other two can be computed (see Exercise 2). Curto et al. (1995) deal with the situation by considering ATP as a dependent variable and constraining AMP and ADP by the requirements above.

A problem similar to the relationships between AMP, ADP, and ATP appears if one would like to make statements about metabolites that are not explicitly modeled, but are directly related to a dependent variable. An example is fructose-6-phosphate (F6P), which is essentially always in equilibrium with glucose-6-phosphate (G6P). Instead of modeling F6P as an additional dependent variable – after all, it changes with time just as much as G6P – one may define constraints that can be evoked if the additional metabolite is needed. Such a constraint is

$$[F6P]/[G6P] = 0.3. \quad (8.4)$$

Should the concentration of F6P appear in any of the model equations, it can be replaced with $0.3 \times [G6P]$. The same types of relationships hold between glyceraldehyde

3-phosphate (G3P) and fructose-1,6-diphosphate (FDP), between 3-phosphoglycerate (3PG) and phosphoenolpyruvate (PEP), and between G6P and UDPG:

$$[G3P]/[FDP] = 0.01,$$
$$[PEP]/[3PG] = 0.1, \qquad (8.5)$$
$$[G6P]/[UDPG] = 1.444.$$

Metabolites defined through such relationships can be eliminated from the equations, which reduces the number of variables and differential equations.

With that, we make the following definitions of dependent variables and list their observed concentrations at steady state:

Dependent Variable	Acronym	Symbol	Conc. (mM)
Internal glucose	G_{In}	X_1	0.0345
Glucose-6-phosphate	G6P	X_2	1.011
Fructose-1,6-diphosphate	FDP	X_3	9.144
Phosphoenolpyruvate	PEP	X_4	0.0095
Adenosine triphosphate	ATP	X_5	1.1278

The independent variables include enzyme activities and other quantities that affect the pathway but are not changed by the dynamics of the system. As said before, the ratio of NAD^+ to NADH remains essentially constant, and therefore also constitutes an independent variable. We define the following list:

Independent Variable	Acronym	Symbol	Value
Glucose uptake	V_{In}^{max}	X_6	19.7 mM min^{-1}
Hexokinase	V_{HK}^{max}	X_7	68.5 mM min^{-1}
Phosphofructokinase	V_{PFK}^{max}	X_8	31.7 mM min^{-1}
Glyceraldehyde-3-phosphate dehydrogenase	V_{GAPD}^{max}	X_9	49.9 mM min^{-1}
Pyruvate kinase	V_{PK}^{max}	X_{10}	3,440 mM min^{-1}
Polysaccharide production (glycogen + trehalose)	V_{POL}^{max}	X_{11}	14.31 mM min^{-1}
Glycerol production	V_{GOL}^{max}	X_{12}	203 mM min^{-1}
ATPase	V_{ATPase}^{max}	X_{13}	25.1 mM min^{-1}
NAD^+/NADH ratio	—	X_{14}	0.042 [unitless]

The definitions in these two lists essentially determine the scope of questions that can possibly be analyzed with the model. For instance, by our combining the production of glycogen and trehalose into one process, the model will not be able to make any statements about the relative quantities of these two polysaccharides. In a similar vein, the model cannot describe deviations from the equilibrium between glucose-6-phosphate and fructose-6-phosphate. The two are tied together by an unbendable constraint.

Setting Up GMA Equations

Instead of going directly to the S-system form, it is sometimes easier first to set up GMA equations. Although there is no problem in formulating the S-system equations from the map in symbolic form, the estimation of parameters is facilitated in the GMA form if these parameters are obtained from traditional rate laws. Once GMA parameters are estimated, the S-system equations are readily derived with straightforward mathematical manipulations.

The mass balance equations for the GMA and the S-system form are shown in Table 8.1. In many cases, the processes of production or degradation are identical in the GMA form and the S-system form. Where this is not the case, all GMA terms with a positive sign are aggregated into a single production term in the S-system form, and all GMA terms with a negative sign are aggregated into a single degradation term in the S-system form. For instance, in the fifth equation these relationships read

$$V_5^+ = V_{51}^+ + V_{52}^+,$$
$$V_5^- = V_{51}^- + V_{52}^- + V_{53}^- + V_{54}^-. \tag{8.6}$$

Studying the map, we can determine immediately which dependent and independent variables are to be included in any of the production or degradation terms in the GMA or the S-system model. It is those and only those variables that directly affect the process. They are listed in Table 8.2.

From this table, one readily constructs the GMA and S-system equations. All variables directly involved are included in a given production or degradation term with their own kinetic order exponent, and each term is assigned a rate constant. To facilitate the comparison between the GMA and S-system forms, we use α- and β-parameters for rate constants and the g- and h-parameters for kinetic orders in both models. In the GMA form, this terminology sometimes requires double indices for the rate constants and triple indices for kinetic orders.

Table 8.1. Mass Balance Equations in the GMA Model and the S-System Model

Variable	GMA	S-system
$\dot{X}_1 =$	$V_1^+ - V_1^-$	$V_1^+ - V_1^-$
$\dot{X}_2 =$	$V_2^+ - V_{21}^- - V_{22}^-$	$V_2^+ - V_2^-$
$\dot{X}_3 =$	$V_3^+ - V_{31}^- - V_{32}^-$	$V_3^+ - V_3^-$
$\dot{X}_4 =$	$V_4^+ - V_4^-$	$V_4^+ - V_4^-$
$\dot{X}_5 =$	$V_{51}^+ + V_{52}^+ - V_{51}^- - V_{52}^- - V_{53}^- - V_{54}^-$	$V_5^+ - V_5^-$

Table 8.2. Variables to Be Included in Each Term of the GMA Model and the S-System Model

Term	GMA	S-system	Variables to Be Included
V_1^+	✓	✓	X_2, X_6
V_1^-	✓	✓	X_1, X_5, X_7
V_2^+	✓	✓	X_1, X_5, X_7
V_{21}^-	✓		X_2, X_5, X_8
V_{22}^-	✓		X_2, X_5, X_{11}
V_2^-		✓	X_2, X_5, X_8, X_{11}
V_3^+	✓	✓	X_2, X_5, X_8
V_{31}^-	✓		X_3, X_5, X_9, X_{14}
V_{32}^-	✓		X_3, X_4, X_5, X_{12}
V_3^-		✓	$X_3, X_4, X_5, X_9, X_{12}, X_{14}$
V_4^+	✓	✓	X_3, X_5, X_9, X_{14}
V_4^-	✓	✓	X_3, X_4, X_5, X_{10}
V_{51}^+	✓		X_3, X_5, X_9, X_{14}
V_{52}^+	✓		X_3, X_4, X_5, X_{10}
V_5^+		✓	$X_3, X_4, X_5, X_9, X_{10}, X_{14}$
V_{51}^-	✓		X_1, X_5, X_7
V_{52}^-	✓		X_2, X_{11}
V_{53}^-	✓		X_2, X_5, X_8
V_{54}^-	✓		X_5, X_{13}
V_5^-		✓	$X_1, X_2, X_5, X_7, X_8, X_{11}, X_{13}$

The GMA equations read

$$\dot{X}_1 = \alpha_1 X_2^{g_{12}} X_6^{g_{16}} - \beta_1 X_1^{h_{11}} X_5^{h_{15}} X_7^{h_{17}},$$

$$\dot{X}_2 = \alpha_2 X_1^{g_{21}} X_5^{g_{25}} X_7^{g_{27}} - \beta_{2,1} X_2^{h_{22,1}} X_5^{h_{25,1}} X_8^{h_{28,1}} - \beta_{2,2} X_2^{h_{22,2}} X_5^{h_{25,2}} X_{11}^{h_{2,11,2}},$$

$$\dot{X}_3 = \alpha_3 X_2^{g_{32}} X_5^{g_{35}} X_8^{g_{38}} - \beta_{3,1} X_3^{h_{33,1}} X_5^{h_{35,1}} X_9^{h_{39,1}} X_{14}^{h_{3,14,1}}$$
$$\quad - \beta_{3,2} X_3^{h_{33,2}} X_4^{h_{34,2}} X_5^{h_{35,2}} X_{12}^{h_{3,12,2}},$$

$$\dot{X}_4 = \alpha_4 X_3^{g_{43}} X_5^{g_{45}} X_9^{g_{49}} X_{14}^{g_{4,14}} - \beta_4 X_3^{h_{43}} X_4^{h_{44}} X_5^{h_{45}} X_{10}^{h_{4,10}},$$

$$\dot{X}_5 = \alpha_{5,1} X_3^{g_{53,1}} X_5^{g_{55,1}} X_9^{g_{59,1}} X_{14}^{g_{5,14,1}} + \alpha_{5,2} X_3^{g_{53,2}} X_4^{g_{54,2}} X_5^{g_{55,2}} X_{10}^{g_{5,10,2}}$$
$$\quad - \beta_{5,1} X_1^{h_{51,1}} X_5^{h_{55,1}} X_7^{h_{57,1}} - \beta_{5,2} X_2^{h_{52,2}} X_{11}^{h_{5,11,2}}$$
$$\quad - \beta_{5,3} X_2^{h_{52,3}} X_5^{h_{55,3}} X_8^{h_{58,3}} - \beta_{5,4} X_5^{h_{55,4}} X_{13}^{h_{5,13,4}}. \tag{8.7}$$

Some of the parameters in these equations are constrained by precursor–product relationships. For instance, we have for the term V_2^+

$$\alpha_2 = \beta_1 = \beta_{51}$$

and

$$g_{21} = h_{11} = h_{51,1},$$
$$g_{25} = h_{15} = h_{55,1}, \qquad (8.8)$$
$$g_{27} = h_{17} = h_{57,1}.$$

Exercise 3 asks you to list all of these relationships.

The S-system equations are obtained in the same fashion. Each dependent or independent variable that directly affects a process is included in the representation of that process with its individual kinetic order. Each production term is assigned a rate constant α, and each degradation term is assigned a rate constant β. The equations thus read

$$\dot{X}_1 = \alpha_1 X_2^{g_{12}} X_6^{g_{16}} - \beta_1 X_1^{h_{11}} X_5^{h_{15}} X_7^{h_{17}},$$
$$\dot{X}_2 = \alpha_2 X_1^{g_{21}} X_5^{g_{25}} X_7^{g_{27}} - \beta_2 X_2^{h_{22}} X_5^{h_{25}} X_8^{h_{28}} X_{11}^{h_{2,11}},$$
$$\dot{X}_3 = \alpha_3 X_2^{g_{32}} X_5^{g_{35}} X_8^{g_{38}} - \beta_3 X_3^{h_{33}} X_4^{h_{34}} X_5^{h_{35}} X_9^{h_{39}} X_{12}^{h_{3,12}} X_{14}^{h_{3,14}}, \qquad (8.9)$$
$$\dot{X}_4 = \alpha_4 X_3^{g_{43}} X_5^{g_{45}} X_9^{g_{49}} X_{14}^{g_{4,14}} - \beta_4 X_3^{h_{43}} X_4^{h_{44}} X_5^{h_{45}} X_{10}^{h_{4,10}},$$
$$\dot{X}_5 = \alpha_5 X_3^{g_{53}} X_4^{g_{54}} X_5^{g_{55}} X_9^{g_{59}} X_{10}^{g_{5,10}} X_{14}^{g_{5,14}} - \beta_5 X_1^{h_{51}} X_2^{h_{52}} X_5^{h_{55}} X_7^{h_{57}} X_8^{h_{58}} X_{11}^{h_{5,11}} X_{13}^{h_{5,13}}.$$

Again, there are precursor–product relationships between V_1^- and V_2^+. Furthermore, several branch point constraints need to be considered. As discussed in Chapter 3, these assure that at steady state the efflux at a branch point is equivalent to the sum of diverging fluxes. In cases where a variable is involved in both diverging fluxes, the aggregated kinetic order is an average of the kinetic orders in the GMA model, weighted with the magnitudes of the diverging fluxes. For example, glucose-6-phosphate (X_2) is the substrate of two diverging pathways, one leading to fructose-1,6-diphosphate (X_3), and the other one to the polysaccharides glycogen and trehalose. Thus, X_2 appears in two degradation terms of the GMA model, which are aggregated in the S-system model. By contrast, ATP (X_5) and phosphofructokinase (X_8) appear only in the branch toward ethanol. Their kinetic orders in the S-system model are computed by simply scaling the GMA parameters by the proportion of the flux going into this branch. Thus, one obtains

$$h_{22} = \frac{h_{22,1} V_{2,1}^- + h_{22,2} V_{2,2}^-}{V_2^-},$$
$$h_{25} = \frac{h_{25,1} V_{2,1}^-}{V_2^-}, \qquad (8.10)$$
$$h_{28} = \frac{h_{28,1} V_{2,1}^-}{V_2^-}.$$

Another example for computing aggregated kinetic orders is the degradation of ATP (X_5). Four terms represent these processes in the GMA model, but only one term appears in the S-system model:

$$h_{55} = \frac{h_{55,1} V_{5,1}^- + h_{55,2} V_{5,2}^- + h_{55,3} V_{5,3}^- + h_{55,4} V_{5,4}^-}{V_5^-}. \tag{8.11}$$

There are numerous such constraints that need to be considered in the construction of the S-system model from the GMA model. Some are demonstrated in the following, and some are left as an exercise.

Parameter Values

For further quantitative analyses, the symbolic models in GMA and S-system form must be specified with numerical values. Curto et al. (1995) achieved this by computing kinetic orders and rate constants that corresponded to traditional rate laws in the papers of Galazzo and Bailey (1990, 1991). We generally follow the same procedure.

Data. Parameter estimation requires experimental measurements and mathematical terms describing the processes in the model. In the present case, the following information is available:

- Steady-state concentrations of dependent variables.
- Levels of independent variables, such as enzyme activities.
- Flux measurements at steady state.
- Traditional rate laws, from which power-law parameters are computed by partial differentiation.

Steady-state values of all dependent variables, levels of all independent variables, and steady-state fluxes were listed with the definitions of dependent and independent variables above.

Sugar transport. The sugar transport into the cell is a function of the external sugar concentration, and it is also regulated by the concentration of glucose-6-phosphate. Experimental measurements at steady state provide a glucose uptake rate of 15.96 mM/min and a concentration [G6P] of 1.011 mM per liter of cell volume. It seems reasonable to assume that the regulated and the unregulated rates relate as

$$V_{In} = V_{In}^0 + K_{G6P} \cdot [G6P]. \tag{8.12}$$

Thus, there are two unknowns, V_{In}^0 and the inhibition constant K_{G6P}, which Galazzo and Bailey (1990, 1991) determined from four measurements under slightly differing conditions. The results suggested that the unregulated transport rate V_{In}^0 and the inhibition constant K_{G6P} are

$$V_{In}^0 = 19.7 \, \text{mM} \, \text{L}^{-1} \, \text{min}^{-1} \tag{8.13}$$

and

$$K_{G6P} = -3.7 \, \text{min}^{-1}. \tag{8.14}$$

Two kinetic orders need to be computed, one with respect to external substrate, X_6, and one with respect to the inhibitor G6P (X_2). The unregulated transport is considered a first-order process, which implies

$$g_{16} = 1. \tag{8.15}$$

The kinetic order with respect to X_2 is computed as usual, namely as

$$g_{12} = \frac{\partial \left(V_{In}^0 + K_{G6P} \cdot [G6P] \right)}{\partial [G6P]} \cdot \frac{[G6P]}{V_{In}} = K_{G6P} \frac{[G6P]}{V_{In}}, \tag{8.16}$$

which is evaluated at the operating point. With $K_{G6P} = -3.7$, $[G6P] = 1.011$, and $V_{In} = 15.96$, the kinetic order is -0.2344.

The rate constant is computed by equating the power-law term to the traditional rate law at steady state and solving for α. We obtain

$$\alpha_1 X_2^{g_{12}} X_6^{g_{16}} = V_{In} \tag{8.17}$$

and

$$\begin{aligned} \alpha_1 &= V_{In} X_2^{-g_{12}} X_6^{-g_{16}} \\ &= 15.96 \times 1.011^{+0.2344} \times 19.7^{-1} \\ &= 0.8122. \end{aligned} \tag{8.18}$$

Hexokinase step. The hexokinase reaction involves two substrates, glucose and ATP, in an ordered single-displacement mechanism (Wilkinson and Rose, 1979). Two options are available for the estimation of kinetic orders. Either they can be obtained directly from rate measurements as functions of substrates, or these data can be fitted first with a traditional rate law that is subsequently reformulated as a power law. The advantage of the first option is that parameters are obtained directly, which eliminates possible errors in choosing or estimating the traditional law. The advantage of the second option is that an appropriate traditional rate law can fill gaps in the data. For instance, we shall see below (see also Exercise 4) that the direct kinetic order estimation of the hexokinase step with respect to ATP is not very reliable, because only one data point is available in the relevant concentration range.

Option 1. The first estimation method consists of directly using data from the literature that characterize the reaction rate as a function of substrate concentrations. Appropriate data are given in Figs. 8.2–8.4. Figure 8.2 shows the glucose

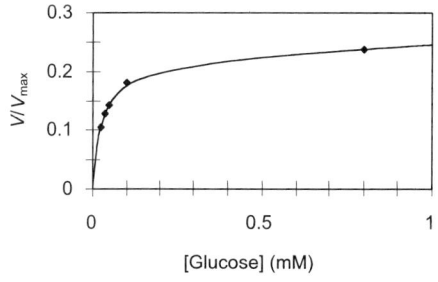

Figure 8.2. Glucose dependence of the hexokinase step; data redrawn from Galazzo and Bailey (1990). The line was fitted by hand.

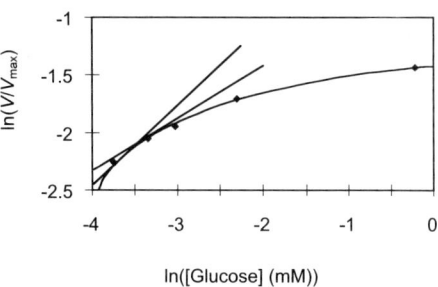

Figure 8.3. Glucose dependence of the hexokinase step in logarithmic coordinates (see also Fig. 8.2). The kinetic order with respect to glucose corresponds to the slope of this function at the steady state (about $-3.4 = \ln 0.0345$).

dependence of the reaction in Cartesian coordinates. Clearly, the reaction rate is a saturated function over the range of substrate concentrations of interest. At pH 4.5, which characterizes the condition of interest, the glucose concentration at steady state is between 0.03 and 0.035 mM (see definition of variables), which lies in the steep part of the saturation curve. To estimate the kinetic order, we transform the data into logarithmic coordinates (Fig. 8.3), where the kinetic order corresponds to the slope at the steady-state point. The natural logarithm in the range of 0.03 to 0.035 mM is about -3.4, and the measured slope is somewhere between 0.5 and 0.75.

Hexokinase also depends on ATP (Fig. 8.4). At steady state, the ATP concentration is about 1.13 mM (see definition of variables), which is slightly outside the measured range. Nonetheless, it appears that the rate is almost saturated at this concentration, and this implies that the kinetic order is slightly positive, but close to zero. An exact value is difficult to estimate because the hand-drawn curve through the data might increase more steeply and saturate earlier (see Exercise 4).

Option 2. Curto et al. (1995) followed a slightly different strategy. As suggested by Wilkinson and Rose (1979), Galazzo and Bailey (1990) had provided a mechanistic formulation of the hexokinase step and estimated its parameters from the data shown in Figs. 8.2 and 8.4. Curto and coworkers used this rate law and estimated kinetic orders for the GMA–S-system model by differentiation. The appropriate mechanistic rate law was given as

$$V_{HK} = \frac{V_{HK}^{max}}{\frac{K_G^s K_{ATP}^m}{G_{In} \cdot ATP} + \frac{K_G^m}{G_{In}} + \frac{K_{ATP}^m}{ATP} + 1} \tag{8.19}$$

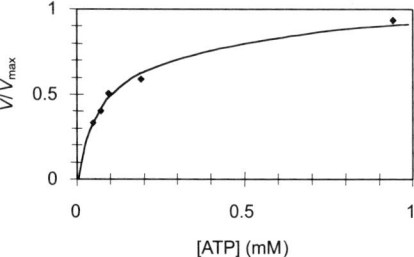

Figure 8.4. ATP dependence of the hexokinase step; data redrawn from Galazzo and Bailey (1990). The line was fitted by hand.

with the parameter values

$$V_{\text{HK}}^{max} = 68.5,$$
$$K_G^s = 0.0062,$$
$$K_G^m = 0.11,$$
$$K_{\text{ATP}}^m = 0.1.$$
(8.20)

Differentiation with respect to the substrates, evaluation at the steady state, multiplication by the substrate concentration, and division by the rate yields the kinetic orders

$$h_{11} = 0.7464 \tag{8.21}$$

and

$$h_{15} = 0.0243. \tag{8.22}$$

Furthermore, the enzyme activity X_7 is entered with the kinetic order

$$h_{17} = 1. \tag{8.23}$$

Equating the power-law term with the traditional rate yields the rate constant

$$\beta_1 = 2.8632 \tag{8.24}$$

(see Exercise 5). Because of their precursor–product relationship, the production term of glucose-6-phosphate, X_2, is thus also identified:

$$g_{21} = 0.7464,$$
$$g_{25} = 0.0243,$$
$$\alpha_2 = 2.8632.$$
(8.25)

Degradation of glucose-6-phosphate. The degradation of glucose-6-phosphate (X_2) occurs through two pathways. One yields fructose-1,6-diphosphate (X_3); X_3 is actually catalyzed from fructose-6-phosphate, but because glucose-6-phosphate and fructose-6-phosphate are in rapid equilibrium, the latter was omitted from the model during the establishment of the equations. The second pathway is catalyzed by glycogen synthetase and leads to UDPG (which is in equilibrium with glucose-6-phosphate) and subsequently to the polysaccharides glycogen and trehalose.

Phosphofructokinase reaction. Phosphofructokinase (PFK) catalyzes fructose-6-phosphate in an allosteric fashion and is modulated by several effectors, the most important of which generally are fructose-2,6-phosphate, AMP, P_i, and pH. Under the conditions studied, the rate is rather insensitive to fructose-2,6-phosphate under wide ranges of concentration. P_i can be subsumed in the rate constant, and this leaves the two substrates, fructose-6-phosphate and ATP, and two effectors, AMP and pH, to be included in the model for the reaction. Evaluating the relevant literature and the simplifications above, Galazzo and Bailey chose the concerted transition model for allosteric enzymes as proposed by Monod et al. (1965) and modified by Hess and Plesser (1979). This rate law reads, in the simplified form used by Galazzo and Bailey,

$$V_{PFK} = V_{PFK}^{max} \cdot \frac{10[F6P]\frac{[ATP]}{0.06} R}{R^2 + L_0(L_1 T)^2}, \qquad (8.26)$$

where

$$\begin{aligned}
V_{PFK}^{max} &= 31.7, \\
R &= 1 + [F6P] + \frac{[ATP]}{0.06} + 10[F6P]\frac{[ATP]}{0.06}, \\
L_0 &= \exp(4.17 \times pH^{in} - 20.4225) - 1658.22, \\
L_1 &= \frac{1 + 0.76[AMP]}{1 + 40[AMP]}, \\
T &= 1 + 0.0005[F6P] + \frac{[ATP]}{0.06} + 0.0005[F6P]\frac{[ATP]}{0.06}.
\end{aligned} \qquad (8.27)$$

This complicated rate law depends on the concentrations of fructose-6-phosphate, ATP, and AMP, and on the intracellular pH, pH^{in}, which under the given conditions is 6.94. Fructose-6-phosphate in itself is not a variable in our model, but it is related to the glucose-6-phosphate by

$$[F6P] = 0.3[G6P]. \qquad (8.28)$$

Since the steady-state concentration of glucose-6-phosphate is known (1.011 mM), this value can be substituted in the rate law. Similarly, AMP is not explicitly modeled, but the constraints involving ATP and ADP allow us to compute the steady-state AMP concentration as 0.8774 mM (see Exercise 2). Finally, the parameter L_0 depends on pH in an exponential fashion that was determined from experimental measurements. The model addresses a pH of 4.5, and this value is to be substituted in the rate law. V_{PFK}^{max} was determined from experimental data. Exercise 6 asks you to compute the traditional rate law at steady state.

From this rate law, the kinetic orders and the rate constant of the GMA model are computed in the usual fashion. Curto et al. (1995) obtained the following values:

$$\begin{aligned}
h_{22,1} &= g_{32} = 0.7318, \\
h_{25,1} &= g_{35} = -0.3941, \\
h_{28,1} &= g_{38} = 1, \\
\beta_{2,1} &= \alpha_3 = 0.5232
\end{aligned} \qquad (8.29)$$

(see Exercise 7). Note that these parameters also appear in the S-system model, namely in the production term of X_3. The degradation term of X_2, by contrast, will be aggregated from this GMA term and the one describing polysaccharide production.

Glycogen synthetase reaction. Glucose-6-phosphate can be converted into the two polysaccharides glycogen and trehalose. Trehalose is usually produced at 10–20% of the glycogen production rate (Kuenzi and Fiechter, 1972), and Galazzo and Bailey assessed the production level for the present scenario as 10%.

The first step toward polysaccharide production is a reversible reaction, which keeps glucose-6-phosphate and UDPG at equilibrium. UDPG is catalyzed by glycogen synthetase in a Michaelis–Menten reaction with respect to the substrate UDPG. Glucose-6-phosphate is a strong allosteric effector of this reaction: Enzyme activity may be stimulated 20 times by its presence (Rothman and Cabib, 1967). This strong effect lets us expect that the kinetic order of this rate with respect to glucose-6-phosphate will be high.

Rothman and Cabib (1967) suggest the following rate law:

$$V_{Gly} = V_{Gly}^{app} \cdot \frac{1}{\frac{K_0^m}{[UDPG]}\left(1 + \frac{K^m}{[G6P]}\right) + 1} \tag{8.30}$$

with parameter values

$$K_0^m = 1 \tag{8.31}$$

and

$$K^m = 1.1 \tag{8.32}$$

and a UDPG concentration of 0.7 mM at steady state. The apparent maximum rate V_{Gly}^{app} is a function of the glucose-6-phosphate concentration, which is appropriately described by a Hill rate law:

$$V_{Gly}^{app} = \frac{14.31[G6P]^{8.25}}{2^{8.25} + [G6P]^{8.25}}. \tag{8.33}$$

The large Hill parameter of 8.25 again reflects the strong effect of glucose-6-phosphate on the reaction. Curto et al. (1995) compute the kinetic order with respect to glucose-6-phosphate as

$$h_{22,2} = 8.6107, \tag{8.34}$$

which, as expected, is very high. The kinetic order with respect to the independent enzyme activity X_{11} is

$$h_{2,11,2} = 1. \tag{8.35}$$

Presumably assuming ATP saturation, Galazzo and Bailey (1990, 1991), as well as Curto et al. (1995), did not include ATP in this term (Cascante, pers. comm.). With a trehalose production rate of 10% of the glycogen production, the rate constant is

$$\beta_{2,2} = 0.0009 \tag{8.36}$$

(see Exercise 8).

Degradation of fructose-1,6-diphosphate. Fructose-1,6-diphosphate constitutes another branch point in the fermentation pathway. It can be converted into two molecules of glycerol or into two molecules of phosphoenolpyruvate. According to Galazzo and Bailey the former reaction, V_{GOL}, is proportional to the pyruvate kinase reaction, V_{PK}, which converts phosphoenolpyruvate into ethanol and is discussed in the next section. The proportionality is

$$V_{GOL}/V_{PK} = 0.05901. \tag{8.37}$$

The parameter values for this rate are

$$h_{33,2} = h_{43} = 0.05,$$
$$h_{34,2} = h_{44} = 0.533,$$
$$h_{35,2} = h_{45} = -0.0822, \tag{8.38}$$
$$h_{3,12,2} = 1,$$
$$\beta_{3,2} = 0.0945/2 = 0.04725$$

(see subsection on pyruvate kinase below). Except for division by 2, the rate constant is the same as in the power-law rate law representing pyruvate kinase, because the proportionality is accounted for in the maximum enzyme activities, which are represented by $X_{12} = 203$ and $X_{10} = 3,440$, respectively, and differ by the factor 0.05901. The rate is divided by 2 because two molecules are produced in the pyruvate kinase step, but only one molecule of fructose-1,6-phosphate is degraded in the glycerol step considered here.

The conversion into phosphoenolpyruvate is catalyzed by glyceraldehyde 3-phosphate dehydrogenase (GAPD). It is subject to crossed product inhibition and competitive inhibition from AMP, ADP, and ATP (Yang and Deal, 1969; Doran, 1985: pp. 122–137). A suitable rate law is

$$V_{GAPD} = V_{GAPD}^{max}\left[1 + \frac{0.0025}{[G3P]} + \frac{0.18}{[NAD^+]} \cdot S + \frac{0.0025 \times 0.18}{[G3P][NAD^+]}\right.$$
$$\left. \times \left(1 + \frac{[NADH]}{0.0003}\right) \cdot S\right]^{-1}, \tag{8.39}$$

where the maximum velocity was obtained from experimental results as $V_{GAPD}^{max} = 49.9$.

As stated earlier, $NAD^+ = 1.919$ and $NADH = 0.081$. Using the moiety conservation relationship $NADH + NAD^+ = \text{constant} = 2$, X_{14} can be expressed in terms of NAD^+ without NADH, or vice versa. Turning the result around yields, for instance,

$$[NAD^+] = \frac{2}{X_{14} + 1}, \tag{8.40}$$

which is substituted in Eq. (8.39) for the computation of the kinetic order with respect to X_{14} (Cascante, pers. comm.).

Glyceraldehyde-3-phosphate is not an explicit variable in the model, but can be computed from the proportionality with fructose-1,6-diphosphate:

$$[G3P] = 0.01[FDP]. \tag{8.41}$$

The quantity S represents the inhibition by adenylates. It is given as

$$S = 1 + \frac{[AMP]}{1.1} + \frac{[ADP]}{1.5} + \frac{[ATP]}{2.5} \tag{8.42}$$

and at steady state has the value

$$S = 1 + \frac{0.877428}{1.1} + \frac{0.994769}{1.5} + \frac{1.127803}{2.5} = 2.9120. \tag{8.43}$$

With these numerical specifications, the rate of GAPD at steady state is

$$V_{GAPD}(\text{steady state}) = 15.057. \tag{8.44}$$

Parameter values for the corresponding GMA terms are computed as usual. For the kinetic order with respect to ATP, one uses a substitution strategy similar to that for NAD$^+$. In this case, one evokes the relationships (8.2) and (8.3). The results are

$$\begin{aligned}
h_{33,1} &= g_{43} = 0.6159, \\
h_{45,1} &= g_{45} = 0.1308, \\
h_{39,1} &= g_{49} = 1, \\
h_{3,14,1} &= g_{4,14} = -0.6088, \\
\beta_{31} &= 0.011, \\
\alpha_4 &= 2\beta_{31} = 0.022
\end{aligned} \tag{8.45}$$

(Exercise 9). It may be surprising that ATP, as an inhibitor, has an associated kinetic order that is positive. From a mathematical point of view, one can see that $S = [ATP] + [ADP] + [AMP]$ appears in the denominator of V_{GAPD}, which leads to an inverse effect. One might also intuitively argue that the higher the ATP concentration, the less ADP is available. ADP is a stronger inhibitor than ATP, which is reflected in their inhibition constants $K_{ADP} = 1.5$ and $K_{ATP} = 2.5$. Thus, a higher ATP concentration has a positive effect on the rate, and the kinetic order is positive.

Pyruvate kinase reaction. The last reaction in the fermentation pathway is the conversion of phosphoenolpyruvate to ethanol. It is catalyzed by pyruvate kinase and follows a complicated rate law that was proposed by Monod et al. (1965) and modified by Hess and Plesser (1979). Its details are presented in Galazzo and Bailey, and it suffices here to present the GMA parameters, which again were obtained by partial differentiation, as has been demonstrated many times. The reaction has two substrates, phosphoenolpyruvate and ADP, and is allosterically activated by fructose-1,6-diphosphate and affected by pH. Curto et al. (1995) compute the appropriate parameter values as

$$\begin{aligned}
h_{43} &= 0.05, \\
h_{44} &= 0.533, \\
h_{45} &= -0.0822, \\
h_{4,10} &= 1, \\
\beta_4 &= 0.0945.
\end{aligned} \tag{8.46}$$

Dynamics of ATP. The only dependent variable that is not yet explicitly specified is ATP, X_5. Actually, some of the contributors to its dynamics have been developed already. Two processes lead to increases in the ATP concentration: V_{GAPD} and V_{PK}. The traditional rate law for V_{GAPD} was developed above. Relevant here is differentiation with respect to ATP, which appears in the expression S of the rate law (see above). The result is

$$g_{53,1} = h_{33,1} = g_{43} = 0.6159,$$
$$g_{55,1} = h_{45,1} = g_{45} = 0.1308,$$
$$g_{59,1} = h_{39,1} = g_{49} = 1, \tag{8.47}$$
$$g_{5,14,1} = h_{3,14,1} = g_{4,14} = -0.6088,$$
$$\alpha_{5,1} = \alpha_4 = 2\beta_{31} = 0.022.$$

As stated above, the pyruvate kinase step produces ATP, and its kinetic orders were computed before. We recall

$$g_{53,2} = h_{43} = 0.05,$$
$$g_{54,2} = h_{44} = 0.533,$$
$$g_{55,2} = h_{45} = -0.0822, \tag{8.48}$$
$$g_{5,10,2} = h_{4,10} = 1,$$
$$\alpha_{5,2} = \beta_4 = 0.0945.$$

ATP is used up in four steps of the pathway: the hexokinase reaction, the second step in the production of polysaccharides, the phosphofructokinase reaction, and an ATPase reaction, which we have not yet encountered. The hexokinase reaction yields the parameter values

$$h_{51,1} = h_{11} = g_{21} = 0.7464,$$
$$h_{55,1} = h_{15} = g_{25} = 0.0243,$$
$$h_{57,1} = h_{17} = g_{27} = 1, \tag{8.49}$$
$$\beta_{5,1} = \beta_1 = \alpha_2 = 2.8632.$$

The polysaccharide production step contributes

$$h_{52,2} = h_{22,2} = 8.6107,$$
$$h_{5,11,2} = h_{2,11,2} = 1, \tag{8.50}$$
$$\beta_{5,2} = \beta_{2,2} = 0.0009.$$

The relevant parameters of the phosphofructokinase reaction are

$$h_{52,3} = h_{22,1} = g_{32} = 0.7318,$$
$$h_{55,3} = h_{25,1} = g_{35} = -0.3941,$$
$$h_{58,3} = h_{28,1} = g_{38} = 1, \tag{8.51}$$
$$\beta_{5,3} = \beta_{2,1} = \alpha_3 = 0.5232.$$

The ATPase reaction is represented by a first-order kinetic of the type

$$V_{\text{ATPase}} = V_{\text{ATPase}}^{max}[\text{ATP}], \tag{8.52}$$

where the maximal enzyme activity V_{ATPase}^{max} is modeled as the independent variables X_{13}. Thus,

$$h_{55,4} = 1,$$
$$h_{5,13,4} = 1. \tag{8.53}$$

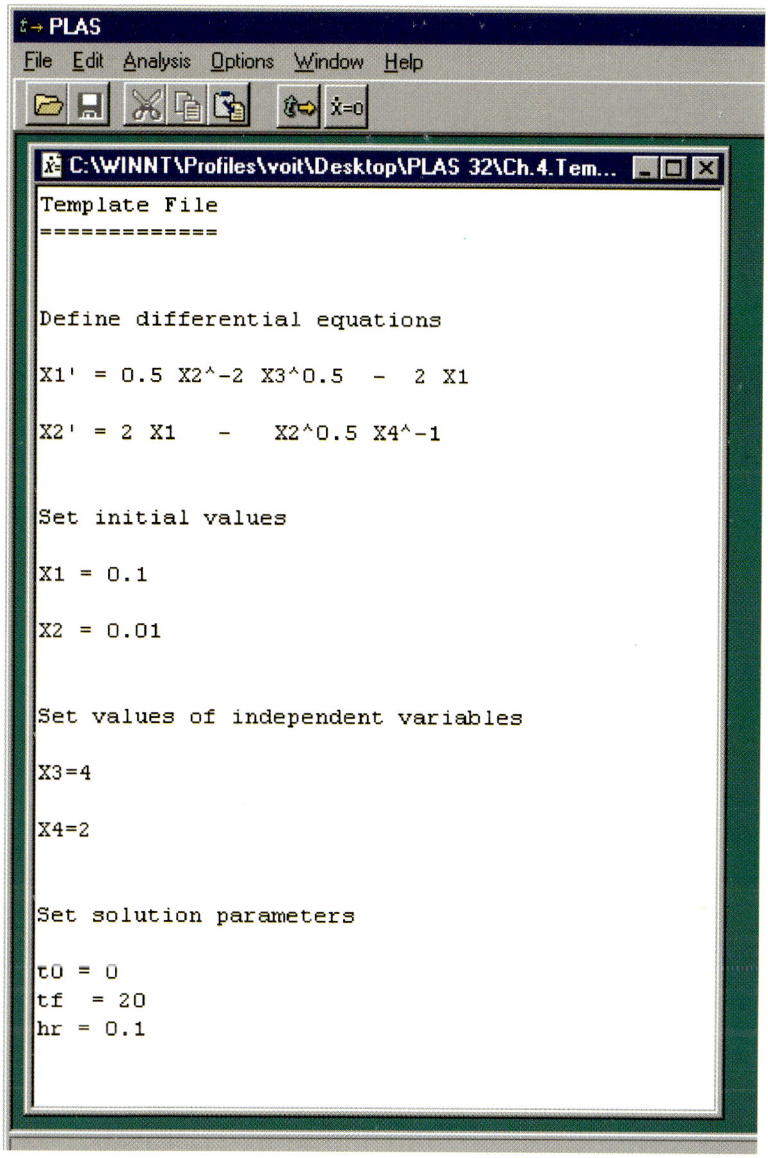

PLAS
File Edit Analysis Options Window Help
ẋ=0
C:\WINNT\Profiles\voit\Desktop\PLAS 32\Ch.4.Tem...
Template File
=============
Define differential equations
X1' = 0.5 X2^-2 X3^0.5 - 2 X1
X2' = 2 X1 - X2^0.5 X4^-1
Set initial values
X1 = 0.1
X2 = 0.01
Set values of independent variables
X3=4
X4=2
Set solution parameters
t0 = 0
tf = 20
hr = 0.1

PLATE I. Template file (*Template.plc*) shows with a simple example how equations, initial values, and other settings are implemented in PLAS. See text, p. 100.

ẋ=0
C:\WINNT\Profiles\voit\Desktop\PLAS 32\Ch.4.Phospho1.plc
File Phospho1.plc
=================
Define differential equations
X1' = a1 X4^g14 X6^g16 - b1 X1^h11 X2^h12 X7^h17
X2' = a2 X1^g21 X2^g22 X5^g25 X7^g27 X10^g210 - b2 X2^h22 X3^h23 X8^h28
X3' = a3 X2^g32 X3^g33 X8^g38 - b3 X3^h33 X9^h39

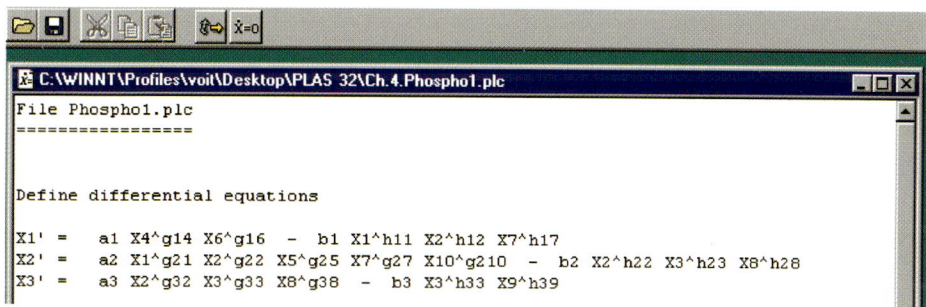

PLATE II. Equations (4.1) as defined and implemented in PLAS. See text, p. 101.

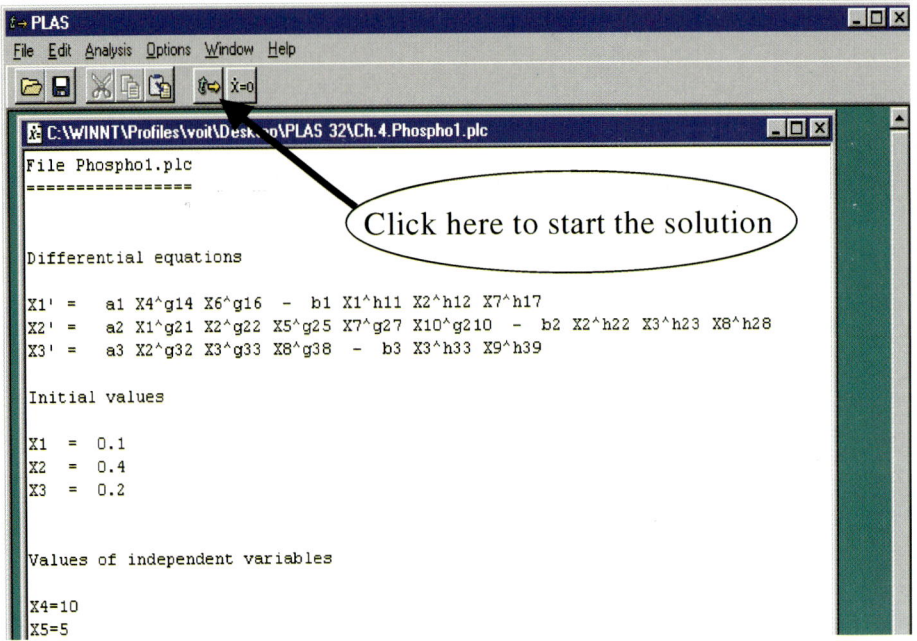

PLATE III. Initiate the solution in PLAS by clicking the button ![]. See text, p. 106.

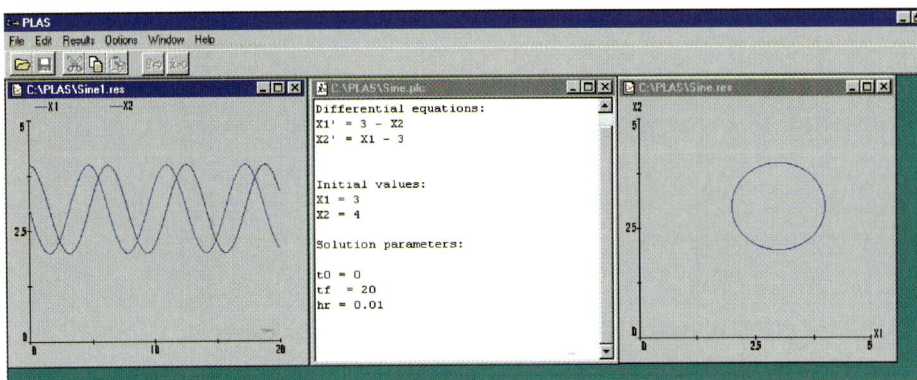

PLATE IV. Analysis of a simple harmonic oscillation. The program and result windows are tiled. The left panel contains a typical time plot, while the right panel shows a phase-plane plot. See text, p. 111.

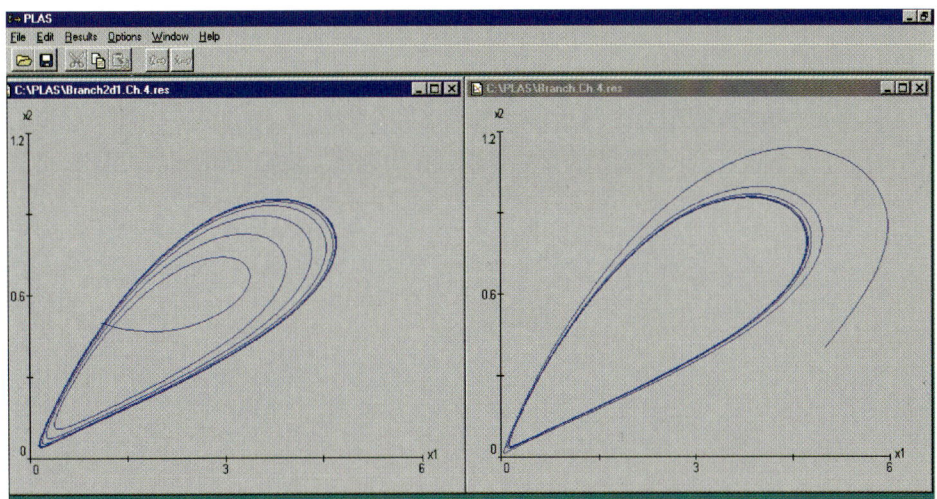

PLATE V. Tiled PLAS output of *Branch16.plc*, showing how a stable limit cycle is approached from either the inside or the outside. See text, p. 114.

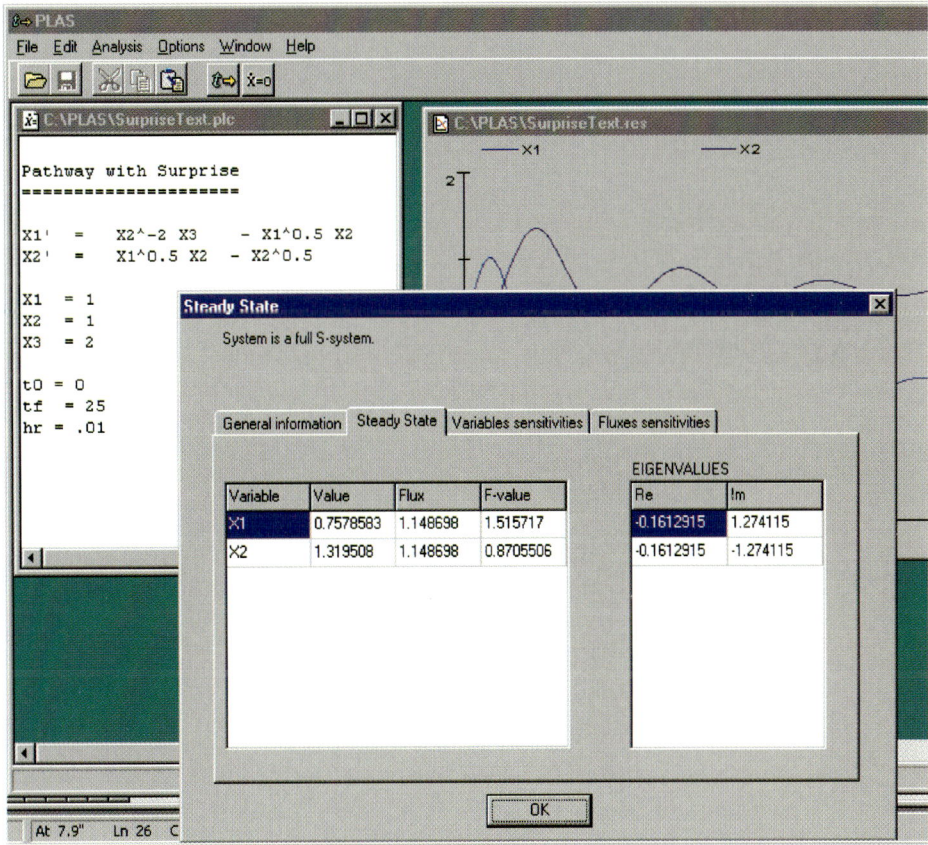

PLATE VI. Steady-state solution of the pathway with surprise [Eq. (4.13)], displayed in PLAS. See text, p. 120.

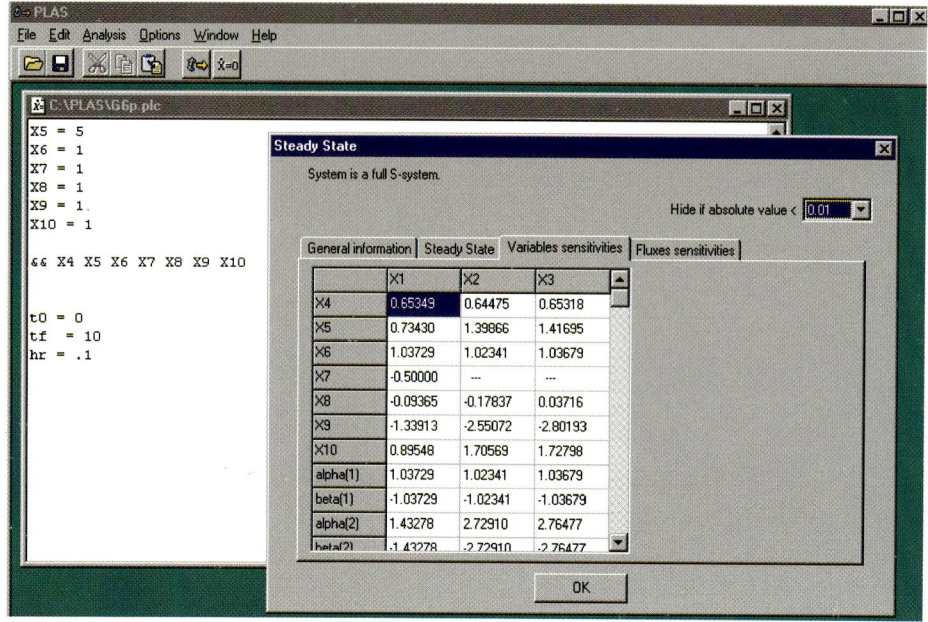

PLATE VII. PLAS displays logarithmic gains and sensitivities. See text, p. 121.

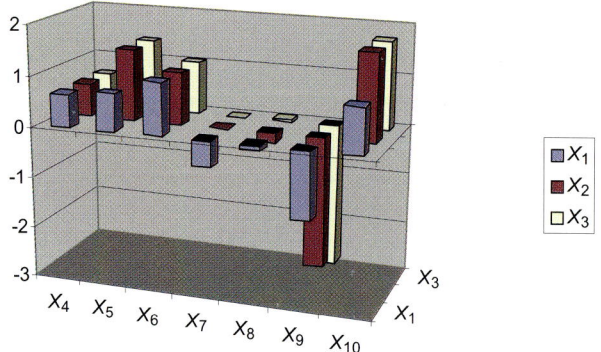

PLATE VIII. Pseudo-three-dimensional representation of gains of the glucose-phosphate pathway, using Excel®. See text, p. 123.

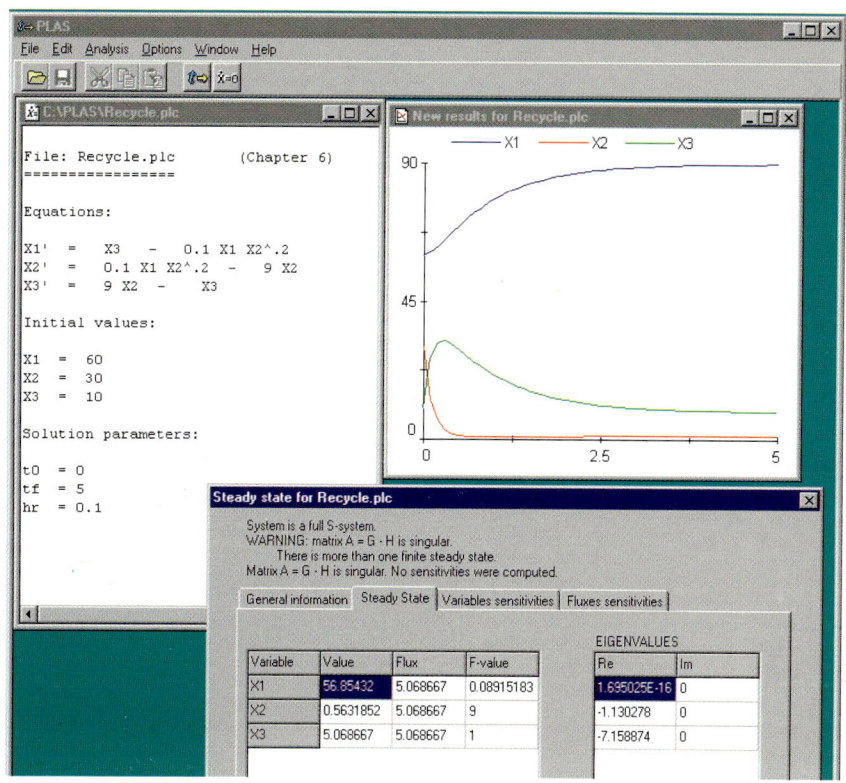

PLATE IX. PLAS computes a steady state, but warns that it is just one of many. See text, p. 199.

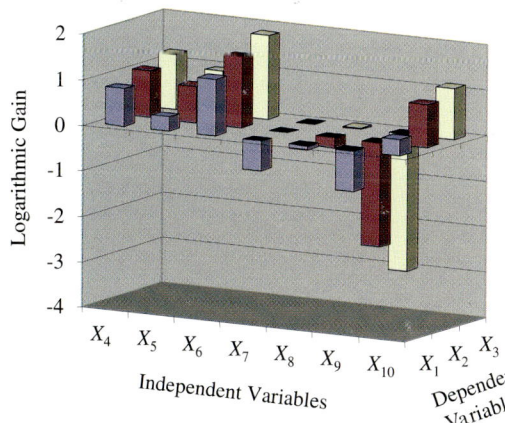

PLATE X. Logarithmic gain profile of a model of glycolysis discussed in Chapter 11. Note the difference in magnitude between the effects of X_8 and X_9. See text, p. 229.

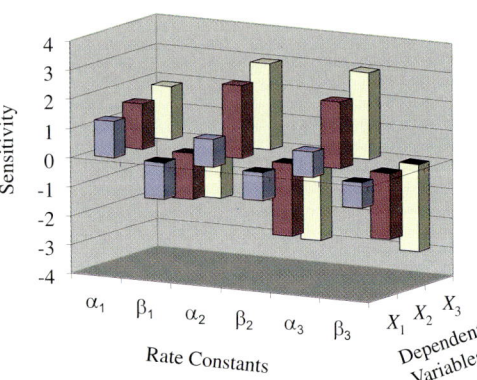

PLATE XI. Unconstrained rate constant sensitivities of the model of glycolysis discussed in Chapter 11. See text, p. 238.

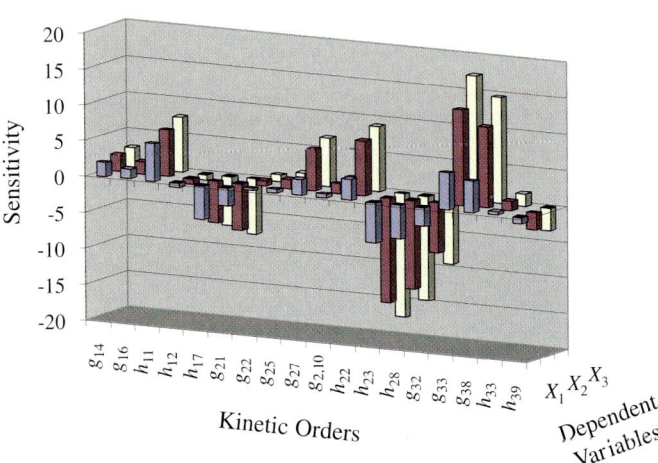

PLATE XII. Unconstrained kinetic order sensitivities of the model of glycolysis discussed in Chapter 11. Unconstrained, some of the kinetic orders appear to be high, but if constraints are taken into account, they are within a reasonable range. For instance, the highest sensitivities shown are associated with h_{2i} and g_{3i} ($i = 2, 3, 8$), which in fact describe the same process and cancel each other. See text, p. 243.

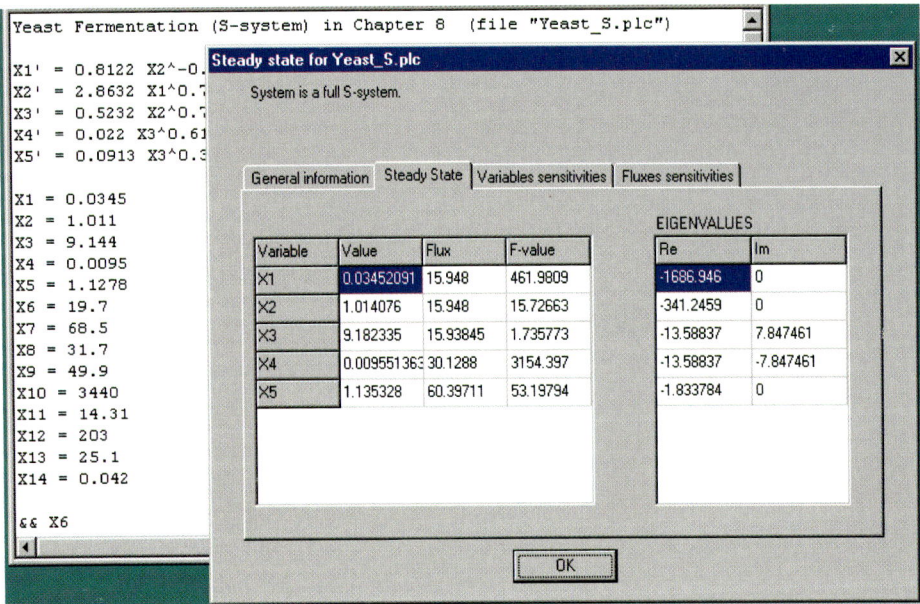

PLATE XIII. PLAS output characterizing the steady state of the S-system model *Yeast_S.plc*. See text, p. 280.

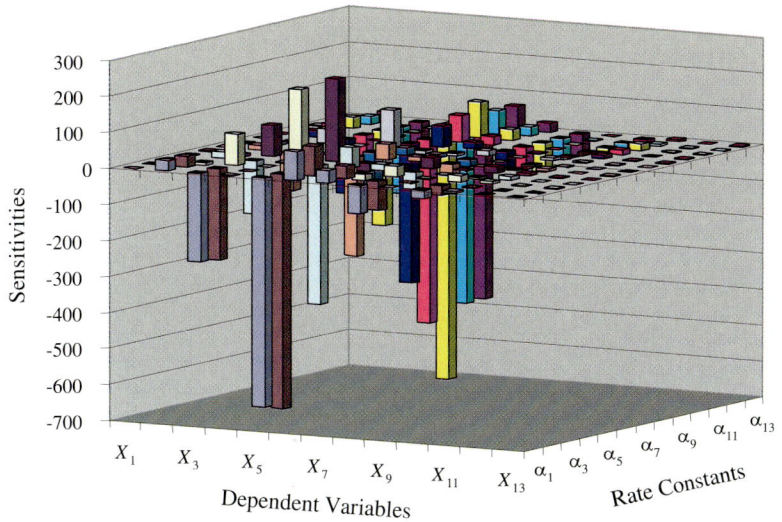

PLATE XIV. Rate constant sensitivities of the original model. See text for interpretation. See text, p. 312.

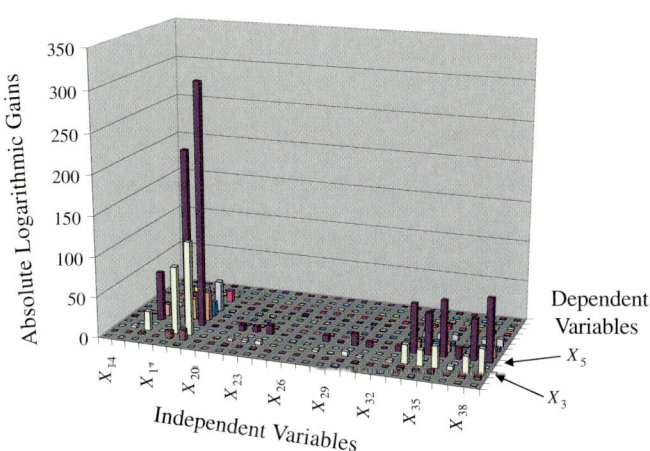

PLATE XV. Logarithmic gains of metabolites with respect to enzyme activities. See text, p. 313.

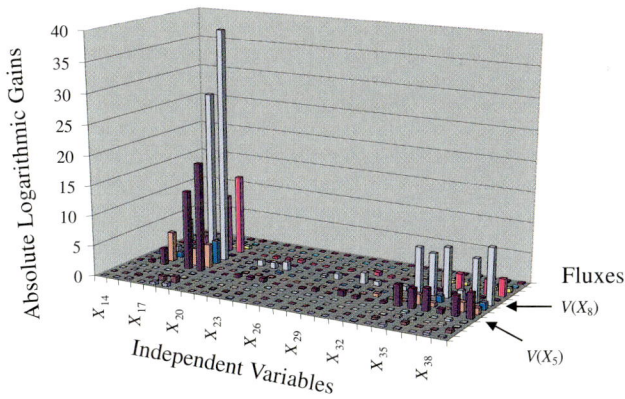

PLATE XVI. Logarithmic gains of fluxes with respect to enzyme activities. See text, p. 313.

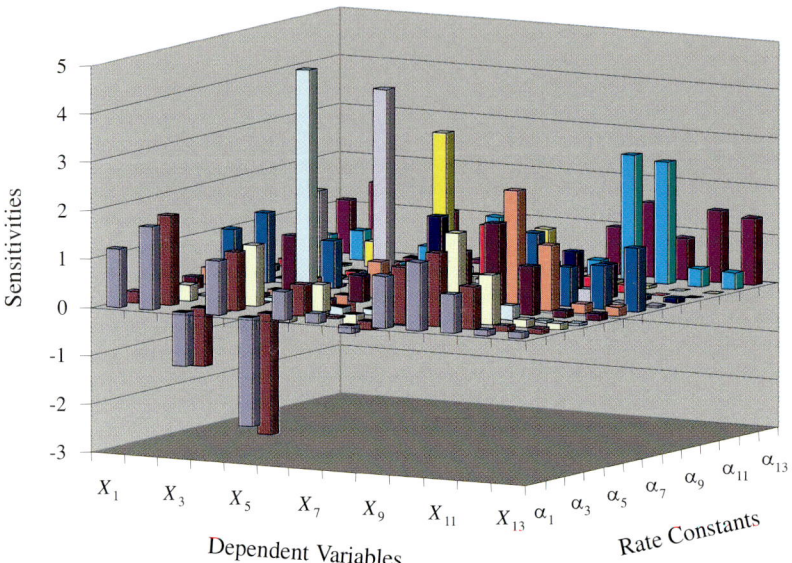

PLATE XVII. Rate constant sensitivities of the modified model. Note the difference in scale of the vertical axis in comparison with Fig. 9.4 (Colorplate VII), indicating the enormous reduction in model sensitivity and consequent increase in robustness. See text, p. 321.

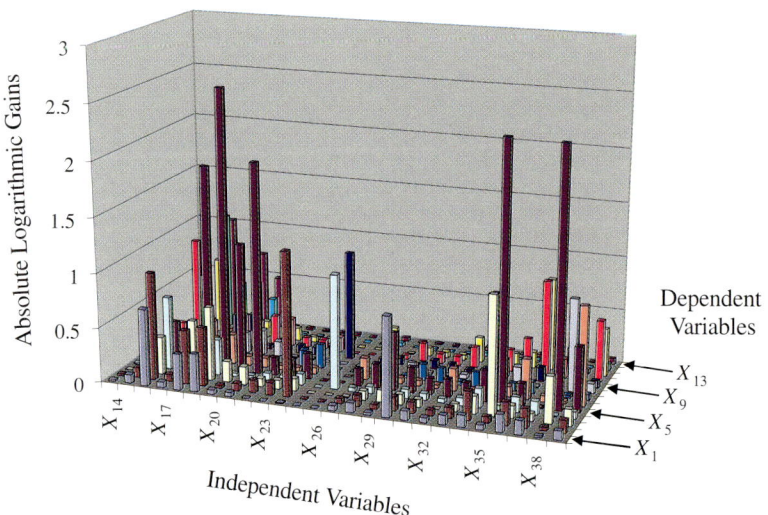

PLATE XVIII. Absolute values of logarithmic gains of the modified model. All gains are now reasonable, the gain profile is much better balanced than in the original model, and the model is generally more robust. Note the difference in scale of the vertical axis in comparison with Fig. 9.5 (Colorplate VII). See text, p. 321.

As a product of two variables, this traditional rate law already is in power-law form, and the rate is

$$\beta_{5,4} = 1. \tag{8.54}$$

We have thus identified the entire GMA model. It reads

$$\begin{aligned}
\dot{X}_1 &= 0.8122 X_2^{-0.2344} X_6 - 2.8632 X_1^{0.7464} X_5^{0.0243} X_7, \\
\dot{X}_2 &= 2.8632 X_1^{0.7464} X_5^{0.0243} X_7 - 0.5232 X_2^{0.7318} X_5^{-0.3941} X_8 \\
&\quad - 0.0009 X_2^{8.6107} X_{11}, \\
\dot{X}_3 &= 0.5232 X_2^{0.7318} X_5^{-0.3941} X_8 - 0.011 X_3^{0.6159} X_5^{0.1308} X_9 X_{14}^{-0.6088} \\
&\quad - 0.04725 X_3^{0.05} X_4^{0.533} X_5^{-0.0822} X_{12}, \\
\dot{X}_4 &= 0.022 X_3^{0.6159} X_5^{0.1308} X_9 X_{14}^{-0.6088} - 0.0945 X_3^{0.05} X_4^{0.533} X_5^{-0.0822} X_{10}, \\
\dot{X}_5 &= 0.022 X_3^{0.6159} X_5^{0.1308} X_9 X_{14}^{-0.6088} + 0.0945 X_3^{0.05} X_4^{0.533} X_5^{-0.0822} X_{10} \\
&\quad - 2.8632 X_1^{0.7464} X_5^{0.0243} X_7 - 0.0009 X_2^{8.6107} X_{11} \\
&\quad - 0.5232 X_2^{0.7318} X_5^{-0.3941} X_8 - X_5 X_{13}.
\end{aligned} \tag{8.55}$$

It is implemented in PLAS as file *Yeast_G.plc*.

S-System Representation

The corresponding S-system is readily obtained from the GMA model by aggregating into one power-law term collections of α-terms or β-terms that appear in any one equation. The equations for X_1 and X_4 are already in S-system form; in the equations of X_2 and X_3, two β-terms have to be aggregated into one, and in the equation of X_5, two α-terms are aggregated and four β terms. The procedure was shown in Chapter 3 and is only demonstrated once here. The remaining aggregations are left as Exercise 10.

Aggregation of the $\beta_{2,1}$- and $\beta_{2,2}$-terms. At steady state, the aggregated β_2-term of the second S-system equation is exactly equivalent to the sum of the $\beta_{2,1}$- and $\beta_{2,2}$-terms of the GMA model. Furthermore, we know which variables are included in this S-system term: all variables that directly affect the degradation of X_2, no matter through which branch. In other words, all variables that are part of any of the corresponding GMA terms are included in the S-system term. Thus, we assert symbolically

$$\beta_2 X_2^{h_{22}} X_5^{h_{25}} X_8^{h_{28}} X_{11}^{h_{2,11}} = \beta_{2,1} X_2^{h_{22,1}} X_5^{h_{25,1}} X_8^{h_{28,1}} + \beta_{2,2} X_2^{h_{22,2}} X_{11}^{h_{2,11,2}}. \tag{8.56}$$

The parameters of the S-system term on the left-hand side are computed from the expression on the right-hand side (RHS) by partial differentiation, as we have done many times. As an example,

$$h_{22} = \frac{\partial \mathrm{RHS}}{\partial X_2} \cdot \frac{X_2}{\mathrm{RHS}}, \tag{8.57}$$

and this expression is evaluated at the steady state. The result is

$$h_{22} = \frac{h_{22,1} \cdot \left(\beta_{2,1} X_2^{h_{22,1}} X_5^{h_{25,1}} X_8^{h_{28,1}}\right) + h_{22,2} \cdot \left(\beta_{2,2} X_2^{h_{22,2}} X_{11}^{h_{2,11,2}}\right)}{\beta_{2,1} X_2^{h_{22,1}} X_5^{h_{25,1}} X_8^{h_{28,1}} + \beta_{2,2} X_2^{h_{22,2}} X_{11}^{h_{2,11,2}}}. \quad (8.58)$$

Numerically, we obtain

$$h_{22} = \frac{0.7318 \times 30.11 + 8.6107 \times 0.014}{30.11 + 0.014} = 0.735. \quad (8.59)$$

In the same way, h_{25}, h_{28}, and $h_{2,11}$ are obtained as -0.394, 0.999, and 0.001, respectively. The rate constant β_2 derives from equating the S-system term and the sum of GMA terms at the steady state and solving for β_2:

$$\beta_2 = \frac{\beta_{2,1} X_2^{h_{22,1}} X_5^{h_{25,1}} X_8^{h_{28,1}} + \beta_{2,2} X_2^{h_{22,2}} X_{11}^{h_{2,11,2}}}{X_2^{h_{22}} X_5^{h_{25}} X_8^{h_{28}} X_{11}^{h_{2,11}}}; \quad (8.60)$$

its numerical value is 0.5239 (see Exercise 11).

S-system model. Once the various terms in the second, third, and fifth equations of the GMA model are aggregated, the S-system model is completely determined. It is emphasized that the aggregation is purely a matter of mathematical manipulations and that, in particular, no further estimations are necessary. One should also note that the strategy of formulating the GMA model first and deriving the S-system model in a second step is not the only option, at least in theory. If appropriate experimental measurements are available, the S-system parameters may be estimated directly. For instance, suppose one measured the overall degradation of glucose-6-phosphate as a function of substrate, ATP, phosphofructokinase, and the activities of enzymes involved in polysaccharide production. Given these data, one could directly estimate the β_2-term of the S-system model, for instance, by plotting the rate against the substrate concentrations in logarithmic coordinates and measuring the slope at the steady-state point. In practice, these measurements are often not available, and if traditional measurements are at hand, the strategy of setting up the GMA model first is the method of choice.

The resulting S-system model reads

$$\begin{aligned}
\dot{X}_1 &= 0.8122 X_2^{-0.2344} X_6 - 2.8632 X_1^{0.7464} X_5^{0.0243} X_7, \\
\dot{X}_2 &= 2.8632 X_1^{0.7464} X_5^{0.0243} X_7 - 0.5239 X_2^{0.735} X_5^{-0.394} X_8^{0.999} X_{11}^{0.001}, \\
\dot{X}_3 &= 0.5232 X_2^{0.7318} X_5^{-0.3941} X_8 \\
&\quad - 0.0148 X_3^{0.584} X_4^{0.03} X_5^{0.119} X_9^{0.944} X_{12}^{0.056} X_{14}^{-0.575}, \\
\dot{X}_4 &= 0.022 X_3^{0.6159} X_5^{0.1308} X_9 X_{14}^{-0.6088} - 0.0945 X_3^{0.05} X_4^{0.533} X_5^{-0.0822} X_{10}, \\
\dot{X}_5 &= 0.0913 X_3^{0.333} X_4^{0.266} X_5^{0.024} X_9^{0.5} X_{10}^{0.5} X_{14}^{-0.304} \\
&\quad - 3.2097 X_1^{0.198} X_2^{0.196} X_5^{0.372} X_7^{0.265} X_8^{0.265} X_{11}^{0.0002} X_{13}^{0.47}.
\end{aligned} \quad (8.61)$$

It is implemented in PLAS as file *Yeast_S.plc*.

Model Analyses

Now that we have achieved our goal of formulating the GMA and the S-system models, we must ask what we can do with them. The sky is the limit at this point, but the most interesting options fall into several categories, which are depicted in Fig. 8.5 (see also Voit and Torres, 1998).

Properties of the model. Before any other analyses are executed, it is mandatory to subject the model to a number of tests that confirm observed properties, indicate desirable features, and generally give us some confidence that the model is consistent and realistic.

The first set of tests is associated with the steady state. Does the model reflect the observed steady state? Is the steady state locally stable? Is the model robust, in the sense that it can tolerate realistic perturbations? Is the model overly sensitive to changes in independent variables or in parameter values?

The second tier of tests addresses the dynamic features. How fast does the system return to its steady state? Do perturbations induce overshoots, undershoots, or oscillations? If so, are they realistic? Have they been observed?

Finally, one may ask about the structure of the system. Why is there a feedforward activation? How does this model compare with a model that does not show this activation? Why is the first step subject to feedback inhibition, and not some later step?

Obviously, a complete analysis of a moderately complicated model requires some effort. Here, we will demonstrate some of the typical steady-state tests and explore some features of the system dynamic, but we will not analyze questions of structure and design. The interested reader may find such analyses (of other models), for instance, in Savageau (1976), Irvine and Savageau (1995ab), Savageau and Sands (1991), and Sorribas and Cascante (1994).

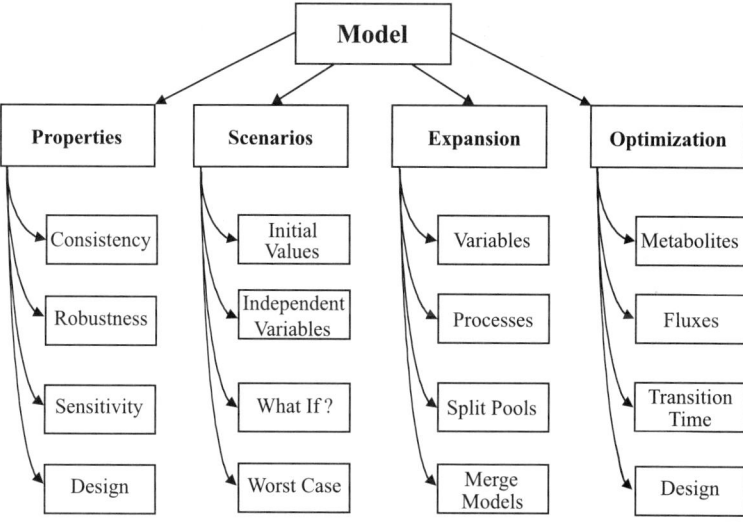

Figure 8.5. Some uses of a model.

Steady state. In principle, the steady state of the S-system could be computed with paper and pencil (see Chapter 6), but we simply implement the system in PLAS and let PLAS do the work. The steady state of the GMA model cannot be computed algebraically, and even in PLAS we must run the dynamical solution to find out to whether the system will reach a steady state and what its characteristics are.

The two models are implemented as *Yeast_S.plc* and *Yeast_G.plc*. Loading *Yeast_S.plc* and clicking the steady-state button demonstrates that the model steady state indeed coincides with the observed steady state, in terms of both metabolites and fluxes. The results are not perfectly precise, but the deviations are minor and due to rounding in the parameter values. The steady-state window in PLAS (Fig. 8.6) also presents the eigenvalues of the system. Most important is that all real parts are negative, which indicates that the steady state is locally stable. Two eigenvalues are complex conjugate, which suggests the potential for oscillatory behavior.

Logarithmic gains. The next check pertains to logarithmic gains and sensitivities. If any of these are in a range of 10 or higher, the model will respond to minor changes or perturbations with quite drastic responses. The logarithmic gains are presented in Table 8.3. The value highest in magnitude is $L(X_4, X_{10}) = -1.77835$, which implies that a 1% change in pyruvate kinase activity leads to a decrease in phosphoenolpyruvate that is still less than 2%. All other responses are even smaller.

Table 8.3 makes it clear that all variables are affected if one of the independent variables is changed. However, the amount of the effect varies by orders of magnitude. The effect of glucose transport (X_6) on phosphoenolpyruvate (X_4) is noticeable, whereas the effect of hexokinase activity (X_7) on phosphoenolpyruvate (X_4) is essentially negligible. The system is thus responsive to perturbations, but it does not react

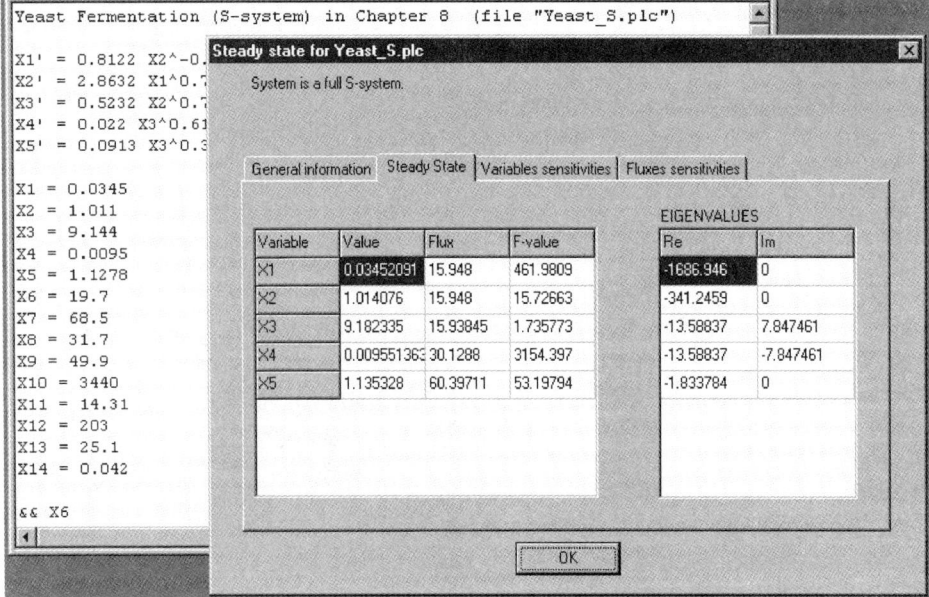

Figure 8.6 (see Color Plate XIII). PLAS output characterizing the steady state of the S-system model *Yeast_S.plc*.

Table 8.3. Logarithmic Gains of S-System Model

	X_1	X_2	X_3	X_4	X_5
X_6	0.907066	1.30747	0.974842	1.30629	0.678832
X_7	−1.33985	0.000214804	−0.00019537	3.7751E-06	0.000528505
X_8	0.286895	−0.937332	0.314442	0.425491	0.229315
X_9	0.00016125	−0.00040912	−1.62386	0.151678	−0.0010066
X_{10}	−0.0177137	0.0449424	0.0505179	−1.77835	0.110577
X_{11}	0.000576826	−0.00167314	−0.00064226	−0.00131272	−0.00157853
X_{12}	0.0173348	−0.0439812	−0.0509291	−0.0973167	−0.108212
X_{13}	0.145591	−0.369387	0.335966	−0.0064918	−0.908843
X_{14}	−0.00031271	0.0007934	0.988585	−0.091827	0.00195209

in a disastrous fashion, where a small perturbation would evoke drastic changes in the balance of metabolites.

Similarly, we can compute the effect of changes in independent variables on any of the fluxes in the pathway. For aggregated fluxes the flux logarithmic gains are easily obtained in PLAS. The results are shown in Table 8.4. Again, the degree of effect varies widely, and again, the system is sufficiently responsive. As for the concentration gains, V_{In} (X_6) has the strongest effect on the system, followed by phosphofructokinase. It is noted that this pattern depends on the experimental conditions. For instance, Cascante et al. (1995b) found a significant effect of ATPase if the cells were immobilized at pH 5.5 (see Exercise 18).

It is also possible to determine the effect of any of the independent variables on non-aggregated fluxes, such as v_{GOL}. For this purpose, we need to express these fluxes in terms of aggregated fluxes. For instance, studying the map or the GMA system, we may define

$$J_1 = v_{PK} = \text{ethanol production},$$
$$J_2 = V_{2,2}^- = v_{POL}, \qquad (8.62)$$
$$J_3 = V_{3,2}^- = v_{GOL},$$

Table 8.4. Logarithmic Flux Gains of S-System Model

	V_1	V_2	V_3	V_4	V_5
X_6	0.693529	0.693529	0.689278	0.689196	0.688388
X_7	−5.0350E-05	−5.0350E-05	−5.1090E-05	−5.1200E-05	−5.1370E-05
X_8	0.219711	0.219711	0.223687	0.223659	0.223393
X_9	9.58966E-05	9.58966E-05	9.73064E-05	−0.00026598	−0.00042271
X_{10}	−0.0105345	−0.0106894	0.0455774	0.0464359	0.0464359
X_{11}	0.000392184	0.000392184	−0.00060230	−0.00060204	−0.00060094
X_{12}	0.0103092	0.0103092	0.0104607	−0.0455213	−0.0454427
X_{13}	0.0865844	0.0865844	0.0878573	0.088045	0.0883377
X_{14}	−0.00018597	−0.00018597	−0.00018871	0.000325008	0.000819765

and assert

$$V_1 = V_2 = J_1/2 + J_2 + J_3/2,$$
$$V_3 = J_1/2 + J_3/2, \qquad (8.63)$$
$$V_4 = J_1.$$

Solving these equations for J_1, J_2, and J_3, one obtains

$$J_1 = V_4,$$
$$J_2 = V_1 - V_3, \qquad (8.64)$$
$$J_3 = 2V_3 - V_4.$$

The computation of gains is now a matter of applying the chain rule. For example, the effect of changes in X_6 on polysaccharide production may be written as $L(J_2, X_6)$. Symbolically, it is computed as

$$L(J_2, X_6) = \frac{\partial J_2}{\partial X_6} \cdot \frac{X_6}{J_2} = \frac{\partial(V_1 - V_3)}{\partial X_6} \cdot \frac{X_6}{V_1 - V_3} = \left(\frac{\partial V_1}{\partial X_6} - \frac{\partial V_3}{\partial X_6}\right)\frac{X_6}{V_1 - V_3}, \qquad (8.65)$$

which, as always, is evaluated at the steady state. The expression can be reformulated in terms of the flux gains in Table 8.4, and the result is

$$L(J_2, X_6) = \frac{V_1}{V_1 - V_3} L(V_1, X_6) - \frac{V_3}{V_1 - V_3} L(V_3, X_6). \qquad (8.66)$$

Using the steady-state values computed in PLAS (see Fig. 8.6), we obtain

$$L(J_2, X_6) = \frac{15.948}{0.0095} \times 0.693529 - \frac{15.9385}{0.0095} \times 0.689278 = 7.8256. \qquad (8.67)$$

Because the difference between the fluxes V_1 and V_3 is small, this gain factor is rather sensitive and must be considered with caution (see Exercise 12). Nonetheless, the gain indicates that changes in the glucose transport step have a significant effect on polysaccharide production. This production step is also fairly sensitive to changed activities of phosphofructokinase. In this case, the gain factor is negative (see Exercises 13 and 14).

Sensitivities. The model should be robust not only with respect to variations in external conditions and other independent variables, but also with respect to its parameter values. As discussed in Chapter 7, this aspect of robustness is assessed with the computation of sensitivities. In principle, this could be achieved algebraically, but of course it is much easier to use PLAS. The result is clear: Out of 55 kinetic order sensitivities, 51 are lower than 1, and the remaining four are between 1 and 5. This is about as good as one can expect. Sensitivities in this range indicate that the system is very robust, but still responsive. Like gains, sensitivities depend on the experimental conditions.

Dynamics. In addition to steady-state analyses, the testing phase of a model must include its dynamics. Questions that we may ask in this context are: How fast does the model return to its steady state after a perturbation? How does the model react to moderate – not just very, very small – perturbations? Are the transient responses between a perturbation and the approach to the steady state reasonable? Does the model change its qualitative behavior, for instance, by starting to oscillate?

With a model of moderate size, the number of possible simulations is unlimited (see also Chapter 4). Sorribas et al. (1995) executed many such simulations and came to the conclusion that the model in every way responds in a reasonable fashion (see Exercise 15). None of the transients were found to be extraordinary, no sustained oscillations were observed, and the time for returning to a steady state was generally short.

Again, such a clean bill of health is not to be taken for granted. Suspended and immobilized cells at pH 5.5 exhibited responses that cast doubt on the model under that condition. In the first case, several metabolites entered into strong oscillations, with ATP even exceeding the total concentration of adenylates. A possible explanation was that the assumption of proportionality between v_{GOL} and v_{PK} might not be justified. In the immobilized cells, phosphoenolpyruvate exhibited enormous overshoots when the ratio of NADH to NAD^+ was decreased. When the pyruvate kinase activity was decreased from 3,440 to 3,000, it took over 8 h for phosphoenolpyruvate to reach its steady-state. The new steady-state level was almost 60 mM, as opposed to the original value of 0.5 mM. The conclusion was that some component of phosphoenolpyruvate turnover was not captured correctly with the model under these altered conditions (see also Exercise 18).

Summary on robustness. Overall, the analysis of stability, gains, sensitivities, and model dynamics demonstrates that the S-system model for suspended cells at pH 4.5 is robust: Small changes in initial values, independent variables, or parameters do not alter the steady state much and evoke dynamic responses that appear reasonable. This robustness of the model is no guarantee that the model is in some sense correct, but it does strengthen our confidence.

Robustness is by no means to be expected routinely. An analogous model for immobilized cells at pH 5.5 was found to have serious problems. Even though it described a situation very similar to the one analyzed here, the model turned out to be so sensitive that it had to be considered unreliable and inappropriate (see Exercise 18). Some of the logarithmic gains were found to be higher than 20, and the logarithmic gain of phosphoenolpyruvate with respect to phosphofructokinase activity was over 60. These results alone do not explain the deficiency of the model under these altered conditions, but they may pinpoint parts of the pathway that require closer scrutiny. In this particular case, the high gain factors were associated mainly with phosphoenolpyruvate, and one may surmise that some part of its dynamics are not modeled right under the altered conditions. The parameter values were measured *in vitro* and based on the same traditional rate laws as in suspended cells. Maybe these assumptions are not justified. Maybe the estimated steady-state concentration of phosphoenolpyruvate is different in suspended and immobilized cells and under different pH conditions. Maybe the kinetic behavior of pyruvate kinase is different under these conditions. These questions are discussed in more detail in Sorribas et al. (1995). Whatever the true explanation may be, it is important to note that the model does not just give dubious answers, but that it does help the researcher to focus on problematic areas of the pathway. A similar situation was discussed in great detail by Shiraishi and Savageau (1992a–d, 1993) and will be analyzed in Chapter 9.

Model Expansion

A model is never complete. Even if the model is robust and produces reliable answers, there are always aspects that can be expanded or questions that require a shift in focus. Expansions usually require the addition of variables and associated equations, the change of an independent into a dependent variable, or the addition of processes or signals. The following chapters discuss some expansions and refinements of models, and we will not discuss this topic further in this chapter (however, see Exercise 16).

Optimization

Optimization of biological systems can be studied from two points of view. The first considers evolution as the driving force behind the improvement and eventual optimization of natural designs. For instance, Savageau (e.g., 1976, 1989, 1996) studied many features of metabolic pathways and expression patterns of genes against various criteria of optimality. Meléndez-Hevia and Torres (1988) and Meléndez-Hevia (1990) studied the naturally optimized features of the pentose phosphate cycle, and Heinrich and coworkers (e.g., Heinrich and Schuster, 1996, 1998; Heinrich et al., 1997) explored how evolution led to the optimization of the catalytic properties of enzymes and enzyme systems. Fell and Small (1986) analyzed the boundary conditions for lipogenesis with methods of operations research. These types of analyses lead to insights into the structural design of metabolic pathways, which will provide very valuable guidance once biotechnology is ready to design new pathways from scratch.

The second approach to optimization in biology targets artificial improvements of systems for some given purpose. Specifically in the area of biotechnology, one of the prominent tasks is the development of microbial strains with increased performance in the production of a desired product. Traditionally, this task has involved a series of iterations of mutagenesis followed by selection. In many cases, this procedure has been very successful, with some new strains producing over hundred times more than the original parent strain. However, progress is slowing down in the optimization of the better-known organisms, and it is necessary to develop methods that can identify new avenues to improving yield.

Because of its great biotechnological importance, the general literature in this area is quite large. In the context of most interest here, one line of research has targeted the network stoichiometry (see also Chapter 2), which can be formulated with matrix equations and is therefore amenable to methods of linear optimization. Examples of this type of approach are Papoutsakis and Meyer (1985), Majewski and Domach (1990), and Savinell (1991) and Savinell and Palsson (1992a–d).

The analysis of the network stoichiometry alone ignores valuable information about the kinetic properties of the system. However, these kinetic properties are obviously characterized by nonlinearities, and this has been a stumbling block in the optimization of full biochemical systems. The problem was overcome to some degree by the realization that steady-state equations of S-systems are linear, thus allowing direct optimization of biochemical systems under steady-state conditions with methods of *linear programming*, for which analytical and numerical methods are well established (Voit, 1992a; Regan et al., 1993).

The philosophy of this optimization strategy is as follows. Suppose, we are interested in maximizing the production of a particular metabolite. The experimental tools we have available are (limited) controls of several input variables, which act either as precursors, exogenous products, or intermediates, or as modulators of some of the processes involved. Of course, there are constraints that must not be violated: Input variables can only be varied within a certain physiological range, internal concentrations must not exceed thresholds of toxicity, and the fluxes cannot be increased indefinitely. Furthermore, for continual production the optimized system must be at a steady state. All these constraints are translated into constraints that can easily be imposed upon an S-system model, and within the S-system methodology the problem can be represented directly as a linear program.

Linear programming. We have seen many times that changes in independent variables cause the system to assume a new steady state. This new steady state is characterized by new metabolite levels and flux levels. The question of interest in a biotechnological setting is: Can we find combinations of values of independent variables such that the new steady state is optimal in some sense. A typical criterion of optimality is an increased flux of producing a desired metabolite. In the fermentation pathway of this case study, examples could be the ethanol flux or the flux of glycerol or polysaccharide production.

Mathematically, the steady state of the S-system model is characterized by a matrix **A** that has as many rows as dependent variables (n) and as many columns as the total number of dependent plus independent variables ($n + m$). If the values of the independent variables are not fixed, the steady-state equation is usually underdetermined and has infinitely many solutions. In linear programming, additional linear constraints are added to such an underdetermined system, and the goal is to find the optimal solution that still satisfies the steady-state equation and all constraints. Many books and papers have been written about linear programming. A standard classic in the field is Luenberger (1984); a short, well-written introduction can be found in Strang (1986).

In our example, the constraints address physiological and biochemical limitations of the system. For instance, the transport of glucose into the cell cannot be arbitrarily increased. The maximum of this flux constitutes a typical constraint. Similarly, metabolite levels must not exceed certain thresholds lest the concentrations be toxic to the cell. Again, these limits pose constraints that the optimal solution must not violate. For a non-S-system model, the symbolic characterization of the steady state and of such constraints is usually so difficult that no good solutions exist. By contrast, the problem is mathematically simple for S-system models, because their steady states are computed from linear equations in logarithmic coordinates. The optimization problem is therefore linear (Voit, 1992a; Regan et al., 1993) and can be implemented straightforwardly in any of numerous software packages.

For S-system models, the linear program has the following form:

(1) maximize ln(flux)

subject to

(2) steady-state equations, expressed in logarithms of variables,

(3) ln(dependent or independent variable) ≤ constant,

(4) ln(dependent or independent variable) ≥ constant,

(5) ln(dependent or independent variable) = constant,

(6) ln(variable) unrestricted,

(7) ln(flux) ≤ constant,

(8) ln(flux) ≥ constant,

(9) ln(flux) unrestricted,

(10) ln(flux1/flux2) ≤ constant.

(8.68)

In this formulation, (1) is a typical *objective function* that is linear in the logarithms of the involved dependent and independent variables. (2) assures that the optimized system is in a steady state, no matter what the altered enzyme concentrations are. (3) and (4) constrain variables to stay within certain limits. (5) forces the variable to be at a given value, whereas (6) is an option that permits any real value for the logarithm of a variable and thus any positive real value for the variable itself. (7)–(9) are the corresponding constraints on fluxes, and (10) requires that the logarithm of the flux ratio flux1/flux2 should stay below certain limit. Numerical examples for these constraints are discussed in the following subsection, where we design linear programs for the optimization of ethanol, glycerol, and carbohydrate production in *Saccharomyces cerevisiae* under conditions of a suspended cell culture at pH 4.5.

Maximization of fluxes

Objective function. The first step in setting up a linear program is the definition of the objective function. As an example, we consider the maximization of ethanol production; other optimization tasks are performed in Torres et al. (1997; see also 1996).

The rate of ethanol production is given as the flux V_4^- [cf. Fig. 8.1 and Eq. (8.61)]:

$$V_4^- = 0.0945 X_3^{0.05} X_4^{0.533} X_5^{-0.0822} X_{10}. \tag{8.69}$$

The goal is to find values for the independent variables such that this flux is maximized. Only X_{10} appears in the flux as an independent variable, but of course this flux is affected by the entire pathway. If the problem were unconstrained, the task would be very simple: Just increase the activity of pyruvate kinase (X_{10}) a lot, and the flux will go up concomitantly (see Exercise 17). In reality, this is impossible, and it is necessary to optimize the flux under constraints whose details need to be determined later.

To make the objective function linear, we subject it to a logarithmic transformation. This is legitimate because a function and its logarithm assume the maximum for the same arguments. Thus, the objective function in logarithmic coordinates $y_i = \ln X_i$ reads

$$\ln 0.0945 + 0.05 y_3 + 0.533 y_4 - 0.0822 y_5 + y_{10}. \tag{8.70}$$

Actually, for our maximization purpose we can omit the logarithm of the rate constant, because it simply shifts the maximal solution by a constant. The objective function for the linear program is thus

$$0.05 y_3 + 0.533 y_4 - 0.0822 y_5 + y_{10}. \tag{8.71}$$

Steady-state constraints. The second step is the formulation of the steady-state constraints, again expressed in terms of the logarithms $y_i = \ln X_i$. For instance, the first S-system equation of the pathway is

$$\dot{X}_1 = 0.8122 X_2^{-0.2344} X_6 - 2.8632 X_1^{0.7464} X_5^{0.0243} X_7. \tag{8.72}$$

At steady state, the derivative is set equal to zero, and the β-term is moved to the left-hand side:

$$2.8632 X_1^{0.7464} X_5^{0.0243} X_7 = 0.8122 X_2^{-0.2344} X_6. \tag{8.73}$$

Next we move X_2 and X_6 with their kinetic orders to the left-hand side and the rate constant $\beta = 2.8632$ to the right-hand side. Taking logarithms on both sides yields

$$0.7464 y_1 + 0.2344 y_2 + 0.0243 y_5 - y_6 + y_7 = -1.260. \tag{8.74}$$

The other four steady-state constraints are derived in the same manner, and the result is

$$\begin{aligned}
&0.7464 y_1 - 0.739 y_2 + 0.418 y_5 + y_7 - 0.999 y_8 - 0.001 y_{11} = -1.699, \\
&0.7318 y_2 - 0.584 y_3 + 0.03 y_4 - 0.513 y_5 + y_8 \\
&\quad - 0.944 y_9 - 0.056 y_{12} + 0.575 y_{14} = -3.5615, \\
&0.5659 y_3 - 0.533 y_4 + 0.213 y_5 + y_9 - y_{10} - 0.608 y_{14} = 1.4564, \\
&-0.198 y_1 - 0.196 y_2 + 0.333 y_3 + 0.266 y_4 - 0.348 y_5 - 0.265 y_7 \\
&\quad - 0.265 y_8 + 0.5 y_9 + 0.5 y_{10} - 0.002 y_{11} - 0.47 y_{13} - 0.304 y_{14} = 33.039.
\end{aligned} \tag{8.75}$$

Constraints on enzyme concentrations. Next, decisions must be made about how much each enzyme should be allowed to change. Considerations here are physiological limitations as well as limits given by current biotechnological capabilities. For instance, one must ask how much an enzyme activity could possibly be increased by recombinant DNA techniques. For this example, we allow most of the enzymes to vary between 1 and 50 times their basal values. However, in order to ensure that metabolism outside this particular pathway remains unperturbed, we decide to keep those enzymes at their basal level that divert flux from the target product. For the same reason, we do not allow the ratio NADH/NAD$^+$ to change. These specifications lead to the following constraints on the independent variables:

$$
\begin{aligned}
&2.980 \leq y_6 = 6.892, \\
&4.226 \leq y_7 \leq 8.138, \\
&3.456 \leq y_8 \leq 7.368, \\
&3.910 \leq y_9 \leq 5.840, \\
&8.143 \leq y_{10} \leq 12.055, \\
&y_{11} = 2.66, \\
&y_{12} = 5.313, \\
&3.222 \leq y_{13} \leq 7.134, \\
&y_{14} = -3.17.
\end{aligned}
\tag{8.76}
$$

Constraints on metabolite concentration. In the same vein, we limit the range of variation of the metabolite concentrations, X_1 to X_5, to 20% about their steady-state levels and set their limits to 0.8 and 1.2 of their basal values. Translated into logarithmic coordinates, the constraints read

$$
\begin{aligned}
&-3.604 \leq y_1 \leq -3.199, \\
&-0.212 \leq y_2 \leq 0.193, \\
&1.989 \leq y_3 \leq 2.395, \\
&-4.879 \leq y_4 \leq -4.474, \\
&-0.109 \leq y_5 \leq 0.295.
\end{aligned}
\tag{8.77}
$$

Mass conservation constraint. In the optimization process, the system deviates from the original steady-state operating point, and as a consequence of the aggregation of fluxes at branch points, some of the stoichiometric relationships that hold in the basal solution are no longer precisely satisfied. In order to avoid significant stoichiometric violations, we add the following constraint that reflects the required dependency between fluxes:

$$V_1^+ / V_4^- > 0.5. \tag{8.78}$$

This constraint ensures that the rate V_4^- (ethanol production) cannot be greater than twice the input flux, in view of the splitting of each fructose diphosphate molecule into two molecules of phosphoenol pyruvate (X_4). Taking logarithms, this flux constraint becomes

$$-0.2344 y_2 - 0.05 y_3 - 0.533 y_4 + 0.0822 y_5 + y_6 - y_{10} > -2.84204967. \tag{8.79}$$

The need for these constraints derives from the aggregation strategy for diverging pathways in S-systems. This may appear to be a disadvantage, and one might think that using a GMA model would be better. However, it is precisely the aggregation that makes optimization with methods of linear programming possible. Without aggregation, the steady state cannot be characterized algebraically, and this prevents us from setting up the constraints in the linear program. These issues are discussed in detail in Torres et al. (1997).

With all constraints defined, the linear program is fully specified, except for an "admissible" starting solution, which is one that is not optimal but does not violate any of the constraints. While such a starting solution is not mandatory in all software algorithms, a good initial guess speeds up the optimization. In our particular case, it is easy to provide such a solution: It is the observed steady-state solution. It may not be optimal with respect to ethanol production, but it is certainly admissible.

Summary of optimization results. Since PLAS does not include an algorithm for dealing with linear programs, we will not go through the details of the optimization procedure. Nonetheless, it is instructive to summarize some results obtained by Torres et al. (1997).

The optimal solution under the given constraints is presented in Table 8.5. This solution corresponds to an ethanol flux that is 3.2 times higher than in the original pathway. To achieve this increase, the independent variables (enzyme activities) need to be increased by factors of between 2 and 4.5.

Changing the lower or upper limits of the permissible enzyme concentrations (thus allowing ranges from 0.01 to 50) does not alter these enhancements. When one raises the limits for the metabolite concentrations, allowing up to 90% variation about the basal levels, the ethanol production can be increased to 5.23 times the original value. In all cases, the yield of ethanol production (expressed as $100\% \times 0.5 \times V_4^- / V_1^+$) is 100%, and inspection of eigenvalues confirms that the optimized S-system is stable.

Table 8.5. Steady-State Solution Optimized for Ethanol Production

Metabolite	Base Value	Optimized Value	Optimized/Base
X_1	0.0345	0.0276	0.8
X_2	1.011	1.213	1.2
X_3	9.144	10.973	1.2
X_4	0.0095	0.0114	1.2
X_5	1.1278	0.902	0.8
Enzyme			
X_6	19.7	62.12	3.15
X_7	68.5	245.82	3.59
X_8	31.7	76.73	2.42
X_9	49.9	147.0	2.95
X_{10}	3,440	9,724.42	2.83
X_{13}	25.1	106.67	4.25
Flux			
$V_{In} = V_1^+$	15.96	48.22	3.02
$V_{POL} = V_{2,2}^-$	0.014	0.067	4.79
$V_{GOL} = 2V_{3,2}^-$	1.777	1.97	1.11
$V_{ethanol} = V_{PK} = V_4^-$	30.11	96.44	3.20

In Torres's analysis, the deviation of the total output flux from the total input flux was 2.1%, which again indicates the relatively small magnitude of error introduced by aggregating fluxes in the S-system representation.

In the optimized solution, all metabolites are either at their upper or their lower limit. This is often – but not always – the case in this type of optimization (cf. Voit, 1992a; Torres et al., 1996, 1997). In other cases, the enzymes may be the limiting quantities, while the metabolites are somewhere in the middle section of their admissible ranges.

The optimized solution prescribes the levels of all controllable enzymes that are predicted to yield a 3.2-fold amplified ethanol flux. This prediction is as reliable as the model. The model is an approximation, and as far as this approximation is an appropriate representation of reality, the predicted yields should not be too far off. At the same time, nobody has confirmed (or refuted) the predictions in Table 8.5 yet.

An obvious question is whether less effort could yield a result that is maybe not quite as good but acceptable. Torres et al. (1997) analyzed this question by allowing only one or a limited set of enzyme activities to be altered. Their results are very interesting: If one, two, three, or even four enzyme activities are optimized, the highest increase in ethanol flux is by a mere factor of 1.16, and if five activities are optimized, the increase is still only by a factor of 1.29 (see Exercise 17). These increases are hardly worth the effort. Similar to observations in a different system (Torres et al., 1996), all (or essentially all) enzymes must be modulated to specific degrees in order to produce worthwhile increases in flux. This result is not entirely surprising in well-studied systems like the present fermentation pathway, because years of experimentation have selected strains that are probably close to optimal already. These experiments would most likely have detected successful strains in which only one or two enzymes are overexpressed. By contrast, selection experiments may not have found combinations of five or more specifically altered enzyme activities.

In comparison with ethanol production, which has been studied for many years, the production of carbohydrates and glycerol in the fermentation pathway of yeast has received less attention. Torres and coworkers optimized the pathway with respect to these fluxes and predicted that under the same constraints as in the ethanol optimization, production of glycerol and carbohydrates can be increased about 10-fold and 100-fold, respectively. As mentioned before, the predictions of these optimizations have not yet been scrutinized in the laboratory. However, it is clear that this approach to flux improvement adds a new screening tool to the repertoire of biotechnology.

EXERCISES

1. Discuss advantages and disadvantages of pooling glucose-6-phosphate and fructose-6-phosphate.
2. Compute the concentrations of AMP and ADP from the constraints in Eqs. (8.2) and (8.3) for the concentrations 0.8, 1, and 1.2 of ATP.
3. List all precursor–product relationships in the GMA equations (8.7) and (8.55).

4. Analyze the reliability of kinetic order estimates of ATP in the hexokinase step.

   4.1. Estimate the kinetic order of the hexokinase step with respect to ATP by measuring the slope of the rate in logarithmic coordinates at the steady-state value of ATP.

   4.2. Suppose you perform an additional rate measurement for [ATP] = 0.9 mM. In the first hypothetical case, the result is 0.8 mM min^{-1}, and in the second case, the result is 0.9 mM^{-1}. Estimate the kinetic order in each case from the slope of the rate. Describe and interpret the results.

   4.3. If you have access to a nonlinear estimation procedure, analyze the cases in part 4.2 by first estimating the traditional rate law and then deriving the GMA parameters. Describe and interpret the results.

5. Confirm the estimates $h_{11} = 0.7464$, $h_{15} = 0.0243$, and $\beta_1 = 2.8632$ by executing the differentiation of V_{HK}.

6. Compute the steady-state rate of the phosphofructokinase reaction.

7. Confirm the parameter values of the power-law term representing the phosphofructokinase reaction.

8. Confirm the parameter values of the power-law term representing the glycogen synthetase reaction.

9. Explain the constraint $\alpha_4 = 2\beta_{31}$ in the power-law representation of the GAPD reaction.

10. Aggregate the terms in the third and fifth equations of the GMA model to obtain the corresponding S-system model.

11. Confirm $h_{25} = -0.394$, $h_{28} = 0.999$, $h_{2,11} = 0.001$, and $\beta_2 = 0.5239$ in the β_2-term of the S-system model.

12. Study the sensitivity of the gain $L(J_2, X_6)$ by rounding the steady-state values for fluxes and flux gains to three, two, and one significant digit. Interpret the results.

13. Compute the gain factor of polysaccharide production with respect to phosphofructokinase activity. Explain why it is negative.

14. Compute all gains of the non-aggregated fluxes J_1, J_2, and J_3.

15. Perform dynamic simulations with the S-system model and the GMA model. Change values of independent variables, choose different initial conditions, and slightly change some of the parameter values. Record and summarize your findings.

16. Add ethanol to the model as dependent variable.

    16.1. Formulate the equation for the new variable, and change the existing equations if necessary.

    16.2. Run some simulations, and discuss the results in terms of steady states and dynamics.

    16.3. Include first-order degradation of ethanol. Execute simulations, and discuss the effect of the rate constant.

    16.4. What kinds of questions can be analyzed with this expanded model that could not be analyzed before?

17. Beginning with the original S-system model, explore to what degree the ethanol flux can be increased by varying V_{In}. Record which constraints are violated if the value

of V_{In} is increased too much. Execute the same type of experiment, varying ATP degradation (X_{13}) or both V_{In} and X_{13}.

18. (Project) Analyze an S-system model of the fermentation pathway for immobilized cells at pH 5.5, which was specified by Curto et al. (1995) as

$$\dot{X}_1 = 1.0848 X_2^{-0.3284} X_6 - 1.3588 X_1^{0.4728} X_5^{0.0275} X_7,$$

$$\dot{X}_2 = 1.3588 X_1^{0.4728} X_5^{0.0275} X_7 - 1.1807 X_2^{0.288} X_5^{-0.324} X_8^{0.753} X_{11}^{0.247},$$

$$\dot{X}_3 = 0.8055 X_2^{0.263} X_5^{-0.4303} X_8$$
$$\quad - 0.048 X_3^{0.377} X_4^{0.002} X_5^{0.056} X_9^{0.93} X_{12}^{0.07} X_{14}^{-0.353},$$

$$\dot{X}_4 = 0.0797 X_3^{0.4061} X_5^{0.1473} X_9 X_{14}^{-0.3798} - 0.03 X_3^{0.0001} X_4^{0.023} X_5^{-1.1451} X_{10},$$

$$\dot{X}_5 = 0.0976 X_3^{0.203} X_4^{0.011} X_5^{-0.497} X_9^{0.5} X_{10}^{0.5} X_{14}^{-0.19}$$
$$\quad - 3.5332 X_1^{0.169} X_2^{0.103} X_5^{0.18} X_7^{0.357} X_8^{0.269} X_{11}^{0.088} X_{13}^{0.286},$$

where the independent variables have the following values: $X_6 = 45.6$, $X_7 = 68.5$, $X_8 = 31.7$, $X_9 = 49.9$, $X_{10} = 3,440$, $X_{11} = 14.31$, $X_{12} = 259.7$, $X_{13} = 14.3$, $X_{14} = 0.011$. Compute steady states, analyze stability, logarithmic gains, sensitivities, and model dynamics. For example, perturb the system by altering the values of some of the dependent or independent variables. Select parameters with particularly low or particularly high sensitivities, change their values by 1%, 5%, 10%, and 20% up or down, and analyze system responses. Record your findings, interpret them, and write a report and an executive summary.

REFERENCES

[43–4], [66], [76], [88], [100–1], [143], [146–7], [153], [165–6], [207], [228], [238–9], [250], [278], [287], [295], [319], [332], [344], [349], [355–9], [374–8], [390–1], [409], [422–4], [439–40], [471], [493], [514].

CHAPTER NINE

Case Study 2 – Diagnosis and Refinement of a Model of the Tricarboxylic Acid Cycle in *Dictyostelium discoideum*

This case study is excerpted from a five-part series of articles by Shiraishi and Savageau (1992a–d, 1993) that used methods of canonical modeling to detect and resolve inconsistencies in models of the tricarboxylic acid cycle in the slime mould *Dictyostelium discoideum*. Shiraishi and Savageau compared features of the S-system methodology with corresponding analyses based on Michaelis–Menten kinetics (Wright et al., 1992; Albe and Wright, 1992) and demonstrated the diagnostic and analytical superiority of the S-system approach over the conventional modeling approach. The results of their analyses and the mutual advantages of the two approaches were subsequently discussed in some detail by Albe and Wright (1994), Wright and Field (1994), and Shiraishi and Savageau (1996).

Background

Dictyostelium discoideum is a free-living cellular slime mold with impressive features and abilities (for a layperson's account see Zimmer, 1998). Its complicated life cycle has two phases (Fig. 9.1). In the vegetative phase, the spore develops into a unicellular *myxamoeba*. It consumes bacteria such as *E. coli*, grows, and multiplies for about ten to twelve hours if the food supply is adequate. Under starving conditions, the amoeboid cells enter the second phase, where, responding chemotactically to excitable waves of cAMP, they aggregate to form a *pseudoplasmodium* that is composed of 10,000 to 15,000 cells. The anterior region of this "slug" contains *prestalk* cells; the posterior region is composed of presumptive spore cells. As the slug migrates, it secretes and deposits a slime sheath of mucopolysaccharides. The cells now begin to differentiate, and significant morphological changes ensue. During the *culmination* process, prestalk cells migrate up toward the tip of the organism and begin to form a centrally located cellulose sheath with highly vacuolated cells and rigid cell walls. The migration of the prestalk cells into the sheath pulls the prespore mass off the substrate and toward the tip of the emerging stalk. This process ultimately results in the formation of the fruiting body, or *sorocarp*. The sorocarp consists of thousands of differentiated spore cells that are supported by the cellulose stalk. The stalk cells eventually die, while the spore cells enter into a state of dormancy, until

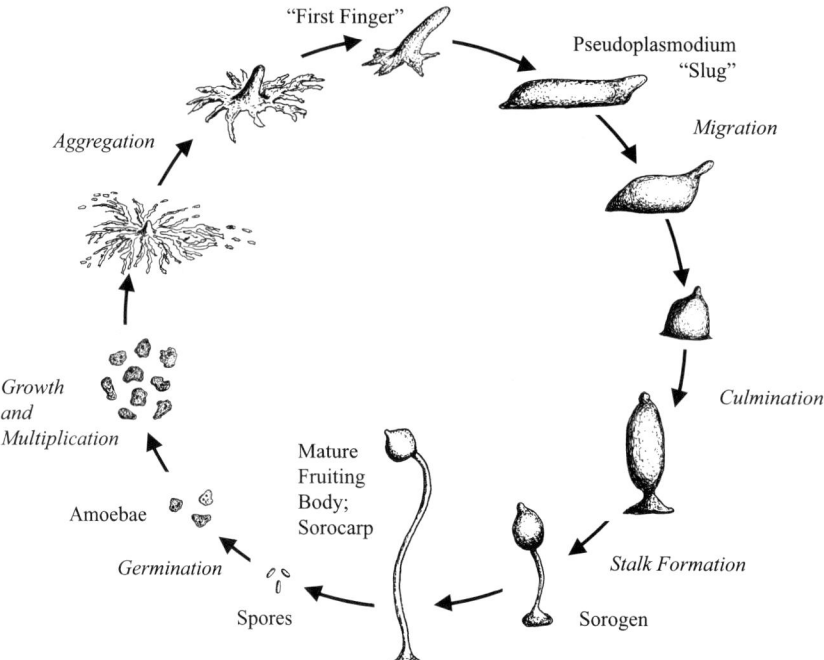

Figure 9.1. Life cycle of *Dictyostelium discoideum*.

the environmental conditions are favorable and a new cycle begins. In the laboratory at 22°C, a complete cycle takes about one day. General descriptions of the development of *Dictyostelium* are found, for instance, in Loomis (1975, 1982), Wilson and Rutherford (1978), Wright and Kelly (1981), and Kimmel (1988). In 1995–1999 alone, several hundred articles about *Dictyostelium* have appeared in the literature.

With respect to kinetic models, *D. discoideum* is intriguing for several reasons. First, the organism is complicated enough to necessitate the use of mathematical models as a framework for understanding its metabolism. Yet, it is convenient from an experimental point of view, and this has resulted in a comprehensive body of data and knowledge about its metabolic pathways. Secondly, during differentiation the cells starve and cease to grow. It is therefore fair to assume that all biochemical processes normally associated with growth are essentially suspended (Wright et al., 1977). Clearly, this greatly simplifies any modeling effort. Third, as one consequence of the cessation of growth, the system is *metabolically closed* (Kelly et al., 1979a). The total amount of carbohydrates per cell is constant, even though their distribution changes during morphogenesis (e.g., White and Sussman, 1961). For instance, glucose is used up while new saccharides like cellulose and trehalose accumulate (Wright et al., 1979). Energy for the differentiation phase is made available through the degradation of its own proteins (Gustafson and Wright, 1972; Wright, 1973). This is accomplished through oxidation in the TCA cycle, which therefore constitutes a crucial component of differentiation. The constancy of total carbohydrates and the balance between protein catabolism on one hand and oxygen consumption and CO_2 production on the other (Kelly et al., 1979b) provide valuable constraints for kinetic models of the system.

The current models of the TCA cycle in *Dictyostelium* have about fifty variables. Models of this degree of complexity don't just appear *de nihilo*. Rather, like *Dictyostelium* itself, they developed from isolated ideas and small models that eventually aggregated, differentiated, and finally came to fruition. It is illuminating to follow some of these developments over the past thirty years, because the types of successive amendments and refinements constitute a fine example of a long-standing modeling effort, its interactions with biochemical experimentation, the permanence of some fundamental questions, and the slight shifting in goals and emphases that reflects trends in biochemical and biomathematical thinking. Since most of these models were created and analyzed by Wright and her coworkers, the general philosophy and nomenclature are consistent throughout the history of these models. This makes it easy to compare the models from the original concepts through numerous amendments up to the most recent models, so to speak in a fast-forward mode.

By modern standards, the first model (Wright, 1968; Wright et al., 1968) was very simple. It described the phosphorylation of uridine 5′-triphosphate and glucose-1-phosphate to uridine diphosphoglucose (UDPG), along with the conversion of UDPG either to end product saccharides or to soluble glycogen. The model also allowed glycogen to be recycled to glucose-1-phosphate. The model consisted of six differential equations. Most processes were modeled as first-order kinetics, but the production of UDPG was represented by a Michaelis–Menten rate law for binary complexes. The model had two purposes. The first, general purpose was to explore the validity of using *in vitro* information for predictions *in vivo*. The second, specific purpose was to analyze the effects of changing the activity of UDPG phosphorylase and the first-order rates of the remaining processes. It is no surprise that the model was not unanimously embraced. Doubts were voiced about the reliability of the data used, about information that was missing or intentionally ignored, and about the appropriateness of the (*in vitro*) Michaelis–Menten rate law for a reaction that was believed *in vivo* to take place on a surface (see discussion following Wright, 1968).

Once the first model was designed, tested, and deemed reasonable, other specific questions were asked. For instance, Wright and Marshall (1971) used the model to explore the effects of changes in enzyme activities (V_{max}-values) on the accumulation of trehalose, which was considered a key marker during differentiation (see also Wilson and Rutherford, 1978).

In a typical model extension, Wright and Gustafson (1972) added to the current system of nine reactions a further reaction of newly recognized importance and fine-tuned the kinetic expressions for four other reactions. As in all models in this series, the kinetic rate laws were of Michaelis–Menten type, though the replacement of maximal activities with time-dependent "activation functions" was considered. The authors also found it useful to develop a specific computer program, METASIM, which replaced an earlier all-purpose differential equation solver provided by IBM.

The paper had three general objectives that today are still at the core of biochemical systems analysis. The first objective was one of intrinsic validity: "to consolidate our knowledge of the metabolism during the differentiation process and make sure it is consistent." The second objective addressed questions of internal control: "to elucidate which factors are most critical to the accumulation of the end products of differentiation." Finally, the third objective was concerned with the interplay between experimentation and mathematical analysis: "to point the way to further biochemical research and experimentation in order to make the mathematical model more explicit."

The next model analysis attempted a more specific ranking of "kinetic positions" of enzymes (Wright and Park, 1975). It was based on simulations of the effects of changed enzyme activities. In canonical modeling, we perform this type of ranking with logarithmic gains and sensitivities, but such an analysis was not executed, presumably because it is very difficult to execute with systems of Michaelis–Menten rate laws (Shiraishi and Savageau, 1992bc). New experimental data suggested some changes in pool sizes of metabolites.

The "fourth expansion" (Wright et al., 1977) strove to "(a) provide a framework within which to judge the relevance of data *in vitro* to conditions operative in the living cells; (b) organize and orient data; hence revealing meaningful experimental approaches; and (c) make specific predictions which can be substantiated." The analysis was primarily a comparison between simulated and observed perturbations in external glucose and their effects. It appears that the authors thought themselves to have reached some sense of closure with this particular model: they called it *fiducial* to indicate "a certain arbitrariness in fixing upon the present model and studying variations from it."

By 1979, the model had reached a size of 20 reactions and 125 parameters, of which 85% were considered factual (Wright et al., 1979). The model was used to simulate permanent alterations and 90-min perturbations in external glucose and uracil conditions. According to the authors, without the model, the experiments would have led to interesting correlations, but could not have been interpreted in a meaningful way. The model, by contrast, allowed interpretations in quantitative, mechanistic terms.

At the same time, the model was expanded in different directions. One extension elucidated a totally new aspect, namely the spatial biochemical heterogeneity of the sorocarp. It had turned out that stalk and spore cells behaved metabolically like two independent cell populations, which suggested that the pathways in the spore and the stalk should be distinguished. This line of analysis almost became a separate theme with variations (Wright et al., 1982; Wright, 1984; Chiew et al., 1985; Wright and Reimers, 1988). Originally, only carbohydrate metabolism was considered, but eventually parts of the citric acid cycle were included, leading to a combined network of 17 pools and 40 reactions.

Following a different route, it was felt that the carbohydrate model should be expanded or complemented with other pathways that were metabolically related. Kelly et al. (1979ab) focused on the tricarboxylic acid (TCA) cycle, which had essentially been ignored before, and "transition models of the citric acid cycle and the hexose–pentose shunt were under construction" (Wright and Kelly, 1981). By this time, the model complexity had grown from 17 parameters in the first model to more than 200 parameters.

The work of Kelly et al. (1979ab) is of particular interest here because of its focus on the TCA cycle. This cycle had been modeled before in other organisms. For instance, a series of papers discussed steady-state and non-steady-state analyses of carbohydrate metabolism in rat liver, including the TCA cycle (Heath, 1968; Heath and Threlfall, 1968; Threlfall and Heath, 1968). Using almost an engineering approach, Sauer et al. (1970) developed a model of the TCA cycle based on a system of linear differential equations to explain radioactive tracer data.

Building upon these ideas, Kelly et al. (1979ab) replaced the linear equations with generalized Michaelis–Menten rate laws and, like Sauer et al. (1970), differentiated between mitochondrial and cytoplasmatic metabolite pools for all metabolites of the TCA cycle. The new focus essentially excluded the carbohydrate pathways that had constituted the core in the earlier models. The TCA model was parameterized primarily from literature data and analyzed with a customized simulation program. The analysis led to predictions about the compartmentalization of enzymes and metabolites between mitochondria and cytoplasm.

The latest TCA model of Wright's group (Wright et al., 1992; Albe and Wright, 1992) is a direct extension of the model of Kelly et al. (1979b). It is based on the same reaction scheme and uses the same types of methods. Its purpose is to provide a framework for characterizing the role of proteins as the major energy source during the differentiation phase of *Dictyostelium*. The model furthermore attempts to reassess one of the original questions, namely, to what degree kinetic measurements obtained *in vitro* are usable for the construction of dynamic models *in vivo* (cf. Wright and Marshall, 1971; Wright et al., 1977; Albe et al., 1989).

Although this model apparently reflects many observations well, a subsequent systematic analysis revealed how treacherous a simulation analysis can be if it is without a solid mathematical foundation. Shiraishi and Savageau (1992a–d, 1993) reanalyzed the model twice, once with the same Michaelis–Menten rate laws that Wright and her coworkers used, and once with a canonical model in the form of an S-system. Close to the operating point, the two models produced essentially identical results, as was to be expected. However, when perturbations were analyzed, the Michaelis–Menten model did not always produce a valid steady state, and conclusions about the presumed steady state were found to be faulty. Because of widely differing dynamic properties among the components of the system, these problems were very difficult to detect with the Michaelis–Menten model, but became directly evident from a systematic S-system analysis of sensitivities and gains. The sensitivity analyses not only indicated that something was wrong with the model, but pinpointed problem portions of the pathway and suggested structurally minor amendments, which nevertheless proved very effective.

This case study summarizes the diagnostic analysis of the model and the remediation of its problems.

Biochemical Map

The tricarboxylic acid cycle in *Dictyostelium* very efficiently produces ATP while decomposing pyruvate to water and CO_2 via acetyl-CoA. Under nutrient-rich conditions, the cycle is fed from ingested proteins that are broken down to amino acids; during periods of starvation, cellular proteins are used up. With reasonable simplifications, the tricarboxylic acid cycle involves 13 dependent metabolites and 26

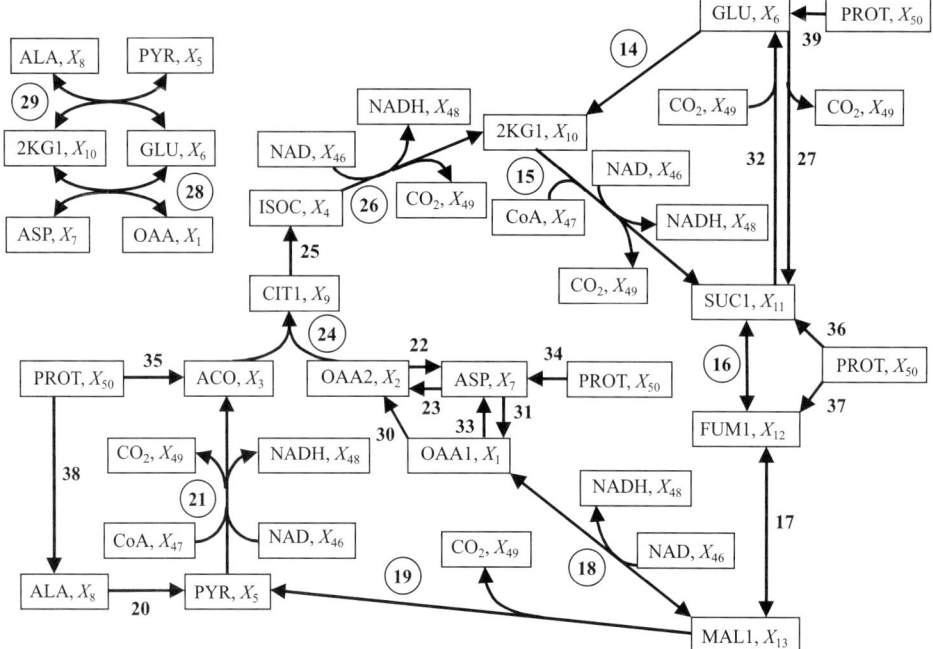

Figure 9.2. Map of the tricarboxylic acid cycle in *Dictyostelium discoideum*, redrawn from Kelly et al. (1979ab), Wright et al. (1992), and Shiraishi and Savageau (1992a). The numbers 1 and 2 were attached to OAA, CIT, 2KG, SUC, FUM, and MAL by Wright and coworkers. They correspond to the subscripts *M* and *C* in the original model of Kelly et al., which differentiated between intramitochondrial and cytoplasmatic concentrations.

enzyme-catalyzed processes. Some of the steps are essentially irreversible, but some are truly reversible.

The map describing the biochemical system is shown in Fig. 9.2. It reflects the intramitochondrial metabolite pools proposed by Kelly et al. (1979ab) and Wright et al. (1992). Following Shiraishi and Savageau (1992a), the 13 dependent metabolites are coded as X_1 through X_{13}, while the 26 enzymes are considered as independent variables and coded as X_{14} through X_{39}. In addition, three cofactors are coded as X_{46}, X_{47}, and X_{48}. The cofactors are modeled as independent variables, because they are assumed to be constant. Finally, the model contains two fixed reservoirs for CO_2 and protein (X_{49} and X_{50}). Although they are shown in Fig. 9.2, they are merged with the rate constants and do not explicitly appear in the equations; steady-state values for these pools are therefore not required. The numbering scheme, in which X_{40} through X_{45} are missing, allows for a later refinement of the model, in which six additional reactions and enzymes are considered.

Our dictionary thus consists of 44 variables, which are listed in Tables 9.1 and 9.2, along with their biochemical names and observed concentrations and fluxes.

Model Equations and Parameter Values

The equations of the S-system model are developed in three steps. First, the map in Fig. 9.2 is expressed in the form of general system equations, or *node equations*. In

DIAGNOSIS AND REFINEMENT

Table 9.1. Dependent Metabolites in the Model with Steady-State Concentrations and Fluxes

Variable Name	Compound (Abbreviation)	Steady-State Concentration X_{is} (mM)	Steady-State Flux V_{is} (mM/min)	Relative Flux V_{is}/X_{is} (min^{-1})
X_1	Oxalacetate 1 (OAA 1)	0.0025	2.19	874
X_2	Oxalacetate 2 (OAA 2)	0.0025	2.19	873
X_3	Acetyl-CoA (ACO)	0.0593	2.00	33.7
X_4	Isocitrate (ISOC)	0.01	2.00	200
X_5	Pyruvate (PYR)	0.291	2.04	7.00
X_6	Glutamate (GLU)	6.03	1.45	0.241
X_7	Aspartate (ASP)	1.85	0.606	0.327
X_8	Alanine (ALA)	4.83	0.947	0.196
X_9	Citrate 1 (CIT 1)	0.025	2.00	80.0
X_{10}	α-Ketoglutarate (2KG1)	0.01	2.65	265
X_{11}	Succinate (SUC)	0.801	3.57	4.46
X_{12}	Fumarate (FUM)	0.04	2.85	71.4
X_{13}	Malate (MAL 1)	0.22	2.85	13.0

a second step, the node equations are translated into symbolic S-system equations, and in the final step, the kinetic orders and rate constants are numerically specified.

The node equations are formulated in the usual fashion, namely by summing all reactions v_{ki} that enter any given pool i and subtracting all reactions v_{ij} that leave this pool. The entering reactions can be collected in an aggregated term V_i^+, and the exiting reactions can be aggregated in V_i^-. The node equations for the tricarboxylic acid cycle read

$$\begin{aligned}
\dot{X}_1 &= (v_{71} + v_{13,1} + v_{10,6}) - (v_{12} + v_{17}) = V_1^+ - V_1^-, \\
\dot{X}_2 &= (v_{12} + v_{72}) - (v_{27} + v_{29}) = V_2^+ - V_2^-, \\
\dot{X}_3 &= (v_{53} + v_{50,3}) - (v_{29}) = V_3^+ - V_3^-, \\
\dot{X}_4 &= (v_{94}) - (v_{4,10}) = V_4^+ - V_4^-, \\
\dot{X}_5 &= (v_{85} + v_{13,5}) - (v_{53} + v_{58}) = V_5^+ - V_5^-, \\
\dot{X}_6 &= (v_{10,6} + v_{11,6} + v_{50,6}) - (v_{58} + v_{6,10} + v_{6,11}) = V_6^+ - V_6^-, \\
\dot{X}_7 &= (v_{17} + v_{27} + v_{50,7}) - (v_{71} + v_{72} + v_{10,6}) = V_7^+ - V_7^-, \\
\dot{X}_8 &= (v_{58} + v_{50,8}) - (v_{85}) = V_8^+ - V_8^-, \\
\dot{X}_9 &= (v_{29}) - (v_{94}) = V_9^+ - V_9^-, \\
\dot{X}_{10} &= (v_{58} + v_{4,10} + v_{6,10}) - (v_{10,6} + v_{10,11}) = V_{10}^+ - V_{10}^-, \\
\dot{X}_{11} &= (v_{6,11} + v_{10,11} + v_{50,11}) - (v_{11,6} + v_{11,12}) = V_{11}^+ - V_{11}^-, \\
\dot{X}_{12} &= (v_{11,12} + v_{50,12}) - (v_{12,13}) = V_{12}^+ - V_{12}^-, \\
\dot{X}_{13} &= (v_{12,13}) - (v_{13,1} + v_{13,5}) = V_{13}^+ - V_{13}^-.
\end{aligned} \quad (9.1)$$

The parentheses indicate how the fluxes v_{ij} have been grouped to form the aggregated fluxes V_i^+ and V_i^-. Note that different strategies of aggregation are already

Table 9.2. Reactions, Enzymes, and Cofactors, Represented as Independent Variables in the Model

Variable Name	Reaction	Abbreviation	EC Number	Concentration (mM)
X_{14}	Glutamate dehydrogenase	GluDH	1.4.1.2	0.977
X_{15}	α-Ketoglutarate dehydrogenase complex	α-KGDH		7.610
X_{16}	Succinate dehydrogenase	SDH	1.3.99.1	3.15
X_{17}	Fumarase		4.2.1.2	25.7
X_{18}	Malate dehydrogenase	MDH	1.1.1.37	77.8
X_{19}	Malic enzyme	ME	1.1.1.40	3.08
X_{20}	Ala → Pyr			0.196
X_{21}	Pyruvate dehydrogenase complex:	PDC		258.0
	Pyruvate dehydrogenase		1.2.4.1	
	Dihydrolipoyl transacetylase		2.3.1.12	
	Dihydrolipoyl dehydrogenase		1.8.1.4	
X_{22}	Oaa 2 → Asp			74.0
X_{23}	Asp → Oaa2			0.1
X_{24}	Citrate synthetase	CS	4.1.3.7	8.24
X_{25}	Aconitase		4.2.1.3	80.0
X_{26}	Isocitrate dehydrogenase	IsocDH	1.1.1.41	271.0
X_{27}	Glu → Suc			0.133
X_{28}	Aspartate transaminase	AspTA	2.6.1.1	9.95
X_{29}	Alanine transaminase	AlaTA	2.6.1.2	26.7
X_{30}	Oaa 1 → Oaa 2			800.0
X_{31}	Asp → Oaa 1			0.1
X_{32}	Suc → Glu			1.00
X_{33}	Oaa 1 → Asp			74.0
X_{34}	Prot → Asp			0.236
X_{35}	Prot → AcCoA			0.46
X_{36}	Prot → Suc			0.360
X_{37}	Prot → Fum			0.08
X_{38}	Prot → Ala			0.45
X_{39}	Prot → Glu			0.414
X_{46}	NAD			0.072
X_{47}	CoA			0.1
X_{48}	NADH			0.18
X_{49}	CO_2			
X_{50}	Pool of proteins			

implemented in these groupings (cf. Chapter 3 and Sorribas and Savageau, 1989c). For instance, aspartate and oxalacetate 1 are converted into each other. Consequently, v_{17} and v_{71} are listed separately and included in V_1^+, V_1^-, V_7^+, and V_7^- (see also Exercise 22). Other fluxes, such as $v_{12,13}$ and $v_{13,1}$, are reversible in principle, but essentially irreversible under the pertinent conditions. They are included only as net forward fluxes. For instance, $v_{12,13}$ only appears in V_{12}^- and V_{13}^+.

DIAGNOSIS AND REFINEMENT

Skipping over the grouping of fluxes in the middle parts of the equations, we see that Eq. (9.1) is in the form of the general system equation

$$\dot{X}_i = V_i^+(X_1, X_2, \ldots, X_{n+m}) - V_i^-(X_1, X_2, \ldots, X_{n+m}), \tag{9.2}$$

which we discussed in Chapter 3, since all fluxes V_i^+ and V_i^- are formally functions of some or all of the dependent or independent variables in the system.

The node equations describe the topology of the map in a succinct form, which however does not say anything about the specific features of the steps involved. For actual computations, the fluxes V_i^+ and V_i^- need to be characterized mathematically from knowledge about the system. All information necessary for the construction of a Michaelis–Menten model can be found in Segel (1975), Kelly et al. (1979ab), Wright et al. (1992), and Albe and Wright (1992). The fluxes take the form below:

V_1^+:

$$v_{71} = X_{31} X_7,$$

$$v_{13,1} = X_{18}(X_{46} X_{13} - X_{48} X_1)$$
$$\times \bigg(0.310 \times 1.33 + 1.33 X_{46} + 0.1 X_{13} + X_{46} X_{13}$$
$$+ 0.270 X_{48} + 0.04 X_1 + X_{48} X_1$$
$$+ \frac{0.27 X_{46} X_{48}}{0.310} + \frac{0.1 X_{13} X_1}{0.27} + \frac{X_{46} X_{13} X_{48}}{0.04} + \frac{X_{13} X_{48} X_1}{3.3}$$
$$+ \frac{X_{46} X_{13} X_1}{0.17} + \frac{X_{46} X_{48} X_1}{0.31} + \frac{0.31 X_{46} X_{13} X_{48} X_1}{0.31 \times 0.27 \times 0.04} \bigg)^{-1}$$
$$\text{(iso-ordered bi bi),} \tag{9.3a}$$

$$v_{10,6} = X_{28} \left(X_7 X_{10} - \frac{X_1 X_6}{9.5} \right)$$
$$\times \bigg(0.33 X_7 + 0.46 X_{10} + X_7 X_{10} + \frac{9.4 X_1}{9.5} + \frac{0.1 X_6}{9.5} + \frac{X_1 X_6}{9.5}$$
$$+ \frac{9.4 X_1 X_7}{0.46 \times 9.5} + \frac{0.46 X_6 X_{10}}{9.4} \bigg) \quad \text{(ping pong bi bi).}$$

V_1^-:

$$v_{12} = X_{30} X_1,$$
$$v_{17} = X_{33} X_1. \tag{9.3b}$$

V_2^+:

$$v_{72} = X_{23} X_7 \quad \text{and} \quad v_{12} \text{ of } V_1^-. \tag{9.3c}$$

V_2^-:

$$v_{27} = X_{22} X_2,$$
$$v_{29} = X_{24} X_2 X_3 \left[0.007 X_3 + 0.01 X_2 \left(1 + \frac{X_{47}}{0.11} \right) + X_2 X_3 \right]^{-1}$$

(Dead-end competitive inhibition ping pong bi bi). (9.3d)

V_3^+ :

$$v_{53} = X_{21}X_5X_{46}X_{47}$$
$$\times \left(0.11X_5X_{47} + 0.01X_5X_{46} + 0.14X_{46}X_{47} + X_5X_{46}X_{47}\right.$$
$$+ \frac{0.14 \times 1.7X_3X_{48}}{0.02} + \frac{0.11X_5X_{47}X_{48}}{0.05} + \frac{0.01X_3X_5X_{46}}{0.02}$$
$$\left. + \frac{0.14 \times 1.7X_3X_5X_{48}}{0.18 \times 0.02}\right)^{-1} \quad \text{(multisite ping pong with [P]=0),}$$

$$v_{50,3} = X_{35}. \tag{9.3e}$$

V_3^- :

$$v_{29} \text{ of } V_2^-. \tag{9.3f}$$

V_4^+ :

$$v_{94} = X_{25}X_9. \tag{9.3g}$$

V_4^- :

$$v_{4,10} = X_{26}X_4X_{46}\left[0.34X_4 + 0.13X_{46}\left(1 + \frac{X_{48}}{0.02}\right) + X_4X_{46}\right]^{-1}$$

(dead-end competitive inhibition ping pong bi bi). (9.3h)

V_5^+ :

$$v_{85} = X_{20}X_8,$$
$$v_{13,5} = \frac{X_{19}X_7X_{13}}{(0.37 + X_{13})(X_7 + 0.1)} \quad \text{(allosteric Michaelis–Menten).} \tag{9.3i}$$

V_5^- :

$$v_{58} = -X_{29}\left(X_8X_{10} - \frac{X_5X_6}{9.5}\right)$$
$$\times \left(0.19X_8 + 0.43X_{10} + X_8X_{10} + \frac{15X_5}{9.5} + \frac{0.87X_6}{9.5} + \frac{X_5X_6}{9.5}\right.$$
$$\left. + \frac{15X_5X_8}{0.43 \times 9.5} + \frac{0.43X_6X_{10}}{15}\right)^{-1} \quad \text{and} \quad v_{53} \text{ of } V_3^+$$

(dead-end competitive inhibition ping pong bi bi). (9.3j)

V_6^+ :

$$v_{11,6} = X_{32}X_{11}, \quad v_{50,6} = X_{39}, \quad \text{and} \quad v_{10,6} \text{ of } V_1^+. \tag{9.3k}$$

V_6^- :

$$v_{6,10} = X_{14}X_6X_{46}\left[0.2X_6 + 2X_{46}\left(1 + \frac{X_{48}}{0.025}\right) + X_6X_{46}\right]^{-1}$$

(ping pong bi bi),

$$v_{6,11} = X_{27}X_6, \quad \text{and} \quad v_{58} \text{ of } V_5^-. \tag{9.3l}$$

V_7^+ :

$$v_{50,7} = X_{34}, \quad v_{17} \text{ of } V_1^-, \quad v_{27} \text{ of } V_2^-. \tag{9.3m}$$

DIAGNOSIS AND REFINEMENT

V_7^- :

$v_{10,6}$ and v_{71} of V_1^+, v_{72} of V_2^+. (9.3n)

V_8^+ :

$v_{50,8} = X_{38}$, and v_{58} of V_5^-. (9.3o)

V_8^- :

v_{85} of V_5^+. (9.3p)

V_9^+ :

v_{29} of V_2^-. (9.3q)

V_9^- :

v_{94} of V_4^+.

V_{10}^+ :

$v_{4,10}$ of V_4^-, v_{58} of V_5^-, and $v_{6,10}$ of V_6^-. (9.3r)

V_{10}^- :

$$v_{10,11} = X_{15} X_{10} X_{46} X_{47} \left(0.07 X_{10} X_{47} + 0.002 X_{10} X_{46} + X_{46} X_{47} \right.$$

$$+ X_{10} X_{46} X_{47} + 1.5 X_{11} X_{48} + \frac{0.07 X_{10} X_{47} X_{48}}{0.018}$$

$$\left. + 0.002 X_{11} X_{10} X_{46} + \frac{1.5 X_{11} X_{10} X_{48}}{0.75} \right)^{-1}$$

and $v_{10,6}$ of V_1^+ (multisite ping pong with $[P]=0$). (9.3s)

V_{11}^+ :

$v_{50,11} = X_{36}$, $v_{6,11}$ of V_6^- and $v_{10,11}$ of V_{10}^-. (9.3t)

V_{11}^- :

$$v_{11,12} = \frac{X_{16}(X_{11} - X_{12}/10)}{0.1 + X_{11} + X_{12}/10} \quad \text{and} \quad v_{11,6} \text{ of } V_6^+ \quad \text{(uni uni)}.$$ (9.3u)

V_{12}^+ :

$v_{50,12} = X_{37}$ and $v_{11,12}$ of V_{11}^- (9.3v)

V_{12}^- :

$$v_{12,13} = \frac{X_{17}(X_{12} - X_{13}/10)}{0.1 + X_{12} + X_{13}/10}.$$ (9.3w)

V_{13}^+ :

$v_{12,13}$ of V_{12}^-. (9.3x)

V_{13}^- :

$v_{13,1}$ of V_1^+ and $v_{13,5}$ of V_5^+. (9.3y)

The corresponding S-system model is derived from these flux descriptions in a straightforward fashion, as shown in Chapter 3 (see also Exercises 1–4). Each influx V_i^+ or efflux V_i^- consists of a product of power-law functions that contains a rate constant and those dependent and independent variables that directly affect that flux. Each variable enters the product with its own kinetic order. The numerical value of the kinetic order is computed from the traditional rate law through partial differentiation, as shown in Chapter 5 and the Appendix.

As an example, consider the production of fumarate (X_{12}), which is represented by the aggregated flux V_{12}^+. Three processes contribute to this flux, namely, the reversible reactions with succinate and malate, and the disassembly of protein. The first of these reactions is coded in the node equations as $v_{11,12}$, the second as $v_{13,12}$, and the third as $v_{50,12}$. Since the flow of material is essentially irreversible from succinate to fumarate to malate, the reversibility of $v_{13,12}$ is taken into account by the net rate $v_{12,13}$ in V_{12}^- and V_{13}^+.

The kinetic order with respect to X_{11} is computed through partial differentiation of V_{12}^+, which yields

$$g_{12,11} = \frac{\partial(v_{11,12} + v_{50,12})}{\partial X_{11}} \cdot \frac{X_{11}}{v_{11,12} + v_{50,12}}. \tag{9.4}$$

For $v_{11,12}$ and $v_{50,12}$, we use the definitions (9.3) given above in the formulation of V_{11}^- and V_{12}^+. Since $v_{50,12}$ is independent of X_{11}, it only appears in the denominator of the last term. Thus, the partial derivative at steady state becomes

$$\frac{\partial v_{11,12}}{\partial X_{11}} = \frac{\partial}{\partial X_{11}} \left(\frac{X_{16} X_{11} - 0.1 X_{16} X_{12}}{X_{11} + 0.1 + 0.1 X_{12}} \right)$$

$$= \frac{3.15(X_{11} + 0.104) - 3.15 X_{11} + 0.0126}{(X_{11} + 0.104)^2} = 0.4154, \tag{9.5}$$

and the desired kinetic order is

$$g_{12,11} = 0.4154 \times \frac{0.801}{2.85} = 0.117. \tag{9.6}$$

The kinetic orders with respect to X_{12} and X_{16} are obtained in the same fashion, except that $v_{11,12}$ is differentiated with respect to X_{12} and X_{16}, respectively. The result is

$$g_{12,12} = \left[\frac{\partial}{\partial X_{12}} \left(\frac{3.15 \times 0.801 - 0.315 X_{12}}{0.901 + 0.1 X_{12}} \right) \right] \cdot \frac{X_{12}}{V_{12}^+}$$

$$= -0.655 \times \frac{0.04}{2.85} = -0.0919, \tag{9.7}$$

$$g_{12,16} = \frac{X_{11} - 0.1 X_{12}}{0.1 + X_{11} + 0.1 X_{12}} \cdot \frac{X_{16}}{V_{12}^+} = 0.881 \times \frac{3.15}{2.85} = 0.973. \tag{9.8}$$

The kinetic order with respect to protein degradation is computed through partial differentiation of $v_{50,12}$, which is simply represented by X_{37}. Because $v_{11,12}$ is independent of X_{37}, we obtain

$$g_{12,37} = \frac{\partial(X_{37} + v_{11,12})}{\partial X_{37}} \cdot \frac{X_{37}}{X_{37} + v_{11,12}} = \frac{X_{37}}{X_{37} + v_{11,12}}, \tag{9.9}$$

DIAGNOSIS AND REFINEMENT

which, as always, is evaluated at the steady state. Using the definition of $v_{11,12}$ given above in the description of V_{11}^- and substituting steady-state values, the result is

$$g_{12,37} = \frac{0.08}{2.854} = 0.028. \tag{9.10}$$

Finally, the rate constant is calculated by dividing the conventional rate law at steady state by the corresponding product of power-law functions (cf. Chapter 5 and Appendix),

$$\alpha_{12} = \frac{V_{12}^+}{X_{11}^{0.117} X_{12}^{-0.00919} X_{16}^{0.973} X_{37}^{0.028}} = 0.998. \tag{9.11}$$

As a second example, the aggregated degradation of succinate depends on X_{11}, X_{12}, X_{16}, and X_{32}. Its kinetic orders are computed from V_{11}^- as

$$h_{11,11} = \frac{\partial V_{11}^-}{\partial X_{11}} \cdot \frac{X_{11}}{V_{11}^-} = 0.317,$$

$$h_{11,12} = \frac{\partial V_{11}^-}{\partial X_{12}} \cdot \frac{X_{12}}{V_{11}^-} = -0.00733,$$

$$h_{11,16} = \frac{\partial V_{11}^-}{\partial X_{16}} \cdot \frac{X_{16}}{V_{11}^-} = 0.776, \tag{9.12}$$

$$h_{11,32} = \frac{\partial V_{11}^-}{\partial X_{32}} \cdot \frac{X_{32}}{V_{11}^-} = 0.224,$$

where all quantities are evaluated at the steady state, i.e., with the values for the metabolites and enzymes given in Tables 9.1 and 9.2. The appropriate rate constant is

$$\beta_{11} = \frac{V_{11}^-}{X_{11}^{h_{11,11}} X_{12}^{h_{11,12}} X_{16}^{h_{11,16}} X_{32}^{h_{11,32}}} = 1.536. \tag{9.13}$$

In this fashion, all kinetic orders and rate constants are obtained from the above rate laws (see also Exercises 5 and 6). The S-system model consists of the following influxes and effluxes:

$$V_1^+ = 0.8282 X_1^{-0.038} X_6^{-0.0204} X_7^{0.106} X_{10}^{0.114} X_{13}^{0.7}$$
$$\times X_{18}^{0.807} X_{28}^{0.108} X_{31}^{0.0848} X_{46}^{0.599} X_{48}^{-0.181}, \tag{9.14a}$$

$$V_1^- = 1.3423 X_1 X_{30}^{0.915} X_{33}^{0.0847}, \tag{9.14b}$$

$$V_2^+ = 1.3401 X_1^{0.915} X_7^{0.0848} X_{23}^{0.0848} X_{30}^{0.915}, \tag{9.14c}$$

$$V_2^- = 17.166 X_2^{0.706} X_3^{0.0716} X_{22}^{0.0848} X_{24}^{0.915} X_{47}^{-0.0341}, \tag{9.14d}$$

$$V_3^+ = 0.0654 X_3^{-0.73} X_5^{0.281} X_{21}^{0.77} X_{35}^{0.23} X_{46}^{0.761} X_{47}^{0.731} X_{48}^{-0.754}, \tag{9.14e}$$

$$V_3^- = 16.242 X_2^{0.679} X_3^{0.0782} X_{24} X_{47}^{-0.0372}, \tag{9.14f}$$

$$V_4^+ = X_9 X_{25}, \tag{9.14g}$$

$$V_4^- = 0.152 X_4^{0.958} X_{26} X_{46}^{0.0348} X_{48}^{-0.862}, \tag{9.14h}$$

$$V_5^+ = 1.875 X_7^{0.0274} X_8^{0.465} X_{13}^{0.336} X_{19}^{0.535} X_{20}^{0.465}, \tag{9.14i}$$

$$V_5^- = 0.01923 X_3^{-0.717} X_5^{0.413} X_6^{0.306} X_8^{-0.29}$$
$$\times X_{10}^{-0.0883} X_{21}^{0.756} X_{29}^{0.244} X_{46}^{0.748} X_{47}^{0.718} X_{48}^{-0.741}, \tag{9.14j}$$

$$V_6^+ = 2.968 X_1^{-0.0184} X_6^{-0.0307} X_7^{0.0323} X_{10}^{0.172} X_{11}^{0.552} X_{28}^{0.163} X_{32}^{0.552} X_{39}^{0.285}, \tag{9.14k}$$

$$V_6^- = 0.3276 X_5^{0.193} X_6^{1.027} X_8^{-0.408} X_{10}^{-0.124} X_{14}^{0.104}$$
$$\times X_{27}^{0.554} X_{29}^{0.342} X_{46}^{0.0443} X_{48}^{-0.0381}, \tag{9.14l}$$

$$V_7^+ = 2.98 X_1^{0.305} X_2^{0.306} X_{22}^{0.306} X_{33}^{0.305} X_{34}^{0.389}, \tag{9.14m}$$

$$V_7^- = 3.864 X_1^{-0.0441} X_6^{-0.0734} X_7^{0.688} X_{10}^{0.411} X_{23}^{0.306} X_{28}^{0.389} X_{31}^{0.306}, \tag{9.14n}$$

$$V_8^+ = 0.1215 X_5^{0.295} X_6^{0.658} X_8^{-0.625} X_{10}^{-0.19} X_{29}^{0.525} X_{38}^{0.475}, \tag{9.14o}$$

$$V_8^- = X_8 X_{20}, \tag{9.14p}$$

$$V_9^+ = 16.242 X_2^{0.679} X_3^{0.0782} X_{24} X_{47}^{-0.0372}, \tag{9.14q}$$

$$V_9^- = X_9 X_{25}, \tag{9.14r}$$

$$V_{10}^+ = 0.156 X_4^{0.724} X_5^{0.106} X_6^{0.259} X_8^{-0.223} X_{10}^{-0.0679}$$
$$\times X_{14}^{0.0568} X_{26}^{0.756} X_{29}^{0.188} X_{46}^{0.0506} X_{48}^{-0.672}, \tag{9.14s}$$

$$V_{10}^- = 0.8063 X_1^{-0.0101} X_6^{-0.0168} X_7^{0.0177} X_{10}^{0.99}$$
$$\times X_{11}^{-0.879} X_{15}^{0.911} X_{28}^{0.0891} X_{46}^{0.882} X_{47}^{0.879} X_{48}^{-0.881}, \tag{9.14t}$$

$$V_{11}^+ = 1.504 X_6^{0.225} X_{10}^{0.663} X_{11}^{-0.651} X_{15}^{0.674} X_{27}^{0.225}$$
$$\times X_{36}^{0.101} X_{46}^{0.653} X_{47}^{0.651} X_{48}^{-0.653}, \tag{9.14u}$$

$$V_{11}^- = 1.536 X_{11}^{0.317} X_{12}^{-0.00733} X_{16}^{0.776} X_{32}^{0.224}, \tag{9.14v}$$

$$V_{12}^+ = 0.998 X_{11}^{0.117} X_{12}^{-0.00919} X_{16}^{0.973} X_{37}^{0.028}, \tag{9.14w}$$

$$V_{12}^- = 8.289 X_{12}^{1.98} X_{13}^{-1.36} X_{17}, \tag{9.14x}$$

$$V_{13}^+ = 8.289 X_{12}^{1.98} X_{13}^{-1.36} X_{17}, \tag{9.14y}$$

$$V_{13}^- = 0.9387 X_1^{-0.0197} X_7^{0.0196} X_{13}^{0.775} X_{18}^{0.618} X_{19}^{0.382} X_{46}^{0.458} X_{48}^{-0.139}. \tag{9.14z}$$

The S-system model is now completely determined, with the exception of initial values that are needed for dynamic simulations. These are typically specified to be somewhere in the vicinity of the steady state. The system is implemented in PLAS as *TCA.plc*.

1. Consistency and robustness analysis. A quick numerical check with PLAS confirms that the S-system model has the expected steady state with the metabolite concentrations given in Table 9.1. At this point, the S-system model and the traditional kinetic model are exactly equivalent. The dependent variables have exactly the same values in the two models, and even the slopes with respect to time are the same, so that very small perturbations are guaranteed to evoke similar responses.

As soon as the system is moved further away from the steady state, the two models may respond differently – how differently can only be determined through numerical analysis.

Although differences between models based on Michaelis–Menten kinetics and those formulated as S-systems are to be expected, the more important question is the appropriateness of each, neither, or both models to represent the actual biochemical system. For instance, are the models stable? Are they consistent? Do they show the correct responses, at least qualitatively? Shiraishi and Savageau analyze these questions in great detail for Michaelis–Menten models as well as for S-system models. We limit our discussions to models in S-system form.

There are two prominent sources for potential inconsistencies. One reason for inaccurate results may be that faulty assumptions were made in the construction of the map or in the implementation of the model equations. Another reason is the estimation of parameter values from *in vitro* experiments; we cannot be sure whether the same results would have been obtained *in vivo*. If true dynamic observations were available, such as transient responses of the tricarboxylic acid cycle to various inputs, the model responses could be compared with these data. As so often, such data are not at hand.

A minimal test for the validity of the model consists in an analysis of stability and of log gains and sensitivities. In some sense, these system properties constitute necessary (but not sufficient) conditions for the robustness of the model. Instability and unusually large sensitivities and gains help us detect if the model is too sensitive with respect to normal perturbations, while stability and reasonable sensitivities and gains are supportive of the robustness of the model, even though they do not guarantee its correctness. Even if some gains or sensitivities turn out to be unreasonable, there is still good news: the results of the sensitivity analysis don't just provide the information that something is wrong with the model, but they actually locate potential trouble spots in it.

Stability of the S-system model can be checked analytically (cf. Chapter 6 and Savageau, 1975, 1976; Voit, 1991) or numerically in PLAS from the steady-state–stability dialog box (see Table 9.3). All real parts of eigenvalues are negative, which indicates that the system is locally stable and will return to the steady state after small perturbations. This is good in that an unstable model would not be an acceptable representation of an actual system that obviously tolerates disturbances every day. In theory, stability is only guaranteed for infinitesimally small perturbations, but in most biochemically relevant cases, that translates into real perturbations of moderate magnitude (cf. Chapter 6). There is one pair of eigenvalues with non-zero imaginary parts, which implies the potential of oscillations.

The substantial range of magnitudes of the real parts of the eigenvalues (from about -10^{-3} to -10^{+2}) is an indication that the dependent variables respond to perturbations at rather different time scales: Some can be expected to return very quickly to their original steady-state value, while others may exhibit an extended transition phase, during which the system is kept away from its normal steady state.

Dynamic simulations confirm these predictions. For instance, if a bolus of fumarate (X_{12}) is added to the system at time zero, the responses among the dependent

Table 9.3. Eigenvalues of the Model of Tricarboxylic Acid Cycle, Obtained from PLAS

Real Part	Imaginary Part
−908.3154	0
−628.031	0
−240.4053	42.76885
−240.4053	−42.76885
−160.5042	0
−66.99394	0
−26.73265	0
−9.126683	0
−1.647335	0
−1.131236	0
−0.2117159	0
−0.02673667	0
−0.003610466	0

metabolites are drastically different (Fig. 9.3; cf. Shiraishi and Savageau, 1992c: p. 22931). The metabolites X_1, X_2, X_4, and X_9 rapidly overshoot and return to their steady-state values within a minute. X_{10}, X_{11}, and X_{13} require several minutes, and X_3 and X_5 return to their steady-state values only after about 1,500 min. X_7 and X_8 are almost unaffected.

This is what happens (cf. Shiraishi and Savageau, 1992c, 1993): The bolus of fumarate (X_{12}) is rapidly converted into malate (X_{13}), most of which is subsequently channeled into oxalacetate (X_1). The excess oxalacetate is propagated through the cycle of citrate (X_9), isocitrate (X_4), α-ketoglutarate (X_{10}), and succinate (X_{11}). Since oxalacetate inhibits aspartate transaminase, the excess also causes increased levels of α-ketoglutarate. Finally, the excess amount of oxalacetate stimulates citrate synthetase, which leads to rapid production of citrate (X_9) and subsequent depletion of acetyl-CoA. Acetyl-CoA, in turn, relieves the inhibition of pyruvate decarboxylase and causes the concentration of pyruvate (X_5) to drop. After about three minutes, the material has been redistributed in the system so that most metabolite concentrations are within 1% of their nominal steady state value. However, the concentration of acetyl-CoA remains depleted by about 6%, and that of pyruvate by about 15%. This apparent, but not real, quasi-steady state in fact constitutes a very long transient period of almost constant concentrations, and it takes about 1,500 min before acetyl-CoA and pyruvate also return to their nominal steady-state values. In other words, the system needs more than a day to recover from a typical perturbation in one of its metabolites (see also Exercises 7 and 8).

Turnover times for the dependent metabolites can also be explored with the analysis of *apparent first-order rate constants*, which are defined as the relative steady-state fluxes, V_{iS}/X_{iS}, and listed in PLAS as *F-values* (see also Savageau, 1976: p. 186). The unit of these quantities is min^{-1}, and therefore small values indicate

DIAGNOSIS AND REFINEMENT

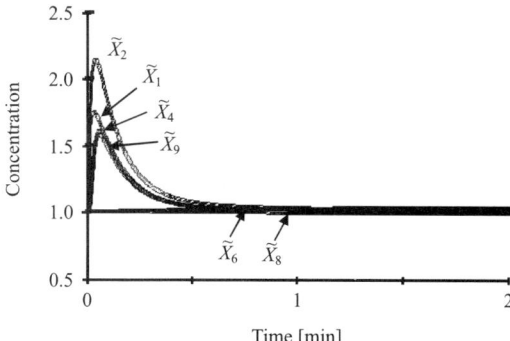

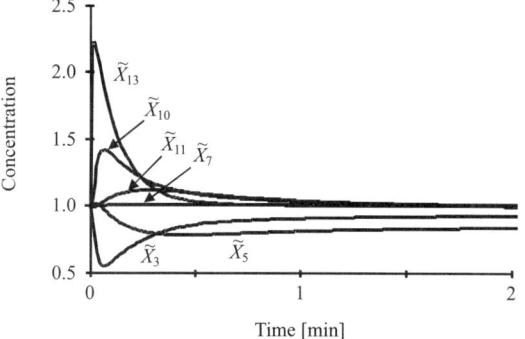

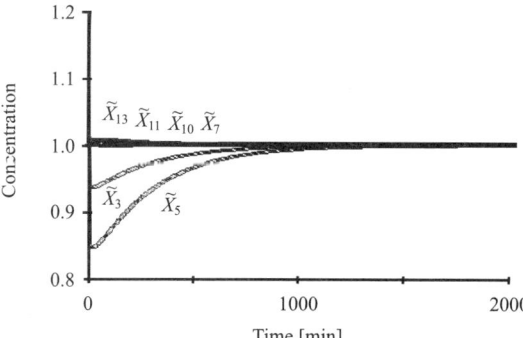

Figure 9.3. Dynamic response of the model to a 10-fold bolus increase in fumarate. Some metabolites quickly recover from the perturbance (top panel). Other metabolites, such as X_{10} and X_{11}, return to their steady-state values within a few minutes (center panel). Yet others, especially X_3 and X_5, seem to approach a new steady state, but in fact have a very long transition period (bottom panel; note the difference in scales). The tildes indicate that all variables are scaled with respect to their steady-state concentrations; in $TCA.plc$ they are denoted by z_i's. See text for explanations.

long turnover times. The numerical values for the model are given in Table 9.4. They suggest that glutamate, aspartate, and alanine have turnover times that are two or three orders of magnitude longer than those of other metabolites. This is consistent with the results of the eigenvalue analysis. In biochemical terms, this result is explained in the following way. Once the aspartate pool becomes expanded, it is slow to return to the original value. As a result, oxalacetate remains elevated and stimulates citrate synthetase, thereby depleting the pool of acetyl-CoA. Acetyl-CoA strongly inhibits pyruvate decarboxylase, thus effecting a long-lasting drop in pyruvate. Furthermore, glutamate is elevated and alanine depleted, and both return only slowly to their pre-disturbance levels, while pyruvate is used to replenish the alanine pool.

Table 9.4. Apparent First-Order Rate Constants (*F*-Values) for the Movement of Material throughout the System.

Metabolite	Variable	F-Value
Oxalacetate 1	X_1	876
Oxalacetate 2	X_2	876
Acetyl-CoA	X_3	33.7
Isocitrate	X_4	200
Pyruvate	X_5	7.0
Glutamate	X_6	0.24
Aspartate	X_7	0.33
Alanine	X_8	0.20
Citrate	X_9	80.0
α-Ketoglutarate	X_{10}	264
Succinate	X_{11}	4.5
Fumarate	X_{12}	71.2
Malate	X_{13}	13.0

Note: Small values of these quantities, which are defined as steady-state flux divided by steady-state concentration, indicate long turnover times.

The overall result of this brief analysis of the steady state, its stability, and time scales is that the model is *self-consistent*. It exhibits the correct steady state, and this steady state is locally stable. However, the different magnitudes of eigenvalues and turnover times are a warning sign. The dynamic simulations indicate that pyruvate and acetyl-CoA are especially slow to respond to perturbations.

Another test of model robustness is the response to changes in malate dehydrogenase and malic enzyme (cf. Albe and Wright, 1992; Shiraishi and Savageau, 1992c; 1993). This test is easily implemented in PLAS by changing the value of the corresponding enzyme activity. For instance, to simulate the scenario of 98% activity in malate dehydrogenase, we simply multiply X_{18} by 0.98 (see also Exercise 9). The responses are discouraging. Pyruvate assumes a new steady-state value of 24.4, which is more than 80 times the original value. Alanine increases to almost twice its original value and acetyl-CoA to more than 5 times the original value. Furthermore, it takes more than 10,000 time units to reach 99% of the new steady-state value of pyruvate. This means that a minute (2%) drop in enzyme activity would lead to a vastly altered steady state, and that the transition would take over a week.

To show how difficult to predict a nonlinear system can be – even though this exercise is not very relevant biochemically – we reduce the activity of malate dehydrogenase further. At 95% activity, the predicted steady-state value of pyruvate is 22,066, and at 93%, it reaches over 2 million. However, on reducing the activity 1% further to 92%, the steady-state value all of a sudden becomes 2.483, and with further decreases it begins to rise again. The steady-state values of pyruvate at 90%, 80%, 70%, 60%, and 50% activity are 3, 4.8, 2.4, 1.1, and 2.4×10^{65}, respectively.

Concomitant with this progression, but at a different rate, acetyl-CoA increases to 4.4 at 95%, 420 at 90%, and about 8.5 million at 80%. At 70% and 60% it decreases to 6.4 and 2.75, but at 50% it assumes a value of 1.3×10^{24}. Alanine increases to 20 at 95%, 92 at 90%, 2,500 at 80%, 105,000 at 70%, 7.8 million at 60%, and 1.3×10^9 at 50%. In stark contrast, some of the other metabolites are almost unaffected throughout the progressive reduction in malate dehydrogenase activity. With decreasing enzyme activity, the maximal real eigenvalue approaches 0, which indicates ever increasing transition times up to the point where the system no longer reaches the steady state.

The analysis of sensitivities could in principle be done analytically (cf. Chapter 7), but considering the size of the model, a strictly numerical exploration is the method of choice. Since there are thirteen dependent variables and steady-state fluxes and more than fifty parameters, a full sensitivity analysis produces almost 4,000 metabolite and flux sensitivities, and this number does not even include the more than 650 logarithmic gains. Studying huge tables of numbers does not always convey an overall impression, and it is therefore recommended to enter the sensitivities from the steady-state dialog boxes in PLAS in a graphics program and to present the results in some graphical fashion.

One should note the ease with which these analyses can be executed in PLAS. As Shiraishi and Savageau (1992c) point out, other modeling approaches do not even permit direct computations of the steady state, either analytically or numerically. For instance, the steady state of a corresponding Michaelis–Menten model has to be computed with iterative methods, and gains must be based on finite differences. The computation of sensitivities is utterly impractical, since one would have to solve the steady-state equations iteratively for each variation in each one of the 55 parameters.

Many graphical representations are possible. Maybe most appealing is a pseudo-three-dimensional plot in which the x-axis shows metabolites or fluxes, the y-axis the rate constant or kinetic order parameters, and the z-axis the value of the sensitivity of a metabolite or flux with respect to a parameter. Depending on the system, this representation provides an immediate intuitive impression of problematic areas in the model. In order to keep the graphs simple, parameters with very low associated sensitivities may be omitted, which is facilitated by the **hide if** < ... option in the sensitivity dialog box in PLAS.

The pseudo-three-dimensional graph can be condensed when one sums over the x-values or along the y-values of the graph. In the first case, one obtains the overall effect of changes in one particular parameter, whereas in the latter case one obtains the overall sensitivity of one particular metabolite or flux. For an order of magnitude analysis, one does not want large positive sensitivities to cancel large negative sensitivities. Therefore the sums of sensitivities are computed for absolute values. Nonetheless, one can indicate what portion of the sum of absolute values is attributed to positive and which to negative sensitivities (cf. Figs. 1–4 in Shiraishi and Savageau, 1992b; Exercise 10).

The rate constant sensitivities of the model are shown in Fig. 9.4. Many of them are very high, and one must wonder whether this is another reflection on the robustness of the model. Although high sensitivities are a warning sign, we must remember

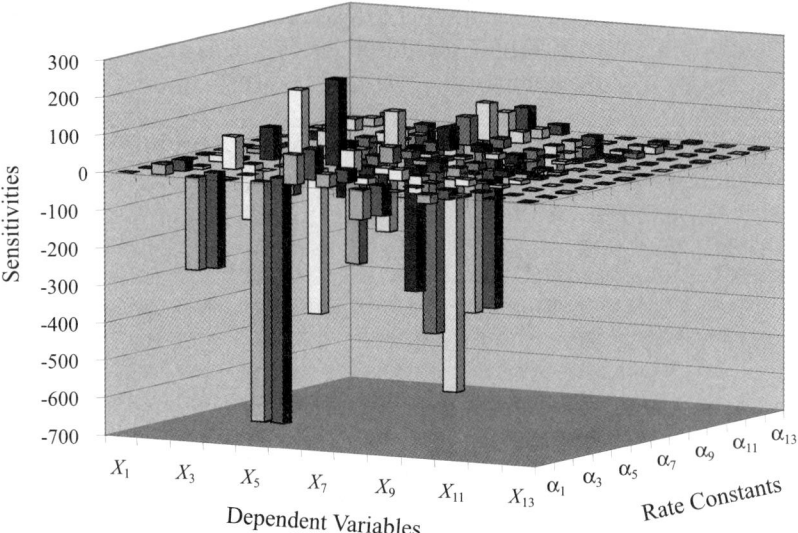

Figure 9.4 (see Color Plate XIV). Rate constant sensitivities of the original model. See text for interpretation.

that the sensitivities in this graph are unconstrained (cf. Chapter 7). For instance, consider the sensitivity of X_5 to changes in α_4, which is listed with an incredibly high negative value of -384.856, which would imply an almost 400-fold amplification of any perturbation in α_4. We note that α_4 is the rate constant of the conversion of citrate into isocitrate by aconitase, and this process is also represented by the beta term of X_9. Because the sensitivity of X_5 to a change in β_9 is equal to -1 times the sensitivity of X_5 to a change in α_9 (cf. Chapter 7), the change in the rate of the aconitase step is measured by the sum $S(X_5, \alpha_4) + (-1)S(X_5, \alpha_9)$. Indeed, these two sensitivities are exactly the same, and the sum is zero. Thus, while each individual rate constant sensitivity is very high, the constrained sensitivity is zero.

As a slightly more complicated example, consider the response of X_5 to a change in α_5, which is quantified by the sensitivity $S(X_5, \alpha_5) = 227.9045$. The process V_5^+ consists of two components, namely v_{85} and $v_{13,5}$, which at the nominal steady state have values of 0.947 and 1.090, respectively. Now assume that the rate v_{85} is changed by 1%. This change corresponds to a percent change of $0.947/(0.947 + 1.090) = 0.4654$ in α_5. This change, in turn, would lead to an increase of about $0.465 \times 227.9045 = 106$ percent in X_5. Changing v_{85} would also change β_8, and because v_{85} constitutes the only process of degradation of X_8, this change would evoke a percent change in X_5 that is given by the sensitivity $S(X_5, \beta_8) = -S(X_5, \alpha_8) = -106.786$. Thus, the total effect of a change in v_{85} on X_5 is characterized by the sum of the weighted sensitivity with respect to α_5 and the (raw) sensitivity with respect to β_8, and this sum is about -0.81%. As explained in Chapter 7, PLAS can be manipulated to produce constrained as well as unconstrained sensitivities (Exercise 11).

Since in actuality a change in a process is most likely related to the activity of the catalyzing enzyme, and since these activities are represented as independent variables, it makes sense to study logarithmic gains instead of – or in addition to – rate constant sensitivities. Logarithmic gain profiles with respect to enzyme activities are presented

DIAGNOSIS AND REFINEMENT 313

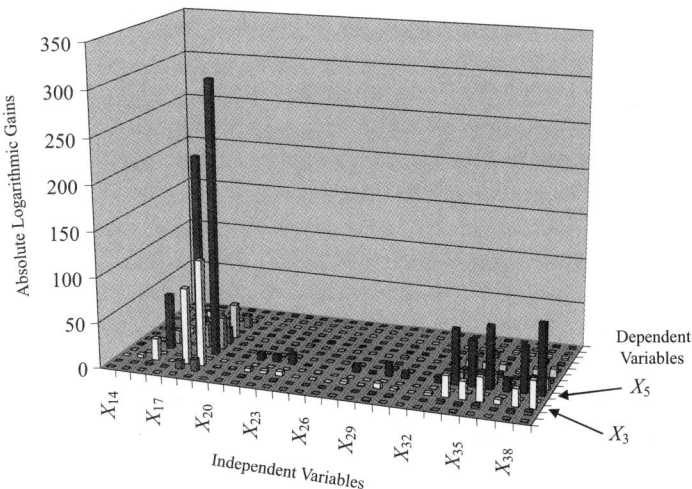

Figure 9.5 (see Color Plate XV). Logarithmic gains of metabolites with respect to enzyme activities.

in Figs. 9.5 and 9.6 for metabolites and fluxes. Similar profiles can be shown for cofactors (Exercise 12).

The gain profiles in Figs. 9.5 and 9.6 illustrate that some of the high rate constant sensitivities (see Fig. 9.4) may be due to constraints, but that the system has some problems nevertheless. The log gains automatically allow for dependences, and yet some very mild changes in enzyme activities lead to enormous amplifications. Most strongly affected are the metabolites X_3 and X_5 and the fluxes $V(X_5)$, $V(X_8)$, and $V(X_{10})$. The log gain of malic enzyme (X_{19}) with respect to pyruvate (X_5) is over 300, which suggests that a moderate 10% increase in enzyme activity would lead to a more than 30-fold increase in the steady-state value of pyruvate! Similarly, pyruvate is very sensitive to changes in the activity of malate dehydrogenase (X_{18}). The products of pyruvate metabolism, especially acetyl-CoA (X_3), alanine, and α-ketoglutarate,

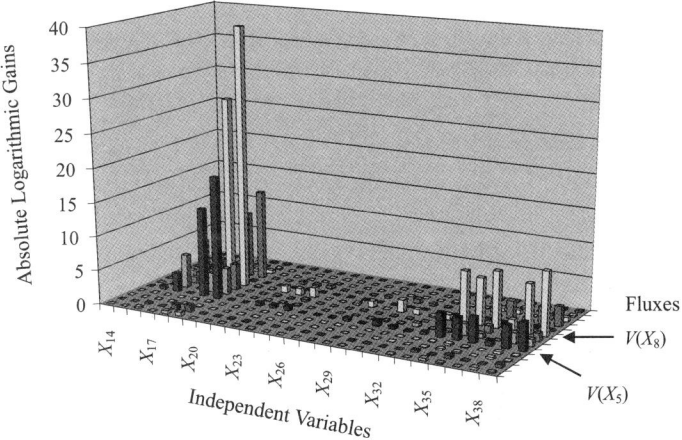

Figure 9.6 (see Color Plate XVI). Logarithmic gains of fluxes with respect to enzyme activities.

are strongly affected as well. Clearly, the distribution of fluxes at the malate branch point is critical and not well modeled.

In contrast to the fluxes associated with malate and pyruvate, the cycle flux is essentially unaffected by changes in most of the independent variables, as can be seen from the log gains of the fluxes $V(X_3)$, $V(X_4)$, $V(X_9)$, $V(X_{12})$, and $V(X_{13})$. The only effective way of altering this cycle flux would be to bring carbon into the cycle, which would correspond to an increase in the activities of the steps $X_{34}, \ldots, X_{39}$ that supply amino acids from protein degradation. This rigidity of the model and its inability to propagate signals in an efficient manner are surprising and appear to be rather limiting for a pathway that is crucial for survival and development of the organism (Shiraishi and Savageau, 1992c).

Kinetic order sensitivities also point to pyruvate (X_5), followed by acetyl-CoA (X_3), and to the fluxes through pyruvate (V_5) and α-ketoglutarate (V_{10}). Of the 1,885 kinetic order sensitivities of fluxes in the system, 66 are in the 100-fold range, 320 in the 10-fold range, 705 in the 1-fold range, 584 in the 0.1-fold range, and the remaining 209 in the 0.01-fold range or below. Of the 1,885 kinetic order sensitivities of metabolites, 45 are in the 1000-fold range, 238 in the 100-fold range, 479 in the 10-fold range, 593 in the 1-fold range, 368 in the 0.1-fold range, and the remaining 162 in the 0.01-fold range or below. Most of the high sensitivities relate to pyruvate metabolism. The highest sensitivity in this category is the response of pyruvate to a change in the kinetic order $h_{10,15}$. An increase in this parameter, which represents the action of the α-ketoglutarate dehydrogenase complex, causes an increase in pyruvate concentration that is an astounding 4,000-fold greater than the stimulus. While these sensitivities again are unconstrained, the results corroborate those of the analyses of gains and rate constant sensitivities, and warn us that the model is not robust. It has severe problems, especially in the area of pyruvate metabolism.

Before we analyze the problems associated with pyruvate metabolism further, we should note that the sensitivity and gain analysis indicates appropriate system behavior in other parts of the pathway, where sensitivities and log gains are attenuating or only mildly amplifying.

If most gains associated with a particular metabolite are zero or very small, the metabolite is so insensitive that it cannot be controlled. As an example, citrate (X_9) and isocitrate (X_4) are largely unaffected by variations in almost any enzyme concentrations. These metabolites precede irreversible steps in the unbranched part of the cycle, and their concentrations remain unchanged along with the flux through the pathway. Only if the immediately related enzymes aconitase (X_{25}) or isocitrate dehydrogenase (X_{26}) are altered will the concentrations of citrate and isocitrate be affected. However, the corresponding gains are negative and of small magnitude and cause the substrate pool to change in inverse proportion [$L(X_4, X_{26}) = L(X_9, X_{25}) = -1$]. Whereas aconitase ($X_{25}$) and isocitrate dehydrogenase (X_{26}) moderately affect their substrates citrate and isocitrate, they have essentially no effect on other metabolites: all corresponding log gains are zero or very small. This can be explained by the fact that these enzymes occupy positions within an unbranched pathway with constant influx (Shiraishi and Savageau, 1992d; see also Curto et al., 1998a).

DIAGNOSIS AND REFINEMENT

Finally, the sensitivity and gain analysis shows that not only enzymes but also cofactors can have a strong effect on the system. For instance, the concentration of NAD affects pyruvate with a log gain of -157 (see Exercise 12). This suggests that variations in such cofactors should not be ignored in models of the tricarboxylic acid cycle.

Modification of the Model

The problems in the current model are related to pyruvate metabolism and, in particular, to the distribution of fluxes at the malate branch point, which is known to be a crucial control point (e.g., see Canela et al., 1987; Albe and Wright, 1992). Furthermore, the turnover times of the pools of glutamate, aspartate, and alanine are suspiciously long, as is seen mathematically from the large ratios of pool sizes to fluxes. This result is interesting in that Kelly et al. (1979b) found these pools difficult to measure because of compartmentation within *Dictyostelium*. They performed several types of experiments, which produced comparable results, and subsequently put most weight on measurements of CO_2 losses from the system, because of their accuracy and ease of determination. These measurements produce accurate estimates for the net flux through the cycle, but provide no information about the utilization of amino acids for protein synthesis.

Inspection of the map or a semiformal network analysis (Shiraishi and Savageau, 1992c) suggests that some of the problems with the model could be caused by the fact that there is a constant supply of amino acids deriving from protein catabolism (reactions 34 to 39), which can only be balanced by losses of CO_2 from the system. A promising modification, therefore, seems to be to allow for the re-incorporation of amino acids in proteins. This appears to be reasonable, given that some protein turnover is a necessary consequence of any differentiation and development process. This modification is reflected in the map simply by adding arrows from amino acids to protein, which complement the existing arrows of amino acid supply from proteins (Fig. 9.7). The mathematical structure of the model changes slightly, because one has to account for six additional rates, which are represented by the new independent variables X_{40} to X_{45} that are defined in Table 9.4. Furthermore, the formerly unidirectional steps of amino acid supply from proteins must be revisited.

During aggregation and culmination in the life cycle of *Dictyostelium*, significant metabolic changes occur. Of particular interest in the present context is that the protein and amino acid pools drop, and that there is very active proteolysis and decreased protein replacement between the early slug stage and culmination (e.g., Gregg and Bronsweig, 1956; Wright and Anderson, 1960ab). Experimental data suggest that about 50% of cell protein is utilized during a 24 h period of development, which roughly corresponds to 2%/h (Liddel and Wright, 1961; White and Sussman, 1961). At the same time, an average of 7% of cellular protein is newly synthesized every hour during preculmination (Wright and Anderson, 1960b; Franke and Sussman, 1973).

These data permit an assessment of the fluxes of amino acid supply and re-incorporation in proteins (Shiraishi and Savageau, 1993). The total protein degradation to amino acids must be 9%/h to allow for the 7% that is reutilized for new

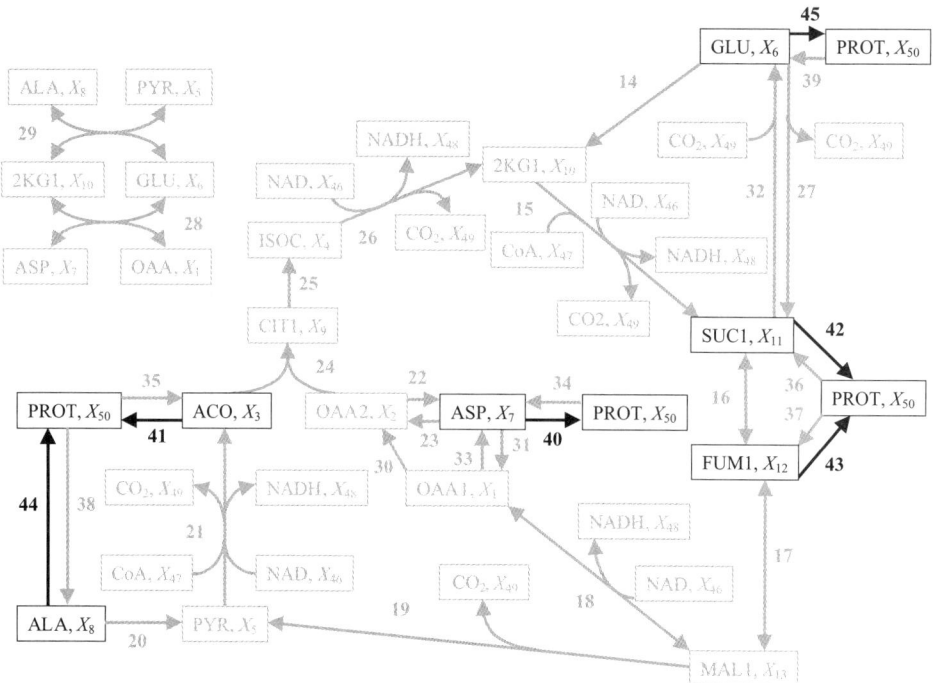

Figure 9.7. Addition of amino acid incorporation in proteins. In comparison with the original map (Fig. 9.2), the unidirectional fluxes from the protein pool into the system are replaced by bi-directional fluxes (highlighted).

protein synthesis and the 2% lost in energy production. The ratio of reverse to forward fluxes thus is $v_R/v_F = 7/9$, which corresponds to about 78% reutilization (see also Exercise 13). One could be more precise by assuming an exponential dynamic for reutilization, which would yield an approximate reutilization rate of 70%. Since the data can be assumed to have an error margin of about 10%, the difference in reutilization is of minor import (Shiraishi and Savageau, 1993).

A second guideline for adjusting parameter values in the modified model is the requirement of *external equivalence* (cf. Savageau, 1985b; Irvine and Savageau, 1985ab; see also Chapter 4): the *net* fluxes, v_{net}, in the modified model should be the same as in the previous model, since those were experimentally measured. These requirements are sufficient to estimate new parameter values for the affected fluxes, since we obtain two linear equations in two unknowns,

$$v_{net} = v_F - v_R \quad \text{and} \quad v_R = \frac{7}{9} v_F, \tag{9.15}$$

which have the solution

$$v_F = \frac{9}{2} v_{net} \quad \text{and} \quad v_R = \frac{7}{2} v_{net}. \tag{9.16}$$

As an example, consider the fluxes between aspartate and the protein pool. The net flux in the original model is $v_{net} = v_{50,7\,old} = X_{34\,old} = 0.236$ mM/min. In the modified model, this net flux is the difference between aspartate supply and aspartate reutilization. The flux of aspartate supply corresponds to $\frac{9}{2}$ of the net flux in the original model, which yields $v_F = v_{50,7} = X_{34} = \frac{9}{2} \times 0.236 = 1.06$ mM/min. Assuming

DIAGNOSIS AND REFINEMENT

Table 9.5. Revised and Additional Steps of Amino Acid Supply from and Incorporation in Proteins, Represented as Independent Variables in the Modified Model

Variable Name	Reaction	Concentration
X_{34}	Protein → Asp	0.106
X_{35}	Protein → AcCoA	0.207
X_{36}	Protein → Suc	0.162
X_{37}	Protein → Fum	0.036
X_{38}	Protein → Ala	0.203
X_{39}	Protein → Glu	0.186
X_{40}	Asp → Protein	0.446
X_{41}	Acetyl-CoA → Protein	27.2
X_{42}	Suc → Protein	1.57
X_{43}	Fum → Protein	7.0
X_{44}	Ala → Protein	0.326
X_{45}	Glu → Protein	0.240

a first-order process, aspartate reutilization takes the symbolic form $v_{7,50} = X_{40} X_7$. According to the considerations above, the magnitude of this flux is $\frac{7}{2}$ of the net flux in the original model, which implies $v_R = v_{7,50} = X_{40} X_7 = 0.826$ mM/min. Even though the modifications change the structure of the model, the observed steady-state concentrations and net fluxes should remain the same. External equivalence implies that the steady-state concentration of aspartate remains the same, namely $X_7 = 1.85$ mM/min. Using this piece of information, one determines the rate of aspartate reutilization as $X_{40} = 0.826/1.85 = 0.446$ min^{-1}. With these values for forward and reverse fluxes, the net flux from protein to aspartate is unchanged at 1.062 mM/min -0.826 mM/min $= 0.236$ mM/min. The computation of the remaining fluxes of supply and reutilization of amino acids is left as Exercise 14.

All independent variables representing protein catabolism, X_{34} to X_{39}, are adjusted in this manner; their updated values are given in the top half of Table 9.5. The numerical values of the new independent variables X_{40} to X_{45}, representing amino acid reutilization, are presented in the bottom half of Table 9.5. The remaining independent variables, X_{14} to X_{33}, are unchanged from the previous model and have the same concentrations as displayed in Table 9.2.

The additional reactions in the system are coded as

$$v_{3,50} = X_3 X_{41}, \quad v_{6,50} = X_6 X_{45}, \quad v_{7,50} = X_7 X_{40}, \quad (9.17)$$
$$v_{8,50} = X_8 X_{44}, \quad v_{11,50} = X_{11} X_{42}, \quad v_{12,50} = X_{12} X_{43}.$$

They are included in the node equations (9.1) of the system, which otherwise are unchanged. Specifically, the utilization rates $v_{i,50}$ are included in the degradation terms V_i^- of the affected variables X_3, X_6, X_7, X_8, X_{11}, and X_{12}. For instance, the node equation for X_{11} in the modified model reads

$$\dot{X}_{11} = (v_{6,11} + v_{10,11} + v_{50,11}) - (v_{11,6} + v_{11,12} + v_{11,50}) = V_{11}^+ - V_{11}^-. \quad (9.18)$$

The inclusion of the reverse reactions typically effects changes in all parameter values of the utilization terms of the corresponding S-system equation. As always, the kinetic orders associated with the power-law representation of a utilization rate are each computed as the partial derivative of this rate with respect to one of the variables, multiplied by this variable, and divided by the rate. Thus, even though reactions like $v_{i,50}$ may not explicitly include all variables in V_i^-, all kinetic orders and the rate constant in the utilization term V_i^- will normally be affected.

As an illustration, the utilization rate of X_{11} in the modified model is

$$\beta_{11} X_{11}^{h_{11,11}} X_{12}^{h_{11,12}} X_{16}^{h_{11,16}} X_{32}^{h_{11,32}} X_{42}^{h_{11,42}}. \tag{9.19}$$

Its kinetic orders and rate constant are computed from V_{11}^- as

$$h_{11,11} = \frac{\partial V_{11}^-}{\partial X_{11}} \cdot \frac{X_{11}}{V_{11}^-} = 0.495,$$

$$h_{11,12} = \frac{\partial V_{11}^-}{\partial X_{12}} \cdot \frac{X_{12}}{V_{11}^-} = -0.00542,$$

$$h_{11,16} = \frac{\partial V_{11}^-}{\partial X_{16}} \cdot \frac{X_{16}}{V_{11}^-} = 0.574, \tag{9.20}$$

$$h_{11,32} = \frac{\partial V_{11}^-}{\partial X_{32}} \cdot \frac{X_{32}}{V_{11}^-} = 0.166,$$

$$h_{11,42} = \frac{\partial V_{11}^-}{\partial X_{42}} \cdot \frac{X_{42}}{V_{11}^-} = 0.261.$$

and

$$\beta_{11} = V_{11}^- / X_{11}^{h_{11,11}} X_{12}^{h_{11,12}} X_{16}^{h_{11,16}} X_{32}^{h_{11,32}} X_{42}^{h_{11,42}} = 2.4373, \tag{9.21}$$

which is quite different from the corresponding equations (9.12) and (9.13) in the original model.

The inclusion of amino acid reutilization affects the S-system equations of the variables X_3, X_6, X_7, X_8, X_{11}, and X_{12}. The new influx and efflux terms are

$$V_3^+ = 0.3231 X_3^{-0.405} X_5^{0.156} X_{21}^{0.427} X_{35}^{0.573} X_{46}^{0.422} X_{47}^{0.405} X_{48}^{-0.418}, \tag{9.22a}$$

$$V_3^- = 9.6952 X_2^{0.376} X_3^{0.489} X_{24}^{0.554} X_{41}^{0.446} X_{47}^{-0.00206}, \tag{9.22b}$$

$$V_6^+ = 2.459 X_1^{-0.00921} X_6^{-0.0154} X_7^{0.0162} X_{10}^{0.086} X_{11}^{0.276} X_{28}^{0.0813} X_{32}^{0.276} X_{39}^{0.6413}, \tag{9.22c}$$

$$V_6^- = 1.1528 X_5^{0.0963} X_6^{1.01} X_8^{-0.204} X_{10}^{-0.062} X_{14}^{0.0518} X_{27}^{0.277}$$
$$\times X_{29}^{0.171} X_{45}^{0.5} X_{46}^{0.0222} X_{48}^{-0.0191}, \tag{9.22d}$$

$$V_7^+ = 2.1167 X_1^{0.129} X_2^{0.129} X_{22}^{0.129} X_{33}^{0.129} X_{34}^{0.741}, \tag{9.22e}$$

$$V_7^- = 3.4893 X_1^{-0.0187} X_6^{-0.0311} X_7^{0.868} X_{10}^{0.174} X_{23}^{0.129} X_{28}^{0.165} X_{31}^{0.129} X_{40}^{0.577}, \tag{9.22f}$$

$$V_8^+ = 0.5724 X_5^{0.111} X_6^{0.247} X_8^{-0.234} X_{10}^{-0.0713} X_{29}^{0.197} X_{38}^{0.803}, \tag{9.22g}$$

$$V_8^- = 1.9369 X_8 X_{20}^{0.375} X_{44}^{0.625}, \tag{9.22h}$$

$$V_{11}^+ = 2.0031 X_6^{0.166} X_{10}^{0.491} X_{11}^{-0.481} X_{15}^{0.499} X_{27}^{0.166} X_{36}^{0.335} X_{46}^{0.483} X_{47}^{0.481} X_{48}^{-0.483},$$
(9.22i)

$$V_{11}^- = 2.4373 X_{11}^{0.495} X_{12}^{-0.00542} X_{16}^{0.574} X_{32}^{0.166} X_{42}^{0.261},$$
(9.22j)

$$V_{12}^+ = 1.271 X_{11}^{0.106} X_{12}^{-0.00836} X_{16}^{0.885} X_{37}^{0.115},$$
(9.22k)

$$V_{12}^- = 9.1694 X_{12}^{1.89} X_{13}^{-1.24} X_{17}^{0.911} X_{43}^{0.0893}.$$
(9.22l)

All other influx and efflux terms of the modified S-system model are the same as in the original model (9.14) (see also Exercise 15). The model is implemented in PLAS as *TCA_Mod.plc*.

2. Analysis of the modified model. The first check again is the numerical computation of the non-zero steady state. Within the accuracy of the parameter values, PLAS produces the same solution as before, which is what we should expect from the two externally equivalent models. The absolute and relative fluxes for metabolites with unaltered utilization terms V_i^- are unaffected, whereas the fluxes through metabolites with altered utilization increase in all cases (see Table 9.6), reflecting the increased turnover associated with reutilization of amino acids.

The analysis of eigenvalues confirms that the steady state is locally stable. It also shows that the smallest real eigenvalue is now -0.34, in contrast to -0.003 in the previous model. Interpretation of these values in terms of time scales suggests that the modified system will respond to perturbations on the order of 100 times more quickly than the slowest variables in the previous model. This is confirmed by dynamic simulations, according to which the metabolites X_3 and X_5 approach their pre-perturbation steady-state values within a few minutes, which appears to be

Table 9.6. Comparison of Relative Fluxes in the Original and the Modified Models

Variable Name	Metabolite	Steady-State Flux V_{is} (mM/min)		Relative Flux V_{is}/X_{is} (min^{-1})	
		Original Model	Modified Model	Original Model	Modified Model
X_1	Oxalacetate 1	2.19	2.19	874	874
X_2	Oxalacetate 2	2.19	2.19	873	873
X_3	Acetyl-CoA	**2.00**	**3.61**	**33.7**	**60.9**
X_4	Isocitrate	2.00	2.00	200	200
X_5	Pyruvate	2.04	2.04	7.0	7.0
X_6	Glutamate	**1.45**	**2.90**	**0.241**	**0.481**
X_7	Aspartate	**0.606**	**1.43**	**0.327**	**0.773**
X_8	Alanine	**0.947**	**2.52**	**0.196**	**0.522**
X_9	Citrate 1	2.00	2.00	80	80
X_{10}	α-Ketoglutarate	2.65	2.65	265	265
X_{11}	Succinate	**3.57**	**4.83**	**4.46**	**6.03**
X_{12}	Fumarate	**2.85**	**3.13**	**71.4**	**78.3**
X_{13}	Malate	2.85	2.85	13	13

Bold entries indicate fluxes affected by model modification.

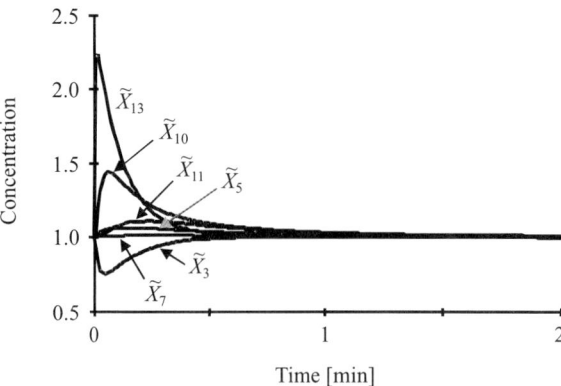

Figure 9.8. Dynamic response of the modified model to a 10-fold bolus increase in fumarate. In comparison with the original model (see Fig. 9.3), even the slow metabolites X_3 and X_5 return fairly quickly to the steady state. The tildes indicate that all variables are scaled with respect to their steady-state concentrations; in *TCA_Mod.plc* they are denoted by z_i's. See text for further discussion.

biologically reasonable, as opposed to more than a day in the previous model (Fig. 9.8). Further explorations are left as Exercise 16.

It is useful again to test the responses to reduced activities in malate dehydrogenase. At a level of 98% activity, the steady state is almost unchanged, and pyruvate increases from 0.291 to its new value of 0.302 in about 10 min. With progressive reductions down to 50% activity, pyruvate, acetyl-CoA, and alanine change in a monotonic fashion. Even at half the original activity of malate dehydrogenase, the pyruvate concentration increases only to 1.05, acetyl-CoA less than doubles, and alanine increases by little more than 10%. The dominant real eigenvalue is essentially unchanged, which indicates reasonable transition times. In light of the earlier results, these improvements are astounding. One might add that the corresponding modified Michaelis–Menten model is not as well behaved. In fact, at half the original enzyme activity, the model becomes unstable, and the pyruvate concentration approaches infinity (Shiraishi and Savageau, 1993).

Another critical test of the modified model is the analysis of sensitivities and gains, since these properties provided strong indication that the previous model did not appropriately describe pyruvate metabolism. The techniques are exactly the same as before; selected results are presented in Figs. 9.9 and 9.10. Inspection of the vertical axes in the pseudo-three-dimensional graphs confirms that the modified system is several orders of magnitude less sensitive than the previous model. All rate constant sensitivities, even in unconstrained form, are below 5 in magnitude. All logarithmic gains of metabolites and fluxes are below 3 in magnitude, and most are in the unit range or below, which suggests that the modified model is very robust (Exercise 17).

The cycle flux, represented by $V(X_3)$, $V(X_4)$, $V(X_9)$, $V(X_{12})$, and $V(X_{13})$, is now better controllable, since most independent variables have some measurable effect. This is in contrast to the original model, whose ability to propagate signals was very restricted.

DIAGNOSIS AND REFINEMENT 321

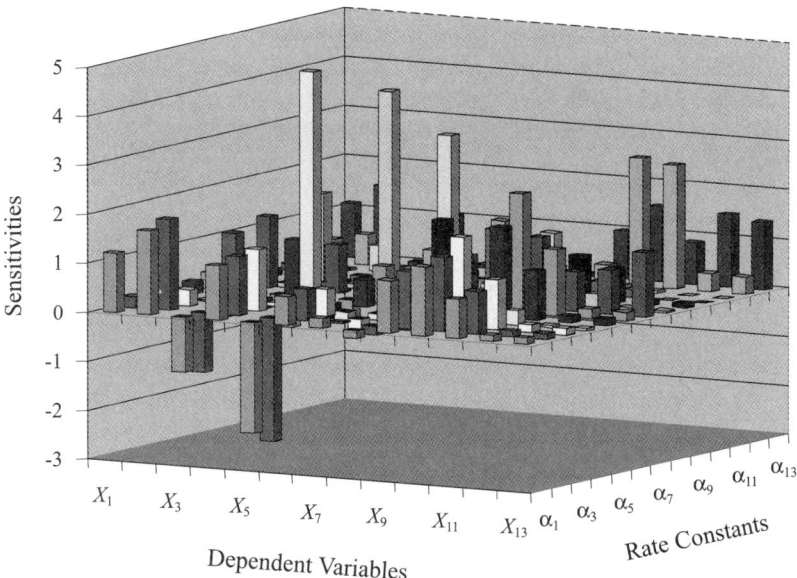

Figure 9.9 (see Color Plate XVII). Rate constant sensitivities of the modified model. Note the difference in scale of the vertical axis in comparison with Fig. 9.4 (Colorplate VII), indicating the enormous reduction in model sensitivity and consequent increase in robustness.

Significant improvements are also evident from an analysis of the kinetic order sensitivities. By far the most sensitivities are below 1, which indicates that typical perturbations are easily tolerated. This is readily demonstrated with the hide if <... option in the sensitivity dialog boxes in PLAS. Setting the threshold at 5 or even 1, most sensitivities and gains disappear. Even unconstrained, the highest sensitivities

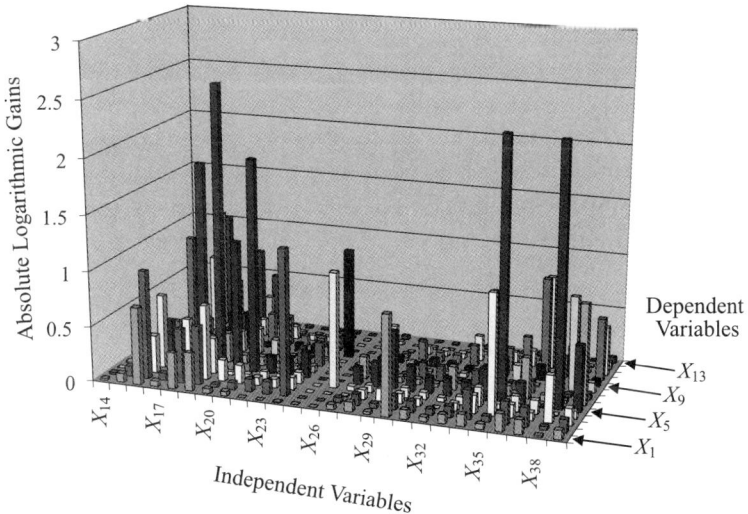

Figure 9.10 (see Color Plate XVIII). Absolute values of logarithmic gains of the modified model. All gains are now reasonable, the gain profile is much better balanced than in the original model, and the model is generally more robust. Note the difference in scale of the vertical axis in comparison with Fig. 9.5 (Colorplate VII).

for fluxes in the modified model are about 20, whereas some of them reached several thousand in the original model. Thus, the abnormally high sensitivities have been reduced by more than two orders of magnitude.

When constraints are taken into account, even the moderately high sensitivities are greatly reduced. For instance, the sensitivities of X_2 and of $V(X_2)$ with respect to h_{11} and $h_{1,30}$ have magnitudes of about 10. However, g_{21} and $g_{2,30}$ refer to the same process, and on taking the precursor–product relationship into account, the sensitivities essentially cancel each other (see also Exercise 19).

The model now appears to be much more reasonable, and the improvements in system properties that are associated with pyruvate metabolism are stunning. Nevertheless, the highest sensitivities and gains are still associated with pyruvate, even though they are now in an acceptable range. Assuming that the modified model structure is reliable, these sensitivities indicate the crucial role of pyruvate metabolism within the TCA cycle.

The impacts of changes in NAD are now also better under control: the effect of NAD on the concentration of pyruvate is reduced from −156 to −3.23. The largest flux gain with respect to NAD is found in the flux through pyruvate, with an (attenuation) value of −0.27, whereas previously the flux through the alanine pool was affected by NAD with a (strongly amplifying) negative gain of −20.

The new independent variables, which take account of amino acid reutilization, have flux gains that are below 1, but rank among the highest in the system. This indicates that amino acid reutilization directly or indirectly plays a significant role in the dynamical redistribution of material throughout the biochemical system.

In the original model, isocitrate and citrate were essentially affected only by their producing and degrading steps, which made these metabolites difficult to control. In the modified model, this situation is also improved, and isocitrate and citrate are now reasonably strongly affected by several enzymes, in particular, succinate dehydrogenase, malic enzyme, and malate dehydrogenase, which show log gains between −0.33 and +0.67. Isocitrate also responds to changes in NADH with a log gain of 0.92.

Outside the comparison of numerical values, one notices that the sensitivities and gains in the modified model are generally much better distributed throughout the system. There are not just a few metabolites or enzymes that would overpower the responses of the system, but control of the pathway is shared among most of the constituents.

Are We All Done?

There is hardly a doubt that the modified model behaves much better than the previous model. Considering the small conceptual changes from the first model (namely, a provision for the reutilization of amino acids), the improvements are amazing. The immediate question is now whether the modified model could be further improved through slightly altered assumptions, updated *in vivo* parameter values, or the addition of dependent and independent variables and processes. Of course, the answer is likely to be yes. However, except for the use of more accurate parameter values, further modifications can be expected to make the model more complicated, and only an assessment of the goals of the model can determine whether the added complexity is justified. Still, it may be a good idea to explore possible minor alterations that are suggested by the results of the current analysis and that may have the potential of further fine-tuning the responses and predictions of the present model.

A critical part of the pathway is still pyruvate metabolism. While greatly improved over the original model, the sensitivity and gain profiles associated with pyruvate still indicate the potential for further improvement. For instance, the totality of influences over pyruvate is about five times stronger than the totality of influences over metabolites in the oxalacetate branch. Furthermore, redistribution of fluxes at the malate branch is still limited. Shiraishi and Savageau (1993) note that this branch point presents a crucial problem in the corresponding Michaelis–Menten model. If the level of malate dehydrogenase drops by 10 or 20%, pyruvate accumulates dangerously fast, and if the level drops to less than 50%, the Michaelis–Menten system is no longer stable. This flaw in the traditional model raises fundamental questions about the validity of its structure and the common practice of using information obtained *in vitro* for models describing a pathway *in vivo* (Shiraishi and Savageau, 1993; see also Albe et al., 1989; Savageau, 1992b, 1995b).

What are our options for further exploring this critical portion of the tricarboxylic acid cycle? The first action might be a re-examination of the kinetic properties that determine the parameters in our models. Of course there could be inaccuracies anywhere in the system, but it might be most efficacious to focus our effort to the most influential sites within the pathway, which are characterized by high sensitivities and gains. For example, the analysis of log gains tells us that succinate dehydrogenase (X_{16}) has a strong overall effect, which is consistent with analyses by Canela et al. (1987), Rigoulet et al. (1988), and Albe and Wright (1992). The equilibrium constants for this enzyme and for the two subsequent enzymes in the pathway, fumarase (X_{17}) and malate dehydrogenase (X_{18}), are reported by Wright et al. (1992) as 10.0, 10.0, and 1.0, whereas typical textbook values are 1.0, 4.5, and 1.1×10^{-5} (Mahler and Cordes, 1966; see also Exercise 21). Furthermore, the low K_{eq} for malate dehydrogenase implies that this reaction can only proceed in the forward direction, because it is closely coupled to citrate synthetase, which has a large K_{eq} (Srere, 1975).

Gain analysis also indicates that NAD plays a significant role. In the present model, NAD is being considered an independent variable, which introduces some rigidity. Considering it a dependent variable instead, or using a buffer box (Voit and Ferreira, 1998; see also Chapter 4), could relieve this rigidity and allow NAD to exert more efficient control over pyruvate, whose concentration is affected by NAD with a relatively strong log gain of -3.2.

Neither the re-examination of kinetic properties of X_{16} to X_{18} nor the floating of NAD addresses directly the redistribution of fluxes at the malate branch point, which, according to Albe and Wright (1992), is a key control point in the TCA cycle of *Dictyostelium*. As a third option for exploring refinements Shiraishi and Savageau (1993) therefore suggest revisiting the control mechanisms at this branch point. Although the proposals are not yet experimentally tested, the model implies that there might be additional reactions removing pyruvate, some inhibition acting on malic enzyme, or a mechanism for diverting malate into the oxalacetate branch. Since a similar redistribution problem at the oxalacetate branch has been successfully solved by *Dictyostelium*, it would not be too surprising if analogous mechanisms governed the malate branch point.

EXERCISES

1. Formulate symbolic GMA and S-system representations for the original and the modified models.
2. Discuss the advantages and disadvantages of considering NAD, NADH, and CoA as independent variables, as opposed to simply including them in the rate constants, as was done, e.g., in some of the steady-state computations in Chapter 6.
3. What would it entail to make NAD, NADH, and CoA dependent variables? Explain additional data needs and structural changes in the equations. Would there be differences between GMA and S-system models with respect to these three cofactors? If so, describe them.
4. What would it entail to make the protein pool a dependent variable? Explain the additional data needs and structural changes in the equations. Differentiate, if necessary, between the original and the modified models.
5. Derive the kinetic orders and rate constants associated with pyruvate. Confirm the numerical values given in the text.
6. Derive the kinetic orders and rate constants associated with succinate in the original and the modified models. Confirm the numerical values given in the text.
7. Simulate the injection of a bolus of malate in the original and the modified models. Before you implement the simulation in PLAS, predict whether the sensitivity and gain profiles will change. Check your prediction. Pay attention to responses at different time scales. Compare the results with those obtained from simulating a bolus of fumarate.
8. Simulate the injection of a bolus of pyruvate in the original and the modified models. Pay attention to responses at different time scales. Compare the results with those obtained from simulating boli of fumarate and malate.
9. Test the robustness of the original and the modified models with respect to (persistent) changes in malic enzyme. Before you implement the simulation in PLAS, predict whether the sensitivity and gain profiles will change. Check your prediction. Study steady-state concentrations and fluxes, eigenvalues, and transition times.
10. Produce two-dimensional sensitivity profiles for rate constants in the original model. Sum absolute values along the x-axis or along the y-axis of the pseudo-three-dimensional graph. Indicate what portion of each sum is attributed to positive and what portion to negative sensitivities (cf. Figs. 1–4 in Shiraishi and Savageau, 1992b).
11. Manipulate PLAS so that it produces constrained rate constant sensitivities in the original model.
12. Show graphically the logarithmic gains of metabolites and fluxes with respect to changes in cofactors in the original model. Discuss gains with respect to CO_2 and protein content.

DIAGNOSIS AND REFINEMENT

13. It was assumed in the model modification that 7% of cellular protein is newly synthesized every hour throughout development. What are the consequences for the modified model if only 5% is synthesized, as is the case in bacteria (Wright and Anderson, 1960b). Compute the forward and reverse fluxes of aspartate supply and reutilization under this assumption.

14. Derive values for new fluxes in the modified model [see Eq. (9.17) and Table 9.5].

15. 15.1. Explain why the kinetic orders in the synthesis term of glutamate change from the original to the modified model. For instance, why does the kinetic order with respect to oxalacetate 1 change, even though the relationships between oxalacetate 1 and glutamate are not directly affected by the modification?

    15.2. Explore situations where the kinetic orders do not change. Find an example.

16. Compare the dynamic responses of the original and the modified models to various perturbations in initial values. Record which responses are fairly similar and which are drastically different. Provide biochemical explanations.

17. Display two- and pseudo-three-dimensional profiles of the logarithmic flux gains of the modified model.

18. Analyze the relatively high kinetic order sensitivities of fluxes associated with $h_{12,12}$. Begin with the unconstrained sensitivities, list all constraints, and assess the constrained effects of changes in this process.

19. Analyze the relatively high kinetic order sensitivities associated with V_5^-. Begin with the unconstrained sensitivities, list all constraints, and assess the constrained effects of changes in this process.

20. Display graphically the logarithmic gains of metabolites and fluxes with respect to changes in cofactors in the modified model. Compare with the original model.

21. (Project) Wright et al. (1992) suggest 10.0, 10.0, and 1.0, respectively, as the equilibrium constants for succinate dehydrogenase, fumarase, and malate dehydrogenase. In contrast, Mahler and Cordes (1966) reported the values 1.0, 4.5, and 1.1×10^{-5}. Using the latter values, derive S-system equations for the modified model, implement the new model in PLAS, and assess the impact of the alterations.

22. (Project) The original and the modified models both account for bi-directional processes between aspartate and oxalacetate 1 and 2. Both fluxes toward aspartate are 74, while the reverse fluxes are 0.1, which is almost three orders of magnitude lower. Predict what happens if the reverse fluxes Asp → Oaa 1 and Asp → Oaa 2 are ignored. Implement this simplification in PLAS, and compare the results with your predictions.

23. (Project) Formulate the modified model as a GMA model. Compare gain and sensitivity profiles, as well as representative dynamical simulation results. Summarize your findings in a non-technical report with a technical appendix.

REFERENCES

[4–6], [41], [51], [68], [97], [125], [131], [140–1], [165–6], [191–2], [197], [220], [222–3], [227], [291], [309], [318–19], [330], [339], [343], [365], [374–9], [396–7], [413], [439], [455], [487], [494], [497–512], [523].

CHAPTER TEN

Case Study 3 – A Sequence of Models Describing Purine Metabolism

Modeling is an iterative process. Based on data, assumptions, and simplifications, equations are set up and analyzed, simulations are executed, and predictions are made that characterize system responses not observed at the time of model design. Almost always the comparison between model output, whether reflecting a known experimental situation or a prediction, yields that the model differs in some ways from the observation. This is to be expected, because the model is an abstraction that is not intended to capture all details of the phenomenon. Nonetheless, in many cases the differences are so significant that the model must be deemed inappropriate. Of course, much thought has gone into the original model, and there is no need to start over again. Instead, the next iteration begins, in which the model is amended, adjusted, fine-tuned, or expanded. The evaluation of this new model has two components: comparison with the observed data and comparison with the previous model. The latter is important because it indicates whether the alterations had any effect at all, and if so, whether they pointed in the right direction.

Clearly, there are no general guidelines for how exactly to alter an existing model if it shows weaknesses, but the present case study may serve as a good example. It begins with a seemingly reasonable model of purine metabolism and goes through standard analyses only to find that there are severe and irreconcilable differences between model structure and reality. Because of the canonical structure of the model, problematic parts of the model are readily identified, and subsequent amendments lead to a second model that ameliorates the former problems. However, the model is still not satisfactory, and yet another iteration begins. This process continues until a reasonable model is obtained.

The progression through several models, the rationale behind alterations from one model to the next, and the diagnostic analyses of each model follow the dissertation research of Dr. Raul Curto (1996), who has generously agreed to have some of his published and unpublished work included in this case study. The main results of his and his collaborators' work (Curto et al., 1997, 1998ab) were recently published in detail. In particular, the complete process of estimating parameters was documented (Curto, 1996; Curto et al., 1998a) and will therefore not be included in this chapter, because it would distract from the main focus of the case study, which is the cycling between model design, model analysis, comparison with observations, and refinement. In his dissertation research, Curto went through about twenty models of increasing sophistication. The case study will not describe all intermediates, but select key models that structurally – and not just numerically – differ from their predecessors.

The pathway of interest is purine metabolism. Purine metabolism provides the organism with building blocks for the synthesis of DNA and RNA and is intimately connected with the dynamics of other key compounds, such as ATP. Numerous diseases are associated with purine metabolism, some causing elevated concentrations of uric acid in blood, which in turn may result in gout, and others with devastating effects leading to mental retardation, self-mutilation, and even death. Several of these diseases are caused by partial or total enzyme deficiencies or by the overexpression of key enzymes. The literature on purine metabolism and its clinical manifestations is huge. Good introductions are found in several editions of the standard work *The Metabolic Basis of Inherited Disease* (Stanbury et al., 1983).

Purine metabolism has many features that make it amenable to biomathematical analysis. While quite complex in itself, the pathway is almost closed: Only a few external compounds are used as substrates, and there are only a few end products. Furthermore, a large body of kinetic information is available. Finally, the complicated control structure of inhibitions and activations prohibits intuitive analyses and makes mathematical approaches mandatory.

Purine metabolism has been modeled several times with slightly different objectives. In one of the earlier models, Starmer et al. (1975) designed a purine model to study the distribution of dietary nitrogen throughout the purine pathway. Franco and Canela (1984) designed a kinetic model based on data from different species and tissues, with the aim of demonstrating the usefulness of computer simulations of complex metabolic systems. Heinmets (1989) developed a model focusing on nucleic acid synthesis from nucleotides and deoxynucleotides, and Bartel and Holzhütter (1990) constructed a model of purine metabolism in rat liver. The goal of our case study is to design an integrated model that is based on all available data and that may aid our understanding of the complex dynamics of purine metabolism in the human body. It follows very closely the work of Curto and his collaborators (Curto, 1996; Curto et al., 1997, 1998ab).

A good model of purine metabolism can be the basis for different types of investigation. First, the model may demonstrate that essential parts of the pathway are not completely understood or misunderstood. Such an insight would be gained from a model that allegedly contained all relevant components and processes, yet failed to describe biochemical or clinical observations. Second, the model may elucidate disease patterns. For instance, simulations of lowered or elevated enzyme activities may not only confirm or predict ultimate disease outcomes such as hyperuricemia, but also suggest which intermediate metabolites might be affected and to what degree. As a third example, a reliable model of the pathway may serve as a screening tool for drugs aimed at ameliorating symptoms or diseases associated with purine metabolism.

We begin with a summary of biochemical information, which is presented as a biochemical map. This summary will not discuss every reaction and every modulation, since all this is presented in detail in the work of Curto (1996) and Curto et al. (1997, 1998ab). The map is then translated into GMA and S-system equations, and, again, we will skip many intermediate steps and just show the results. In Curto's original analysis, parameter estimation was probably the modeling step that required most effort, which is not surprising in that the final S-system model has over 100 kinetic orders and 32 rate constants. We will simply list Curto's results. Once the model is numerically identified, we will execute standard analyses of validation, which assess steady-state characteristics and dynamic features. For the former, we will check stability and develop sensitivity and gain profiles. For the latter, we will simulate different perturbations and disease patterns in PLAS. During each iteration of the modeling process, we will study how discrepancies between model results and observations are used to expand and refine the model.

BIOCHEMISTRY OF PURINE METABOLISM

Before we set up model equations, it is useful to review the reactions that drive purine synthesis. Reactions that are deemed most relevant are summarized in Fig. 10.1; biochemical details of the pathway can be found, for instance, in Stanbury et al. (1983) and Curto et al. (1998a).

The first key metabolite of purine biosynthesis is 5-phosphoribosyl-α-1-pyrophosphate (PRPP), which results from a transfer of a pyrophosphate group of ATP to ribose-5-phosphate. This reaction is catalyzed by phosphoribosylpyrophosphate synthetase and modulated by various metabolites of purine metabolism. A linear chain of reactions converts PRPP into inosine monophosphate (IMP), which constitutes the central branch point of purine metabolism. IMP is converted into AMP or GMP. As far as guanosine, adenosine, and their immediate derivatives are not used in the synthesis of nucleic acids or as energy storage, they can be converted into hypoxanthine (HX) and xanthine (Xa), the latter of which is oxidized to uric acid (UA).

In addition to this *neobiosynthesis* of ribonucleotides, two very important salvage pathways help maintain the pool of IMP and, thus, of adenosine and guanosine compounds. In these salvage pathways, adenine phosphoribosyltransferase (APRT) and hypoxanthine-guanine phosphoribosyltransferase (HGPRT) condense free bases with PRPP, thereby forming ribonucleotides in one step. This salvaging mechanism appears to be crucial, since enzyme deficiencies that compromise the reactions involved can have devastating effects which may result in severe hyperuricemia, mental retardation, self-mutilation, and death. The entire pathway of purine neobiosynthesis and salvage is tightly controlled by manifold feedbacks and cross-activations, and many of the reactions are reversible.

FIRST MODEL

For the first model, we construct a map that contains those metabolites that we consider most important (see Fig. 10.1). A key element is the primary substrate,

CASE STUDY 3 – A SEQUENCE OF MODELS DESCRIBING PURINE METABOLISM

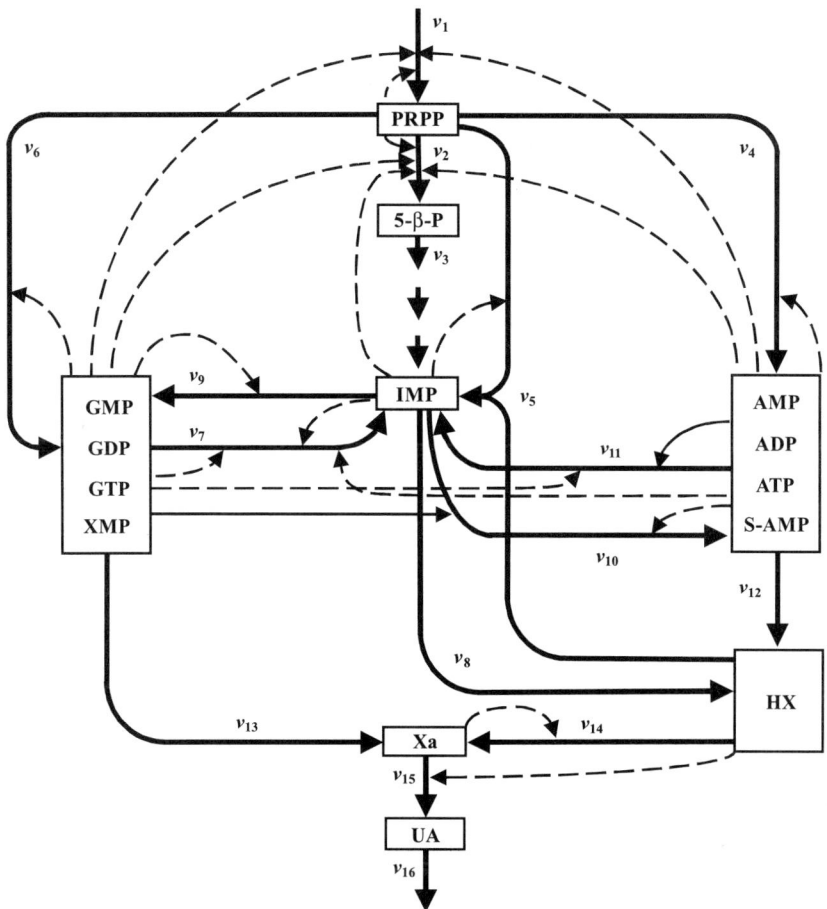

Figure 10.1. First model of purine metabolism (see text for details).

phosphoribosyl pyrophosphate (PRPP). We don't consider the details of its production, but take into account the catalyzing enzyme – PRPP synthetase (PRPPS) – and the fact that the production is subject to feedback inhibition. PRPP is utilized in four ways. For purine synthesis *de novo*, PRPP is converted into β-5-phosphoribosyl amine by the enzyme amidophosphoribosyltransferase. PRPP furthermore contributes significantly to three salvage pathways. These involve phosphoribosyltransferase reactions in which PRPP condenses with free bases to form ribonucleotides (Wyngaarden and Kelley, 1983: p. 1066). Two phosphoribosyltransferases are known. Adenine phosphoribosyltransferase (APRT) uses primarily adenine and some related compounds, and hypoxanthine-guanine phosphoribosyltransferase (HGPRT) acts upon hypoxanthine and guanine, and to a minor degree on xanthine and some other compounds. These salvage reactions are essentially irreversible. Deficiencies in APRT and HGPRT can cause severe diseases. Both the synthesis and utilization of PRPP are heavily controlled by feedback regulation.

Amidophosphoribosyltransferase catalyzes the first step in a sequence that eventually yields inosine monophosphate (IMP). Since there are no branch points or control signals downstream of β-5-phosphoribosyl amine, further intermediates are not included in the map and the model. This omission simplifies the model and does not affect the steady state, but it may have an effect on the dynamic responses of the model (see Exercise 29). IMP is the central metabolite of purine metabolism. It is in this location that the organism can channel purine rings into adenylates and guanylates or toward hypoxanthine, from where they may be incorporated in RNA or DNA, recycled, or disposed of. All fluxes into or out of the IMP pool are very tightly controlled. If adenylates are not needed, they are converted into hypoxanthine, from where they may be recycled or further degraded to xanthine, which is ultimately oxidized to uric acid. Excess guanylates are catalyzed by guanase to xanthine and oxidized to uric acid (Exercise 1).

Most of the variables are self-explanatory from Fig. 10.1, except for some that represent pools of several metabolites. One of these pools includes AMP, ADP, ATP, and adenylosuccinate (S-AMP). These compounds are converted into each other in numerous ways, and representing the dynamics within the pool would complicate the model very significantly. The same arguments suggest the pooling of GMP, GDP, GTP, and XMP.

It is important to reflect on the merits and disadvantages of such pooling. Clearly, pooling simplifies the model considerably, since eight metabolites are represented by two variables. Without the simplification, each variable would have to be represented with its own equation, requiring six additional sets of parameter values and initial values. However, pooling precludes our asking important questions. For instance, there is no possibility for this model to help us assess imbalances between ATP and ADP in patients with compromised purine metabolism. For this first model, we accept this limitation, hoping that the assumptions are reasonable and valid on one hand, and yield mathematical tractability and usefulness on the other (see Exercise 2).

While the map looks reasonable and a corresponding model could readily be implemented in PLAS, the structure of the system turns out to be fundamentally flawed. Closer analysis shows that, independent of the rate laws and parameter values, the model cannot maintain a steady state and at the same time balance the purine ring moieties (Curto, 1996). To see this, we list and compare those fluxes that produce and degrade purine rings. The only purine-producing reaction is characterized by the step v_3, and the only loss of rings occurs through v_{16}. However, these two cannot be balanced at steady state. Setting the system equations equal to zero, we obtain, even without further specifying rate laws,

IMP: $(v_3 + v_5 + v_7 + v_{11}) - (v_8 + v_9 + v_{10}) = 0$,
Adenylates: $(v_4 + v_{10}) - (v_{11} + v_{12}) = 0$,
Guanylates: $(v_6 + v_9) - (v_7 + v_{13}) = 0$,
HX: $(v_8 + v_{12}) - (v_5 + v_{14}) = 0$,
Xa: $(v_{13} + v_{14}) - v_{15} = 0$,
UA: $v_{15} - v_{16} = 0$.

These are algebraic equations that we may add up. The result is

$$v_3 + v_4 + v_6 = v_{16},$$

which implies that the purine ring input (v_3) cannot be the same as the loss of purine rings (v_{16}), since v_4 and v_6 are always positive. The root of the problem is that the model does not allow for the fact that the step v_4 recycles one molecule of adenine and that v_6 recycles one molecule of guanine. To balance influx and efflux of the purine ring moiety, v_4 and v_6 would have to be added to v_{16}.

SECOND MODEL

We could simply implement the recycling mechanisms, but use the opportunity for some other amendments. Since we haven't analyzed the details of the first model, these amendments may appear to have no rational foundation. However, they resulted from Curto's (1996) careful evaluations of the first model and a number of minor revisions. We consider the following alterations:

1. There is no real reason to retain β-5-phosphoribosyl-1-amine. The metabolite itself is not of much interest. Also, subsequent intermediates are omitted anyway. Finally, later steps in this pathway from PRPP to IMP are not modulated, so that we are not risking the loss of potential control (see also Exercise 29).
2. It appears useful to include pools for deoxynucleotides that correspond to the nucleotide pools.
3. Step v_8 actually consists of two reactions. 5′-Nucleotidase transforms IMP into inosine, which is transformed to hypoxanthine by purine nucleoside phosphorylase. Similarly, v_{12} is composed of three steps. First, unspecific phosphatases convert AMP unto adenosine. Second, adenosine is deaminated to inosine. Finally, purine nucleoside phosphorylase catalyzes the conversion of inosine to hypoxanthine. Analogous mechanisms act on the corresponding deoxynucleotides. Here, adenosine deaminase converts deoxyadenosine to deoxyinosine. Since IMP is so strongly controlled, its dynamics should be modeled as accurately as possible. Thus, to reflect these steps, pools of deoxynocleotides are included, a pool containing guanine, guanosine, and deoxyguanosine is included, adenosine and S-adenosyl-L-methionine (SAM) are added to the pool of adenylates, and inosine and deoxyinosine are added to the hypoxanthine pool. The inclusion of adenosine with the adenylates is justified because adenosine and AMP are readily converted into each other (e.g., Bontemps et al., 1983). SAM is the starting point for the polyamine pathway, generating adenine. SAM and adenosine are readily converted into each other. The aggregation of inosine with deoxyinosine and hypoxanthine is justified for similar reasons.
4. Again because of the central position of IMP, the fluxes leaving this pool may have to be represented in better detail. IMP is directly converted into adenylosuccinate (S-AMP) or xanthine monophosphate (XMP), so these metabolites should be represented. As a first option, we add them to the pools of adenylates and guanylates, respectively. This strategy may have to be revisited, since the conversion of XMP into GMP is essentially irreversible.

Setting Up Equations

The map of the second model (without modulations) is given in Fig. 10.2. The describing system contains 11 dependent variables, X_1–X_{11}. We code them as shown in Table 10.1.

In addition to the coding of variables, it is useful to identify the enzyme-catalyzed reactions and non-enzymatic steps that constitute the pathway. They are listed in Table 10.2 along with abbreviated names and E.C. numbers.

Processes. Even though the pathway is quite complex, the simple rules for translating a map into canonical equations are the same as for much simpler chains of

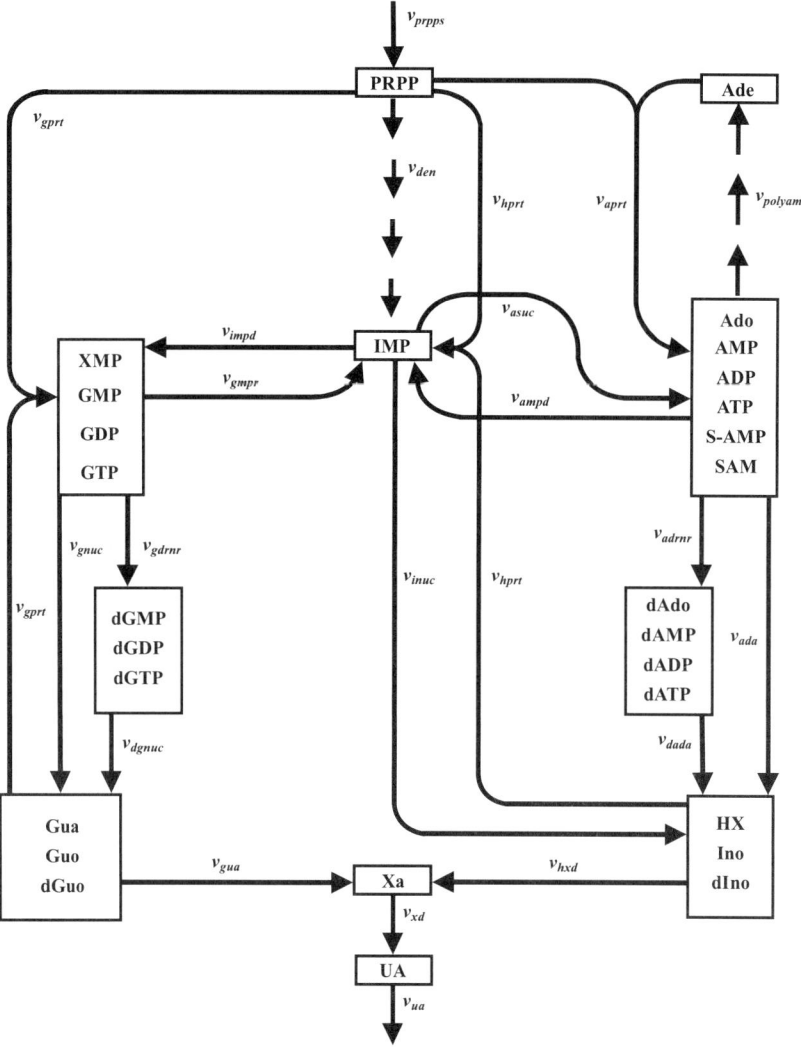

Figure 10.2. Second model of purine metabolism without representation of activations and inhibitions (see text for details).

Table 10.1. Variables of Second Model of Purine Metabolism

Variable	Abbreviation	Metabolite
X_1	PRPP	Phosphoribosylpyrophosphate
X_2	IMP	Inosine monophosphate
X_3	S-AMP	Adenylosuccinate
	Ado	Adenosine
	AMP	Adenosine monophosphate
	ADP	Adenosine diphosphate
	ATP	Adenosine triphosphate
	SAM	S-adenosyl-L-methionine
X_4	XMP	Xanthosine monophosphate
	GMP	Guanosine monophosphate
	GDP	Guanosine diphosphate
	GTP	Guanosine triphosphate
X_5	dAdo	Deoxyadenosine
	dAMP	Deoxyadenosine monophosphate
	dADP	Deoxyadenosine diphosphate
	dATP	Deoxyadenosine triphosphate
X_6	dGMP	Deoxyguanosine monophosphate
	dGDP	Deoxyguanosine diphosphate
	dGTP	Deoxyguanosine triphosphate
X_7	Ade	Adenine
X_8	HX	Hypoxanthine
	Ino	Inosine
	dIno	Deoxyinosine
X_9	Gua	Guanine
	Guo	Guanosine
	dGuo	Deoxyguanosine
X_{10}	Xa	Xanthine
X_{11}	UA	Uric acid

reactions. For reasons of parameter estimation, it is useful first to design a canonical model in GMA form and then to derive the S-system form from it. This procedure has been executed many times before, and we leave most of it as an exercise. As just one illustration, we formulate the equations for the key metabolite PRPP.

The dynamics of PRPP is affected by five reactions. One describes the synthesis of PRPP from R5P, and the remaining four are pathways that use PRPP as a substrate. The synthesizing reaction has R5P as substrate and is modulated by numerous inhibitors. One is PRPP itself, and four others of relevance are AMP, ADP, GDP, and GTP. The latter are not represented individually in our model, and we have to use the aggregated pools X_3 and X_4 as substitutes and adjust the kinetic orders accordingly. The term representing PRPP synthesis thus includes X_1, X_3, and X_4 as variables:

$$v_{prpps} = v_{prpps}(X_1, X_3, X_4).$$

Table 10.2. Enzymatic Reactions and Steps of Second Model of Purine Metabolism

Flux	Enzyme Name or Step	E.C.
v_{prpps}	Phosphoribosylpyrophosphate synthetase (PRPPS)	2.7.6.1.
v_{gprt}	Hypoxanthine-guanine phosphoribosyltransferase (HGPRT)	2.4.2.8.
v_{hprt}	Hypoxanthine-guanine phosphoribosyltransferase (HGPRT)	2.4.2.8.
v_{aprt}	Adenine phosphoribosyltransferase (APRT)	2.4.2.7.
v_{den}	*De novo* synthesis (amidophosphoribosyltransferase; ATASE)	2.4.2.14.
v_{asuc}	Adenylosuccinate synthetase (ASUC)	6.3.4.4.
v_{impd}	IMP dehydrogenase (IMPD)	1.1.1.205.
v_{ampd}	AMP deaminase (AMPD)	3.5.4.6.
v_{gmpr}	GMP reductase (GMPR)	1.6.6.8.
v_{polyam}	Polyamine pathway (S-adenosylmethionine decarboxylase; SAMD)	4.1.1.50.
v_{inuc}	5'-Nucleotidase (5NUC)	3.1.3.5.
v_{gnuc}	5'-Nucleotidase (5NUC)	3.1.3.5.
v_{dgnuc}	5'(3') Nucleotidase (3NUC)	3.1.3.31.
v_{ada}	Adenosine deaminase (ADA)	3.5.4.4.
v_{dada}	Adenosine deaminase (ADA)	3.5.4.4.
v_{adrnr}	Diribonucleotide reductase (DRNR)	1.17.4.1.
v_{gdrnr}	Diribonucleotide reductase (DRNR)	1.17.4.1.
v_{gua}	Guanine hydrolase (GUA)	3.5.4.3.
v_{hxd}	Xanthine oxidase or xanthine dehydrogenase (XD)	1.2.1.37.
v_{xd}	Xanthine oxidase or xanthine dehydrogenase (XD)	1.2.1.37.
v_{ua}	Uric acid excretion	Non-enzymatic

All kinetic orders in the term reflect feedback inhibition by different products and are negative. Since there is no branching, the corresponding power-law term is the same for the GMA and the S-system. The synthesis term thus reads, in symbolic terms,

$$v_{prpps} = V_1^+ = \alpha_1 X_1^{g_{11}} X_3^{g_{13}} X_4^{g_{14}}.$$

PRPP is degraded, transformed, or otherwise depleted via four pathways. The main pathway of *de novo purine synthesis* leads through a linear chain of reactions to IMP. In our model, this chain is represented as a single step (v_{den}), and the enzyme responsible for its dynamic is amidophosphoribosyltransferase. The step is allometrically inhibited by adenyl and guanyl ribonucleotides and by IMP in a synergistic fashion (Wyngaarden and Kelley, 1983: p. 1070). The inhibition is competitive with respect to PRPP. The step is furthermore activated by PRPP. The rate thus is a function of the first four variables:

$$v_{den} = v_{den}(X_1, X_2, X_3, X_4).$$

Three salvage pathways, catalyzed by adenine phosphoribosyltransferase (APRT) and by hypoxanthine-guanine phosphoribosyltransferase (HGPRT), use PRPP to form AMP, GMP, or IMP. In all three cases, the product has an inhibitory effect on the reaction, and there is also discussion about the relevance of potential reversibility of the reactions (see, e.g., Henderson et al., 1968; Giacomello and Salerno, 1978). Both product inhibition and reversibility suggest the inclusion of the product in the rate law. In symbolic terms, the three rates are

$$v_{aprt} = v_{aprt}(X_1, X_3, X_7),$$
$$v_{gprt} = v_{gprt}(X_1, X_4, X_9),$$
$$v_{hprt} = v_{hprt}(X_1, X_2, X_8).$$

The symbolic analogues of the four rates in the GMA model can be directly formulated from the given information about dependences of some of the system variables (Exercise 3). In the S-system model, the utilization of PRPP is aggregated into one term, which includes as variables the main substrate PRPP, the substrates Ade, Gua, and HX of the salvage pathways, and the inhibitors AMP, GMP, and IMP. The kinetic orders associated with substrates are positive, and those associated with inhibitors are negative. The formulation of the degradation term of PRPP in symbolic S-system form is left as the second part of Exercise 3.

Every rate of synthesis or degradation is symbolically formulated in this fashion. First, its dependences on the dependent or independent variables are identified. This information leads straightforwardly to the corresponding GMA term. In a second step, the aggregated S-system term is formulated from the collection of GMA terms.

Parameter estimation. As mentioned in the beginning, the focus of this chapter is not the estimation of parameter values, even though this step is crucial and required a very significant effort in the original analysis. The interested reader is encouraged to study the estimation process by reading the original documents (Curto, 1996; Curto et al., 1998a). A few examples were also given in Chapter 5. It suffices here to outline the strategy of parameter estimation and to discuss some of the idiosyncrasies of this particular pathway.

The experimental information available for the pathway consists of three mayor types of information:

1. Steady-state concentrations.
2. Steady-state fluxes.
3. Kinetic characteristics, such as Michaelis constants and maximal rates.

In addition, there is some clinical information pertaining to diseases of purine metabolism.

Steady-state concentrations. It is not surprising that there is no unique set of steady-state concentrations for the metabolites. The pathway is a key component of metabolism, and because it is directly associated with adenylates and thus with the availability of easily accessible energy, its numerical characteristics not only depend on the species and tissue but also on the physiological state of the organism. For example, the concentration of IMP is on the order of 10 μM in erythrocytes and brain cells and on the order of 100 μM in kidney and liver. If fructose is increased, its concentration in the liver may increase to 1,000 μM. Other concentrations are rather stable. For instance, under physiological conditions, there is not much fluctuation in the total concentration of the adenylate pool.

An exploration of the literature shows that the steady-state concentrations of most of the metabolites involved generally remain within about an order of magnitude.

Table 10.3. Summary of Average Steady-State Concentrations

Variable	Abbreviation	Individual Concentration (mM)	Total Concentration (mM)
X_1	PRPP	5	5
X_2	IMP	100	100
X_3	S-AMP	0.2	
	Ado	0.5	
	AMP	200	2,500
	ADP	400	
	ATP	1,900	
	SAM	4	
X_4	XMP	25	
	GMP	25	425
	GDP	75	
	GTP	300	
X_5	dAdo	<0.1	
	dAMP	0.5	
	dADP	1.4	6
	dATP	4	
X_6	dGMP	0.1	
	dGDP	0.5	3
	dGTP	2.4	
X_7	Ade	10	10
X_8	HX	6.9	
	Ino	3	10
	dIno	0.1	
X_9	Gua	0.5	
	Guo	4.4	5
	dGuo	0.1	
X_{10}	Xa	5	5
X_{11}	UA	100	100

Curto et al. (1998a) present a detailed account of the pertinent literature, and it suffices here to state the steady-state concentrations that will be used for the model; they are listed in Table 10.3.

Mathematically precise concentration values, though difficult to establish, are mandatory to get the numerical modeling process started. Of course, the numerical values are subject to scrutiny and adjustment if needed. Refinements will be guided by sensitivity and dynamical analyses of the model, which will give us some impression of the importance of the accuracy or uncertainty of the chosen values. If the model responds very strongly to small alterations, better accuracy is required than if the model turns out to be rather insensitive. As an example, we will analyze the effect of changing the concentration of adenine and its associated flux. For the present model, we choose [Ade] = 10, $v_{aprt} = 10$, which reflects measurements in

CASE STUDY 3 – A SEQUENCE OF MODELS DESCRIBING PURINE METABOLISM

human brain. Later, we will reduce these quantities to 1, which is supported by data from other tissues (van Acker et al., 1977; Simmonds et al., 1989).

Steady-state fluxes. Specific information about fluxes is often difficult to obtain, especially if one is interested in direct measurements in the intact human body. It is therefore necessary to deduce information from other types of data. For instance, the amount of a chemical compound that is metabolized by the human body during a given time period is sometimes known. Assuming an average body weight (BW) of 70 kg, one can estimate several of the required fluxes, which are then expressed, for instance, in units of $\mu\text{mol min}^{-1} \text{kg}^{-1}$.

The model contains 21 fluxes. Direct measurements of only four of these are available. They are, in alphabetical order,

$$v_{aprt} = 10 \ \mu\text{mol min}^{-1}\text{kg}^{-1},$$

$$v_{hxd} = 0.05 \ \mu\text{mol min}^{-1}\text{kg}^{-1},$$

$$v_{ua} = 2.27 \ \mu\text{mol min}^{-1}\text{kg}^{-1},$$

$$v_{xd} = 0.03 \ \mu\text{mol min}^{-1}\text{kg}^{-1}.$$

At steady state, the totality of all fluxes entering a pool must equal the totality of all fluxes leaving that pool. Thus, the 11 equations of the dependent variables are set equal to zero, which allows us to formulate 11 of the fluxes as linear combinations of the remaining 10. The steady-state equations are

X_1: $\quad v_{prpps} = v_{gprt} + v_{hprt} + v_{aprt} + v_{den},$

X_2: $\quad v_{den} + v_{gmpr} + v_{ampd} + v_{hprt} = v_{impd} + v_{asuc} + v_{inuc},$

X_3: $\quad v_{aprt} + v_{asuc} = v_{ampd} + v_{polyam} + v_{adrnr} + v_{ada},$

X_4: $\quad v_{gprt} + v_{impd} = v_{gmpr} + v_{gdrnr} + v_{gnuc},$

X_5: $\quad v_{adrnr} = v_{dada},$

X_6: $\quad v_{gdrnr} = v_{dgnuc},$

X_7: $\quad v_{polyam} = v_{aprt},$

X_8: $\quad v_{inuc} + v_{ada} + v_{dada} = v_{hprt} + v_{hxd},$

X_9: $\quad v_{gnuc} + v_{dgnuc} = v_{gprt} + v_{gua},$

X_{10}: $\quad v_{hxd} + v_{gua} = v_{xd},$

X_{11}: $\quad v_{xd} = v_{ua}.$

For a complete kinetic characterization, six more constraints between fluxes are needed, and these can be extracted from other knowledge about the system. Curto (1996) obtained the following results:

$$v_{hxd} + v_{hprt} \approx 4.9 \ \mu\text{mol min}^{-1}\text{kg}^{-1},$$

$$v_{hprt} = v_{gprt},$$

$$v_{hprt} \approx 3v_{hxd},$$

$$v_{gdrnr} + v_{gnuc} = 9v_{gmpr},$$

$$v_{ampd} \approx 3v_{ada},$$

$$v_{asuc} = 5v_{impd}.$$

Table 10.4. Steady-State Flux Rates, Listed in Alphabetical Order

$v_{ada} = 2.1$	$v_{adrnr} = 0.2$	$v_{ampd} = 5.69$	$v_{aprt} = 10$
$v_{asuc} = 8$	$v_{dada} = 0.2$	$v_{den} = 2.3$	$v_{dgnuc} = 0.1$
$v_{gdrnr} = 0.1$	$v_{gmpr} = 0.5$	$v_{gnuc} = 4.7$	$v_{gprt} = 3.7$
$v_{gua} = 1.1$	$v_{hprt} = 3.7$	$v_{hxd} = 1.23$	$v_{impd} = 1.6$
$v_{inuc} = 2.68$	$v_{polyam} = 1.01$	$v_{prpps} = 20.79$	$v_{ua} = 2.3$
$v_{xd} = 2.3$			

Note: Units are μmol min^{-1} kg^{-1}.

These estimates and constraints are sufficient to compute the flux values at steady state; they are presented in Table 10.4. Although these values have been calculated for our specific modeling purposes, they are interesting in themselves because they provide some insight in the distribution of fluxes in human purine metabolism. Overall, there is quite a spread, with the influx to the pathway exceeding the smallest fluxes by over 200-fold.

Kinetic characteristics. The steady-state concentrations and fluxes are not sufficient to estimate a full kinetic model. For this purpose, more detailed kinetic information is needed. In some cases, direct measurements allow us to estimate kinetic orders. Two examples for this situation were given in Chapter 5. They described the dependence of adenine excretion and uric acid excretion as functions of their substrate concentrations. In most cases, however, such information is unavailable, and the best (and maybe only) option is the transformation of a traditional rate law into the power-law form. This transformation is simply a matter of power-law approximation, as was discussed in Chapter 5. As mentioned at the beginning of the chapter, parameter estimation is not the focus of this case, and we simply re-state results as they were derived by Curto (1996) and Curto et al. (1998a).

Model Equations and Analysis

Curto (unpublished) computed all kinetic orders for this model from traditional rate laws and from other types of biochemical and clinical information. In a first step, he computed the kinetic orders of the GMA model. These kinetic orders are coded with a subscript that consists of the reaction name and the index of the variable that exerts the effect. For instance, f_{prpps4} refers to the effect of X_4 on v_{prpps} (PRPP synthetase step). The results are given in Table 10.5 in alphabetical order. The rate constants are computed from the fact that the power-law term and the traditional rate law are equal at the steady state (see Appendix). For instance, the rate constant α_{prrps} is computed as

$$\alpha_{prpps} = v_{prpps} \cdot \left(X_1^{f_{prpps1}} X_3^{f_{prpps3}} X_4^{f_{prpps4}} \right)^{-1},$$

which is evaluated by plugging in steady-state values (see Exercise 4).

The corresponding S-system model is obtained in two steps also. First, the aggregated kinetic orders are computed as averages of the GMA kinetic orders, weighted

CASE STUDY 3 – A SEQUENCE OF MODELS DESCRIBING PURINE METABOLISM

Table 10.5. Kinetic Orders of Second Model of Purine Metabolism (GMA Form).

$f_{ada3} = 0.97$	$f_{adrnr3} = 0.1$	$f_{adrnr5} = -0.3$	$f_{adrnr6} = 0.87$
$f_{ampd3} = 2.7$	$f_{ampd4} = -0.04$	$f_{aprt1} = 0.5$	$f_{aprt3} = -0.8$
$f_{aprt7} = 0.2$	$f_{asuc2} = 0.4$	$f_{asuc3} = -0.21$	$f_{asuc4} = 0.2$
$f_{dada5} = 1$	$f_{den1} = 2$	$f_{den2} = -0.06$	$f_{den3} = -0.25$
$f_{den4} = -0.2$	$f_{dgnuc6} = 1$	$f_{gdrnr4} = 0.4$	$f_{gdrnr5} = -1.2$
$f_{gdrnr6} = -0.39$	$f_{gmpr2} = -0.15$	$f_{gmpr3} = -0.07$	$f_{gmpr4} = -0.06$
$f_{gnuc4} = 0.9$	$f_{gprt1} = 1.2$	$f_{gprt4} = -1.2$	$f_{gprt9} = 0.42$
$f_{gua9} = 0.5$	$f_{hprt1} = 1.1$	$f_{hprt2} = -0.89$	$f_{hprt8} = 0.48$
$f_{hxd8} = 0.45$	$f_{hxd10} = -0.22$	$f_{impd2} = 0.15$	$f_{impd4} = -0.12$
$f_{inuc2} = 0.8$	$f_{polyam3} = 0.9$	$f_{prpps1} = -0.03$	$f_{prpps3} = -0.45$
$f_{prpps4} = -0.04$	$f_{ua11} = 2.21$	$f_{xd8} = -0.55$	$f_{xd10} = 0.78$

with the magnitudes of the involved fluxes. For example, the aggregated kinetic order for the degradation of X_3 is composed of four terms:

$$h_{33} = \frac{f_{polyam3}v_{polyam} + f_{ampd3}v_{ampd} + f_{ada3}v_{ada} + f_{adrnr3}v_{adrnr}}{v_{polyam} + v_{ampd} + v_{ada} + v_{adrnr}}.$$

In a second step, the rate constants are computed in the usual way.

The S-system describing the map in Fig. 10.2 has the following form:

$$\dot{X}_1 = 935.725 X_1^{-0.03} X_3^{-0.45} X_4^{-0.04}$$
$$- 862.225 X_1^{0.9} X_2^{-0.166} X_3^{-0.453} X_4^{-0.237} X_7^{0.106} X_8^{0.0858} X_9^{0.0751},$$
$$\dot{X}_2 = 0.001150 X_1^{0.711} X_2^{-0.287} X_3^{1.21} X_4^{-0.0589} X_8^{0.146}$$
$$- 2.2336 X_2^{0.452} X_3^{-0.138} X_4^{0.115},$$
$$\dot{X}_3 = 190.153 X_1^{0.289} X_2^{0.168} X_3^{-0.552} X_4^{0.0842} X_7^{0.116}$$
$$- 0.000260225 X_3^{1.44} X_4^{-0.012} X_5^{-0.00316} X_6^{0.00916},$$
$$\dot{X}_4 = 138.151 X_1^{0.838} X_2^{0.0453} X_4^{-0.874} X_9^{0.293}$$
$$- 0.0493729 X_2^{-0.0142} X_3^{-0.00660} X_4^{0.8} X_5^{-0.0226} X_6^{-0.00736},$$
$$\dot{X}_5 = 0.0601985 X_3^{0.1} X_5^{-0.3} X_6^{0.87} - 0.0333333 X_5,$$
$$\dot{X}_6 = 0.117086 X_4^{0.4} X_5^{-1.2} X_6^{-0.39} - 0.0333333 X_6,$$
$$\dot{X}_7 = 0.00962159 X_3^{0.9} - 1{,}622.78 X_1^{0.5} X_3^{-0.8} X_7^{0.2},$$
$$\dot{X}_8 = 0.0249405 X_2^{0.424} X_3^{0.416} X_5^{0.0408}$$
$$- 10.4238 X_1^{0.831} X_2^{-0.672} X_8^{0.473} X_{10}^{-0.0539},$$
$$\dot{X}_9 = 0.0226833 X_4^{0.881} X_6^{0.0208} - 144.479 X_1^{0.925} X_4^{-0.925} X_9^{0.438},$$
$$\dot{X}_{10} = 1.09662 X_8^{0.235} X_9^{0.239} X_{10}^{-0.115} - 2.32558 X_8^{-0.55} X_{10}^{0.78},$$
$$\dot{X}_{11} = 2.32558 X_8^{-0.55} X_{10}^{0.78} - 8.74436 \times 10^{-5} X_{11}^{2.21}.$$

The system is implemented in the file *Purine2S.plc*.

The first check is the computation of the steady state in PLAS. Both the concentrations and the fluxes are correct within the accuracy of the parameter estimation. Furthermore, all eigenvalues have negative real parts, which confirms that the model is locally stable. There are two pairs of complex conjugate eigenvalues, which imply the potential for oscillatory behavior.

The next step of analysis may either be an analysis of sensitivities or the exploration of the system dynamics with simulations of typical scenarios. We decide to begin with some simulations in PLAS.

Simulations. A typical situation is a temporary increase in one of the system variables. For instance, we may double the initial value of PRPP. The response is very reasonable, and within a short time all variables have returned to the original steady state. Numerous biochemical and clinical observations suggest that the adenylates remain essentially unaffected by temporary perturbations. Indeed, the response in this simulation is a mere rise from 2,500 to 2,501.23 before the adenylates return to the steady state. The responses to a 10-fold initial concentration of PRPP are stronger, with the adenine concentration (X_7) temporarily dropping from 10 to 2 and the guanine pool (X_9) dropping from 5 to 0.3, before they return. These drops may be a cause of concern, but then again, PRPP was increased to the 10-fold nominal level. Overall, the responses to temporary increases in PRPP and other metabolites appear to be qualitatively consistent with biochemical and clinical findings, but they are not always very accurate quantitatively (see Exercises 5 and 6).

A different type of simulation investigates structural changes. An important scenario is a permanently elevated level of PRPP, which in reality may have numerous reasons. The most common may be overproduction of PRPP due to increased activity of PRPP synthetase. This case is readily implemented by enlarging the value of the corresponding rate constant, α_1. Doubling the rate leads to a number of responses, some of which are very reasonable, while others clearly are not. For instance, PRPP increases by 70%, the adenylate pool increases by about 50%, and uric acid increases by 40%. The rate of conversion of xanthine into uric acid, represented by v_{xd}, doubles. The results reflect clinical observations quite well. At the same time, *de novo* synthesis increases 40-fold, IMP increases from 100 to about 650, and the hypoxanthine pool increases from 10 to almost 650. While *de novo* synthesis and these metabolite concentrations are known to increase in patients with PRPP synthetase superactivity, the extent is vastly overpredicted by the model.

Yet another observation we may test is the observation that removal of the inhibition of the enzyme AMPD by guanylates results in an increase in uric acid and the flux v_{xd} (van den Berghe and Hers, 1980). This test is a little bit more complicated, since the inhibition is part of two terms and thus represented by two parameters, h_{34} and g_{24}. The first kinetic order represents a singular effect and is easily changed from -0.012 to 0. However, the kinetic order g_{24} is a conglomerate of three effects. One describes the reversible flux between GMP (which is part of X_4) and IMP (X_2), the second one describes the inhibition of interest here, and the third one describes the inhibition of the initial enzyme of *de novo* synthesis, amidophosphoribosyltransferase,

CASE STUDY 3 – A SEQUENCE OF MODELS DESCRIBING PURINE METABOLISM

which is represented by v_{den}. Since V_2^+ also includes the salvage pathway catalyzed by v_{hprt}, the kinetic order is formally given as

$$g_{24} = \left(\frac{\partial v_{gmpr}}{\partial X_4} + \frac{\partial v_{ampd}}{\partial X_4} + \frac{\partial v_{hprt}}{\partial X_4} + \frac{\partial v_{den}}{\partial X_4} \right) \frac{X_4}{V_2^+},$$

which is to be evaluated at the steady-state operating point. The individual components are given as

$$v_{ampd} = \alpha_{ampd} X_3^{f_{ampd3}} X_4^{f_{ampd4}},$$

$$v_{den} = \alpha_{den} X_1^{f_{den1}} X_2^{f_{den2}} X_3^{f_{den3}} X_4^{f_{den4}},$$

$$v_{gmpr} = \alpha_{gmpr} X_2^{f_{gmpr2}} X_3^{f_{gmpr3}} X_4^{f_{gmpr4}},$$

$$v_{hprt} = \alpha_{hprt} X_1^{f_{hprt1}} X_2^{f_{hprt2}} X_8^{f_{hprt8}}.$$

For the computation of g_{24}, only steady-state flux values and the kinetic orders with respect to X_4 are needed (see Exercise 7). The former are listed in Table 10.3, and the latter are $f_{ampd4} = -0.04$, $f_{den4} = -0.2$, $f_{gmpr4} = -0.06$, and $f_{hprt4} = 0$ (Curto, unpublished; however see Curto et al., 1998a for a similar representation). With these numerical values, the kinetic order is

$$g_{24} = \frac{-0.06 \times 0.5 - 0.04 \times 5.69 - 0.2 \times 2.3}{0.5 + 5.69 + 2.3 + 3.7} = -0.589.$$

Eliminating the inhibitory effect on v_{ampd} changes the kinetic order to

$$g_{24}^* = \frac{-0.06 \times 0.5 - 0.2 \times 2.3}{0.5 + 5.69 + 2.3 + 3.7} = -0.40.$$

We reset h_{34} and g_{24} thus and execute the simulation in PLAS, with the following result. The system response is qualitatively correct but quantitatively too weak: the concentration of uric acid increases by merely 4%, and the conversion of xanthine to uric acid by less than 10%. [Confirm this result (Exercise 8).]

We will skip Curto's analyses of partial or almost complete deficiency in the activity of hypoxanthine-guanine phosphoribosyltransferase (HGPRT) or adenine phosphoribosyltransferase (APRT) at this point and only mention that they are not represented well by the model. HGPRT deficiency leads to a disease complex known as Lesh–Nyhan syndrome that overshadows all other problems associated with purine metabolism because of the severity of its symptoms. With the enzyme failing, the important salvage of free guanine bases is not functional, and the organism develops symptoms ranging from hyperuricemia to mental retardation and death. Since the model has exhibited some other flaws already, we will not parameterize it at a severely HGPRT- or APRT-deficient state.

In summary, the model exhibits reasonable trends, but has some flaws when it comes to quantitative predictions.

Analysis of sensitivities. In addition to dynamical explorations, we may check the sensitivities of the model. Particularly high sensitivities may point out problematic parts of the model and suggest means of amelioration.

The model does not have independent variables, and thus, there are no logarithmic gains to be computed. Most of the sensitivities with respect to rate constants are unremarkable. For the first six variables, the highest sensitivities have a magnitude of about 3, which means that the corresponding variables change by 3% or less if one of the rate constants is changed by 1%. The situation is different for X_7 through X_{10}. Here, alterations in α_3 or β_3 are magnified between 10-fold and almost 20-fold. For instance, the sensitivity $S(X_7, \alpha_3)$ is almost 18, and $S(X_7, \alpha_7)$ is about 14. Not only adenine (X_7) is affected. While not quite as sensitive, X_8, X_9, and X_{10} respond with about 10-fold magnification to alterations in rate constants α_3 and/or α_4.

The sensitivities with respect to kinetic orders leave no doubt that something is not quite right with this model. Numerous sensitivities are in the range between 10 and 20, which alone would be cause for concern. However, some of the unconstrained sensitivities associated with adenylate metabolism are between 100 and 200 in magnitude, which is a clear warning sign.

In summary, the sensitivity analysis suggests problems in the model structure or numerical specification. The most severe problems are associated with the adenylate pool and its dynamics. Refined analysis of adenylate kinetics is thus the starting point for the next iteration of the modeling process.

Modifications

1. Polyamine pathway. The sensitivity profile suggests to scrutinize the polyamine pathway more closely. In the current model, SAM is pooled with the adenylates, and that might be one of the causes for the high sensitivities. In reality, ATP and methionine form SAM + P_i + PP_i in a reaction catalyzed by methionine adenosyltransferase. In a second step, SAM is decarboxylated. In the current model, the kinetic order for the degradation of SAM, $f_{polyam3}$, is based on the decarboxylase reaction and has a value of 0.9. It may be more appropriate instead to base the kinetic order on the methionine adenosyltransferase reaction. Assuming a Michaelis–Menten reaction with a K_m of 450 for ATP (Oden and Clarke, 1983) and a steady-state concentration of 1,900, the kinetic order becomes $f_{polyam3} = K_m/(K_m+[ATP]) = 0.2$ (cf. Chapter 5). In the GMA model, the corresponding rate constants have to be adjusted, and in the S-system model, the aggregated kinetic order must be recomputed. For the S-system model, the following adjustments reflect the change in $f_{polyam3}$. The kinetic order g_{73} simply changes from 0.9 to 0.2. In order to regain the same (observed) flux value, the rate constant is recomputed as $\alpha_7 = 2.3$. The kinetic order h_{33} is aggregated from several processes, as shown above. Changing $f_{polyam3}$ to 0.2 results in a new kinetic order $h_{33} = 1.03$ and an associated rate constant $\beta_3 = 0.0062$.

The result of the adjustment turns out to be unremarkable. The sensitivities are somewhat better behaved, but the system responses are essentially unaffected.

2. Adenylate turnover. Another possibility of amending the problematic features of adenylate metabolism is to revisit the magnitude of the flux between adenylates and adenine. The literature information about this flux is somewhat uncertain (cf. Ayvazian and Skupp, 1965; Page et al., 1991). So far, the model has assumed that v_{aprt} (=10) is about three times as large as v_{hprt} and v_{gprt} (=3.7). The pertinent literature would allow a lower ratio, and we set $v_{aprt} = 5$. To regain a steady state, v_{polyam} must have the same value, and v_{prpps} becomes 14.7. The result of this alteration is negligible, except that the sensitivities slightly decrease.

3. Combination of modifications 1 and 2. The two previous amendments both improved the sensitivity profiles to a minor degree, and it might be useful to test the combined effect. Since the system is highly nonlinear, synergisms between different effectors cannot be excluded. However, in this particular case, the combined effect is as unremarkable as the individual effects. Similarly, further lowering the flux value of v_{aprt} to 1, which could possibly be justified with data from the literature, does not yield significant improvements. It thus appears that minor variations on the second model will not lead to satisfactory answers, and we decide to split SAM from the pool of adenylates.

THIRD MODEL

Since the pooling of S-adenosylmethionine (SAM) with other adenylates in the previous models seemed to create problems, we decide to define SAM as an additional dependent variable. In this amended model, SAM is synthesized from ATP and methionine by the enzyme methionine adenosyl transferase; the step is coded as v_{mat}. SAM can be degraded via two pathways, the (formerly modeled) polyamine pathway (v_{polyam}) yielding adenine (Ade), or the transmethylation pathway (v_{trans}) which ultimately produces adenosine (Ado). Since Ado is still part of the pool of adenylates, the latter pathway thus amounts to some recycling of SAM.

The amendments just incorporated did not result from new biochemical discoveries. The pathways governing the dynamics of SAM had been known before, but it had been assumed in the former models that the details of these steps were of negligible importance for the overall dynamics of purine metabolism. The simulations and sensitivity analyses caused us to doubt these assumptions, and the canonical structure of the model suggested the alterations introduced here.

Investigation of the kinetic and dynamic properties of the pathways associated with SAM yields that most of SAM is channeled into the transmethylation pathway, producing adenosine, while only small amounts are converted into adenine (Barber et al., 1986). This pathway is inhibited by SAM.

A second set of alterations is implemented in this third model. Analysis of the second model showed high parameter sensitivities with respect to $f_{prpps,adenylates}$, the kinetic order representing the effect of adenylates on PRPP synthetase. While this kinetic order was determined from the best available data, the high sensitivity indicates some problem with the dynamics of PRPP. A known process that as yet had been ignored, because it was assumed to be irrelevant for our system, is the synthesis

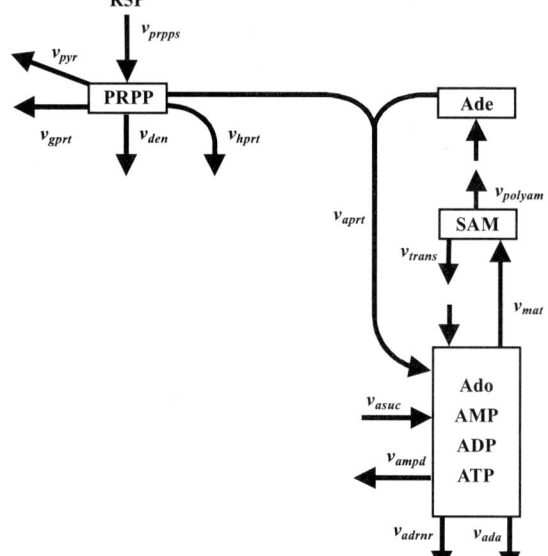

Figure 10.3. Amendments leading to the third model of purine metabolism.

of pyrimidines from PRPP. We add this process and code it as the flux v_{pyr}. The kinetic order of PRPP in v_{pyr} is difficult to estimate, and it is assumed that the production of pyrimidines is roughly similar to that of purines. Since the first enzymes of both pyrimidine and purine synthesis are phosphoribosyltransferases, it is assumed that the corresponding kinetic orders are similar. As required for the increased use of PRPP while maintaining the same steady state, the flux v_{prpps} increases to 20.79.

The amended portions of the model are shown in Fig. 10.3. All definitions of variables are the same as in the previous model, except that SAM is taken out of the pool X_3 and now coded as the new dependent variable X_{12}. The kinetic orders associated with SAM are

$$f_{polyam,SAM} = 0.9, \quad f_{mat,adenylates} = 0.2, \quad f_{mat,SAM} = -0.6,$$
$$f_{trans,adenylates} = 0.33.$$

Furthermore, the use of PRPP for pyrimidine production is reflected in the parameter $f_{pyr,PRPP} = 1.27$. With this information, GMA and S-system models can be constructed as before (Exercise 9).

Implementation in PLAS confirms that the correct steady state is obtained, and stability analysis confirms that it is stable. Furthermore, the sensitivities are reduced to a more reasonable level, even though some of them are still high (Exercise 9).

With the steady-state features improved, we can check to what degree model results are consistent with clinical and biochemical observations. Summarily, the following results are obtained (Curto, 1996; Exercise 10):

1. The model captures reasonably well the observed constancy of the adenine and guanine nucleotide pools under various conditions and perturbations.

2. The model captures reasonably well several types of perturbations associated with hypoxanthine.
3. Several observations (e.g., Ayvazian and Skupp, 1965; Jerushalmy et al., 1973; Jacobs et al., 1988) in humans suggest that more adenine is incorporated in adenylates than guanine is incorporated in guanylates. These observations are not well modeled.
4. Removing the inhibition of AMPD by guanylates and by P_i results in increases in UA and v_{xd}. The model reflects these observations reasonably well.
5. The model predicts well the increases in v_{den} and the concentration of UA in patients with increased PRPP synthetase activity.
6. The model does not represent severe deficiencies in HGPRT and APRT well.
7. Observations regarding xanthinuria and uric acid overproduction are not modeled very well.

FOURTH MODEL

A remaining problem with the models developed so far is an unphysiological accumulation of adenine, xanthine, and hypoxanthine in the simulation of HGPRT or xanthine dehydrogenase deficiency (tests 6 and 7 above). The exact values of these increased metabolites depend on the parameter specifications but are always too high. Because such accumulation seems unreasonable, the possibility of purine base excretion is introduced. The new fluxes (v_{ade}, v_x, and v_{hx}) appear to be not very significant in healthy subjects, but noticeable in patients with some purine metabolic diseases. As suggested in the literature (e.g., Ayvazian and Skupp, 1965; Bontemps et al., 1983), we quantify the fluxes as $v_{ade} = 0.01$, $v_x = 0.03$, and $v_{hx} = 0.05$ μmol min^{-1} kg^{-1} and the new kinetic orders as $f_{ade,ade} = 0.5$, $f_{hx,hx} = 2.21$, and $f_{x,x} = 2.21$ (Exercise 11). Directly applicable data are unavailable, and these parameter values are educated guesses that result from surrogate information about other, similar processes. This situation may make you uncomfortable, but it is a common occurrence that can only be ameliorated by sensitivity studies, simulations, and the search for additional data. The introduction of v_{ade} affects the steady state of the adenylates only slightly, and insistence on the original steady state requires very minor adjustments in other fluxes of adenylate metabolism.

Since these alterations are somewhat peripheral to the system, the steady state and its stability and robustness don't change much. However, the model captures better the consequences of minor deficiencies in HGPRT and APRT, and the accumulation of purine bases now corresponds to observations.

"FINAL" MODEL

At this point we don't know it yet, but this will be our "final" model. The quotation marks, of course, indicate that a model is seldom really final and definite, but the model proposed in this section constitutes the conclusion of a major phase of model development.

As mentioned in the previous section, there are uncertainties about the degradation and excretion of adenine, hypoxanthine, and xanthine. Curto (unpublished) reanalyzed the corresponding parameters with sensitivity analyses. Furthermore, he questioned whether hypoxanthine and xanthine inhibit each other's degradation *in vivo*. Implementing various combinations of parameter values and comparing results with the few data available, he concluded that inhibition of the two degradation steps seemed minor. Similarly, he analyzed other kinetic orders that were subject to relatively large uncertainties, such as $f_{ampd,adenylates}$ and $f_{prpps,adenylates}$, by implementing them with different values within the observed ranges.

As yet another amendment, Curto (1996) suggested allowing for the incorporation and usage of adenylates and guanylates into and from DNA and RNA. DNA and RNA are assembled from adenylates and guanylates by polymerases. In order to indicate that this assembly requires both substrates, the steps are represented with double-tailed arrows. Similarly, the degradation of DNA and RNA by several exonucleases, endonucleases, and ribonuclease simultaneously generates guanylates and adenylates, and the reactions are represented with double-headed arrows. Since different moieties of guanylates and adenylates are used up or generated, respectively, the steps are designated in the map with individual names.

To avoid problems due to XMP and S-AMP being constituents of the guanylate and adenylate pools, they are taken out of those pools and defined as individual dependent variables. Finally, in order to explore the roles of ribose-5-phosphate and phosphate, these are now defined explicitly as independent variables X_{17} and X_{18}.

Thus, the amended model consists of 16 dependent variables (X_1–X_{16}) and two independent variables (X_{17}–X_{18}). Table 10.6 shows how they are coded. The enzyme catalyzed reactions and non-enzymatic steps that constitute the pathway are listed in Table 10.7 along with abbreviated names, E.C. numbers, and their steady-state flux rates.

With the old definitions and the new amendments implemented, the "final" model is shown in Fig. 10.4.

Processes

As before, it is a good idea to design two models, one in GMA form and the other in S-system form. We illustrate the procedure again with the dynamics of PRPP, which now is affected by six reactions. The synthesizing reaction has R5P as substrate, which is now explicitly included as an independent variable, and is modulated by numerous inhibitors. One is PRPP itself, and four others of relevance are AMP, ADP, GDP, and GTP. The latter are not represented individually in our model, and we have to use the aggregated pools X_4 and X_8 as substitutes. In addition to the inhibitory modulations, P_i is an allosteric activator (Wyngaarden and Kelley, 1983: p. 1069) and therefore is included in the synthesis term of PRPP with a positive kinetic order. The term representing PRPP synthesis thus includes as variables X_1, X_4, X_8, X_{17}, and X_{18}:

$$v_{prpps} = v_{prpps}(X_1, X_4, X_8, X_{17}, X_{18}).$$

CASE STUDY 3 – A SEQUENCE OF MODELS DESCRIBING PURINE METABOLISM

Table 10.6. Variables of the "Final" Model of Purine Metabolism and Their Steady-State Values

Variable	Abbreviation	Metabolite	Steady-State Concentration (mM)
X_1	PRPP	Phosphoribosylpyrophosphate	5
X_2	IMP	Inosine monophosphate	100
X_3	S-AMP	Adenylosuccinate	0.2
X_4	Ado AMP ADP ATP	Adenosine Adenosine monophosphate Adenosine diphosphate Adenosine triphosphate	2,500
X_5	SAM	S-adenosyl-L-methionine	4
X_6	Ade	Adenine	1
X_7	XMP	Xanthosine monophosphate	25
X_8	GMP GDP GTP	Guanosine monophosphate Guanosine diphosphate Guanosine triphosphate	400
X_9	dAdo dAMP dADP dATP	Deoxyadenosine Deoxyadenosine monophosphate Deoxyadenosine diphosphate Deoxyadenosine triphosphate	6
X_{10}	dGMP dGDP dGTP	Deoxyguanosine monophosphate Deoxyguanosine diphosphate Deoxyguanosine triphosphate	3
X_{11}	RNA	Ribonucleic acid	28,600
X_{12}	DNA	Deoxyribonucleic acid	5,160
X_{13}	HX Ino dIno	Hypoxanthine Inosine Deoxyinosine	10
X_{14}	Xa	Xanthine	5
X_{15}	Gua Guo dGuo	Guanine Guanosine Deoxyguanosine	5
X_{16}	UA	Uric acid	100
X_{17}	R5P	Ribose-5-phosphate	18
X_{18}	P_i	Phosphate	1,400

The kinetic orders associated with X_{17} and X_{18} are positive, whereas all other kinetic orders in the term are negative. Since there is no branching, the corresponding power-law term is the same for the GMA and the S-system. In either case, the synthesis term thus reads, in symbolic terms,

$$v_{prpps} = V_1^+ = \alpha_1 X_1^{g_{11}} X_4^{g_{14}} X_8^{g_{18}} X_{17}^{g_{1,17}} X_{18}^{g_{1,18}}.$$

Table 10.7. Enzymes and Steps of the "Final" Model of Purine Metabolism, Along with Their Steady-State Rates

Flux	Enzyme Name or Step	E.C. Number	Steady-state Flux Rate (μ mol min^{-1} kg^{-1})
v_{prpps}	Phosphoribosylpyrophosphate synthetase (PRPPS)	2.7.6.1.	20.79
v_{gprt}	Hypoxanthine-guanine phosphoribosyltransferase (HGPRT)	2.4.2.8.	3.7
v_{hprt}	Hypoxanthine-guanine phosphoribosyltransferase (HGPRT)	2.4.2.8.	3.7
v_{aprt}	Adenine phosphoribosyltransferase (APRT)	2.4.2.7.	1
v_{den}	De novo synthesis (Amidophosphoribosyltransferase; ATASE)	2.4.2.14.	2.39
v_{pyr}	Pyrimidine synthesis	Several enzymes	10
v_{asuc}	Adenylosuccinate synthetase (ASUC)	6.3.4.4.	8
v_{asli}	Adenylosuccinate lyase (ASLI)	4.3.2.2.	8
v_{impd}	IMP dehydrogenase (IMPD)	1.1.1.205.	1.6
v_{gmps}	GMP synthetase (GMPS)	6.3.4.1.	1.6
v_{ampd}	AMP deaminase (AMPD)	3.5.4.6.	5.69
v_{gmpr}	GMP reductase (GMPR)	1.6.6.8.	0.5
v_{trans}	Tansmethylation pathway (protein O-methyltransferase; MT)	2.1.1.24.	13.99
v_{mat}	Methionine adenosyltransferase (MAT)	2.5.1.6.	15
v_{polyam}	Polyamine pathway (S-adenosylmethionine decarboxylase; SAMD)	4.1.1.50.	1.01
v_{ade}	Adenine oxidation (xanthine oxidase)	1.2.1.37.	0.01
v_{inuc}	5'-Nucleotidase (5NUC)	3.1.3.5.	2.68
v_{gnuc}	5'-Nucleotidase (5NUC)	3.1.3.5.	4.7
v_{arna}	RNA polymerase (from ATP) (RNAP)	2.7.7.6.	1,980
v_{grna}	RNA polymerase (from GTP) (RNAP)	2.7.7.6.	1,320
v_{rnaa}	RNases (to AMP) (RNAN)	Several enzymes	1,980
v_{rnag}	RNases (to GMP) (RNAN)	Several enzymes	1,320
v_{dgnuc}	5'(3')-Nucleotidase (3NUC)	3.1.3.31.	0.1
v_{ada}	Adenosine deaminase (ADA)	3.5.4.4.	2.1
v_{dada}	Adenosine deaminase (ADA)	3.5.4.4.	0.2
v_{adrnr}	Diribonucleotide reductase (DRNR)	1.17.4.1.	0.2
v_{gdrnr}	Diribonucleotide reductase (DRNR)	1.17.4.1.	0.1
v_{gua}	Guanine hydrolase (GUA)	3.5.4.3.	1.1
v_{adna}	DNA polymerase (from dATP) (DNAP)	2.7.7.7.	10
v_{gdna}	DNA polymerase (from dGTP) (DNAP)	2.7.7.7.	6.8
v_{dnaa}	DNases (to dAMP) (DNAN)	Several enzymes	10
v_{dnag}	DNases (to dGMP) (DNAN)	Several enzymes	6.8
v_{hx}	Hypoxanthine excretion	Non-enzymatic	0.05
v_{hxd}	Xanthine oxidase or xanthine dehydrogenase (XD)	1.2.1.37.	1.23
v_{xd}	Xanthine oxidase or xanthine dehydrogenase (XD)	1.2.1.37.	2.3
v_{x}	Xanthine excretion	Non-enzymatic	0.03
v_{ua}	Uric acid excretion	Non-enzymatic	2.3

CASE STUDY 3 – A SEQUENCE OF MODELS DESCRIBING PURINE METABOLISM

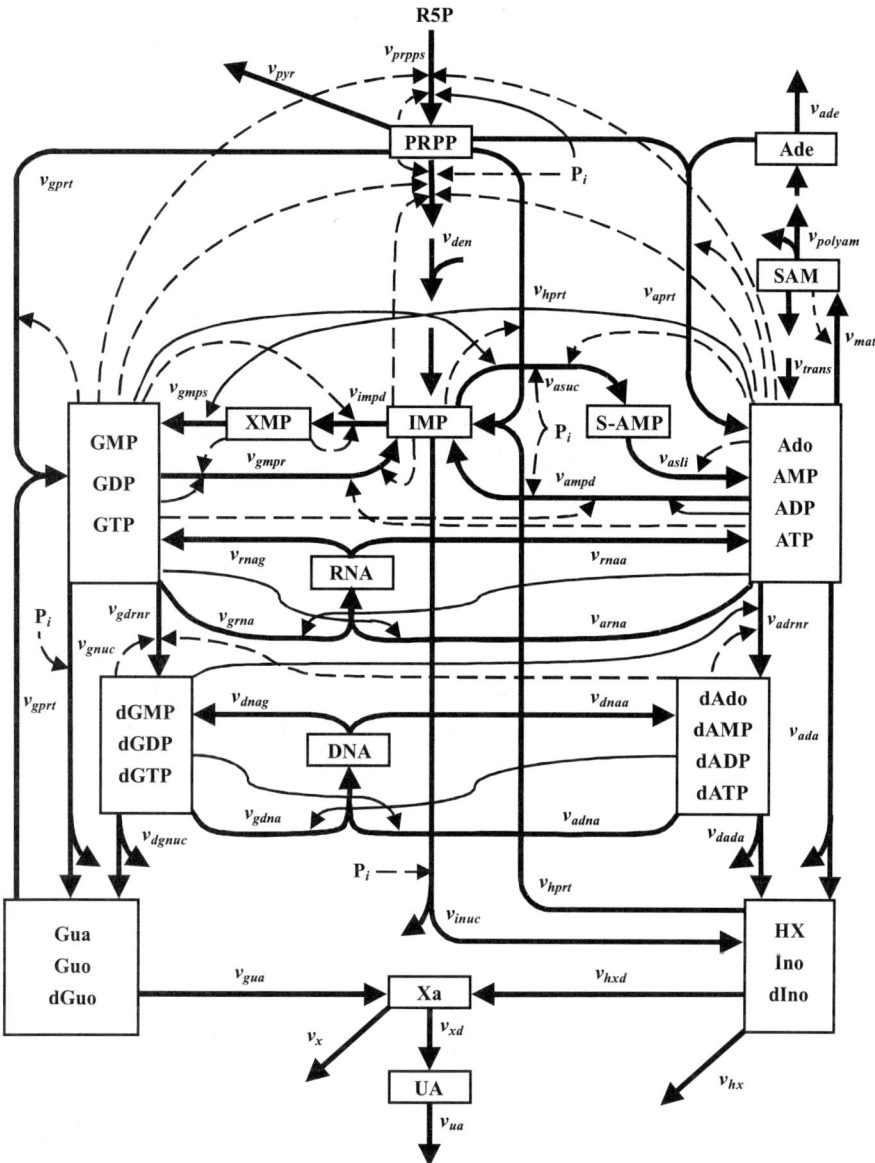

Figure 10.4. Map of the "final" model. Light solid arrows represent activation, while light dashed arrows represent inhibition. Curved heavy arrows entering or leaving the pathway indicate purine ring and ribose moieties that balance the stoichiometry of the system.

The degradation of PRPP in this model is a function of five variables:

$$v_{den} = v_{den}(X_1, X_2, X_4, X_8, X_{18}).$$

The salvage pathways, catalyzed by hypoxanthine-guanine phosphoribosyltransferase (HGPRT) and by adenine phosphoribosyltransferase (APRT), use PRPP to form IMP, GMP, or AMP. These reactions are potentially reversible, which suggests

Table 10.8. Kinetic Orders of the "Final" Model of Purine Metabolism (GMA Form)

$f_{ada4} = 0.97$	$f_{ade6} = 0.55$	$f_{adna9} = 0.42$	$f_{adna10} = 0.33$
$f_{adrnr4} = 0.1$	$f_{adrnr9} = -0.3$	$f_{adrnr10} = 0.87$	$f_{ampd4} = 0.8$
$f_{ampd8} = -0.03$	$f_{ampd18} = -0.1$	$f_{aprt1} = 0.5$	$f_{aprt4} = -0.8$
$f_{aprt6} = 0.75$	$f_{arna4} = 0.05$	$f_{arna8} = 0.13$	$f_{asli3} = 0.99$
$f_{asli4} = -0.95$	$f_{asuc2} = 0.4$	$f_{asuc4} = -0.24$	$f_{asuc8} = 0.2$
$f_{asuc18} = -0.05$	$f_{dada9} = 1$	$f_{den1} = 2$	$f_{den2} = -0.06$
$f_{den4} = -0.25$	$f_{den8} = -0.2$	$f_{den18} = -0.08$	$f_{dgnuc10} = 1$
$f_{dnaa12} = 1$	$f_{dnag12} = 1$	$f_{gdna9} = 0.42$	$f_{gdna10} = 0.33$
$f_{gdrnr8} = 0.4$	$f_{gdrnr9} = -1.2$	$f_{gdrnr10} = -0.39$	$f_{gmpr2} = -0.15$
$f_{gmpr4} = -0.07$	$f_{gmpr7} = -0.76$	$f_{gmpr8} = 0.7$	$f_{gmps4} = 0.12$
$f_{gmps7} = 0.16$	$f_{gnuc8} = 0.9$	$f_{gnuc18} = -0.34$	$f_{gprt1} = 1.2$
$f_{gprt8} = -1.2$	$f_{gprt15} = 0.42$	$f_{grna4} = 0.05$	$f_{grna8} = 0.13$
$f_{gua15} = 0.5$	$f_{hprt1} = 1.1$	$f_{hprt2} = -0.89$	$f_{hprt13} = 0.48$
$f_{hx13} = 1.12$	$f_{hxd13} = 0.65$	$f_{impd2} = 0.15$	$f_{impd7} = -0.09$
$f_{impd8} = -0.03$	$f_{inuc2} = 0.8$	$f_{inuc18} = -0.36$	$f_{mat4} = 0.2$
$f_{mat5} = -0.6$	$f_{polyam5} = 0.9$	$f_{prpps1} = -0.03$	$f_{prpps4} = -0.45$
$f_{prpps8} = -0.04$	$f_{prpps17} = 0.65$	$f_{prpps18} = 0.7$	$f_{pyr1} = 1.27$
$f_{rnaa11} = 1$	$f_{rnag11} = 1$	$f_{trans5} = 0.33$	$f_{ua16} = 2.21$
$f_{x14} = 2.0$	$f_{xd14} = 0.55$		

the inclusion of the products in the rate laws. In symbolic terms, the three rates are

$$v_{gprt} = v_{gprt}(X_1, X_8, X_{15}),$$
$$v_{hprt} = v_{hprt}(X_1, X_2, X_{13}),$$
$$v_{aprt} = v_{aprt}(X_1, X_4, X_6).$$

Finally, PRPP may be used for pyrimidine synthesis, which is coded as v_{pyr}. The rate law for this process is simply a function of its substrate:

$$v_{pyr} = v_{pyr}(X_1).$$

Thus, the procedure is exactly the same as before, but the result is somewhat different, since the new model is more detailed.

The kinetic orders and rate constants of the corresponding GMA system equations are given in Tables 10.8 and 10.9 (see also Exercise 13) in alphabetical order. The PLAS file is named *PurineGMA.plc*.

Implementation of the system in PLAS confirms the steady state, at least approximately. Since there is no analytical solution, PLAS needs to search for a numerical surrogate, which usually is very close to the true steady-state solution. PLAS begins the search with the initial values, but if these are too different from the steady state, a numerical solution may not be found. In such a case, the steady-state solution must be obtained by selecting different initial values or by solving the differential equations for a sufficiently long time period (see Exercise 14).

As a second check, we display the sensitivities (see Exercise 15). Most of them are small in magnitude, as we would hope. However, there are some huge values among the rate constant and the kinetic order sensitivities. Closer inspection shows that they

Table 10.9. Rate Constants of the "Final" Model of Purine Metabolism (GMA Form)

$\alpha_{ada} = 0.001062$	$\alpha_{ade} = 0.01$	$\alpha_{adna} = 3.2789$	$\alpha_{adrnr} = 0.0602$
$\alpha_{ampd} = 0.02688$	$\alpha_{aprt} = 233.8$	$\alpha_{arna} = 614.5$	$\alpha_{asli} = 66544.7$
$\alpha_{asuc} = 3.5932$	$\alpha_{dada} = 0.03333$	$\alpha_{den} = 5.2728$	$\alpha_{dgnuc} = 0.03333$
$\alpha_{dnaa} = 0.001938$	$\alpha_{dnag} = 0.001318$	$\alpha_{gdna} = 2.2296$	$\alpha_{gdrnr} = 0.1199$
$\alpha_{gmpr} = 0.3005$	$\alpha_{gmps} = 0.3738$	$\alpha_{gnuc} = 0.2511$	$\alpha_{gprt} = 361.69$
$\alpha_{grna} = 409.6$	$\alpha_{gua} = 0.4919$	$\alpha_{hprt} = 12.569$	$\alpha_{hx} = 0.003793$
$\alpha_{hxd} = 0.2754$	$\alpha_{impd} = 1.2823$	$\alpha_{inuc} = 0.9135$	$\alpha_{mat} = 7.2067$
$\alpha_{polyam} = 0.29$	$\alpha_{prpps} = 0.898$	$\alpha_{pyr} = 1.2951$	$\alpha_{rnaa} = 0.06923$
$\alpha_{rnag} = 0.04615$	$\alpha_{trans} = 8.8539$	$\alpha_{ua} = 0.00008744$	$\alpha_{x} = 0.0012$
$\alpha_{xd} = 0.949$			

are all related to the dynamics of RNA and, to a lesser degree, DNA. Values of this magnitude are of real concern. Numerical simulations confirm the strong effect of changing one of the rate constants or kinetic orders in question. What could be the reason? Stepping back, we realize that the troubling parameters are not independent. For instance, the polymerase that catalyzes the assembly of RNA uses both adenylates and guanylates. Thus, if we change a parameter in the step v_{arna}, we must make the corresponding change in the step v_{grna}. The same arguments hold for the assembly of DNA and the disassembly of RNA and DNA.

As a specific example, consider a 1% increase in the rate constant of v_{rnaa}. According to the sensitivity profile (row *alpha(4,3)*, if the v_{rnaa} term is the third one listed in the fourth equation in PLAS), the expected changes in X_1, X_2, and X_{11} are a 162.5% decrease and 263% and 11.7% increases, respectively. These imply very strong effects, which, however, would represent the unrealistic situation that we could change the production of adenylates from RNA without concomitantly changing the degradation of RNA toward adenylates and guanylates. The corresponding 1% change in v_{rnag} evokes effects on X_1, X_2, and X_{11} of the following sizes: −56.4%, −169.8%, and 55.6% (see row labeled *alpha(8,3)*). Finally, the two degradation processes of RNA have the following effects (see row labeled *beta(11)*): 218.9%, 93.3%, and −68.4%. Thus, the overall effect on X_1 is −162.5% −56.4% + 218.9%, which is 0! The same result is obtained for X_2, and the overall percentage change in X_{11} is 1. Considering the appropriate constraints, the system is very well behaved. (See also Exercise 16.)

The S-system model is computed from the GMA model by aggregation of terms at branch points. For instance, the utilization of PRPP is aggregated into one term, which includes as variables the main substrate PRPP, the substrates Ade, Gua, and HX of the salvage pathways, and the inhibitors AMP, GMP, IMP, and P_i. The kinetic orders associated with substrates are positive, and those associated with inhibitors are negative. The formulation of the degradation term of PRPP in symbolic S-system form is left as Exercise 17.

Every rate of synthesis or degradation is symbolically formulated in this fashion. The parameter estimation, of course, uses information from the previous models, and adds specifics deriving from the latest amendments. Details of this estimation process are found in Curto et al. (1998a). The "final" S-system model in parametric form reads

PRPP:
$$\dot{X}_1 = 0.898 X_1^{-0.03} X_4^{-0.45} X_8^{-0.04} X_{17}^{0.65} X_{18}^{0.7}$$
$$- 31.335 X_1^{1.27} X_2^{-0.165} X_4^{-0.067} X_6^{0.036} X_8^{-0.237} X_{13}^{0.085} X_{15}^{0.075} X_{18}^{-0.0092}.$$

IMP:
$$\dot{X}_2 = 1.629 X_1^{0.720} X_2^{-0.28} X_4^{0.319} X_7^{-0.03} X_8^{-0.024} X_{13}^{0.144} X_{18}^{-0.06}$$
$$- 5.584 X_2^{0.454} X_4^{-0.156} X_7^{-0.012} X_8^{0.126} X_{18}^{-0.11}.$$

S-AMP:
$$\dot{X}_3 = 3.5932 X_2^{0.4} X_4^{-0.24} X_8^{0.2} X_{18}^{-0.05} - 66544.7 X_3^{0.99} X_4^{-0.95}.$$

Adenylates (AMP, ADP, ATP, Ado):
$$\dot{X}_4 = 0.082082 X_1^{0.00025} X_3^{0.004} X_4^{-0.00418} X_5^{0.0023} X_6^{0.00037} X_{11}^{0.988}$$
$$- 614.27 X_4^{0.054} X_5^{-0.0045} X_8^{0.128} X_9^{-0.00003} X_{10}^{0.00087} X_{18}^{-0.00028}.$$

SAM:
$$\dot{X}_5 = 7.2067 X_4^{0.2} X_5^{-0.6} - 9.006 X_5^{0.368}.$$

Ade:
$$\dot{X}_6 = 0.29 X_5^{0.9} - 220.14 X_1^{0.495} X_4^{-0.79} X_6^{0.748}.$$

XMP:
$$\dot{X}_7 = 1.2823 X_2^{0.15} X_7^{-0.09} X_8^{-0.03} - 0.3738 X_4^{0.12} X_7^{0.16}.$$

Guanylates (GMP, GDP, GTP):
$$\dot{X}_8 = 0.048842 X_1^{0.0034} X_4^{0.00014} X_7^{0.00019} X_8^{-0.0034} X_{11}^{0.996} X_{15}^{0.001}$$
$$- 408.11 X_2^{-0.000057} X_4^{0.0498} X_7^{-0.00029} X_8^{0.133} X_9^{-0.00009} X_{10}^{-0.000029} X_{18}^{-0.001}.$$

d-Adenylates (dAMP, dADP, dATP, dAdo):
$$\dot{X}_9 = 0.00229 X_4^{0.002} X_9^{-0.0058} X_{10}^{0.017} X_{12}^{0.98} - 3.3 X_9^{0.431} X_{10}^{0.324}.$$

d-Guanylates (dGMP, dGDP, dGTP):
$$\dot{X}_{10} = 0.00151 X_8^{0.0058} X_9^{-0.017} X_{10}^{-0.0058} X_{12}^{0.986} - 2.2655 X_9^{0.413} X_{10}^{0.34}.$$

RNA:
$$\dot{X}_{11} = 1024.195 X_4^{0.05} X_8^{0.13} - 0.115389 X_{11}.$$

DNA:
$$\dot{X}_{12} = 5.5062 X_9^{0.42} X_{10}^{0.33} - 0.0032545 X_{12}.$$

HX, Ino, dIno:
$$\dot{X}_{13} = 0.1033 X_2^{0.43} X_4^{0.409} X_9^{0.04} X_{18}^{-0.19} - 8.2827 X_1^{0.817} X_2^{-0.66} X_{13}^{0.528}.$$

Xa:
$$\dot{X}_{14} = 0.7234 X_{13}^{0.343} X_{15}^{0.236} - 0.9325 X_{14}^{0.569}.$$

Gua, Guo, dGuo:

$$\dot{X}_{15} = 0.2615\,X_8^{0.881}\,X_{10}^{0.02}\,X_{18}^{-0.33} - 136.6\,X_1^{0.925}\,X_8^{-0.925}\,X_{15}^{0.438}.$$

UA:

$$\dot{X}_{16} = 0.949\,X_{14}^{0.55} - 0.00008744\,X_{16}^{2.21}.$$

Ribose-5-phosphate (R5P):

$$X_{17} = 18 \quad \text{(independent)}.$$

Phosphate (P_i):

$$X_{18} = 1{,}400 \quad \text{(independent)}.$$

Implementation in PLAS (file *PurineSfin.plc*) confirms the steady state. In contrast to the GMA model, the solution is not obtained with a search algorithm, but from the linear steady-state equations of the analytical solution. Consequently, all differences between the observed and the computed steady state are due to rounding while computing rate constants. As in the GMA model, the parameter sensitivities fall into two groups: The first group consists of those associated with the metabolites of primary interest. Most of them are very reasonable, with magnitudes of just a few percent. In stark contrast, the second group contains very high sensitivities that are associated with RNA and DNA and the pools directly connected to them. In this case, the sensitivities refer to the aggregated quantities. There is no need to compute the deaggregated sensitivities, since these are exactly the same as for the GMA model, which we have computed before (Exercise 18).

Simulations

Changes in input. As a first simple test, we can explore the effect of increased input. This can be accomplished in two ways. First, we can check in PLAS the logarithmic gains of the S-system model with respect to X_{17}. Interestingly, all metabolites and fluxes have positive gains, which means that (moderate) increases in input lead to overall elevations in concentrations. None of the gains is very large, though. For example, the gains of PRPP (X_1), IMP (X_2), and uric acid (X_{16}) are listed as 0.38, 0.95, and 0.29, respectively. The variable affected most, with a gain of 1.73, is hypoxanthine. This strong response suggests that increases in PRPP are leading to increased salvage via the HGPRT reaction.

As a second test, we initialize the system with a doubled value of R5P. The steady-state values of X_1, X_2, X_{13}, and X_{16} become 6.5, 193, 33.2, and 122.6, respectively, which is in line with the predictions based on logarithmic gains. The dynamical solution shows the transient responses. The first aspect worth noticing is that it takes the system about two days (t0 = 0, tf = 2,880) to come close to the new steady state. While this may sound slow, it is actually consistent with clinical observations related to treating hyperuricemia with allopurinol, which show transient responses at a one- to two-day time scale (Wyngaarden and Kelley, 1983: p. 1103).

The shape of the responses varies among the metabolites. PRPP increases within 12 min to about 160% of the original value, before very slowly decreasing to the new steady state of 6.5 (see Fig. 10.5). IMP follows PRPP rather quickly, increasing

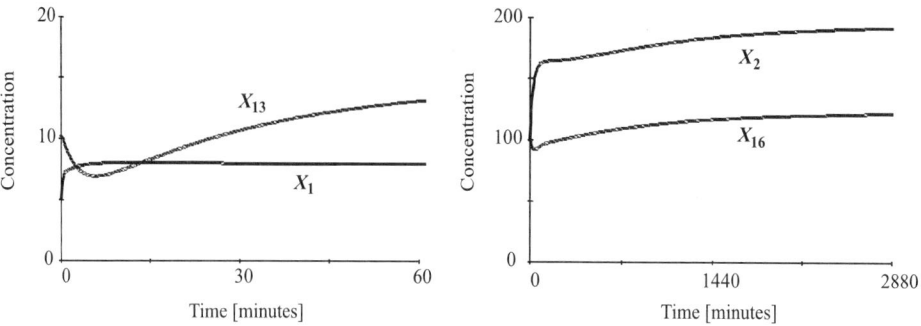

Figure 10.5. Selected responses to doubling the input R5P. Note different scales.

within 1h to about 160. Subsequently, the IMP concentration continues to grow, but at a much slower rate (see Fig. 10.5). The response in hypoxanthine is different. The immediate response is a *decrease* in concentration to about two-thirds. After 26 min, hypoxanthine is back at the original level but continues to increase to more than three times the old value. The uric acid concentration initially falls to about 90%, before regaining its original value after about four hours ($t = 240$) and increasing to the new 23% elevated steady state (see Fig. 10.5).

Although individual fluxes like v_{den} are not immediately reported in the S-system model because they are aggregated with other fluxes, they are readily made explicit by inclusion as a transformation in PLAS. For instance, to see v_{den}, just add the line

$$\text{vden} = 5.2728\ X1^{\wedge}2\ X2^{\wedge} - 0.06\ X4^{\wedge} - 0.25\ X8^{\wedge} - 0.2\ X18^{\wedge} - 0.08$$

which corresponds to the definition in the GMA model and is characterized by rate constant α_{den} and the five kinetic orders $f_{den,i}$ listed in Tables 10.8 and 10.9 (see Exercise 19).

Temporary perturbations. As a representative example, we initialize the GMA system and the S-system at the observed steady state, except that we set PRPP = 50, which corresponds to a tenfold concentration. In both models, the excess is quickly absorbed. After less than one time unit, PRPP has already returned to a value below 6, and after 10 units, PRPP is essentially back at the original steady state of 5. One immediate recipient of the excess PRPP is IMP. Its concentration rises quickly before returning rather slowly to the steady state. The pools of adenylates and guanylates are essentially unaffected, which confirms numerous clinical observations. Because of the salvage pathway catalyzed by HGPRT, hypoxanthine is used with PRPP to produce IMP. Its concentration quickly falls to about two-thirds of its original size before returning slowly to the original value. The uric acid concentration is almost unaffected by this temporary perturbation. (See also Exercises 20 and 21.)

As a second example, consider an exogenous increase in the hypoxanthine pool from 10 to 1000. Again, this is readily implemented by simply initializing the system correspondingly and studying the responses. Qualitatively, the GMA and the S-system model react similarly, but numerically there are some differences in the magnitude of the responses. For instance, the flux v_{den} decreases from 2.39 to about 0.56 in the

GMA model, whereas it only decreases to 1.3 in the S-system model. The simplest way to see this flux in either model is to define it in PLAS as a transformation, as was done above.

It is difficult to say which model is more accurate in this case. On one hand, there are clinical data (King et al., 1983) that seem to favor the GMA result. On the other hand, theoretical studies (Voit and Savageau, 1987; Sorribas and Savageau, 1989c) have shown that the S-system model may be expected to be more accurate, at least in comparison with traditional rate laws like that of Michaelis and Menten. Caution is necessary in assessing this case, no matter which model is chosen, since the hypoxanthine concentration is increased 100-fold, which constitutes a rather large deviation from the operating point. The validity of both models is guaranteed only in the vicinity of the operating point, and although experience has shown that the "vicinity" may sometimes span several orders of magnitude, one must not forget the underlying theoretical caveat.

Exercise 22 discusses a further set of perturbation experiments.

Changes in enzyme activities

PRPPS. A well-studied example in this category is *superactivity* of the first enzyme of the pathway, PRPP synthetase (PRPPS). Clinical studies have shown that increases in PRPPS lead to elevated concentrations of PRPP (between 13 and 29), HX (30), Xa (10), and UA (up to 300), while adenylates and guanylates remain fairly unaffected (e.g., Becker et al., 1989b; Jimenez et al., 1989). Also, the flux v_{den} increases (to between about 5 and 12; Becker et al., 1989b). Superactivity is implemented by doubling the rate constant α_1, and v_{den} is explicitly included as a transformation variable, as above (see also Exercise 23). The results are fairly similar between the two models, and they are also similar to results obtained from a Michaelis–Menten model (Curto et al., 1997). PRPP increases to about 7.5, HX to between 40 and 64, Xa to about 15, and UA to about 135. The flux v_{den} increases to about 4.4.

Of course, it is possible to study the transient responses. They are quite unremarkable and also similar for the S-system and the GMA model. These responses, however, are not very meaningful. PRPPS superactivity is a disease of a permanent nature. An individual exhibits either normal activity or superactivity, but it is not to be expected that permanent superactivity will develop all of a sudden.

AMPD. The enzyme AMPD is normally inhibited by guanylates and P_i. However, a disorder has been reported where the enzyme is resistant to inhibition (van den Berghe and Hers, 1980). This lack of inhibition results in increased degradation of xanthine and subsequently increased uric acid levels. The situation is easily implemented. For instance, in the GMA model the inhibitory effects appear in the equations of IMP and the adenylates as the kinetic orders $f_{ampd8} = -0.03$ and $f_{ampd18} = -0.1$. Setting the appropriate kinetic orders to zero, the resulting uric acid concentration is increased by 17% and v_{xd} by 41% (Exercise 24).

To implement the same structural change in the S-system, the aggregated kinetic orders g_{28}, $g_{2,18}$, h_{48}, and $h_{4,18}$ are adjusted by subtracting the contribution of the inhibitory effect. Assuming the overall measured flux v_{ampd} is unchanged, the new kinetic order g'_{28}, for instance, is

$$g'_{28} = g_{28} - \frac{f_{ampd8} v_{ampd}}{\text{steady-state flux through IMP}} = -0.024 - \frac{-0.03 \times 5.69}{12.29} = -0.01$$

(Exercise 25). Similarly, the other kinetic orders become

$g'_{2,18} = -0.0137,$

$h'_{48} = 1.28085,$

$h'_{4,18} = 0.$

It is no coincidence that $h'_{4,18}$ becomes zero, because the inhibition of AMPD is the only effect of P_i on the degradation of adenylates in the original model.

The results are very similar to those obtained with the GMA model, which is to be expected, because the elimination of AMPD inhibition does not move the system far away from the original steady state. The uric acid concentration increases by about 15%, and v_{xd} by 35%.

HGPRT. The most devastating inherited disease associated with purine metabolism is HGPRT deficiency (e.g., Kelley and Wyngaarden, 1983). Biochemically, the recycling of hypoxanthine and guanine is compromised, which results, among many metabolic imbalances, in strongly elevated serum levels of uric acid. The problems of hyperuricemia are in severe cases compounded by imbalances in guanylates and other metabolites.

From a modeling point of view, a severe enzyme deficiency must be considered with caution. Foremost, one has to realize that the simulation of HGPRT deficiency addresses patients that are born with the disease. It does not happen that a normal individual gradually develops the disease. Thus, we are dealing with a pathway that operates at either one of two very different operating points, the normal point or the enzyme-deficient point. We shall see in the following that the same GMA and the S-system models simultaneously capture both the normal and the diseased states, at least to some degree. However, such a demand is in general too stringent as a criterion for the quality of a metabolic model. If the same model *structure*, parameterized once at the normal state and re-parameterized at the disease state, is able to capture physiological variations around these points, then the model can be deemed successful.

At a less philosophical and more numerical level, there is concern that a severely enzyme-deficient pathway deviates too much from its original operating point to yield valid responses. The idea of knocking out 99% of an enzyme activity may be too drastic a deviation. While there is no proof that the model predictions are reliable, it turns out in this case that the new steady-state concentrations are not as far from the original operating point as the reduction in enzyme activity might suggest. This encourages us to continue cautiously with the analysis.

Finally, one must question whether patients with severe HGPRT deficiency can be expected to have the same characteristics as normal subjects, outside the deficiency itself. The best way of avoiding this potential problem would be to use only data on concentrations and fluxes in HGPRT-deficient patients. This luxury is not available, however, and if we want to explore the effects of this disease, we have no choice but to return to information about normal subjects, keeping in mind the above caveats.

As in the simulation of PRPPS superactivity, the transients between the original and a new steady state are irrelevant, because individuals do not spontaneously develop HGPRT deficiency. Nevertheless, a full dynamical model is of interest if we want to study perturbations in diseased patients or, maybe more importantly, drug treatments.

The implementation of HGPRT deficiency proceeds on the same principles as in the previous cases of PRPPS superactivity and inhibition-resistant AMPD. For the GMA model, the simplest approach may be the definition of a deficiency term $1 - D$ that is included as a factor in all affected rates. D is set to 0 for normal subjects (no deficiency) and to a value of (e.g.) 0.8 for 80% deficiency. The term $1 - D$ appears in the equations of PRPP, IMP, the pool of guanylates, and the pool containing hypoxanthine.

The implementation in the S-system model is somewhat more involved, but because it has been shown (Voit and Savageau, 1987; Sorribas and Savageau, 1989c) that the aggregation in the S-system form improves accuracy and thereby widens the range of valid approximation, the effort may be worthwhile. Also, the S-system allows the immediate display of steady-state solutions, including stability and sensitivities, while it turns out that the numerical solver in PLAS does not always find a numerical solution to the GMA model, thus requiring us to solve the differential equations. Since the dynamics of the system is not very fast, it is necessary to specify rather long time intervals, such as $t0 = 0$, $tf = 10,000$, and this costs computation time. While one might argue that a minute or two of solution time is negligible in light of the complexity of the overall project, these minutes accumulate in a comprehensive exploration of the system.

Each new aggregated kinetic order is computed from the corresponding old kinetic order by subtracting the missing contribution of HGPRT and dividing by the new steady-state flux, which also is decreased by the deficiency. For instance, the new kinetic order g'_{21} is

$$g'_{21} = \frac{g_{21} - D\frac{f_{hprt1} v_{hprt}}{SSF_2}}{1 - D\frac{v_{hprt}}{SSF_2}} = \frac{0.72 - 0.33116 D}{1 - 0.30106 D},$$

where SSF_2 denotes the steady-state flux through X_2. In this fashion, one obtains

$$h'_{11} = (1.27 - 0.40933 D)/(1 - 0.35594 D),$$
$$h'_{12} = (-0.165 + 0.15894 D)/(1 - 0.17797 D),$$
$$h'_{18} = (-0.237 + 0.21356 D)/(1 - 0.17797 D),$$
$$h'_{1,13} = (0.085 - 0.085 D)/(1 - 0.17797 D),$$
$$h'_{1,15} = (0.075 - 0.075 D)/(1 - 0.17797 D),$$

$$g'_{22} = (-0.28 + 0.26794 D)/(1 - 0.30106 D),$$
$$g'_{2,13} = (0.144 - 0.144 D)/(1 - 0.30106 D),$$
$$g'_{81} = (0.0034 - 0.0034 D)/(1 - 0.0027916 D),$$
$$g'_{88} = (-0.0034 + 0.0034 D)/(1 - 0.0027916 D),$$
$$g'_{8,15} = (0.001 - 0.001 D)/(1 - 0.0027916 D),$$
$$h'_{13,1} = (0.817 - 0.817 D)/(1 - 0.74297 D),$$
$$h'_{13,2} = (-0.66 + 0.66 D)/(1 - 0.74297 D),$$
$$h'_{13,13} = (0.528 - 0.35671 D)/(1 - 0.74297 D),$$
$$h'_{15,1} = (0.925 - 0.925 D)/(1 - 0.77083 D),$$
$$h'_{15,1} = (-0.925 + 0.925 D)/(1 - 0.77083 D),$$
$$h'_{15,15} = (0.438 - 0.32375 D)/(1 - 0.77083 D).$$

Notice that h_{11} contains two components that are affected: one representing v_{hprt} and the other v_{gprt}.

The affected rate constants are recomputed with the new kinetic orders and adjusted fluxes at the original steady state. It would be better to compute these from observed data on concentrations and fluxes in HGPRT-deficient patients, but this information is not available. For instance, α'_2 becomes

$$\alpha'_2 = \frac{SSF_2 - D \cdot v_{hprt}}{X_{1S}^{g'_{21}} X_{2S}^{g'_{22}} X_{4S}^{g_{24}} X_{7S}^{g_{27}} X_{8S}^{g_{28}} X_{13S}^{g'_{2,13}} X_{18}^{g_{2,18}}}$$
$$= \frac{12.29 - 3.7 D}{5^{g'_{21}} 100^{g'_{22}} 2500^{0.319} 25^{-0.03} 400^{-0.024} 10^{g'_{2,13}} 1400^{-0.06}}.$$

The remaining rate constants are

$$\beta'_1 = \frac{20.79 - 2 \times 3.7 D}{5^{h'_{11}} 100^{h'_{12}} 2500^{-0.067} 10^{0.036} 400^{h'_{18}} 10^{h'_{1,13}} 5^{h'_{1,15}} 1400^{-0.0092}},$$
$$\alpha'_8 = \frac{1325.4 - 3.7 D}{5^{g'_{81}} 2500^{0.00014} 25^{0.00019} 400^{g'_{88}} 28600^{0.996} 5^{g'_{8,15}}},$$
$$\beta'_{13} = \frac{4.98 - 3.7 D}{5^{h'_{13,1}} 100^{h'_{13,2}} 10^{h'_{13,13}}},$$
$$\beta'_{15} = \frac{4.8 - 3.7 D}{5^{h'_{15,1}} 400^{h'_{15,8}} 5^{h'_{15,15}}}.$$

These definitions of kinetic orders and rate constants need to be executed only once. Implementing them in this semi-symbolic form in PLAS allows us to model different degrees of deficiency by simply changing the deficiency factor D. Clearly, setting $D = 0$ returns the original system of normal purine metabolism. The models are found in files *PurineGHGPRT.plc* and *PurineSHGPRT.plc*.

As an example, we simulate a severe 99% HGPRT deficiency with the GMA and the S-system models and report those results for which there are corresponding

Table 10.10. Comparison of Representative Concentrations and Fluxes in Normal Subjects and Patients with Severe HGPRT Deficiency (Data and Model Predictions)

Variable	Control: Steady-State Value	99% HGPRT Deficiency			
		GMA	S-system	Clinical Observation	Reference
X_1	5	7.1	7.7	Moderate increase	[23]
X_{13}	10	70.0	46.8	50	[133]
X_{14}	5	22.6	23.2	15	[133]
X_{16}	100	145.6	146.5	150	[98]
v_{den}	2.39	6.3	7.4	7–10	[509]
				48	[80]
v_{xd}	2.3	5.3	5.4	7–14	[350]
v_{hx}	0.05	0.44	0.28	0.45	[133]
				Only moderate increase	[186: p. 1125ff.]
v_x	0.03	0.61	0.64	0.27	[133]
				Only moderate increase	[186: p. 1125ff.]
v_{ua}	2.3	5.3	5.4	7–14	[350]

observations. The comparison is given in Table 10.10. The results indicate that the two models reflect clinical observations qualitatively and in most cases semi-quantitatively. In assessing the results, one must remember that clinical data are often the result of very small sample sizes and sometimes measurements on a single patient. Nonetheless, the consistency between models and data is encouraging.

It is generally accepted that purine synthesis is increased in HGPRT-deficient patients. One explanation is the elevated concentration of PRPP, which some consider critical in such patients (cf. Kelley and Wyngaarden, 1983: p. 1127). A second proposed explanation is that IMP and GMP concentrations might be low, which would lead to decreased inhibition of purine synthesis and subsequent higher levels. *In vivo* measurements apparently are difficult to obtain, but the few available data seem to indicate that GMP and IMP levels are relatively normal. The guanylates are low in both models, but it is not possible to separate out GMP specifically.

Interestingly, the GMA and S-system models give different answers with respect to IMP: In the GMA model, the steady-state concentration of IMP is increased to about 114, while it is decreased to about 72 in the S-system model. Further exploration shows the following: For mild deficiencies (e.g., 20%), both models predict IMP between 97 and 98. For increased severity of the deficiency (increasing values of D), the steady-state concentration of IMP decreases monotonically in the S-system model until it reaches 71 for total deficiency ($D = 1$). In the GMA model, by contrast, IMP decreases for values of D between 0 and 0.8; in the latter case the steady-state concentration of IMP is minimal at about 95.8. For more severe deficiencies, the values climb up to 123.5 (at $D = 1$: total deficiency). It cannot be decided which of the two models exhibits the better results for severe deficiencies, but the discrepancy is a warning sign that one or both models are leaving the range of valid approximation.

Many details about HGPRT deficiency are not known, and sometimes there are differing reports. For instance, it is unclear whether this deficiency is accompanied by changes in APRT activity. In some cell systems and patients it is increased, in some unaffected, in some decreased (cf. Kelley and Wyngaarden, 1983: p. 1125 f.). In the S-system and GMA models, the activity is essentially unaltered. Some further aspects of these deficiencies are discussed in Curto et al. (1998b).

Drug treatment. An increased or decreased enzyme activity simulates a (permanent) disease state, and one might ask what drug intervention could be considered to alleviate problems associated with the disease. One of the common drugs is allopurinol (e.g., Stanbury et al., 1983: p. 34; Wyngaarden and Kelley, 1983: p. 1102), which is an analogue of hypoxanthine. It effectively inhibits xanthine oxidase and is a successful agent for controlling uric acid production. *In vivo*, allopurinol is rapidly metabolized to oxipurinol, which also inhibits xanthine oxidase.

There are various levels of sophistication at which allopurinol treatment can be modeled. The simplest is the simulation of a constant treatment, which effectively lowers the activity of xanthine oxidase. Since this enzyme catalyzes two steps, namely the degradation of hypoxanthine and of xanthine, the decreased activity has to be implemented in several places that are associated with v_{hxd} and v_{xd}. For instance, using the GMA model, we introduce a new independent variable AP, which is defined as the effect of allopurinol on the activity of xanthine oxidase. AP is included as $1 - \text{AP}$ in the equations of X_{13}, X_{14}, and X_{16}. These equations read:

$$X13' = 0.001062 X4^\wedge.97 + 0.03333 X9 + 0.9135 X2^\wedge.8 X18^\wedge - 0.36 \gg$$
$$\gg -12.569 X1^\wedge 1.1 X2^\wedge - .89 X13^\wedge.48 - .003793 X13^\wedge 1.12 \gg$$
$$\gg -0.2754(1 - \text{AP})X13^\wedge.65$$
$$X14' = 0.2754(1 - \text{AP})X13^\wedge.65 + 0.4919 X15^\wedge.5 \gg$$
$$\gg -0.949(1 - \text{AP})X14^\wedge.55 - 0.0012 X14^\wedge 2$$
$$X16' = .949(1 - \text{AP})X14^\wedge.55 - .00008744 X16^\wedge 2.21$$

An additional command like

$$\text{AP} = 0.6$$

signifies 60% inhibition and 40% remaining activity.

Let's suppose we are studying an individual with decreased renal clearance of uric acid. This disease state is readily modeled by decreasing the rate v_{ua}. Summarizing findings of several authors (cf. Wyngaarden and Kelley, 1983: p. 1089), the kidneys in these patients clear uric acid at a rate of about 70–85%. Thus, we include the factor 0.75 in the degradation term $0.000087144 X_{16}^{2.21}$ of the 16th equation. Without allopurinol, the uric acid concentration is about 114; all other metabolite concentrations are unaffected. Now suppose the patient is treated with allopurinol, which we assume to inhibit at a level of 50%. Implementing this in PLAS, the uric acid concentration, which was formerly elevated to about 114 because of doubled PRPPS activity, is now about 104. We also see one of the immediate side effects of allopurinol: As observed (cf. Wyngaarden and Kelley, 1983: p. 1103), the hypoxanthine and xanthine pools are elevated – in this particular simulation to about 17 and 12, respectively.

At a higher level of sophistication, one could allow for the fact that allopurinol also inhibits amidophosphoribosyltransferase. Furthermore, using an input module, one could model the fact that allopurinol has a half-life of only 2 to 3 h and that the oxipurinol formed from allopurinol has a half-life of about 13.5 to 28 h (Elion et al., 1966; Hande et al., 1978).

CONCLUSIONS

This case study describes the *de novo* development of a model and its subsequent refinements. It reflects a situation that is very typical, even though it is seldom presented in the literature. Who wants to see preliminary models that don't work – especially if there are proven amendments? In fact, the paucity of reports of sequential model developments is something of a disservice, especially to the novice, since even a modeling genius rarely gets it right the first time. Besides, there is no absolute "getting it right." New demands on the model, new data, new scenarios all may suggest taking the best available model and making it better.

The central result of the present analysis consists of two models, one in GMA form and the other in S-system form. Close to the operating point, which we, according to good tradition, set at the steady state, the two models respond very similarly, even though their mathematical structures differ because of the aggregation strategy underlying the S-system. For large deviations from the operating point, the models exhibit some differences, which often – but not always – are minor. For instance, in the case of severe HGPRT deficiency, the steady-state concentration of IMP increases in the GMA model and decreases in the S-system model. Without definite experimental data, it is impossible to decide whether one or the other model, or both models, are incorrect. Theory is no help, because both models are designed as local representations and guaranteed only for small variations about the steady state. The GMA model seems to have more appeal with many biochemists, since it is easier to identify a particular flux in a GMA model than in an S-system model, where several fluxes may be aggregated. However, in favor of the S-system model it has been shown that – at least in small model systems of Michaelis–Menten rate laws – the S-system representation is more accurate than the GMA representation (Voit and Savageau, 1987; Sorribas and Savageau, 1989c). Furthermore, every piece of information available from the GMA model can also be retrieved from the S-system model. Finally, the S-system representation allows the computation of the steady state without time-consuming dynamical solutions or a search algorithm.

This case study has executed analyses that exemplify how biochemical system information is on the brink of becoming relevant in a clinical setting. If models like the ones proposed here are sufficiently tested and fine-tuned, they may serve as screening tools for drugs or enzyme replacement therapies; they may yield insights that are otherwise unobtainable; and they can be expected eventually to reduce the number of animal experiments.

EXERCISES

1. Formulate a symbolic S-system model (without the specification of parameter values) for the map in Fig. 10.1.
2. Discuss how the model would need to be changed if one wanted to study imbalances between ADP and ATP.
3. Formulate the four rate laws of PRPP degradation as symbolic GMA and S-system terms in the map of Fig. 10.2. Specify all relationships between the parameters in the GMA system and the S-system terms.
4. Use Tables 10.4 and 10.5 to construct the second model in GMA form. Check its steady state in PLAS.
5. Check system responses of the second model (S-system form) to temporary increases in metabolites other than PRPP. Report the maximum and minimum values of all dependent variables during the transition from the perturbed initial state to the steady state. Summarize and interpret the findings.
6. Check system responses of the second model (GMA form) to temporary increases in metabolites other than PRPP. Compare with results from Exercise 5.
7. Show that for the computation of g_{24} only steady-state flux values and the kinetic orders with respect to X_4, but not with respect to other variables, are needed.
8. Analyzing the second model, we reset h_{34} and g_{24}, with the result that the system response was qualitatively correct but quantitatively too weak: the concentration of uric acid increased by merely 4%, and the conversion of xanthine to uric acid by less than 10%. Confirm this in PLAS.
9. Construct the GMA and S-system equations for the third model. Implement the models in PLAS. Confirm steady states and stability. Analyze sensitivities. Why doesn't PLAS produce logarithmic gains?
10. Quantify Curto's results 1, 3, 4, and 5 about the third model, which are presented in the text.
11. Formulate the rate laws in GMA and S-system form that differentiate the fourth from the third model.
12. There is discussion about whether v_{pyr} in some animal species is composed of individual pathways for pyrimidine, tryptophan, and histidine synthesis. If so, the synthesis of tryptophan and histidine is believed to be small in comparison with the synthesis of pyrimidine. How would the "final" model change if v_{pyr} were split up into individual pathways for pyrimidine, tryptophan, and histidine synthesis?
13. Construct the GMA equations from Tables 10.8 and 10.9, and compare with the PLAS file *PurineGMA.plc*.
14. Confirm the steady state of the "final" GMA model and its stability. First, specify the observed steady-state concentrations as initial values. In a second analysis, use

CASE STUDY 3 – A SEQUENCE OF MODELS DESCRIBING PURINE METABOLISM

the same model, but change the initial values. Check whether PLAS still finds the steady state. If not, solve the differential equations. For how long do you have to run the solution before the system reaches the steady state?

15. Analyze the sensitivities of the "final" GMA model. Is there a (simple) relationship between the magnitude of a parameter value and its corresponding sensitivities?

16. Analyze the large fluxes associated with RNA.

    16.1. Assess the overall effects of changing the kinetic orders in the production of adenylates from RNA.

    16.2. Suppose the polymerase reaction, which governs the assembly of RNA, is saturated with respect to adenylates and guanylates. In other words, RNA is produced as needed, independent of the current concentrations of adenylates and guanylates, as long as these are within their physiological ranges. Change the GMA model correspondingly, and check the raw and constrained sensitivities of the system. How would one have to make this change in the corresponding S-system?

17. Formulate symbolically the degradation term of PRPP in the final S-system model. Characterize constraints on the parameters involved.

18. Discuss in words or prove mathematically that the deaggregated sensitivities of the S-system model are exactly the same as the corresponding quantities of the GMA model.

19. Study the responses of individual and aggregated fluxes to changes in input R5P. Which fluxes are affected most? Which least? Distinguish between increases and decreases in flux rate, if applicable.

20. Compare the responses of the S-system and GMA models to an initial PRPP concentration of 50. Summarize your findings in an executive summary.

21. Study the responses of the system (in S-system and GMA form) to endogenous supplies of IMP, Ade, and Xa. Summarize your findings in a concise report.

22. Numerous clinical studies suggest that the adenine and guanine nucleotide pools are rather insensitive to temporary fluctuations in other metabolites. Investigate these observations with the S-system or GMA model.

23. Discuss to what degree doubling V_{max} in a Michaelis–Menten model corresponds to doubling the rate constant in a power-law model.

24. Lack of inhibition of the enzyme AMPD results in increased degradation of xanthine and subsequently increased uric acid levels. Confirm with PLAS the results given in the text.

25. Confirm the formula for $g'_{2,8}$ in the S-system simulation of an abnormality in which the enzyme AMPD is resistant to inhibition by guanylates and P_i. Develop the corresponding formulas for $g'_{2,18}$, $h'_{4,8}$, and $h'_{4,18}$, and confirm the numerical results given in the text. How would you model partial resistance to inhibition? Why is the assumption important that the overall flux v_{ampd} is unchanged? What needs to be done if the flux is different?

26. Model the effects of a partial deficiency in the activity of the enzyme xanthine dehydrogenase, which catalyzes the degradation of xanthine to uric acid.

27. Several "uricosuric" drugs are on the market that increase the excretion of uric acid (Wyngaarden and Kelley, 1983: p. 1101). Model their action.

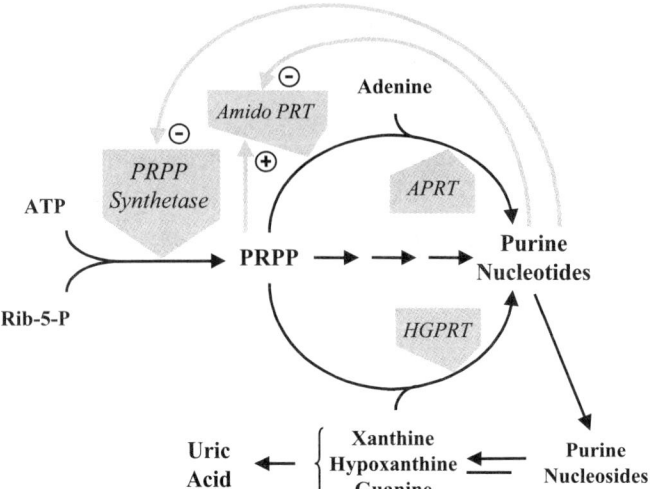

Figure 10.6. Simplified representation of purine metabolism. Redrawn from Becker et al. (1989a).

28. According to Delbarre et al. (1968), the xanthine oxidase inhibitor "thiopurinol... reduces uric acid synthesis without a concomitant increase in oxypurine excretion." How could this be modeled?

29. (Project) Study the effects of including one, two, or more intermediates between PRPP and IMP in the "final" GMA or S-system model. Select steady-state values of 10 and fluxes of the magnitude v_{den}. Analyze steady states, stability, sensitivities, and dynamic responses.

30. (Project) Model different degrees of APRT deficiency. Clinically, this deficiency is almost asymptomatic, except that the excretion of adenine metabolites (Ade, hydroxyadenine, and 2,8-dihydroxyadenine) is increased to about 0.5 μmol min^{-1} kg^{-1} (van Acker et al., 1977; van Acker and Simonds, 1991). Implement a severe deficiency, and compare observations with model results.

31. (Project) Model allopurinol treatment in HGPRT-deficient patients. Summarize your findings in a technical and a non-technical report.

32. (Project) Becker et al. (1989a) used the schematic representation in Fig. 10.6 to study the effects of PRPPS in otherwise normal subjects and in HGPRT-deficient patients. Formulate a symbolic model in GMA or S-system form. Estimate parameters from the information given in this chapter. Compare results.

REFERENCES

[7], [16], [18], [23–4], [34], [65], [67–70], [81–2], [95], [99], [116], [133–4], [142], [148], [168–70], [190], [198], [263], [271], [364], [382], [396], [399–400], [429–31], [466], [513], [525].

CHAPTER ELEVEN

Case Study 4 – Algebraic Analysis of the Initial Steps of the Glycolytic–Glycogenolytic Pathway in Perfused Rat Liver

Some standard procedures are often included in a biochemical systems analysis, no matter what the ultimate goal of the analysis is. This case study illustrates these procedures. It focuses on parameter estimation, which is never as clear-cut as a mathematician would hope, and on steady-state analysis by algebraic means. The pathway of this case study consists of the initial reactions of glycolysis and glycogenolysis. Dr. Marta Cascante of the University of Barcelona, Spain, suggested this example, generously collected much of the material, and executed many of the algebraic analyses. The experimental data were obtained from a number of sources, but in particular from the original work of Dr. Néstor Torres of the University of La Laguna, Tenerife (Spain), who also contributed greatly to this chapter by supplying information and feedback not readily available otherwise. Background information on this pathway can be found in most biochemical textbooks, in the reviews of Scrutton and Utter (1968) and of Newsholme and Start (1973), and in Torres's publications (Torres, 1986, 1994ad; Torres et al., 1986).

CONSTRUCTION OF THE MAP

Glycolysis and glycogenolysis are central steps of carbohydrate metabolism. They yield the common intermediate glucose-6-phosphate, which can subsequently be used by the organism as a major source of energy. Under adequate nutritional conditions, the concentration of glucose in human blood varies only slightly, and even fasting for several days does not lead to significant drops in glucose concentration. Under conditions of low glucose input, the organism must access other energy supplies. For extended periods of low glucose input, such as fasting for several days, higher animals retrieve energy from lipids, which have a high energy storage potential. For faster demands, energy can be provided by the utilization of glycogen, which is more readily accessible than lipids. Glycogen is stored in the liver and other tissues in large quantities, reaching 0.5% to 1% in skeletal muscle and making up 2% to 8% of wet liver weight. Glycogen is constantly formed and degraded, and the liver of an animal starved for just one day loses almost all glycogen. Subsequent feeding of glycogenic substances leads to rapid replenishment (White et al., 1968: pp. 443–445).

The main route of glycogen utilization is glycogenolysis via glycogen phosphorolysis to glucose-1-phosphate and subsequent conversion to glucose-6-phosphate via the phosphoglucomutase reaction. Glucose-6-phosphate can be used for energy production or formation of glucose. It is clear that a detailed, quantitative understanding of the kinetic and dynamic properties of glucose, glycogen, and glucose-6-phosphate is central to any assessments of the energy economy of organisms and of diseases like diabetes that are associated with the balance and imbalance of carbohydrates.

A comprehensive analysis of all facets of glycolysis and glycogenolysis would be an overwhelmingly complicated endeavor, because ubiquitous compounds like glucose, fructose, ATP, and NAD^+ are involved, which are substrates or modulators of a wide range of other pathways. The first task is therefore the identification of those compounds and reactions that are to be considered in the analysis. They are shown in Fig. 11.1. As mentioned many times by now, the identification of constituents and the subsequent construction of the map are most crucial steps. They determine how realistic and reliable the model can possibly be, but also acknowledge the need for simplification, which ignores many of the known factors and compounds. There is no question that it is possible to design alternate models of glycolysis and glycogenolysis that are all reasonable and relevant. In the present case, the goal of the analysis is the demonstration of some of the algebraic methods that were introduced in earlier chapters, and for this purpose, the small size of the system in Fig. 11.1 is the deciding factor in the design of the model.

By our definition, the map in Fig. 11.1 is now our reality. Compounds, factors, and modulators not shown are assumed to be constant or irrelevant. In a real application, one could imagine a later refinement or expansion of the model, but for now, the map shows everything the analysis is able to consider.

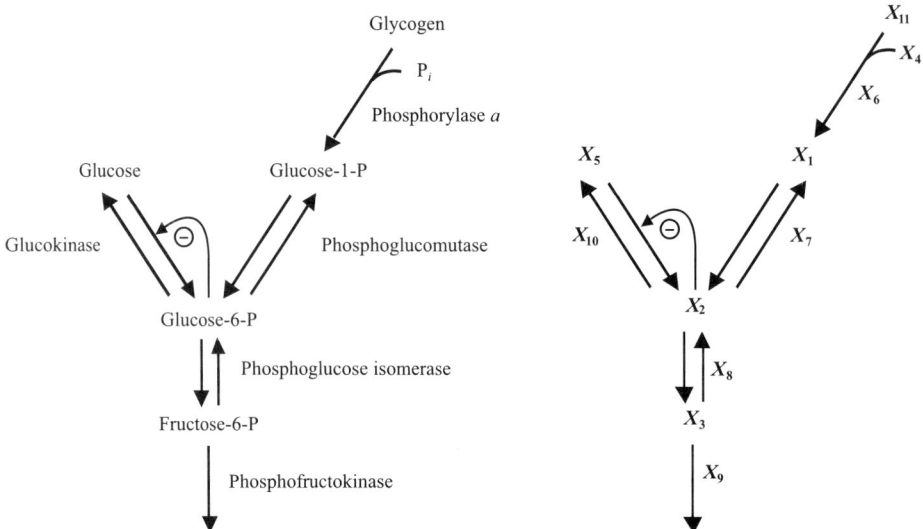

Figure 11.1. Initial steps of glycolysis and glycogenolysis, in terms of biochemical names and in terms of model variables.

CASE STUDY 4 – ALGEBRAIC ANALYSIS

As always, the first step in translating the map into a model is the identification of dependent and independent variables. We define:

Dependent Variables:

X_1 = glucose-1-P (G1P),
X_2 = glucose-6-P (G1P),
X_3 = fructose-6-P (F6P).

Independent Variables:

$X_4 = P_i$
X_5 = glucose,
X_6 = phosphorylase a,
X_7 = phosphoglucomutase,
X_8 = phosphoglucose isomerase,
X_9 = phosphofructokinase,
X_{10} = glucokinase,
X_{11} = glycogen.

Three comments on these definitions are in order:

1. Glucose (X_5) is defined as an independent variable. This is justified because the glucokinase reaction is essentially irreversible, with a high equilibrium constant in the range of several thousand (cf. Bassham and Krause, 1969; Newsholme and Start, 1973: p. 263; see also Exercise 1.1).
2. Glycogen is also modeled as an independent variable, because the model describes a situation analyzed by Torres (1986, 1994a) where glycogen is experimentally held saturated (see also Exercise 1.2).
3. 15 μM fructose-2,6-biphosphate was added to the medium. This metabolite is a strong effector of phosphofructokinase that is produced *in vivo* while the system is working. The addition guarantees operation of the pathway at a steady state (cf. Torres, 1986, 1994a).

The biochemical map in terms of dependent and independent variables is shown in the right panel of Fig. 11.1. If there were doubts about the consistency or validity of this map or about the inclusion or omission of variables, they would have to be removed now. Otherwise, these doubts would vitiate all further analyses and results.

The model equations are set up in two steps. First, we write down the general mass balance equations, using generic symbols like V_1^+ and V_1^- for fluxes and indicating which dependent and independent variables affect each one of these fluxes. Since only three of the variables are dependent, we have three equations with a total of eleven variables. The mass balance equations read

$$\dot{X}_1 = V_1^+(X_4, X_6, X_{11}) - V_1^-(X_1, X_2, X_7),$$
$$\dot{X}_2 = V_2^+(X_1, X_2, X_5, X_7, X_{10}) - V_2^-(X_2, X_3, X_8), \qquad (11.1)$$
$$\dot{X}_3 = V_3^+(X_2, X_3, X_8) - V_3^-(X_3, X_9).$$

In the second step, these equations are directly translated into S-system equations, following the recipes in Chapter 3: Each variable in each flux term appears in the corresponding product of power-law functions with an associated kinetic order parameter g_{ij} or h_{ij} whose two indices coincide with the numbers of the affected and the affecting variables, respectively. Furthermore, each flux term receives a rate constant α_i or β_i whose index coincides with the number of the equation. The result is

$$\begin{aligned}
\dot{X}_1 &= \alpha_1 X_4^{g_{14}} X_6^{g_{16}} X_{11}^{g_{1,11}} - \beta_1 X_1^{h_{11}} X_2^{h_{12}} X_7^{h_{17}}, \\
\dot{X}_2 &= \alpha_2 X_1^{g_{21}} X_2^{g_{22}} X_5^{g_{25}} X_7^{g_{27}} X_{10}^{g_{2,10}} - \beta_2 X_2^{h_{22}} X_3^{h_{23}} X_8^{h_{28}}, \\
\dot{X}_3 &= \alpha_3 X_2^{g_{32}} X_3^{g_{33}} X_8^{g_{38}} - \beta_3 X_3^{h_{33}} X_9^{h_{39}}.
\end{aligned} \quad (11.2)$$

Since glycogen is experimentally held saturated, small changes in its amount have no effect on the phosphorylase a reaction in the synthesis of glucose-6-phosphate (V_1^+). Consequently, the parameter $g_{1,11}$ is zero, which suggests omission of the term $X_{11}^{g_{1,11}}$ from the flux V_1^+. Other than that, we cannot say much about the numerical values of the parameters in Eq. (11.2).

We have not explicitly formulated the flux V_5^-, since X_5 is an independent variable. However, this is useful for later purposes and easy to do, by pretending for a moment that X_5 is a dependent variable. The appropriate definition is

$$V_5^- = \beta_5 X_2^{h_{52}} X_5^{h_{55}} X_{10}^{h_{5,10}}. \quad (11.3)$$

PARAMETER VALUES

Data

We consider a typical situation where the primary data consist of steady-state concentrations and steady-state fluxes and where additional information is available about the kinetic properties of some of the reactions. Most kinetic properties of the pathway are obtained from Newsholme and Start (1973) and from Torres (1994a), but several pieces of information are not available in these sources or seem inappropriate and therefore have to be retrieved from other literature reports.

As is typical in the estimation of parameters for the model, it is close to impossible to establish one definite set of parameter values. Even intrinsic features of enzymes like equilibrium constants, Michaelis constants, and maximal activities are often reported in the literature with significant discrepancies, and it is not unusual that the necessary information must be pieced together from different experimental or physiological conditions or even different species. This creates unavoidable uncertainties in the model definition and the results of the analysis, and about the only remedy is a careful assessment of sensitivities and gains and of the overall robustness and reasonableness of the model and its responses to simulation experiments.

The first set of data pertains to steady-state concentrations and enzyme activities. Torres (1994a) determined values for X_1, X_2, and X_3 as 0.067, 0.465, and

0.15 mM, respectively. He also measured the involved fluxes and listed them as

(flux through G1P) = 0.0187 μmol/min per milligram protein,
(flux through G6P) = 0.0113 μmol/min per milligram protein,
(flux through F6P) = 0.0300 μmol/min per milligram protein.

For purposes of parameter estimation, we will have to compare these fluxes with maximal enzyme activities. These are obtained in most cases from Newsholme and Start (1973), who list fluxes in *micromoles per minute per gram fresh weight*. It is thus advisable to convert Torres's units into units used in the literature. In Torres's experiments, 1 g of fresh liver was homogenized in 3 ml of buffer, yielding a homogenate volume of 3.5 ml (Torres, personal communication). Like calf, cow, and pig (cf. Geigy, 1960), rat liver is assumed to have an approximate protein content of 20%. Using these values and converting milligrams into grams, we obtain, for instance, for the flux through G6P

$$0.0113 \ \mu\text{mol/min per milligram protein}$$
$$= 0.0113 \times 1000 \times 20\%/3.5$$
$$= 0.646 \ \mu\text{mol/min per gram fresh liver weight.}$$

The conversion factor thus is CF = 57.14, and the fluxes in the new units read

$V_1^+ = 1.068 \ \mu$mol/min per gram fresh liver weight,
$V_5^- = 0.646 \ \mu$mol/min per gram fresh liver weight,
$V_3^- = 1.714 \ \mu$mol/min per gram fresh liver weight.

The remaining fluxes are mathematically determined by these three fluxes because of precursor–product and branch point relationships at steady state:

$$V_1^- = V_1^+,$$
$$V_2^+ = V_1^- + V_5^-,$$
$$V_2^- = V_3^+ = V_3^-.$$

The independent variables of the model describe the concentrations of glucose, P_i, and glycogen, as well as the activities of the enzymes involved. An appropriate glucose concentration for Torres's system is 5 mM, which is consistent with normal mammalian metabolism (e.g., Newsholme and Start, 1973: pp. 264–268; Torres et al., 1986). The phosphate concentration in liver is usually high, and Torres uses a value of 10 mM, as also suggested by Newsholme and Start (1973: p. 152). Torres fixed the glycogen concentration experimentally at 50 mM. However, since the model describes a situation of glycogen saturation, the numerical value of X_{11} (glycogen) does not enter the model [see discussion above after Eq. (11.2)].

For consistency, the enzyme activities reported by Torres (1994a) are also multiplied with the conversion factor CF. After the conversion, the activities of phosphofructokinase and glucokinase are consistent with values reported by Scrutton and Utter (1968) and Newsholme and Start (1973: p. 264). However, the measured maximal activities of phosphorylase a, phosphoglucomutase, and phosphoglucose isomerase appear to be appreciably lower than comparable values in the literature, and we decided to use literature values instead. Precise numerical values for the enzyme activities are not absolutely necessary, since the rate constants are free parameters that adjust the power-law terms to the observed steady-state fluxes. We could thus specify the enzyme activities as 1 in arbitrary units and compensate with the definition of the rate constants.

Summarizing direct measurements and applying CF where appropriate, Torres's (1994a) measurements and auxiliary literature values suggest the following definitions:

Dependent Variables:

[G1P]: $X_{1S} = 0.067$ mM,

[G6P]: $X_{2S} = 0.465$ mM,

[F6P]: $X_{3S} = 0.150$ mM.

Independent Variables:

[P$_i$]: $X_4 = 10.00$ mM,

[glucose]: $X_5 = 5.00$ mM,

phosphorylase a: $X_6 = 3.00$ μmol/min per gram fresh liver weight,

phosphoglucomutase: $X_7 = 40.00$ μmol/min per gram fresh liver weight,

phosphoglucose isomerase: $X_8 = 136.00$ μmol/min per gram fresh liver weight,

phosphofructokinase: $X_9 = 2.86$ μmol/min per gram fresh liver weight,

glucokinase: $X_{10} = 4.00$ μmol/min per gram fresh liver weight,

[glycogen]: $X_{11} = 50.00$ mM   (not needed for the model).

Steady-State Fluxes:

$V_1^+ = V_1^- = 1.068$ μmol/min per gram fresh liver weight,

$V_2^+ = V_2^- = 1.714$ μmol/min per gram fresh liver weight,

$V_3^+ = V_3^- = 1.714$ μmol/min per gram fresh liver weight,

$V_5^- = 0.646$ μmol/min per gram fresh liver weight.

Constraints

The data are used to determine kinetic order and rate constant parameters. Before we actually do this, it is advisable to identify constraints in the system, of which there

CASE STUDY 4 – ALGEBRAIC ANALYSIS

are two. The first one is the precursor–product relationship between X_2 and X_3, and the second one is the branch point constraint at X_2.

The precursor–product relationship between X_2 and X_3 requires that all parameters of V_2^- must equal those of V_3^+, since V_2^- and V_3^+ actually represent the same process. This necessitates

$$\alpha_3 = \beta_2, \quad g_{32} = h_{22}, \quad g_{33} = h_{23}, \quad g_{38} = h_{28}. \tag{11.4}$$

The branch point constraint at X_2 requires that the sum of the two fluxes entering X_2 from X_1 and X_5, namely V_1^- and V_5^-, must equal the influx V_2^+. The branch point constraint can be expressed as

$$V_2^+ = V_1^- + V_5^- \tag{11.5}$$

at steady state, i.e., as

$$\alpha_2 X_1^{g_{21}} X_2^{g_{22}} X_5^{g_{25}} X_7^{g_{27}} X_{10}^{g_{2\,10}} = \beta_1 X_1^{h_{11}} X_2^{h_{12}} X_7^{h_{17}} + \beta_5 X_2^{h_{52}} X_5^{h_{55}} X_{10}^{h_{5\,10}}. \tag{11.6}$$

Recalling the discussion about branch points from Chapter 3, the kinetic orders on the left-hand side are computed through partial differentiation, because this term can be considered a power-law representation of the right-hand side. This representation is exact at an operating point, which we choose to be the steady state, and sufficiently accurate in the vicinity of this point.

For instance, we compute g_{21} as

$$g_{21} = \frac{\partial V_2^+}{\partial X_1} \cdot \frac{X_1}{V_2^+}$$

$$= \frac{\partial \left(\beta_1 X_1^{h_{11}} X_2^{h_{12}} X_7^{h_{17}}\right)}{\partial X_1} \cdot \frac{X_1}{V_2^+} + \frac{\partial \left(\beta_5 X_2^{h_{52}} X_5^{h_{bb}} X_{10}^{h_{5,10}}\right)}{\partial X_1} \cdot \frac{X_1}{V_2^+}, \tag{11.7}$$

which is to be evaluated at the steady state. The first term in Eq. (11.7) is

$$\frac{\partial \left(\beta_1 X_1^{h_{11}} X_2^{h_{12}} X_7^{h_{17}}\right)}{\partial X_1} \cdot \frac{X_1}{V_2^+} = \beta_1 h_{11} X_1^{h_{11}-1} X_2^{h_{12}} X_7^{h_{17}} \cdot \frac{X_1}{V_2^+}, \tag{11.8}$$

which is equivalent to $h_{11} V_1^- / V_2^+$. The second differential in Eq. (11.7) is zero, because $\beta_5 X_2^{h_{52}} X_5^{h_{55}} X_{10}^{h_{5,10}}$ is independent of X_1. Thus,

$$g_{21} = h_{11} \frac{V_1^-}{V_2^+} \tag{11.9}$$

which is evaluated at the steady state. g_{25}, g_{27}, and $g_{2,10}$ are derived in the same fashion, and the result is

$$g_{25} = h_{55} \frac{V_5^-}{V_2^+}, \quad g_{27} = h_{17} \frac{V_1^-}{V_2^+}, \quad g_{2,10} = h_{5,10} \frac{V_5^-}{V_2^+}. \tag{11.10}$$

The derivation of g_{22} is slightly more complicated, because both V_1^- and V_5^- depend on X_2. In this case, the constrained kinetic order is equivalent to the weighted average of the two independent kinetic orders:

$$g_{22} = h_{12} \frac{V_1^-}{V_2^+} + h_{52} \frac{V_5^-}{V_2^+}. \tag{11.11}$$

The proof of this result is left as Exercise 2.

Kinetic Orders

The independent kinetic orders are represented by the parameters g_{14}, g_{16}, h_{11}, h_{12}, h_{17}, h_{22}, h_{23}, h_{28}, h_{33}, h_{39}, h_{52}, h_{55}, and $h_{5,10}$. All estimations are executed with methods described in Chapter 5: Parameter Estimation.

1. g_{14} and g_{16} quantify the phosphorylase *a* reaction. g_{14} is determined with the method of Groen et al. (1982a) as

$$g_{14} = \frac{1}{1 - \Gamma/K_{eq}} \left(1 - \frac{V_1^+}{V_{max}}\right). \tag{11.12}$$

According to Newsholme and Start (1973: p. 152), the equilibrium constant for phosphorylase *a* is $K_{eq} = 0.3$; V_{max} is low in comparison with other enzymes of glycolysis and similar to that of phosphofructokinase, which is at the order of 2 μmol min^{-1} g^{-1} (p. 154) or somewhat higher (pp. 264 and 283; Scrutton and Utter, 1968: p. 254). In our system, the activity of phosphofructokinase was measured as 2.86, and we thus use the value $V_{max} = 3$ for the phosphorylase *a* reaction. The mass action ratio in our system is $\Gamma = [G1P]/[P_i] = 0.0067$. With these values and $V_1^+ = 1.068$, the kinetic order is

$$g_{14} = 0.66.$$

Scrutton and Utter (1968: p. 262) list the maximal activity of phosphorylase *a* (in the direction of glycogen synthesis) as 37, which is at odds with the assessment by Newsholme and Start (1973) and the value we use. The potential effects of this discrepancy should be evaluated with sensitivity analyses and dynamic simulations characterizing the robustness of the model (see Exercise 1.3).

As the standard default, we assume that the reaction rate is proportional to enzyme activity. Thus,

$$g_{16} = 1.$$

2. h_{11}, h_{12}, and h_{17} quantify the phosphoglucomutase reaction. Suppose $V_{max(F)}$ and $V_{max(R)}$ denote the maximal rates of the forward and reverse reactions, respectively. The kinetic orders h_{11} and h_{12} are again computed from the formulas of Groen et al. (1982a) as

$$\begin{aligned} h_{11} &= \frac{1}{1 - \Gamma/K_{eq}} \left(1 - \frac{V_1^-}{V_{max(F)}}\right), \\ h_{12} &= \frac{-\Gamma/K_{eq}}{1 - \Gamma/K_{eq}} \left(1 + \frac{V_1^-}{V_{max(R)}}\right). \end{aligned} \tag{11.13}$$

CASE STUDY 4 – ALGEBRAIC ANALYSIS

The value of the equilibrium constant is obtained from Mahler and Cordes (1969) and Reich and Sel'kov (1981) as $K_{eq} = 19$. We assume $V_{max(F)} = V_{max(R)} = V_{max}$, which has a value of 37 μmol min^{-1} g^{-1} (Newsholme and Start, 1973: p. 154) to 43 μmol min^{-1} g^{-1} (Scrutton and Utter, 1968: p. 262). Γ is determined to be $\Gamma = [G6P]/[G1P] = 0.465/0.067 = 6.94$. With $V_{max} = 37$ μmol min^{-1} g^{-1} and $V_1^- = 1.068$ μmol min^{-1} g^{-1}, the desired kinetic orders are

$h_{11} = 1.53,$
$h_{12} = -0.59.$

Assuming proportionality between rate and enzyme activity, we set $h_{17} = 1$.

3. h_{52}, h_{55}, and $h_{5,10}$ represent input to the system via the glucokinase reaction. Similarly to the previous cases, the dependence on the substrate concentration is given as h_{55} and takes the form

$$h_{55} = \frac{1}{1 - \Gamma/K_{eq}} \left(1 - \frac{V_5^-}{V_{max(F)}}\right). \tag{11.14}$$

According to Bassham and Krause (1969), the equilibrium constant of this reaction has a large value: $K_{eq} = 1,550$. Newsholme and Start (1973: p. 263) list a range of 3,900 to 5,500 and a mass action ratio of 0.016. The extreme difference between equilibrium constant and mass action ratio allows us to simplify Eq. (11.14) to

$$h_{55} = 1 - \frac{V_5^-}{V_{max}} \tag{11.15}$$

(see Exercise 1 and Chapter 5 for discussion of this simplification).

The maximal activity was measured as 4 (cf. Torres, 1994a), which is essentially the same as the value 4.3 reported by Scrutton and Utter (1968: p. 260) and by Newsholme and Start (1973: p. 266). Using the steady-state flux $V_5^- = 0.646$, the appropriate value for the kinetic order is

$h_{55} = 0.84.$

There is discussion about whether glucose-6-phosphate inhibits the glucokinase step (Newsholme and Start, 1973: pp. 261–267; Hers and Hue, 1983). It is clear that if there is feedback at all, the inhibition is very weak. The kinetic order is therefore assumed to be negative with a low magnitude. We follow the suggestion of Cascante et al. (1991b) and specify

$h_{52} = -0.1.$

Torres (1994a) selected a somewhat stronger inhibition value of -0.25. Again assuming proportionality between rate and enzyme activity, we set

$h_{5,10} = 1.$

4. The kinetic orders $g_{21}, g_{22}, g_{25}, g_{27}$, and $g_{2,10}$ are computed from the branch point constraints discussed in the previous section. They take the values

$$g_{21} = 1.53 \times \frac{1.068}{1.714} = 0.95,$$

$$g_{22} = -0.59 \times \frac{1.068}{1.714} - 0.1 \times \frac{0.646}{1.714} = -0.41,$$

$$g_{25} = 0.84 \times \frac{0.646}{1.714} = 0.32,$$

$$g_{27} = 1 \times \frac{1.068}{1.714} = 0.62,$$

$$g_{2,10} = 1 \times \frac{0.646}{1.714} = 0.38.$$

5. h_{22}, h_{23}, h_{28} characterize the phosphoglucose isomerase reaction. Analogous to the previous estimation, h_{22} and h_{23} are given as

$$\begin{aligned} h_{22} &= \frac{1}{1 - \Gamma/K_{eq}} \left(1 - \frac{V_3^+}{V_{max(F)}}\right), \\ h_{23} &= \frac{-\Gamma/K_{eq}}{1 - \Gamma/K_{eq}} \left(1 + \frac{V_3^+}{V_{max(R)}}\right). \end{aligned} \tag{11.16}$$

The reaction takes place close to equilibrium. Using the concentrations of glucose-6-phosphate and fructose-6-phosphate at steady state, one obtains the corresponding mass action ratio as 0.323. According to Newsholme and Start (1973: pp. 263–264), the mass action ratio is 0.31 and the maximum velocity of the reaction is $V_{max} = 136 \ \mu\text{mol min}^{-1} \text{ g}^{-1}$ in rat liver and averages at 280 $\mu\text{mol min}^{-1} \text{ g}^{-1}$ in "all vertebrate species." The range for the equilibrium constant is listed as 0.36–0.47, which brackets the value $K_{eq} = 0.43$ given by Slein (1950) and Bassham and Krause (1969). Thus,

$$\begin{aligned} h_{22} &= g_{32} = 3.97, \\ h_{23} &= g_{33} = -3.06, \\ h_{28} &= g_{38} = 1 \end{aligned}$$

(see also Exercise 23).

6. h_{33} represents the phosphofructokinase reaction. Under experimental conditions of fructose-2,6-bisphosphate saturation, the activity of phosphofructokinase increases hyperbolically with respect to fructose-6-phosphate (Newsholme and Start, 1973: pp. 106–107; Hers and Van Schaftingen, 1982; Hers and Hue, 1983). This behavior has been confirmed for an experimental set-up similar to the present (Torres et al., 1986; Torres, 1994a). Furthermore, the concentration of fructose-6-phosphate is above K_M (Torres, 1986), which implies that the corresponding kinetic order is between 0 and 0.5. We set

$$h_{33} = 0.3$$

CASE STUDY 4 – ALGEBRAIC ANALYSIS

(cf. Torres, 1994a). Analyzing the sensitivity of the system with respect to this kinetic order will indicate whether the uncertainty in this parameter value is a cause for concern. As before, we assume proportionality between rate and enzyme activity and set $h_{39} = 1$.

Rate Constants

All kinetic orders are now available, and the only estimations left address the rate constants. These are determined by equating the measured fluxes with the corresponding power-law terms at steady state, and the result is

$$\alpha_1 = V_1^+ / X_4^{g_{14}} X_6^{g_{16}} X_{11}^{g_{111}} = 0.077884314,$$
$$\beta_1 = V_1^- / X_1^{h_{11}} X_2^{h_{12}} X_7^{h_{17}} = 1.062708258,$$
$$\alpha_2 = V_2^+ / X_1^{g_{21}} X_2^{g_{22}} X_5^{g_{25}} X_7^{g_{27}} X_{10}^{g_{210}} = 0.585012402,$$
$$\beta_2 = V_2^- / X_2^{h_{22}} X_3^{h_{23}} X_8^{h_{28}} = 7.93456 \times 10^{-4}, \qquad (11.17)$$
$$\alpha_3 = V_3^+ / X_2^{g_{32}} X_3^{g_{33}} X_8^{g_{38}} = 7.93456 \times 10^{-4},$$
$$\beta_3 = V_3^- / X_3^{h_{33}} X_9^{h_{39}} = 1.05880847,$$
$$\beta_5 = V_5^- / X_2^{h_{52}} X_5^{h_{55}} X_{10}^{h_{510}} = 0.038706421.$$

S-System Equations

Substituting the numerical values in Eq. (11.2), the S-system model becomes

$$\dot{X}_1 = 0.077884314 X_4^{0.66} X_6 - 1.062708258 X_1^{1.53} X_2^{-0.59} X_7,$$
$$\dot{X}_2 = 0.585012402 X_1^{0.95} X_2^{-0.41} X_5^{0.32} X_7^{0.62} X_{10}^{0.38}$$
$$\quad - 7.93456 \times 10^{-4} X_2^{3.97} X_3^{-3.06} X_8, \qquad (11.18)$$
$$\dot{X}_3 = 7.93456 \times 10^{-4} X_2^{3.97} X_3^{-3.06} X_8 - 1.05880847 X_3^{0.3} X_9.$$

The system is implemented in PLAS as *Glycolysis.plc*.

CONSISTENCY AND ROBUSTNESS

Steady-State Solution

It is always advisable to check whether the system is indeed defined correctly. Outside an item by item comparison, we may check the steady-state solution, which should return the experimentally measured concentrations and fluxes that were used for the estimation of parameter values. The $\dot{x}=0$ command in PLAS produces the following concentrations, fluxes, and eigenvalues:

$$X_1 = 0.067, \quad V_1 = 1.068, \quad \text{Re}_1 = -25.08371, \quad \text{Im}_1 = 0,$$
$$X_2 = 0.465, \quad V_2 = 1.714, \quad \text{Re}_2 = -1.037407, \quad \text{Im}_2 = 0,$$
$$X_3 = 0.15, \quad V_3 = 1.714, \quad \text{Re}_3 = -52.80591, \quad \text{Im}_3 = 0.$$

The concentration and flux values echo the experimental steady state. Of course, these numbers are no proof that we entered everything correctly, but they provide good circumstantial evidence. It is also useful to check the value of the flux V_5^-. This is easily done in PLAS by defining it as an additional variable. The command reads

$$V5 = 0.038706421\ X2\wedge(-0.1)\ X5\wedge 0.84\ X10$$

Initializing the system at the steady state, we obtain the flux value by solving the system for just a few time steps. The reported value is 0.646, which is the correct result.

Since we are analyzing a real-world system, we can expect the steady state to be stable. The steady-state display in PLAS shows that all eigenvalues have negative real parts and that all imaginary parts are zero. This indicates local stability and does not suggest oscillations (cf. Chapter 6).

Logarithmic Gains and Sensitivities

To obtain the logarithmic gains in PLAS, we identify the independent variables with the command

&& X4 X5 X6 X7 X8 X9 X10

Calling up the steady-state solution with the button, the logarithmic gains are displayed under the *Sensitivities* tab. They are shown in Table 11.1. The numerical values of the gains indicate that the system is responsive to changes in independent variables, but that the responses are quite moderate. The highest gain in magnitude, $L(X_3, X_9)$, is about -3, which means that a 1% increase in phosphofructokinase activity (X_9) evokes a 3% decrease in fructose-6-phosphate (X_3); see also Exercises 3 and 4.

As a test of these gains, we may increase the phosphofructokinase activity by 10%, which is accomplished by setting $X_9 = 2.86 \times 1.1$. The resulting steady state becomes $X_1 = 0.0615973$, $X_2 = 0.37391$, $X_3 = 0.112693$, and the relative decrease in X_3 is $(0.15 - 0.112693)/0.15 = 24.9\%$, which is indeed of the right magnitude. The relative change in X_1 is an 8.1% decrease, which is close to the 8.8% decrease predicted by the logarithmic gain. As the logarithmic gains indicate, the three dependent variables react differently to the same change in an independent variable. In a later

Table 11.1. Logarithmic Gains of Concentrations and Fluxes

	X_1	X_2	X_3	$V(X_1)$	$V(X_2)$	$V(X_3)$
X_4	0.8283062	1.029336	1.21621	0.66	0.364863	0.364863
X_5	0.3099501	0.8037688	0.9496912	0	0.2849073	0.2849073
X_6	1.255009	1.5596	1.842742	1	0.5528227	0.5528227
X_7	−0.6544811	−0.002298359	−0.00271562	0	−0.000814686	−0.000814686
X_8	−0.08648161	−0.2242659	0.03263826	0	0.009791477	0.009791477
X_9	−0.8821124	−2.287512	−3.000423	0	0.09987306	0.09987306
X_{10}	0.3680657	0.9544755	1.127758	0	0.3383275	0.3383275

section, we will show that the steady-state solution is a linear function of the independent variables in logarithmic coordinates and that the slope coefficients indeed are the logarithmic gains (see also Chapter 7). Thus, expressed in terms of changes in logarithms, the gain predictions are precise, and they also give us a quick impression of the relative changes that are to be expected in the concentration variables themselves.

The sensitivities with respect to rate constants exhibit a similar pattern, with values that are at most about 3 in magnitude (see Exercise 5). The sensitivities with respect to kinetic orders, however, are not as moderate (see Exercise 6). In particular, the model is very sensitive in its parameters of the phosphoglucose isomerase step. For instance, the sensitivity of X_3 with respect to h_{23} is about -17. If we increase just h_{23} by 1%, X_3 falls from 0.15 to 0.124107, and an increase by 10% leads to a nonsensical steady state. Does that mean the model is unreliable? Not really, since we have ignored that h_{23} and g_{33} are actually the same parameter. Just changing one of the two would describe an unrealistic alteration. Indeed, if both are simultaneously increased by 10%, the resulting steady state is only mildly affected. For instance, X_3 increases by less than 2%, from 0.15 to 0.152841.

Perturbations

The sensitivity analysis implied robustness with respect to small changes in independent variables or parameters, and the goal now is to test moderate changes and their dynamic effects. The range of possible simulations of small or large perturbations is unlimited. We focus on three representative scenarios:

1. Perturbation in one of the dependent variables.
2. Response to an altered independent variable.
3. Change in a parameter value.

Perturbation in one of the dependent variables. First we investigate how the system responds to the one-time addition of exogenous glucose-1-phosphate. This scenario may be implemented in PLAS in different ways. The default option is simply to initiate the system at the former steady state, except that X_1 is augmented by the exogenous amount. For instance, to double the amount, we initiate with $X_1 = 0.134$. The results are a fast drop in X_1, slight overshoots in X_2 and X_3, and an eventual return to the original steady state. As an alternative, we may simulate a more realistic scenario. We initiate the system at its steady state, as what one might call a "pre-experiment" of 1 min. The actual experiment, namely addition of a bolus of glucose-1-phosphate, begins at minute 1. This strategy may be implemented in PLAS by adding to the first equation a term

   +Bolus

and including the commands

   Bolus = 0
   @1Bolus = 67
   @1.001Bolus = 0

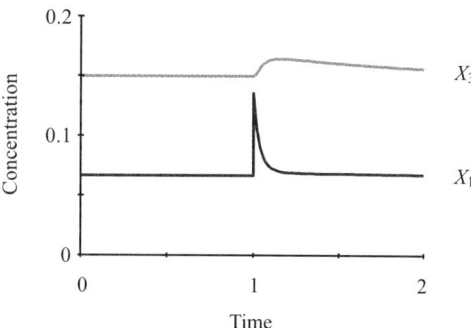

Figure 11.2. System response to a bolus injection of glucose-1-phosphate at time $t = 1$. X_1 rapidly drops back to the original steady state, while X_3 exhibits a slower response with slight overshoot. X_2 is not shown, because its values are outside the scale. Other settings: $t0 = 0$, $tf = 2$, $hr = 0.001$.

The size and duration of the bolus are chosen so that their product reflects the total amount of added glucose-1-phosphate. The specifications above $[67 \times (1.001 - 1) = 0.067]$ again reflect a doubling of the steady-state amount. The results (shown in Fig. 11.2) are not 100% equivalent with the results above but very similar (see Exercises 7 and 8).

Response to an altered independent variable. The first simulation has shown that the system responds to a one-time perturbation in a dependent variable with a fast increase in concentrations and subsequent return to the previous steady state. The response to a persistent change in an independent variable is quite different. Suppose we want to model the system response to a doubled glucose input. As in the previous simulation, we initialize the simulation at the steady state for 1 min before the change. At time 1, we increase the independent variable X_5 (permanently) from 5 to 10, while we leave all dependent variables and the other independent variables unaltered. Thus, we specify

@ 1 X5 = 10

As a final time, we select $t = 4$. All dependent variables start to increase at time 1, with quite a noticeable rate of change in the beginning and an apparent saturation toward later times (see Fig. 11.3). There are no overshoots, and it seems that the system will not return to the original steady state. Will it reach a new steady state? We have two ways to check. We may extend the solution to a larger time to see whether the system will eventually settle down. Or we may define $X_5 = 10$ in PLAS and

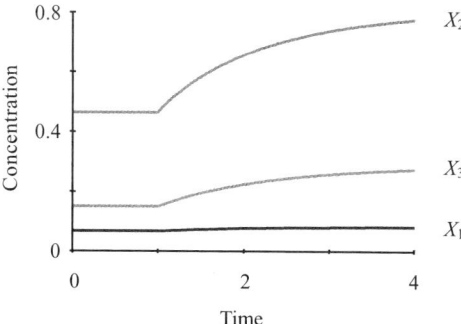

Figure 11.3. System response to a doubled glucose input at time 1.

CASE STUDY 4 – ALGEBRAIC ANALYSIS

use the steady-state command, which produces $X_{1S} = 0.0830575$, $X_{2S} = 0.81173$, $X_{3S} = 0.289719$, where the subscript S indicates that these are steady-state values. Both options confirm that the system has a new steady state. (For the steady-state computation, PLAS ignores the statement @ 1 X5=10 and shows the original steady state based on the initialization with $X_5 = 5$.) Exercise 9 analyzes more changes in independent variables.

Issues of robustness: Changes in parameter values. As an illustration, we test the ramifications of the assumption $h_{52} = -0.1$, which was made to reflect the slight inhibitory effect of glucose-6-phosphate on the phosphorylation of glucose. The chosen parameter value also affects the parameter g_{22} through the constraint at the branch point X_2. What are the consequences of the assumption concerning h_{52}? Because X_5 is not a dependent variables, h_{52} does not explicitly appear in the system equations. As a substitute, we check the sensitivity of the steady state with respect to g_{22} and find that all sensitivities of dependent variables and fluxes with respect to g_{22} are below 1:

$$S(X_1, g_{22}) = 0.3040846,$$
$$S(X_2, g_{22}) = 0.7885583,$$
$$S(X_3, g_{22}) = 0.9317192.$$

Thus, any small increase in g_{22} will lead to relative changes in the dependent variables that are slightly attenuating the effect.

As an alternative, we may just use different values of h_{52} and study the effect. Using Eq. (11.11) with $V_1^+ = 1.068$, $V_2^+ = 1.714$, $V_5^- = 0.646$, and $h_{12} = -0.59$, we obtain the following. If we select $h_{52} = -0.5$, thereby reflecting stronger inhibition, then g_{22} increases in magnitude from -0.41 to -0.56. If $h_{52} = 0$, representing no inhibition at all, then $g_{22} = -0.37$.

Changing the value of g_{22}, of course, alters the value of the flux V_2^+, which would lead to a disparity with the observed flux value. This is compensated by adjusting the rate constant according to the formula

$$\tilde{\alpha}_2 = \alpha_2 X_{2S}^{g_{22}(old) - g_{22}(new)} \tag{11.19}$$

(the proof of this assertion is left as Exercise 10). For example, for the case $h_{52} = -0.5$, the new rate constant is

$$\tilde{\alpha}_2 = \alpha_2 X_{2S}^{-0.41+0.56} = 0.585012402 \times 0.465^{0.15} = 0.521534468. \tag{11.20}$$

Just to make sure, we change g_{22} and α_2 in the model *Gluco1.plc*, and check that the steady state is correct. We also call up the eigenvalues and find that again all real parts are negative. The compensating adjustment of α_2 follows the philosophy of *controlled mathematical comparisons*, which maintains that comparisons of two alternative system designs are only fair if the two systems are externally as equivalent as possible. The foremost criterion of external equivalence is usually that the two systems have the same steady state. For more details on controlled mathematical comparisons, see Irvine and Savageau (1985ab), Savageau (1985b), Savageau and Sands (1991), Hlavacek and Savageau (1995–1997), and Chapter 12.

Similar results are obtained for $g_{22} = -0.37$ and, correspondingly, $\alpha_2 = 0.603208$. Is the system always stable, no matter what g_{22}? Out of curiosity, we try $g_{22} = 0$ and adjust the rate constant so that the steady state is correct. Checking the eigenvalues, we find indeed that the system has become unstable. Further exploration demonstrates that the system loses stability somewhere between $g_{22} = -0.01$ and $g_{22} = -0.02$ (confirm this in PLAS). What does this imply about h_{52}? Solving Eq. (11.11) for h_{52} and assuming that everything else remains unchanged, we obtain for $g_{22} = -0.01$

$$h_{52} = \frac{V_2^+}{V_5^-}\left(g_{22} - \frac{V_1^-}{V_2^+}h_{12}\right) = 1.002. \qquad (11.21)$$

The system becomes unstable only if the effect of glucose-6-phosphate on phosphorylation is rather strongly *activating* instead of *inhibiting*.

In addition to considerations of stability, one could study the effect of changes in h_{52} on other systems responses. For instance, one may study the responses to doubling glucose input. For $X_5 = 10$, the systems with different g_{22} and adjusted rate constants do reach different steady states, but for reasonable inhibition values of h_{52}, the system responses are quite similar (see Exercises 11–13).

SYMBOLIC STEADY-STATE ANALYSIS

Steady-State Equations

For more theoretical analyses, such as a general investigation of logarithmic gains and sensitivities, we need equations characterizing the steady state in an explicit algebraic form. These equations can be studied with respect to their structure and algebraic features, or they can be used for numerical evaluations, upon substitution of numerical values. None of these analyses require much computing power. Indeed, my sons in middle and high school executed all of the following matrix computations on their pocket calculators.

To compute the steady state, we set the three symbolic system equations equal to zero and divide each equation by its α_i rate constant. The result is

$$0 = X_4^{g_{14}} X_6^{g_{16}} - \frac{\beta_1}{\alpha_1} X_1^{h_{11}} X_2^{h_{12}} X_7^{h_{17}},$$

$$0 = X_1^{g_{21}} X_2^{g_{22}} X_5^{g_{25}} X_7^{g_{27}} X_{10}^{g_{2\,10}} - \frac{\beta_2}{\alpha_2} X_2^{h_{22}} X_3^{h_{23}} X_8^{h_{28}}, \qquad (11.22)$$

$$0 = X_2^{g_{32}} X_3^{g_{33}} X_8^{g_{38}} - \frac{\beta_3}{\alpha_3} X_3^{h_{33}} X_9^{h_{39}}.$$

Next, we add the negative term in each equation to both sides and define $y_i = \ln X_i$ for $i = 1, 2, \ldots, 10$ and $b_i = \ln(\beta_i/\alpha_i)$ for $i = 1, 2, 3$. When we recall that power

CASE STUDY 4 – ALGEBRAIC ANALYSIS

functions can be expressed in terms of exponential and logarithmic functions as $X_i^{g_{ij}} = \exp(g_{ij} \ln X_j)$, we obtain

$$\begin{aligned}
&\exp(b_1)\exp(h_{11}y_1)\exp(h_{12}y_2)\exp(h_{17}y_7) = \exp(g_{14}y_4)\exp(g_{16}y_6), \\
&\exp(b_2)\exp(h_{22}y_2)\exp(h_{23}y_3)\exp(h_{28}y_8) \\
&\quad = \exp(g_{21}y_1)\exp(g_{22}y_2)\exp(g_{25}y_5)\exp(g_{27}y_7)\exp(g_{2\,10}y_{10}), \\
&\exp(b_3)\exp(h_{33}y_3)\exp(h_{39}y_9) = \exp(g_{32}y_2)\exp(g_{33}y_3)\exp(g_{38}y_8).
\end{aligned} \quad (11.23)$$

These equations become much simpler when we take logarithms on both sides, which reduces the products to sums and the exponential functions to their arguments. The linear steady-state equations are

$$\begin{aligned}
&b_1 + h_{11}y_1 + h_{12}y_2 + h_{17}y_7 = g_{14}y_4 + g_{16}y_6, \\
&b_2 + h_{22}y_2 + h_{23}y_3 + h_{28}y_8 = g_{21}y_1 + g_{22}y_2 + g_{25}y_5 \\
&\qquad\qquad\qquad\qquad\qquad\qquad\qquad + g_{27}y_7 + g_{2\,10}y_{10}, \\
&b_3 + h_{33}y_3 + h_{39}y_9 = g_{32}y_2 + g_{33}y_3 + g_{38}y_8.
\end{aligned} \quad (11.24)$$

These linear equations are equivalently represented in the form of a matrix equation. Since the system contains independent variables, we can split up this matrix equation into parts associated exclusively with dependent variables and parts relating dependent to independent variables. Recalling results from Chapter 6, the general form of this matrix equation is

$$\mathbf{A}_D \vec{y}_D + \mathbf{A}_I \vec{y}_I = \vec{b}. \quad (11.25)$$

Recall that $\mathbf{A}_D$ is the matrix that contains as elements a_{ij} the differences between the g_{ij}'s and h_{ij}'s that correspond to dependent variables and that $\mathbf{A}_I$ contains as elements the differences between kinetic orders with respect to independent variables. In our particular example, Eq. (11.25) takes the form

$$\begin{pmatrix} -h_{11} & -h_{12} & 0 \\ g_{21} & g_{22}-h_{22} & -h_{23} \\ 0 & g_{32} & g_{33}-h_{33} \end{pmatrix} \begin{pmatrix} y_1 \\ y_2 \\ y_3 \end{pmatrix}$$

$$+ \begin{pmatrix} g_{14} & 0 & g_{16} & -h_{17} & 0 & 0 & 0 \\ 0 & g_{25} & 0 & g_{27} & -h_{28} & 0 & g_{2,10} \\ 0 & 0 & 0 & 0 & g_{38} & -h_{39} & 0 \end{pmatrix} \begin{pmatrix} y_4 \\ y_5 \\ y_6 \\ y_7 \\ y_8 \\ y_9 \\ y_{10} \end{pmatrix} = \begin{pmatrix} b_1 \\ b_2 \\ b_3 \end{pmatrix}.$$

(11.26)

We easily convince ourselves that this is the appropriate matrix equation. For instance, the first linear equation represented in this matrix equation is obtained by multiplying the first rows of the two matrices with the corresponding y-vectors, the result of which equals b_1 (see Appendix). Executing this operation, we obtain

$$-h_{11}y_1 - h_{12}y_2 + g_{14}y_4 + g_{16}y_6 - h_{17}y_7 = b_1, \tag{11.27}$$

which is equivalent to the first equation in (11.24).

As discussed in Chapter 6 and the Appendix, we can use the matrix inverse $\mathbf{A}_D^{-1}$ to express the solution in terms of coefficients and independent variables. The solution is formally given as

$$\vec{y}_D = \mathbf{A}_D^{-1}\vec{b} - \mathbf{A}_D^{-1}\mathbf{A}_I\vec{y}_I. \tag{11.28}$$

Evaluation of $\mathbf{A}_D^{-1}$ requires us to compute the 3×3 determinant $|\mathbf{A}_D|$ and the 2×2 determinants $|\mathbf{A}_{ij}|$ which are obtained from the matrix $\mathbf{A}_D$ by eliminating the ith row and the jth column. These determinants, divided by $|\mathbf{A}_D|$ and multiplied by $+1$ or -1, depending on their position, become the entries (j, i) of $\mathbf{A}_D^{-1}$ (see Appendix).

All required determinants are computed in a straightforward fashion, as shown in the Appendix. The symbolic results, along with numerical values for our example, are

$$\mathbf{A}_D = \begin{pmatrix} -1.53 & 0.59 & 0 \\ 0.95 & -4.38 & 3.06 \\ 0 & 3.97 & -3.36 \end{pmatrix}, \tag{11.29a}$$

$$\mathbf{A}_I = \begin{pmatrix} 0.66 & 0 & 1 & -1 & 0 & 0 & 0 \\ 0 & 0.32 & 0 & 0.62 & -1 & 0 & 0.38 \\ 0 & 0 & 0 & 0 & 1 & -1 & 0 \end{pmatrix}, \tag{11.29b}$$

$$|\mathbf{A}_D| = -h_{11}(g_{22} - h_{22})(g_{33} - h_{33}) - h_{11}g_{32}h_{23} + h_{12}g_{21}(g_{33} - h_{33}) = -2.047, \tag{11.29c}$$

$$|\mathbf{A}_{11}| = (g_{22} - h_{22})(g_{33} - h_{33}) + g_{32}h_{23} = 2.569, \tag{11.29d}$$

$$|\mathbf{A}_{12}| = g_{21}(g_{33} - h_{33}) = -3.192, \tag{11.29e}$$

$$|\mathbf{A}_{13}| = g_{21}g_{32} = 3.772, \tag{11.29f}$$

$$|\mathbf{A}_{21}| = -h_{12}(g_{33} - h_{33}) = -1.982, \tag{11.29g}$$

$$|\mathbf{A}_{22}| = -h_{11}(g_{33} - h_{33}) = 5.141, \tag{11.29h}$$

$$|\mathbf{A}_{23}| = -h_{11}g_{32} = -6.074, \tag{11.29i}$$

$$|\mathbf{A}_{31}| = h_{12}h_{23} = 1.805, \tag{11.29j}$$

$$|\mathbf{A}_{32}| = h_{11}h_{23} = -4.682, \tag{11.29k}$$

$$|\mathbf{A}_{33}| = -h_{11}(g_{22} - h_{22}) + h_{12}g_{21} = 6.141, \tag{11.29l}$$

$$\mathbf{A}_D^{-1} = \frac{1}{|\mathbf{A}_D|}\begin{pmatrix} +|\mathbf{A}_{11}| & -|\mathbf{A}_{21}| & +|\mathbf{A}_{31}| \\ -|\mathbf{A}_{12}| & +|\mathbf{A}_{22}| & -|\mathbf{A}_{32}| \\ +|\mathbf{A}_{13}| & -|\mathbf{A}_{23}| & +|\mathbf{A}_{33}| \end{pmatrix} = \begin{pmatrix} -1.255 & -0.969 & -0.882 \\ -1.560 & -2.512 & -2.288 \\ -1.843 & -2.968 & -3.000 \end{pmatrix}. \tag{11.29m}$$

CASE STUDY 4 – ALGEBRAIC ANALYSIS

Substitution of numerical values in Eq. (11.28) yields

$$\vec{y}_D = \begin{pmatrix} -1.255 & -0.969 & -0.882 \\ -1.560 & -2.512 & -2.288 \\ -1.843 & -2.968 & -3.000 \end{pmatrix} \begin{pmatrix} 2.613 \\ -6.603 \\ 7.196 \end{pmatrix}$$

$$- \begin{pmatrix} -1.255 & -0.969 & -0.882 \\ -1.560 & -2.512 & -2.288 \\ -1.843 & -2.968 & -3.000 \end{pmatrix}$$

$$\times \begin{pmatrix} 0.66 & 0 & 1 & -1 & 0 & 0 & 0 \\ 0 & 0.32 & 0 & 0.62 & -1 & 0 & 0.38 \\ 0 & 0 & 0 & 0 & 1 & -1 & 0 \end{pmatrix} \begin{pmatrix} \ln 10 \\ \ln 5 \\ \ln 3 \\ \ln 40 \\ \ln 136 \\ \ln 2.86 \\ \ln 4 \end{pmatrix}$$

$$= \begin{pmatrix} -3.231 \\ -3.951 \\ -6.810 \end{pmatrix}$$

$$+ \begin{pmatrix} 0.828 & 0.310 & 1.255 & -0.654 & -0.086 & -0.882 & 0.368 \\ 1.029 & 0.804 & 1.560 & -0.002 & -0.224 & -2.288 & 0.954 \\ 1.216 & 0.950 & 1.843 & -0.003 & 0.033 & -3.000 & 1.128 \end{pmatrix} \begin{pmatrix} 2.303 \\ 1.609 \\ 1.099 \\ 3.689 \\ 4.913 \\ 1.051 \\ 1.386 \end{pmatrix}$$

$$= \begin{pmatrix} -2.702 \\ -0.765 \\ -1.896 \end{pmatrix}. \tag{11.30}$$

The steady-state solution in Cartesian coordinates, $X_1 = 0.067$, $X_2 = 0.465$, $X_3 = 0.15$, is obtained through exponentiation of each number in the last vector. Like $\vec{y}_D$, the solution can also be written as a vector:

$$\vec{X}_D = \begin{pmatrix} 0.067 \\ 0.465 \\ 0.150 \end{pmatrix}.$$

This solution, of course, is nothing new *per se*. What's new is that we can now use the algebraic solution for structural analyses of the model that are independent of particular parameter values. Detailed examples for such analyses can be found throughout the literature, e.g., in Savageau (1976), Irvine and Savageau (1985ab), Sorribas and Cascante (1994), Hlavacek and Savageau (1995, 1996, 1997), and Puigjaner et al. (1995).

Logarithmic Gains of Metabolites

The explicit steady-state solution allows us to analyze gains. These are obtained directly in symbolic or numerical form, since we found in Chapter 7 that all logarithmic gains of metabolites are given as

$$L(X_j, X_k) = (-1) \cdot \left[\text{element } (j, k) \text{ of } \mathbf{A}_D^{-1}\mathbf{A}_I\right] \tag{11.31}$$

or, summarized in the form of a log gain matrix, as

$$\mathbf{L}(\mathbf{X}_D, \mathbf{X}_I) = -\mathbf{A}_D^{-1}\mathbf{A}_I. \tag{11.32}$$

For our numerical example, the log gains of metabolites are thus

$$\mathbf{L}(\mathbf{X}_D, \mathbf{X}_I) = -\begin{pmatrix} -1.255 & -0.969 & -0.882 \\ -1.560 & -2.512 & -2.288 \\ -1.843 & -2.968 & -3.000 \end{pmatrix}$$

$$\times \begin{pmatrix} 0.66 & 0 & 1 & -1 & 0 & 0 & 0 \\ 0 & 0.32 & 0 & 0.62 & -1 & 0 & 0.38 \\ 0 & 0 & 0 & 0 & 1 & -1 & 0 \end{pmatrix}$$

$$\begin{array}{c} \\ = \begin{array}{c} X_1 \\ X_2 \\ X_3 \end{array} \end{array} \begin{array}{cccccc} X_4 & X_5 & X_6 & X_7 & X_8 & X_9 & X_{10} \\ \begin{pmatrix} 0.828 & 0.310 & 1.255 & -0.654 & -0.086 & -0.882 & 0.368 \\ 1.029 & 0.804 & 1.560 & -0.002 & -0.224 & -2.288 & 0.954 \\ 1.216 & 0.950 & 1.843 & -0.003 & 0.033 & -3.000 & 1.128 \end{pmatrix}\end{array}.$$

We have included in this representation the names of the dependent and independent variables to facilitate the recognition of particular gains. For instance, if the concentration of P_i (independent variable X_4) is increased by 5%, glucose-6-phosphate (dependent variable X_2) will increase by about $1.029 \times 5\% = 5.145\%$, whereas a 5% increase in the concentration of phosphoglucomutase (independent variable X_7) results in about a $-0.654 \times 5\% = -3.27\%$ increase, which means a +3.27% decrease, in glucose-1-phosphate (dependent variable X_1) and has almost no effect on glucose-6-phosphate (X_2) and fructose-6-phosphate (X_3), because the corresponding gains are close to zero. Exact predictions of changes in steady-state values are obtained by studying changes in logarithmic coordinates (see Chapter 7).

It is important to realize that the matrix $\mathbf{A}_D$ is independent of all properties associated with independent variables. Thus, if a different experimental set-up affects only the kinetic orders of some of the independent variables, the effects of these changes are easy to analyze, because the hard part in the above computation was the inversion of $\mathbf{A}_D$. For instance, if through the introduction of a mutation the kinetic order of the phosphorylase reaction with respect to P_i, g_{14}, is halved from 0.66 to 0.33 while the rate of the reaction is doubled ($\alpha_1 = 0.155768628$), the systemic effects only appear in one element of $\mathbf{A}_I$ and in b_1. In particular, the logarithmic gain matrix only

CASE STUDY 4 – ALGEBRAIC ANALYSIS

changes in the first column, whereas the rest remains unaffected:

$$L(X_D, X_I) = - \begin{pmatrix} -1.255 & -0.969 & -0.882 \\ -1.560 & -2.512 & -2.288 \\ -1.843 & -2.968 & -3.000 \end{pmatrix}$$

$$\times \begin{pmatrix} 0.33 & 0 & 1 & -1 & 0 & 0 & 0 \\ 0 & 0.32 & 0 & 0.62 & -1 & 0 & 0.38 \\ 0 & 0 & 0 & 0 & 1 & -1 & 0 \end{pmatrix}$$

$$= \begin{matrix} & X_4 & X_5 & X_6 & X_7 & X_8 & X_9 & X_{10} \\ X_1 \\ X_2 \\ X_3 \end{matrix} \begin{pmatrix} 0.414 & 0.310 & 1.255 & -0.654 & -0.086 & -0.882 & 0.368 \\ 0.515 & 0.804 & 1.560 & -0.002 & -0.224 & -2.288 & 0.954 \\ 0.608 & 0.950 & 1.843 & -0.003 & 0.033 & -3.000 & 1.128 \end{pmatrix}.$$

(11.33)

Logarithmic Gains of Fluxes

Changes in independent variables usually result in changes in fluxes. The magnitudes of these changes are characterized by logarithmic gains of fluxes. For instance, the expression $L(V_2^+, X_4)$ represents the effect of phosphate $(P_i; X_4)$ on the synthesis V_2^+ of glucose-6-phosphate (X_2). As was discussed in Chapter 7, gains of fluxes can be computed from the matrices of kinetic order parameters g_{ij} that are associated strictly with the dependent variables and of the logarithmic gains of metabolites, $L(X_j, X_k)$. Specifically, the matrix of flux gains can be expressed as

$$L(V^+, X) = \begin{pmatrix} g_{1,n+1} & g_{1,n+2} & \cdots & g_{1,n+m} \\ g_{2,n+1} & g_{2,n+2} & \cdots & g_{2,n+m} \\ \vdots & \vdots & & \vdots \\ g_{n,n+1} & g_{n,n+2} & \cdots & g_{n,n+m} \end{pmatrix} + \begin{pmatrix} g_{11} & g_{12} & \cdots & g_{1n} \\ g_{21} & g_{22} & \cdots & g_{2n} \\ \vdots & \vdots & & \vdots \\ g_{n1} & g_{n2} & \cdots & g_{nn} \end{pmatrix} L(X_D, X_I)$$

(11.34)

[cf. Eqs. (7.29), (7.30)].

Since we have already computed $L(X_D, X_I)$ in the previous section, the determination of flux gains is readily accomplished. We obtain through direct substitution

$$L(V^+, X) = \begin{pmatrix} 0.66 & 0 & 1 & 0 & 0 & 0 & 0 \\ 0 & 0.32 & 0 & 0.62 & 0 & 0 & 0.38 \\ 0 & 0 & 0 & 0 & 1 & 0 & 0 \end{pmatrix} + \begin{pmatrix} 0 & 0 & 0 \\ 0.95 & -0.41 & 0 \\ 0 & 3.97 & -3.06 \end{pmatrix}$$

$$\times \begin{pmatrix} 0.828 & 0.310 & 1.255 & -0.654 & -0.086 & -0.882 & 0.368 \\ 1.029 & 0.804 & 1.560 & -0.002 & -0.224 & -2.288 & 0.954 \\ 1.216 & 0.950 & 1.843 & -0.003 & 0.033 & -3.000 & 1.128 \end{pmatrix},$$

$$L(V^+, X) = \begin{matrix} & X_4 & X_5 & X_6 & X_7 & X_8 & X_9 & X_{10} \\ V_1^+ \\ V_2^+ \\ V_3^+ \end{matrix} \begin{pmatrix} 0.66 & 0 & 1 & 0 & 0 & 0 & 0 \\ 0.365 & 0.285 & 0.553 & -0.0008 & 0.01 & 0.1 & 0.338 \\ 0.365 & 0.285 & 0.553 & -0.0008 & 0.01 & 0.1 & 0.338 \end{pmatrix}.$$

(11.35)

Note that the first matrix is *not* identical to the matrix $\mathbf{A}_I$ in Eq. (11.29b). In this particular case it has a similar appearance because many h_{ij} associated with independent variables are zero. For the computation of flux gains, only the g_{ij} are to be considered.

The effects on the fluxes V_2^+ and V_3^+ are identical. The reason is that the flux V_2^+ equals the flux V_2^-, because we are dealing with a steady state, and that V_2^- equals V_3^+, because both terms represent the same process. Comparison with results obtained in PLAS shows that the numerical values are slightly off, which is due to rounding in the matrices to three decimal places. These numerical differences are not significant, since usually just the magnitudes of the gains and their relative sizes characterize the responsiveness and robustness of the model.

Another example of the use of the explicit representation of the steady state is the exploration of different input conditions. It is not difficult to see that only one or a few components in $\vec{y}_I$ change, whereas everything else remains unchanged, in the steady-state matrix equation and the logarithmic gain matrix.

For new input conditions, the steady-state equation is formally still

$$\vec{y}_D = \begin{pmatrix} -3.231 \\ -3.951 \\ -6.810 \end{pmatrix} + \mathbf{L}(\mathbf{X}_D, \mathbf{X}_I)\vec{y}_I, \tag{11.36}$$

where the first vector is the same product $\mathbf{A}_D^{-1}\vec{b}$ as before, $\mathbf{L}(\mathbf{X}_D, \mathbf{X}_I)$ is the old logarithmic gain matrix, and $\vec{y}_I$ is the new input vector. As a numerical example, let's suppose the glucose concentration, represented by X_5, is doubled from 5 to 10 mM. The only symbolic change occurs in the second component of the vector $\vec{y}_I$, and we obtain

$$\vec{y}_D = \mathbf{A}_D^{-1}\vec{b} + \mathbf{L}(\mathbf{X}_D, \mathbf{X}_I)\begin{pmatrix} \ln 10 \\ \ln \mathbf{10} \\ \ln 3 \\ \ln 40 \\ \ln 136 \\ \ln 2.86 \\ \ln 4 \end{pmatrix}$$

$$= \begin{pmatrix} -3.231 \\ -3.951 \\ -6.810 \end{pmatrix}$$

$$+ \begin{pmatrix} 0.828 & 0.310 & 1.255 & -0.654 & -0.086 & -0.882 & 0.368 \\ 1.029 & 0.804 & 1.560 & -0.002 & -0.224 & -2.288 & 0.954 \\ 1.216 & 0.950 & 1.843 & -0.003 & 0.033 & -3.000 & 1.128 \end{pmatrix} \begin{pmatrix} \ln 10 \\ \ln \mathbf{10} \\ \ln 3 \\ \ln 40 \\ \ln 136 \\ \ln 2.86 \\ \ln 4 \end{pmatrix}. \tag{11.37}$$

Matrix multiplication and addition of the two vectors produces the new steady state as

$$\vec{y}_D = \begin{pmatrix} -2.487 \\ -0.208 \\ -1.238 \end{pmatrix}, \quad \vec{X}_D = \begin{pmatrix} 0.083 \\ 0.812 \\ 0.290 \end{pmatrix}.$$

All the steady-state values are different from the original scenario, illustrating the connectivity of the system. The effect of the change in X_5 is absolutely and relatively different for the three dependent variables.

We can easily simulate the doubling of X_5 with PLAS and check how accurate these predictions are. A simulation also reveals how fast the new steady state is reached. While theoretically the new steady state is reached only after an infinite time period, a simulation can characterize the 90%, 95%, and 99% marks, which usually cannot be obtained from paper-and-pencil analyses. We have different options of implementation and show one that may be most representative. We solve the original system from $t = -1$ to $t = 10$ and initialize with the original steady-state values. The first part of this solution, from $t = -1$ to $t = 0$, is simply a check that the system is indeed at a steady state. Upon completion of this preparation phase, i.e., at time 0, the actual experiment starts. It is specified in PLAS as

@ 0 X5 = 10

Thus, for the duration of the actual experiment, i.e., between $t = 0$ and $t = 10$, the concentration of X_5 is doubled. As the step size we may choose $hr = 1$, which corresponds to having the metabolite concentrations reported every minute. A selection of results is given in Table 11.2, showing that the system approaches a solution similar to that predicted by the gain analysis. Is the difference due to insufficient time? Will the solution finally reach the predicted values? We can test these questions by extending the solution or by clicking the steady-state button. The numerically computed steady state is given in the last column of Table 11.2. It is still slightly different from the solution based on gains, but the differences are due to numerical inaccuracies in inverting the matrix and solving for the steady state in PLAS.

If X_5 is of particular interest, we can even express the steady-state solution as a *function* of the glucose input. For this purpose, we split up the above steady-state equation into those parts that directly depend on X_5 and those that do not. The logarithm of X_5 appears as the second element in the vector of the equation, but

Table 11.2. Simulated Responses after Doubling X_5

Variable	$t = 0$	Response						Steady State
		1	2	3	4	5	10	
X_1	0.067	0.0763	0.0800	0.0816	0.0823	0.0827	0.0830	0.0830575
X_2	0.465	0.6576	0.7387	0.7764	0.7945	0.8033	0.8115	0.81173
X_3	0.15	0.2242	0.2583	0.2745	0.2823	0.2861	0.2896	0.289719

nowhere else:

$$\vec{y}_D = \mathbf{A}_D^{-1}\vec{b} + \mathbf{L}(\mathbf{X}_D, \mathbf{X}_I) \begin{pmatrix} \ln 10 \\ \ln X_5 \\ \ln 3 \\ \ln 40 \\ \ln 136 \\ \ln 2.86 \\ \ln 4 \end{pmatrix}. \tag{11.38}$$

Clearly, the first term on the right-hand side is unaffected, and in the multiplication of the logarithmic gain matrix with the vector, only the second column is affected by X_5. Thus, we may write

$$\vec{y}_D = \begin{pmatrix} -3.231 \\ -3.951 \\ -6.810 \end{pmatrix}$$

$$+ \begin{pmatrix} 0.828 & 1.255 & -0.654 & -0.086 & -0.882 & 0.368 \\ 1.029 & 1.560 & -0.002 & -0.224 & -2.288 & 0.954 \\ 1.216 & 1.843 & -0.003 & 0.033 & -3.000 & 1.128 \end{pmatrix} \begin{pmatrix} \ln 10 \\ \ln 3 \\ \ln 40 \\ \ln 136 \\ \ln 2.86 \\ \ln 4 \end{pmatrix}$$

$$+ \begin{pmatrix} 0.310 \\ 0.804 \\ 0.950 \end{pmatrix} \ln X_5 \tag{11.39}$$

(see Exercise 16). Evaluating the matrix multiplication and adding the result to the first vector on the right-hand side, we obtain

$$\vec{y}_D = \begin{pmatrix} -3.201 \\ -2.058 \\ -3.424 \end{pmatrix} + \begin{pmatrix} 0.310 \\ 0.804 \\ 0.950 \end{pmatrix} \ln X_5. \tag{11.40}$$

The steady-state concentrations in Cartesian coordinates are computed by exponentiation. For instance, we find

$$X_{2S} = \exp(-2.058) X_5^{0.804} = 0.01277 X_5^{0.804}. \tag{11.41}$$

As a check, for $X_5 = 5$ the steady-state concentration of X_2 is 0.4658, which, within the accuracy of our computation, confirms earlier results. Equation (11.41) not only enables us to compute steady states for any glucose input X_5, but we now know the mathematical form of the general solution: All dependent variables at steady state are power functions of X_5, and each has its own parameters (see Exercise 25).

Transition Times

The transition time is the average time a molecule remains in the pathway (cf. Chapter 6). It characterizes in a simple manner an important aspect of the dynamics of the pathway, without requiring numerical solutions to differential equations. In a biotechnological setting, one is often interested in the productivity of a pathway and thus in minimizing the (*lag*) time it takes the system to reach a new steady state. For such a task, the transition time can be a powerful criterion of optimality (e.g., Torres, 1994d).

The transition time is defined for our system as

$$\tau = \frac{\sigma}{V_3^+}, \tag{11.42}$$

where

$$\sigma = X_1 + X_2 + X_3, \tag{11.43}$$

which is evaluated at the steady state (Easterby, 1981). In our example, the transition time is 0.398 min.

Using the results of the previous section, we can express σ as a function of the independent variable X_5. Simple substitution of numerical results yields

$$\sigma = \exp(-3.201)X_5^{0.310} + \exp(-2.058)X_5^{0.804} + \exp(-3.424)X_5^{0.950}. \tag{11.44}$$

To compute the transition time, we must express V_3^+ in terms of X_5. Since V_3^+ and V_3^- are equivalent at steady state, we use V_3^-, which has only two instead of three terms:

$$V_3^- = 1.0588\left[\exp(-3.424)X_5^{0.95}\right]^{0.3} \cdot X_9. \tag{11.45}$$

The independent variable X_9 has a value of 2.86, and we obtain

$$\tau = \frac{\exp(-3.201)X_5^{0.310} + \exp(-2.058)X_5^{0.804} \exp(-3.424)X_5^{0.950}}{1.084 X_5^{0.285}}$$

$$= 0.0376 X_5^{0.025} + 0.1178 X_5^{0.519} + 0.0301 X_5^{0.665}. \tag{11.46}$$

This equation expresses the transition time as a function of glucose input. The function turns out to be a fairly complicated sum of power-law functions (see Exercise 17).

With the definition of the appropriate logarithmic gain, one can explore how a change in an independent variable affects the transition time. In analogy with other logarithmic gains, the gain of the transition time is defined as

$$L(\tau, X_k) = \frac{1}{\tau} \cdot \frac{\partial \tau}{\partial y_k} \tag{11.47}$$

(see also Cascante et al., 1991b; Torres, 1994d; and Exercise 18).

As shown in Chapter 7, the vector of logarithmic gains of transition times can be computed from gain matrices of metabolites and fluxes, according to the formula

$$L(\tau, X_I) = -L(V_3^+, X_I) + \left(\frac{X_1}{\sigma}, \frac{X_2}{\sigma}, \frac{X_3}{\sigma}\right) L(X_D, X_I)$$

$$= -\begin{bmatrix} & X_4 & X_5 & X_6 & X_7 & X_8 & X_9 & X_{10} \\ V_3^+ & (0.365 & 0.285 & 0.553 & -0.0008 & 0.01 & 0.1 & 0.338) \end{bmatrix}$$

$$+ \left(\frac{0.067}{0.682} \quad \frac{0.465}{0.682} \quad \frac{0.15}{0.682}\right)$$

$$\times \begin{bmatrix} & X_4 & X_5 & X_6 & X_7 & X_8 & X_9 & X_{10} \\ X_1 & \begin{pmatrix} 0.828 & 0.310 & 1.255 & -0.654 & -0.086 & -0.882 & 0.368 \\ X_2 & 1.029 & 0.804 & 1.560 & -0.002 & -0.224 & -2.288 & 0.954 \\ X_3 & 1.216 & 0.950 & 1.843 & -0.003 & 0.033 & -3.000 & 1.128 \end{pmatrix} \end{bmatrix}$$

$$= \begin{bmatrix} & X_4 & X_5 & X_6 & X_7 & X_8 & X_9 & X_{10} \\ \tau & (0.686 & 0.503 & 1.039 & -0.065 & -0.164 & -2.406 & 0.597) \end{bmatrix}.$$

(11.48)

For instance, a 5% increase in the concentration of $P_i(X_4)$ extends the transition time by about 3.43%, from 0.3979 to 0.4115 min. This result can be checked in PLAS by first computing the original transition τ_{ori}; secondly the new transition time τ_{new}, which is obtained from the steady-state solution upon increasing P_i by 5%; and finally the relative change as $(\tau_{ori} - \tau_{new})/\tau_{ori}$. The result is about 3.4%, which is quite similar to the prediction above. As is typical, different independent variables affect the transition time to a different degree. For example, a 5% increase in phosphoglucomutase activity (X_7) evokes a very slight decrease of about 0.3%.

It is interesting to note that phosphofructokinase (X_9) has the strongest effect on the transition time, whereas the effect of glucokinase (X_{10}) is small. This is opposite to their effects on the magnitude of the output flux $V_3^+ = V_3^-$, as indicated by their logarithmic flux gains [see. Eq. (11.35)]. The same phenomenon was described by Cascante et al. (1991b) for a similar glycolytic pathway without glycogenolysis. Torres (1994d) confirmed experimentally the accelerating effect of increased phosphofructokinase activity.

Again, the gain of the transition time can be expressed as a function of one of the independent variables. For instance, the gain in terms of X_5 takes the form

$$L(\tau, X_5) = -0.285 + \frac{1}{\sigma}(0.01262 X_5^{0.310} + 0.10268 X_5^{0.804} + 0.02949 X_5^{0.950})$$

(11.49)

(see Exercise 19). And again, the gain is a combination of power-law functions in the independent variable.

Rate Constant Sensitivities

The matrix A_D^{-1} can be interpreted as a matrix containing all parameter sensitivities with respect to rate constants, since we showed in Chapter 7 that

$$A_D^{-1} = M.$$

(11.50)

CASE STUDY 4 – ALGEBRAIC ANALYSIS

Each element M_{ij} of $\mathbf{M}$ is equivalent to a sensitivity with respect to β_j and to the negative of the corresponding sensitivity with respect to α_j, under the assumption that all rate constants are independent of each other.

When the assumption of independence is not satisfied, we have to be careful in interpreting the numerical values of the elements of $\mathbf{M}$ as sensitivities. For instance, because of the precursor–product relationship between X_2 and X_3, α_3 is normally equal to β_2, as we discussed during the estimation of parameter values. If we ignore these constraints, we claim, in effect, that we can change the degradation reaction of glucose-6-phosphate (X_2) without altering the production of fructose-6-phosphate (X_3). However, since all gain and sensitivity computations are based on linear algebra, the sensitivities can be computed from the matrix $\mathbf{M}$ even if constraints exist, as was shown in Chapter 7. We illustrate sensitivity computations with three examples, one not affected by constraints, and the other two allowing for the precursor–product relationship between X_2 and X_3 and for the conservation of flux at the branch point X_2, respectively.

Example 1: Unconstrained parameters. Suppose we are interested in effects of kinetic alterations in the phosphofructokinase reaction. Specifically, let's study the effects of small changes in β_3. This parameter is not constrained, and we can read its effect directly off the third column of matrix $\mathbf{M}$, which is equal to $\mathbf{A}_D^{-1}$ and thus is [cf. (11.29) and (11.30)]

$$\mathbf{M} = \begin{pmatrix} -1.255 & -0.969 & -0.882 \\ -1.560 & -2.512 & -2.288 \\ -1.843 & -2.968 & -3.000 \end{pmatrix}. \tag{11.51}$$

The sensitivities of the dependent metabolites with respect to changes in rate constant β_3 are shown in the third column of $\mathbf{M}$. They are

$S(X_1, \beta_3) = -0.882,$
$S(X_2, \beta_3) = -2.288,$
$S(X_3, \beta_3) = -3.000.$

(Check these in PLAS.) If β_3 is increased by 3%, then X_1, X_2, and X_3 are predicted to decrease by about 2.6%, 7%, and 9%, respectively.

Example 2: Precursor–product relationships. Now let's investigate a change of rate in the phosphoglucose isomerase reaction. In this case, β_2 and α_3 are simultaneously affected to the same degree. The corresponding unconstrained sensitivities, obtained from $\mathbf{M}$, are

$S(X_1, \beta_2)_U = -0.969,\quad S(X_1, \alpha_3)_U = 0.882,$
$S(X_2, \beta_2)_U = -2.512,\quad S(X_2, \alpha_3)_U = 2.288,$
$S(X_3, \beta_2)_U = -2.968,\quad S(X_3, \alpha_3)_U = 3.000,$

where the subscript U indicates that we are dealing with unconstrained sensitivities. The effects of a simultaneous change of 2% in β_2 and α_3 on each dependent variable are given as 2% times the sum of the appropriate sensitivities; thus,

$$S(X_1, \beta_2 \text{ and } \alpha_3) = 2\% \cdot [S(X_1, \beta_2)_U + S(X_1, \alpha_3)_U] = -0.174\%,$$
$$S(X_2, \beta_2 \text{ and } \alpha_3) = 2\% \cdot [S(X_2, \beta_2)_U + S(X_2, \alpha_3)_U] = -0.448\%, \quad (11.52)$$
$$S(X_2, \beta_2 \text{ and } \alpha_3) = 2\% \cdot [S(X_3, \beta_2)_U + S(X_3, \alpha_3)_U] = +0.064\%.$$

The predicted sensitivities can be compared with simulation results in PLAS. First, we check the original steady state as $(0.067, 0.465, 0.15)$, then the steady state upon increasing both β_2 and α_3 by 2%, which yields $(0.0668854, 0.46294, 0.150097)$. The relative changes are computed as the differences, divided by the original values. The simulated changes are

(relative change in X_1) = -0.171%,
(relative change in X_2) = -0.443%,
(relative change in X_3) = $+0.065\%$,

which is consistent with the predictions based on sensitivities.

Why are we allowed simply to add the unconstrained sensitivities? To justify this step, let's go back to the original definition of the sensitivity of a metabolite with respect to a rate constant, which is given as

$$S(X_i, \alpha_j) = \frac{\partial \ln X_i}{\partial \ln \alpha_j} = \frac{\partial y_i}{\partial \ln \alpha_j} \quad (11.53)$$

and

$$S(X_i, \beta_j) = \frac{\partial \ln X_i}{\partial \ln \beta_j} = \frac{\partial y_i}{\partial \ln \beta_j}, \quad (11.54)$$

where X_i is a dependent variable. Computation of rate constant sensitivities requires the steady-state solution in algebraic form. Studying the steady-state equation (11.28), we notice that β_2 and α_3 only appear in the vector $\vec{b}$, and it is not difficult to see that the entire second part of the equation, $\mathbf{A}_D^{-1}\mathbf{A}_I\vec{y}_I$, which represents the effects of independent variables, disappears during differentiation with respect to β_2 and α_3. The constraint $\beta_2 = \alpha_3$ is made explicit when we substitute $\ln \beta_i - \ln \alpha_i$ for b_i in the first part of the steady-state equation and replace α_3 with β_2:

$$\mathbf{A}_D^{-1}\vec{b} = \begin{pmatrix} M_{11} & M_{12} & M_{13} \\ M_{21} & M_{22} & M_{23} \\ M_{31} & M_{32} & M_{33} \end{pmatrix} \begin{pmatrix} \ln \beta_1 - \ln \alpha_1 \\ \ln \beta_2 - \ln \alpha_2 \\ \ln \beta_3 - \ln \beta_2 \end{pmatrix}. \quad (11.55)$$

For simplicity in notation, we have also substituted $\mathbf{M}$ for $\mathbf{A}_D^{-1}$. When we differentiate one of the dependent variables (in logarithmic form), say y_1, with respect to $\ln \beta_2$, we use the first of the three steady-state equations, namely

$$y_1 = M_{11}(\ln \beta_1 - \ln \alpha_1) + M_{12}(\ln \beta_2 - \ln \alpha_2) + M_{13}(\ln \beta_3 - \ln \beta_2)$$
$$\times (\text{—the first element of } \mathbf{A}_D^{-1}\mathbf{A}_I\vec{y}_I, \text{ which is independent of } \beta_2). \quad (11.56)$$

CASE STUDY 4 – ALGEBRAIC ANALYSIS

Differentiation yields

$$\frac{\partial y_1}{\partial \ln \beta_2} = M_{12} + (-1)M_{13}, \quad (11.57)$$

which justifies our previous computation.

PLAS can be tricked into computing these constrained sensitivities. To this end, the numerical values of β_2 and α_3 are replaced with the same symbolic name, e.g., *beta2*, and this new symbol is numerically specified and declared as an independent variable with the statements

beta2 = 0.000793456

&& beta2

PLAS now treats the rate constant *beta2* like any other independent variable, and computation of its logarithmic gain corresponds to the constrained sensitivity. (Check this in PLAS.)

Example 3: Branch point constraints. The situation is somewhat more complicated for sensitivities at branch points. For instance, the synthesis of glucose-6-phosphate (X_2) is driven by the conversion of glucose-1-phosphate (X_1) and of glucose (X_5), which in our experimental set-up is held saturated. As was stated during the estimation of parameter values, the fluxes at the branch point must satisfy the constraint

$$V_2^+ = V_1^- + V_5^-. \quad (11.58)$$

In contrast to the previous example, β_1 and α_2 are not equal, but α_2 is the average of β_1 and β_5 weighted with the magnitudes of the two fluxes V_1^- and V_5^-. This weighting translates directly into the corresponding sensitivities. To see why that is so, let's look at the sensitivity $S(X_1, \beta_1)$, which is defined as

$$S(X_1, \beta_1) = \frac{\partial \ln X_1}{\partial \ln \beta_1} = \frac{\partial y_1}{\partial \ln \beta_1}. \quad (11.59)$$

This statement is true, but does not explicitly show that β_1 is related to α_2 and that we have to take account of this dependence. The situation becomes clearer when we consider, as in the previous Example 2, the steady-state equation

$$y_1 = M_{11}(\ln \beta_1 - \ln \alpha_1) + M_{12}(\ln \beta_2 - \ln \alpha_2) + M_{13}(\ln \beta_3 - \ln \beta_2)$$
$$\times \left(- \text{ the first element of } \mathbf{A}_D^{-1}\mathbf{A}_I \vec{y}_I, \text{ which is independent of } \beta_1 \text{ and } \alpha_2 \right). \quad (11.60)$$

This time, we must focus on β_1 and α_2. Differentiation with respect to β_1 consists of two parts: One deals with the expression $M_{11} \ln \beta_1$, and the other one with $-M_{12} \ln \alpha_2$, requiring application of the chain rule. Specifically, we compute

$$S(X_1, \beta_1) = \frac{\partial y_1}{\partial \ln \beta_1} + \frac{\partial y_1}{\partial \ln \alpha_2} \cdot \frac{\partial \ln \alpha_2}{\partial \ln \beta_1}$$
$$= \frac{\partial y_1}{\partial \beta_1}\beta_1 + \frac{\partial y_1}{\partial \alpha_2}\alpha_2 \frac{\partial \alpha_2}{\partial \beta_1} \cdot \frac{\beta_1}{\alpha_2}$$
$$= \frac{\partial y_1}{\partial \beta_1}\beta_1 + \frac{\partial y_1}{\partial \alpha_2}\beta_1 \frac{\partial \alpha_2}{\partial \beta_1}. \quad (11.61)$$

Comparison with the elements in **M** helps us identify the partial derivatives of y_1 as

$$\frac{\partial y_1}{\partial \beta_1} \cdot \beta_1 = M_{11} \quad \text{and} \quad \frac{\partial y_1}{\partial \alpha_2} \beta_1 = -M_{12}\frac{\beta_1}{\alpha_2}. \tag{11.62}$$

The elements M_{11} and M_{12} are known, unconstrained sensitivities, and the only task left is the analysis of $\partial \alpha_2/\partial \beta_1$. For this purpose, we utilize the fact that, because of Eq. (11.58), α_2 can be expressed in terms of β_1 and β_5 as

$$\alpha_2 X_1^{g_{21}} X_2^{g_{22}} X_5^{g_{25}} X_7^{g_{27}} X_{10}^{g_{2,10}} = \beta_1 X_1^{b_{11}} X_2^{b_{12}} X_7^{b_{17}} + \beta_5 X_2^{b_{52}} X_5^{b_{55}} X_{10}^{b_{5,10}} \tag{11.63}$$

and thus as

$$\alpha_2 = \beta_1 \frac{X_1^{b_{11}} X_2^{b_{12}} X_7^{b_{17}}}{X_1^{g_{21}} X_2^{g_{22}} X_5^{g_{25}} X_7^{g_{27}} X_{10}^{g_{2,10}}} + \beta_5 \frac{X_2^{b_{52}} X_5^{b_{55}} X_{10}^{b_{5,10}}}{X_1^{g_{21}} X_2^{g_{22}} X_5^{g_{25}} X_7^{g_{27}} X_{10}^{g_{2,10}}}. \tag{11.64}$$

Differentiation of α_2 with respect to β_1 simply yields the first ratio of power-law terms, which needs to be evaluated at the steady state; the second ratio is independent of β_1 and drops out. Closer inspection shows that this ratio is actually

$$\frac{V_1^-}{\beta_1} \bigg/ \frac{V_2^+}{\alpha_2} = \frac{\alpha_2 V_1^-}{\beta_1 V_2^+}.$$

Since all these computations address the steady state, steady-state values are substituted for all X's, and we obtain

$$\begin{aligned} S(X_1, \beta_1) &= M_{11} + \frac{\alpha_2 V_1^-}{\beta_1 V_2^+}(-1)M_{12}\frac{\beta_1}{\alpha_2} \\ &= M_{11} - \frac{V_1^-}{V_2^+}M_{12} \\ &= -1.255 + \frac{1.068}{1.714} \times 0.969 \\ &= -0.651 \end{aligned} \tag{11.65}$$

Again, PLAS can be forced to compute these types of constrained sensitivities. As was to be expected, the implementation is somewhat more convoluted than in Example 2. First, the numerical value of β_1 is replaced with a symbolic name, such as *beta1*, which is declared as an independent variable. Secondly, the numerical value of α_2 is replaced with the expression in Eq. (11.64), where steady-state values are substituted for all X-values. Also, the numerical value of β_5 is substituted, while β_1 is replaced with the new independent variable *beta1*. The altered PLAS command lines are

```
X1' = .077884314 X4^0.66 X6 - beta1 X1^1.53 X2^(-0.59) X7
X2' = 0.34131412*(beta1*1.004979487
      + .646) X1^0.95 X2^ - 0.41 X5^0.32   >>
   >>  X7^0.62 X10^0.38 - .000793456 X2^3.97 X3^(-3.06) X8
X3' = .000793456 X2^3.97 X3^(-3.06) X8 - 1.05880847 X3^0.3 X9

beta1 = 1.062708258
&& beta1
```

The logarithmic gains with respect to *beta1* produce the constrained sensitivities. The first of these is $S(X_1, \beta_1) = -0.651475$, which is consistent with the above value, within the accuracy of our matrix computations. The sensitivities of X_2 and X_3 with respect to β_1 are computed with the same methods, either by hand or with PLAS (see Exercise 20).

For a complete characterization of the sensitivities with respect to rate constants, the *independent* rate constants are distinguished from the *dependent* rate constants. In our particular case, α_2 and α_3 are dependent, because they can be uniquely expressed in terms of the other system characteristics by evoking precursor–product and branch point constraints. This distinction is not unique, and as an alternative one could, for instance, consider α_2 as independent and β_1 as dependent. The constrained sensitivities of metabolites with respect to a given set of independent rate constants can be collected succinctly in a sensitivity matrix (see Exercise 21).

In many cases, the actual values of sensitivities are not as important as their relative magnitudes. Small magnitudes imply robustness of the model, whereas large magnitudes are often a warning sign that the model is too sensitive to withstand normal perturbations. For instance, imagine the sensitivity of a variable X with respect to a parameter p had a value of 100. A seemingly insignificant change in p of just 3% would result in a 300% change in X, thus increasing X to four times the original value. Analyses of this type were illustrated in previous Case Studies.

Kinetic Order Sensitivities

Sensitivities of metabolites with respect to kinetic orders require a fair amount of computation, at least if constraints must be considered. None of this is difficult in principle (see Chapter 7), but there is a lot of it. There are situations that require sensitivities in symbolic form (e.g., cf. Savageau, 1976: Ch.9). In these cases a symbolic algebra program comes in very handy (cf. Weinberger, 1991; Sorribas, 1996). In most other cases, one is well advised to use numerical instead of symbolic solutions, and for this purpose the most convenient tool is PLAS, which readily produces at least the unconstrained kinetic order sensitivities. The constrained sensitivities have to be computed from the unconstrained sensitivities by means of the constraint equations, which at this point is not possible in PLAS (see Exercise 22).

As stated for sensitivities with respect to rate constants, the absolute values of kinetic order sensitivities are sometimes not as important as their relative magnitudes. The analysis of such *sensitivity profiles* often offers insight into the control structure of the system and into weaknesses in the model design.

CONCLUDING REMARKS

This Case Study has emphasized several aspects of biochemical modeling that come up very frequently. The first again addressed model design and estimation. As is typical, far-reaching decisions about the inclusion and exclusion of metabolites and processes had to be made, and even seemingly clear-cut features, such as the maximal enzyme activities, turned out to be subject to sometimes great uncertainty. It was shown how to deal with these uncertainties: One may characterize the sensitivities and the robustness of the model to changes in inputs or parameter values, or one may design different models and compare the results. Though we did not discuss the second option, a comparison is possible between our results and those obtained by Torres (1994a), who dealt with the same system, but made different assumptions in the parameter estimation and selected different parameter values in some instances. Obviously, a single deviation in a key parameter changes every numerical output, but it is worth noting that qualitatively and quantitatively Torres obtained fairly similar results, which again reflects the robustness of the chosen model structure. For instance, Torres showed that the model is rather insensitive with respect to rate constants, and confirmed strong effects of the same kinetic orders that were identified here.

The second aspect of this case was the algebraic treatment of the steady-state features of a model. While it seems that numerical analyses in PLAS would be sufficient, there are occasions where a full or partial algebraic characterization may provide insights that can hardly be obtained with a numerical analysis. Examples are structural analyses that comprehensively elucidate the role of particular processes or signals of interest. For instance, Irvine and Savageau (1985ab) used primarily algebraic methods to identify the role of some feedback signals in a model of the immune response. Our case study developed a full algebraic representation of the steady state and its sensitivities. This representation not only allowed us to compute new steady states quickly under varying input conditions, it also led to sensitivity profiles that indicate which parts of the pathway can be effectively altered with reasonable effort. This type of analysis may explain disease patterns, since alterations in very sensitive parts of the pathway may be expected to lead to more severe symptoms than alterations in very insensitive parts. Sensitivity profiles may also guide strategies for optimizing pathways in a biotechnological setting. While not always entirely reliable, sensitivity profiles very often suggest to the bioengineer which enzyme activities are to be altered in order to improve the yield of a desired product (cf. Voit, 1992a; Regan et al., 1993; Hatzimanikatis and Bailey, 1996; Hatzimanikatis et al., 1996ab; Torres et al., 1996, 1997, 1998). Our analysis furthermore demonstrated how derived features like the transition time can be expressed as functions of one or several of the independent variables. Such representations are useful for evaluating the effects of changes in inputs on overall productivity and for subsequent optimization (cf. Torres, 1994d).

EXERCISES

1. 1.1. Discuss the implications that a small equilibrium constant in the glucokinase reaction would have. Distinguish numerical and structural ramifications.
   1.2. Discuss the implications of considering glycogen at saturated levels. Distinguish numerical and structural ramifications.
   1.3. Scrutton and Utter (1968) list the maximal activity of phosphorylase *a* as 37. Recompute parameters as necessary for that case. Discuss the implications.

2. Show that the constrained kinetic order g_{22} is equivalent to the weighted average of the two independent kinetic orders h_{12} and h_{52}, according to the formula

$$g_{22} = \frac{V_1^-}{V_2^+}h_{12} + \frac{V_5^-}{V_2^+}h_{52}.$$

3. Confirm with paper-and-pencil computations that the logarithmic gains of X_1 and X_2 with respect to X_9 are different in sign from the gain with respect to X_3. State in words what this suggests.

4. Check algebraically whether the logarithmic gains of the flux through X_1 with respect to X_5, X_7, X_8, X_9, and X_{10} are exactly 0 or only approximately equal to 0.

5. Compute in PLAS the sensitivities of the system with respect to rate constants. Which rate constant is most influential? Why do some rate constants exhibit the same sensitivity?

6. Compute in PLAS the unconstrained sensitivities of the system with respect to kinetic orders.

7. Compare the two methods of adding exogenous glucose-1-phosphate to the system. Interpret the slight differences in results.

8. Study the effects of other perturbations of dependent variables. Is it possible to destabilize the system?

9. Analyze other changes in independent variables. For instance, study the effects of $X_7 = 1$ or 100 and of $X_9 = 0.1$ or 10. Study simultaneous changes in the two independent variables. Record and interpret the findings. Can you find values for the independent variables that destabilize the system?

10. Show that changing g_{22} requires the rate constant α_2 to be adjusted in the following fashion:

$$\tilde{\alpha}_2 = \alpha_2 X_{2S}^{g_{22}(old)-g_{22}(new)}.$$

Implement this adjustment in PLAS in such a fashion that you only have to enter the new kinetic order and PLAS automatically resets the rate constant. Discuss whether (and if so, which) other adjustments are necessary for fair comparisons with the original system. Test your computations with $h_{52} = -0.25$, as was suggested by Torres (1994a).

11. Study dynamic responses of systems in which h_{52} is changed to either 0 or -0.5. Compare the responses with those of the original system. For example, study the effect of altered glucose inputs in the three systems.

12. The estimation of h_{33} is somewhat uncertain. Explore the impact of setting the value of this kinetic order equal to 0.1 or 0.5.

13. Study system responses for a series of altered values of h_{33}. Find useful ways, such as graphs, of presenting the results.

14. Use Eqs. (11.26)–(11.30) to compute the steady state of the system if glucose-6-phosphate does not inhibit the glucosekinase reaction. Execute the computation with and without adjustment of the rate constant. Check the results in PLAS.

15. Compute the logarithmic concentration and flux gains of the system if glucose-6-phosphate does not inhibit the glucosekinase reaction. Execute the computation with and without adjustment of the rate constant. Check the results in PLAS.

16. Confirm that it is legitimate to split the steady-state Eq. (11.39) so that the terms with $\ln X_5$ are separated from the rest.
17. Plot the transition time as a function of glucose input.
18. Express the logarithmic gain of the transition time in terms of X_k rather than y_k.
19. Confirm the formula for $L(\tau, X_5)$. Substitute formally the expression of σ as a function of X_5, and show that the gain in some sense consists of a constant plus a weighted sum of power-law functions in X_5. What are the weights? What does the constant represent?
20. Compute with algebraic means the sensitivities of X_2 and X_3 with respect to β_1. Check the accuracy of results with PLAS.
21. Compute the sensitivities with respect to β_5. Hint: Use the unconstrained sensitivities with respect to α_2, and work out arguments analogous to those following Eq. (11.64). Put all constrained rate constant sensitivities into a sensitivity matrix that shows how each of the dependent variables responds to a change in one of the independent rate constants.
22. Compute by hand the sensitivities of X_1, X_2, and X_3 with respect to h_{22} and h_{17}.
23. The numerical specifications for the phosphoglucose isomerase step vary slightly among different conditions. Using a mass action ratio of 0.31 or 0.32, different equilibrium constants between 0.36 and 0.47, and different maximal rates between 136 and 280 μmol min^{-1} g^{-1}, recompute S-system parameters as necessary. Which measurement is most influential? What is the significance of the altered values in terms of steady states and model dynamics?
24. (Project) For the estimation of h_{33}, we used observations by Hers and Hue (1983) and by Torres et al. (1986) suggesting the hyperbolic dependence of the phosphofructokinase activity on fructose-6-phosphate, and set the value for this kinetic order to 0.3. Explore the ramifications of changing this numerical value.
25. (Project) Express the steady-state solution as a function of P_i. Test whether the result that all dependent variables at steady state are power functions of X_5 extends to other independent variables.
26. (Project) Explore the responses of the model to two simultaneous perturbations, namely the decrease in glucose concentration and the increase in the activity of phosphorylase a. According to Newsholme and Start (1973), these conditions cause the liver to degrade mainly glycogen. Decreases in glucose concentration negatively affect the fluxes, while increases in phosphorylase a have the opposite effect. Characterize the relative effect of the two perturbations. Explore conditions under which the two effects compensate each other.

REFERENCES

[19], [47], [79], [112], [126], [136–8], [151–2], [155–7], [165–6], [227], [257], [283], [287–8], [319], [330], [363], [384], [389–90], [415–16], [419], [421–5], [439–40], [482], [486].

CHAPTER TWELVE

Epilogue – Canonical Modeling Beyond Biochemistry

Biochemical systems stimulated the original interest in canonical modeling. They were selected for a number of good reasons. Without doubt, biochemical systems are ubiquitous in the living world and relevant to many branches of modern biology, medicine, and biotechnology. Furthermore, even in the early stages of canonical modeling, a huge amount of kinetic information about enzymes and metabolic pathways was available, and continually improving methods of biochemistry and molecular biology have brought forth an explosion in knowledge about the constituents of biochemical systems. Finally, a good reason for analyzing biochemical systems is that examples are known for about every degree of complexity. This is of significance, because any new mathematical theory or methodological framework needs to evolve from simple phenomena and methods to increasingly more complicated and relevant applications. At the beginning, the new theory and its associated methods of analysis require examples with few variables. These examples are often quite trivial. In the context of biochemical systems, for instance, one finds a large literature on linear pathways with only a few metabolites. If these first analyses are successful, the theory needs applications that gradually increase in complexity.

This can clearly be seen in the development of canonical models. The size and complexity of pathways that can be modeled and numerically analyzed with current methods of canonical modeling have about reached a point where relevant, quantitative predictions can be made. For instance, we are approaching a level of expertise in biotechnology where we can begin to manipulate fermentation processes in a systematic fashion that is guided by biomathematical analysis. The same may be said about the development and screening of drugs. The red-blood-cell model (Ni and Savageau, 1996ab) and the model of purine metabolism (Curto et al., 1997, 1998ab) have the potential of forming the base for targeted pharmaceutical research. With the collective experience with such systems growing, it will become easier in the future to develop models of moderate or large size for new pathways and for specific therapeutic or biotechnological purposes. Thus, biochemical systems were the first target of canonical modeling, and they will very likely remain among the most fruitful applications in the foreseeable future.

Nevertheless, biochemical systems have not been the only focus or source of insights. Other applications have broadened the realm of canonical modeling and stimulated the exploration of novel types of methods and techniques. Some of these applications are still related to biochemistry, while others are quite removed from it. For instance, canonical models were employed to study tRNA proofreading mechanisms in *E. coli* (Okamoto and Savageau, 1984ab; Okamoto et al., 1991). These models differ in emphasis from most systems discussed so far, yet are part of biochemistry. Much progress has been made in the deciphering of rules governing the regulation of gene expression (see below). These genetic control systems are different from metabolic pathways, but are intimately related to biochemical processes. Applying methods of biochemical system analysis to other biological phenomena, Irvine and Savageau (1985ab) studied regulatory questions about the immune response, and Torres (1996) analyzed the flow of magnesium through a tropical ecosystem. Applications of a truly non-biochemical nature include studies in forest management, as well as in applied mathematics and statistics.

This chapter reviews some of the themes that are not biochemical but have been approached with methods of canonical modeling and biochemical systems theory. Most of these themes have not only resulted in new subject-oriented insights, but also contributed to the repertoire of methods. For instance, Irvine and Savageau's (1985ab) immunological analysis defined the techniques with which *controlled mathematical comparisons* are executed (see below). In addition to the more application-oriented studies, quite a substantial body of work has addressed the mathematical properties of canonical models, and this research, in turn, has led to applications in non-biological areas such as computational statistics and numerical analysis.

BIOLOGICAL APPLICATIONS

Gene Regulation and Demand Theory

Whereas activators and inhibitors exert immediate control over the short-term functioning of a metabolic pathway, longer-term control rests with the regulation of gene expression. The amount of biological detail on gene regulation is overwhelming, but from a structural point of view, there are only a few generic types of regulatory *circuitry*. In the *classical* circuit, expression of a regulator is almost unaffected by the concentration of effector proteins; effector and regulator are *uncoupled*. In the *autogenous* circuit, by contrast, regulator and effector proteins form bi-functional proteins and their expression is perfectly coupled. Other types of coupling have been observed between these two extremes. In addition to the type of circuitry, the regulatory mechanism itself can be implemented through *induction* or *activation* (positive mode) or *repression* (negative mode). In the former case, the transcriptional units are normally turned off, and induction or activation is required to initiate gene expression. In the latter case, the transcriptional units are normally turned on, and the appearance of the repressor slows down or turns off gene expression.

Over the past twenty-five years, Savageau has been studying extensively the various mechanisms of gene regulation with methods of canonical modeling (Savageau,

1974ab, 1976, 1977, 1979c, 1983abc, 1985c, 1989, 1996, 1998ab, 1999; Savageau and Sands, 1991; Hlavacek and Savageau, 1995, 1996, 1997). As an introduction to the topic, the reader may begin with Chapter 14 of Savageau's (1976) book, Savageau (1977, 1989), or the concise review in Neidhardt and Savageau (1996). Savageau and his collaborators pursued two lines of research that complemented each other. One is of a qualitative nature, leading to a conceptual framework called *demand theory*, and the other is formulated in terms of S-system models that address the structural features of the regulatory circuitry itself.

Demand theory. The goal of demand theory is to explain the observation that some genes are controlled positively and others negatively. Specifically, Savageau (1977) set out to answer the following questions: Are the differences in molecular design significant? Are they simply historical accidents that represent functionally equivalent solutions to the same regulatory problem? Alternatively, have they been selected to meet specific needs and, if so, can we determine the functional implications inherent in each design and the nature of the selective forces that have given rise to them?

At first glance, the positive or negative mode of regulation seems to be coincidental, and models of the two modes yield almost indistinguishable responses. Critical differences emerge only in their reactions to mutations in the regulatory mechanism itself. Using arguments of selection and fitness, demand theory predicts a mode of control that gives the organism a selective advantage. This prediction can be summarized in a simple rule: When mutations follow the "natural mutational tendency (entropy)" (Neidhardt and Savageau, 1996), repressor-controlled systems are likely to become constitutively expressed, whereas activator-controlled systems are likely to become super-repressed (Savageau, 1977).

Demand theory furthermore proposes that the mode of gene regulation depends on the demand for gene expression. Specifically, a gene is positively controlled if the gene is normally expressed at the high end of its regulatable region and is thus in high demand under the typical environmental conditions of the organism. A gene is negatively controlled in the opposite situation, namely, if it is in low demand under typical conditions and is normally expressed at the low end of its regulatable region (e.g., Savageau, 1977, 1979c; Neidhardt and Savageau, 1996). As an example, Neidhardt and Savageau compare the utilization of arginine and catabolism of maltose by *E. coli* in the colon of warm-blooded animals. Arginine is abundant in this environment, and since the demand for the gene product is low, the bacterium can afford to repress the gene encoding the biosynthetic pathway for arginine. As predicted by demand theory, the mode of regulation is negative through repressor control. By contrast, the maltose catabolic system is positively controlled by an activator-mediated system, and demand theory predicts that the demand for expression must be high. Indeed, the disaccharide maltose is abundant in the colon. The simple correlation between demand for gene expression and mode of gene control, which is postulated by demand theory, has been confirmed for dozens of operons.

Circuitries of gene regulation. In contrast to the qualitative demand theory, the exploration of advantages and disadvantages of a particular regulatory circuitry requires the full spectrum of quantitative methods provided by canonical modeling. The generic approach for these types of analysis has been the *method of controlled mathematical comparisons*, which allows the objective comparison of alternative designs; mathematical details of this method are described in a later section (see also Chapter 11).

The first step in a comparison of alternative candidate models is the establishment of quality criteria that define a "good" system. Some criteria are almost universal, while others are subject specific. For instance, most systems are expected to be stable at their steady states and able to tolerate moderate perturbations. We have discussed these issues many times in this book in terms of eigenvalues, parameter sensitivities, and logarithmic gains. For regulatory gene circuits, Hlavacek and Savageau (1995, 1996, 1997) added the subject specific criteria of *decisiveness, efficiency, responsiveness*, and *selectivity*. In a decisive circuit, there should be a sharp threshold in substrate concentration that separates induction from non-induction. Efficiency refers to the gain in product, which should offset the cost of induction. The criterion of responsiveness favors speed of response to environmental changes, whereas the criterion of selectivity favors a system that limits the accumulation of regulator during induction.

Criteria of this nature are used as measuring sticks against which alternative circuits are evaluated. Since they are selected independently of the mathematical structure of the candidate models and solely on biological considerations, they provide an objective means of determining what type of regulatory circuit is best suited under a given set of environmental and organismal conditions.

Hlavacek and Savageau (1995) analyzed perfectly coupled and completely uncoupled circuits in this fashion and were led to two definite predictions:

1. If the capacity of induction is small, the circuit is expected to be repressor-mediated and perfectly coupled, or to be activator-mediated and completely uncoupled.
2. If the capacity is large, the circuit is expected to be repressor-mediated and completely uncoupled, or to be activator-mediated and perfectly coupled.

In a similar vein, Hlavacek and Savageau (1996, 1997) compared directly coupled, uncoupled, and inversely coupled circuits, in which, respectively, regulator gene expression increases, remains constant, or decreases with effector gene expression.

Growth Dynamics

Growth phenomena have been tantalizing human curiosity for a very long time. There is evidence that Babylonian and eastern European peoples assessed sizes and trends in the growth of populations many thousand years ago (see Savageau, 1979a). The study of growth is so interesting because, on one hand, the size of a growing organism or population typically follows a rather simple time course, and on the other hand, growth is the consequence of millions of processes that occur at different levels of biological organization and at different time scales. The puzzle thus is this: How can so many processes and subsystems be orchestrated to result in such simple growth functions?

EPILOGUE – CANONICAL MODELING BEYOND BIOCHEMISTRY

In the context of canonical modeling, growth has been approached from different angles. Savageau (1979ab, 1980) argued the following. The uncounted processes leading to the growth of an organism or population form a huge system that – at least in a thought experiment – can be formulated as an S-system model. The dynamics of the variables involved have vastly different time scales. For instance, biochemical processes take place on the order of seconds or minutes, whereas aging processes may span the entire life. Very fast processes are essentially always in steady state, because they react rapidly to perturbations, at least in comparison with the slower processes. Other processes, by contrast, are so slow that their dynamics are flat: A rise in global temperature is so insignificant over the life span of an individual organism that one may legitimately consider it constant. Since the fast and the slow processes are either at steady state or constant, the describing equations can be set equal to zero and solved, and the associated variables in effect become parameters. What is left between the very fast and the very slow phenomena comprises those processes that take place at the time scale of interest. These *temporally dominant* processes constitute a model of greatly reduced size, which, nonetheless, still has the canonical structure of an S-system. Savageau showed that many famous "growth laws" in the literature emerge naturally from this procedure.

A fair number of canonical analyses has been devoted to the growth of trees and forests. Some of these analyses try to connect the growth dynamics with biochemical and physiological processes, while others target individual trees as the smallest units of the model. Torres (1996) proposed a model of magnesium flow through a tropical forest. Except for the fact that the pools in his model are entities like plants, litter, carnivores, and soil, the philosophy and techniques of his analysis are very similar to the typical biochemical analyses we have encountered in this book. The S-system equations for the model are set up from a map, parameter values are estimated from information found in the literature, and the analyses focus on stability, sensitivities, gain profiles, and some dynamics. The results identify and rank the role of the various forest components in the magnesium cycle. For instance, they point to the soil as a magnesium donor with high capacity.

Voit and Sands (1996ab) addressed the question of biomass allocation in growing trees. Using as variables different tree compartments, and as fluxes processes like carbon uptake from the soil and transport between compartments, they demonstrated that the partitioning of biomass is a function of nitrogen availability as well as tree age. While set up in a general fashion, the model intrinsically explained many observations that had been stated in the literature in a fragmented *ad hoc* fashion.

Martin (1997) used canonical S-system methods for an innovative condensation of a complex simulation model for forest dynamics. The starting point was an earlier detailed computer model that allowed the consideration of very many parameters and the simulation of numerous scenarios of forest management. An S-system model was used as an approximation of this original model under purposely limited sets of conditions. Martin showed that the overall dynamics of the simplifying S-system was very adequate and that it permitted the extrapolation of results to scenarios that could not be modeled with the original model because of severe data gaps.

At the population level of forests, Voit (1988a, 1990a) used S-system methods to explain the phenomenon of self-thinning and the so-called $\frac{3}{2}$ *rule* in even-aged tree stands. This rule, which is widely supported by observations (e.g., White, 1980, 1981), states that the relationship between the number of trees per unit area and the average tree size with time approaches a power-law function with exponent $\frac{3}{2}$. Voit gave a systems-based explanation of the phenomenon and discussed the generality and limitations of the $\frac{3}{2}$ rule. Using a simplified version of Voit's model, Torsella and Bin Razali (1991) analyzed actual data and predicted the effect of tree spacing in managed loblolly pine forests.

Johnson (1991) analyzed the *sweep* of trees, which is a measure of crookedness, with respect to maximum lumber yield. In the process, Johnson extended S-system methodology to include a stochastic term of Wiener type and provided a technique of conditional least squares parameter estimation based on quasi-linearization and Kalman filtering.

MATHEMATICAL RESEARCH

The never-changing structure of canonical models, combined with their ability to capture nonlinear phenomena, has always been intriguing. Some consequences of the special structure of S-systems were recognized in the first analyses and, indeed, were essential criteria of selection among the numerous possible nonlinear modeling structures. For instance, the fact that the steady-state equations of S-systems become linear upon taking logarithms has always been a cornerstone of canonical analysis (Savageau, 1969ab).

The mathematical research on canonical models falls into several overlapping categories. The first deals with the numerical analysis of the differential equations as well as the steady-state equations. The second category addresses the accuracy and general structural features of canonical models. The third category contains analytic methods that are made possible by the linear structure of the steady-state equations. The fourth category deals with algebraic analyses and methods that transform power-law models into alternative, equivalent forms, which all have their own advantages. Finally, considerable effort has been devoted to topics of computational and applied statistics. While these are seemingly unrelated to biochemical modeling, recent insights show close conceptual connections.

Accuracy and Model Structure

It has been discussed numerous times that every model is, by design, an approximation that intentionally or unintentionally ignores some details of reality and accepts errors in data representation. This raises the immediate question of how accurate a given model actually is. The problem in answering this question is that the true underlying functions are seldom known, so that a straightforward comparison between alternative models and reality is not possible. In rare cases, models can be assessed against directly corresponding measurements of processes. Usually, however, only input–output data are available, and a model evaluation based on such data is not always reliable.

A particular question in this context has been the relative accuracy of S-systems and GMA systems. The former have clear advantages for steady-state analysis, whereas the latter are sometimes closer to biochemical intuition. Though no final answer is available, several articles have elucidated different aspects of the two alternative representations. Voit and Savageau (1987) compared the accuracy of S-systems and GMA systems with respect to Michaelis–Menten models and demonstrated that, maybe contrary to intuition, S-systems provided the more accurate models in almost all representative test cases studied. Sorribas and Savageau (1989abc) confirmed this result with a careful analysis of systems with flux reversal that operated close to the thermodynamic equilibrium. Curto et al. (1998a; see also Chapter 9) developed parallel models in Michaelis–Menten, GMA, and S-system form for purine metabolism and found that close to the steady state all three yielded about the same results. Severe enzyme deficiencies were apparently better modeled with a GMA model. However, the relevance of this observation must be questioned, since a normal individual and an enzyme-deficient subject constitute two entirely different systems without transitions between them.

The accuracy of a model depends on its structure and its parameter values. Issues of parameter estimations were discussed in great detail in Chapter 5, but a few words should be said here about investigations of the structure of a model. In most cases, the model is constructed from the bottom up, beginning with variables and underlying processes, and connecting them in accordance with a known or assumed structure. However, not all interactions between variables may be known, and some effort has been devoted over the years to exploring the existence of interactions, and thus the structure of the model. These analyses have been greatly facilitated by the fact that the structure of a canonical model translates directly into the values of some of its parameter values. For instance, if a kinetic order is changed from -0.5 to 0, a former inhibiting influence has been eliminated, and the structure of the model is altered. Some approaches to assessing the structure of a model are closely linked with the method of mathematically controlled comparisons, which is reviewed in a later section. A review of structural analyses was presented in Voit (1996c), and it may suffice here to provide representative examples.

Savageau (1976) analyzed in a symbolic fashion all possible patterns of feedback inhibition of a linear pathway. Based on several criteria of biological superiority, the analysis demonstrated why feedback of the initial substrate, exerted by the end product, is so frequently seen in nature, and why other imaginable patterns of feedback regulation are not observed. Irvine and Savageau (1985ab) compared alternative models of the immune response and characterized the potential role of a hypothesized mechanism of feedback regulation. Sorribas and Cascante (1994) proposed a method of structure identification based on logarithmic gain profiles and perturbation experiments. Using a didactic test system of moderate size, the authors were able to identify from "experiments" the "unknown" regulatory structure of the system. Hatzimanikatis et al. (1996b) used methods of mixed-integer linear programming to optimize the structure of a metabolic model under predefined constraints.

Salvador (1996, 1997, 2000ab) took yet a different approach to assessing the appropriateness of a model structure. He argued that typical sensitivity analyses are based on first-order approximation and that their linear character misses synergisms that might exist between different parameters. In other words, one cannot necessarily expect that the system response to two simultaneous perturbations in different parameters corresponds to the sum of responses to individual perturbations. Salvador therefore proposed to study *synergisms* between stimuli, which were defined in direct generalization of sensitivities as second-order approximations. The profiles of these synergisms were used for a classification of system responses in terms of their additive and multiplicative nature.

Using canonical modeling methods, Savageau (1993b, 1995a, 1998c) analyzed the consequences of the common assumption that all reactions occur under homogenous, well-mixed conditions. Relaxing this assumption is of great relevance, since it is becoming increasingly clear that many enzyme-catalyzed processes take place on surfaces or in channels, and that the restriction to two- or one-dimensional spaces drastically affects the dynamics of the reaction. Savageau showed that without the assumption of homogeneity the Michaelis–Menten mechanism becomes mathematically more complicated. The resulting rate laws are best represented with *fractal kinetics*, which differ from the traditional Michaelis–Menten rate law, but are nonetheless direct special cases of canonical models.

Numerical Methods

While nothing can beat the rigor and crispness of a mathematical proof or analytical solution to a problem, numerical solutions are an absolute necessity for realistic modeling. Analytical solutions often simply do not exist or are so cumbersome that they are almost useless for practical purposes. A good example is the analytical solution of a simple two-variable S-system that includes most growth functions (Voit and Savageau, 1984). Not only the numerical features, but even the structure of this solution depends on the numerical values of the kinetic orders involved. If these are integers, the solution proceeds in one direction, but if they are just a tad off an integer, the solution proceeds in a different direction. As another example, consider the computation of the steady state of an S-system with ten or twenty dependent variables. In principle, the solution could be obtained with paper and pencil, using methods of linear algebra, as they were explained in Chapter 6. However, who would want to do that by hand? Since systems of moderate and large size were envisioned from the beginning, the development of efficient numerical methods has been an ongoing endeavor throughout the history of canonical models.

Of greatest importance has always been an algorithm for integrating the differential equations. It was realized early on (e.g., Savageau, 1970; 1976: p. 139 f.) that it saved computer time to transform the variables logarithmically before numerically integrating the differential equations. Most solvers for canonical models have been using this principle. The breakthrough in the development of numerical solutions for canonical models came with the discovery that derivatives of products of power-law functions can be formulated symbolically in a very efficient, recursive fashion (Irvine, 1988; Irvine and Savageau, 1990). This recursion made it feasible to compute Taylor approximations of essentially any order. Combined with methods of optimizing the size of the next solution step, Irvine and Savageau developed an algorithm that at the time ran faster on a PC than the best all-purpose mainframe algorithms available. The algorithm was included in a fully interactive package, ESSYNS (Voit et al., 1989). PLAS was based upon the algorithmic concepts in ESSYNS but further improved the handling of data and results and fully exploited similar patterns of kinetic orders in different terms (Ferreira, 1992). It also embedded the algorithm in a user-friendly Windows environment and added the uncounted conveniences that we currently enjoy.

Shiraishi and Fujiwara (1996) expanded the principles of the algorithm of Irvine and Savageau to solve two-point boundary value problems with high accuracy. Combining this algorithm with the method of recasting (Savageau and Voit, 1987; see also below) and with a shooting method based on a Newton–Raphson search method, Shiraishi and Fujiwara showed that not only canonical models but also boundary value problems containing other types of functions can be solved very efficiently. As an example, they analyzed an immobilized enzyme reaction of Michaelis–Menten type.

Ni and Savageau (1996ab) developed a very interesting computer program that scans S-systems for putative errors in model structure and parameter values. The program makes heavy use of the S-system structure and of the fact that steady states, stability, and sensitivities can be formulated symbolically in a generally applicable fashion. It takes account of internal and external equivalence and allows the user to define criteria of model quality. In one of their applications to red-blood-cell metabolism, the algorithm screened over 1,200 possible model structures and identified 26 as the most likely to succeed.

The computation of steady states in S-systems follows well-known methods of linear algebra (cf. Chapter 6), but the same is not true for GMA systems. Setting the GMA differential equations equal to zero results in multiple products of power-law functions that cannot be linearized by taking logarithms. Because of the importance of GMA systems, approaches have been developed to characterize their steady states with numerical methods (Savageau, 1993a; Mueller et al., 1998; see also Hasegawa and Shiraishi, 1996).

Outside number-crunching routines, some authors have written symbolic or semi-symbolic programs in languages like Macsyma and Mathematica that allow analyses without the specification of parameter values (e.g., Weinberger, 1991; Sorribas, 1996).

Consequences of the Explicit Linear Nature of the Steady-State Equations in S-Systems

The method of controlled mathematical comparisons. One of the fundamental questions of biomathematical research is why a natural system is designed the way it is. There are always possible alternatives, and it is not a trivial matter to discover why nature has selected one design over all its competitors. For instance, why are some genetic regulatory systems activator-controlled and others repressor-controlled? Some comments about the systems themselves were given in the section on gene regulation. It is important here to realize that controlled comparisons are very much dependent on the structure of canonical models and, in particular, the explicit form of the steady state. Examples of controlled mathematical comparisons include Savageau, 1985b; Irvine and Savageau, 1985ab; Hlavacek and Savageau, 1995, 1996, 1997).

The key concept of controlled comparisons is the *internal* and *external equivalence* of two or more competing systems. One system may have the design observed in nature, and the others may have designs that apparently are just as reasonable. In the simplest case, two systems are compared that differ in a single process. To assess the role of this process, all other parts of the two systems are made the same, constituting internal equivalence. Furthermore, one tries to make the two systems as similar to the outside observer as possible, constituting external equivalence. After all, one can usually observe the steady state of the natural system, so that all designs that can allegedly explain the system must satisfy this steady state. Since analyses of this type are usually executed not with numerical models but with S-system models in symbolic form, it is almost mandatory that the steady state can be characterized symbolically. The same is true for stability and, to some degree, for sensitivities and gains. In typical controlled comparisons, the range of stability is considered a criterion for fitness, because one can argue that a system with a wider margin of stability can tolerate larger perturbations. Similarly, a controlled comparison of sensitivities and gains suggests which of the competitors is most robust.

Optimization. As a direct offshoot of the linearity of the steady-state equations, it was realized that these linear equations could be used for purposes of optimization (Voit, 1992a; Regan et al., 1993, Torres et al., 1996–1998). Consider a situation where some metabolic production process is modeled as an S-system and operates at a steady state. A typical example is a batch fermentation process. The numerical characteristics of this steady state depend on all components of the system and, in particular, on the values of the independent variables, which often are under the control of the experimenter. Now suppose that the feature of interest is the accumulation of some end product or some output flux. Changing any given independent variable will presumably affect the value of this product or flux, though not necessarily in the desired direction. The logarithmic gain profile gives some indication of such an effect. However, a change in an independent variable will have an effect not only on the metabolite or flux of interest, but on the entire system of reactions. Therefore, one cannot arbitrarily increase or decrease one independent variable to obtain the desired effect, lest the system lose fitness and be no longer functional. In mathematical terms, the optimization is constrained by limits on the concentrations of some or all of the metabolites and some or all of the fluxes. Since all fluxes are represented as products of power-law functions, they become linear in logarithmic coordinates, just like the steady-state equations. This is very important in that it allows the optimization problem to be formulated entirely with linear equations or inequalities. Specifically, the optimization becomes a linear program (e.g., Luenberger, 1984; Strang, 1986), as we discussed in Chapter 8.

This approach to optimizing a nonlinear system is mathematically as valid as the S-system representation itself. If the S-system is an approximate model, questions of accuracy and validity must be considered, since the validity of the S-system is mathematically guaranteed only in a vicinity of the (steady-state) operating point. Of course, optimization is not of interest close to the existing point, but has as a goal to move one of the metabolites or fluxes away from that operating point. Thus, the accuracy of the predicted optimal solution depends on the accuracy of the S-system approximation. Comparative analyses suggest that the two types of accuracy are sufficient in many relevant cases (Torres et al., 1996–1998).

The original optimization method of S-system models was subsequently expanded to allow for the optimization of the transition time (Torres, 1994d) and the optimization of the model structure (Hatzimanikatis et al., 1996ab). Petkov and Maranas (1997) furthermore studied the effects of uncertainties in the optimization of metabolic pathways.

A rather different approach to optimization was taken by Chaudhuri and Johnson (1990) and Ganguly and Chaudhuri (1995), who investigated optimal control and harvesting strategies for fisheries, and by Johnson (1985, 1988), Brown (1991), and Brown and Johnson (1992), who analyzed various aspects of optimal control in agricultural systems, such as the optimal feeding of turkeys.

Structural stability. Chapter 6 discussed the assessment of stability in S-systems. For most biochemical systems of practical relevance, stability is a prerequisite, and an unstable system is deemed unrealistic. From a mathematical point of view, the transition from stability to instability can be very interesting, because the formerly stable point may be replaced by a *stable oscillation*, whose center is an unstable steady state. The characterization of such transitions and of *critical points* where the transition happens is the object of *structural stability analysis*.

An example of a stable oscillation is the heartbeat of a healthy animal, which can be visualized in an electrocardiogram that shows (complicated) oscillations about some base value. If the organism is stimulated, the shape of the oscillation may change in frequency or amplitude or both, but after a while, the heartbeat will normally return to the healthy oscillation. This phenomenon is also characterized by the term *limit cycle*, which derives from the representation of the oscillation in the phase plane. In the phase plane, one dependent variable is plotted against another dependent variable, and in the case of a limit cycle a (*cyclical*) pattern emerges that repeats itself after each oscillation (see also Chapter 4). Mathematicians have been interested for about a century in characterizing limit cycles directly from the differential equations of the model. Lewis (1991) demonstrated that models in S-system form are exceptionally easy to characterize with respect to the most common type of limit cycle behavior. For two-variable systems the analysis becomes so simple that it can easily be executed by hand. Given a two-variable S-system and Lewis's criterion, one can read off from the equations whether the model has the potential for a stable or unstable limit cycle. Building upon these insights, Lewis (1992) succeeded in fully characterizing the entire spectrum of possible behaviors of two-variable S-systems according to their structure, which is a rare achievement for nonlinear dynamical systems.

Recasting

It has been obvious since the beginnings of canonical modeling that the power-law structure contains numerous famous phenomena as trivial special cases. For instance, exponential growth and decay are readily formulated as S-systems with one positive or negative linear term, respectively. Somewhat less obvious are other growth functions, such as the logistic growth and Gompertz's law, whose differential forms are widely known (Savageau, 1979ab, 1980). What came as a surprise is that virtually any function that is the solution of some ordinary differential equation can also be written equivalently as the solution of a GMA or S-system (Voit and Savageau, 1986; Savageau and Voit, 1987; Voit, 1988b, 1990b; see also Kerner, 1981, and Peschel and Mende, 1986, for the recasting of equations into other canonical forms). The term "equivalent" signifies that these functions are not *approximated* by GMA or S-systems, but that they are reproduced *exactly*, with perfect accuracy. This discovery proved that essentially all smooth nonlinearities are special cases of canonical models in both GMA and S-system form. It also proved that these canonical models are rich enough to capture about every process of relevance, including weird-looking oscillations and deterministic chaos. Finally, it proved that there will never be a complete analytical solution of general GMA or S-systems, since it is known that some differential equations have no analytical solutions in terms of explicit algebraic functions.

The recasting method led to a number of disparate research themes. Two of these are *classification* and *computational statistics*. Another theme that was briefly mentioned in Chapter 4 is the combination of recasting and modeling (see also Voit, 1990c).

Classification. If a wide variety of different-looking functions can be subsumed under one umbrella structure, an obvious question for a mathematician is whether all – or at least some – functions can be classified as special cases of one pattern. One could argue that one would formulate all functions as differential equations and recast these as S-systems, and that henceforth every function could be categorized simply by its S-system parameters. For the limited set of growth functions, such a classification was successful (Savageau, 1979a, 1980; see also Voit, 1985, 1990b; Voit et al., 1985). One problem with this approach is that the recasting process is by no means unique. In fact, it was shown that the same original function can be represented as an S-system in many different ways (e.g., see Voit 1991: Chapter 12). Even worse, these different representations may have different dimensions, which means that it might be possible to represent the same function as a system of three or eight S-system differential equations. To some degree, these different representations can be transformed into each other, and Fairén and Hernández-Bermejo have presented theorems and methods for determining minimal representations (Fairén and Hernández-Bermejo, 1996; Hernández-Bermejo and Fairén, 1997; see also Voit, 1992b, 1996c). Whether their classification is optimal in practical applications remains to be seen.

The classification of growth functions in S-system format has been used in different contexts. It was also shown that essentially all continuous statistical distribution functions can be formulated as S-systems, though a classification in this form has not provided significant benefits.

Statistical Issues

Savageau (1982) showed that many of the typical statistical densities or distributions can be recast as one- or two-variable S-systems. After more progress in recasting, Rust spearheaded an effort at recasting all distribution functions, including those that otherwise were very difficult to evaluate with any method, such as the noncentral F- and t-distributions (e.g., Rust and Voit, 1990; Voit and Rust, 1990, 1992; Lee and Mahir, 1994ab). He also suggested several new approaches to statistical computation with S-systems. These included the assessments of power (Rust and Voit, 1990) and the development of generalized distributions, such as the normal and the noncentral t-distribution, which allowed for heavier tails and fractional degrees of freedom (Rust and Voit, 1989; Rust, 1991).

Voit (1992c) pursued a different approach by arguing that cumulative distribution functions always rise from 0 to 1 in a monotonic fashion, which is typically S-shaped. Since a single S-system equation models such a sigmoid shape, he explored how closely traditional distributions could be approximated by a single S-system equation, which was subsequently dubbed the *S-distribution*. It was shown that, indeed, most traditional continuous and discrete distributions are well approximated by the S-distribution and that the S-distribution provides a natural method of comparing and classifying traditional distributions. These observations, combined with some desirable statistical properties, suggested the S-distribution as a new and very flexible tool for representing experimental data (Voit and Yu, 1994; Yu and Voit, 1995, 1996; Voit and Schwacke, 1998, 2000; Voit, 2000).

Several studies followed up on these ideas. It was shown that the S-distribution is particularly well suited for the evaluation in trends in distributions over time. For instance, Balthis et al. (1996) demonstrated how the distribution of contaminants in fish changed over the lifetime of the fish. Sorribas et al. (1999) characterized growth patterns in populations of children in the Catalonia region of Spain, and Voit and Sorribas (2000) assessed similar patterns in tree populations. Because of its ability to model different types of skewness, the S-distribution was also found to be an intriguing tool for random number generation and, thus, Monte Carlo simulations (e.g., Voit et al., 1995).

Finally, closing the circle, there has been recent interest in the connections between statistics and S-system modeling. Arguing similarly to Savageau, who explained growth functions as temporally dominant processes of complex underlying systems, survival functions associated with S-distributions were explained as the processes dominating very complex systems that determine death or survival of individuals in a population (Voit, 1998). It was shown that the linear–logistic model and the Cox model of statistics and epidemiology, whose validity could not be explained before, are in fact direct consequences of a generic disease model in S-system form (Voit and Knapp, 1997). The closely related area of environmental risk assessment uses a diverse spectrum of concepts and models, yet it was demonstrated that most of them are special canonical models, which suggests that canonical models may provide a general modeling framework for risk assessment (Voit and Schubauer-Berigan, 1998;

Voit, 2000bc). Maybe the closest connection between metabolic modeling and statistic is seen in the analysis of metabolic systems with statistically distributed parameters. Not much has been done in this area, but it is clear that questions of susceptibility to drugs or environmental stressors need to deal with such questions. For the simplest cases of a single distributed input, Voit (1996a) showed that a metabolic model can be formulated as a complicated function that transforms the statistical distribution of input into some distribution of an output metabolite of interest.

REFERENCES

[14], [36–7], [50], [67–9], [84], [89], [103], [135], [137–8], [149], [155–7], [164–7], [171–3], [194], [212–13], [217–18], [226], [230], [251], [254], [258–9], [266–8], [279], [281], [287], [299–303], [304–5], [310–12], [316–17], [319–28], [330–2], [340–2], [344–9], [352], [373], [389–90], [392], [394–6], [407], [419–20], [422–4], [426], [433–42], [445–52], [454], [456–61], [464–70], [473], [482], [488–9], [520–1].

Appendix

The level of mathematical literacy among biochemists varies dramatically. At one end of the spectrum there are some who are very sophisticated and use complex mathematical concepts in their daily research, while their colleagues at the other end may dread mathematics as soon as it goes beyond high school algebra. This book is intended as an introduction to biochemical modeling for those with an interest in the topic, whether they have a solid mathematical background or not. As a compromise between not boring the mathematically astute and not alienating the less trained, relevant college mathematics has been taken out of the body of the book and put in this appendix.

The appendix begins with a listing of topics with which the reader should be familiar. This is followed by a review of some mathematical terminology and some tools that are not necessarily a part of a biochemist's training but that are the basis for understanding this text. The newly introduced terms and tools are illustrated, where feasible, with examples of some biochemical relevance. There is absolutely no claim of completeness in any of the topics presented. For example, from the huge body of knowledge about matrices, we just choose a few techniques that are most useful in the given context. We will not prove theorems or develop techniques from first principles. Instead, mathematics is used as a tool whose validity we will not question. It is hoped that this goal-oriented hands-on approach will appeal especially to those whose mathematical foundation is not particularly solid. What's primarily required of the reader is an interest in learning the concepts of biochemical systems analysis and a willingness to pursue new ways of thinking.

REQUIRED PRIOR KNOWLEDGE

The reader should be familiar with the mathematical notations introduced in high school, with indexed variables, and with the notation of sums and products. We will use some Greek letters and, on rare occasion, basic terms associated with sets and elements.

The reader should have a good working knowledge of the linear, exponential, natural logarithmic, and power-law functions, as well as simple polynomials and rational functions. Relationships between the exponential and logarithmic functions should be firmly understood. The text assumes familiarity with the notion and computation of ordinary derivatives of these functions. Except for one brief background section in Chapter 3, knowledge about integration is not required. While not absolutely necessary, a general understanding of the method of linear regression is helpful in Chapter 7.

OVERVIEW OF CONCEPTS DESCRIBED IN THIS APPENDIX

1. *Functions of a single variable:* General features, Taylor approximation, linearization, power-law approximation.
2. *Functions of several variables:* General features, partial derivatives, power-law approximation.
3. *Linear algebra:* Vectors and matrices, determinants, systems of linear equations, inverse matrices.

FUNCTIONS OF A SINGLE VARIABLE

General Features

Suppose P and Q are two sets; that is, they are collections of "things," any sort of entities, or as we say in general, *elements*. The elements could be people, metabolites, real numbers, or any other entities. The sets P and Q may contain the same types of elements or entirely different elements. The elements in P have the generic name p and the elements in Q the generic name q. Symbolically, we write $p \in P$ and $q \in Q$ and say "p is an element of P and q is an element of Q."

A *function* associates in a unique fashion an element q to an element p. For instance, let P be the set of all people and Q the set of all women. We define f as the function that associates to each individual his or her natural mother. There is exactly one natural mother for everybody, so all elements $p \in P$ are accounted for; at the same time, a mother can have several children, which translates into f associating several p with the same q. Furthermore, not every woman is a mother, and thus, not every $q \in Q$ is "reached" by the "motherhood function." Thus, it is not necessary that different elements of P be associated with different elements of Q or that each q has a counterpart in P. The only requirement is that for each $p \in P$ there is one and only one corresponding $q \in Q$.

If Q is the set of positive numbers, then the rule of associating the age to a person is a function. Another example of a function is the operation that associates to every real number its negative. In this case, P and Q both are the set of all real numbers, $\mathbb{R}$, all elements of P are accounted for, and this time all elements of Q are reached by the function. In contrast, the association between student and teacher is not a function, since most students have several teachers and most teachers have many students. The rule $1/p$ also is not a function for all real numbers, since no real number q is associated with $p = 0$. However, $1/p$ is a function for all real numbers except 0.

APPENDIX

There are different notations for a function. Since we almost always deal with sets of numbers, we use the traditional notation, $q = f(p)$, which is shorthand for "q is a function of p, and the name of the function is f." The set P of which p is an element is called the *domain* of the function. The set Q of which q is an element is called the *counterdomain*. The set of all elements $q \in Q$ that are actually reached by the function constitutes its *range*. p is called the *independent variable* or the *argument*, and q the *dependent variable*, since its value depends on the value of p.

Instead of one independent and one dependent variable, a function can have several independent and dependent variables. For instance, the ordinary sum is a function that associates one dependent variable, z, with two independent variables, x and y. We write a function of two arguments in symbolic form as $z = f(x, y)$. It is also possible that a function has one independent and two dependent variables. For instance, the independent variable could refer to a particular person and the dependent variables to the person's height and weight.

It is clear that dependent variables are crucial players in biochemical systems analysis. They are the components of a system that respond to influences from the inside or outside in complicated ways that we try to understand. Depending on the conditions, dependent variables can increase or decrease, oscillate, or even disappear. Biochemical systems analysis is also interested in independent variables. These have a different character, since they don't change during an experiment and are often under the experimenter's control. An example is the pH of a buffer. Dependent and independent variables are discussed in detail in Chapter 1.

When a dependent variable X_i changes during an experiment, we say in mathematical terminology that it *depends on time*, or *is a function of time*. Roughly speaking, a function provides us with a rule that associates one and only one value of X_i with each point in a given set of (time) points. As an example, consider data published by Berry et al. (1990), who incubated hepatocytes from fasted rats with various concentrations of glucose in the absence of inhibitors. Given a particular glucose concentration, they studied the change in lactate concentration over time. Incubated in 80 mM glucose, the hepatocytes produced the following concentration of lactate per vessel:

Incubation period (min)	*Lactate/Vessel (µmol)*
5	0.5
10	1.3
20	3.4
40	6.2
60	6.8
80	6.8
100	6.2

The results show that lactate production is a function of incubation time, since for each incubation time there is one and only one lactate concentration, as it is mathematically required from a valid function. For a different glucose concentration the produced lactate is different, but again, there is only one concentration at any given time.

Repeated experiments may be plotted on the same graph, thus giving the impression that several lactate concentrations are measured at one time point. While this summary of several experiments is a legitimate graphical representation, it does not constitute a function in the strict mathematical sense. Note in the above example that the same lactate concentration was measured at 40 and 100 min. There is nothing mathematically wrong with this, since the only requirement is the uniqueness of a measurement for each one point in time.

Time can be represented as absolute time, such as "February 25, 1995, 3:15:00 p.m.," or through relative time, such as "10 min," meaning "10 minutes after the beginning of the experiment." In unambiguous cases the unit is often omitted, and we allow expressions like "at $t = 10$," meaning "exactly 10 time units after starting the experiment." To address the value of X_i exactly 10 time units after starting the experiment, we write X_i at $t = 10$, $X_i(t = 10)$, or $X_i(10)$.

For the above experiment, one may call the lactate concentration X_1, and can formulate the experimental results as

$X_1(5 \text{ min}) = 0.5 \ \mu\text{mol},$
$X_1(10 \text{ min}) = 1.3 \ \mu\text{mol},$
$X_1(20 \text{ min}) = 3.4 \ \mu\text{mol},$
$X_1(40 \text{ min}) = 6.2 \ \mu\text{mol},$
$X_1(60 \text{ min}) = 6.8 \ \mu\text{mol},$
$X_1(80 \text{ min}) = 6.8 \ \mu\text{mol},$
$X_1(100 \text{ min}) = 6.2 \ \mu\text{mol},$

and thus define a function of concentration versus time.

Instead of listing all time points, a function can often be expressed in a more succinct form. For k time points $t_1, t_2, t_3, \ldots, t_k$, we can formulate the time dependence of X_i with an expression like

X_i at all time points t_j,

(where j can be any number between 1 and k, 1 and k included), or, more succinctly,

$X_i(t_j), \quad j = 1, 2, \ldots, k.$

Note that this notation involves two indices, i and j, that have not much to do with each other: i identifies a dependent variable, while j represents all time points of interest. As long as the correct representation is assured, we can code an index with a different symbol, without any change in meaning. For example, the previous notation signals exactly the same as

$X_i(t_h), \quad h = 1, 2, \ldots, k.$

If X_i is defined at a limited number of time points, as in the examples above, we say that $X_i(t)$ is a *discrete* function. In contrast, it is also possible mathematically to define X_i for *all* time points of a certain time period, even though we cannot measure or observe all these X_i. For example, we can define X_i to represent the concentration of a metabolite during the time interval between 1:00 p.m. and 8:00 p.m., or, in generic terms, between the time points t_b (begin) and t_e (end). In this case, we write

$$X_i(t), \quad t \in [t_b, t_e].$$

The expression $t \in [t_b, t_e]$ means *for each time point t that is (an element of the time interval) between t_b and t_e*. In contrast to the previous expression, $X_i(t_j)$ ($j = 1, 2, \ldots, n$), X_i is now defined for all time points of the given interval, not just for a limited number of selected time points. The typical graph for such a function is some line that may go up or down and associates one and only one value of X_i with each time point. The notation $t_b \leq t \leq t_e$ is used equivalently for the expression $t \in [t_b, t_e]$.

Even if the system variable X_i is defined for all time points of an entire interval, X_i may exhibit *jumps*. For example, if X_i represents light intensity and the light switch is turned on and off, the graph of X_i versus time exhibits jumps.

If the values of X_i do not jump within the time interval $[t_b, t_e]$, X_i is called *continuous*. Almost all functions used in this book are continuous and, in addition, they are *smooth*. Smoothness in this context means that the graph of the function has no corners and that it has a unique slope at each point. As a rule of thumb, one of my professors in college, Dr. P. Dombrowski, said that "a continuous functions has no holes" and that "a smooth function, in addition, allows one to ride a bicycle along its graph without stopping." A continuous graph with corners is not smooth, because it requires the bicyclist to come to a stop and to reposition the bicycle in the new direction.

The shape of the graph of X_i versus t may be simple or very complicated. Simple functions of great biochemical interest are *strictly monotone functions*. These functions either constantly increase or constantly decrease but never remain constant or change from increasing to decreasing or from decreasing to increasing. Examples of strictly monotone functions include growth functions and survival functions. Strictly monotone functions never have a peak or a valley, never overshoot or undershoot, and never oscillate. They never assume the same value twice. Oscillations are the paradigm of non-monotonic functions.

Taylor Approximation

Chapter 3 discusses at some length why we need approximations and for what qualities we should look in the selection of an appropriate approximation. This section of the appendix focuses on the mathematical details, without providing much rationale.

In loose terminology, an approximation is some mathematical construct that more or less behaves like the "original": some sort of a model that mimics reality. For more mathematical rigor, we define an approximation as a function that assumes exactly the same value as the approximated function for at least one point, the so-called *operating point*, and that has values very similar to the approximated function close to this operating point. The approximation may be given as an explicit function, such as a linear function, or in the form of a differential equation or a set of differential equations.

If the approximating function consists of a straight line, we called it a *linear approximation* or a *linearization*, and if it is curved we call it a *nonlinear approximation*.

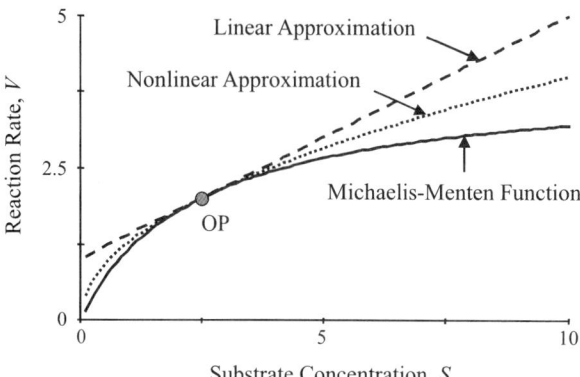

Figure A.1. Michaelis–Menten rate law ($V_{max} = 4$, $K_M = 2.5$) with linear and nonlinear (power-law) approximations. The operating point (OP) is chosen for a substrate concentration S equaling K_M and a corresponding rate V of $V_{max}/2 = 2$.

Figure A.1 shows the Michaelis–Menten rate law approximated with a linear and a nonlinear (power-law) function. All three coincide at the operating point (OP), which in the illustration is chosen to be at K_M.

One of the most important theorems in mathematics is attributed to Brook Taylor (1685–1731). It states that any smooth function (one that has sufficiently many continuous derivatives) can be approximated by a polynomial and prescribes how the coefficients of this polynomial are to be determined. If the polynomial has infinitely many terms, it is an exact representation of the original function, if the latter is sufficiently well behaved. To indicate the infinite number of polynomial terms, some texts call this structure a *Taylor series*, rather than a polynomial. If the polynomial contains a finite number of terms, it may differ from the original. In this case, it is called a *truncated Taylor series* or a *(truncated) Taylor polynomial*. How much the original and the truncated Taylor polynomial differ depends on the particular function and the number of polynomial terms. The coefficients of the polynomial are computed through differentiation of the original function.

Specifically, if the function $f(x)$ has at least n continuous derivatives, it is approximated at the operating point p by

$$f(x) \approx f(p) + f'(p)(x-p) + \frac{1}{2!} f''(p)(x-p)^2 + \frac{1}{3!} f'''(p)(x-p)^3 + \cdots$$
$$+ \frac{1}{(n-1)!} f^{(n-1)}(p)(x-p)^{n-1}. \tag{A.1}$$

The expressions $2!$, $3!$, $(n-1)!$ are called *factorials*. They are merely a shorthand for products of all positive integers up to the given number, e.g., $6! = 1 \times 2 \times 3 \times 4 \times 5 \times 6$. The symbols f', f'', f''', and $f^{(n-1)}$ denote the first, second, third, and $(n-1)$st derivative of $f(x)$, respectively. It is not difficult to see that the original function f and the Taylor approximation are exactly the same at the operating point p: Just substitute p for x.

Taylor does not prescribe how one should pick n, the number of terms in the polynomial. Usually, the choice is a compromise between accuracy and computational effort. In most cases the overall accuracy improves when we specify a larger number n, but it is obvious that we have to have more computing to do with each added derivative.

Linearization

Probably the most widely used Taylor approximation is *linearization*, which corresponds to $n = 1$:

$$f(x) \approx f(p) + f'(p)(x - p). \tag{A.2}$$

The right-hand side indeed constitutes a linear function that goes through the point $(p, f(p))$ and has the same slope as the original $f(x)$ at this point. The first fact becomes evident when we evaluate the approximation for $x = p$, which results in $f(p)$. The second fact can be seen when we compare the right-hand side with the general linear function $y = mx + b$, where m is the slope. In our case, m corresponds to $f'(p)$, which, as the derivative, is equal to the slope of the original function $f(x)$ at the operating point p. The y-intercept b is $f(p) - pf'(p)$.

The replacement of a complicated function $f(x)$ with a simple linear function may not appear to be much of an approximation, since it often deviates quite rapidly from the original function (see Fig. A.1). However, it is often more efficient to piece together many linearizations, computed at different operating points, than to compute many of the higher derivatives. Furthermore, the crude approximation turns out to be sufficient in many applications. Linearization is a cornerstone of mathematical analysis in the engineering sciences.

There is a distinct difference between linear approximation and linear regression. Linear approximation doesn't use data, but requires a function (known or unknown) that is being approximated. It replaces this function with a straight line, and the two are exactly the same at the operating point. Linear regression doesn't use an original function or an operating point. Instead it requires data points, and averages a trend in these points by providing a straight line that overall deviates from the data as little as possible.

Power-Law Approximation

While simple and successful in engineering, linearization is not often satisfactory in biochemistry, because the original functions are too strongly curved to be accurately represented by a straight line. A compromise is possible, though. Experience has shown that linearization provides good representations when one first moves from the usual straight Cartesian coordinates into logarithmic coordinates. In the case of a Michaelis–Menten rate law this step would mean expressing the logarithm of the rate of product formation as a function of the logarithm of substrate concentration. Once this *transformation* is executed, the new function is linearized as shown above. Upon this linearization in the logarithmic coordinate system, one returns to the straight Cartesian coordinates, where the result is nonlinear.

This may appear to be a cumbersome and baroque procedure. Fact is that this type of approximation is very useful in that it provides a good nonlinear representation that, to some degree, can be analyzed with linear methods. Furthermore, there is a shortcut that circumvents the transformations between Cartesian and logarithmic coordinate systems and directly yields the final result. It has been shown that this type of linearization in logarithmic coordinates always leads to power-law functions (cf. Savageau, 1976; Voit, 1991: Chapter 2). Thus, the bottom line is

$$f(x) \approx \alpha X^g. \tag{A.3}$$

Where is the operating point? Where are the derivatives? All features of the linear Taylor approximation are actually there. They are not explicitly visible but contribute to the definition of the parameters α and g. In the logarithmic coordinate system, g is the slope of the linearization, which corresponds to the slope $f'(p)$ in the Cartesian linearization. α is a factor that assures that the approximation is exact at the operating point p. The parameters of the power-law approximation are defined as follows. The exponent is given as

$$g = \frac{df}{dx} \cdot \frac{x}{f(x)}, \tag{A.4}$$

which is evaluated at the operating point p. Interpreted differently, g is the derivative of the logarithm of f with respect to the logarithm of x. That is, if we plotted the logarithm of f versus the logarithm of x, the tangent at the operating point p would have the slope g.

The multiplier α is defined as

$$\alpha = f(p) p^{-g}. \tag{A.5}$$

As an example, let's compute the power-law approximation to the function

$$f(x) = 1 - \exp(-x), \tag{A.6}$$

which is zero for $x = 0$ and approaches 1 in a monotone fashion for increasing values of x. For the power-law approximation, we first compute the exponent g as

$$g = \frac{df}{dx} \cdot \frac{x}{f(x)} = \exp(-x) \cdot \frac{x}{1 - \exp(-x)}, \tag{A.7}$$

which is to be evaluated at an operating point of our choice. For example, if the operating point is chosen as 0.5, we obtain $g = 0.7708$. In the second step, we compute the corresponding multiplier as

$$\alpha = f(p) p^{-g} = 0.3935 \times 0.5^{-0.7708} = 0.6713. \tag{A.8}$$

If we choose $p = 0.75$, the parameters of the power-law approximation are $g = 0.6714$ and $\alpha = 0.64$. Thus, the mathematical structure of the approximation is independent of the operating point, but its numerical values are not. The parameters of the power-law approximations are functions of the operating point.

Power-law approximations combine many of the convenient properties of the linear approximation with the curvilinear nature of responses in biochemical systems. There is no proof that this approximation is optimal, but there is a lot of circumstantial evidence attesting to its biochemical validity. Some of this evidence is presented in Chapter 3.

A very interesting fact about the linear and power-law approximation technique is that we can symbolically formulate the approximating functions even if the original, approximated function is not known. Whatever the original function, the linearization always takes the general form

$$y = mx + b, \tag{A.9}$$

and the power-law approximation always takes the form

$$z = \alpha x^g. \tag{A.10}$$

Of course, if the original function is unknown, we don't know the values of the parameters m, b, α, and g, yet we do know the mathematical structure of the approximating functions. This is of great importance, because it allows us to formulate approximate models symbolically for biochemical processes that are not understood mechanistically in terms of substrates, enzymes, modulators, and effectors. Chapter 5 uses these facts for the estimation of parameter values of approximate models from experimental data.

FUNCTIONS OF SEVERAL VARIABLES

General Features

All functions discussed in the previous section had only one argument. Studying the responses of a dependent variable X_i, we frequently find that this variable does not just change with time but depends on other dependent and independent variables. It is only logical to express these dependences by means of functions. For instance, in an earlier example of lactate production in hepatocytes, the amount of lactate produced depends on the concentration of glucose: For each (fixed) glucose concentration, the production of lactate is a different function of time, as shown in Fig. A.2. Focusing on glucose concentration rather than on incubation time as the independent variable, we can say that lactate production is a function of the glucose concentration. We can eliminate time as a variable, for instance, by collecting all data exactly one hour after incubation. Upon this shift in emphasis, the data of Berry et al. (1990) can be written as

Glucose Concentration (mM)	Lactate /Vessel after 1 h of Incubation (μmol)
10	0.8
20	3.0
40	5.3
80	6.8

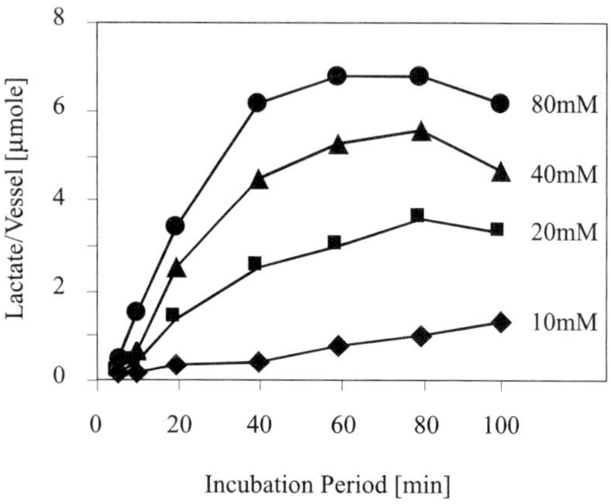

Figure A.2. Effects of glucose concentration (10 to 80 mM) on lactate formation in hepatocyes. Redrawn from Berry et al. (1990).

If we code lactate as X_1 and glucose as X_2, we can refer to $X_1(X_2)$ or make mathematical statements such as

$$X_1(10 \text{ mM}) = 0.8 \ \mu\text{mol}.$$

In addition to glucose, hepatocytes require sodium fluoride, and one could obtain data on lactate production as a function of sodium fluoride concentration. Like the time dependence of X_1 or to the functional relationships between lactate (X_1) and glucose (X_2), the functional dependence of lactate on sodium fluoride (X_3) can be abbreviated as $X_1(X_3)$.

In complicated pathways, a dependent variable X_i may be a function of many other variables, and in such a case all variables that affect X_i are listed in parentheses, separated by commas. In the above example of lactate production in incubated hepatocytes, the simultaneous dependence on glucose and sodium fluoride can be written as $X_1(X_2, X_3)$ or, explicitly including time dependence, as $X_1(t, X_2, X_3)$. Of course, these expressions themselves are almost devoid of information, since they do not specify *how* X_1 depends on t, X_2, and X_3. This type of information is provided by an assignment in form of an equation.

For example, the dependent variable X_4 may be a function of three other variables X_1, X_2, and X_3, which could be defined by the assignment

$$X_4 = X_4(X_1, X_2, X_3) = X_1 \exp(X_2 + X_3). \tag{A.11}$$

In this particular case, X_4 is equivalent to X_1 multiplied by the exponential of the sum of X_2 and X_3.

As second example, suppose X_6 is the product of a constant c, the variable X_1, and the variable X_8 squared:

$$X_6(X_1, X_8) = cX_1 X_8^2. \tag{A.12}$$

The graph of X_6 as a function of X_1 and X_8 is a surface in three-dimensional space, since for each *pair* of values, one for X_1 and one for X_8, there is exactly one value of X_6. Two axes represent the variables X_1 and X_8, and a third axis provides the scale

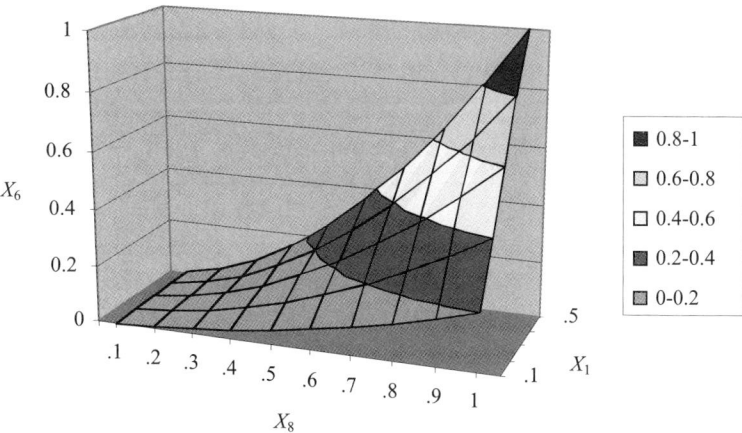

Figure A.3. X_6 is a function of two independent variables: $X_6(X_1, X_8) = 2X_1 X_8^2$. The graph is shown for $X_1 \in [0.1, 0.5]$ and $X_8 \in [0.1, 1]$.

for X_6. Of course, one cannot really show a three-dimensional object on a flat piece of paper, but the *pseudo-three-dimensional* plot of Fig. A.3 gives an impression of what X_6 looks like for X_1 between 0 and 0.5 and X_8 between 0 and 1, if $c = 2$.

In addition to a three-dimensional representation, one can acquire some feel for the function by studying slices of the three-dimensional graph. One fixes one of the two variables, X_1 or X_8, and plots how X_6 changes as a function of just the other variable. In Fig. A.3, these slices would follow the grid lines. When we fix $X_8 = 1$, then X_6 reduces to the simple linear function $X_8 = cX_1$. For $X_8 = 0.5$, X_6 is still a linear function, but the slope now is $cX_8^2 = 0.25c$. For $X_8 = 0$, X_6 is constant at the value 0. When we fix X_1, X_6 is a parabola of the form kX_8^2, where k is equal to c times the fixed value of X_1. For any fixed X_8, the graphs of X_6 versus X_1 are slices of the three-dimensional graph of $X_6(X_1, X_8)$ that are parallel to the X_8-axis, and for any fixed X_1, the graphs of X_6 versus X_8 are slices parallel to the X_1-axis.

To plot X_4 of the earlier example as a function of the three variables X_1, X_2, and X_3 is impossible, since the graph is an object in four-dimensional space. However, we can again develop some impression of the function by studying two-dimensional or three-dimensional "slices." For three-dimensional representations one fixes one of the three variables X_1, X_2, X_3 and studies X_4 as a function of the remaining two, and for two-dimensional graphs one fixes two of the three variables, X_1, X_2, X_3 and studies X_4 as a function of the remaining one.

When X_i is a function of other dependent variables, and these are functions of time, then X_i is a function of functions of time and, therefore, itself a function of time. Consequently, if all functional relationships between X_i and the relevant dependent variables and all functional relationships between the relevant dependent variables and time are known, X_i can be plotted as a simple graph against time. This graph shows the temporal behavior of X_i and only implicitly represents the dependence of X_i on other system variables. Thus, we can analyze different aspects of the same dependent variable by studying its temporal behavior or its dependence on other variables. These two aspects complement each other and are of equal importance.

As an example, suppose X_1 is a function of X_2, say $X_1(X_2) = \exp(X_2)$, and X_2 is a function of time, say $X_2(t) = t - 1$. Then $X_1(t) = X_1(X_2(t)) = \exp(t - 1)$. Both X_1 and X_2, can be studied as functions of time, and in addition, a graph of X_1 versus X_2 shows directly how X_1 depends on X_2, namely in an exponential fashion.

Partial Derivatives

In some cases, we may be interested in the change of a function of several variables with respect to a change in one of these variables. For instance, we may ask: How does the function $V(X_1, X_2, X_3, \ldots, X_n)$ change in response to a change in X_3? This type of change is formally computed by evaluating the ordinary derivative of V with respect to X_3, while considering all other X's as constants. As an example, suppose V is a product of power functions:

$$V(X_1, X_2, X_3, \ldots, X_n) = \alpha X_1^{g_1} X_2^{g_2} X_3^{g_3} \cdots X_n^{g_n}. \tag{A.13}$$

The rate of change of V with respect to X_3 is thus

$$\begin{aligned} \frac{\partial V}{\partial X_3} &= \left(\alpha X_1^{g_1} X_2^{g_2} X_4^{g_4} \cdots X_n^{g_n} \right) \cdot g_3 X_3^{g_3 - 1} \\ &= \alpha g_3 X_1^{g_1} X_2^{g_2} X_3^{g_3 - 1} X_4^{g_4} \cdots X_n^{g_n}. \end{aligned} \tag{A.14}$$

Rates of change of this type are referred to as *partial derivatives*, where the term *partial* indicates that the change in V is computed only with respect to one of its arguments and not with respect to several or all variables simultaneously. Partial derivatives are notationally distinguished from ordinary derivatives through the use of the "curly d" symbol ∂.

It is useful to define for variables of several variables a measure of overall change, not just in one direction, such as X_3 in the above example, but in any general direction that is composed of simultaneous changes in several variables. This type of change is expressed in the *total differential* and defined for a function $V(X_1, X_2, \ldots, X_n)$ as

$$dV = \frac{\partial V}{\partial X_1} dX_1 + \frac{\partial V}{\partial X_2} dX_2 + \frac{\partial V}{\partial X_3} dX_3 + \cdots + \frac{\partial V}{\partial X_n} dX_n. \tag{A.15}$$

One may intuitively grasp the meaning of this definition by considering the partial derivatives as slopes in the directions of the axes of the coordinate system, which are represented by $X_1, X_2, \ldots, X_n$, and the terms $dX_1, dX_2, \ldots, dX_n$ as the lengths of the steps taken in these directions. The concept of the total differential has nothing to do with "infinitesimally small quantities." Instead, it describes the linear component of the multivariate, nonlinear function V (cf. Courant, 1972: p. 61). This linear component approximates the nonlinear function V well if the steps dX_i are small.

Power-Law Approximation in Several Variables

Because of their greater complexity, functions of several variables need to be approximated more often than functions of a single argument. The philosophy and much of the mathematical machinery are the same, except that we are now dealing with partial instead of ordinary derivatives. For the theoretical foundation, the starting point again is Taylor's theorem, which this time is formulated for functions in several variables. A prominent special case again is linearization, where only the first derivatives are retained. There are several first derivatives now, one for each variable.

As said in the beginning of the Appendix, it is not our goal to strive for completeness, and therefore we limit our discussion to the multivariate power-law approximation, which again corresponds to (Taylor) linearization in logarithmic coordinates. We will discuss a function of n variables, but as an intuitive example for the following, imagine a function of just two variables, such as a process rate that depends on a substrate and an inhibitor.

Suppose the function V depends on n variables X_i and is written as $V(X_1, X_2, X_3, X_4, \ldots, X_n)$. To compute the power-law approximation, we again choose an operating point P. In contrast to the earlier one-variable procedure, the operating point now has n coordinates $p_1, p_2, \ldots, p_n$. As before, we move to logarithmic coordinates for each X-direction and for V; linearize in each direction, according to Taylor's theorem; and return to the straight, Cartesian coordinates. As in the one-dimensional case, we can circumvent almost all detail with shortcuts. The result is analogous to the one-variable case: the function V can be approximated by a product of power-law functions of the form

$$V = \alpha \cdot X_1^{g_1} X_2^{g_2} X_3^{g_3} X_4^{g_4} \cdots X_n^{g_n}. \tag{A.16}$$

In strict analogy to previous results, the parameters g_i are the slopes of the approximating function (in logarithmic coordinates), and the factor α assures that the approximated function and the approximation exactly coincide at the operating point of our choice.

Since we have several variables now, the parameters g_i are computed through *partial* differentiation of V with respect to one of the X's. They are given as

$$g_i = \frac{\partial V}{\partial X_i} \cdot \frac{X_i}{V}, \tag{A.17}$$

which are evaluated at the operating point P with the coordinates $p_1, p_2, \ldots, p_n$. The factor α is computed once all g_i are determined. It is given as

$$\alpha = V(p_1, p_2, \ldots, p_n) p_1^{-g_1} p_2^{-g_2} \cdots p_n^{-g_n}. \tag{A.18}$$

As an example, consider the function

$$X_4 = X_4(X_1, X_2, X_3) = X_1 \exp(X_2 + 2X_3). \tag{A.19}$$

Without transporting the function into a logarithmic coordinate system, we directly compute the exponents as partial derivatives, multiplied by the appropriate X and divided by the approximated function. Thus,

$$g_1 = \frac{\partial X_4}{\partial X_1} \cdot \frac{X_1}{X_4} = \exp(X_2 + 2X_3)\frac{X_1}{X_1 \exp(X_2 + 2X_3)} = 1, \tag{A.20}$$

$$g_2 = \frac{\partial X_4}{\partial X_2} \cdot \frac{X_2}{X_4} = X_1 \times 1 \times \exp(X_2 + 2X_3)\frac{X_2}{X_1 \exp(X_2 + 2X_3)} = X_2, \tag{A.21}$$

$$g_3 = \frac{\partial X_4}{\partial X_3} \cdot \frac{X_3}{X_4} = X_1 \times 2 \times \exp(X_2 + 2X_3)\frac{X_3}{X_1 \exp(X_2 + 2X_3)} = 2X_3, \tag{A.22}$$

which are to be evaluated at the operating point of choice. In this particular case, g_1 does not depend on the operating point; it is always equal to 1. This is so because the function X_4 happens to be a product of X_1 and some other term, and X_1 in itself is already a (very simple) power-law function. g_2 and g_3 in this example are equal to p_2 and $2p_3$, where p_2 and p_3 are the second and third coordinates of the operating point. Other examples are found throughout the book, and especially in Chapter 5.

Because of its conceptual importance, it is useful to reiterate comments about the linear and power-law approximation techniques in one variable, namely, that we can symbolically formulate an approximating function even if the original, approximated function is not known. The analogue is true in this multivariate case: for a function V of n variables $X_1, X_2, \ldots, X_n$, the power-law approximation always takes the form

$$V = \alpha \cdot X_1^{g_1} X_2^{g_2} X_3^{g_3} X_4^{g_4} \cdots X_n^{g_n}. \tag{A.23}$$

This fact allows us to construct mathematical models symbolically for very complex biochemical systems. Again, if the original function is unknown, we don't know the values of the parameters α and g_i, yet we do know the mathematical structure of the approximating functions. Chapter 5 uses this fact for the estimation of parameters from experimental data.

LINEAR ALGEBRA

Even though the biochemical models used in this book are nonlinear, important analyses lead to systems of linear equations. In two or three variables, these systems can be solved with the method of substitution in which, for example, two equations are expressed in terms of the first variable, X_1, and equated, whereby X_1 is eliminated. For systems of more than three or four variables, these methods are cumbersome and prone to errors. The mathematical discipline of linear algebra has addressed these problems by developing convenient notations and methods that exclusively deal with systems of linear equations. We will make use of these methods in symbolic analyses of steady states and therefore introduce them here as far as we shall need them. Readers not interested in symbolic analyses may be able to get by without this section and without Chapters 6 and 7. However, the concepts here are essential for a deeper understanding of steady-state and sensitivity analyses and, to some degree, of parameter estimation. The notion of vectors and matrices is considered by some as something unnecessarily complicated, but in fact, these concepts are convenient tools that help us with the bookkeeping in large systems.

Vectors and Matrices

A *vector* is a collection of numbers or symbols that can naturally be lined up. For instance, we can address the four quantities y_1, y_2, y_3, and y_4 simultaneously as the vector

$$\begin{pmatrix} y_1 \\ y_2 \\ y_3 \\ y_4 \end{pmatrix}. \tag{A.24}$$

We say that this vector has the *components* y_1, y_2, y_3, and y_4, and denote it as y or $\vec{y}$:

$$\vec{y} = \begin{pmatrix} y_1 \\ y_2 \\ y_3 \\ y_4 \end{pmatrix}. \tag{A.25}$$

When we write the components in a column, $\vec{y}$ is called a *column vector*. It is sometimes convenient to write the components of a vector in a row instead of a column. This is indicated by the superscript tr, which stands for *transposed*:

$$\vec{y}^{tr} = \begin{pmatrix} y_1 \\ y_2 \\ y_3 \\ y_4 \end{pmatrix}^{tr} = (y_1, y_2, y_3, y_4). \tag{A.26}$$

Transposing a column vector yields a *row vector*, and transposing a row vector returns a column vector; for instance:

$$(y_1, y_2, y_3, y_4)^{tr} = \begin{pmatrix} y_1 \\ y_2 \\ y_3 \\ y_4 \end{pmatrix}. \tag{A.27}$$

Quantities a_{ij} with two indices can be organized in the form of a vector, but it is more convenient to order them with respect to i and j. For this purpose, we define a two-dimensional array, which is called a *matrix*, and denote it with a bold capital letter, such as **A**. The first row of this matrix contains all quantities a_{ij} whose first index is 1, the second row contains all quantities a_{ij} whose first index is 2, and so on. At the same time, the first column of this matrix contains all quantities a_{ij} whose second index is 1, the second column contains all quantities a_{ij} whose second index is 2, and so on. For $n = 4$, the matrix reads

$$\mathbf{A} = \begin{pmatrix} a_{11} & a_{12} & a_{13} & a_{14} \\ a_{21} & a_{22} & a_{23} & a_{24} \\ a_{31} & a_{32} & a_{33} & a_{34} \\ a_{41} & a_{42} & a_{43} & a_{44} \end{pmatrix}. \tag{A.28}$$

If a matrix contains n rows and m columns, we call it an $n \times m$ matrix. In Eq. (A.28), $\mathbf{A}$ is a 4×4 matrix, but in general, the numbers of rows and columns need not be the same. In a formal sense, column vectors are matrices with n rows and just one column ($n \times 1$ matrices), and row vectors are matrices with just one row and n columns ($1 \times n$ matrices).

As with vectors, we may form the transpose of a matrix, in which rows become columns and vice versa, and denote it with the superscript tr. For the 4×4 matrix $\mathbf{A}$, the transpose is

$$\mathbf{A}^{\mathrm{tr}} = \begin{pmatrix} a_{11} & a_{21} & a_{31} & a_{41} \\ a_{12} & a_{22} & a_{32} & a_{42} \\ a_{13} & a_{23} & a_{33} & a_{43} \\ a_{14} & a_{24} & a_{34} & a_{44} \end{pmatrix}. \tag{A.29}$$

The transpose of a transposed matrix again is the original matrix.

In addition to vectors and matrices, we need an operation that connects matrices with vectors in a manner that is convenient for the analysis of systems of linear equations. Most relevant for us is *matrix multiplication*. It applies to matrices and vectors. Specifically, one defines the product of a matrix with a vector,

$$\mathbf{A}\vec{y} = \begin{pmatrix} a_{11} & a_{12} & a_{13} & a_{14} \\ a_{21} & a_{22} & a_{23} & a_{24} \\ a_{31} & a_{32} & a_{33} & a_{34} \\ a_{41} & a_{42} & a_{43} & a_{44} \end{pmatrix} \begin{pmatrix} y_1 \\ y_2 \\ y_3 \\ y_4 \end{pmatrix}, \tag{A.30}$$

as the column vector

$$\begin{pmatrix} a_{11}y_1 + a_{12}y_2 + a_{13}y_3 + a_{14}y_4 \\ a_{21}y_1 + a_{22}y_2 + a_{23}y_3 + a_{24}y_4 \\ a_{31}y_1 + a_{32}y_2 + a_{33}y_3 + a_{34}y_4 \\ a_{41}y_1 + a_{42}y_2 + a_{43}y_3 + a_{44}y_4 \end{pmatrix}. \tag{A.31}$$

This vector is obtained in the following way:

1. Start with the first row of the matrix.
   - Multiply the first element with the first component of the vector.
   - Add to the result the product of the second element of the first row with the second component of the vector.
   - Add to the result the product of the third element of the first row with the third component of the vector.

- Add to the result the product of the fourth element of the first row with the fourth component of the vector.
2. Continue with the second, third, and fourth rows of the matrix in the same manner.

Note that the result is actually a vector, because each row contains only one component, which happens to be expressed in the form of a sum.

With this notation, we can formulate the *matrix equation*

$$\mathbf{A}\vec{y} = \vec{b}, \tag{A.32}$$

which equates two column vectors, namely, (i) the product of the matrix $\mathbf{A}$ with $\vec{y}$, and (ii) $\vec{b}$. After execution of the matrix multiplication, this equation reads

$$\begin{pmatrix} a_{11}y_1 + a_{12}y_2 + a_{13}y_3 + a_{14}y_4 \\ a_{21}y_1 + a_{22}y_2 + a_{23}y_3 + a_{24}y_4 \\ a_{31}y_1 + a_{32}y_2 + a_{33}y_3 + a_{34}y_4 \\ a_{41}y_1 + a_{42}y_2 + a_{43}y_3 + a_{44}y_4 \end{pmatrix} = \begin{pmatrix} b_1 \\ b_2 \\ b_3 \\ b_4 \end{pmatrix}. \tag{A.33}$$

Equation (A.33) represents a system of four linear equations, the first of which is

$$a_{11}y_1 + a_{12}y_2 + a_{13}y_3 + a_{14}y_4 = b_1. \tag{A.34}$$

Note that for the multiplication of a vector $\vec{y}$ with a matrix $\mathbf{A}$, $\mathbf{A}$ must have exactly as many columns as $\vec{y}$ has components, whereas the number of rows in $\mathbf{A}$ is irrelevant. For instance, the set of three equations in four variables

$$\begin{aligned} a_{11}y_1 + a_{12}y_2 + a_{13}y_3 + a_{14}y_4 &= b_1, \\ a_{21}y_1 + a_{22}y_2 + a_{23}y_3 + a_{24}y_4 &= b_2, \\ a_{31}y_1 + a_{32}y_2 + a_{33}y_3 + a_{34}y_4 &= b_3 \end{aligned} \tag{A.35}$$

can be written as the matrix equation

$$\mathbf{A}\vec{y} = \vec{b}, \tag{A.36}$$

where now $\mathbf{A}$ is a 4×3 matrix and the solution vector $\vec{b}$ has only three components:

$$\begin{pmatrix} a_{11} & a_{12} & a_{13} & a_{14} \\ a_{21} & a_{22} & a_{23} & a_{24} \\ a_{31} & a_{32} & a_{33} & a_{34} \end{pmatrix} \begin{pmatrix} y_1 \\ y_2 \\ y_3 \\ y_4 \end{pmatrix} = \begin{pmatrix} a_{11}y_1 + a_{12}y_2 + a_{13}y_3 + a_{14}y_4 \\ a_{21}y_1 + a_{22}y_2 + a_{23}y_3 + a_{24}y_4 \\ a_{31}y_1 + a_{32}y_2 + a_{33}y_3 + a_{34}y_4 \end{pmatrix} = \begin{pmatrix} b_1 \\ b_2 \\ b_3 \end{pmatrix}. \tag{A.37}$$

Analogously, a set of four linear equations in three variables corresponds to a matrix equation in which the matrix is 4×3, the variable vector has three components, and the solution vector has four components.

The multiplication of two matrices $\mathbf{A}$ and $\mathbf{B}$ is very similar to the multiplication of a vector and a matrix. We just consider $\mathbf{B}$ as a collection of column vectors, perform the matrix–vector multiplication for each column vector, and write the resulting column vectors in one matrix.

For instance, if **A** is a 4×2 matrix and **B** is a 2×3 matrix, then their product is

$$\begin{pmatrix} a_{11} & a_{12} \\ a_{21} & a_{22} \\ a_{31} & a_{32} \\ a_{41} & a_{42} \end{pmatrix} \begin{pmatrix} b_{11} & b_{12} & b_{13} \\ b_{21} & b_{22} & b_{23} \end{pmatrix}$$

$$= \begin{pmatrix} a_{11}b_{11} + a_{12}b_{21} & a_{11}b_{12} + a_{12}b_{22} & a_{11}b_{13} + a_{12}b_{23} \\ a_{21}b_{11} + a_{22}b_{21} & a_{21}b_{12} + a_{22}b_{22} & a_{21}b_{13} + a_{22}b_{23} \\ a_{31}b_{11} + a_{32}b_{21} & a_{31}b_{12} + a_{32}b_{22} & a_{31}b_{13} + a_{32}b_{23} \\ a_{41}b_{11} + a_{42}b_{21} & a_{41}b_{12} + a_{42}b_{22} & a_{41}b_{13} + a_{42}b_{23} \end{pmatrix} \quad (A.38)$$

$$= \begin{pmatrix} c_{11} & c_{12} & c_{13} \\ c_{21} & c_{22} & c_{23} \\ c_{31} & c_{32} & c_{33} \\ c_{41} & c_{42} & c_{43} \end{pmatrix},$$

which is a 4×3 matrix.

It is easy to see that the dimensions of the involved matrices or vectors have to match up. In Eq. (A.37), the vector $\vec{y}$ must have four components; otherwise the association between the entries a_{ij} and the variables y_i is not defined; in Eq. (A.38), **B** must have two rows. Generally, we must require that in the matrix equation

$$\mathbf{AB} = \mathbf{C} \quad (A.39)$$

the matrix **A** have as many columns as **B** has rows. If **A** is an $n \times m$ matrix and **B** is an $m \times k$ matrix, then their product **C** is a $n \times k$ matrix. This statement covers the case that **A** is a row vector with m components and **B** is a matrix with m rows, such as

$$\begin{pmatrix} a_{11} & a_{12} & a_{13} & a_{14} \end{pmatrix} \begin{pmatrix} b_{11} & b_{12} & b_{13} \\ b_{21} & b_{22} & b_{23} \\ b_{31} & b_{32} & b_{33} \\ b_{41} & b_{42} & b_{43} \end{pmatrix}, \quad (A.40)$$

or that **B** is a column vector with m components:

$$\begin{pmatrix} a_{11} & a_{12} & a_{13} & a_{14} \end{pmatrix} \begin{pmatrix} b_{11} \\ b_{21} \\ b_{31} \\ b_{41} \end{pmatrix}. \quad (A.41)$$

In the first case, the result is a row vector with three components, and in the latter case, the result is a 1×1 matrix or, in other words, just an ordinary number, namely

$$c_{11} = a_{11}b_{11} + a_{12}b_{21} + a_{13}b_{31} + a_{14}b_{41}. \quad (A.42)$$

Determinants

A *determinant* is a number that is associated with a *square* $n \times n$ matrix **A**. It can be positive, zero, or negative. For matrices in which the numbers of rows and columns

APPENDIX

differ, no determinant is defined. The determinant is denoted as det **A** or as |**A**|, and can be written explicitly as

$$\det \mathbf{A} = |\mathbf{A}| = \begin{vmatrix} a_{11} & a_{12} & a_{13} & \cdots & a_{1n} \\ a_{21} & a_{22} & a_{23} & \cdots & a_{2n} \\ a_{31} & a_{32} & a_{33} & \cdots & a_{3n} \\ \vdots & \vdots & \vdots & & \vdots \\ a_{n1} & a_{n2} & a_{n3} & \cdots & a_{nn} \end{vmatrix}. \tag{A.43}$$

Note the slight difference in notation: A matrix is written as an array within parentheses, while a determinant is bordered by straight vertical lines.

A determinant characterizes a matrix in a way that is difficult to explain intuitively. However, for our purposes it is sufficient to know that determinants, among other features, are directly related to the solutions of matrix equations. Because of their importance in many branches of mathematics, there is a vast literature on determinants. We discuss only two issues: How to compute determinants and how to use them to solve linear equations.

The computation of a determinant depends on its size. For 2×2 and 3×3 determinants we use shortcuts. The strategy for larger determinants is to reduce them in size until one reaches 3×3 determinants.

Computation of a 2×2 determinant. The determinant associated with the matrix

$$\mathbf{A} = \begin{pmatrix} a_{11} & a_{12} \\ a_{21} & a_{22} \end{pmatrix} \tag{A.44}$$

is defined as

$$|\mathbf{A}| = \begin{vmatrix} a_{11} & a_{12} \\ a_{21} & a_{22} \end{vmatrix} = a_{11}a_{22} - a_{12}a_{21}. \tag{A.45}$$

That is, one multiplies along the diagonals and subtracts the results. Note that the result, like any determinant, is a number.

For example, Chapter 6 discusses that the steady state of a biochemical system is characterized by a matrix equation. For the pathway in Fig. 6.1 and the parameter values given in Eq. (6.2), the characterizing matrix is

$$\mathbf{A} = \begin{pmatrix} -1 & -2 \\ 1 & -0.5 \end{pmatrix}, \tag{A.46}$$

and its determinant is

$$|\mathbf{A}| = \begin{vmatrix} -1 & -2 \\ 1 & -0.5 \end{vmatrix} = 0.5 + 2 = 2.5. \tag{A.47}$$

Computation of a 3×3 determinant. The determinant associated with the matrix

$$\mathbf{A} = \begin{pmatrix} a_{11} & a_{12} & a_{13} \\ a_{21} & a_{22} & a_{23} \\ a_{31} & a_{32} & a_{33} \end{pmatrix} \tag{A.48}$$

is

$$|\mathbf{A}| = \begin{vmatrix} a_{11} & a_{12} & a_{13} \\ a_{21} & a_{22} & a_{23} \\ a_{31} & a_{32} & a_{33} \end{vmatrix}$$
$$= a_{11}a_{22}a_{33} + a_{12}a_{23}a_{31} + a_{13}a_{21}a_{32} - a_{11}a_{23}a_{32} - a_{12}a_{21}a_{33} - a_{13}a_{22}a_{31}.$$
(A.49)

One easy way to remember which multiplications have to be performed is the rule of Sarrus (see, e.g., Bronstein and Semendjajew, 1973). It says: Write down the determinant twice, side by side, as shown below:

$$\begin{matrix} a_{11} & a_{12} & a_{13} & a_{11} & a_{12} & a_{13} \\ a_{21} & a_{22} & a_{23} & a_{21} & a_{22} & a_{23} \\ a_{31} & a_{32} & a_{33} & a_{31} & a_{32} & a_{33} \end{matrix}$$
(A.50)

The elements on each of the diagonals from a_{11}, a_{12}, and a_{13} down to the *right* (solid lines) are multiplied, and the products are given a *positive* sign. The elements on each of the diagonals from a_{11}, a_{12}, and a_{13} down to the *left* (dashed lines) are multiplied, and the products are given a *negative* sign.

For example, the determinant of

$$\mathbf{A} = \begin{pmatrix} 1 & 2 & 1 \\ 0 & 1 & 0 \\ 3 & 1 & -4 \end{pmatrix}$$
(A.51)

is

$$\det \mathbf{A} = 1 \times 1 \times (-4) + 2 \times 0 \times 3 + 1 \times 0 \times 1 - 1 \times 0 \times 1$$
$$- 2 \times 0 \times (-4) - 1 \times 1 \times 3 = -7.$$
(A.52)

Computation of $n \times n$ determinants ($n > 3$). Determinants associated with $n \times n$ matrices of more than three rows and columns are slightly more complicated to compute. Several strategies are available, and each of them may be best in a given situation. In order to keep things simple, we just show one standard method, known as *expansion by the first row* and credited to Pierre Simon Laplace (1749–1827). This is what needs to be done:

1. Associate alternately plus signs and minus signs to the elements in the first row, starting with a plus sign:

$$a_{11}^+ \quad a_{12}^- \quad a_{13}^+ \quad a_{14}^- \quad \cdots.$$

2. Define n new determinants associated with $(n-1) \times (n-1)$ matrices $\mathbf{A}_{1j}$ that are generated from $\mathbf{A}$ by deleting the first row and one column at a time. For instance,

APPENDIX

A_{11} is obtained from A when the first row and the first column are eliminated, and A_{13} is

$$|A_{13}| = \begin{vmatrix} a_{21} & a_{22} & a_{24} & \cdots & a_{2n} \\ a_{31} & a_{32} & a_{34} & \cdots & a_{3n} \\ a_{41} & a_{42} & a_{44} & \cdots & a_{4n} \\ \vdots & \vdots & \vdots & & \vdots \\ a_{n1} & a_{n2} & a_{n4} & \cdots & a_{nn} \end{vmatrix}. \tag{A.53}$$

These reduced determinants are sometimes called *minors* of $|A|$ (e.g., Kreyszig, 1993: p. 372). Notice that the last values of the first and the second index are still n, since the first row and the third column have been eliminated.

3. Multiply each element a_{1j} of the first row of A by the sign that was determined in step 1, and by the determinant $|A_{1j}|$. The result is the desired determinant, expressed in terms of determinants of size $(n-1) \times (n-1)$:

$$|A| = a_{11}|A_{11}| - a_{12}|A_{12}| + a_{13}|A_{13}| - a_{14}|A_{14}| + \cdots - \cdots + \cdots. \tag{A.54}$$

If the original matrix A is 4×4, then the matrices A_{1j} are 3×3, and we finish up the computation with the method of Sarrus. If the original matrix A is 5×5, then the matrices A_{1j} are 4×4, and each of them must be expanded by its first row. Needless to say, for large n, the computation by hand becomes tedious, but computer programs exist to compute determinants of virtually any size.

Example. Suppose

$$A = \begin{pmatrix} a_{11} & a_{12} & a_{13} & a_{14} \\ a_{21} & a_{22} & a_{23} & a_{24} \\ a_{31} & a_{32} & a_{33} & a_{34} \\ a_{41} & a_{42} & a_{43} & a_{44} \end{pmatrix} = \begin{pmatrix} 1 & 2 & 0 & -2 \\ 1 & 0 & 0 & 1 \\ 4 & -0.5 & 2 & 0 \\ -1 & 0 & 3 & 1 \end{pmatrix}. \tag{A.55}$$

Then

$$\begin{aligned} \det A &= a_{11}|A_{11}| - a_{12}|A_{12}| + a_{13}|A_{13}| - a_{14}|A_{14}| \\ &= a_{11}(a_{22}a_{33}a_{44} + a_{23}a_{34}a_{42} + a_{24}a_{32}a_{43} - a_{22}a_{34}a_{43} \\ &\quad - a_{23}a_{32}a_{44} - a_{24}a_{33}a_{42}) - a_{12}(a_{21}a_{33}a_{44} + a_{23}a_{34}a_{41} + a_{24}a_{31}a_{43} \\ &\quad - a_{21}a_{34}a_{43} - a_{23}a_{31}a_{44} - a_{24}a_{33}a_{41}) + a_{13}(a_{21}a_{32}a_{44} + a_{22}a_{34}a_{41} \\ &\quad + a_{24}a_{31}a_{42} - a_{21}a_{34}a_{42} - a_{22}a_{31}a_{44} - a_{24}a_{32}a_{41}) - a_{14}(a_{21}a_{32}a_{43} \\ &\quad + a_{22}a_{33}a_{41} + a_{23}a_{31}a_{42} - a_{21}a_{33}a_{42} - a_{22}a_{31}a_{43} - a_{23}a_{32}a_{41}) \\ &= 1 \times (0 + 0 - 0.5 \times 3 - 0 - 0 - 0) - 2 \times (2 + 0 + 4 \times 3 - 0 - 0 + 1) \\ &\quad + 0 + 2 \times (-0.5 \times 3 + 0 + 0 - 0 - 0 - 0) \\ &= -1.5 - 30 - 3 = -34.5. \end{aligned} \tag{A.56}$$

Remarks

1. The *absolute* value of a determinant does not change when the first row is exchanged with another row; however, if an even-numbered row becomes the first row, we have to multiply the resulting determinant by -1. It is often advantageous

for expansion to write the row with the most zeros as the first row, because then many terms in the expansion are zero.
2. If two rows or two columns of $\mathbf{A}$ are equal, then det $\mathbf{A} = 0$.
3. det $\mathbf{A}^{tr}$ = det $\mathbf{A}$. Consequently, if one column contains many zeros, it is advantageous to compute det $\mathbf{A}^{tr}$ by transposing $\mathbf{A}$ and interchanging rows so that many zeros appears as the first row of $\mathbf{A}^{tr}$.

Example. Consider the matrix

$$\mathbf{A} = \begin{pmatrix} a_{11} & a_{12} & a_{13} & a_{14} \\ a_{21} & a_{22} & a_{23} & a_{24} \\ a_{31} & a_{32} & a_{33} & a_{34} \\ a_{41} & a_{42} & a_{43} & a_{44} \end{pmatrix} = \begin{pmatrix} 1 & 2 & 3 & -2 \\ 1 & 0 & 0 & 1 \\ 4 & 0 & 2 & 0 \\ -1 & 0 & 3 & 1 \end{pmatrix}. \tag{A.57}$$

Using the transpose

$$\mathbf{A}^{tr} = \begin{pmatrix} 1 & 1 & 4 & -1 \\ 2 & 0 & 0 & 0 \\ 3 & 0 & 2 & 3 \\ -2 & 1 & 0 & 1 \end{pmatrix}, \tag{A.58}$$

and interchanging the first and the second row, so that we have

$$\det \mathbf{A} = -\det \begin{pmatrix} 2 & 0 & 0 & 0 \\ 1 & 1 & 4 & -1 \\ 3 & 0 & 2 & 3 \\ -2 & 1 & 0 & 1 \end{pmatrix}, \tag{A.59}$$

the expansion simply becomes

$$\det \mathbf{A} = -2 \begin{vmatrix} 1 & 4 & -1 \\ 0 & 2 & 3 \\ 1 & 0 & 1 \end{vmatrix}$$

$$= -2(2 + 12 + 0 - 0 - 0 + 2) = -32. \tag{A.60}$$

Systems of Linear Equations

The goal of this subsection is to solve matrix equations of the form

$$\mathbf{A}\vec{y} = \vec{b}. \tag{A.61}$$

In the context of biochemical systems analysis, these types of equations are very important for the characterization of steady states (see Chapters 6 and 7). In a nutshell, the metabolite concentrations in a biochemical system do not change if all fluxes entering any given metabolite pool are in balance with all fluxes exiting this pool. This condition of the system is called a *steady state* or, sometimes, an *equilibrium*. If a biochemical system of n dependent variables is represented as a canonical model in S-system form (see Chapter 3), the steady state is characterized by a set of n linear equations of the type

$$a_{11}y_1 + a_{12}y_2 + \cdots + a_{1n}y_n = b_1,$$
$$a_{21}y_1 + a_{22}y_2 + \cdots + a_{2n}y_n = b_2,$$
$$\vdots$$
$$a_{n1}y_1 + a_{n2}y_2 + \cdots + a_{nn}y_n = b_n.$$

(A.62)

These equations can be written compactly as the matrix equation (A.61), and both representations contain exactly the same information. The first task of almost any biochemical systems analysis is the characterization of the system's steady state. This means finding numerical values for $y_1, y_2, \ldots, y_n$ that satisfy Eq. (A.62) or, equivalently, Eq. (A.61). The latter is actually easier, and therefore solving matrix equations is a crucial part of biochemical systems analysis.

Having made this statement, one must note that many computer programs are available for obtaining solutions of systems of linear equations if the model is fully parameterized. In particular, PLAS computes numerical steady-state solutions very efficiently. Hand computations are not irrelevant, though. Many biochemical systems analyses that address questions about the structure of a model are based on such computations (e.g., Savageau, 1976: Chapters 9–16; Irvine and Savageau, 1985ab; Lewis, 1991; Sorribas and Cascante, 1994; Hlavacek and Savageau, 1995, 1996, 1997; Ni and Savageau, 1996b).

Gabriel Cramer (1704–1752) developed a method that uses determinants to solve matrix equations of the form (A.61). It is known as *Cramer's rule* and says that each component y_i of the solution vector $\vec{y}$ is given as the ratio of two determinants: det **A**, and the determinant of a matrix $\mathbf{A}_i$, which is obtained from **A** by replacing the ith column with the vector $\vec{b}$:

$$y_i = \frac{\det \mathbf{A}_i}{\det \mathbf{A}}$$

$$= \frac{\begin{vmatrix} a_{11} & a_{12} & a_{13} & \cdots & b_1 & \cdots & a_{1n} \\ a_{21} & a_{22} & a_{23} & \cdots & b_2 & \cdots & a_{2n} \\ a_{31} & a_{32} & a_{33} & \cdots & b_3 & \cdots & a_{3n} \\ \vdots & \vdots & \vdots & & \vdots & \vdots & \vdots \\ a_{n1} & a_{n2} & a_{n3} & \cdots & b_n & \cdots & a_{nn} \end{vmatrix}}{\begin{vmatrix} a_{11} & a_{12} & a_{13} & \cdots & a_{1i} & \cdots & a_{1n} \\ a_{21} & a_{22} & a_{23} & \cdots & a_{2i} & \cdots & a_{2n} \\ a_{31} & a_{32} & a_{33} & \cdots & a_{3i} & \cdots & a_{3n} \\ \vdots & \vdots & \vdots & & \vdots & \vdots & \vdots \\ a_{n1} & a_{n2} & a_{n3} & \cdots & a_{ni} & \cdots & a_{nn} \end{vmatrix}},$$

(A.63)

which is defined as long as $\det \mathbf{A} \neq 0$.

If $\det \mathbf{A} = 0$, Cramer's rule does not work. In this case, the matrix equation either has no solution at all or it has no unique solution but infinitely many. We need to be able to identify these cases, and we do that by realizing that the determinant vanishes, but beyond this realization we shall not have to deal with these solution families. We therefore do not discuss methods for evaluating matrix equations where the system determinant is zero, and refer the reader to the book by Savageau (1976), which deals with the issues in the context of canonical S-system models.

As an example for Cramer's method, let's solve the system

$$\begin{pmatrix} 0.5 & 0 & 1 & 1 \\ -0.5 & 0.5 & 0 & 0 \\ 0 & -0.5 & 1 & 0 \\ 0 & -0.5 & 0 & 1 \end{pmatrix} \cdot \vec{y} = \begin{pmatrix} 0 \\ \ln 0.5 \\ \ln 1.6 \\ \ln 1.6 \end{pmatrix}, \quad (A.64)$$

which comes up in the body of Chapter 6 as a steady-state equation in matrix form. The determinant associated with the system matrix $\mathbf{A}$ is computed via expansion by the first row:

$$\det \mathbf{A} = 0.5 \times \begin{vmatrix} 0.5 & 0 & 0 \\ -0.5 & 1 & 0 \\ -0.5 & 0 & 1 \end{vmatrix} - 0 + 1 \times \begin{vmatrix} -0.5 & 0.5 & 0 \\ 0 & -0.5 & 0 \\ 0 & -0.5 & 1 \end{vmatrix}$$

$$- 1 \times \begin{vmatrix} -0.5 & 0.5 & 0 \\ 0 & -0.5 & 1 \\ 0 & -0.5 & 0 \end{vmatrix}$$

$$= 0.5 \times 0.5 + 0.25 - (-0.25) = 0.75. \quad (A.65)$$

The determinant is not zero, so we can proceed with Cramer's rule and compute the determinants $|\mathbf{A}_1|, \cdots, |\mathbf{A}_4|$, in which one column at a time is replaced with the solution vector $\vec{b}$. For simplicity of notation, we call the non-zero components of this vector $a = \ln 0.5$, $b = \ln 1.6$, and $c = \ln 1.6$. Even though b and c have the same value, we use different symbols to make the computation of the determinants as lucid as possible. Each of the determinants is 4×4 and thus needs to expanded:

$$\det \mathbf{A}_1 = \begin{vmatrix} 0 & 0 & 1 & 1 \\ a & 0.5 & 0 & 0 \\ b & -0.5 & 1 & 0 \\ c & -0.5 & 0 & 1 \end{vmatrix} = 1 \times \begin{vmatrix} a & 0.5 & 0 \\ b & -0.5 & 0 \\ c & -0.5 & 1 \end{vmatrix} - 1 \times \begin{vmatrix} a & 0.5 & 0 \\ b & -0.5 & 1 \\ c & -0.5 & 0 \end{vmatrix}$$

$$= -0.5a - 0.5b - 0.5c - 0.5a$$
$$= -\ln 0.5 - \ln 1.6 = 0.22314, \quad (A.66a)$$

$$\det \mathbf{A}_2 = \begin{vmatrix} 0.5 & 0 & 1 & 1 \\ -0.5 & a & 0 & 0 \\ 0 & b & 1 & 0 \\ 0 & c & 0 & 1 \end{vmatrix}$$

$$= 0.5 \times \begin{vmatrix} a & 0 & 0 \\ b & 1 & 0 \\ c & 0 & 1 \end{vmatrix} + 1 \times \begin{vmatrix} -0.5 & a & 0 \\ 0 & b & 0 \\ 0 & c & 1 \end{vmatrix} - 1 \times \begin{vmatrix} -0.5 & a & 0 \\ 0 & b & 1 \\ 0 & c & 0 \end{vmatrix}$$

$$= 0.5a - 0.5b - 0.5c$$
$$= \ln 0.5 - \ln 1.6 = -0.816577, \quad (A.66b)$$

APPENDIX

$$\det \mathbf{A}_3 = \begin{vmatrix} 0.5 & 0 & 0 & 1 \\ -0.5 & 0.5 & a & 0 \\ 0 & -0.5 & b & 0 \\ 0 & -0.5 & c & 1 \end{vmatrix} = 0.5 \times \begin{vmatrix} 0.5 & a & 0 \\ -0.5 & b & 0 \\ -0.5 & c & 1 \end{vmatrix} - 1 \times \begin{vmatrix} -0.5 & 0.5 & a \\ 0 & -0.5 & b \\ 0 & -0.5 & c \end{vmatrix}$$

$$= 0.5 \cdot (0.5b + 0.5a) - (0.25c - 0.25b)$$

$$= 0.25(\ln 0.5 + \ln 1.6) = -0.055786,$$

$$\det \mathbf{A}_4 = \begin{vmatrix} 0.5 & 0 & 1 & 0 \\ -0.5 & 0.5 & 0 & a \\ 0 & -0.5 & 1 & b \\ 0 & -0.5 & 0 & c \end{vmatrix} = 0.5 \times \begin{vmatrix} 0.5 & 0 & a \\ -0.5 & 1 & b \\ -0.5 & 0 & c \end{vmatrix} + 1 \times \begin{vmatrix} -0.5 & 0.5 & a \\ 0 & -0.5 & b \\ 0 & -0.5 & c \end{vmatrix}$$

$$= 0.5(0.5c + 0.5a) + (0.25c - 0.25b)$$

$$= 0.25(\ln 0.5 + \ln 1.6) = -0.055786. \qquad (A.66c)$$

Division of these determinants by $|\mathbf{A}|$ yields the solution

$$y_1 = \frac{\det \mathbf{A}_1}{\det \mathbf{A}} = 0.29752,$$

$$y_2 = \frac{\det \mathbf{A}_2}{\det \mathbf{A}} = -1.0888,$$

$$y_3 = \frac{\det \mathbf{A}_3}{\det \mathbf{A}} = -0.074381,$$

$$y_4 = \frac{\det \mathbf{A}_4}{\det \mathbf{A}} = -0.074381.$$

(A.67)

Substitution of these values in Eq. (A.64) confirms the result: all the linear equations are satisfied (see also Eq. (6.38)).

Inverse Matrices

Except for 0, any real number has a unique *reciprocal*, that is, a number which multiplied by the first number results in 1. A similar statement is true for matrices: any $n \times n$ matrix whose determinant is not zero has a unique *inverse*. The inverse matrix of $\mathbf{A}$ is denoted by $\mathbf{A}^{-1}$. When we multiply such an inverse matrix by the original matrix, the result is the *identity* (unit) matrix $\mathbf{I}$:

$$\mathbf{A}\mathbf{A}^{-1} = \mathbf{A}^{-1}\mathbf{A} = \mathbf{I}. \qquad (A.68)$$

The identity matrix is always square and has all ones in the diagonal and zeros everywhere else. For $n = 4$, the identity matrix is

$$\mathbf{I} = \begin{pmatrix} 1 & 0 & 0 & 0 \\ 0 & 1 & 0 & 0 \\ 0 & 0 & 1 & 0 \\ 0 & 0 & 0 & 1 \end{pmatrix}. \qquad (A.69)$$

If a matrix is multiplied by the identity matrix of appropriate size, the matrix remains unchanged. If a vector is multiplied by the identity matrix of appropriate size, the vector remains unchanged.

What are inverse matrices good for? Consider again the matrix equation (A.61), which in the present context characterizes the steady state of a biochemical system. Cramer's rule allowed us to compute one component of the solution vector at a time. Matrix inversion yields the entire vector at once. This can be seen when we multiply both sides of the matrix equation by $\mathbf{A}^{-1}$:

$$\mathbf{A}^{-1}\mathbf{A}\vec{y} = \mathbf{A}^{-1}\vec{b}. \tag{A.70}$$

The left-hand side reduces to $\mathbf{I}\vec{y}$, which equals $\vec{y}$, and represents the desired solution. Thus, if we know $\mathbf{A}^{-1}$, simple multiplication by the vector $\vec{b}$ yields the solution of the matrix equation and, equivalently, the solution of a system of linear equations.

The computation of the inverse of a matrix is based on determinants and somewhat resembles a multidimensional Cramer's rule scheme. In general, the computation requires three steps:

1. Compute the determinant $|\mathbf{A}|$. If this determinant is zero or undefined, no inverse can be computed.
2. Compute all determinants of matrices $\mathbf{A}_{ij}$ that are obtained from $\mathbf{A}$ when the ith row and the jth column are eliminated. These matrices $\mathbf{A}_{ij}$ have $n-1$ rows and columns, and there are n^2 many of them.
3. Compute

$$\mathbf{A}^{-1} = \frac{1}{|\mathbf{A}|} \begin{pmatrix} +|\mathbf{A}_{11}| & -|\mathbf{A}_{21}| & \cdots & \pm|\mathbf{A}_{n1}| \\ -|\mathbf{A}_{12}| & +|\mathbf{A}_{22}| & \cdots & \mp|\mathbf{A}_{n2}| \\ \vdots & \vdots & & \vdots \\ \mp|\mathbf{A}_{1n}| & \pm|\mathbf{A}_{2n}| & \cdots & \mp|\mathbf{A}_{nn}| \end{pmatrix}. \tag{A.71}$$

Note in this computation that the determinants $|\mathbf{A}_{ij}|$ are placed in the matrix on the right-hand side in an unexpected way: For instance, $|\mathbf{A}_{21}|$ appears where we might expect $|\mathbf{A}_{12}|$ and vice versa. Also note that a plus or a minus sign is associated to each minor determinant, depending on its "checkerboard color." We have used $\pm$ and $\mp$ symbols to indicate that plus signs or minus signs may be in these positions, depending on whether the original matrix $\mathbf{A}$ has an even or an odd number of rows and columns.

As an example, if $\mathbf{A}$ has three rows and columns and takes the form

$$\mathbf{A} = \begin{pmatrix} a & b & c \\ d & e & f \\ g & h & i \end{pmatrix}, \tag{A.72}$$

its inverse is

$$\mathbf{A}^{-1} = \frac{1}{|\mathbf{A}|} \begin{pmatrix} +|\mathbf{A}_{11}| & -|\mathbf{A}_{21}| & +|\mathbf{A}_{31}| \\ -|\mathbf{A}_{12}| & +|\mathbf{A}_{22}| & -|\mathbf{A}_{32}| \\ +|\mathbf{A}_{13}| & -|\mathbf{A}_{23}| & +|\mathbf{A}_{33}| \end{pmatrix}, \tag{A.73}$$

where

$$|A_{11}| = \begin{vmatrix} e & f \\ h & i \end{vmatrix}, \quad |A_{21}| = \begin{vmatrix} b & c \\ h & i \end{vmatrix}, \quad |A_{31}| = \begin{vmatrix} b & c \\ e & f \end{vmatrix},$$

$$|A_{12}| = \begin{vmatrix} d & f \\ g & i \end{vmatrix}, \quad |A_{22}| = \begin{vmatrix} a & c \\ g & i \end{vmatrix}, \quad |A_{32}| = \begin{vmatrix} a & c \\ d & f \end{vmatrix}, \quad (A.74)$$

$$|A_{13}| = \begin{vmatrix} d & e \\ g & h \end{vmatrix}, \quad |A_{23}| = \begin{vmatrix} a & b \\ g & h \end{vmatrix}, \quad |A_{33}| = \begin{vmatrix} a & b \\ d & e \end{vmatrix}.$$

We use the concept of an inverse matrix for symbolic analyses of steady states in Chapters 6 and 7.

EXERCISES

1. Which of the following associations are functions? Explain why.

    1.1. Associate to each integer number its successor. For instance, associate 4 to 3, 5 to 4, etc.

    1.2. Associate to each integer number its successor and its predecessor. For instance, associate 2 and 4 to 3, 3 and 5 to 4, etc.

    1.3. The equation $x^2 + y^2 = 4$ describes the circle with radius 2 at the origin of the coordinate system. Is y a function of x? Is x a function of y?

2. Find at least two examples each of functions that are

    2.1. strictly monotone and continuous;

    2.2. strictly monotone and discrete;

    2.3. not strictly monotone and continuous;

    2.4. not strictly monotone and discrete.

3. Show that the exponential function $\exp(t)$ generates similar values as the following Taylor polynomials:

    3.1. $X(t) = 1 + t + 0.5t^2$.

    3.2. $Y(t) = 1 + t + 0.5t^2 + t^3/6$.

    3.3. $Z(t) = 1 + t + 0.5t^2 + t^3/6 + t^4/24$.

    Compare $\exp(t)$ with these functions for $t = 0, 1, 2, 10, -1, -2,$ and -10. Discuss how closely these representations approximate the exponential function.

4. Compute Taylor polynomials up to the fifth term for the following functions:

    4.1. $V(t) = \sin t$.

    4.2. $W(t) = \exp(t + 2)$.

    4.3. $Y(t) = t^4$.

    4.4. $X(t) = 4 + t + 1.2t^3 + 2t^4$.

    4.5. $Z(t) = \alpha t^g$ for α, g any real numbers. Hint: Distinguish different cases, such as $g = 0$, $g = 1$, g an integer, etc.

5. Show that for arguments between -1 and 1 the natural logarithm $\ln(1+t)$ generates similar values as the following polynomials:
   5.1. $X(t) = t - 0.5t^2$.
   5.2. $Y(t) = t - 0.5t^2 + t^3/3$.
   5.3. $Z(t) = t - 0.5t^2 + t^3/3 - t^4/4$.

   Compare $\ln(1+t)$ with these truncated Taylor series for $t = 0, \pm 0.1, \pm 0.2, \pm 0.5,$ and 1.

6. Show that for arguments greater than or equal to 0.5, the natural logarithm $\ln t$ generates similar values as the following function:

$$X(t) = \left(\frac{t-1}{t}\right) + \frac{1}{2}\left(\frac{t-1}{t}\right)^2 + \frac{1}{3}\left(\frac{t-1}{t}\right)^3 + \frac{1}{4}\left(\frac{t-1}{t}\right)^4.$$

   Compare $\ln t$ with this truncated Taylor series for $t = 0.5, 1, 2,$ and 10.

7. Compute linear approximations and power-law approximations for the following functions. Select in each case $t = 0.1, 1,$ and 10 as operating points.
   7.1. $y = \exp(t+1)$.
   7.2. $y = \ln(t+1)$.
   7.3. $y = \exp(2\ln t)$.
   7.4. $y = \ln[\exp(t)\exp(t^2)]$.

8. Compute linear approximations and power-law approximations for the following functions. Select in each case $t = 0.1, 1,$ and 10 as operating points.
   8.1. $y = 5t$.
   8.2. $y = t + 2t^2 - 0.5t^3$.
   8.3. $y = (2t)/(t+1)$.
   8.4. $y = (2t^3)/(t^3+1)$.

9. For the following functions compute the rate of change (partial derivative) with respect to X_1:
   9.1. $V(X_1, X_2) = X_1 + X_2$.
   9.2. $V(X_1, X_2) = \exp(X_1 + X_2)$.
   9.3. $V(X_1, X_2, X_3) = \exp(X_2 + X_3)$.
   9.4. $V(X_1, X_2, X_3) = X_1 \exp(X_2 + X_3)$.

10. Compute power-law approximations for the following functions. Select several different operating points for each case.
    10.1. $V(X_1, X_2) = X_1 + X_2$.
    10.2. $V(X_1, X_2) = \exp(X_1 + X_2)$.
    10.3. $V(X_1, X_2, X_3) = X_1 \exp(X_2 + X_3)$.
    10.4. $V(X_1, X_2, X_3) = X_1 X_2 X_3$.
    10.5. $V(X_1, X_2) = 2X_1/(1 + X_1 + X_2)$.

11. Compute the determinants of the following matrices:
    11.1. $\begin{pmatrix} 1 & 1 \\ 1 & 0 \end{pmatrix}$.

APPENDIX 441

11.2. $\begin{pmatrix} 2 & 2 \\ 2 & 0 \end{pmatrix}.$

11.3. $\begin{pmatrix} 1 & 0.5 \\ 0.1 & 10 \end{pmatrix}.$

11.4. $\begin{pmatrix} 1 & 2 & 2 \\ 0 & 2 & 0 \\ 1 & 1 & 1 \end{pmatrix}.$

11.5. $\begin{pmatrix} 1 & 1 & 3 & 0.5 \\ 2 & 0 & 1 & 4 \\ 3 & 0 & 2 & 2 \\ 0.5 & 0 & 1 & 0.1 \end{pmatrix}.$

11.6. $\begin{pmatrix} 2 & 1 & 2 & 2 \\ 1 & 4 & 4 & 1 \\ 0 & 3 & 0 & 1 \\ 1 & 0 & 2 & 2 \end{pmatrix}.$

12. Write the following sets of equations in the form of a matrix equations and solve them:

    12.1.
    $$2y_1 + 4y_2 + y_3 = 2,$$
    $$y_1 - 0.5y_2 + y_3 = 1,$$
    $$4y_1 + 4y_2 + 4y_3 = 4.$$

    12.2.
    $$y_1 + 4y_2 + 1 = 0,$$
    $$2y_1 + y_3 + y_4 = 2,$$
    $$2y_1 + 4y_2 - 1 = 1,$$
    $$y_1 + y_2 - y_3 + y_4 = 0.$$

    12.3.
    $$y_1 + y_2 - 1 = 0,$$
    $$y_1 + y_3 - 1 = 0,$$
    $$y_1 + y_4 + 1 = 0,$$
    $$y_2 + y_3 - y_4 + 1 = 0.$$

13. Compute the inverses of the following matrices:

    13.1. $\begin{pmatrix} 1 & 1 \\ 1 & 0 \end{pmatrix}.$

    13.2. $\begin{pmatrix} 1 & 2 & 2 \\ 0 & 2 & 0 \\ 1 & 1 & 1 \end{pmatrix}.$

13.3. $\begin{pmatrix} 1 & 0 & 0 & 0 \\ 0 & 1 & 0 & 0 \\ 0 & 0 & 1 & 0 \\ 0 & 0 & 0 & 1 \end{pmatrix}.$

13.4. $\begin{pmatrix} 2 & 1 & 2 & 2 \\ 1 & 4 & 4 & 1 \\ 0 & 3 & 0 & 1 \\ 1 & 0 & 2 & 2 \end{pmatrix}.$

Theoretical puzzles

14. Using expansion, show that the determinant of a matrix is always zero if one row or one column contains only zeros.
15. Show with examples (or prove mathematically) that the determinant of a matrix is zero if two rows are identical.
16. Test with examples whether the products **AB** and **BA** of two square matrices are necessarily the same or not.
17. Show that the rule for computing 2×2 determinants in fact can be considered an expansion, such as we used for larger determinants.
18. Show that expansion of a 3×3 determinant yields the same result as the rule of Sarrus.

REFERENCES

[28], [35], [62], [155–7], [165–6], [205], [217], [259], [319], [390], [439].

Hints and Partial Solutions

CHAPTER ONE

Exercise 1.1
Dependent:

$X_1 =$ lysine,
$X_2 = \alpha$-ketoglutarate,
$X_3 =$ saccharopine.

Independent:

$X_4 =$ enzyme (lysine-oxoglutarate reductase).

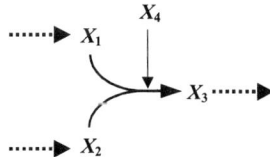

The dotted arrows are optional, indicating whether the system is open or closed. In the case of an enzyme deficiency, the numerical value of the independent variable X_4 (the enzyme activity) would be reduced by some percentage.

Exercise 2.1
Proper.

Exercise 2.5
Improper: dashed arrow doesn't make sense.

Exercise 2.9
Proper.

Exercise 2.13
Improper: arrow from X_3 to X_4 should be a flow arrow, as opposed to a signal arrow.

Exercise 2.17
Proper.

Exercise 3.2
(a) Open system; material continues to flow through it.
(b) No material leaves the system; influx material eventually accumulates in pool X_9 without end.
(c) Closed system; if X_1 is dependent, all material accumulates in pool X_9, and the dynamics then stops; if X_1 is independent, material continues to accumulate in pool X_9.

Exercise 3.4
(a) X_3 inhibits the conversion of X_1 into X_2; the more X_3 there is, the slower the process.
(b) X_3 activates the conversion of X_1 into X_2; the more X_3 there is, the faster the process.

CHAPTER TWO

Exercise 1
If it has not happened already, the solver has to be set to BDF Gear (under *Options*), because the equations are not in power-law form.

V_{max} and K_M affect the speed with which the substrate is converted into product. Qualitatively, the curves are rather similar, even though the shapes are somewhat affected.

Special cases result if the kinetic parameters are zero. Biologically, this may not be very relevant, but one can check out the results anyway. For $V_{max} = 0$, the right-hand sides of the differential equations are 0, indicating that there is no change in S and P. Indeed, the variables don't change over time. For $K_M = 0$, the variable S drops out of the right-hand sides, and the right-hand sides are constant at $-V_{max}$ and V_{max}, respectively. The decrease in S and increase in P are linear.

Exercise 4
The decay rate k affects the speed with which X_1 disappears. For $k = 0.2$, it takes between 11 and 12 time units until 90% of X_1 are used up. This is most easily seen from the *Table* option under *Results*, where the value of X_1 at time 11 is given as above 1 [which is 10% of the initial value $X_1(0) = 10$] and the value at time 12 is below 1. This result on time is independent of the initial value of X_1. Confirm this, for instance, by setting $X_1(0) = 20$.

Changing the value of *hr* does not internally affect the accuracy of the solution. However, this parameter dictates at which time points the solution is *displayed*. The default graphics setting connects the solution points with straight lines, and if *hr* is large, the graph has a ragged appearance.

CHAPTER THREE

Exercise 1
An S-system model is always a special case of a GMA model with the same number of variables. There is no limitation as to how many power-law terms may be on the right-hand side of a GMA system, so the one positive and one negative term of an S-system qualify for a GMA system. The opposite is not true. For instance, if the GMA system

HINTS AND PARTIAL SOLUTIONS

has two equations with three terms each, and if these terms differ in their kinetic orders, the system cannot in general be written as an S-system with two variables. However, it is noted that such a system can be written exactly as an S-system with more than two variables. How this is accomplished is beyond the scope of this chapter but can be found in Savageau and Voit (1987).

Exercise 3
Suppose one α-term consists exclusively of the rate constant. This means that this term does not depend on any of the variables, but that material is pumped into the system at a constant rate, no matter what the dynamics of the system is doing otherwise. If a β-term consists exclusively of the rate constant, the system is "leaking" material at a constant rate. If the rate is high, all material may be lost. One or more of the variables will eventually hit zero, and the dynamics comes to a stop. If the solver is set to BDF (Gear), some variables may actually become negative. Biochemically speaking this wouldn't make sense, but negative values could be of relevance in other applications.

Exercise 5
The S-system and the GMA representation are exactly the same.

Exercise 10
g_{13} is the only negative kinetic order. All other parameters are positive.

Exercise 13
The equations for X_1 through X_4 are the same in both cases, namely

$$\dot{X}_1 = \alpha_1 X_5^{g_{15}} - \beta_1 X_1^{h_{11}},$$
$$\dot{X}_2 = \alpha_2 X_1^{g_{21}} X_4^{g_{24}} - \beta_2 X_2^{h_{22}},$$
$$\dot{X}_3 = \alpha_3 X_1^{g_{31}} X_2^{g_{32}} - \beta_3 X_3^{h_{33}},$$
$$\dot{X}_4 = \alpha_4 X_3^{g_{43}} - \beta_4 X_4^{h_{44}}.$$

In the top panel, X_5 is an independent variables, and one adds to the differential equations the algebraic statement

$$X_5 = \text{constant}.$$

In the bottom panel, X_5 is a dependent variable, and one adds the differential equation

$$\dot{X}_5 = -\beta_5 X_5^{h_{55}}.$$

Steady-state flux constraints need to be considered for all variables. The flux between X_5 and X_1 must equal the sum of the two fluxes from X_1 to X_2 and X_3, the two fluxes entering X_2 must equal the efflux to X_3, and the two fluxes entering X_3 must equal the efflux toward X_4.

The dynamics of the two systems is different. In the first case, X_5 is fed into the system at a constant rate. Since no material leaves the system, the total mass in the system keeps increasing. In the second case, the dependent variable X_5 is being used up. Since, again, no material is leaving, the original amount of material is being redistributed, with X_1 and X_5 emptying and all material cycling between X_2, X_3, and X_4. It is instructive to choose parameter values and to implement and analyze the systems in PLAS.

Exercise 15

Suppose the fluxes are converging as shown below:

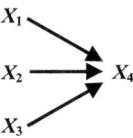

The branch point constraint at the steady state is

$$\alpha_4 X_1^{g_{41}} X_2^{g_{42}} X_3^{g_{43}} = \beta_1 X_1^{h_{11}} + \beta_2 X_2^{h_{22}} + \beta_3 X_3^{h_{33}} = V.$$

The kinetic orders are obtained by partial differentiation. For instance,

$$g_{41} = \frac{\partial V}{\partial X_1} \frac{X_1}{V} \qquad \text{(at steady state)}$$

$$= \left(h_{11} \beta_1 X_1^{h_{11}-1}\right) \frac{X_1}{V} \qquad \text{(at steady state)}$$

$$= \frac{h_{11} \beta_1 X_{1S}^{h_{11}}}{\beta_1 X_{1S}^{h_{11}} + \beta_2 X_{2S}^{h_{22}} + \beta_3 X_{3S}^{h_{33}}},$$

where X_{1S}, X_{2S}, and X_{3S} denote steady-state values. The rate constant α_4 is obtained as

$$\alpha_4 = \frac{\beta_1 X_{1S}^{h_{11}} + \beta_2 X_{2S}^{h_{22}} + \beta_3 X_{3S}^{h_{33}}}{X_{1S}^{g_{41}} X_{2S}^{g_{42}} X_{3S}^{g_{43}}}.$$

Exercise 17

The same S-system can represent different pathways. For instance, the equation

$$\dot{X}_1 = \alpha_1 X_0^{g_{10}} - \beta_1 X_1^{h_{11}}$$

may represent the two pathways (a) and (b) below:

(a) $X_0 \longrightarrow X_1 \longrightarrow$

(b) $X_0 \longrightarrow X_1 \longrightarrow$

Exercise 18

The pathway is represented by the system

$$\dot{X}_1 = \alpha_1 X_4^{g_{14}} - \beta_1 X_1^{h_{11}},$$

$$\dot{X}_2 = \alpha_2 X_5^{g_{25}} - \beta_2 X_2^{h_{22}},$$

$$\dot{X}_3 = \alpha_3 X_1^{g_{31}} X_2^{g_{32}} - \beta_3 X_3^{h_{33}},$$

$X_4, X_5 = \text{constant.}$

HINTS AND PARTIAL SOLUTIONS

The constraint equations for the converging branch point are constructed by equating all steady-state fluxes that enter a pool with all steady-state fluxes leaving this pool. In this case, the two steady-state fluxes leaving X_1 and X_2 must equal the flux entering X_3. Thus, the constraint equation reads

$$V_{1S}^- + V_{2S}^- = V_{3S}^+,$$

which is equivalent to

$$\beta_1 X_{1S}^{h_{11}} + \beta_2 X_{2S}^{h_{22}} = \alpha_3 X_{1S}^{g_{31}} X_{2S}^{g_{32}},$$

where the subscript S reminds us that all evaluations are executed at the steady state or another operating point of our choice.

The constraint relationships between parameters are derived from considering V_{3S}^+ as the power-law approximation of $V_{1S}^- + V_{2S}^-$ and computing its parameters through partial differentiation, as shown in the chapter and in the Appendix. The result is

$$g_{31} = \frac{\partial (V_1^- + V_2^-)}{\partial X_1} \frac{X_1}{V_1^- + V_2^-} \quad \text{at steady state}$$
$$= h_{11} V_{1S}^- / (V_{1S}^- + V_{2S}^-)$$
$$= \beta_1 h_{11} X_{1S}^{h_{11}} / (\beta_1 X_{1S}^{h_{11}} + \beta_2 X_{2S}^{h_{22}}),$$
$$g_{32} = h_{22} V_{2S}^- / (V_{1S}^- + V_{2S}^-)$$
$$= \beta_2 h_{22} X_{2S}^{h_{22}} / (\beta_1 X_{1S}^{h_{11}} + \beta_2 X_{2S}^{h_{22}}),$$
$$\alpha_3 = (V_{1S}^- + V_{2S}^-) / (X_{1S}^{g_{31}} X_{2S}^{g_{32}})$$
$$= (\beta_1 X_{1S}^{h_{11}} + \beta_2 X_{2S}^{h_{22}}) / (X_{1S}^{g_{31}} X_{2S}^{g_{32}})$$

The kinetic orders g_{31} and g_{32} can be interpreted as the proportions of the individual fluxes V_{1S}^- and V_{2S}^- with respect to the overall flux V_{3S}^+ that are weighted with the kinetic orders h_{11} and h_{22}.

Exercise 20 (Partial solution; from Sorribas and Savageau, 1989c)
For the reversible strategy, the parameter values of the second equation are $\alpha_2 = 82.9$, $\beta_2 = 83.3$, $g_{21} = 0.0888$, $g_{22} = -0.521$, $g_{23} = 0.516$, $h_{21} = -0.267$, $h_{22} = 0.395$, and $h_{23} = -0.0476$. For the irreversible strategy, the parameter values of the second equation are $\alpha_2 = 3.33$, $\beta_2 = 3.49$, $g_{21} = 3.2$, $g_{22} = -3.1$, $h_{22} = 5.14$, and $h_{23} = -5.07$.

CHAPTER FOUR

Exercise 3
Reduced enzyme activities may be implemented as lowered values of the corresponding independent variable, in this case phosphofructokinase (X_9). One has to be careful, though, with extreme deficiencies, since these may be accompanied with an entirely different set of steady-state values. This topic is discussed in detail in Chapter 10.

Selected results are

X_9	X_{1S}	X_{2S}	X_{3S}
1	0.05	0.5	0.16
0.9	0.058	0.659	0.217
0.75	0.074	1.05	0.361
0.5	0.127	2.950	1.124

In the last scenario, $X_9 = 0.5$, the system needs several thousand time units to reach the new steady state.

For low values of X_9, e.g., $X_9 = 0.1$ or $X_9 = 0.001$, the approximation apparently is no longer valid. The system has an unreasonable steady state (see comments above), and two of the eigenvalues approach 0, indicating a gradual loss of stability.

Exercise 6
In this particular case, the GMA system description is identical with the S-system description (4.9). The reason is that the kinetic orders of the two degradation processes of X_2, one ($2\,X_2^{0.5}$) for the production of X_3 and the other ($8\,X_2^{0.5}$) for the production of X_4, are the same (0.5). Their sum is a simple power-law function, namely $10\,X_2^{0.5}$, which appears as the aggregated degradation term of X_2. For most branched systems, the S-system and the GMA representation are different.

In the given case, the results of S-system and GMA analyses are obviously identical.

Exercise 10
If all rate constants are *multiplied* by the same positive number, the steady state is maintained, but the dynamics are faster (if the multiplier is greater than 1) or slower (if the multiplier is less than 1). If the multiplier is negative, the system dynamics runs backwards in time. If the rate constants are multiplied with different factors, or if the same constant is *added* to each rate constant, about anything can happen.

Exercise 14
The pathway loses stability for values of g_{13} between -11.3 and -11.4. This is most easily checked by clicking the [x=0] button and seeing whether all eigenvalues have negative real parts. One can also look at the dynamical solutions. If there is a stable steady state, the oscillations will completely dampen. By contrast, in the unstable case the oscillations approach some stable (limit cycle) pattern. This may best be visualized in the phase plane by initializing the system outside and then inside the limit cycle.

To see the differences between borderline cases like $g_{13} = -11.3$ and $g_{13} = -11.4$ in the time course plot, one has to extend the solution to a large final value t_f.

Exercise 16.1
Introduce a new variable $\tau = t/2$. To express the system in terms of this new time variable, we have to use the chain rule of differentiation, which in this case says

$$\frac{dX_i}{d\tau} = \frac{dX_i}{dt} \cdot \frac{dt}{d\tau} = \frac{dX_i}{dt} \times 2.$$

HINTS AND PARTIAL SOLUTIONS

Replacing dX_i/dt with the right-hand side of the corresponding S-system equation shows that all rate constants are multiplied by 2: the dynamics are twice as fast as before.

Exercise C.2
If X_1 is an independent variable, it is supplied at a constant rate. Thus, if material leaving X_3 and X_4 is continuously captured, the capture pools grow without end. By contrast, if X_1 is dependent, the diagram indicates that no input is replenishing the X_1-pool. Consequently, there is a finite amount of material, which will end up in capture pools fed by X_3 and X_4. How much will end up in which capture pool depends on the parameter values of the system. However it is distributed, the total amount originally given in X_1, X_2, X_3, and X_4 in the end equals the total amount in the two capture pools.

To check these statements in PLAS, define additional variables X_5 and X_6, whose right-hand sides are equivalent to the β-terms of X_3 and X_4, respectively. If X_1 is dependent, it requires an equation without an α-term, but with an appropriate β-term. If X_1 is independent, it is not represented by a differential equation, and only its constant value has to be specified.

Exercise C.5
As an example, one may use the branched pathway of Fig. 4.10, which is represented as an S-system in Eq. (4.9). To implement a circadian input, the independent variable X_5 is replaced with a variable that exhibits sinusoidal oscillations. Such a variable is given in Eq. (4.8). Thus, simply change the variable names in Eq. (4.8) to X_5 and X_6, adjust the center of the oscillation to 0.5, and put the two systems together:

$$\dot{X}_1 = 10 X_3^{g_{13}} X_5 - 5 X_1^{0.5}, \quad X_1(0) = 1.1,$$
$$\dot{X}_2 = 5 X_1^{0.5} - 10 X_2^{0.5}, \quad X_2(0) = 0.5,$$
$$\dot{X}_3 = 2 X_2^{0.5} - 1.25 X_3^{0.5}, \quad X_3(0) = 0.9,$$
$$\dot{X}_4 = 8 X_2^{0.5} - 5 X_4^{0.5}, \quad X_4(0) = 0.75,$$
$$\dot{X}_5 = 0.5 - X_6, \quad X_5(0) = 0.5,$$
$$\dot{X}_6 = X_5 - 0.5, \quad X_6(0) = 0.75.$$

The amplitude of the input oscillation is determined by the initial value $X_6(0)$. To change the frequency of the sine input in relation to the time scale of the pathway model, the four rate constants in the equations of X_5 and X_6 may be multiplied by the same factor.

CHAPTER FIVE

Exercise 1
The figure below shows a Microsoft Excel® plot of all data in logarithmic coordinates, including corresponding values of the computed regression line. The slope (kinetic order) computed in Excel® is about 2.3.

Kinetic order of UA in UA Excretion

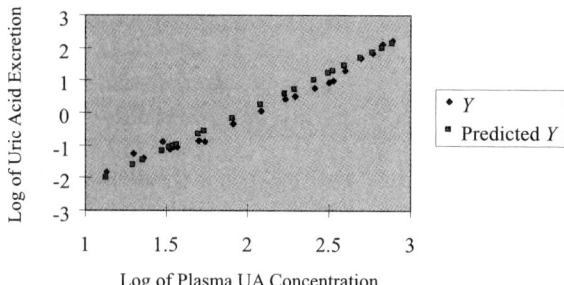

It appears that one straight line captures all data with sufficient accuracy. A formal analysis would consist of fitting data for the normal and the diseased subjects separately and testing for significant differences between the two regression lines with an F-statistic (e.g., Neter and Wasserman, 1974).

Exercise 6

6.1 Simple algebraic rewriting of the formula shows that Stitt's formula is an approximation of the logarithmic derivative, which defines the kinetic order:

$$g_{ik} = \frac{X_2(V_1 - V_2)}{V_2(X_1 - X_2)} \approx \frac{\partial V}{\partial X} \cdot \frac{X}{V}.$$

The formula is valid if the differences in fluxes and concentrations are (close to infinitesimally) small.

6.2 In Example 2, extracting data points from the plot, one obtains the values $\ln X_1 = -1.1$, $\ln X_2 = -0.96$, $\ln X_3 = -0.95$, $\ln X_4 = -0.8$, $\ln V_1 = 1$, $\ln V_2 = 2.25$, $\ln V_3 = 1.7$, $\ln V_4 = 2.15$, which corresponds to $X_1 = 0.333$, $X_2 = 0.383$, $X_3 = 0.387$, $X_4 = 0.45$, $V_1 = 2.718$, $V_2 = 9.49$, $V_3 = 5.474$, and $V_4 = 8.585$. According to Stitt's formula, one computes, for instance, the kinetic order for the data pair (X_1, X_3) as

$$g = \frac{0.387}{0.333}\left(1 - \frac{5.474}{2.718}\right) \bigg/ \left[\frac{5.474}{2.718}\left(1 - \frac{0.387}{0.333}\right)\right] = 3.608.$$

In Example 3, the first two data points are relevant. The have the numerical values $\ln X_1 = 0.42$, $\ln X_2 = 0.95$, $\ln V_1 = 1.84$, $\ln V_2 = 1.81$, and thus $X_1 = 1.522$, $X_2 = 2.586$, $V_1 = 6.234$, and $V_2 = 6.050$. Accordingly,

$$g = \frac{2.586}{1.522}\left(1 - \frac{6.050}{6.234}\right) \bigg/ \left[\frac{6.050}{6.234}\left(1 - \frac{2.586}{1.522}\right)\right] = -0.0737.$$

By slightly changing the numerical values of the data points, study how sensitive these results are with respect to inaccuracies in measured concentrations and fluxes.

Exercise 9

Details of this system are described in Chapter 9.

9.1 The power-law representation for citrate production is

$$v_1 = 16.242 \, \text{OAA}^{0.679} \, \text{ACO}^{0.0782} \, \text{CS} \, \text{CoA}^{-0.0372}.$$

9.2 The power-law representation for succinate degradation is derived from the aggregation of two processes, namely

$$v_2 = \text{SUC} \cdot \text{ENZ}$$

and v_3, as given. The aggregated power-law representation is

$$v = 1.536 \, \text{SUC}^{0.317} \, \text{FUM}^{-0.00733} \, \text{SDH}^{0.776} \, \text{ENZ}^{0.224}.$$

Exercise 10 (first equation, first kinetic order)

The kinetic order g_{ji} describes the effect of X_i on the production of X_j. It is computed by partial differentiation of v_{net} with respect to X_i and multiplying the result with X_i/v_{net}:

$$\frac{\partial v_{net}}{\partial X_i} = \frac{V_{max(F)}/K_{Mi}}{1 + X_i/K_{Mi} + X_j/K_{Mj}} - \frac{(V_{max(F)}/K_{Mi})(X_i - X_j/K_{eq})K_{Mi}^{-1}}{(1 + X_i/K_{Mi} + X_j/K_{Mj})^2},$$

$$g_{ji} = \frac{\partial v_{net}}{\partial X_i} \cdot \frac{X_i}{v_{net}} = \frac{X_i}{X_i - X_j/K_{eq}} - \frac{X_i/K_{Mi}}{1 + X_i/K_{Mi} + X_j/K_{Mj}}.$$

Dividing the first term on the right-hand side by X_i yields

$$\frac{X_i}{X_i - X_j/K_{eq}} = \frac{1}{1 - (X_j/X_i)/K_{eq}} = \frac{1}{1 - \Gamma/K_{eq}}.$$

Inspecting the functional form given for the forward reaction, one confirms directly that the second term is equal to $v_F/V_{max(F)}$, and the overall result is Eq. (5.36).

Exercise 12

There are several ways of estimating parameters. For instance, one may plot X_1 and X_2 versus time, hand-fit the time courses, and measure slopes at all time points. On substituting the slopes for the derivatives $\dot{X}_1$ and $\dot{X}_2$ and taking logarithms, the differential equations, evaluated at all time points, become a set of linear equations that can be solved by regular linear regression.

The resulting parameter values depend somewhat on the measured slopes. Reasonable values are

$$g_{11} = -0.52, \quad g_{12} = 0.47,$$
$$h_{21} = 1.12, \quad h_{22} = 0.49,$$
$$\alpha_1 = 0.04, \quad \beta_2 = 0.18.$$

As an alternative, the parameter values may be estimated from the differential equations directly or upon replacement of derivatives with slopes, as above, but without the logarithmic transformation. The results in each case can be expected to differ slightly.

CHAPTER SIX

Exercise 1.2

The steady-state equations in logarithmic variables are

$$-0.5 y_1 + 2 y_2 - y_3 = 0,$$
$$1.5 y_1 - 2.5 y_2 = \ln 2,$$
$$-2 y_1 + 0.5 y_2 - 0.5 y_3 = -\ln 4.$$

The equations can be solved by substitution or with matrix methods, as shown in the Appendix. The solution to this set is $y_1 = 0.743865267$, $y_2 = 0.169060288$, $y_3 = -0.033812038$. The corresponding solution in X-variables is $X_1 = 2.104053$, $X_2 = 1.184191529$, $X_3 = 0.9667532$.

Exercise 1.4

The steady-state equations in logarithmic variables are

$$-0.5y_1 - 0.8y_3 - 0.1y_4 + y_5 = -\ln 2,$$
$$1.5y_1 - 0.7y_2 = \ln 4,$$
$$-y_1 + 1.5y_2 - 2y_3 = -\ln 2.5,$$
$$-y_1 + 1.5y_2 - 2y_4 = -\ln 1.5.$$

The solution to this set is $y_1 = 1.561242971$, $y_2 = 1.365100346$, $y_3 = 0.701348954$, $y_4 = 0.445936321$. The corresponding solution in X-variables is $X_1 = 4.76474$, $X_2 = 3.916116$, $X_3 = 2.016471$, $X_4 = 1.561952$. The variable y_5 is the logarithm of the independent variable; thus, $y_5 = \ln 2$.

Exercise 3.1

$$|A| = 1 \times 4 \times 12 + 2 \times 6 \times 4 + 3 \times 2 \times 8 - 3 \times 4 \times 4$$
$$- 1 \times 6 \times 8 - 2 \times 2 \times 12 = 0.$$

Exercise 3.3

$|C| = 1.$

Exercise 5

The right-hand side of each linear steady-state equation has the form $\ln(\beta_i/\alpha_i)$. If α_i or β_i is zero, the logarithm is undefined.

Exercise 7

No. The simplest counterexample is the comparison of two S-systems with the same kinetic orders. If all rate constants in system 2 are twice those of system 1, the steady-state equations are identical, but the dynamics are twice as fast in system 2; thus, the two systems are different. One can construct any number of counterexamples that differ much more in structure than this trivial example. See, for instance, Exercise 17 of Chapter 3.

Exercise 11.1

If all variables have steady-state values of 1, all products of power-law functions, excluding the rate constants, have a value of 1. To equate the two products of power-law functions in each equation, its two rate constants must be the same. However, the rate constants may differ from one equation to the next.

Furthermore, consider the system

$$\dot{X}_1 = 2X_1 X_2 - 2X_1,$$
$$\dot{X}_2 = 2X_2 - 2X_2^2,$$

in which all rate constants are the same. To be in steady state, the system must satisfy $X_2 = 1$, which is most easily seen from the second equation. However, given $X_2 = 1$,

HINTS AND PARTIAL SOLUTIONS

the variable X_1 can take any positive value or even be zero. Thus, the assertion in this problem is not true in general.

Exercise 14.2
In this unusual case, all three equations lead to the same steady-state equation, and this equation simply says that X_1 has to equal X_2. The variables $X_1 = X_2$ may take any steady-state value, including zero, as long as they are equal. Furthermore, X_3 does not appear anywhere on the right-hand side. Thus, X_3 may take any steady-state value, including zero. Explore the dynamics of this system in PLAS using different initial values for X_1, X_2, and X_3. Explain.

Exercise 17
The determinant for the system is

$$\text{DET} = \begin{vmatrix} F_1 a_{11} - \lambda & F_1 a_{12} \\ F_2 a_{21} & F_2 a_{22} - \lambda \end{vmatrix}$$
$$= F_1 a_{11} F_2 a_{22} - F_1 a_{11} \lambda - F_2 a_{22} \lambda + \lambda^2 - F_1 a_{12} F_2 a_{21}.$$

The coefficients of the characteristic polynomial are $\phi_1 = -F_1 a_{11} - F_2 a_{22}$ and $\phi_2 = F_1 F_2 (a_{11} a_{22} - a_{12} a_{21})$. The only remaining non-zero quantity of the Routh array is $\Delta_{11} = \phi_2$. For stability, ϕ_1 and Δ_{11} must be positive. Since F_1 and F_2 are always positive, the desired inequalities follow directly.

CHAPTER SEVEN

Exercise 3

$$\mathbf{A} = \begin{pmatrix} -1 & -2 & 0.5 & 0 \\ 1 & -0.5 & 0 & 1 \end{pmatrix}, \quad \mathbf{A}_D^{-1} = \frac{1}{2.5}\begin{pmatrix} -0.5 & 2 \\ -1 & -1 \end{pmatrix}, \quad \mathbf{A}_I = \begin{pmatrix} 0.5 & 0 \\ 0 & 1 \end{pmatrix},$$

$$\mathbf{A}_D^{-1}\mathbf{A}_I = \begin{pmatrix} -0.1 & 0.8 \\ -0.2 & -0.4 \end{pmatrix}.$$

Exercise 9

$$\vec{y}_D = \begin{pmatrix} -1.247664886 \\ 0.27725894 \end{pmatrix}, \quad S(X, \beta) = \begin{pmatrix} -0.2 & 0.8 \\ -0.4 & -0.4 \end{pmatrix}, \quad \vec{b} = \begin{pmatrix} \ln 4 \\ -\ln 2 \end{pmatrix},$$

$$L(X_D, X_I) = \begin{pmatrix} 0.1 & -0.8 \\ 0.2 & 0.4 \end{pmatrix}, \quad \vec{y}_I = \begin{pmatrix} \ln 4 \\ \ln 2 \end{pmatrix}.$$

Substituting these quantities in Eq. (7.53) confirms equality.

Exercise 14

$$S(V, \alpha) = \begin{pmatrix} 0.2 & -0.8 \\ 0.2 & 0.2 \end{pmatrix}, \quad S(X, \alpha) = \begin{pmatrix} 0.2 & -0.8 \\ 0.4 & 0.4 \end{pmatrix}.$$

For example, $l = 1$, $i = 1$:

$$0.2 = 1 + (0)(0.2) + (-2)(0.4).$$

For $l = 1$, $i = 2$,

$$-0.8 = (0)(-0.8) + (-2)(0.4).$$

Check the results for $l = 2$.

Exercise 17 [Eq. (7.93)]

$$S(X, \alpha) = \begin{pmatrix} 0.2 & -0.8 \\ 0.4 & 0.4 \end{pmatrix}, \quad S(X, \beta) = \begin{pmatrix} -0.2 & 0.8 \\ -0.4 & -0.4 \end{pmatrix}, \quad g = \begin{pmatrix} 0 & -2 \\ 1 & 0 \end{pmatrix},$$

$$h = \begin{pmatrix} 1 & 0 \\ 0 & 0.5 \end{pmatrix}.$$

For $l = 1$, $k = 2$,

$$[(0.2)(-2) + (-0.2)(0)] + [(-0.8)(0) + (0.8)(0.5)] = 0.$$

For $l = 2$, $k = 1$,

$$[(0.4)(0) + (-0.4)(1)] + [(0.4)(1) + (-0.4)(0)] = 0.$$

Exercise 18

$$S(X, \alpha) \begin{pmatrix} g_{11} & g_{12} & \cdots & g_{1n} \\ g_{21} & g_{22} & \cdots & g_{2n} \\ \vdots & \vdots & & \vdots \\ g_{n1} & g_{n2} & \cdots & g_{nn} \end{pmatrix} + S(X, \beta) \begin{pmatrix} h_{11} & h_{12} & \cdots & h_{1n} \\ h_{21} & h_{22} & \cdots & h_{2n} \\ \vdots & \vdots & & \vdots \\ h_{n1} & h_{n2} & \cdots & h_{nn} \end{pmatrix}$$

$$= \begin{pmatrix} -1 & 0 & \cdots & 0 \\ 0 & -1 & \cdots & 0 \\ \vdots & \vdots & & \vdots \\ 0 & 0 & \cdots & -1 \end{pmatrix}.$$

CHAPTER EIGHT

Exercise 2

Call $M = [\text{AMP}]$, $D = [\text{ADP}]$, and $T = [\text{ATP}]$. The constraints can be written as $D + M + T = 3$ and $M = D^2/T$. Substituting the latter into the first, one obtains the quadratic equation $D^2 + DT + (T^2 - 3T) = 0$. The equation has two solutions, only one of which is positive:

$$D = -0.5T + 0.5\sqrt{T^2 - 4T^2 + 12T}.$$

For $T = 1$, the square root term is 3, which yields $D = 1$. Consequently, $M = D^2/T = 1$. Thus, all adenylates are present at the same concentration of 1 mM. For $T = 0.8$, the solution is $D = 0.98564$, $M = 1.21436$. For $T = 1.2$, the solution is $D = 0.98745$, $M = 0.81255$.

HINTS AND PARTIAL SOLUTIONS

Exercise 6

$R = 1 + 0.3033 + 18.7967 + 57.0116 = 77.1116,$

$T = 1 + 0.00015 + 18.7967 + 0.00285 = 19.7997,$

$L_0 = 5000.5341 - 1658.22 = 3342.3141,$

$L_1 = 0.046178,$

$$V_{steady\text{-}state} = 31.7 \times \frac{10 \times 0.3033 \times 18.7967 \times 77.1116}{77.1116^2 + 3342.3141 \times 0.8360} = 15.944.$$

The result is within rounding of the value of 15.946 used before.

Exercise 11

The kinetic orders of the S-system term are computed from the corresponding terms of the GMA system. According to Table 8.1, the terms of interest are $V_2^- = V_{21}^- + V_{22}^-$ with $V_2^- = 15.96$, $V_{21}^- = 15.946$, and $V_{22}^- = 0.014$. Substituting power-law terms leads to the constraint in Eq. (8.56):

$$\beta_2 X_2^{h_{22}} X_5^{h_{25}} X_8^{h_{28}} X_{11}^{h_{2,11}} = \beta_{2,1} X_2^{h_{22,1}} X_5^{h_{25,1}} X_8^{h_{28,1}} + \beta_{2,2} X_2^{h_{22,2}} X_{11}^{h_{2,11,2}}.$$

Partial differentiation of the right-hand side (RHS) yields

$$h_{25} = \frac{\partial \text{RHS}}{\partial X_5} \cdot \frac{X_5}{\text{RHS}}$$

$$= \frac{h_{25,1} \cdot \left(\beta_{2,1} X_2^{h_{22,1}} X_5^{h_{25,1}} X_8^{h_{28,1}} \right)}{\beta_{2,1} X_2^{h_{22,1}} X_5^{h_{25,1}} X_8^{h_{28,1}} + \beta_{2,2} X_2^{h_{22,2}} X_{11}^{h_{2,11,2}}} = \frac{h_{25,1} V_{2,1}^-}{V_2^-},$$

which is evaluated at the steady state. The result is

$$h_{25} = \frac{-0.3941 \times 15.946}{15.96} = -0.394.$$

Similarly, by direct derivation or using the result in Eq. (8.10),

$$h_{28} = \frac{h_{28,1} V_{2,1}^-}{V_2^-} = \frac{1 \times 15.946}{15.96} = 0.999,$$

$$h_{2,11} = \frac{h_{2,11,2} V_{2,2}^-}{V_2^-} = \frac{1 \times 0.014}{15.96} = 0.001.$$

The rate constant is computed from the constraint above [Eq. (8.56)] as

$$\beta_2 = \frac{V_{2,1}^- + V_{2,2}^-}{X_2^{h_{22}} X_5^{h_{25}} X_8^{h_{28}} X_{11}^{h_{2,11}}} = \frac{15.96}{30.46} = 0.5239,$$

where steady-state values are used for X_2, X_5, X_8, and X_{11}.

Exercise 13

The computation follows the procedure shown in the text [Eqs. (8.62)–(8.67)]. The effect of a change in phosphofructokinase activity (X_8) on polysaccharide production is written as $L(J_2, X_8)$ and computed at the steady state as

$$L(J_2, X_8) = \frac{\partial J_2}{\partial X_8} \cdot \frac{X_8}{J_2} = \frac{\partial (V_1 - V_3)}{\partial X_8} \cdot \frac{X_8}{V_1 - V_3} = \left(\frac{\partial V_1}{\partial X_8} - \frac{\partial V_3}{\partial X_8}\right) \frac{X_8}{V_1 - V_3}$$

$$= \frac{V_1}{V_1 - V_3} L(V_1, X_8) - \frac{V_3}{V_1 - V_3} L(V_3, X_8).$$

Substituting the steady-state values displayed in Fig. 8.6, one obtains

$$L(J_2, X_8) = \frac{15.948}{0.00955} \times 0.219711 - \frac{15.93845}{0.00955} \times 0.223687 = -6.416. \quad (8.67)$$

The gain factor is negative, because increases in phosphofructokinase activity cause more material to be channeled toward fructose-1.6-phosphate, thereby reducing the production of polysaccharides.

CHAPTER NINE

Exercise 2
With respect to dynamical solutions or computations of steady states, it is immaterial whether NAD, NADH, and CoA are kept explicit or subsumed in the corresponding rate constants. Subsuming them would reduce the number of symbols. However, this simplification is not very significant in light of the advantages of making NAD, NADH, and CoA explicit. These advantages include the logistic ease of changing their numerical values from one experiment to another without the need of recomputing several rate constants, which contain NAD, NADH, and CoA with different powers. Furthermore, implementation as independent variables allows us directly to compute logarithmic gains without worries about precursor–product or branch point constraints that have to be considered in the assessment of rate constant sensitivities.

Exercise 5 (Production of pyruvate)
Pyruvate is produced from alanine (X_8) and malate (X_{13}); the two rates are $v_{85} = X_{20} X_8$ and $v_{13,5} = X_{19} X_7 X_{13}/[(0.37 + X_{13})(X_7 + 0.1)]$. The kinetic orders are computed by partial differentiation. For instance, for X_{20}, one computes

$$g_{5,20} = \frac{\partial (v_{85} + v_{13,5})}{\partial X_{20}} \cdot \frac{X_{20}}{v_{85} + v_{13,5}}$$

and evaluates this quantity at the steady-state operating point. Thus,

$$g_{5,20} = X_8 \frac{X_{20}}{v_{85} + v_{13,5}} = 4.83 \times \frac{0.196}{0.94668 + 1.0896} = 0.465.$$

Similarly,

$$g_{57} = \frac{X_{13} X_{19}(0.37 + X_{13})(X_7 + 0.1) - X_7 X_{13} X_{19}(0.37 + X_{13})}{[(0.37 + X_{13})(X_7 + 0.1)]^2} \cdot \frac{X_7}{v_{85} + v_{13,5}}$$

$$= 0.0302023 \frac{1.85}{2.03628} = 0.0274,$$

HINTS AND PARTIAL SOLUTIONS

$$g_{58} = X_{20}\frac{X_8}{v_{85} + v_{13,5}} = g_{5,20} = 0.465,$$

$$g_{5,13} = \frac{X_7 X_{19}(0.37 + X_{13})(X_7 + 0.1) - X_7 X_{13} X_{19}(0.1 + X_7)}{[(0.37 + X_{13})(X_7 + 0.1)]^2} \cdot \frac{X_{13}}{v_{85} + v_{13,5}}$$

$$= 3.10589 \frac{0.22}{2.03628} = 0.336,$$

$$g_{5,19} = \frac{X_7 X_{13}}{(0.37 + X_{13})(X_7 + 0.1)} \cdot \frac{X_{19}}{v_{85} + v_{13,5}} = 0.535.$$

The rate constant is computed from the constraint that $V_5^+ = v_{85} + v_{13,5}$ at steady state. Thus,

$$\alpha_5 = \frac{v_{85} + v_{13,5}}{X_7^{0.0274} X_8^{0.465} X_{13}^{0.336} X_{19}^{0.535} X_{20}^{0.465}} = 1.875.$$

The parameter values of the degradation term V_5^- are computed analogously.

Exercise 8
A straightforward implementation consists of (i) adding a term + bolus to the equation of X_5 (pyruvate); (ii) adding statements such as bolus $= 0$, @ 1 bolus $= 10$, and @ 1.1 bolus $= 0$; and (iii) initializing the system at the original steady state. One may study the system responses in terms of the original variables or the scaled variables $\tilde{X}_i$. The latter option may be advantageous, because with it one does not have to compare vastly different scales.

Solving the original model from 0 to 20 shows the expected sharp increase (bolus) in pyruvate, which is gradually degraded. At $t = 20$, the system appears to have reached a new steady state. However, this is not so: the system will return to the original steady state (check this by clicking), but it takes a very long time to do so.

Most variables temporarily increase in response to the bolus. Outside pyruvate, acetyl-CoA (X_3) shows the highest increase of about 65% before it very slowly returns to the steady state. Oxalacetate 2 (X_2), glutamate (X_6), and aspartate (X_7) initially decrease, before reapproaching the steady state. As with other perturbations of the original system, the time scales of the dynamics differ drastically among the dependent variables.

The modified model behaves much better than the original model. Within 10 time units, the bolus of pyruvate is metabolized, and the system has essentially returned to its original steady state.

Exercise 13
Assuming that protein loss for energy production is unchanged at 2%, the ratio of reverse to forward fluxes is now $v_R/v_F = \frac{5}{7}$, which corresponds to about 71% reutilization. Expressed in terms of net flux, we obtain $v_F = \frac{7}{2} v_{net}$ and $v_R = \frac{5}{2} v_{net}$.

The net flux between aspartate and the protein pool in the original model is $v_{net} = v_{50,7\,old} = X_{34\,old} = 0.236$ mM/min. In the modified model, the flux of aspartate supply corresponds to $\frac{7}{2}$ of the net flux in the original model, which yields $v_F = v_{50,7} = X_{34} = \frac{7}{2} \times 0.236 = 0.826$ mM/min and $v_R = v_{7,5} = X_{40} X_7 = \frac{5}{2} \times 0.236 = 0.590$ mM/min. Since $X_7 = 1.85$ mM/min is unchanged, the rate of aspartate reutilization is $X_{40} = 0.59/1.85 = 0.319$ min^{-1}. With these values for the forward and reverse fluxes, the net

Table H1.

10% Increase in	Response of Original Model	Response of Modified Model
X_1	Fast	Fast
X_2	Fast	Fast
X_3	Slow (X_3, X_5 not back @ 20)	Back @ 2
X_4	Fast	Fast
X_5	Very slow (apparent steady state @ 20)	Back @ 10
X_6	Very slow (X_5 almost zero)	Back @ 20
X_7	Very slow (X_5 almost zero)	Back @ 10
X_8	Very slow (X_5 exceeds steady-state value by more than 70 times	Back @ 20 (X_5 exceeds steady-state value by more than 18 times)
X_9	Fast	Fast
X_{10}	Fast	Fast
X_{11}	Very slow (apparent steady state @ 20)	Back @ 10
X_{12}	Fast	Fast
X_{13}	Very slow (apparent steady state @ 20)	Back @ 5

flux from protein to aspartate is unchanged at 0.826 mM/min −0.590 mM/min = 0.236 mM/min.

Exercise 16

There are unlimited possibilities for exploring the effects of perturbations in initial values. As one example, Table H1 shows some results of increasing any one initial value by a factor of 10. As the table indicates, the responses vary widely. Some perturbations are mitigated very quickly, while others cause some of the system variables to reach rather high or low values before eventually returning to the original steady state, which, within reason, is independent of the initial values (check this by clicking $\boxed{\dot{x}=0}$). The results in the table are qualitative, indicating whether the system absorbs the perturbation quickly or slowly, where "slowly" means that the values of some variables are still noticeably different from the steady-state values at the arbitrarily picked final time 20. The shorthand "Back @ 10" means that the system has essentially returned to the steady state at time 10. It is obvious that X_3 and X_5 are the most critical variables, as they were in simulations described in the text.

CHAPTER TEN

Exercise 2

As the current models are set up, ADP and ATP are part of the same pool. To study imbalances between the two, they need to be represented in separate pools. This requires individual differential equations for ADP and ATP (and the rest of the adenylate pool) that reflect all processes leading to synthesis and degradation of ADP and ATP, respectively. For normal subjects, the steady-state sizes of these pools reflect normal physiology. Imbalances may occur when some of the processes (potentially *anywhere* in the system) are altered. Specifics are very difficult to predict without a formal analysis or sufficiently many simulations.

Exercise 4 (Purine model 2 in form of a GMA system; PLAS notation)

$X1' = 0.9\ X1^{\wedge} - .03\ X4^{\wedge} - .45\ X8^{\wedge} - 0.04\ X17^{\wedge}.65\ X18^{\wedge}.7 \gg$
$\gg \quad -5.2728\ X1^{\wedge}2\ X2^{\wedge} - 0.06\ X4^{\wedge} - 0.25\ X8^{\wedge} - 0.2\ X18^{\wedge} \gg$
$\gg \quad -361.69\ X1^{\wedge}1.2\ X8^{\wedge} - 0.08 - 233.8\ X1^{\wedge}.5\ X4^{\wedge} \gg$
$\gg \quad -0.8\ X6^{\wedge}.75 - 1.2\ X15^{\wedge}.42 - 12.569\ X1^{\wedge}1.1\ X2^{\wedge} \gg$
$\gg \quad -.89\ X13^{\wedge}.48 - 1.2951\ X1^{\wedge}1.27$

$X2' = 5.2728\ X1^{\wedge}2\ X2^{\wedge} - 0.06\ X4^{\wedge} - 0.25\ X8^{\wedge} - 0.2\ X18^{\wedge} - 0.08 \gg$
$\gg \quad +12.569\ X1^{\wedge}1.1\ X2^{\wedge} - .89\ X13^{\wedge}.48 + 0.02688\ X4^{\wedge}.8\ X8^{\wedge} \gg$
$\gg \quad -0.03\ X18^{\wedge} - 0.1 + 0.3005\ X2^{\wedge} - .15\ X4^{\wedge} - .07\ X7^{\wedge} \gg$
$\gg \quad -0.76\ X8^{\wedge}.7 - 3.5932\ X2^{\wedge}.4\ X4^{\wedge} - 0.24\ X8^{\wedge}0.2\ X18^{\wedge} \gg$
$\gg \quad -0.05 - 1.2823\ X2^{\wedge}.15\ X7^{\wedge} - 0.09\ X8^{\wedge} - 0.03 \gg$
$\gg \quad -0.9135\ X2^{\wedge}.8\ X18^{\wedge} - 0.36$

$X3' = 3.5932\ X2^{\wedge}.4\ X4^{\wedge} - .24\ X8^{\wedge}.2\ X18^{\wedge} - .05 \gg$
$\gg \quad -66544.7\ X3^{\wedge}.99\ X4^{\wedge} - 0.95$

$X4' = 66544.7\ X3^{\wedge}.99\ X4^{\wedge} - 0.95 + 233.8\ X1^{\wedge}.5\ X4^{\wedge} - 0.8\ X6^{\wedge}.75 \gg$
$\gg \quad +0.06923\ X11 + 8.8539\ X5^{\wedge}.33 - 7.2067\ X4^{\wedge}.2\ X5^{\wedge} - .6 \gg$
$\gg \quad -0.02688\ X4^{\wedge}.8\ X8^{\wedge} - 0.03\ X18^{\wedge} - 0.1 \gg$
$\gg \quad -614.5\ X4^{\wedge}0.05\ X8^{\wedge}0.13 - 0.001062\ X4^{\wedge}.97 \gg$
$\gg \quad -0.0602\ X4^{\wedge}.1\ X9^{\wedge} - .3\ X10^{\wedge}0.87$

$X5' = 7.2067\ X4^{\wedge}.2\ X5^{\wedge} - 0.6 - 8.8539\ X5^{\wedge}.33 - 0.29\ X5^{\wedge}.9$

$X6' = 0.29\ X5^{\wedge}.9 - 233.8\ X1^{\wedge}.5\ X4^{\wedge} - 0.8\ X6^{\wedge}.75 - 0.01\ X6^{\wedge}0.55$

$X7' = 1.2823\ X2^{\wedge}.15\ X7^{\wedge} - 0.09\ X8^{\wedge} - 0.03 \gg$
$\gg \quad -0.3738\ X4^{\wedge}0.12\ X7^{\wedge}0.16$

$X8' = 0.3738\ X4^{\wedge}0.12\ X7^{\wedge}0.16 + 361.69\ X1^{\wedge}1.2\ X8^{\wedge} \gg$
$\gg \quad -1.2\ X15^{\wedge}.42 + 0.04615\ X11 - 0.3005\ X2^{\wedge} - .15\ X4^{\wedge} \gg$
$\gg \quad -.07\ X7^{\wedge} - 0.76\ X8^{\wedge}.7 - 409.6\ X4^{\wedge}0.05\ X8^{\wedge}.13 \gg$
$\gg \quad -0.2511\ X8^{\wedge}0.9\ X18^{\wedge} - 0.34 - .1199\ X8^{\wedge}.4\ X9^{\wedge} \gg$
$\gg \quad -1.2\ X10^{\wedge} - .39$

$X9' = 0.0602\ X4^{\wedge}.1\ X9^{\wedge} - .3\ X10^{\wedge}0.87 + 0.001938\ X12 \gg$
$\gg \quad -3.2789\ X9^{\wedge}.42\ X10^{\wedge}.33 - 0.03333\ X9$

$X10' = .1199\ X8^{\wedge}.4\ X9^{\wedge} - 1.2\ X10^{\wedge} - .39 + .001318\ X12 \gg$
$\gg \quad -2.2296\ X9^{\wedge}.42\ X10^{\wedge}.33 - 0.03333\ X10$

$X11' = 614.5\ X4^{\wedge}0.05\ X8^{\wedge}0.13 + 409.6\ X4^{\wedge}0.05\ X8^{\wedge}.13 \gg$
$\gg \quad -0.06923\ X11 - 0.04615\ X11$

$X12' = 3.2789\ X9^{\wedge}.42\ X10^{\wedge}.33 + 2.2296\ X9^{\wedge}.42\ X10^{\wedge}.33 \gg$
$\gg \quad -.001318\ X12 - .001938\ X12$

$X13' = 0.001062\ X4^{\wedge}.97 + 0.03333\ X9 + 0.9135\ X2^{\wedge}.8\ X18^{\wedge} - 0.36 \gg$
$\gg \quad -12.569\ X1^{\wedge}1.1\ X2^{\wedge} - .89\ X13^{\wedge}.48 - 0.003793\ X13^{\wedge}1.12 \gg$
$\gg \quad -0.2754\ X13^{\wedge}.65$

$X14' = 0.2754\ X13^{\wedge}.65 + 0.4919\ X15^{\wedge}.5 - 0.949\ X14^{\wedge}.55 \gg$
$\gg \quad -0.0012\ X14^{\wedge}2$

```
X15'= 0.2511 X8^0.9 X18^ − 0.34 + 0.03333 X10 >>
>>     − 361.69 X1^1.2 X8^ − 1.2 X15^.42 − 0.4919 X15^.5
X16'= .949 X14^.55 − .00008744 X16^2.21
```

X1 = 5
X2 = 100
X3 = 0.2
X4 = 2500
X5 = 4
X6 = 1
X7 = 25
X8 = 400
X9 = 6
X10 = 3
X11 = 28600
X12 = 5160
X13 = 10
X14 = 5
X15 = 5
X16 = 100
X17 = 18
X18 = 1400

t0 = 0
tf = 1000
hr = 1

Exercise 7

The kinetic order g_{24} is defined as

$$g_{24} = \left(\frac{\partial v_{gmpr}}{\partial X_4} + \frac{\partial v_{ampd}}{\partial X_4} + \frac{\partial v_{hprt}}{\partial X_4} + \frac{\partial v_{den}}{\partial X_4}\right) \frac{X_4}{V_2^+}.$$

All component rates were previously defined as products of power-law functions, namely, $v_{ampd} = \alpha_{ampd} X_3^{f_{ampd3}} X_4^{f_{ampd4}}$, $v_{den} = \alpha_{den} X_1^{f_{den1}} X_2^{f_{den2}} X_3^{f_{den3}} X_4^{f_{den4}}$, $v_{gmpr} = \alpha_{gmpr} X_2^{f_{gmpr2}} X_3^{f_{gmpr3}} X_4^{f_{gmpr4}}$, and $v_{hprt} = \alpha_{hprt} X_1^{f_{hprt1}} X_2^{f_{hprt2}} X_8^{f_{hprt8}}$. Computation of the partial derivative and multiplication with X_4/V_2^+ in each case leads to an expression of the form $f \cdot v/V_2^+$. For instance, in the case of v_{den} one obtains

$$\frac{\partial v_{den}}{\partial X_4} \cdot \frac{X_4}{V_2^+} = f_{den4} \cdot \alpha_{den} X_1^{f_{den1}} X_2^{f_{den2}} X_3^{f_{den3}} X_4^{f_{den4}-1} \cdot \frac{X_4}{V_2^+} = f_{den4} \frac{v_{den}}{V_2^+}.$$

Thus, outside the known rates v_{ampd}, v_{den}, v_{gmpr}, v_{hprt} (and V_2^+, which actually is the sum of the four), only the kinetic orders with respect to X_4 are needed for the computation.

Exercise 12

Splitting up v_{pyr} is irrelevant for the S-system model, as long as the same total amount is being degraded. Subsequently, the relative amounts used for tryptophan or histidine are immaterial. The corresponding GMA model would be slightly different, because two β-terms would describe the synthesis of tryptophan and histidine. The relative amounts for tryptophan and histidine synthesis would be reflected in the steady-state values of the two fluxes; their sum would be equivalent with the current v_{pyr}.

Exercise 14 (Partial solution)

Quasi steady state of GMA, computed by solving the system from t0 = 0 to tf=10,000:

10000	5.017591	98.26205	0.1981791	2475.284	3.991847
	0.9846734	24.7933	410.1931	6.005045	3.02478
	28680.16	5175.295	9.515787	5.057937	5.504281
	100.286				

At tf=10,000, X_9 and are X_{12} still (slightly) moving, so a steady state has not quite been reached within the accuracy of the computer.

PLAS finds the steady-state solution if the initial values are not too far from it.

Exercise 17

The degradation term of PRPP in the final S-system model is

$$V_1^- = \beta_1 X_1^{b_{11}} X_2^{b_{12}} X_4^{b_{14}} X_6^{b_{16}} X_8^{b_{18}} X_{13}^{b_{1,13}} X_{15}^{b_{1,15}} X_{18}^{b_{1,18}}.$$

At steady state, this degradation term must equal the sum of the rates v_{pyr}, v_{den}, v_{aprt}, v_{gprt}, and v_{hprt}. The equivalence is ensured through constraints on the kinetic orders, which are sums of (relative) partial derivatives, as shown many times (e.g., see solution to Exercise 7), and on the numerical value of the rate constant β_1.

Exercise 21 (Partial solution)

The supply of IMP or other metabolites can be realized in a biochemical model in different ways. One may implement a one-time bolus or increase the rate constant of the appropriate production term, which corresponds to the addition of some constant influx. In the first case, the system returns to the original steady state, as long as the bolus is of reasonable size. The features to be studied in this case are the magnitudes of overshoots and undershoots, as well as the duration of the transient behavior, before the system has essentially returned to the steady state. In the second case, the steady state will most likely be different. In this situation, one may study how much the new steady state differs from the original state, how long it takes the system to reach the new steady state, and what the transients look like.

As an example, we study an additional supply of IMP in the final S-system model. In the first case, we add a bolus of 100 during the time period [0.5, 0.6] and study the dynamics over the time period [0, 100]. As expected, the steady state is not affected (check this by clicking). In response to the bolus, X_2 rapidly rises by about 10% to 110, before it gradually falls back to its original value. This process is rather slow, and at $t = 100$, X_2 is still more than 1% above the steady state. Most of the system variables are almost unaffected by the surplus in IMP. This is in line with clinical and biochemical observations that IMP is very tightly controlled. The only variable reacting noticeably is hypoxanthine (X_{13}), which temporarily increases by over 10% about 10 time units after the bolus.

To realize a constant oversupply of IMP, one may change α_2. Trying different values, one finds that multiplication of α_2 by 1.03 leads to a steady-state value of IMP that is about 10% increased. As in the former case of a bolus, most system variables are rather immune to this permanent change. Noticeable exceptions are hypoxanthine (30% increased) and xanthine (20% increased). These two metabolites are immediate recipients of the excess in IMP. Interestingly, the uric acid level increases only by about 5%. Further analysis shows that the excess amounts of hypoxanthine and xanthine are handled by increased rates of elimination. Specifically, v_x and v_{bx} increase from 0.03 to 0.04 and from 0.05 to 0.065, respectively. By contrast, the rate of the salvage pathway (v_{hprt}) is only very slightly increased (from 3.7 to 3.75). The new steady state is approached in a gradual fashion without interesting transient behavior.

Exercise 27

The effect of uricosuric drugs is rather readily implemented. Since only the excretion of uric acid is affected, the necessary alterations are limited to v_{ua}. In the S-system model, this process is represented by the β-term of the variable X_{16} (uric acid), which has the form $0.00008744 X_{16}^{2.21}$. An increase in uric acid excretion may be implemented in three ways: an increase in the rate constant, an increase in the kinetic order, or a combination of both. Which implementation is most realistic depends on the mechanism of the drug.

CHAPTER ELEVEN

Exercise 1.1

The most direct consequence is a lower value for the kinetic order h_{55}. The sensitivities with respect to this kinetic order are among the highest, and changes in h_{55} might change the sensitivity profile. As long as the reaction is still far from thermodynamic equilibrium, this might be the main consequence. However, if the reaction is close to the equilibrium, the reverse reaction must be considered, and it might even be necessary to consider glucose as a dependent, rather than an independent variable. This would require major redesigning of the model, beginning with the addition of a new equation.

Exercise 5

Sensitivities with respect to rate constants:

	X_1	X_2	X_3
α_1	1.255009	1.5596	1.842742
β_1	−1.255009	−1.5596	−1.842742
α_2	0.968594	2.511778	2.967785
β_2	−0.968594	−2.511778	−2.967785
α_3	0.8821124	2.287512	3.000423
β_3	−0.8821124	−2.287512	−3.000423

HINTS AND PARTIAL SOLUTIONS

	$V(X_1)$	$V(X_2)$	$V(X_3)$
α_1	1	0.5528227	0.5528227
β_1	0	-0.5528227	-0.5528227
α_2	0	0.8903355	0.8903355
β_2	0	0.1096645	-0.8903355
α_3	0	-0.09987306	0.9001269
β_3	0	0.09987306	0.09987306

There is no clear winner among the rate constants; all affect the system in similar ways. The sensitivities of metabolites with respect to the two rate constants of each equation are equal in magnitude and opposite in sign. This is no coincidence. Reasons are given in Chapter 7.

Exercise 6
The sensitivities of metabolites with respect to kinetic orders are given in Table H2. The sensitivities of fluxes with respect to kinetic orders are given in Table H3.

Exercise 12 (Partial solution)
As a first, quick test, look at the parameter sensitivities of the dependent variables with respect to h_{33}. For $h_{33} = 0.3$, they are

$$S(X_1, h_{33}) = 0.5020419,$$
$$S(X_2, h_{33}) = 1.301905,$$
$$S(X_3, h_{33}) = 1.707649.$$

Table H2.

	X_1	X_2	X_3
$g(1, 4)$	1.907245	2.370134	2.800427
$g(1, 6)$	1.378769	1.713396	2.024459
$h(1, 1)$	5.190324	6.450018	7.621003
$h(1, 2)$	-0.56698	-0.7045862	-0.8325021
$h(1, 7)$	-4.629578	-5.753178	-6.797654
$g(2, 1)$	-2.487262	-6.450018	-7.621003
$g(2, 2)$	0.3040846	0.7885583	0.9317192
$g(2, 5)$	0.4988454	1.293616	1.528469
$g(2, 7)$	2.215276	5.7447	6.787636
$g(2, 10)$	0.5102474	1.323184	1.563405
$h(2, 2)$	2.944429	7.635552	9.021769
$h(2, 3)$	-5.622869	-14.58134	-17.22855
$h(2, 8)$	-4.758368	-12.3395	-14.5797
$g(3, 2)$	-2.681533	-6.953806	-9.120986
$g(3, 3)$	5.120827	13.27943	17.41802
$g(3, 8)$	4.333514	11.23776	14.74004
$h(3, 3)$	0.5020419	1.301905	1.707649
$h(3, 9)$	-0.9269428	-2.403767	-3.15291

Table H3.

	$V(X_1)$	$V(X_2)$	$V(X_3)$
$g(1,4)$	1.907245	2.370134	2.800427
$g(1,6)$	1.378769	1.713396	2.024459
$h(1,1)$	5.190324	6.450018	7.621003
$h(1,2)$	−0.56698	−0.7045862	−0.8325021
$h(1,7)$	−4.629578	−5.753178	−6.797654
$g(2,1)$	−2.487262	−6.450018	−7.621003
$g(2,2)$	0.3040846	0.7885583	0.9317192
$g(2,5)$	0.4988454	1.293616	1.528469
$g(2,7)$	2.215276	5.7447	6.787636
$g(2,10)$	0.5102474	1.323184	1.563405
$h(2,2)$	2.944429	7.635552	9.021769
$h(2,3)$	−5.622869	−14.58134	−17.22855
$h(2,8)$	−4.758368	−12.3395	−14.5797
$g(3,2)$	−2.681533	−6.953806	−9.120986
$g(3,3)$	5.120827	13.27943	17.41802
$g(3,8)$	4.333514	11.23776	14.74004
$h(3,3)$	0.5020419	1.301905	1.707649
$h(3,9)$	−0.9269428	−2.403767	−3.15291

Thus, any small increase in h_{33} will lead to comparable relative changes in the dependent variables, with an attenuation of the effect in X_1 and slight amplifications in X_2 and X_3. To explore more substantial changes, set $h_{33} = 0.5$. The steady state becomes

$$X_{1S} = 0.0825881,$$
$$X_{2S} = 0.799886,$$
$$X_{3S} = 0.305551.$$

All three steady-state concentrations are affected; the direction of change and, to some degree, the magnitudes of change are in line with the trends predicted by the sensitivity analysis. For fair (controlled) comparisons, one should adjust β_3. Beyond the steady-state values, one should study stability and execute representative perturbation experiments.

Exercise 19

The first weight is the product of $L(X_1, X_5)$ (which is 0.310) with the factor in the steady-state value of X_1, expressed as a function of X_5, which is $\exp(-3.201)$ [see Eqs. (11.40) and (11.41)]. The powers of X_5 are the log gains with respect to X_5. They enter the equation through steady-state expressions like $X_{1S} = \exp(-3.201) X_5^{0.310}$. The constant is the log gain of V_3^+ with respect to X_5.

Exercise 20

$$S(X_2, \beta_1) = M_{21} - \frac{V_1^-}{V_2^+} M_{22} = 0.00524,$$

$$S(X_3, \beta_1) = M_{31} - \frac{V_1^-}{V_2^+} M_{32} = 0.00637.$$

HINTS AND PARTIAL SOLUTIONS

PLAS computes these sensitivities as $S(X_2, \beta_1) = 0.00549782$ and $S(X_3, \beta_1) = 0.00649594$.

Exercise 22

h_{22}

With respect to X_1, the unconstrained sensitivity with respect to h_{22} is reported in PLAS as 2.944. If h_{22} is changed, g_{32} is changed to the same degree. The unconstrained sensitivity with respect to g_{32} is -2.682. The sum of the two yields an overall sensitivity of 0.262. Similarly, the overall sensitivities with respect to X_2 and X_3 are 0.682 and -0.099. To visualize these effects in PLAS, one may increase both, h_{22} and g_{32}, by 1%. The resulting steady state is (0.06717481, 0.4681528, 0.1498526), which is approximately 0.26%, 0.68%, and -0.1% changed in comparison with the original steady state.

h_{17}

The computation of constrained sensitivities with respect to h_{17} is more complicated, since the corresponding kinetic order of the production term V_2^+, namely g_{27}, is part of an aggregated power-law term. The value of g_{27} was calculated in the text as $0.62h_{17}$.

The computation of the constrained sensitivity of X_1 with respect to h_{17} begins with a combination of the steady-state equations (11.26) and (11.30), which may be formulated as

$$\vec{y}_D = \begin{pmatrix} -1.255 & -0.969 & -0.882 \\ -1.560 & -2.512 & -2.288 \\ -1.843 & -2.968 & -3.000 \end{pmatrix} \begin{pmatrix} 2.613 \\ -6.603 \\ 7.196 \end{pmatrix}$$

$$- \begin{pmatrix} -1.255 & -0.969 & -0.882 \\ -1.560 & -2.512 & -2.288 \\ -1.843 & -2.968 & -3.000 \end{pmatrix}$$

$$\times \begin{pmatrix} g_{14} & 0 & g_{16} & -h_{17} & 0 & 0 & 0 \\ 0 & g_{25} & 0 & g_{27} & -h_{28} & 0 & g_{2,10} \\ 0 & 0 & 0 & 0 & g_{38} & -h_{39} & 0 \end{pmatrix} \begin{pmatrix} \ln 10 \\ \ln 5 \\ \ln 3 \\ \ln 40 \\ \ln 136 \\ \ln 2.86 \\ \ln 4 \end{pmatrix}.$$

This combination of numerical and symbolic entries was chosen to make the role of h_{17} and g_{27} explicit. When y_1 (the first component of $\vec{y}_D$) is differentiated with respect to h_{17}, the first term (matrix × vector) drops out, and in the second term only the component $-(-1.255) \times (-h_{17}) \times \ln 40$ is of relevance, since everything else is independent of h_{17}. Differentiating this component with respect to h_{17} yields the value -4.6295. As discussed in Chapter 7, this value is to be multiplied by the value of the kinetic order, which here happens to be 1, and indeed, the result -4.6295 is the unconstrained sensitivity reported by PLAS. Similarly, the term of interest for the sensitivity with respect to g_{27} is $-(-0.969) \times (g_{27}) \times \ln 40$, which yields an unconstrained sensitivity of 3.5745. Multiplication by $g_{27} = 0.62$ yields the unconstrained sensitivity value 2.216 reported in PLAS.

A 1% change in h_{17} evokes a simultaneous change of 0.62% in g_{27}, because $g_{27} = 0.62h_{17}$. Thus, the overall effect of a 1% change in h_{17} is $-4.6295\% + 0.62 \times 3.5745\% = -2.41\%$. One can test this result by increasing h_{17} and g_{27} simultaneously by, say, 0.1%.

PLAS computes the new steady-state value of X_1 as 0.06683844, which is indeed about 0.241% less than the original value.

Exercise 23 (General comments)
Trying different conditions shows that the kinetic orders can be quite different, which indicates that the parameters are sensitive to experimental measurements. Furthermore, the portion of the power-law term that is directly affected, namely $X_2^{h_{22}} X_3^{h_{23}}$, varies widely between the original case and the alternative systems. However, the overall effect on the system is not as great as one might expect. First, the rate constant β_2 is to be adjusted, so that the observed steady state is regained. Second, if h_{22}, h_{23}, and β_2 are changed, the corresponding parameters in V_3^+ must be changed accordingly. While some of the individual parameter sensitivities are high, the overall effect of the changes is moderate. This should be assessed with simulations exploring transient and long-term responses.

APPENDIX

Exercise 2
2.1 Examples include the logarithmic and exponential functions.
2.2 An example is $y(n) = n^2$ for $n = 0, 1, 2, \ldots$.
2.3 An example is $y(x) = x^2$ for $x \in (-\infty, +\infty)$.
2.4 An example is $y(n) = n^2$ for $n = 0, \pm 1, \pm 2, \ldots$.

Exercise 4
4.1 Calling the operating point p, one obtains

$$V(t) \approx \sin p + \cos p(t-p) + \frac{-\sin p}{1 \times 2}(t-p)^2 + \frac{-\cos p}{1 \times 2 \times 3}(t-p)^3$$
$$+ \frac{\sin p}{1 \times 2 \times 3 \times 4}(t-p)^4 + \frac{\cos p}{1 \times 2 \times 3 \times 4 \times 5}(t-p)^5.$$

If the operating point is chosen as 0, the result becomes much simpler, since $\sin p = 0$ and $\cos p = 1$. Thus,

$$V(t) \approx t - \frac{1}{6}t^3 + \frac{1}{120}t^5.$$

4.3 $Y(t)$ is already a (very simple) polynomial, and the best polynomial approximation is this function itself. This is readily confirmed by computing the Taylor series $T(t)$ at some operating point p:

$$T(t) = p^4 + 4p^3(t-p) + \frac{12p^2}{2}(t-p)^2 + \frac{24p}{6}(t-p)^3 + \frac{24}{24}(t-p)^4$$
$$= p^4 + 4p^3 t - 4p^4 + 6p^2 t^2 - 12p^3 t + 6p^4 + 4pt^3 - 12p^2 t^2 + 12p^3 t$$
$$- 4p^4 + t^4 - 4pt^3 + 6p^2 t^2 - 4p^3 t + p^4$$
$$= t^4.$$

4.4 Although one could go through the formal procedures, inspection shows that this function already is a polynomial of degree 4. Indeed, computing the Taylor series, one finds that the fifth-order term drops out, and the result is exactly the same as the given $X(t)$.

HINTS AND PARTIAL SOLUTIONS 467

Exercise 7
7.1 The linear approximation at the operating point p is given as
$$L(t) = \exp(p+1) + \exp(p+1) \cdot (t-p)$$
$$= (1 + t - p)\exp(p+1).$$

The power-law approximation at the operating point p is
$$P(t) = \alpha t^g,$$

where $g = \exp(p+1) \cdot p / \exp(p+1) = p$ and $\alpha = \exp(p+1) \cdot p^{-p}$.

7.3 y can actually be written in a much simpler form, namely, $y = t^2$ (convince yourself that this is true). With that, we already have a power-law function. The linear approximation is
$$L(t) = p^2 + 2p(t-p) = 2pt - p^2.$$

Exercise 9
9.2 $\partial V/\partial X_1 = \exp(X_1 + X_2)$.
9.3 $\partial V/\partial X_1 = 0$.

Exercise 10
10.2 The kinetic orders are computed through partial differentiation and evaluation at the operating point of choice. For instance,
$$g_1 = \frac{\partial V}{\partial X_1} \cdot \frac{X_1}{V}$$
$$= \exp(X_1 + X_2) \cdot \frac{X_1}{\exp(X_1 + X_2)} = X_1 \quad \text{at the operating point}$$

Similarly, $g_2 = X_2$ at the operating point. The rate constant α is obtained in a second step by equating $\alpha X_1^{g_1} X_2^{g_2}$ with V. If the operating point is (P_1, P_2), the power-law approximation PL is
$$PL(X_1, X_2) = \exp(P_1 + P_2) P_1^{-P_1} P_2^{-P_2} X_1^{P_1} \cdot X_2^{P_2}.$$

10.3 With respect to X_2 and X_3, the power-law approximation is the same as in Exercise 10.2. With respect to X_1, V already is in (very simple) power-law form. Thus, the power-law approximation PL is
$$PL(X_1, X_2, X_3) = \exp(P_2 + P_3) \cdot P_2^{-P_2} P_3^{-P_3} X_1 X_2^{P_2} X_3^{P_3}.$$

Exercise 11
11.2 -4.
11.4 $2 - 4 = -2$.
11.5 The easiest way to compute this determinant is probably expansion by the second column, since this column contains three zeros. According to remarks 3 and 1 in the text, we thus take the transpose and exchange the first and second rows, thereby

picking up a minus sign. Thus, we compute the determinant

$$-\begin{vmatrix} 1 & 0 & 0 & 0 \\ 1 & 2 & 3 & 0.5 \\ 3 & 1 & 2 & 1 \\ 0.5 & 4 & 2 & 0.1 \end{vmatrix} = -1 \times \begin{vmatrix} 2 & 3 & 0.5 \\ 1 & 2 & 1 \\ 4 & 2 & 0.1 \end{vmatrix}.$$

The determinant on the right-hand side is computed with Sarrus's rule. The result is

$$\det = -1(0.4 + 12 + 1 - 4 - 0.3 - 4) = -5.1.$$

Exercise 12

12.3 The matrix equation is

$$\begin{pmatrix} 1 & 1 & 0 & 0 \\ 1 & 0 & 1 & 0 \\ 1 & 0 & 0 & 1 \\ 0 & 1 & 1 & -1 \end{pmatrix} \begin{pmatrix} y_1 \\ y_2 \\ y_3 \\ y_4 \end{pmatrix} = \begin{pmatrix} 1 \\ 1 \\ -1 \\ -1 \end{pmatrix}.$$

The system determinant is computed through expansion by the first row. The result is $\det \mathbf{A} = 1$. According to Cramer's rule, one obtains

$$y_1 = \frac{\begin{vmatrix} 1 & 1 & 0 & 0 \\ 1 & 0 & 1 & 0 \\ -1 & 0 & 0 & 1 \\ -1 & 1 & 1 & -1 \end{vmatrix}}{\det \mathbf{A}} = 4.$$

The remaining components of the solution could be obtained in the same way, but it is easier simply to plug in y_1 into the equations and to obtain $y_2 = y_3 = -3$, $y_4 = -5$.

Exercise 16

The products $\mathbf{AB}$ and $\mathbf{BA}$ of two square matrices may be the same, but they are not necessarily the same. If one of the matrices is a multiple of a identity matrix, the products are the same. For instance, suppose

$$\mathbf{A} = \begin{pmatrix} 2 & 0 \\ 0 & 2 \end{pmatrix}, \quad \mathbf{B} = \begin{pmatrix} 1 & 2 \\ 3 & 4 \end{pmatrix}, \quad \mathbf{C} = \begin{pmatrix} 2 & 1 \\ 0 & 2 \end{pmatrix}.$$

Direct multiplication shows that $\mathbf{AB} = \mathbf{BA}$ and $\mathbf{AC} = \mathbf{CA}$. However, $\mathbf{BC} \neq \mathbf{CB}$.

Exercise 18

Consider the determinant

$$D = \begin{vmatrix} a & b & c \\ d & e & f \\ g & h & i \end{vmatrix}.$$

According to Sarrus's rule, its value is $aei + bfg + cdh - ceg - afh - bdi$. Expansion by the first row yields $a(ei - fh) - b(di - fg) + c(dh - eg)$. Multiplying out produces directly the result of Sarrus's method.

References

[1] Achs, M.J., L. Garfinkel, and D. Garfinkel, 1991: A computer model of pancreatic islet glycolysis. *J. Theor. Biol.* **150**(1): 109–135.

[2] Adolph, E.F., 1949: Quantitative relations in the physiological constitutions of mammals. *Science* **109**: 579–585.

[3] Agresar, G., and M.A. Savageau, 1998: A numerical method for solving large stiff systems of differential equations in power-law form. PowBioSys Symposium, Oeiras, Portugal (Abstract).

[4] Albe, K.R., M.H. Butler, and B.E. Wright, 1989: Cellular concentrations of enzymes and their substrates. *J. Theor. Biol.* **143**: 163–195.

[5] Albe, K.R., and B.E. Wright, 1992: Systems analysis of the tricarboxylic acid cycle in *Dictyostelium discoideum*. II. Control analysis. *J. Biol. Chem.* **267**: 3106–3114.

[6] Albe, K.R., and B.E. Wright, 1994: Carbohydrate metabolism in *Dictyostelium discoideum*: I. Model construction. *J. Theor. Biol.* **169**: 243–251.

[7] Ayvazian, J.H., and S. Skupp, 1965: The study of purine utilization and excretion in a xanthinuric man, *J. Clin. Invest.* **44**: 1248–1260.

[8] Bairoch A., 1993: The ENZYME data bank. *Nucleic Acids Res.* **21**(3): 3155–3156.

[9] Bairoch A., 1999: The ENZYME data bank in 1999. *Nucleic Acids Res.* **27**(1): 310–311.

[10] Bairoch, A., and R. Apweiler, 1997a: The SWISS-PROT protein sequence database: its relevance to human molecular medical research. *J. Molec. Med.* **75**(5): 312–316.

[11] Bairoch, A., and R. Apweiler, 1997b: The SWISS-PROT protein sequence data bank and its supplement TrEMBL. *Nucl. Acids Res.* **25**(1): 31–36.

[12] Bairoch, A., and R. Apweiler, 1999: The SWISS-PROT protein sequence data bank and its supplement TrEMBL in 1999. *Nucl. Acids Res.* **27**(1): 49–54.

[13] Bairoch, A., and B. Boeckmann, 1991: The SWISS-PROT protein sequence data bank. *Nucleic Acids Res.* **19**, Suppl: 2247–2249.

[14] Balthis, W.L., E.O. Voit, and G.M. Meaburn, 1996: Setting prediction limits for mercury concentrations in fish having high bioaccumulation potential. *Environmetrics* **7**: 429–439.

[15] Bangerter, J.K., 1998: *Uncertainty and Variability Analysis in the Estimation of Human Exposure to Mercury from Seafood Consumption Using Two-Dimensional Monte Carlo Simulations*. Ph.D. Dissertation, Medical University of South Carolina, Charleston, SC.

[16] Barber, J.R., B.H. Morimoto, L.S. Brunauer, and S. Clarke, 1986: Metabolism of S-adenosyl-L-methionine in intact human erythrocytes. *Biochim. Biophys. Acta* **887**: 361–372.

[17] Barker, W.C., J.S. Garavelli, P.B. McGarvey, C.R. Marzec, B.C. Orcutt, G.Y. Srinivasarao, L.-S.L. Yeh, R.S. Ledley, H.-W. Mewes, F. Pfeiffer, A. Tsugita, and C. Wu, 1999: The PIR-international protein sequence database. *Nucl. Acids Res.* **27**(1): 39–43.

[18] Bartel, T., and H. Holzhütter, 1990: Mathematical modelling of the purine metabolism of the rat liver. *Biochim. Biophys. Acta* **1035**: 331–339.

[19] Bassham, J.A., and G.H. Krause, 1969: Free energy changes and metabolic regulation in steady-state photosynthetic carbon reduction. *Biochim. Biophys. Acta* **189**(2): 207–221.

[20] Batschelet, E., 1979: *Introduction to Mathematics for Life Scientists* (3rd edition). Springer-Verlag, Berlin.

[21] Battey, J., E. Jordan, D. Cox, and W. Dove, 1999: An action plan for mouse genomics. *Nat. Genet.* **21**: 73–75.

[22] Baxevanis, A.D., and B.F.F. Ouellette, 1998: *Bioinformatics. A Practical Guide to the Analysis of Genes and Proteins*. Wiley-Interscience, New York.

[23] Becker, M.A., M. Kim, and K. Husain, 1989a: PRPP and purine nucleotide metabolism in human lymphoblasts with both PRPP synthetase superactivity and HGPRT deficiency. *Adv. Exp. Med. Biol.* **253B**: 13–20.

[24] Becker, M.A., J.G. Puig, F.A. Mateos, M.L. Jimenez, M. Kim, and H.A. Simmonds, 1989b: Neurodevelopmental impairment and deranged PRPP and purine nucleotide synthesis in inherited superactivity of PRPPsynthetase. *Adv. Exp. Med. Biol.* **253A**: 15–22.

[25] Benson, D.A., M.S. Boguski, D.J. Lipman, J. Ostell, and B.F.F. Ouellette, 1998: GenBank. *Nucl. Acids Res.* **26**(1): 1–7.

[26] Benson, D.A., M.S. Boguski, D.J. Lipman, J. Ostell, B.F.F. Ouellette, B.A. Rapp, and D.L. Wheeler, 1999: GenBank. *Nucl. Acids Res.* **27**(1): 12–17.

[27] Berg, P.H., E.O. Voit, and R. White, 1996: A pharmacodynamic model for the action of the antibiotic Imipenem on *Pseudomonas in vitro*. *Bull. Math. Biol.* **58**(5): 923–938.

[28] Bernstein, F.C., T.F. Koetzle, G.J. Williams, E.F. Meyer, M.D. Brice, J.R. Rogers, O. Kennard, T. Shimanouchi, and M. Tasumi, 1977: The protein data bank: a computer-based archival file for macromolecular structures. *J. Mol. Biol.* **112**: 535–542.

[29] Berry, M.N., R.B. Gregory, A.R. Grivell, D.C. Henly, J.W. Phillips, P.G. Wallace, and G.R. Welch, 1990: Constraints in the application of control analysis to the study of metabolism in hepatocytes, in: Cornish-Bowden, A., and M.L. Cárdenas (Eds.), *Control of Metabolic Processes*, Plenum Press, pp. 343–350.

[30] Bish, D.R., and M.L. Mavrovouniotis, 1998: Enzymatic reaction rate limits with constraints on equilibrium constants and experimental parameters. *BioSystems* **47**: 37–60.

[31] BMDP® Statistical Software, 1998: Los Angeles, CA.

[32] Boguski, M. S., 1998: Bioinformatics – a new era. *Trends Guide to Bioinformatics, Elsevier Trend Journals, Supplement*, pp. 1–3.

[33] Bohnensack, R., and W. Halangk, 1986: Control of respiration and of motility in ejaculated bull spermatozoa. *Biochim. Biophys. Acta* **850**(1): 72–79.

[34] Bontemps, F., G. van den Berghe, and H.G. Hers, 1983: Evidence for a substrate cycle between AMP and adenosine in isolated hepatocytes. *Proc. Natl. Acad. Sci. USA* **80**: 2829–2833.

[35] Bronstein, I.N., and K.A. Semendjajew, 1973: *Taschenbuch der Mathematik* (12th edition), B.G. Teubner Verlagsgesellschaft, Leipzig.

[36] Brown, A.B., 1991: *Determining the Economically Optimum Slaughter Age and Time Paths for Dietary Energy and Protein*. Ph.D. Dissertation, Department of Agricultural and Resource Economics, North Carolina State University, Raleigh, NC.

[37] Brown, A.B., and T. Johnson, 1992: Use of an S-system in optimal control of a biological system. Departmental Report, Department of Agricultural and Resource Economics, North Carolina State University, Raleigh, NC. Presented at the Second S-System Symposium, Tampa, FL, 1992.

[38] Brown, P.O., and D. Botstein, 1999: Exploring the new world of the genome with DNA microarrays. *Nat. Genet.* 21 (Supplement 1): 33–37.

[39] Brownstein, M.J., J.M. Trent, and M.S. Boguski, 1998: Functional genomics. *Trends Guide to Bioinformatics, Elsevier Trend Journals, Supplement*, pp. 27–29.

[40] Burks, C., 1999: Molecular biology database list. *Nucl. Acids Res.* 27(1): 1–9.

[41] Canela, E.I., I. Ginesla, and R. Franco, 1987: Simulation of the purine nucleotide cycle as an anaplerotic process in skeletal muscle. *Arch. Biochem. Biophys.* 254(1): 142–155.

[42] Carnap, R., 1966. *Philosophical Foundations of Physics*. Basic Books, New York.

[43] Cascante, M., R. Curto, and A. Sorribas, 1995a: Testing the robustness of the steady-state characteristics of a metabolic pathway: parameter sensitivity as a basic feature for model validation. *J. Biol. Systems* 3: 105–113.

[44] Cascante, M., R. Curto, and A. Sorribas, 1995b: Comparative characterization of the fermentation pathway of *Saccharomyces cerevisiae* using biochemical systems theory and metabolic control analysis. Steady-state analysis. *Math. Biosci.* 130: 51–69.

[45] Cascante, M., R. Franco, and E. Canela, 1991a: Sensitivity analysis: a common foundation of theories for the quantitative study of metabolic control, in: Voit, E.O. (Ed.), *Canonical Nonlinear Modeling. S-System Approach to Understanding Complexity*, Van Nostrand Reinhold, New York, Ch. 4.

[46] Cascante, M., A. Sorribas, and E.I. Canela, 1994: Enzyme–enzyme interactions and metabolite channeling: alternative mechanisms and their evolutionary significance. *Biochem. J.* 298: 313–320.

[47] Cascante, M., N.V. Torres, R. Franco, E. Meléndez-Hevia, and E.I. Canela, 1991b: Control analysis of transition times. *Molec. Cell. Biochem.* 101: 83–91.

[48] *C. elegans* Sequencing Consortium, 1998: Genome sequence of the nematode *C. elegans*: a platform for investigating biology. *Science* 282: 2012–2018.

[49] Chance, B., D. Garfinkel, J.J. Higgins, and B. Hess, 1960: Metabolic control mechanisms. *J. Biol. Chem.* 235(8): 2426–2439.

[50] Chaudhuri, K., and T. Johnson, 1990: Bioeconomic dynamics of a fishery modeled as an S-system. *Math. Biosci.* 99: 231–249.

[51] Chiew, Y.Y., J.M. Reimers, and B.E. Wright, 1985: Steady state models of spore cell metabolism in *Dictyostelium discoideum*. *J. Biol. Chem.* 260: 15325–15331.

[52] Chock, P.B., S.G. Rhee, and E.R. Stadtman, 1980: Interconvertible enzyme cascades in cellular regulation. *Ann. Rev. Biochem*, 49: 813–843.

[53] Cleland, W.W., 1967: Enzyme kinetics. *Ann. Rev. Biochem.* 36: 77–112.

[54] Cole, S.T., R. Brosch, J. Parkhill, T. Garnier, C. Churcher, D. Harris, S.V. Gordon, K. Eiglmeier, S. Gas, C.E. Barry 3rd., et al., 1998: Deciphering the biology of Mycobacterium tuberculosis from the complete genome sequence. *Nature* 393: 537–544.

[55] Collado-Vides, J., 1989: A transformational-grammar approach to the study of the regulation of gene expression. *J. Theor. Biol.* 136: 403–425.

[56] Collado-Vides, J., R.M. Guièrrez-Ríos, and G. Bel-Enguix, 1998: Networks of transcriptional regulation encoded in a grammatical model. *BioSystems* 47: 103–118.

[57] Collins, F.S., A. Patrinos, E. Jordan, A. Chakravarti, R. Gesteland, L. Walters, and the members of the DOE and NIH planning groups, 1998: New goals for the US Human Genome Project: 1998–2003. *Science* **282**: 682–689.

[58] Cornish-Bowden, A., 1976: The effect of natural selection on enzymic catalysis. *J. Mol. Biol.* **101**: 1–9.

[59] Cornish-Bowden, A., 1989: Metabolic control theory and biochemical systems theory: different objectives, different assumptions, different results. *J. Theor. Biol.* **136**: 365–377.

[60] Cornish-Bowden, A., and M.L. Cárdenas (Eds.), 1990: *Control of Metabolic Processes*. NATO ASI Series A, Vol. 190, Plenum Press, New York.

[61] Corton, J.C., S.P. Anderson, A.J. Stauber, D.B. Janszen, J.S. Kimbell, and R.B. Conolly, 1999: Entering the Era of Toxicogenomics with DNA Microarrays. *CIIT Activities*, http://www.ciit.org/ACT99/ACTIVITIESFEB99/feb99.html.

[62] Courant, R., 1972: *Vorlesungen über Differential- und Integralrechnung. Zweiter Band*. Springer-Verlag, Berlin.

[63] Crowley, P.H., 1975: Natural selection and the Michaelis constant. *J. Theor. Biol.* **50**: 461–475.

[64] Crummenerl, E., 1969: *Bi uns doahäim*. Gebr. Zimmermann Verlag, Balve/Westf., Germany.

[65] Curto, R., 1996: *The Usefulness of Mathematical Models in Biotechnological and Clinical Studies: Analysis of Ethanol Fermentation in* Saccharomyces cerevisiae *and Purine Metabolism in Humans*. Tesis Doctoral, Universitat de Barcelona.

[66] Curto, R., A. Sorribas, and M. Cascante, 1995: Comparative characterization of the fermentation pathway of *Saccharomyces cerevisiae* using biochemical systems theory and metabolic control analysis. Model definition and nomenclature. *Math. Biosci.* **130**: 25–50.

[67] Curto, R., E.O. Voit, A. Sorribas, and M. Cascante, 1997: Validation and steady-state analysis of a power-law model of purine metabolism. *Biochem. J.* **324**: 761–775.

[68] Curto, R., E.O. Voit, A. Sorribas, and M. Cascante, 1998a: Mathematical models of purine metabolism in man. *Math. Biosci.* **151**: 1–49.

[69] Curto, R., E.O. Voit, A. Sorribas, and M. Cascante, 1998b: Analysis of abnormalities in purine metabolism leading to gout and to neurological dysfunctions in man. *Biochem. J.* **329**: 477–487.

[70] Delbarre, F., C. Auscher, A. Degery, H. Brouilhet, and J.-L. Olivier, 1968: Le traitment de la dyspurine goutteuse par la mercaptopyrazolopyrimidine (MPP:thiopurinol). *Presse Medicale* **76**(49): 2329.

[71] Delgado, J., and J.C. Liao, 1992a: Determination of flux control coefficients from transient metabolite concentrations. *Biochem. J.* **282**: 919–927.

[72] Delgado, J., and J.C. Liao, 1992b: Metabolic control analysis using transient metabolite concentrations. Determination of metabolite concentration control coefficients. *Biochem. J.* **285**: 954–972.

[73] DeRisi, J.L., V.R. Iyer, and P.O. Brown, 1997. Exploring the metabolic and genetic control of gene expression on a genomic scale. *Science* **278**: 680–686.

[74] de Verdier, C.H., A. Ericson, F. Niklasson, and M. Westman, 1977: Adenine metabolism in man. *Scand. J. Clin. Lab. Invest.* **37**: 567–575.

[75] Dong, H., R.J. O'Brien, E.T. Fung, A.A. Lanahan, P.F. Worley, and R.L. Huganir, 1997: GRIP: a synaptic PDZ domain-containing protein that interacts with AMPA receptors. *Nature* **386**: 279–284.

[76] Doran, P.M., 1985: *Effects of Immobilization on the Metabolism of Yeast (Fermentation, Cells, Ethanol Production)*. Ph.D. Dissertation, California Institute of Technology.

[77] Duggan, D.J., M. Bittner, Y. Chen, P. Meltzer, and J.M. Trent, 1999: Expression profiling using cDNA microarrays. *Nat. Genet.* **21** (Supplement), 10–14.

[78] Dykhuizen, D.E., A.M. Dean, and D.L. Hartl, 1987: Metabolic flux and fitness. *Genetics* **115**: 25–31.

[79] Easterby, J.S., 1981: A generalized theory of the transition time for sequential enzyme reactions. *Biochem. J.* **285**: 965–972.

[80] Edelstein-Keshet, L., 1988: *Mathematical Models in Biology*. McGraw-Hill, Inc., New York, NY.

[81] Edwards, N.L., D. Recker, and I.H. Fox, 1979: Overproduction of uric acid in hypoxanthine-guanine phosphoribosyltransferase deficiency. *J. Clin. Invest.* **63**: 922–930.

[82] Elion, G.B., A. Kovensky, G.H. Hitchings, E. Metz, and R.W. Rundles, 1966: Metabolic studies of allopurinol, an inhibitor of xanthine oxidase. *Biochem. Pharmacol.* **15**: 863–880.

[83] Environmental Protection Agency (EPA USA), 1992: *Guidelines for Exposure Assessment*. Federal Register **57**(104), 22888–22938.

[84] Fairén, V., and B. Hernández-Bermejo, 1996: Mass action law conjugate representation for general chemical mechanisms. *J. Phys. Chem.* **100**: 19023–19028.

[85] Faux, M.C., and J.D. Scott, 1996: Molecular glue: kinase anchoring and scaffold proteins. *Cell* **85**(1): 9–12.

[86] Fell, D.A., 1992: Metabolic control analysis – a survey of its theoretical and experimental development. *Biochem. J.* **286**: 313–330.

[87] Fell, D.A., 1997: *Understanding the Control of Metabolism*. Portland Press, London.

[88] Fell, D.A., and J.R. Small, 1986: Fat synthesis in adipose tissue. An examination of stoichiometric constraints. *Biochem. J.* **238**: 781–786.

[89] Ferreira, A.E.N., 1992: Linked lists representation of S-systems: an alternative way to implement numerical algorithms. Second S-System Symposium, Tampa, FL (Abstract).

[90] Fersht, A.R., 1974: Catalysis, binding and enzyme–substrate complementarity. *Proc. Roy. Soc. Ser. B* **187**, 397–407.

[91] Fleischmann R.D., M.D. Adams, O. White, R.A. Clayton, E.F. Kirkness, A.R. Kerlavage, C.J. Bult, J.F. Tomb, B.A. Dougherty, J.M. Merrick, et al., 1995: Whole-genome random sequencing and assembly of *Haemophilus influenzae* Rd. *Science* **269**: 496–512.

[92] Flint, H.J., R.W. Tateson, I.B. Bartelmess, D.J. Porteous, W. Donachies, and H. Kacser, 1981: Control of the flux in the arginine pathway of *Neurospora crassa*. *Biochem. J.* **200**: 231–246.

[93] FlyBase Consortium, 1999: The FlyBase database of the Drosophila genome projects and community literature. *Nucl. Acids Res.* **27**(1): 85–88.

[94] Fodor, S., 1997: Massively parallel genomics. *Science* **277**: 393–395.

[95] Franco, R., and E.I. Canela, 1984: Computer simulation of purine metabolism. *Eur. J. Biochem.* **144**: 305–315.

[96] Frank, P.M., 1978: *Introduction to System Sensitivity Theory*. Academic Press, New York.

[97] Franke, J., and M. Sussman, 1973: Accumulation of uridine diphosphoglucose pyrophosphorylase in *Dictyostelium discoideum* via preferential synthesis. *J. Mol. Biol.* **81**(2): 173–185.

[98] Fujibuchi, W., S. Goto, H. Migimatsu, I. Uchiyama, A. Ogiwara, Y. Akiyama, and M. Kanehisa, 1998: DBGET/LinkDB: an integrated database retrieval system, in: Altman, R.B., A.K. Dunker, L. Hunter, and T.E. Klein (Eds.), *Pacific Symp. Biocomputing '98*, pp. 683–694.

[99] Fujimori, S., T. Tagaya, N. Yamaoka, N. Kamatani, and L. Akaoka, 1991: Molecular analysis of hypoxanthine-guanine phosphoribosyltransferase deficiency in Japanese patients. *Adv. Exp. Med. Biol.* **309B**: 101–104.

[100] Galazzo, J.L., and J.E. Bailey, 1990: Fermentation pathway kinetics and metabolic flux control in suspended and immobilized *Saccharomyces cerevisiae*. *Enzyme Microb. Technol.* **12**: 162–172.

[101] Galazzo, J.L., and J.E. Bailey, 1991: Errata. *Enzyme Microb. Technol.* **13**: 363.

[102] Gallagher, R., and T. Appenzeller, 1999: Beyond reductionism. *Science* **284**: 79.

[103] Ganguly, S., and K.S. Chaudhuri, 1995: Regulation of a single-species fishery by taxation. *Ecological Modeling* **82(1)**: 51–60.

[104] Garfinkel, D., 1968: The role of computer simulation in biochemistry. *Comp. & Biomed. Res.* **2(1)**: 31–44.

[105] Garfinkel. D., 1980: Computer modeling, complex biological systems, and their simplifications. *Amer. J. Phys.* **239(1)**: R1–R6.

[106] Garfinkel, D., 1985: Computer-based modeling of biological systems which are inherently complex: problems, strategies, and methods. *Biomed. Biochim. Acta* **44(6)**: 823–829.

[107] Garfinkel, D., M.J. Achs, and L. Dzubow, 1974: Simulation of biological systems at the level of biochemistry and physiology. *Fed. Proc.* **33(2)**: 176–182.

[108] Garfinkel, D., R.A. Frenkel, and L. Garfinkel, 1968: Simulation of the detailed regulation of glycolysis in a heart supernatant preparation. *Comp. & Biomed. Res.* **2(1)**: 68–91.

[109] Garfinkel, D., L. Garfinkel, M. Pring, S.B.Green, and B. Chance, 1970: Computer applications to biochemical kinetics. *Ann. Rev. Biochem.* **39**: 473–498.

[110] Gauss®, Aptech Systems, Inc., P.O. Box 6487, Kent, WA 98064.

[111] Gavalas, G.R., 1968: *Nonlinear Differential Equations of Chemically Reacting Systems*. Springer-Verlag, Berlin.

[112] Geigy, A.G., 1960: *Documenta Geigy: Wissenschaftliche Tabellen* (6th edition). Pharmazeutische Abteilung der Firma J.R., Geigy A.G., Basel.

[113] Gerber, G., H. Preissler, R. Heinrich, and S.M. Rapoport, 1974: Hexokinase of human erythrocytes. Purification, kinetic model and its application to the conditions in the cell. *Eur. J. Biochem* **45(1)**: 39–52.

[114] George, D.G., W.C. Barker, H.-W. Mewes, F. Pfeiffer, and A. Tsugita, 1996a: The PIR-international protein sequence database. *Nucl. Acids Res.* **24(1)**: 17–20.

[115] George, D.G., L.T. Hunt, and W.C. Barker, 1996b: PIR-international protein sequence database. *Methods in Enzymology* **266**: 41–59.

[116] Giacomello, A., and C. Salerno, 1978: Human hypoxanthine-guanine phosphoribosyltransferase. *J. Biol. Chem.* **253**: 6038–6044.

[117] Giersch, C., 1994: Determining elasticities from multiple measurements of steady-state flux rates and metabolite concentrations: theory. *J. Theor. Biol.* **169**: 89–99.

[118] Giersch, C., 1995: Determining elasticities from multiple measurements of steady-state flux rates and metabolite concentrations: application of the multiple modulation method to a reconstituted pathway. *Eur. J. Biochem.* **227**: 194–201.

REFERENCES

[119] Goffeau, A., B.G. Barrell, H. Bussey, R.W. Davis, B. Dujon, H. Feldmann, F. Galibert, J.D. Hoheisel, C. Jacq, M. Johnston, et al., 1996: Life with 6000 genes. *Science* **274**: 546–567.

[120] Goldberg, D.E., 1989: *Genetic Algorithms in Search, Optimization, and Machine Learning*. Addison-Wesley, Reading, MA.

[121] Goldenfeld, N., and L.P. Kadanoff, 1999: Simple lessons from complexity. *Science* **284**: 87–89.

[122] Goto, S., H. Bono, H. Ogata, W. Fujibuchi, T. Nishioka, K. Sato, and M. Kanehisa, 1997: Organising and computing metabolic pathway data in terms of binary relations, in: Altman, R.B., A.K. Dunker, L. Hunter, and T.E. Klein (Eds.), *Pacific Symp. Biocomputing '97*, pp. 175–186.

[123] Goto, S., T. Nishioka, and M. Kanehisa, 1998: LIGAND: chemical database for enzyme reactions. *Bioinformatics* **14**(7), 591–599.

[124] Goto, S., T. Nishioka, and M. Kanehisa, 1999: LIGAND database for enzyme, compounds, and reactions. *Nucleic Acids Res.* **27**(1): 377–379.

[125] Gregg, J.H., and R.D. Bronsweig, 1956: Dry weight loss during culmination of the slime mold, *Dictyostelium discoideum*. *J. Cell. Comp. Physiol.* **47**: 483–488.

[126] Groen, A.K., R. van der Meer, H.V. Westerhoff, R.J.A. Wanders, T.P.M. Akerboom, and J.M. Tager, 1982a: Control of metabolic fluxes, in: H.Sies (Ed.), *Metabolic Compartmentation*, Academic Press, New York, pp. 9–37.

[127] Groen, A.K., C.W.T. van Roermund, R.C. Vervoorn, and J.M. Tager, 1986: Control of gluconeogenesis in rat liver cells. Flux control coefficients of the enzymes in the gluconeogenic pathway in the absence and presence of glucagon. *Biochem. J.* **237**(2): 379–389.

[128] Groen, A.K., R.C. Vervoorn, R. Van der Meer, and J.M. Tager, 1983: Control of gluconeogenesis in rat liver cells. I. Kinetics of the individual enzymes and the effect of glucagon. *J. Biol. Chem.* **258**(23): 14346–14353.

[129] Groen, A.K., R.J.A. Wanders, H.V. Westerhoff, R. van der Meer, and J.M. Tager, 1982b: Quantification of the contribution of various steps to the control of mitochondrial respiration. *J. Biol. Chem.* **257**(6): 2754–2757.

[130] Guckenheimer, J., and P. Holmes, 1983: *Nonlinear Oscillations, Dynamical Systems, and Bifurcations of Vector Fields*. Springer-Verlag, New York.

[131] Gustafson, G.L., and B.E. Wright, 1972: Analysis of approaches used in studying differentiation of the cellular slime mold. *CRC Critical Reviews in Microbiology* **1**: 453–478.

[132] Haldane, J.B.S., 1930: *Enzymes*. Longmans, Green, London [reprinted by M.I.T. Press, Cambridge, MA, 1965].

[133] Hande, K., E. Reed, and B. Chabner, 1978: Allopurinol kinetics. *Clin. Pharmacol. Ther.* **23**(5): 598–605.

[134] Harkness, R.A., 1989: Lesch–Nyhan syndrome: reduced amino acid concentrations in CSF and brain, *Adv. Exp. Med. Biol.* **253A**: 159–163.

[135] Hasegawa, T., and F. Shiraishi, 1996: Application of biochemical systems theory to determination of intrinsic kinetic parameters of an immobilized enzyme reaction following Michaelis–Menten kinetics, in: Yamakawa, T., and G. Matsumoto (Eds.), *Methodologies for the Conception, Design, and Application of Intelligent Systems*, World Scientific, Singapore, pp. 183–186.

[136] Hatzimanikatis, V., and J.E. Bailey, 1996: MCA has more to say. *J. Theor. Biol.* **182**: 233–242.

[137] Hatzimanikatis, V., C.A. Floudas, and J.E. Bailey, 1996a: Optimization of regulatory architectures in metabolic reaction networks. *Biotechnol. Bioeng.* **52(4)**: 485–500.

[138] Hatzimanikatis, V., C.A. Floudas, and J.E. Bailey, 1996b: Analysis and design of metabolic reaction networks via mixed-integer linear optimization. *AIChE J.* **42(5)**: 1277–1292.

[139] Hayt, W. H., and J.E. Kemmerly, 1978: *Engineering Circuit Analysis*. McGraw-Hill, New York.

[140] Heath, D.F., 1968: The redistribution of carbon label by the reactions involved in glycolysis, gluconeogenesis, and the tricarboxylic acid cycle in rat liver. *Biochem. J.* **110**: 313–335.

[141] Heath, D.F., and C.J. Threlfall, 1968: The interaction of glycolysis, gluconeogenesis and the tricarboxylic acid cycle *in vivo*. *Biochem. J.* **110**: 337–362.

[142] Heinmets, F., 1989: Supercomputer analysis of purine and pyrimidine metabolism leading to DNA-synthesis. *Cell Biophys.* **14**: 283–323.

[143] Heinrich, R., F. Montero, E. Klipp, T.G. Wadell, and E. Meléndez-Hevia, 1997: Theoretical approaches to the evolutionary optimisation of glycolysis: thermodynamic and kinetic constraints. *Eur. J. Biochem.* **243**: 191–201.

[144] Heinrich, R., S.M. Rapoport, and T.A. Rapoport, 1977: Metabolic regulation and mathematical models. *Prog. Biophys. Mol. Biol.* **32**: 1–82.

[145] Heinrich, R., and T.A. Rapoport, 1974: A linear steady-state treatment of enzymatic chains: general properties, control and effector strength. *Eur. J. Biochem.* **42**: 89–95.

[146] Heinrich, R., and S. Schuster, 1996: *The Regulation of Cellular Systems*. Chapman and Hall, New York.

[147] Heinrich, R., and S. Schuster, 1998: The modeling of metabolic systems. Structure, control, and optimality. *BioSystems* **47**: 61–77.

[148] Henderson, J.F., L.W. Brox, W.N. Kelley, F.M. Rosenbloom, and J.E. Seegmiller, 1968: Kinetic studies of hypoxanthine-guanine phosphoribosyltransferase, *J. Biol. Chem.* **243**: 2514–2522.

[149] Hernández-Bermejo, B., and V. Fairén, 1997: Lotka–Volterra representation of general nonlinear systems. *Math. Biosci.* **140**: 1–32.

[150] Hernández-Bermejo, B., V. Fairén, and A. Sorribas, 1999: Power-law modeling based on least-squares minimization criteria. *Math. Biosci.*, **161(1–2)**: 83–94.

[151] Hers, H.G., and L. Hue, 1983: Gluconeogenesis and related aspects of glycolysis. *Ann. Rev. Biochem.* **52**: 617–653.

[152] Hers, H.G., and E. Van Schaftingen, 1982: Fructose-2,6-bisphosphate two years after its discovery. *Biochem. J.* **206**: 1–12.

[153] Hess, B., and T. Plesser, 1979: Temporal and spatial order in biochemical systems. *Ann. New York Acad. Sci.* **316**: 203–213.

[154] Hill, A.V., 1910: Possible effects of the aggregation of the molecules of haemoglobin on its dissociation curves. *J. Physiol.* **40**: iv–viii.

[155] Hlavacek, W.S., and M.A. Savageau, 1995: Subunit structure of regulator proteins influences the design of gene circuitry: analysis of perfectly coupled and completely uncoupled circuits. *J. Molec. Biol.* **248**: 739–755.

[156] Hlavacek, W.S., and M.A. Savageau, 1996: Rules for coupled expression of regulator and effector genes in inducible circuits. *J. Molec. Biol.* **255(1)**: 121–139.

[157] Hlavacek, W.S., and M.A. Savageau, 1997: Completely uncoupled and perfectly coupled gene expression in repressible systems. *J. Molec. Biol.* **266**: 538–558.

[158] Hochachka, P.W., and G.N. Somero, 1984: *Biochemical Adaptation*. Princeton University Press, Princeton, NJ.
[159] Hodges, P.E., A.H.Z. McKee, B.P.Davis, W.E. Payne, and J.I. Garrels, 1999: Yeast Protein Database (YPD): a model for the organization and presentation of genome-wide functional data. *Nucleic Acids Res.* **27**(1): 69–73.
[160] Hofestädt, R., 1993: A simulation shell to model metabolic pathways. *Systems Anal. Modelling Simulation* **11**: 253–262.
[161] Hofestädt, R., and U. Scholz, 1998: Information processing for the analysis of metabolic pathways and inborn errors. *BioSystems* **47**: 91–102.
[162] Hofmeyer, J.H., H. Kacser, and K.J. van der Merwe, 1986: Metabolic control analysis of moiety-conserved cycles. *Eur. J. Biochem.* **155**(3): 631–641.
[163] Huxley, J.S., 1932: *Problems in Relative Growth*. Dial, New York.
[164] Irvine, D.H., 1988: Efficient solution of nonlinear models expressed in S-system canonical form. *Math. Comput. Modelling* **11**, 123–128.
[165] Irvine, D.H., and M.A. Savageau, 1985a: Network regulation of the immune response: alternative control points for suppressor modulation of effector lymphocytes. *J. Immunol.* **134**: 2100–2116.
[166] Irvine, D.H., and M.A. Savageau, 1985b: Network regulation of the immune response: modulation of suppressor lymphocytes by alternative signals including contrasuppression. *J. Immunol.* **134**: 2117–2130.
[167] Irvine, D.H., and M.A. Savageau, 1990. Efficient solution of nonlinear ordinary differential equations expressed in S-system canonical form. *SIAM J. Numer. Anal.* **27**: 704–735.
[168] Jacobs, A.E.M., A. Oosterhof, and J.H. Veerkamp, 1988: Purine and pyrimidine metabolism in human muscle and cultured muscle cells. *Biochim. Biophys. Acta* **970**: 130–136.
[169] Jerushalmy, Z., O. Sperling, H. Pinkhas, M. Krinska, and A. Vries, 1973: Enzymes of purine metabolism in platelets: phosphoribosylpyrophosphate synthetase and purine phosphoribosyltransferases. *Adv. Exp. Med. Biol.* **41**: 159–162.
[170] Jimenez, M.L., J.G. Puig, F.A. Mateos, T.H. Ramos, J.S. Melian, V.G. Nieto, and M.A. Becker, 1989: Increased purine nucleotide degradation in the central nervous system (CNS) in PRPPS synthetase superactivity, *Adv. Exp. Med. Biol.* **253A**: 9–13.
[171] Johnson, T., 1985: Modeling growth for economic analysis of dynamic agricultural systems. Departmental Report, Department of Agricultural and Resource Economics, North Carolina State University, Raleigh, NC.
[172] Johnson, T., 1988: Estimation and simulation of S-systems. *Math. Computer Modeling* **11**: 134–139.
[173] Johnson, T., 1991: Estimating parameters of S-systems, in: Voit, E.O. (Ed.), *Canonical Nonlinear Modeling. S-System Approach to Understanding Complexity*, Van Nostrand Reinhold, New York, Ch. 11.
[174] Jollow, D., and D. McMillan, 1998: Ethnic variation and genetic susceptibility: glucose-6-phosphate dehydrogenase deficiency, in: Mendelsohn, M.L., L. Mohr, and J.P. Peeters (Eds.), *Biomarkers: Medical and Workplace Applications*, Joseph Henry Press, Washington, DC, pp. 227–239.
[175] Jury, E.I., 1971: The inners approach to some problems of system theory. *IEEE Trans. Automatic Contr.*, AC-16, pp. 233–240.
[176] Kacser, H., 1991: A superior theory? *J. Theor. Biol.* **149**: 141–144.

[177] Kacser, H., and L. Acerenza, 1993: A universal method for increase in metabolite production. *Eur. J. Biochem.* **216**: 361–367.

[178] Kacser, H., and J.A. Burns, 1973: The control of flux. *Symp. Soc. Exp. Biol.* **27**: 65–104.

[179] Kacser, H., and J.A. Burns, 1979: Molecular democracy; who shares the controls? *Biochem. Soc. Trans.* **7**: 1149–1160.

[180] Kacser, H., and J.A. Burns, 1981: The molecular basis of dominance. *Genetics* **97**: 639–666.

[181] Kacser, H., H.M. Sauro, and L. Acerenza, 1990: Enzyme-enzyme interactions and control analysis: 1. The case of non-additivity: monomer-oligomer associations. *Eur. J. Biochem.* **187**(3): 481–491.

[182] Kahn, P., 1995: From genome to proteome: looking at a cell's proteins. *Science* **270**: 369–370.

[183] Kanehisa, M., 1997: A database of post-genome analysis. *Trends Genet.* **13**: 375–376.

[184] Kanehisa, M., 1998: Databases of biological information. *Trends Guide to Bioinformatics, Elsevier Trend Journals, Supplement*, pp. 24–26.

[185] Karp, P.D., and S. Paley, 1996: Integrated access to metabolic and genomic data. *J. Comput. Biol.* **3**(1): 191–212.

[186] Karp, P.D., M. Riley, S.M. Paley, A. Pelligrini-Toole, and M. Krummenacker, 1999: EcoCyc: encyclopedia of *Escherichia coli* genes and metabolism. *Nucl. Acids Res.* **27**(1): 55–58.

[187] Karp, P.D., M. Riley, S.M. Paley, and A. Pelligrini-Toole, 1996: EcoCyc: an encyclopedia of *Escherichia coli* genes and metabolism. *Nucl. Acids Res.* **24**(1): 32–39.

[188] Karp, P.D., M. Riley, M. Saier, I.T. Paulsen, S.M. Paley, and A. Pelligrini-Toole, 2000: The EcoCyc and MetaCyc databases. *Nucl. Acids Res.* **28**(1), 56–59.

[189] Kauffman, S.A., 1993: *The Origins of Order. Self-Organization and Selection in Evolution*. Oxford University Press, New York.

[190] Kelley, W.N., and J.B. Wyngaarden, 1983: Clinical syndromes associated with hypoxanthine-guanine phosphoribosyltransferase deficiency, in: Stanbury, J.B., J.B. Wyngaarden, D.S. Fredrickson, J.L. Goldstein, and M.S. Brown (Eds.), *The Metabolic Basis of Inherited Disease*, McGraw-Hill, New York, pp. 1115–1143.

[191] Kelly, P.J., J.K. Kelleher, and B.E. Wright, 1979a: The tricarboxylic acid cycle in *Dictyostelium discoideum*. Metabolic concentrations, oxygen uptake, and ^{14}C-labelled amino acid labelling patterns. *Biochem. J.* **184**: 581–588.

[192] Kelly, P.J., J.K. Kelleher, and B.E. Wright, 1979b: The tricarboxylic acid cycle in *Dictyostelium discoideum*. A model of the cycle at preculmination and aggregation. *Biochem. J.* **184**, 589–597.

[193] Kermack, W.O., and A.G. McKendrick, 1927: Contributions to the mathematical theory of epidemics. *J. Roy. Statist. Soc.* **115**, 700–721.

[194] Kerner, E.H., 1981: Universal formats for nonlinear ordinary differential systems. *J. Math. Phys.* **22**(7): 1366–1371.

[195] Kholodenko, B.N., M. Cascante, and H.V. Westerhoff, 1995: Control theory of metabolic channelling. *Molec. Cell. Biochem.* **143**(2): 151–168.

[196] Kholodenko, B.N., S. Schuster, J. Garcia, H.V. Westerhoff, and M. Cascante, 1998: Control analysis of metabolic systems involving quasi-equilibrium reactions. *Biochim. Biophys Acta* **1379**(3): 337–352.

[197] Kimmel, A.R. (Ed.), 1988: *Molecular Biology of Dictyostelium Development*. A.R. Liss, New York.

[198] King, M.E., J.M. Honeysett, and S.B. Howell, 1983: Regulation of *de novo* purine synthesis in human bone marrow mononuclear cells by hypoxanthine. *J. Clin. Invest.* **72**: 965–970.

[199] Koch, C., and G. Laurent, 1999: Complexity and the nervous system. *Science* **284**: 96–98.

[200] Kohn, M.C., L.E. Menten, and D. Garfinkel, 1979: A convenient computer program for fitting enzymatic rate laws to steady-state data. *Computers Biomed. Res.* **12**(5): 461–469.

[201] Kohn, M.C., and D.R. Lemieux, 1991: Identification of regulatory properties of metabolic networks by graph theoretical methods. *J. Theor. Biol.* **150**: 3–25.

[202] Kopelman, R., 1986: Rate processes on fractals: theory, simulations, experiments. *J. Statist. Phys.* **42**: 185–200.

[203] Kopelman, R., 1991: Reaction kinetics in restricted spaces. *Israel J. Chem.* **31**: 147–157.

[204] Koshland, D.E., and K.E. Neet, 1968: The catalytic and regulatory properties of enzymes. *Ann. Rev. Biochem.* **37**: 359–410.

[205] Kreyszig, E., 1993: *Advanced Engineering Mathematics* (7th edition). Wiley, New York.

[206] Kruckeberg, A., H.E. Neuhaus, R. Feil, L. Gottlieb, and M. Stitt, 1989: Decreased-activity mutants of phosphoglucose isomerase in the cytosol and chloroplast of *Clarkia xantiana*. *Biochem. J.* **261**: 457–467.

[207] Kuenzi, M., and A. Fiechter, 1972: Regulation of carbohydrate composition of *Saccharomyces cerevisiae* under growth limitation. *Arch. Mikrobiol.* **84**(3): 254–265.

[208] Kuhn, T.S., 1962: *The Structure of Scientific Revolutions*. University of Chicago Press, Chicago.

[209] Lashkari, D.A., J.L. DeRisi, J.H. McCusker, A.F. Namath, C. Gentile, S.Y. Hwang, P.O. Brown, and R.W. Davis, 1997: Yeast microarrays for genome wide parallel genetic and gene expression analysis. *Proc. Natl. Acad. Sci. USA* **94**(24): 13057–13062.

[210] Laszlo, E., 1972: *The Systems View of the World*. George Braziller, New York.

[211] Lathem, W., and G.P. Rodnan, 1962: Impairment of uric acid excretion in gout. *J. Clin. Invest.* **41**: 1955–1963.

[212] Lee, M.H., and B.R. Ahmad Mahir, 1994a: Penggunaan pendekatan sistem-S dan ESSYNS dalam analisis taburan normal. *Pertanika J. Sci. Technol.* **2**(2): 165–173.

[213] Lee, M.H., and B.R. Ahmad Mahir, 1994b: Analisis berangka bagi taburan khi kuasa dua melalui persamaan terbitan sistem-S. *Sains Malaysiana* **23**(4): 129–147.

[214] Lehninger, A.L., 1970: *Biochemistry: The Molecular Basis of Cell Structure and Function*. Worth Publishers, New York.

[215] Lehninger, A.L., D.L. Nelson, and M.M. Cox, 1993: *Principles of Biochemistry: With an Extended Discussion*. Worth Publishers, New York.

[216] Leicester, H.M., 1974: *Development of Biochemical Concepts from Ancient to Modern Times*. Harvard University Press, Cambridge, MA.

[217] Lewis, D.C., 1991: A qualitative analysis of S-systems: Hopf bifurcations, in: Voit, E.O. (Ed.), *Canonical Nonlinear Modeling. S-System Approach to Understanding Complexity*, Van Nostrand Reinhold, New York, Ch. 16.

[218] Lewis, D.C., 1992: A classification of the equilibria of two-dimensional S-systems. Second S-System Symposium, Tampa, FL (Abstract).

[219] Liang, S., S. Fuhrman, and R. Somogyi, 1998: REVEAL, a general reverse engineering algorithm for inference of genetic network architecture, in: *Pacific Symp. Biocomputing*, pp. 18–29.

[220] Liddel, G.U., and B.E. Wright, 1961: The effect of glucose on respiration of the differentiating slime mold. *Developm. Biol.* **3**: 265–276.

[221] Lipshutz, R.J., S.P.A. Fodor, T.R. Gingeras, and D.J. Lockhart, 1999: High density synthetic oligonucleotide arrays. *Nat. Genet.* **21** (Supplement): **20**–24.

[222] Loomis, W.F., 1975: *Dictyostelium discoideum – A Developmental System*. Academic Press, New York.

[223] Loomis, W.F. (Ed.), 1982: *The Development of Dictyostelium discoideum*. Academic Press, New York.

[224] Lotka, A.J., 1924: *Elements of Physical Biology*. Williams and Wilkins, Baltimore (reprinted as *Elements of Mathematical Biology*, Dover, New York, 1956).

[225] Lowry, O.H., and J.V. Passonneau, 1964: The relationships between substrates and enzymes of glycolysis in brain. *J. Biol. Chem.* **239**: 31–42.

[226] Luenberger, D.G., 1984: *Linear and Nonlinear Programming*. Addison-Wesley, Reading, MA.

[227] Mahler, H.R., and E.H. Cordes, 1966: *Biological Chemistry*. Harper, New York, 1966 and 1969.

[228] Majewski, R.A., and M.M. Domach, 1990: Simple constrained-optimization view of acetate overflow in *E. coli*. *Bioetchn. Bioeng.* **35**: 732–738.

[229] Maple®, Waterloo Maple Inc., 57 Erb Street W., Waterloo, Ontario, N2L 6C2, Canada.

[230] Martin, P.-G., 1997: The use of canonical S-system modelling for condensation of complex dynamic models. *Ecol. Model.* **103**: 43–70.

[231] Martini, G., and M.V. Ursini, 1996: A new lease of life for an old enzyme. *BioEssays* **18**(8): 631–637.

[232] MathCad®, MathSoft, Inc., 101 Main Street, Cambridge, MA 02142, USA.

[233] Mathematica®, Wolfram Research, Inc., 100 Trade Center Drive, Champaign, IL 61820, USA.

[234] Maurer, H., 1969: *Theoretische Grundlagen der Programmiersprachen – Theorie der Syntax*. Bibliographisches Institut, Mannheim, Wien, Zürich.

[235] Mavrovouniotis, M.L., 1988: *Computer-Aided Design of Biochemical Pathways*. Ph.D. Dissertation, Massachusetts Institute of Technology.

[236] Mavrovouniotis, M.L., G. Stephanopoulos, and G. Stephanopoulos, 1990a: Estimation of upper-bounds for the rates of enzymatic reactions. *Chem. Eng. Commun.* **93**: 211–236.

[237] Mavrovouniotis, M.L., G. Stephanopoulos, and G Stephanopoulos, 1990b: Computer-aided synthesis of biochemical pathways. *Biotechn. Bioeng.* **36**: 1119–1132.

[238] Meléndez-Hevia, E., 1990: The game of the pentose phosphate cycle: a mathematical approach to study the optimization in design of metabolic pathways during evolution. *Biomed. Biochim. Acta* **49**: 903–916.

[239] Meléndez-Hevia, E., and N.V. Torres, 1988: Economy of design in metabolic pathways: further remarks on the game of the pentose phosphate cycle. *J. Theor. Biol.* **132**: 97–111.

[240] Meléndez-Hevia, E., N.V. Torres, and J. Sicilia, 1990: A generalization of metabolic control analysis to conditions of no proportionality between activity and concentration of enzymes. *J. Theor. Biol.* **142**(4): 443–451.

[241] Mendelson, E., 1964: *Introduction to Mathematical Logic*. Van Nostrand Reinhold, New York.
[242] Menten, L.E., M.C. Kohn, and D. Garfinkel, 1981: A convenient computer program for estimation of enzyme and metabolite concentrations in multienzyme systems. *Comput. Biomed. Res.* **14**(1): 91–102.
[243] Mewes, H.-W., K. Heumann, A. Kaps, K. Mayer, F. Pfeiffer, S. Stocker, and D. Frishman, 1999: MIPS: a database for genomes and protein sequences. *Nucl. Acids Res.* **27**(1), 44–48.
[244] Michaelis, L., and M.L. Menten, 1913: Die Kinetik der Invertinwirkung. *Biochem. Zeitschrift*, **49**: 333–369.
[245] Michal, G., 1993: *Biochemical Pathways* (wall chart). Boehringer Mannheim.
[246] Michal, G., 1998a: *Biochemical Pathways*. Spektrum Akademie Verlag, Heidelberg.
[247] Michal, G., 1998b: On representation of metabolic pathways. *BioSystems* **47**: 1–7.
[248] Middleton, R., and H. Kacser, 1983: Enzyme variation, metabolic flux and fitness: alcohol dehydrogenase in *Drosophila melanogaster*. *Genetics* **105**: 633–650.
[249] Miller, G.A., 1956: The magical number seven, plus or minus two. Some limits to our capacity for processing information. *Psychol. Rev.* **63**: 81–97.
[250] Monod, J., J. Wyman, and J.-P. Changeaux, 1965: On the nature of allosteric transitions: a plausible model. *J. Mol. Biol.* **12**: 88–118.
[251] Mueller, K.M., S.A. Burns, and M.A. Savageau, 1998: A comparison of the monomial method and the S-system method for solving systems of algebraic equations. *Appl. Math. Comput.* **90**(2–3): 167–180.
[252] Murray, J.D., 1990: *Mathematical Biology* (2nd printing). Springer-Verlag, Berlin.
[253] Needham, J., 1942: *Biochemistry and Morphogenesis*. Cambridge University Press, Cambridge.
[254] Neidhardt, F.C., and M.A. Savageau, 1996: Regulation beyond the operon, in: Neidhardt, F.C., R. Curtiss III, J.L. Ingraham, E.C.C. Lin, K.B. Low, B. Magasanik, W.S. Reznikoff, M. Riley, M. Schaechter, and H.E. Umbarger (Eds.), *Escherichia coli and Salmonella: Cellular and Molecular Biology*, Vol. 1 (2nd edition), Amer. Soc. Microbiology, Washington, DC, pp. 1310–1324.
[255] Neter, J., and W. Wasserman, 1974: *Applied Linear Statistical Models*. Richard D. Irwin, Homewood, IL.
[256] Neuhaus, H.E., and M. Stitt, 1990: Control analysis of phosphosynthate partitioning – impact of reduced activity of the ADP-glucose pyrophosphorylase or plastid phosphoglucomutase on the fluxes to starch and sucrose in *Arabidopsis thaliana* (L) Heynh., *Planta* **182**(3): 445–454.
[257] Newsholme, E.A., and C. Start, 1973: *Regulation in Metabolism*. Wiley, London.
[258] Ni, T.-C., and M.A. Savageau, 1996a: Application of biochemical systems theory to metabolism in human red blood cells. *J. Biol. Chem.* **271**(14): 7927–7941.
[259] Ni, T.-C., and M.A. Savageau, 1996b: Model assessment and refinement using strategies from biochemical systems theory: application to metabolism in human red blood cells. *J. Theor. Biol.* **179**: 329–368.
[260] Nishizuka, T. (Ed.), 1980: *Metabolic Maps*. Biochemical Society of Japan.
[261] Nishizuka, T. (Ed.), 1997: *Cell Functions and Metabolic Maps*. Biochemical Society of Japan.
[262] Norman, G.R., and D.L. Streiner, 1994: *Biostatistics: The Bare Essentials*. Mosby, St. Louis.

[263] Oden, K.L., and S. Clarke, 1983: S-adenosyl-L-metionine synthase from human erythrocytes: role in the regulation of cellular S-adenosylmethionine levels, *Biochemistry* **22**: 2978–2986.

[264] Ogata, H., S. Goto, W. Fujibuchi, and M. Kanehisa, 1998: Computation with the KEGG pathway database. *BioSystems* **47**: 119–128.

[265] Ogata, H., S. Goto, W. Fujibuchi, H. Bono, and M. Kanehisa, 1999: KEGG: Kyoto Encyclopedia of Genes and Genomes. *Nucl. Acids Res.* **27**(1): 29–34.

[266] Okamoto, M., and M.A. Savageau, 1984a: Integrated function of a kinetic proofreading mechanism: steady-state analysis testing internal consistency of data obtained *in vivo* and *in vitro* and predicting parameter values. *Biochemistry* **23**: 1701–1709.

[267] Okamoto, M., and M.A. Savageau, 1984b: Integrated function of a kinetic proofreading mechanism: dynamic analysis separating the effects of speed and substrate competition on accuracy. *Biochemistry* **23**: 1710–1715.

[268] Okamoto, M., J.-J. Yoshii, and K. Hayashi, 1991: GTP-dependent kinetic proofreading, in: Voit, E.O. (Ed.): *Canonical Nonlinear Modeling. S-System Approach to Understanding Complexity.* Van Nostrand Reinhold, New York, Ch. 9.

[269] Olsen, L.F., and H. Degn, 1977: Chaos in an enzyme reaction. *Nature* **267**: 177–178.

[270] Ovadi, J., 1991: Physiological significance of metabolic channelling. *J. Theor. Biol.* **152**: 1–22.

[271] Page, T., W.L. Nyhan, A.L. Yu, and J. Yu, 1991: A syndrome of megaloblastic anemia, inmunodeficiency, and exesive nucleotide degradation. *Adv. Exp. Med. Biol.* **309B**: 345–348.

[272] Palisade Corporation, 1997: *@Risk: Advanced Risk Analysis for Spreadsheets*. Newfield, NY.

[273] Palsson, B.Ø., R. Jamier, and E.N. Lightfoot, 1984: Mathematical modelling of dynamics and control in metabolic networks. II. Simple dimeric enzymes. *J. Theor. Biol.* **111**: 303–321.

[274] Palsson, B.Ø., and E.N. Lightfoot, 1984: Mathematical modelling of dynamics and control in metabolic networks. I. On Michaelis–Menten kinetics. *J. Theor. Biol.* **111**: 273–302.

[275] Palsson, B.Ø., H. Palsson, and E.N. Lightfoot, 1985: Mathematical modelling of dynamics and control in metabolic networks. III. Linear reaction sequences. *J. Theor. Biol.* **113**, 231–259.

[276] Pandolfi, P.P., F. Sonati, R. Rivi, P. Mason, F. Grosveld, and L. Luzzatto, 1996: Targeted disruption of the housekeeping gene encoding glucose-6-phosphate dehydrogenase (G6PD): G6PD is dispensable for pentose synthesis but essential for defense against oxidative stress. *EMBO J.* **19**: 5209–5215.

[277] Panetta, J.C., 1998: A mathematical model of drug resistance: heterogeneous tumors. *Math. Biosci.* **147**: 41–61.

[278] Papoutsakis, E.T., and C.L. Meyer, 1985: Equations and calculations of product yields and preferred pathways for butanediol and mixed-acid fermentations. *Biotechnol. Bioeng.* **27**: 50–66.

[279] Peschel, M., and W. Mende, 1986: *The Predator-Prey Model: Do We Live in a Volterra World?* Akademie-Verlag, Berlin.

[280] Peterson, J.L., 1981: *Petri Net Theory and the Modeling of Systems*. Prentice-Hall, Englewood Cliffs, NJ.

[281] Petkov, S.B., and C.D. Maranas, 1997: Quantitative assessment of uncertainty in the optimization of metabolic pathways. *Biotechn. Bioeng.* **56**(2): 145–161.

[282] Pickard, W.F., 1983: Three interpretations of the self-thinning rule. *Ann. Bot.* **51**: 749–757.
[283] Puigjaner, J., M. Cascante, and A. Sorribas, 1995: Assessing optimal designs in metabolic pathways. *J. Biol. Systems* **3**(1): 197–206.
[284] Rabitz, H.K., M. Kramer, and D. Dacol, 1983: Sensitivity analysis in chemical kinetics. *Ann. Rev. Phys. Chem.* **34**: 1419–1461.
[285] Reder, C., 1998: Metabolic control theory: a structural approach. *J. Theor. Biol.* **135**: 175–201.
[286] Reddy, V.N., M.N. Liebman, and M.L. Mavrovouniotis, 1996: Qualitative analysis of biochemical reaction systems. *Comput. Biol. Med.* **26**(1): 9–24.
[287] Regan, L., I.D.L. Bogle, and P. Dunhill, 1993: Simulation and optimization of metabolic pathways. *Computers Chem. Engng.* **17**(5/6): 627–637.
[288] Reich, J.G., and E.E. Sel'kov, 1981: *Energy Metabolism of the Cell. A Theoretical Treatise.* Academic Press, London.
[289] Reisig, W., 1985: *Petri Nets: An Introduction.* Springer-Verlag, New York.
[290] Ribeiro Filho, J.L., P.C. Treleaven, and C. Alippi, 1994: Genetic-algorithm programming environments. *IEEE Computer*, June, pp. 28–43.
[291] Rigoulet, M., N. Averet, J.P. Mazat, B. Guerin, and F. Cohadon, 1988: Redistribution of the flux-control coefficients in mitochondrial oxidative phosphorylations in the course of brain edema. *Biochim. Biophys. Acta* **932**(1): 116–123.
[292] Riley, M., 1998: Genes and proteins of *Escherichia coli* K-12 (GenProtEC). *Nucl. Acids Res.* **26**(1): 54.
[293] Rolleston, F.S., 1972: A theoretical background to the use of measured concentrations of intermediates in the study of the control of intermediary metabolism. *Curr. Topics Cell. Regul.* **5**: 47–75.
[294] Roman, G.C., and D. Garfinkel, 1978: BIOSSIM – a structured machine-independent biological simulation language. *Comput. Biomed. Res.* **11**(1): 3–15.
[295] Rothman, L., and E. Cabib, 1967: Allosteric properties of yeast glycogen synthetase. I. General kinetic study. *Biochemistry* **6**(7): 2098–2112.
[296] Routh, E.J., 1930: *Advanced Part of Dynamics of a System of Rigid Bodies* (6th edition), Vol. II. Macmillan, London.
[297] Rubin, C.S., and O.M. Rosen, 1975: Protein phosphorylation. *Ann. Rev. Biochem* **44**: 831–887.
[298] Rugen, P., and B. Callahan, 1996: An overview of Monte Carlo, a fifty year perspective. *Human and Ecological Risk Assessment* **2**(4): 671–680.
[299] Rust, P.F., 1991: Modeling approximately normal distributions with S-systems, in: *Proceedings of the Annual Meeting of the American Statistical Association, Statistical Computing Section*, pp. 162–166.
[300] Rust, P.F., and E.O. Voit, 1989: S-system analysis of the noncentral t distribution with fractional degrees of freedom, in: *Proceedings of the Annual Meeting of the American Statistical Association, Statistical Computing Section*, pp. 84–87.
[301] Rust, P.F., and E.O. Voit, 1990: Statistical densities, cumulatives, quantiles, and power obtained by S-system differential equations. *J. Amer. Statist. Assoc.* **85**(410): 572–578.
[302] Salvador, A., 1996: Steady-state synergisms in kinetic models: estimation and applications, in: Yamakawa, T., and G. Matsumoto (Eds.), *Methodologies for the Conception, Design, and Application of Intelligent Systems*, World Scientific, Singapore, pp. 143–146.

[303] Salvador, A., 1997: *Development of Methodology and Software for Analysis of Kinetic Models of Metabolic Processes. Application to the Mitochondrial Metabolism of Lipid Hydroperoxides*. Doctoral Thesis, Universidade de Lisboa, Lisboa, Portugal.

[304] Salvador, A., 2000a: Synergism analysis of metabolic processes: I. Conceptual framework. *Math. Biosci.* **163**(2): 105–129.

[305] Salvador, A., 2000b: Synergism analysis of metabolic processes: II. Tensor formulation and treatment of stoichiometric constraints. *Math. Biosci.* **163**(2): 131–158.

[306] Sands, P.J., and E.O. Voit, 1996: Flux-based estimation of parameters in S-systems. *Ecol. Modeling* **93**: 75–88.

[307] Sargent, R., and E. Wainwright (Eds.), 1996: *Crystal Ball: Forecasting & Risk Analysis for Spreadsheet Users*, Version 4.0. CG Press, Broomfield, CO.

[308] SAS®, 1990: SAS Institute, Inc., Cary, NC.

[309] Sauer, F., J.D. Erfle, and M.R. Binns, 1970: Turnover rates and intracellular pool size distribution of citrate cycle intermediates in normal, diabetic and fat-fed rats estimated by computer analysis from specific activity decay data of ^{14}C-labeled citrate cycle acids. *Eur. J. Biochem.* **17**: 350–363.

[310] Savageau, M.A., 1969a: Biochemical Systems Analysis, I. Some mathematical properties of the rate law for the component enzymatic reactions. *J. Theor. Biol.* **25**: 365–369.

[311] Savageau, M.A., 1969b: Biochemical Systems Analysis, II. The steady-state solutions for an n-pool system using a power-law approximation. *J. Theor. Biol.* **25**: 370–379.

[312] Savageau, M.A., 1970: Biochemical Systems Analysis, III. Dynamic solutions using a power-law approximation. *J. Theor. Biol.* **26**: 215–226.

[313] Savageau, M.A., 1971a: Concepts relating the behavior of biochemical systems to their underlying molecular properties. *Arch. Biochem. Biophys.* **145**: 612–621.

[314] Savageau, M.A., 1971b: Parameter sensitivity as a Criterion for evaluating and comparing the performance of biochemical systems. *Nature* **229**(5286): 542–544.

[315] Savageau, M.A., 1972: The behavior of intact biochemical control systems. *Curr. Topics Cell. Regulation* **6**: 63–129.

[316] Savageau, M.A., 1974a: Comparison of classical and autogenous systems of regulation in inducible operons. *Nature (London)* **252**: 546–549.

[317] Savageau, M.A., 1974b: Genetic regulatory mechanisms and the ecological niche of *Escherichia coli*. *Proc. Natl. Acad. Sci. USA* **71**: 2354–2455.

[318] Savageau, M.A., 1975: Optimal design of feedback control by inhibition: dynamic considerations. *J. Mol. Evol.* **5**: 199–222.

[319] Savageau, M.A., 1976: *Biochemical Systems Analysis. A Study of Function and Design in Molecular Biology*. Addison-Wesley, Reading, MA.

[320] Savageau, M.A., 1977: Design of molecular control mechanisms and the demand for gene expression. *Proc. Natl. Acad. Sci. USA* **74**: 5647–5651.

[321] Savageau, M.A., 1979a: Growth of complex systems can be related to the properties of their underlying determinants. *Proc. Nat. Acad. Sci. USA* **76**: 5413–5417.

[322] Savageau, M.A., 1979b: Allometric morphogenesis of complex systems: Derivation of the basic equations from first principles. *Proc. Nat. Acad. Sci. USA.* **76**: 6023–6025.

[323] Savageau, M.A., 1979c: Autogenous and classical regulation of gene expression: a general theory and experimental evidence, in: Goldberger, R.F., P. Berg, R.T. Schimke, K. Moldave, P. Leder, and L.E. Hood (Eds.), *Biological Regulation and Development*, Vol. 1, Plenum, New York, pp. 57–108.

[324] Savageau, M.A., 1980: Growth equations: a general equation and a survey of special cases. *Math. Biosci.* **48**: 267–278.

[325] Savageau, M.A., 1982: A suprasystem of probability distributions. *Biometr. J.* **24**: 323–330.

[326] Savageau, M.A., 1983a: *Escherichia coli* habitats, cell types, and molecular mechanisms of gene control. *Am. Nat.* **122**: 732–744.

[327] Savageau, M.A., 1983b: Regulation of differentiated cell-specific functions. *Proc. Natl. Acad. Sci. USA* **80**: 1411–1415.

[328] Savageau, M.A., 1983c: Models of gene function: general methods of kinetic analysis and specific ecological correlates, in: Blanch, H.W., E.T. Papoutsakis, and G.N. Stephanopoulos (Eds.), *Foundations of Biochemical Engineering Kinetics and Thermodynamics in Biological Systems*, American Chemical Society, Washington, DC, pp. 3–25.

[329] Savageau, M.A., 1985a: Mathematics of organizationally complex systems. *Biomed. Biochim. Acta* **44**: 839–844.

[330] Savageau, M.A., 1985b: A theory of alternative designs for biochemical control systems. *Biomed. Biochim. Acta* **44**: 875–880.

[331] Savageau, M.A., 1985c: Coupled circuits of gene regulation, in: Calendar, R., and L. Gold (Eds.), *Sequence Specificity in Transcription and Translation*, Alan R. Liss, New York, pp. 633–642.

[332] Savageau, M.A., 1989: Are there rules governing patterns of regulation? in: Goodwin, B.C., and P.T. Saunders (Eds.), *Theoretical Biology – Epigenetic and Evolutionary Order*, Edinburgh University Press, Edinburgh, U.K., pp. 42–66.

[333] Savageau, M.A., 1991a: Biochemical systems theory: operational differences among variant representations and their significance. *J. Theor. Biol.* **151**: 509–530.

[334] Savageau, M.A., 1991b: Metabolite channeling: implications for the regulation of metabolism and for quantitative description of reactions *in vivo*. *J. Theor. Biol.* **152**: 85–92.

[335] Savageau, M.A., 1991c: The challenge of reconstruction. *The New Biologist* **3**: 101–102.

[336] Savageau, M.A., 1991d: Reconstructionist molecular biology. *The New Biologist* **3**: 190–197.

[337] Savageau, M.A., 1991e: 20 years of S-systems, in: Voit, E.O. (Ed.): *Canonical Nonlinear Modeling. S-System Approach to Understanding Complexity*, Van Nostrand Reinhold, New York, Ch. 1.

[338] Savageau, M.A., 1992a: Dominance according to metabolic control analysis: major achievement or house of cards? *J. Theor. Biol.* **154**: 131–136.

[339] Savageau, M.A., 1992b: Critique of the enzymologist's test tube, in: Bittar, E.E. (Ed.), *Fundamentals of Medical Cell Biology*, Vol. 3A, JAI Press Inc., Greenwich, CT, pp. 45–108.

[340] Savageau, M.A., 1993a: Finding multiple roots of nonlinear algebraic equations using S-system methodology. *Appl. Math. Comput.* **55**: 187–199.

[341] Savageau, M.A., 1993b: Influence of fractal kinetics on molecular recognition. *J. Molec. Recognition* **6**: 149–157.

[342] Savageau, M.A., 1995a: Michaelis–Menten mechanism reconsidered: implications of fractal kinetics. *J. Theor. Biol.* **176**: 115–124.

[343] Savageau, M.A., 1995b: Enzyme kinetics *in vitro* and *in vivo*: Michaelis–Menten revisited, in: Bittar, E.E. (Ed.), *Principles of Medical Biology*, Vol. 4, JAI Press, Greenwich, CT, pp. 93–146.

[344] Savageau, M.A., 1996: A kinetic formalism for integrative molecular biology: manifestation in biochemical systems theory and use in elucidating design principles for

gene circuits, in: Collado-Vides, J., B. Magasanik, and T.F. Smith (Eds.), *Integrative Approaches to Molecular Biology*, MIT Press, Cambridge, MA, pp. 115–146.

[345] Savageau, M.A., 1998a: Demand theory for gene regulation: quantitative development of the theory. *Genetics* **149**: 1665–1676.

[346] Savageau, M.A., 1998b: Demand theory for gene regulation: quantitative application to the lactose and maltose operons of *Escherichia coli, Genetics* **149**: 1665–1676.

[347] Savageau, M.A., 1998c: Development of fractal kinetic theory for enzyme-catalyzed reactions and implications for the design of biochemical pathways. *BioSystems* **47**: 9–36.

[348] Savageau, M.A., 1999: Design of gene circuitry by natural selection: analysis of the lactose catabolic system in *Escherichia coli*. *Biochem. Soc. Trans.* **27(2)**: 264–270.

[349] Savageau, M.A., and P.J. Sands, 1991: Completely uncoupled and perfectly coupled circuits for inducible gene regulation, in: Voit, E.O. (Ed.): *Canonical Nonlinear Modeling. S-System Approach to Understanding Complexity*, Van Nostrand Reinhold, New York, Ch. 8.

[350] Savageau, M.A., and A. Sorribas, 1989: Constraints among molecular and systemic properties: Implications for Physiological Genetics. *J. Theor. Biol.* **141**: 93–115.

[351] Savageau, M.A., and E.O. Voit, 1982: Power-law approach to modeling biological systems; I. Theory. *J. Ferment. Technol.* **60(3)**: 221–228.

[352] Savageau, M.A., and E.O. Voit, 1987: Recasting nonlinear differential equations as S-systems: a canonical nonlinear form. *Math. Biosci.* **87**: 83–115.

[353] Savageau, M.A., E.O. Voit, and D.H. Irvine, 1987a: Biochemical systems theory and metabolic control theory. I. Fundamental similarities and differences. *Math. Biosci.* **86**: 127–145.

[354] Savageau, M.A., E.O. Voit, and D.H. Irvine, 1987b: Biochemical systems theory and metabolic control theory. II. The role of summation and connectivity relationships. *Math. Biosci.* **86**: 147–169.

[355] Savinell, J.M., 1991: *Analysis of Stoichiometry in Metabolic Networks*. Ph.D. Dissertation, University of Michigan.

[356] Savinell, J.M., and B.Ø. Palsson, 1992a: Network analysis of intermediary metabolism using linear optimization. I. Development of mathematical formalism. *J. Theor. Biol.* **154(4)**: 421–454.

[357] Savinell, J.M., and B.Ø. Palsson, 1992b: Network analysis of intermediary metabolism using linear optimization. II. Interpretation of *hybridoma* cell metabolism. *J. Theor. Biol.* **154(4)**: 455–473.

[358] Savinell, J.M., and B.Ø. Palsson, 1992c: Optimal selection of metabolic fluxes for *in vivo* measurement. I. Development of mathematical methods. *J. Theor. Biol.* **155(2)**: 201–214.

[359] Savinell, J.M., and B.Ø. Palsson, 1992d: Optimal selection of metabolic fluxes for *in vivo* measurement. II. Application to *Escherichia coli* and *hybridoma* cell metabolism. *J. Theor. Biol.* **155(2)**: 215–242.

[360] Schaeffer H.J., A.D. Catling, S.T. Eblen, L.S. Collier, A. Krauss, and M.J. Weber, 1998: MP1: a MEK binding partner that enhances enzymatic activation of the MAP kinase cascade. *Science* **281**: 1668–1671.

[361] Schena, M., D. Shalon, R. Heller, A. Chai, P.O. Brown, and R.W. Davis, 1996: Parallel human genome analysis: microarray-based expression monitoring of 1000 genes. *Proc. Natl. Acad. Sci.* **93**: 10614–10619.

[362] Schomburg, D. (Ed.), 1990: *Enzyme Handbook 1-12*. Springer-Verlag, Heidelberg.
[363] Scrutton, M.C., and M. F. Utter, 1968: The regulation of glycolysis and gluconeogenesis in animal tissues. *Ann. Rev. Biochem.* **37**: 249–303.
[364] Seegmiller, J.E., and F.M. Rosenbloom, 1967: Enzyme defect associated with a sex-linked human neurological disorder and excessive purine synthesis. *Science* **155**: 1682–1684.
[365] Segel, I.H., 1975: *Enzyme Kinetics*. Wiley, New York.
[366] Segel, L.A., 1991: *Biological Kinetics*. Cambridge University Press, Cambridge.
[367] Selkov E., S. Basmanova, T. Gaasterland, I. Goryanin, Y. Gretchkin., N. Maltsev, V. Nenashev, R. Overbeek, E. Panyushkina., L. Pronevitch, E. Selkov Jr., and I. Yunus, 1996: The metabolic pathway collection from ERM: the enzyme and metabolic pathway database. *Nucleic Acids Res.* **24**(1): 26–29.
[368] Selkov E., M. Galimova, I. Goryanin, Y. Gretchkin, N. Ivanova, Y. Komarov, N. Maltsev, N. Mikhailova, V. Nenashev, R. Overbeek, E. Panyushkina, L. Pronevitch, and E. Selkov, Jr., 1997: The metabolic pathway collection: an update. *Nucleic Acids Res.* **25**(1): 37–38.
[369] Selkov, E. Jr., Y. Gretchkin, N. Mikhailova, and E. Selkov, 1998: MPW: the Metabolic Pathways Database. *Nucleic Acids Res.* **26**(1): 43–45.
[370] Seressiotis, A., and J.E. Bailey, 1988: MPS: an artificially intelligent software system for the analysis and synthesis of metabolic pathways. *Biotechn. Bioeng.* **31**: 587–602.
[371] Service, R.F., 1999: Exploring the systems of life. *Science* **284**: 80–83.
[372] Shannon, C.E., 1948: A mathematical theory of communication. *Bell Syst. Tech. J.* **27**: 379–423, 623–656.
[373] Shiraishi, F., and S. Fujiwara, 1996: An efficient method for solving two-point boundary value problems with extremely high accuracy. *J. Chem. Engng. Japan* **29**(1): 88–94.
[374] Shiraishi, F., and M.A. Savageau, 1992a: The tricarboxylic acid cycle in *Dictyostelium discoideum*. I. Formulation of alternative kinetic representations. *J. Biol. Chem.* **267**: 22912–22918.
[375] Shiraishi, F., and M.A. Savageau, 1992b: The tricarboxylic acid cycle in *Dictyostelium discoideum*. II. Evaluation of model consistency and robustness. *J. Biol. Chem.* **267**: 22919–22925.
[376] Shiraishi, F., and M.A. Savageau, 1992c: The tricarboxylic acid cycle in *Dictyostelium discoideum*. III. Analysis of steady state and dynamic behaviour. *J. Biol. Chem.* **267**: 22926–22933.
[377] Shiraishi, F., and M.A. Savageau, 1992d: The tricarboxylic acid cycle in *Dictyostelium discoideum*. IV. Resolution of discrepancies between alterntative methods of analysis. *J. Biol. Chem.* **267**: 22934–22943.
[378] Shiraishi, F., and M.A. Savageau, 1993: The tricarboxylic acid cycle in *Dictyostelium discoideum*. V. Systemic effects of including protein turnover in the current model. *J. Biol. Chem.* **268**: 16917–16928.
[379] Shiraishi, F., and M.A. Savageau, 1996: The tricarboxylic acid cycle in *Dictyostelium discoideum* – two methods of analysis applied to the same model. *J. Theor. Biol.* **178**: 219–222.
[380] Shott, S., 1990: *Statistics for Health Professionals*. W.B. Saunders, Philadelphia.
[381] Šiljak, D.D., 1969: *Nonlinear Systems. The Parameter Analysis and Design*. Wiley, New York.
[382] Simmonds, H.A., A.S. Sahota, and K.J. Van Acker, 1989: Adenine phosphoribosyltransferase deficiency and 2,8-dihydroxyadenine lithiasis, in: Scriver, C.R.,

A.L. Beaudet, W.S. Sly, and D. Valle (Eds.), *The Metabolic Basis of Inherited Disease.*, McGraw-Hill, pp. 1029–1044.

[383] Simpson, G.G., 1975: Meanings of reductionism. *Science* **188**: 836–838.

[384] Slein, M.W., 1950: Phosphomannose isomerase. *J. Biol. Chem.* **186**: 753–761.

[385] Small, J.R., 1988: *Theoretical Aspects of Metabolic Control*. Ph.D. Dissertation, Oxford Polytechnic.

[386] Smith, G.K., R.G. Knowles, C.I. Pogson, M. Salter, M. Hanlon, and R. Mullin, 1990: Flux control coefficients of glycinamide ribonucleotide transformylase for *de novo* purine biosynthesis, in: Cornish-Bowden, A., and M.L. Cárdenas (Eds.), *Control of Metabolic Processes*, NATO ASI Series A, Vol. 190, Plenum Press, New York, pp. 385–387.

[387] Somogyi, R., and C.A. Sniegoski, 1996: Modeling the complexity of genetic networks: Understanding multigenic and pleiotropic regulation. *Complexity* **1**: 45–63.

[388] Somogyi, Z., and S. Liang, 1998: Modeling the normal and neoplastic cell cycle with "realistic Boolean genetic networks": their application for understanding carcinogenesis and assessing therapeutic strategies, in: *Pacific Symp. Biocomputing*, pp. 66–76.

[389] Sorribas, A., 1996: Exploring the properties of signal transduction pathways by mathematical controlled comparisons based on power-law models, in: Yamakawa, T., and G. Matsumoto (Eds.), *Methodologies for the Conception, Design, and Application of Intelligent Systems*, World Scientific, Singapore, pp. 179–182.

[390] Sorribas, A., and M. Cascante, 1994: Structure identifiability in metabolic pathways: parameter estimation in models based on the power-law formalism. *Biochem. J.* **298**: 303–311.

[391] Sorribas, A., R. Curto, and M. Cascante, 1995: Comparative characterization of the fermentation pathway of *Saccharomyces cerevisiae* using biochemical systems theory and metabolic control analysis. Model validation and dynamic behavior. *Math. Biosci.* **130**: 71–84.

[392] Sorribas, A., J. March, and E.O. Voit, 2000: Estimating age-related trends in cross-sectional studies using S-distributions. *Statist. Med.*, **10**(5): 697–713.

[393] Sorribas, A., S. Samitier, E.I. Canela, and M. Cascante, 1993: Metabolic pathway characterization from transient response data obtained in situ: parameter estimation in S-system models. *J. Theor. Biol.* **162**: 81–102.

[394] Sorribas, A., and M.A. Savageau, 1989a: A comparison of variant theories of intact biochemical systems. 1. Enzyme–enzyme interactions and biochemical systems theory. *Math. Biosci.* **94**: 161–193.

[395] Sorribas, A., and M.A. Savageau, 1989b: A comparison of variant theories of intact biochemical systems. 2. Flux oriented and metabolic control theories. *Math. Biosci.* **94**: 195–238.

[396] Sorribas, A., and M.A. Savageau, 1989c: Strategies for representing metabolic pathways within biochemical systems theory: reversible pathways. *Math. Biosci.* **94**: 239–269.

[397] Srere, P.A., 1975: The enzymology of the formation and breakdown of citrate. *Adv. Enzymol. Related Areas Molec. Biol.* **43**: 57–101.

[398] Srinivas, M., and L.M. Patnaik, 1994: Genetic algorithms: a survey. *IEEE Computer*, June, pp. 17–26.

[399] Stanbury, J.B., J.B. Wyngaarden, D.S. Fredrickson, J.L. Goldstein, and M.S. Brown (Eds.), 1983: *The Metabolic Basis of Inherited Disease* (5th edition). McGraw-Hill, New York.

[400] Starmer, C.F., O. Sperling, and J. B. Wyngaarden, 1975: A kinetic model for the intramolecular distribution of ^{15}N in uric acid in patients with primary gout fed ^{15}N-glycine. *Math. Biosci.* **25**: 105–123.

[401] Statistica®. StatSoft, Tulsa, OK, 1984–1995.

[402] Steinbuch, K., 1977: Denken in Modellen, in: G. Schaefer (Ed.), *Leitthemen 77/2*, Westermann, Braunschweig, Germany.

[403] Stitt, M., 1989: Control analysis of photosynthetic sucrose synthesis: assignment of elasticity coefficients and flux-control coefficients to the cytosolic fructose 1,6-biphosphatase and sucrose phosphate synthase. *Phil. Trans. Royal Soc. London B* **323**: 327–338.

[404] Stitt, M., W.P. Quick, U. Schurr, E.-D. Schulze, S.R. Rodermel, and L. Bogorad, 1991: Decreased ribulose-1,5-bisphosphate carboxylase-oxygenase in transgenic tobacco transformed with antisense rbcS. 2. Flux-control coefficients for photosynthesis in varying light, CO_2, and air humidity. *Planta* **183(4)**: 555–566.

[405] Stoesser, G., P. Sterk, M.A. Tuli, P.J. Stoehr, and G.N. Cameron, 1997: The EMBL nucleotide sequence database. *Nucl. Acids. Res.* **25(1)**: 7–14.

[406] Stoesser, G., M.A. Tuli, R. Lopez, and P. Sterk, 1999: The EMBL nucleotide sequence database. *Nucl. Acids. Res.* **27(1)**: 18–24.

[407] Strang, G., 1986: *Introduction to Applied Mathematics*. Wellesley-Cambridge Press, Wellesley, MA.

[408] Stryer, L., 1985: *Biochemistry*, 4th edition. W.H. Freeman, New York.

[409] Su, S., and P.J. Russel, 1968: Adenylate kinase from bakers' yeast. 3. Equilibria: equilibrium exchange and mechanism. *J. Biol. Chem.* **243**: 3826–3833.

[410] Sugawara, H., S. Miyazaki, T. Gojobori, and Y. Tateno, 1999: DNA Data Bank of Japan dealing with large-scale data submission. *Nucl. Acids. Res.* **27(1)**: 25–28.

[411] Thompson, D.W., 1917: *Growth and Form*. Cambridge University Press, Cambridge.

[412] Thornton, J.M., 1998: The future of bioinformatics. *Trends Guide to Bioinformatics, Elsevier Trend Journals, Supplement*, pp. 30–31.

[413] Threlfall, C.J., and D.F. Heath, 1968: Compartmentation between glycolysis and gluconeogenesis in rat liver. *Biochem. J.* **110**: 303–312.

[414] Tominaga, D. and M. Okamoto, 1998: Optimization method for nonlinear system model in power-law formalism. PowBioSys Symposium, Oerias, Portugal (Abstract).

[415] Torres, N.V., 1986: Doctoral Thesis. University of La Laguna, Tenerife.

[416] Torres, N.V., 1994a: Modelization and experimental studies on the control of the glycolytic–glucogenolytic pathway in rat liver. *Molec. Cell. Biochem.* **132**: 117–126.

[417] Torres, N.V., 1994b: Modelling approach to control of carbohydrate metabolism during citric acid accumulation by *Aspergillus niger*. I. Model definition and stability analysis. *Biotechn. Bioeng.* **44**: 104–111.

[418] Torres, N.V., 1994c: Modelling approach to control of carbohydrate metabolism during citric acid accumulation by *Aspergillus niger*. II. Steady-state analysis and optimization. *Biotechn. Bioeng.* **44**: 112–118.

[419] Torres, N.V., 1994d: Application of the transition time of metabolic systems as a criterion for optimization of metabolic processes. *Biotechnol. Bioeng.* **44**: 291–296.

[420] Torres, N.V., 1996: S-system modelling approach to ecosystem: Application to a study of magnesium flow in a tropical forest. *Ecol. Model.* **89**: 109–120.

[421] Torres, N.V., F. Mateo, E. Meléndez-Hevia, and H. Kacser, 1986: Kinetics of metabolic pathways. *Biochem. J.* **234**: 169–174.

[422] Torres, N. V., E.O. Voit, and C. H. Alcón, 1996: Optimization of nonlinear biotechnological processes with linear programming. Application to citric acid production in *Aspergillus niger. Biotechnol. Bioeng.* **49**: 247–258.

[423] Torres, N.V., E.O. Voit, C. Glez-Alcón, and F. Rodriguez, 1997: An indirect optimization method for biochemical systems: description of method and application to the maximization of the rate of ethanol, glycerol, and carbohydrate production in *Saccharomyces cerevisiae. Biotech. Bioeng.* **55**(5): 758–772.

[424] Torres, N.V., E.O. Voit, C. Glez-Alcón, and F. Rodríguez, 1998: A novel approach to design of overexpression strategy for metabolic engineering. Application to the carbohydrate metabolism in the citric acid producing mould *Aspergillus niger. Food Technol.* **36**: 177–184.

[425] Torres, N.V., E.O. Voit, and F. Rodríguez-Acosta, 2000: Optimization of biotechnological processes with canonical models. Methodology, applications and quality assessment of the approach. Submitted.

[426] Torsella, J., and A.M. Bin Razali, 1991: An analysis of forestry data, in: Voit, E.O. (Ed.), *Canonical Nonlinear Modeling. S-System Approach to Understanding Complexity*, Van Nostrand Reinhold, New York, Ch. 10.

[427] Turing, A., 1936: On computable numbers, with an application to the Entscheidungsproblem. *Proc. London Math. Soc. Ser. 2* **42**: 230–265.

[428] Vallino, J.J., and G. Stephanopoulos, 1993: Metabolic flux distribution in *Corynebacterium glutamicum* during growth and lysine overproduction. *Biotechnol. Bioeng.* **41**: 633–646.

[429] Van Acker, K.J., H.A. Simonds, C. Potter, and J.S. Cameron, 1977: Complete deficiency of adenine phosphoribosyltransferase. *N. Engl. J. Med.* **297**: 127–132.

[430] Van Acker, K.J., and H.A. Simonds, 1991: Long-term evolution of type 1 adenine phosphoribosyltransferase (APRT) deficiency, *Adv. Exp. Med. Biol.* **309B**: 91–94.

[431] van den Berghe, G., and H.-G. Hers, 1980: Abnormal AMP deaminase in primary gout. *The Lancet* **2**: 1090.

[432] Varma, A., and B.Ø. Palsson, 1995: Parametric sensitivity of stoichiometric flux balance models applied to wild-type *Escherichia coli* metabolism. *Biotechnol. Bioeng.* **45**: 69–79.

[433] Voit, E.O., 1985: Cell cycles and growth laws. The CCC-model. *J. Theor. Biol.* **114**: 589–599.

[434] Voit, E.O., 1988a: Dynamics of self-thinning plant stands. *Ann. Bot.* **62**: 67–78.

[435] Voit, E.O., 1988b: Recasting nonlinear models as S-systems. *Math. Comput. Modelling* **11**: 140–145.

[436] Voit, E.O., 1990a: Generic modeling of population dynamics with S-systems, in: O. Arino, D.E. Axelrod, and M. Kimmel (Eds.), *Proceedings of the 2nd International Conference on Mathematical Population Dynamics,* Marcel Dekker, pp. 261–280.

[437] Voit, E.O., 1990b: S-system analysis of endemic infections. *Comput. Math. Appl.* **20**(4–6): 161–173.

[438] Voit, E.O., 1990c: Canonical nonlinear simulation of complex systems, in: B. Schmidt (Ed.), *Proceedings of the 1990 European Simulation Multiconference*, Nürnberg, W. Germany, pp. 34–39.

[439] Voit, E.O. (Ed.), 1991: *Canonical Nonlinear Modeling. S-System Approach to Understanding Complexity*. Van Nostrand Reinhold, New York.

[440] Voit, E.O., 1992a: Optimization in integrated biochemical systems. *Biotechnol. Bioeng.* **40**: 572–582.

[441] Voit, E.O., 1992b: Symmetries of S-systems. *Math. Biosci.* **109**: 19–37.
[442] Voit, E.O., 1992c: The S-distribution. A tool for approximation and classification of univariate, unimodal probability distributions. *Biometrical J.* **34**(7): 855–878.
[443] Voit, E.O., 1992d: Selecting a model for integrated biomedical systems, in: Eisenfeld, J., M. Witten, and D.S. Levine (Eds.), *Biomedical Modeling and Simulation*, Elsevier Science, Amsterdam.
[444] Voit, E.O., 1993: S-system modeling of complex systems with chaotic input. *Environmetrics* **4**(2): 153–186.
[445] Voit, E.O., 1996a: Dynamic trends in distributions, *Biometr. J.* **38**(5): 587–603.
[446] Voit, E.O., 1996b: Structure analysis and optimal design with canonical models, in: *Tutorials of the 4th International Conference on Soft Computing*, Fuzzy Logic System Institute, 680–41 Kawazu, Iizuka, Fukuoka 820, Japan, pp. 33–64.
[447] Voit, E.O., 1996c: How many variables? Some comments on the dimensionality of nonlinear systems, in: V. Lakshmikantham (Ed.), *Proceedings of the World Congress of Nonlinear Analysts, Tampa, FL, 1992*, Walter de Gruyter, Berlin, New York.
[448] Voit, E.O., 1998: Canonical modeling: a link between environmental models and statistics, *Austrian J. Statist.* **27**: 109–121.
[449] Voit, E.O., 2000a: A maximum likelihood estimator for shape parameters of S-distributions. *Biometrical Journal* (in press).
[450] Voit, E.O., 2000b: Utility of Biochemical Systems Theory for the analysis of metabolic effects from low-dose chemical exposure. *Risk Analysis* (in press).
[451] Voit, E.O., 2000c: Canonical modeling: a review of concepts with emphasis on environmental health. *Environmental Health Perspectives* (submitted).
[452] Voit, E.O., H.J. Anton, and J. Blecker, 1985: Regenerative growth curves. *Math. Biosci.* **73**: 253–269.
[453] Voit, E.O., W. L. Balthis, and R. A. Holser, 1993: Conditional Monte Carlo modeling with S-systems, in: McAleer, M., and A. Jakeman (Eds.), *Proceedings of the International Congress on Modelling and Simulation*, Perth, W. Australia, pp. 1223–1234.
[454] Voit, E.O., W. L. Balthis, and R. A. Holser, 1995: Hierarchical Monte Carlo modeling with S-distributions: concepts and illustrative analysis of mercury contamination in king mackerel. *Environ. Int.* **21**: 627–635.
[455] Voit, E.O., and A.E.N. Ferreira, 1998: Buffering in models of integrated biochemical systems. *J. Theor. Biol.* **191**: 429–438.
[456] Voit, E.O., D.H. Irvine, and M.A. Savageau, 1989: *The User's Guide to ESSYNS*, Medical University of South Carolina Press.
[457] Voit, E.O., and R.G. Knapp, 1997: Derivation of the linear–logistic model and Cox's proportional hazard model from a canonical system description, *Statist. Med.* **16**: 1705–1729.
[458] Voit, E.O., and P.F. Rust, 1990: Evaluation of the noncentral t distribution with S-systems. *Biometrical J.* **32**(6): 681–695.
[459] Voit, E.O., and P.F. Rust, 1992: Tutorial: S-system analysis of continuous univariate probability distributions. *J. Statis. Comput. Simul.* **42**: 187–249.
[460] Voit, E.O., and P.J. Sands, 1996a: Modeling forest growth. I. Canonical approach. *Ecol. Model.* **86**: 51–71.
[461] Voit, E.O., and P.J. Sands, 1996b: Modeling forest growth. II. Biomass partitioning in Scots pine. *Ecol. Model.* **86**: 73–89.
[462] Voit, E.O., and M.A. Savageau, 1982a: Power-law approach to modeling biological systems; II. Application to ethanol production. *J. Ferment. Technol.* **60**(3): 229–232.

[463] Voit, E.O., and M.A. Savageau, 1982b: Power-law approach to modeling biological systems; III. Methods of analysis. *J. Ferment. Technol.* 60(3): 233–241.

[464] Voit, E.O., and M.A. Savageau, 1984: Analytical solutions to a generalized growth equation. *J. Math. Anal. Appl.* 103(2): 380–386.

[465] Voit, E.O., and M.A. Savageau, 1986: Equivalence between S-systems and Volterra-systems. *Math. Biosci.* 78: 47–55.

[466] Voit, E.O., and M.A. Savageau, 1987: Accuracy of alternative representations for integrated biochemical systems. *Biochemistry* 26: 6869–6880.

[467] Voit, E.O., and M. K. Schubauer-Berigan, 1998: The role of canonical modeling as an unifying framework for ecological and human risk assessment, in: Newman, M.C., and C.L. Strojan (Eds.), *Risk Assessment: Logic and Measurement*, Ann Arbor Press, Chelsea, MI, pp. 101–139.

[468] Voit, E.O., and L.H. Schwacke, 1998: Scalability properties of the S-distribution, *Biometr. J.* 40: 665–684.

[469] Voit, E.O., and L.H. Schwacke, 2000: Random number generation from right-skewed, symmetric, and left-skewed distribution. *Risk Anal.,* 20(1): 59–71.

[470] Voit, E.O., and A. Sorribas, 2000: Computer modeling of dynamically changing distributions of random variables, *Math. Comput. Modelling,* 31: 217–225.

[471] Voit, E.O., and N.V. Torres, 1998: Canonical modeling of complex pathways in biotechnology. *Recent Res. Devel. Biotech. Bioeng.* 1: 321–341.

[472] Voit, E.O., and P.N. Yi, 1990: Comparison of alternative isoeffect curves in radiotherapy. *Bull. Math. Biol.* 52(5): 657–675.

[473] Voit, E.O., and S. Yu, 1994: The S-distribution. Approximation of discrete distributions. *Biometrical J.* 36: 205–219.

[474] von Bertalanffy, L., 1968: *General Systems Theory.* George Braziller, New York.

[475] von Neumann, J., 1951: *The General and Logical Theory of Automata.* Collected Works, Vol. 5. Macmillan, New York (reprinted 1963).

[476] von Neumann, J., and O. Morgenstern, 1944: *The Theory of Games and Economic Behavior.* Princeton University Press, Princeton, NJ.

[477] von Weizsäcker, C.F., 1979: *Die Einheit der Natur* (5th edition). Hanser Verlag, München.

[478] Wallace, J., 1986: *Interpretation of Diagnostic Tests* (4th edition). Little, Brown, Boston.

[479] Walsh, K., and D.E. Koshland, Jr., 1985: Characterization of rate-controlling steps *in vivo* by use of an adjustable expression vector. *Proc. Natl. Acad. Sci. USA* 82: 3577–3581.

[480] Walsh, K., M. Schena, A. Flint, and D.E. Koshland, Jr., 1987: Compensatory regulation in metabolic pathways – responses to increases and decreases in citrate synthase levels. *Biochem. Soc. Symp.* 54: 183–195.

[481] Wanders, R. J. A., C.W.T. van Roemund, and A. J. Meijer, 1984: Analysis of the control of citrulline synthesis in isolated rat-liver mitochondria. *Eur. J. Biochem.* 142(2): 247–254.

[482] Weinberger, V., 1991: Symbolic analysis of S-systems with MACSYMA, in: Voit, E.O. (Ed.), *Canonical Nonlinear Modeling. S-System Approach to Understanding Complexity*, Van Nostrand Reinhold, New York, Ch. 6.

[483] Welch, G.R., 1977: On the role of organized multienzyme systems in cellular metabolism: a general synthesis. *Prog. Biophys. Mol. Biol.* 32: 103–191.

REFERENCES

[484] Wen, X., S. Fuhrman, G.S. Michaels, D.B. Carr, S. Smith, J.L. Barker, and R. Somogyi, 1998: Large-scale temporal gene expression mapping of central nervous system development. *Proc. Natl. Acad. Sci. USA* **95**(1): 334–339.

[485] Weng, G., U.S. Bhalla, and R. Iyengar, 1999: Complexity in biological signaling systems. *Science* **284**: 92–96.

[486] White, A., P. Handler, and E.L. Smith, 1968: *Principles of Biochemistry* (4th edition). McGraw-Hill, New York.

[487] White, G.J., and M. Sussman, 1961: Metabolism of major cell components during slime mold morphogenesis. *Biochim. Biophys. Acta* **53**: 285–293.

[488] White, J., 1980: Demographic factors in populations of plants, in: Solbrig, O.T. (Ed.), *Demography and Evolution of Plant Populations*, Blackwell Scientific Publications, Oxford, pp. 21–48.

[489] White, J., 1981: The allometric interpretation of the self-thinning rule. *J. Theor. Biol.* **89**: 475–500.

[490] Whitmarsh, A.J., J. Cavanagh, C. Tournier, J. Yasuda, and R.J. Davis, 1998: A mammalian scaffold complex that selectively mediates MAP kinase activation. *Science* **281**: 1671–1674.

[491] Wiener, N., 1948: *Cybernetics, or Control and Communication in the Animal and the Machine*. Wiley, New York.

[492] Wilkins, M.R., C. Pasquali, R.D. Appel, K. Ou, O. Golaz, J.-C. Sanchez, J.X. Yan, A.A. Gooley, G. Hughes, I. Humphery-Smith, K.L. Williams, and D.F. Hochstrasser, 1996: From proteins to proteomes: large scale protein identification by two-dimensional electrophoresis and amino acid analysis. *Bio/Technology* **14**: 61–65.

[493] Wilkinson, K.D., and I.A. Rose, 1979: Isotope trapping studies of yeast hexokinase during steady state catalysis. A combined rapid quench and isotope trapping technique. *J. Biol. Chem.* **254**(24): 12567–12572.

[494] Wilson, J.B., and C.L. Rutherford, 1978: ATP, trehalose, glucose, and ammonium ion localization in the two cell types of *Dictyostelium discoideum*. *J. Cell Physiol.* **94**: 37–46.

[495] Wodicka, L., H. Dong, M. Mittmann, M.-H. Ho, and D. J. Lockhart, 1997: Genome-wide expression monitoring in *Saccharomyces cerevisiae*. *Nat. Biotechnol.* **15**(13): 1359–1367.

[496] Woolfolk, C.A., and E.R. Stadtman, 1967: Regulation of glutamine synthetase. III. Cumulative feedback inhibition of glutamine synthetase from *Escherichia coli*. *Arch. Biochem. Biophys.* **118**: 736–755.

[497] Wright, B.E., 1968: An analysis of metabolism underlying differentiation in *Dictyostelium discoideum*. *J. Cell Physiol.* **72** (Suppl.1): 145–160.

[498] Wright, B.E., 1973: *Critical Variables in Differentiation*. Prentice Hall, Englewood Cliffs, NJ.

[499] Wright, B.E., 1984: Constraints on the models of carbohydrate metabolism in the two cell types of *Dictyostelium discoideum*. *J. Theor. Biol.* **110**: 445–460.

[500] Wright, B.E., and M.L. Anderson, 1960a: Protein and amino acid turnover during differentiation in the slime mold. I. Utilization of endogenous amino acids and proteins. *Biochim. Biophys. Acta* **43**: 62–66.

[501] Wright, B.E., and M.L. Anderson, 1960b: Protein and amino acid turnover during differentiation in the slime mold. II. Incorporation of [^{35}S]methionine into the amino acid pool and into protein. *Biochim. Biophys. Acta* **43**: 67–78.

[502] Wright, B.E., M.H. Butler, and K.R. Albe, 1992: Systems analysis of the tricarboxylic acid cycle in *Dictyostelium discoideum*. I. The basis for model construction. *J. Biol. Chem.* **267**: 3101–3105.

[503] Wright, B.E., and R.J. Field, 1994: The tricarboxylic acid cycle in *Dictyostelium discoideum*: two methods of analysis using the same data. *J. Biol. Chem.* **269**(31): 19931–19992.

[504] Wright, B.E., and G.L. Gustafson, 1972: Expansion of the kinetic model of differentiation in *Dictyostelium discoideum*. *J. Biol. Chem.* **247**(24): 7875–7884.

[505] Wright, B.E., and P.J. Kelly, 1981: Kinetic models of metabolism in intact cells, tissues, and organisms. *Curr. Top. Cell. Regul.* **19**: 103–158.

[506] Wright, B.E., and R. Marshall, 1971: Trehalose synthesis during differentiation in *Dictyostelium discoideum*. *J. Biol. Chem.* **246**: 5335–5339.

[507] Wright, B.E., and D.J.M. Park, 1975: An analysis of the kinetic positions held by five enzymes of carbohydrate metabolism in *Dictyostelium discoideum*. *J. Biol. Chem.* **250**: 2219–2226.

[508] Wright, B.E., and J.M. Reimers, 1988: Steady-state models of glucose perturbed *Dictyostelium discoideum*. *J. Biol. Chem.* **263**: 14906–14912.

[509] Wright, B.E., W. Simon, and B.T. Walsh, 1968: A kinetic model of metabolism essential to differentiation in *Dictyostelium discoideum*. *Proc. Natl. Acad. Sci. USA* **60**: 644–651.

[510] Wright, B.E., A. Tai, and K.A. Killick, 1977: Fourth expansion and glucose perturbation of the *Dictyostelium* kinetic model. *Eur. J. Biochem.* **74**: 217–225.

[511] Wright, B.E., A. Tai, K.A. Killick, and D.A. Thomas, 1979: The effects of exogenous glucose, uracil, and inorganic phosphate on differentiation in *Dictyostelium discoideum*. *Arch. Biochem. Biophys.* **192**: 489–499.

[512] Wright, B.E., D.A. Thomas, and D.A. Ingalls, 1982: Metabolic compartments in *Dictyostelium discoideum*. *J. Biol. Chem.* **257**: 7587–7594.

[513] Wyngaarden, J.B., and W.N. Kelley, 1983: Gout, in: Stanbury, J.B., J.B. Wyngaarden, D.S. Fredrickson, J.L. Goldstein, and M.S. Brown (Eds.), *The Metabolic Basis of Inherited Disease*, McGraw-Hill, New York, pp. 1043–1114.

[514] Yang, S.T., and W.C. Deal, Jr., 1969: Metabolic control and structure of glycolytic enzymes. VI. Competitive inhibition of yeast glyceraldehyde 3-phosphate dehydrogenase by cyclic adenosine monophosphate, adenosine triphosphate, and other adenine-containing compounds. *Biochemistry* **8**(7): 2806–2813.

[515] Yao, L., and W.A. Sethares, 1994: Nonlinear parameter estimation via the genetic algorithm. *IEEE Trans. Signal Process.* **42**(4): 927–935.

[516] Yates, F.E., 1977: Explanation in science: is there a general theory of systems? *Am. J. Physiol.* **233**(2): R169–R170.

[517] Yates, F.E., 1978: Complexity and the limits to knowledge. *Am. J. Physiol.* **235**(4): R201–R204.

[518] Yates, F.E., 1992: Fractal applications in biology: scaling time in biochemical networks. *Methods Enzymol.* **210**: 636–675.

[519] Yeargers, E.K., R.W. Shonkwiler, and J.V. Herod, 1996: *An Introduction to the Mathematics of Biology: With Computer Algebra Models*. Birkhäuser, Boston.

[520] Yu, S., and E.O. Voit, 1995: A simple, flexible failure model. *Biometrical J.* **37**: 595–609.

[521] Yu, S., and E.O. Voit, 1996: A graphical classification of survival distributions, in: Jewell, N.P., A.C. Kimber, M.-L. T. Lee, and G.A. Whitmore (Eds.), *Lifetime Data:*

Models in Reliability and Survival Analysis, Kluwer Academic, Dordrecht, pp. 385–392.

[522] Zhang, Z., E.O. Voit, and L.H. Schwacke, 1996: Parameter estimation and sensitivity analysis of S-systems using a genetic algorithm, in: Yamakawa, T., and G. Matsumoto (Eds.), *Methodologies for the Conception, Design, and Application of Intelligent Systems*, World Scientific, Singapore.

[523] Zimmer, C., 1998: The slime alternative. *Discover*, September, pp. 86–93.

[524] Zimmer, C., 1999: Life after chaos. *Science* 284: 84–86.

[525] Zoref-Shani, E., G. Kessler-Icekson, L. Wasserman, and O. Sperling, 1984: Characterization of purine nucleotide metabolism in primary rat cardiomyocyte cultures, *Biochim. Biophys. Acta* 804: 161–168.

Web Sites of Interest

Affymetrix:
http://www.affymetrix.com/

ANGIS: Australian National Genomic Information Service:
http://morgan.angis.su.oz.au/

Brown Laboratory at Stanford:
http://cmgm.stanford.edu/pbrown/mguide/index.html and
http://cmgm.stanford.edu/pbrown/array.html

CBC Bibliography Database:
http://www.cbc.umn.edu/ResearchProjects/BIBLIOGRAPHY/bibliography.html

DDBJ: DNA Data Bank of Japan:
http://www.ddbj.nig.ac.jp/

EC Enzyme database:
http://nucleus.agron.missouri.edu/enzyme.html

EcoCyc:
http://ecocyc.PangeaSystems.com/ecocyc/

EBI: European Bioinformatics Institute:
http://www.ebi.ac.uk/ebi_home.html

EMBL: European Molecular Biology Laboratory:
http://www.embl-heidelberg.de/
http://www.embl-heidelberg.de/Services/index.html

EMP: Enzymology Database:
http://wit.mcs.anl.gov/EMP/

ERM: Enzyme Reaction Mechanism:
http://sirius.shome.eu.org/~miliusha/

ExPASy [Swiss Institute of Bioinformatics (SIB)]:
http://www.expasy.ch/

FlyNets (Drosophila pathways):
http://gifts.univ-mrs.fr/GIFTS_home_page.html
http://gifts.univ-mrs.fr/FlyNets/FlyNets_home_page.html

GenBank:
http://www.ncbi.nlm.nih.gov/

GenProtEC:
http://dbase.mbl.edu/genprotec/start

KEGG: Kyoto Encyclopedia of Genes and Genomes:
http://www.genome.ad.jp/kegg/kegg2.html

LIGAND:
http://www.genome.ad.jp/dbget-bin/www_bfind?ligand

Lawrence Berkeley National Laboratory:
http://www-gsd.lbl.gov/

List of biochemical databases:
http://www.oup.co.uk/nar/Volume_27/Issue_01/summary/gkc105_gml.html
http://www.cc.um.edu.my/biocomp/pathwaydb.html

MIPS: Database for genomes and protein sequences:
http://www.mips.biochem.mpg.de

MPD: Metabolic Pathway Database:
http://wit.mcs.anl.gov/MPW/

MSD: Macromolecular Structure Database:
http://msd.ebi.ac.uk/

National Human Genome Research Institute:
http://www.nhgri.nih.gov/
http://www.nhgri.nih.gov/DIR/LCG/15K/HTML/microlinks.html

PDB: Protein Data Bank:
http://www.rcsb.org/

PIR: Protein Information Resource:
http://www.oup.co.uk/nar/Volume_25/Issue_01/html/gka023_gml.html

Prokaryotic Database:
http://www.bic.nus.edu.sg/proka.html

RasMol & Chime Molecular Visualization:
http://www.umass.edu/microbio/rasmol/

RCSB: Research Collaboratory for Structural Bioinformatics:
http://www.rcsb.org/

SIB: Swiss Institute of Bioinformatics:
http://www.isb-sib.ch/

REFERENCES

SWISS-PROT (Geneva Bioinformatics):
http://www.genebio.com/sprot.html

UM-BBD (University of Minnesota Biocatalysis/Biodegradation Database):
http://www.labmed.umn.edu/umbbd/index.html

University of Washington Resource Facility for Kinetic Analysis:
http://weber.u.washington.edu/~rfka/

WIT (What is There?):
http://wit.mcs.anl.gov/WIT2

WormPD:
http://www.proteome.com/

Yeast Protein Database (YPD):
http://www.proteome.com/

Author Index

Acerenza, L., 62, 155, 157, 175, 186, 190, 246, 478
Achs, M.J., 38, 469, 474
Adams, M.D., 8, 473
Adolph, E.F., 56, 469
Agresar, G., 129, 469
Ahmad Mahir, B.R., 411, 479
Akaoka, L., 359, 474
Akerboom, T.P.M., 162, 163, 372, 475
Akiyama, Y., 16, 474
Albe, K.R., 171, 293, 297, 298, 301, 310, 315, 323, 325, 469, 494
Alcón, C.H., 83, 223, 286, 290, 396, 408, 409, 490
Alippi, C., 174, 483
Anderson, M.L., 315, 325, 493
Anderson, S.P., 17, 472
Anton, H.J., 410, 491
Appel, R.D., 9, 493
Appenzeller, T., 5, 474
Apweiler, R., 18, 469
Auscher, C., 364, 472
Averet, N., 323, 483
Ayvazian, J.H., 343, 345, 469

Bailey, J.E., 61, 187, 260, 261, 268, 269, 270, 273, 396, 405, 409, 474, 475, 476, 487
Bairoch, A., 15, 18, 469
Balthis, W.L., 140, 411, 469, 491
Bangerter, J.K., 140, 469
Barber, J.R., 343, 470
Barker, J.L., 63, 493
Barker, W.C., 18, 470, 474
Barrell, B.G., 8, 9, 475
Barry, C.E., 3rd., 8, 471
Bartel, T., 327, 470
Bartelmess, I.B., 149, 473
Basmanova, S., 15, 487

Bassham, J.A., 367, 373, 374, 470
Batschelet, E., 56, 470
Battey, J., 8, 470
Baxevanis, A.D., 15, 470
Becker, M.A., 355, 359, 364, 470, 477
Bel-Enguix, G., 62, 471
Benson, D.A., 18, 470
Berg, P.H., 174, 183, 470
Bernstein, F.C., 18, 470
Berry, M.N., 415, 421, 422, 470
Bhalla, U.S., 3, 171, 493
Bin Razali, A.M., 175, 176, 189, 404, 490
Binns, M.R., 297, 484
Bish, D.R., 171, 470
Bittner, M., 18, 473
Blecker, J., 410, 491
Boeckmann, B., 18, 469
Bogle, I.D.L., 285, 286, 396, 408, 483
Bogorad, L., 149, 489
Boguski, M.S., 8, 9, 18, 470, 471
Bohnensack, R., 187, 247, 470
Bono, H., 15, 16, 62, 475, 482
Bontemps, F., 331, 345, 470
Botstein, D., 17, 471
Brice, M.D., 18, 470
Bronstein, I.N., 432, 470
Bronsweig, R.D., 315, 475
Brosch, R., 8, 471
Brouilhet, H., 364, 472
Brown, A.B., 409, 471
Brown, M.S., 12, 13, 17, 327, 328, 360, 488
Brown, P.O., 17, 18, 471, 472, 479, 486
Brownstein, M.J., 8, 9, 471
Brox, L.W., 334, 476
Brunauer, L.S., 343, 470
Bult, C.J., 8, 473
Burks, C., 15, 471

Burns, J.A., 60, 155, 157, 175, 186, 190, 245, 246, 478
Burns, S.A., 407, 481
Bussey, H., 8, 9, 475
Butler, M.H., 171, 293, 297, 298, 301, 323, 325, 469, 494

Cabib, E., 273, 483
Callahan, B., 139, 483
Cameron, G.N., 18, 489
Cameron, J.S., 337, 364, 490
Canela, E.I., 171, 184, 216, 232, 251, 315, 323, 327, 373, 389, 390, 471, 473, 488
Cárdenas, M.L., 38, 60, 83, 245, 472
Carnap, R., 7, 19, 40, 471
Carr, D.B., 63, 493
Cascante, M., 59, 60, 83, 113, 144, 145, 146, 148, 149, 150, 171, 184, 185, 216, 223, 232, 246, 251, 260, 261, 268, 270, 272, 273, 275, 279, 281, 283, 284, 314, 326, 327, 328, 335, 336, 338, 341, 352, 355, 360, 373, 383, 389, 390, 399, 405, 435, 471, 472, 478, 483, 488
Catling, A.D., 171, 486
Cavanagh, J., 171, 493
Chabner, B., 361, 475
Chai, A., 17, 486
Chakravarti, A., 8, 9, 472
Chance, B., 38, 471, 474
Changeaux, J.-P., 271, 275, 481
Chaudhuri, K.S., 409, 471, 474
Chen, Y., 18, 473
Chiew, Y.Y., 296, 471
Chock, P.B., 86, 471
Churcher, C., 8, 471
Clarke, S., 342, 343, 470, 482
Clayton, R.A., 8, 473
Cleland, W.W., 38, 471
Cohadon, F., 323, 483
Cole, S.T., 8, 471
Collado-Vides, J., 62, 471
Collier, L.S., 171, 486
Collins, F.S., 8, 9, 472
Conolly, R.B., 17, 472
Cordes, E.H., 323, 325, 480
Cornish-Bowden, A., 38, 60, 83, 160, 245, 246, 472
Corton, J.C., 17, 472
Courant, R., 424, 472
Cox, D., 8, 470
Cox, M.M., 15, 479
Cramer, Gabriel, 435
Crowley, P.H., 160, 472
Crummenerl, E., 1, 472

Curto, R., 59, 60, 83, 113, 145, 146, 148, 149, 150, 185, 223, 246, 260, 261, 268, 270, 272, 273, 275, 281, 283, 284, 314, 326, 327, 328, 330, 331, 335, 336, 337, 338, 341, 344, 346, 352, 355, 360, 399, 405, 471, 472, 488

Dacol, D., 222, 483
Davis, B.P., 16, 477
Davis, R.J., 171, 493
Davis, R.W., 8, 9, 17, 18, 475, 479, 486
Deal, W.C., Jr., 274, 494
Dean, A.M., 149, 473
Degery, A., 364, 472
Degn, H., 57, 482
Delbarre, F., 364, 472
Delgado, J., 157, 472
DeRisi, J.L., 17, 18, 472, 479
de Verdier, C.H., 146, 149, 150, 472
Domach, M.M., 284, 480
Donachies, W., 149, 473
Dong, H., 18, 171, 472, 493
Doran, P.M., 260, 274, 473
Dougherty, B.A., 8, 473
Dove, W., 8, 470
Duggan, D.J., 18, 473
Dujon, B., 8, 9, 475
Dunhill, P., 285, 286, 396, 408, 483
Dykhuizen, D.E., 149, 473
Dzubow, L., 38, 474

Easterby, J.S., 216, 232, 389, 473
Eblen, S.T., 171, 486
Edelstein-Keshet, L., 212, 473
Edwards, N.L., 359, 473
Eiglmeier, K., 8, 471
Einstein, Albert, 3
Elion, G.B., 361, 473
Environmental Protection Agency (EPA USA), 140, 473
Erfle, J.D., 297, 484
Ericson, A., 146, 149, 150, 472

Fairén, V., 59, 129, 164, 410, 473, 476
Faux, M.C., 171, 473
Feil, R., 149, 151, 152, 479
Feldmann, H., 8, 9, 475
Fell, D.A., 38, 60, 62, 83, 86, 147, 148, 149, 156, 157, 161, 163, 172, 223, 245, 249, 284, 473
Ferreira, A.E.N., 67, 91, 135, 136, 137, 323, 407, 473, 491
Fersht, A.R., 160, 473
Fiechter, A., 272, 479
Field, R.J., 293, 494
Fleischmann, R.D., 8, 473
Flint, A., 149, 492

Flint, H.J., 149, 473
Floudas, C.A., 396, 405, 409, 476
FlyBase Consortium, 16, 473
Fodor, S.P.A., 17, 473, 480
Fox, I.H., 359, 473
Franco, R., 216, 232, 251, 315, 323, 327, 373, 389, 390, 471, 473
Frank, P.M., 222, 473
Franke, J., 315, 473
Fredrickson, D.S., 12, 13, 17, 327, 328, 360, 488
Frenkel, R.A., 38, 474
Frishman, D., 18, 481
Fuhrman, S., 63, 480, 493
Fujibuchi, W., 15, 16, 62, 474, 475, 482
Fujimori, S., 359, 474
Fujiwara, S., 407, 487
Fung, E.T., 171, 472

Gaasterland, T., 15, 487
Galazzo, J.L., 187, 260, 261, 268, 269, 270, 273, 474
Galibert, F., 8, 9, 475
Galilei, Galileo, 55
Galimova, M., 15, 16, 487
Gallagher, R., 5, 474
Ganguly, S., 409, 474
Garavelli, J.S., 18, 470
Garcia, J., 246, 478
Garfinkel, D., 2, 5, 38, 40, 172, 469, 471, 474, 479, 481, 483
Garnier, T., 8, 471
Garrels, J.I., 16, 477
Gas, S., 8, 471
Gauss, 173
Gavalas, G.R., 61, 251, 474
Geigy A.G., 40, 369, 474
Gentile, C., 18, 479
George, D.G., 18, 474
Gerber, G., 40, 474
Gesteland, R., 8, 9, 472
Giacomello, A., 334, 474
Giersch, C., 156, 474
Ginesla, I., 315, 323, 471
Gingeras, T.R., 17, 480
Glez-Alcón, C., 59, 83, 223, 260, 286, 289, 290, 396, 408, 409, 490
Goffeau, A., 8, 9, 475
Gojobori, T., 18, 489
Golaz, O., 9, 493
Goldberg, D.E., 174, 475
Goldenfeld, N., 3, 475
Goldstein, J.L., 12, 13, 17, 327, 328, 360, 488
Gooley, A.A., 9, 493
Gordon, S.V., 8, 471
Goryanin, I., 15, 16, 487

Goto, S., 15, 16, 62, 474, 475, 482
Gottlieb, L., 149, 151, 152, 479
Green, S.B., 38, 474
Gregg, J.H., 315, 475
Gregory, R.B., 415, 421, 422, 470
Gretchkin, Y., 15, 16, 487
Grivell, A.R., 415, 421, 422, 470
Groen, A.K., 148, 150, 151, 152, 162, 163, 372, 475
Grosveld, F., 17, 482
Guckenheimer, J., 208, 209, 475
Guerin, B., 323, 483
Guièrrez-Ríos, R.M., 62, 471
Gustafson, G.L., 294, 295, 475, 494

Halangk, W., 187, 247, 470
Haldane, J.B.S., 162, 475
Hande, K., 361, 475
Handler, P., 365, 493
Hanlon, M., 149, 488
Harkness, R.A., 359, 475
Harris, D., 8, 471
Hartl, D.L., 149, 473
Hasegawa, T., 407, 475
Hatzimanikatis, V., 396, 405, 409, 475, 476
Hayashi, K., 400, 482
Hayt, W.H., 48, 476
Heath, D.F., 297, 476, 489
Heinmets, F., 327, 476
Heinrich, R., 6, 38, 40, 60, 61, 119, 223, 245, 251, 284, 474, 476
Heller, R., 17, 486
Henderson, J.F., 334, 476
Henly, D.C., 415, 421, 422, 470
Hernández-Bermejo, B., 59, 129, 164, 410, 473, 476
Herod, J.V., 131, 494
Hers, H.-G., 331, 340, 345, 355, 374, 398, 470, 476, 490
Hess, B., 38, 272, 275, 471, 476
Heumann, K., 18, 481
Higgins, J.J., 38, 471
Hill, A.V., 16, 38, 74, 476
Hitchings, G.H., 361, 473
Hlavacek, W.S., 58, 114, 380, 383, 401, 402, 408, 435, 476
Ho, M.-H., 18, 493
Hochachka, P.W., 160, 477
Hochstrasser, D.F., 9, 493
Hodges, P.E., 16, 477
Hofestädt, R., 15, 62, 477
Hofmeyer, J.H., 246, 477
Hoheisel, J.D., 8, 9, 475
Holmes, Oliver Wendell, 8
Holmes, P., 208, 209, 475

Holser, R.A., 140, 411, 491
Holzhütter, H., 327, 470
Honeysett, J.M., 355, 479
Howell, S.B., 355, 479
Hue, L., 374, 398, 476
Huganir, R.L., 171, 472
Hughes, G., 9, 493
Humphery-Smith, I., 9, 493
Hunt, L.T., 18, 474
Husain, K., 359, 364, 470
Huxley, J.S., 56, 577
Hwang, S.Y., 18, 479

Ingalls, D.A., 296, 494
Irvine, D.H., 58, 59, 60, 67, 83, 87, 114, 148, 180, 237, 245, 246, 247, 279, 316, 380, 383, 396, 400, 405, 406, 407, 408, 435, 477, 486, 491
Ivanova, N., 15, 16, 487
Iyengar, R., 3, 171, 493
Iyer, V.R., 17, 18, 472

Jacobs, A.E.M., 345, 477
Jacq, C., 8, 9, 475
Jamier, R., 129, 209, 482
Janszen, D.B., 17, 472
Jerushalmy, Z., 345, 477
Jimenez, M.L., 355, 470, 477
Johnson, T., 174, 409, 471, 477
Johnston, M., 8, 9, 475
Jollow, D., 17, 477
Jordan, E., 8, 9, 470, 472
Jury, E.I., 212, 477

Kacser, H., 60, 62, 149, 155, 157, 175, 186, 190, 245, 246, 365, 369, 374, 398, 473, 477, 478, 481, 489
Kadanoff, L.P., 3, 475
Kahn, P., 9, 478
Kamatani, N., 359, 474
Kanehisa, M., 5, 8, 15, 16, 62, 474, 475, 478, 482
Kaps, A., 18, 481
Karp, P.D., 16, 478
Kauffman, S.A., 62, 63, 478
Kelleher, J.K., 294, 296, 297, 298, 301, 315
Kelley, W.N., 329, 334, 346, 353, 356, 359, 360, 361, 363, 476, 478, 494
Kelly, P.J., 294, 296, 297, 298, 301, 315, 494
Kemmerly, J.E., 48, 476
Kennard, O., 18, 470
Kerlavage, A.R., 473
Kermack, W.O., 220, 478
Kerner, E.H., 410, 478
Kessler-Icekson, G., 359, 495
Kholodenko, B.N., 246, 478
Killick, K.A., 294, 296, 297, 494

Kim, M., 355, 359, 364, 470
Kimbell, J.S., 17, 472
Kimmel, A.R., 294, 478
King, M.E., 355, 479
Kirkness, E.F., 473
Klipp, E., 284, 476
Knapp, R.G., 411, 491
Knowles, R.G., 149, 488
Koch, C., 3, 479
Koetzle, T.F., 18, 470
Kohn, M.C., 64, 172, 187, 479, 481
Komarov, Y., 15, 16, 487
Kopelman, R., 171, 479
Koshland, D.E., 38, 479
Koshland, D.E., Jr., 149, 492
Kovensky, A., 361, 473
Kramer, M., 222, 483
Krause, G.H., 367, 373, 374, 470
Krauss, A., 171, 486
Kreyszig, E., 433, 479
Krinska, M., 345, 477
Kruckeberg, A., 149, 151, 152, 479
Krummenacker, M., 16, 478
Kuenzi, M., 272, 479
Kuhn, T.S., 4, 5, 479

Lanahan, A.A., 171, 472
Laplace, Pierre Simon, 2, 432
Lashkari, D.A., 18, 479
Laszlo, E., 3, 5, 479
Lathem, W., 185, 479
Laurent, G., 3, 479
Lavoisier, Antoine, 2
Ledley, R.S., 18, 470
Lee, M.H., 411, 479
Lehninger, A.L., 15, 479
Leicester, H.M., 1, 2, 479
Lemieux, D.R., 64, 187, 479
Lewis, D.C., 409, 479
Liang, S., 62, 63, 480, 489
Liao, J.C., 157, 472
Liddel, G.U., 315, 480
Liebman, M.N., 64, 483
Lightfoot, E.N., 129, 209, 482
Lipman, D.J., 18, 470
Lipshutz, R.J., 17, 480
Lockhart, D.J., 17, 18, 480, 493
Loomis, W.F., 294, 480
Lopez, R., 18, 489
Lotka, A.J., 4, 5, 46, 480
Lowry, O.H., 159, 480
Luenberger, D.G., 62, 285, 408, 480
Luzzatto, L., 17, 482

Mahler, H.R., 323, 325, 480
Majewski, R.A., 284, 480

AUTHOR INDEX

Maltsev, N., 15, 16, 487
Maranas, C.D., 409, 482
March, J., 411, 488
Marshall, R., 295, 297, 494
Martin, P.-G., 403, 480
Martini, G., 17, 480
Marzec, C.R., 18, 470
Mason, P., 17, 482
Mateo, F., 365, 369, 374, 398, 489
Mateos, F.A., 355, 470, 477
Maurer, H., 62, 480
Mavrovouniotis, M.L., 61, 64, 171, 470, 480, 483
Mayer, K., 18, 481
Mazat, J.P., 323, 483
McCusker, J.H., 18, 479
McGarvey, P.B., 18, 470
McKee, A.H.Z., 16, 477
McKendrick, A.G., 220, 478
McMillan, D., 17, 477
Meaburn, G.M., 411, 469
Meijer, A.J., 152, 153, 154, 492
Meléndez-Hevia, E., 216, 232, 246, 284, 365, 369, 373, 374, 389, 390, 398, 471, 476, 480, 489
Melian, J.S., 355, 477
Meltzer, P., 18, 473
Mende, W., 56, 60, 129, 410, 482
Mendelson, E., 62, 481
Menten, L.E., 172, 479, 481
Menten, M.L., 38, 43, 481
Merrick, J.M., 8, 473
Metz, E., 361, 473
Mewes, H.-W., 18, 470, 474, 481
Meyer, C.L., 284, 482
Meyer, E.F., 18, 470
Michaelis, L., 38, 43, 481
Michaels, G.S., 63, 493
Michal, G., 15, 481
Middleton, R., 149, 481
Migimatsu, H., 16, 474
Mikhailova, N., 15, 16, 487
Miller, G.A., 3, 481
Mittmann, M., 18, 493
Miyazaki, S., 18, 489
Monod, J., 271, 275, 481
Montero, F., 284, 476
Morgenstern, O., 5, 492
Morimoto, B.H., 343, 470
Mueller, K.M., 407, 481
Mullin, R., 149, 488
Murray, J.D., 131, 481

Namath, A.F., 18, 479
Needham, J., 56, 481
Neet, K.E., 38, 479
Neidhardt, F.C., 401, 481

Nelson, D.L., 15, 479
Nenashev, V., 15, 16, 487
Neter, J., 186, 450, 481
Neuhaus, H.E., 149, 151, 152, 479
Newsholme, E.A., 365, 367, 368, 369, 370, 372, 373, 374, 398, 481
Newton, 173
Ni, T.-C., 113, 145, 399, 407, 435, 481
Nieto, V.G., 355, 477
Niklasson, F., 146, 149, 150, 472
Nishioka, T., 15, 16, 62, 475
Nishizuka, T., 15, 481
Norman, G.R., 140, 481
Nyhan, W.L., 343, 482

O'Brien, R.J., 171, 472
Oden, K.L., 342, 482
Ogata, H., 15, 16, 62, 475, 482
Ogiwara, A., 16, 474
Okamoto, M., 174, 223, 400, 482, 489
Olivier, J.-L., 364, 472
Olsen, L.F., 57, 482
Oosterhof, A., 345, 477
Orcutt, B.C., 18, 470
Ostell, J., 18, 470
Ou, K., 9, 493
Ouellette, B.F.F., 15, 18, 470
Ovadi, J., 171, 482
Overbeek, R., 15, 16, 487

Page, T., 343, 482
Paley, S., 16, 478
Palsson, B.Ø., 61, 62, 129, 209, 284, 482, 486, 490
Palsson, H., 129, 209, 482
Pandolfi, P.P., 17, 482
Panetta, J.C., 141, 482
Panyushkina, E., 15, 16, 487
Papoutsakis, E.T., 284, 482
Park, D.J.M., 296, 494
Parkhill, J., 8, 471
Pasquali, C., 9, 493
Passonneau, J.V., 159, 480
Patnaik, L.M., 174, 488
Patrinos, A., 8, 9, 472
Payne, W.E., 16, 477
Pelligrini-Toole, A., 16, 478
Peschel, M., 56, 60, 129, 410, 482
Peterson, J.L., 64, 482
Petkov, S.B., 409, 482
Pfeiffer, F., 18, 470, 474, 481
Phillips, J.W., 415, 421, 422, 470
Pickard, W.F., 56, 483
Pinkhas, H., 345, 477
Plesser, T., 272, 275, 476
Pogson, C.I., 149, 488

Porteous, D.J., 149, 473
Potter, C., 337, 364, 490
Preissler, H., 40, 474
Pring, M., 38, 474
Pronevitch, L., 15, 16, 487
Puig, J.G., 355, 470, 477
Puigjaner, J., 60, 246, 383, 483

Quick, W.P., 149, 489

Rabitz, H.K., 222, 483
Ramos, T.H., 355, 477
Rapoport, S.M., 6, 40, 61, 119, 251, 474, 476
Rapoport, T.A., 6, 60, 61, 119, 245, 251, 476
Rapp, B.A., 18, 470
Recker, D., 359, 473
Reddy, V.N., 64, 483
Reder, C., 61, 251, 257, 483
Reed, E., 361, 475
Regan, L., 285, 286, 396, 408, 483
Reich, J.G., 373, 483
Reimers, J.M., 296, 471, 494
Reisig, W., 64, 483
Rhee, S.G., 86, 471
Ribeiro Filho, J.L., 174, 483
Rigoulet, M., 323, 483
Riley, M., 16, 478, 483
Rivi, R., 17, 482
Rodermel, S.R., 149, 489
Rodnan, G.P., 185, 479
Rodríguez-Acosta, F., 490
Rodríguez, F., 59, 83, 223, 260, 286, 289, 290, 396, 408, 409, 490
Rogers, J.R, 18, 470
Rolleston, F.S., 483
Roman, G.C., 38, 483
Rose, I.A., 269, 270, 493
Rosen, O.M., 86, 483
Rosenbloom, F.M., 334, 359, 476, 487
Rothman, L., 273, 483
Routh, E.J., 210, 483
Rubin, C.S., 86, 483
Rugen, P., 139, 483
Rundles, R.W., 361, 473
Russel, P.J., 263, 489
Rust, P.F., 137, 411, 483, 491
Rutherford, C.L., 294, 295, 493

Sahota, A.S., 337, 487
Salerno, C., 334, 474
Salter, M., 149, 488
Salvador, A., 223, 405, 483, 484
Samitier, S., 184, 488
Sanchez, J.-C., 9, 493
Sands, P.J., 87, 158, 173, 175, 179, 183, 210, 251, 279, 380, 401, 403, 484, 486, 491

Sargent, R., 140, 484
Sato, K., 16, 62, 475
Sauer, F., 297, 484
Sauro, H.M., 155, 157, 175, 186, 190, 246, 478
Savageau, M.A., 3, 4, 5, 6, 7, 38, 40, 50, 51, 53, 55, 56, 57, 58, 59, 60, 67, 68, 83, 86, 87, 90, 91, 92, 96, 113, 114, 124, 128, 129, 137, 138, 144, 145, 146, 148, 158, 161, 165, 171, 172, 175, 179, 180, 188, 196, 201, 204, 206, 207, 209, 210, 220, 223, 224, 237, 242, 245, 246, 247, 279, 284, 293, 296, 297, 298, 300, 307, 308, 310, 311, 314, 315, 316, 320, 323, 324, 355, 357, 361, 380, 383, 395, 396, 399, 400, 401, 402, 403, 404, 405, 406, 407, 408, 410, 411, 420, 435, 436, 445, 447, 469, 476, 481, 482, 484, 485, 486, 488, 491
Savinell, J.M., 62, 284, 486
Schaeffer, H.J., 171, 486
Schena, M., 17, 149, 486, 492
Scholz, U., 15, 62, 477
Schomburg, D., 15, 487
Schubauer-Berigan, M.K., 411, 492
Schulze, E.-D., 149, 489
Schurr, U., 149, 489
Schuster, S., 38, 60, 61, 119, 223, 245, 246, 251, 284, 476, 478
Schwacke, L.H., 174, 411, 492, 495
Scott, J.D., 171, 473
Scrutton, M.C., 365, 370, 372, 373, 396, 487
Seegmiller, J.E., 334, 359, 476, 487
Segel, I.H., 301, 487
Segel, L.A., 38, 487
Selkov, E., 15, 16, 487
Selkov, E., Jr., 15, 16, 487
Sel'kov, E.E., 373, 483
Semendjajew, K.A., 432, 470
Seressiotis, A., 61, 487
Service, R.F., 6, 487
Sethares, W.A., 174, 494
Shalon, D., 17, 486
Shannon, C.E., 5, 487
Shimanouchi, T., 18, 470
Shiraishi, F., 38, 113, 129, 145, 161, 165, 171, 172, 188, 209, 223, 246, 284, 293, 296, 297, 298, 308, 310, 311, 314, 315, 316, 320, 323, 324, 407, 475, 487
Shonkwiler, R.W., 131, 494
Shott, S., 140, 487
Sicilia, J., 246, 480
Šiljak, D.D., 222, 487
Simmonds, H.A., 337, 355, 470, 487
Simon, W., 295, 494
Simonds, H.A., 337, 364, 490
Simpson, G.G., 2, 4, 488
Skupp, S., 343, 345, 469
Slein, M.W., 374, 488

AUTHOR INDEX

Small, J.R., 62, 149, 156, 284, 473, 488
Smith, E.L., 365, 493
Smith, G.K., 149, 488
Smith, S., 63, 493
Sniegoski, C.A., 62, 63, 488
Somero, G.N., 160, 477
Somogyi, R., 62, 63, 480, 488, 489, 493
Sonati, F., 17, 482
Sorribas, A., 59, 60, 83, 91, 92, 96, 98, 113, 144, 145, 146, 148, 149, 150, 164, 171, 172, 184, 185, 223, 242, 246, 260, 261, 268, 270, 272, 273, 275, 279, 281, 283, 284, 300, 314, 326, 327, 328, 335, 336, 338, 341, 352, 355, 357, 360, 361, 383, 395, 399, 405, 407, 411, 435, 447, 471, 472, 476, 483, 486, 488, 492
Sperling, O., 327, 345, 359, 477, 488, 495
Srere, P.A., 323, 488
Srinivas, M., 174, 488
Srinivasarao, G.Y., 18, 470
Stadtman, E.R., 39, 86, 471, 493
Stanbury, J.B., 12, 13, 17, 327, 328, 360, 488
Starmer, C.F., 327, 488
Start, C., 365, 367, 368, 369, 370, 372, 373, 374, 398, 481
Stauber, A.J., 17, 472
Steinbuch, K., 7, 489
Stephanopoulos, G., 61, 171, 480, 490
Sterk, P., 18, 489
Stitt, M., 149, 151, 152, 479, 489
Stocker, S., 18, 481
Stoehr, P.J., 18, 489
Stoesser, G., 18, 489
Strang, G., 285, 408, 489
Streiner, D.L., 140, 481
Stryer, L., 15, 489
Su, S., 263, 489
Sugawara, H., 18, 489
Sussman, M., 294, 315, 473, 493

Tagaya, T., 359, 474
Tager, J.M., 148, 150, 151, 152, 162, 163, 372, 475
Tai, A., 294, 296, 297, 494
Tasumi, M., 18, 470
Tateno, Y., 18, 489
Tateson, R.W., 149, 473
Taylor, Brook, 418
Thomas, D.A., 294, 296, 494
Thompson, D.W., 56, 489
Thornton, J.M., 9, 489
Threlfall, C.J., 297, 476, 489
Tomb, J.F., 8, 473
Tominaga, D., 174, 489

Torres, N.V. , 59, 83, 91, 113, 145, 167, 216, 223, 232, 246, 260, 279, 284, 286, 289, 290, 365, 367, 368, 369, 370, 373, 374, 375, 389, 390, 396, 397, 398, 400, 403, 408, 409, 471, 480, 489, 490, 492
Torsella, J., 175, 176, 189, 404, 490
Tournier, C., 171, 493
Treleaven, P.C., 174, 483
Trent, J.M., 8, 9, 18, 471, 473
Tsugita, A., 18, 470, 474
Tuli, M.A., 18, 489
Turing, A., 63, 490

Uchiyama, I., 16, 474
Ulam, Stanislav, 139
Ursini, M.V., 17, 480
Utter, M.F., 365, 370, 372, 373, 396, 487

Vallino, J.J., 61, 490
Van Acker, K.J., 337, 364, 487, 490
van den Berghe, G., 331, 340, 345, 355, 470, 490
Van der Meer, R., 148, 151, 162, 163, 372, 475
van der Merwe, K.J., 246, 477
van Roermund, C.W.T., 150, 151, 152, 153, 154, 163, 475, 492
Van Schaftingen, E., 374, 476
Varma, A., 61, 490
Veerkamp, J.H., 345, 477
Vervoorn, R.C., 150, 151, 152, 163, 475
Voit, E.O., 50, 51, 55, 57, 59, 60, 68, 83, 91, 113, 119, 135, 136, 137, 138, 139, 140, 141, 145, 146, 148, 149, 150, 164, 173, 174, 175, 183, 185, 210, 223, 237, 245, 246, 247, 251, 260, 279, 285, 286, 289, 290, 307, 314, 323, 326, 327, 328, 335, 336, 338, 341, 352, 355, 357, 360, 361, 396, 399, 403, 404, 405, 406, 407, 408, 409, 410, 411, 412, 420, 445, 469, 470, 472, 483, 484, 486, 488, 490, 491, 492, 494, 495
von Bertalanffy, L., 3, 5, 492
von Neumann, J., 5, 63, 139, 492
von Weizsäcker, C.F., 9, 492
Vries, A., 345, 477

Wadell, T.G., 284, 476
Wainwright, E., 140, 484
Wallace, J., 40, 492
Wallace, P.G., 415, 421, 422, 470
Walsh, B.T., 295, 494
Walsh, K., 149, 492
Walters, L., 8, 9, 472
Wanders, R.J.A., 148, 152, 153, 154, 162, 163, 372, 475, 492

Wasserman, L., 359, 495
Wasserman, W., 186, 450, 481
Weber, M.J., 171, 486
Weinberger, V., 98, 395, 407, 492
Welch, G.R., 171, 415, 421, 422, 470, 492
Wen, X., 63, 493
Weng, G., 3, 171, 493
Westerhoff, H.V., 148, 162, 163, 246, 372, 475, 478
Westman, M., 146, 149, 150, 472
Wheeler, D.L., 18, 470
White, A., 365, 493
White, G.J., 294, 315, 493
White, J., 56, 404, 493
White, O., 8, 473
White, R., 174, 183, 470
Whitmarsh, A.J., 171, 493
Wiener, N., 5, 493
Wilkins, M.R., 9, 493
Wilkinson, K.D., 269, 270, 493
Williams, G.J., 18, 470
Williams, K.L., 9, 493
Wilson, J.B., 294, 295, 493
Wodicka, L., 18, 493
Woolfolk, C.A., 39, 493
Worley, P.F., 171, 472

Wright, B.E., 171, 293, 294, 295, 296, 297, 298, 301, 310, 315, 323, 325, 469, 471, 475, 480, 493, 494
Wu, C., 18, 470
Wyman, J., 271, 275, 481
Wyngaarden, J.B., 12, 13, 17, 327, 328, 329, 334, 346, 353, 356, 359, 360, 361, 363, 478, 488, 494

Yamaoka, N., 359, 474
Yan, J.X., 9, 493
Yang, S.T., 274, 494
Yao, L., 174, 494
Yasuda, J., 171, 493
Yates, F.E., 3, 4, 6, 57, 97, 494
Yeargers, E.K., 131, 494
Yeh, L.-S.L., 18, 470
Yi, P.N., 50, 492
Yoshii, J.-J., 400, 482
Yu, A.L., 343, 482
Yu, J., 343, 482
Yu, S., 175, 411, 492, 494
Yunus, I., 15, 487

Zhang, Z., 174, 495
Zimmer, C., 57, 293, 495
Zoref-Shani, E., 359, 495

Subject Index

"&&" declaration (of independent variable), 121, 240, 376, 393, 394
"at" (@) statement, 110

absolute change, 22, 55, 68–69, 71, 225, 230
abstraction, 57, 326
accumulation, 46, 64, 77, 109, 128, 132, 133, 167, 262, 295, 345, 402, 408
accuracy, 58, 59, 67, 83, 92, 106, 107, 148, 149, 157, 158, 226, 315, 319, 336, 340, 357, 388, 395, 398, 404–410, 419, 444, 450, 461
acetate, 249
acetyl-CoA, 188, 297, 308, 309, 310, 311, 313, 314, 320, 457
aconitase, 312, 314
activation and activators, 54, 55, 115, 126, 135, 183, 215, 327, 444 (*see also* cross-activation)
 allosteric, 273, 275, 346
 in anaerobic fermentation pathway, 262
 branched pathway with, 236
 of carbamoyl phosphate synthetase, 153
 of degradation, 220
 feedback, 198, 206, 220
 feedforward, 279
 of gene regulation, 400, 401, 402, 407
 inhibition compared, 4, 220
 kinetic orders, 145
 in maltose catabolic system, 401
 notation, 14, 27, 55, 349
 of pancreatic zymogen, 29–31
 phosphorylase cascade, 86
 of purine metabolism, 334, 349
 of reaction, 14, 27, 28, 86, 112–113
 strength of, 216, 380
 time-dependent, 295
ad hoc models, 58
ad hoc symbols, 78

additive, 3, 219, 406
adenine (Ade), 56, 146, 247, 331, 333, 335, 336, 338, 340, 342–345, 347, 351, 352, 363
 excretion, 146, 149–150, 338, 346, 364
 oxidation, 348
adenine phosphoribosyltransferase (APRT), 328, 329, 334, 341, 348, 349
adenosine (Ado), 56, 328, 333, 336, 343, 347, 352
adenosine deaminase (ADA), 331, 334, 348
adenosine diphosphate (ADP), 27, 56
adenosine monophosphate (AMP), 56, 90, 128, 263, 271, 272, 274, 275, 290, 328, 330, 331, 333, 335, 336, 346, 347, 349, 351, 352, 454
adenosine triphosphate (ATP), 12, 14, 27, 28, 56, 90, 128, 150, 152, 153, 167, 168, 187, 188, 247–249, 262–264, 267, 269, 270–272, 274–278, 283, 290–292, 297, 327, 328, 330, 333, 336, 342, 343, 346–348, 352, 362, 366, 454, 458
adenylates, 56, 128, 263, 274, 283, 330, 331, 335, 340, 342–346, 351, 352, 354–356, 363, 454, 458
adenylate kinase, 263
adenylosuccinate (S-AMP), 56, 330, 331, 333, 336, 338, 346, 347, 352
adenylosuccinate lyase (ASLI), 348
adenylosuccinate synthetase (ASUC), 334, 348
adrenaline, 6, 38
aerobic ATP formation, 247
affinity, 109
Affymetrix, 18
aggregated kinetic order, 85, 87, 267, 338, 339, 342, 355, 357
aggregation of fluxes, 83, 87, 91–92, 119, 288
aggregation of terms, 265, 267, 272, 351, 448

alanine (ALA), 299, 309, 310, 311, 313, 315, 319, 320, 322, 456
alanine transaminase, 300
alcohol, 2
algebra, 62, 167, 173, 191, 200, 209, 252, 391, 395, 406, 407, 413, 414, 426
  Boolean, 63
algebraic analysis (*see* algebraic steady-state evaluation)
algebraic steady-state evaluation, 61
  branch point constraints, 393–395
  examples, 212–216
  exercises, 217–221
  of glycolytic–glycogenolytic pathway in perfused rat liver, 365–398
  kinetic order sensitivities, 395
  of linear pathway with feedback, 84, 212–213
  local stability, 208–216
  log gains of fluxes, 385–388
  log gains of metabolites, 384–385
  matrix notation, 200
  of irregular S-systems, 206–208
  of pathway with feedback and cross-activation, 213–216
  precursor–product relationships, 391–393
  rate constant sensitivities, 390–395
  of regular S-systems, 83, 119, 121, 201–206
  steady-state equations, 193–198, 380–384
  transition time, 216–217, 389–390
  two-variable system, 212
  unconstrained parameters, 391
  zero steady states, 198–200
alginate-immobilized cells, 260
allele, 149
alignments, sequence, 18
allometry/allometric relationships, 55–56, 57, 70–73, 74
allopurinol, 353, 360, 361, 364
allosteric
  activator, 273, 275, 346
  enzymes, 163, 271
  inhibition, 76
  Michaelis–Menten reaction, 302
alphabet, 62
α-ketoglutarate, 31, 32, 165, 299, 308, 310, 313, 314, 319, 443
α-ketoglutarate dehydrogenase, 300, 314
α-term, 52, 54, 67, 70, 73, 77, 78, 201, 210, 219, 229, 240, 277, 445, 449
amendment of models, 32, 81, 128, 295, 297, 326, 331, 343, 344, 346, 351, 361
amidophosphoribosyltransferase (ATASE), 329, 334, 340, 348, 361
AMP deaminase (AMPD), 334, 340, 345, 348, 355–356, 357, 363
amphibolic pathways, 25, 92
amplification/amplifiers, 86, 222, 312, 314, 322
amplitudes, 110, 111, 116, 117, 118, 121, 125, 137, 409, 449
anaerobic fermentation pathway, 260–261
  activators, 262
  aggregation of β-terms, 277–278
  analysis of model, 279–284
  ATP dynamics, 275–277
  constraints, 287–289
  data, 268
  dynamics, 282–283
  enzyme concentrations, 287–288
  exercises, 290–292
  expansion of model, 284
  features, 261–262
  flux maximization, 286–290
  fructose-1,6-diphosphate degradation, 273–275
  glucose-6-phosphate degradation, 271–273
  glycogen synthetase reaction, 272–273
  GMA equations, 265–268
  hexokinase step, 269–271
  linear programming, 285–286
  logarithmic gains, 280–282
  mass conservation, 288–289
  metabolite concentration, 288
  objective function, 286–287
  optimization, 284–286, 289–290
  parameter values, 268–277
  phosphofructokinase reaction, 271–272
  properties of model, 279–284
  pyruvate kinase reaction, 275
  robustness of model, 283–284
  sensitivities, 282
  S-system representation, 277–278
  steady state, 280, 287
  sugar transport, 268–269
  variables defined, 262–265
analytical evaluation, 14, 83
analytical solution, 65–66, 75, 210, 350, 353, 406, 410
animal experiments, 8, 362, 365, 401
antagonistic interactions, 11
apparent first-order rate constant, 308
approximate character, 83
approximation
  in biochemical system models, 37–38, 57, 88, 90, 142, 144, 147, 160, 164, 184, 225, 226, 290, 338, 357, 360, 403, 404, 409, 414, 417–421, 425, 426, 440, 447, 448, 450, 466, 467
  of degradation, 59
  second-order, 406
  theory, 56, 57
arcs, 64

SUBJECT INDEX

arginine, 401
Argonne National Laboratory, 16
aromatic amino acid, 20
arrayer, 17
arrows, 151, 315
   activation, 349
   arched, 135
   bi-directional, 28–29
   curved, 349
   dashed, 443
   dotted, 443
   double-headed, 25, 26, 27, 28–29, 346
   double-tailed, 27, 28–29, 87, 346
   feedback control, 4, 13, 251
   flow, 14, 20, 23, 25, 27, 28–29, 31, 93, 130, 133, 251, 443
   flux, 25, 26, 27
   improper diagrams, 14, 31, 443
   inhibition, 349
   modulation, 26, 27, 31, 155
   signal, 443
artificial intelligence, 61
aspartate (ASP), 32, 299, 300, 309, 310, 315–317, 319, 325, 457, 458
aspartate β-semialdehyde, 32
aspartate transaminase, 300, 308
aspartyl phosphate, 32
Aspergillus niger, 162, 167–171
associative, 3, 23, 414, 439
assumptions, 37, 38, 39, 49, 58, 72, 128, 139, 144, 171, 178, 241, 283, 307, 322, 325, 326, 330, 343, 363, 379, 391, 396, 406
   quasi-steady-state, 37
asymptomatic, 364
ATP turnover, 187, 247, 248, 249
Australian National Genomic Information Service (ANGIS), 18
autogenous circuits, 400
automata theory, 63
auxiliary variables, 102, 138, 220, 370

β-5-phosphoribosyl amine, 32, 329–330, 331
β-terms, 23, 50, 51, 52, 70, 73, 77–80, 82, 108, 194, 201, 210, 219, 229, 237, 240, 244, 247, 265, 267, 287, 446, 450, 462, 463
   aggregation, 277–278
backward reaction, 98
bacteria, 8, 11, 17, 295, 327, 405
baker's yeast, 8
balance, 52, 72, 95, 195, 196, 283, 296, 317, 323, 333, 334, 353, 370, 438
balance equation, 61, 93, 94, 267, 371
baseline, 109, 112, 114, 116, 140, 141, 142, 148, 228, 232
baseline experiment, 52, 115

batch processes, 22, 23, 412
best fit, 151, 174
best-case scenario, 140
bi bi, 301, 302
bicyclic glutamine synthase cascade, 86
bi-directional arrows, 28–29
bi-directional processes, 315, 325
bimolecular reactions, 53, 87, 88–89, 171
biochemical
   interpretation, 8, 9
   map (*see* maps, biochemical)
   parameters, 63 (*see also* parameters)
   reactions, 55
   signals, 9, 14, 17, 47, 53, 63, 86, 93, 187, 248, 284, 314, 320, 330–331, 397, 444
   time scales, 6, 209, 307, 403, 404
biochemical system models, 7
   alternatives to S-system models, 58–64
   approximations, 37–38
   analytical convenience, 58
   change in dependent variables, 39, 41–49, 51, 56
   computer simulations, 38–39
   differential equations, 42, 43, 45, 47, 48, 65–67
   examples, 52, 54–55
   exercises, 64–65, 75
   integrated, 39, 40–41
   kinetic orders, 37, 53–54, 55
   linear relationships, 56
   mechanistic approach, 39
   networks as languages, 62–64
   parameters, 39, 51–55
   power-function representations, 59–60
   rate constants, 51–52
   rate laws in, 37–43, 47
   relative changes in flux, 73–75
   S-system form, 49–51, 55–58, 65–75
   stoichiometric network form, 60–62
   theoretical justification, 57–58, 60, 70–73
   validity, 55–57
   variables, 17, 24, 25, 38, 39, 48, 56
bioinformatics, 8, 9, 16, 18, 19
biomass allocation, 404
biomolecular structures, 18
biomolecular time scales, 6
biotechnological optimization, 83, 260, 285, 390, 397
biotechnology, 18, 22, 62, 64, 223, 284, 287, 290, 400
bi-phasic response, 107
bi-substrate reaction, 46, 47
blood (*see also* res-blood-cell)
   clotting, 86
   drug concentrations in, 133, 134

blood (*cont.*)
  glucose in, 24, 366
  serum, 40, 208, 357
  uric acid in, 327
BMDP AR®, 172, 178
body weight, 139, 150, 337
Boehringer Mannheim's Biochemical Pathways Module, 16
Boehringer Metabolic Map, 15
bolus experiments, 103, 107–108, 109, 125, 128, 130–131, 133, 134, 142, 307, 308, 309, 320, 324, 377, 378, 457, 461, 462
bookkeeping, 3, 47, 94, 427
Boolean algebra, 62–63
Boolean network, 63
Boolean rule, 63
bound form of an enzyme, 21, 90, 171
brain, 3, 159, 335, 336–337
branch
branch point(s), 12, 17, 27, 32, 119
  aggregation of terms at, 351
  constraints, 82–86, 89, 111, 191, 267, 371, 374, 379, 393–396, 446, 447, 456
  convergence, 25, 26, 27, 77, 95, 447
  didactic pathway, 256
  divergence, 25, 26, 27, 77, 95, 267, 323–324
  in fermentation pathway, 273
  fluxes at, 17, 82, 119, 184, 176, 288, 314, 315, 323, 369, 391, 393
  malate, 314, 315, 323–324
  modulated, 87, 89
  pathways with (*see* branched pathways)
  precursor–product relationships at, 176, 369, 371
  of purine metabolism, 328
  sensitivities at, 393
branched electrical circuits, 48
branched pathways, 17, 127, 449
  with activation, 236
  bolus experiments with, 134
  degradation in, 191
  dynamic data, 176–179
  equation construction, 77, 81–86, 89
  with feedback, 111–115, 125, 126, 202
  with inhibition, 81–82, 126
  oscillatory patterns in, 111
  parameter estimation, 176–179
  and productivity of systems, 131–132, 134, 141
  representation of, 12, 13, 448
  stability of, 126
  steady state, 119, 213
buffer/buffering, 57, 135–137, 142, 369
buffer boxes, 91, 134–137, 323, 415
buffer constant, 136

Caenorhabditis elegans, 8, 16
cAMP, 293
canonical modeling, 65, 260, 293, 296, 399
  biological applications, 400–404
  input modules, 133–134
  mathematical research, 404–412
canonical structure, 326, 343
carbamoyl phosphate, 153, 154
carbamoyl phosphate synthetase, 153, 154, 155
carbohydrate metabolism, 167, 168, 296, 297, 365
carbon flow, 14
carbonic acid, 2
carboxypeptidase, 29–30
cardiovascular system, 6, 133
carrying capacity, 123
Cartesian coordinates, 146, 147, 149, 154, 202, 270, 383, 388, 419–420, 425
cascades
  equation construction, 86–87
  with feedback, 86
  mechanism, 29, 86, 179
  parameter estimation, 179–184
cause-and-effect, 3–4
CBC Bibliography Database, 18
cDNA, 17–18
cell cycle, 7, 63
cell population model, 7
cellulose, 293, 294
chain rule, 74, 138, 252, 257, 282, 393, 448
changes (*see also* perturbations)
  in dependent variables, 39, 41–49, 51, 56, 108, 109, 121, 122–123, 246, 377–378
  in independent variables, 108–109, 122, 378–379
  in input, 116, 353–354
  instantaneous temporal, 41, 42, 67
  in modulation, 39
  in parameters, 244–245
  in rate constants, 45
  in values, 109–110, 379–380
channels, 25, 82, 146, 171, 262, 330, 343, 406
chaos, 51, 57, 63, 410
characteristic polynomial, 211, 212, 213, 453
checkerboard position, 438
chemical energy, 14, 28
chemical groundplan, 56
chemical structure, 12, 13, 15
chemotherapeutic, 141
chemnodes, 64
chymotrypsin, 29–30, 160
chymotrypsinogen, 29–30
circadian variation, 141, 449
circuits/circuitry
  autogenous, 400

SUBJECT INDEX

branched electric, 48
classical, 400
gene regulation, 4, 158–159, 400, 401–402
citrate, 188, 308, 309, 312, 314, 322, 323, 450
citrate synthase, 149
citric acid, 167
citric acid cycle, 296
citrulline, 152–155
clamped pathways, 92, 96
clamping experiment, 128
classical circuits, 400
classification of functions, 77, 406, 410
classificatory concepts, 19
clearance, 77, 133, 360
cloning, 17
closed systems, 93–94, 95, 207, 294, 443, 444
CO_2 production, 294, 297, 298, 300, 324
cobalamin, 32
coefficients (in steady-state equation), 197–198, 200, 204, 205, 206, 211, 212, 213, 219
coenzymes, 15, 26, 27
cofactors, 11, 15, 27, 32, 63, 64, 165, 298, 300, 313, 315, 324, 325
combined pool, 128, 296
common names, 19, 20
competition, 57, 189, 407, 408
competitive inhibition, 24–25, 76, 154, 161, 274, 301, 302, 334
complex
  conjugate, 280, 340
  numbers, 124, 209, 210, 213
  oscillations, 51
  plane, 124
  systems, 3, 4, 6, 9, 11, 38, 97, 118, 327, 403, 411, 426
complexes, 7, 39, 171, 295, 300, 314, 341
complexity, 3, 4, 5, 6, 7, 10, 11, 13, 38, 58, 90, 97, 129, 172, 262, 295, 296, 323, 357, 399, 425
computational biology, 8–9
computer-aided analyses, 63, 103–105
computer simulations, 38–39, 97–99
  Bolus experiments, 107–108
  branched pathway with feedback, 111–115
  buffer boxes, 134–137
  and computer-aided analyses, 103–105
  condensation of pools, 129–131
  controlled comparisons, 114–118
  defined, 97
  dynamics of the system, 105–119
  exact analogues of other models, 137–138
  examples, 111–119
  exercises, 125–127, 140–142
  in formal language networks, 62, 63
  of glucose-6-phosphate metabolism, 98–99, 103–105

  input changes, 353–354
  logarithmic gains, 120–122
  model refinements, 127–129
  modification of system characteristics, 109–110
  Monte Carlo simulation, 138–140
  parameter sensitivities, 122–123
  persistent changes, 108–109
  phase-plane plots, 110–111
  productivity, 131–133
  purine metabolism, 340–341, 353–355
  simple pathway with surprise, 115–119
  specification of S-system equations, 99–103
  stability, 123–125
  steady-state analysis, 105, 119–127
  steps in, 98
  temporary perturbations, 354–355
  time-dependent inputs, 133–134
  of transient responses, 105–107
  tricarboxylic acid cycle model, 319–322
condensation of pools, 129–131
connectivity relationships, 61, 246–249
conservation law, 199–200, 207, 246, 247
conservation of flux, 79, 82, 391
conservation of mass, 46, 257
  constraints, 90–91, 288
consistency, 55, 57, 133, 367, 392, 395
  analysis, 246, 260, 306–315, 323, 375–380
  between models and data, 40, 57, 138, 279, 340, 344, 353, 359, 369–370
constrained parameters, 87, 71, 191–192, 241, 243, 245, 287–288
constrained systems, 239–241, 257–258
constraint(s), 95, 163, 184, 215, 216, 240, 258, 263, 264, 285, 291, 455, 457
  in bimolecular reactions, 88–89
  branch point, 82–86, 89, 111, 191, 267, 371, 374, 379, 393–396, 446, 447, 456
  for conserved masses, 90–91, 288
  derivation, 87–91
  enzyme concentration, 287–288
  equations, 72, 79–80, 82–91, 95, 99, 191–192, 287, 370–372, 395, 447
  in glycolytic–glycogenolytic pathway model, 370–372
  precursor–product, 79–80, 82, 83, 99
  steady-state, 72, 191–192, 287
  stoichiometric, 83, 90, 191
contraindication, 1
control, 8, 9, 12, 13, 21, 22, 58, 60, 63, 83, 86, 107, 108, 113, 149, 150, 223, 245, 248, 249, 295, 315, 322, 323, 327, 329, 331, 395, 400, 401, 408, 409, 415
control structure, 8, 12, 13, 327, 395
control variables, 22

controlled mathematical comparisons, 58, 114–118, 127, 379, 400, 405, 407–408
convergence branch point, 25, 26, 27, 77, 95, 447
converging pathway, 77, 87, 95
cooperativity, 187, 248
correlation, 9, 19, 140, 296, 401
cosine waves, 110, 111, 125
counterdomain, 415
counterintuitive, 4, 41
Cox model, 411
Cramer's procedure, 200, 436
Cramer's rule, 201, 202, 203, 204, 207, 208, 236, 238, 435, 436, 438, 468
criterion of optimality, 285, 389
critical point, 409
cross-activation, 213–215, 328
cyanide, 248
cybernetics, 5
cycle (*see also* tricarboxylic acid cycle)
   cell, 7, 63
   citric acid, 296
   flux, 314, 320
   futile, 262
   life , 293, 315
   limit, 114, 118, 119, 125, 126, 219, 409, 448
   phase-plane plots, 118
cycling between model design, 327
cytoplasm, 297

damped oscillation, 114, 116, 448
data (*see also* dynamic data)
   steady-state, 146–158
databases, 15–16, 18–19 (*see also specific databases*)
DBGET/LinkDB, 15–16
decisiveness, 402
deficiency, 17, 31, 41, 283, 327, 328, 329, 341, 345, 356–361, 363, 364, 405, 443, 447
degradation, 74
   activation of, 220, 214
   adenine, 346
   adenylates, 330, 356
   ADP, 458
   aggregation of terms, 265, 267, 272, 448
   approximation of, 59
   ATP, 267–268, 292, 458
   in branched pathways, 191
   buffering and, 135
   DNA/RNA, 56, 346, 351
   drugs, 133
   ethanol, 291
   flux, 85, 191
   functions, 59
   fructose-6-phosphate, 109, 273–275
   glucose-6-phosphate, 167–171, 271–273, 278, 391
   glycogen, 365, 398
   in GMA form, 265, 267, 355, 362, 448
   inhibition, 52, 195, 206, 220, 346, 355
   isocitrate, 165–167, 322
   kinetic order, 55, 127, 169, 304, 339, 342, 448
   metabolite, 77, 78
   phosphoenolpyruvate, 262
   power-law rate law, 167, 168, 451
   processes, 26, 31, 45, 49, 53, 54, 72, 79, 135, 265, 312
   product, 94
   of proteins, 294, 314, 315–316
   PRPP, 12, 334, 335, 349, 351, 362, 363, 461
   of purine, 330
   of pyruvate, 457, 461
   rate constants, 50, 123, 267
   of SAM, 342, 343
   speed, 45, 52, 57, 75, 164, 240
   in S-system form, 265, 267, 277, 351, 362, 363, 448, 461
   in splitting reactions, 90
   substrate, 46, 75, 90
   of succinate, 188, 305, 451
   term, 50, 51, 55, 67–68, 70, 77, 87, 88, 108, 119, 135, 168, 317, 335, 360, 457, 461
   of xanthine, 355, 360, 363
degrees of freedom, 191, 411
delay equations, 131
demand theory, 400, 401
deoxyadenosine (dAdo), 331, 333, 336, 347, 352
deoxyadenosine diphosphate (dADP), 333, 336, 347, 352
deoxyadenosine monophosphate (dAMP), 333, 336, 347, 348, 352
deoxyadenosine triphosphate (dATP), 333, 336, 347, 348, 352
deoxyguanosine (dGuo), 331, 333, 336, 347, 353
deoxyguanosine diphosphate (dGDP), 333, 336, 347, 352
deoxyguanosine monophosphate (dGMP), 333, 336, 347, 348, 352
deoxyguanosine triphosphate (dGTP), 333, 336, 347, 348, 352
deoxyinosine (dINO), 331, 333, 347
deoxynucleotides, 327, 331
dependent variables, 262–264, 268, 275, 284, 285, 291, 306, 307, 311, 324, 332, 337, 343, 346, 362, 368, 376, 379, 381, 384, 385, 387, 388, 391, 392, 397, 398, 406, 409, 415, 416, 421–423, 434, 445, 457, 464
bolus experiments, 108
for branched pathways, 82, 112, 141, 177, 178
in cascades, 86, 180–182

SUBJECT INDEX

change in, 39, 41–49, 51, 56, 105, 108, 109, 121, 122–123, 246, 377–378
closed vs. open systems, 93, 94
conserved masses and, 90
defined, 21
differential equations, 48
enumeration of, 76–77
equation construction, 48, 52, 54, 77–79
flux sensitivity, 244
in glycolysis, 28–29, 370, 377–378
independent variables distinguished, 22, 23
intermediate, 22, 131
for linear pathways, 115
logarithmic gain, 224, 225, 227–231
mapping, 24, 27, 28, 29–30, 31
notation, 59
parameter estimation, 122–123, 143, 144, 172, 173, 175, 177, 178, 180–182, 190
phase-plane plots, 110, 111
for productivity assessment, 131–133
replacement with independent variable, 128
rate constant sensitivities, 237
reduced set of, 257, 258
sensitivity analysis, 224, 225, 227–231, 237, 244, 246, 247, 249, 250, 252–254, 257, 258
S-system, 193, 196, 197, 198, 205, 206, 209, 212, 217, 220
steady-state values, 119, 120, 121, 122, 123, 124, 193, 196, 197, 198, 205, 206, 209, 212, 217, 220
sum of, 102
value, 22, 52, 102, 119, 120, 121, 122, 123, 124, 127
in zymogen activation, 29, 30
derivatives, 38, 42, 44, 56, 65, 110, 136, 138, 147, 159, 161, 165, 189, 196, 198, 217, 220, 25, 227, 230, 234, 235, 242, 244, 251, 253, 254, 255, 258, 260, 287, 328, 406, 418, 419, 420, 451
logarithmic, 156, 191, 252, 450
partial, 88, 89, 166, 167, 170, 225, 230, 241, 304, 318, 394, 414, 424, 425, 440, 460, 461
slopes instead of, 174–184
design principles, 3–4, 7
design process, 8, 9, 16, 17, 262, 395–396
determinants, 193, 200–204, 206–210, 213, 216, 218, 220, 239, 382, 414, 435, 437, 440, 453
expansion, 211, 214, 432, 434, 436, 442, 467–468
linear algebra, 430–434
minors, 433, 438
total, 252–255
determinism, 3
deterministic, 1, 410
detoxification, 136

developmental time scales, 6
diabetes, 6, 366
diagnostic analysis, 297
dichotomous effect, 12
dictionary, 20, 28, 29, 298
Dictyostelium discoideum, 129, 165–167, 171–172, 188, 223, 293–325
didactic pathway, 115, 127, 176, 256, 405
differential equations, 8, 42–48, 51, 56–59, 65–68, 75, 77, 94, 99, 101–103, 107, 123, 127, 135, 137, 138, 172–175, 178, 193–195, 203, 209, 220, 224, 250, 256, 264, 295, 297, 350, 357, 363, 389, 404, 406, 407, 409, 410, 417, 444, 445, 449, 451, 458
differential, total, 155, 424
differentiation, 140, 144, 159, 166, 190, 193, 219, 229, 233, 236, 241, 243–245, 247, 248, 251, 257, 270, 291, 294, 295, 297, 315, 392, 418, 425, 446–448, 455 (*see also* partial differentiation)
digitonin technique, 151, 152
dihydrolipoyl dehydrogenase, 300
dihydrolipoyl transacetylase, 300
dihydroxyadenine, 364
directed connections, 64
directed graph, 61, 63
diribonucleotide reductase (DRNR), 334, 348
disease pattern, 327, 328, 396
disease state, 138, 193, 356, 360
disease, metabolic, 345
dissociation, 39
distribution, 46, 139, 294, 314, 315, 327, 338, 410, 411, 412
divergence branch point, 25, 26, 27, 77, 95, 267, 323–324
diverging pathway, 95, 267, 288
DNA, 63, 145, 327, 330, 347, 348, 352, 353
  chips, 17, 18
  degradation, 56, 346, 351
  microarrays, 17–18
  model, 6
  recombinant, 287–288
  sequencing, 8, 18
  synthesis and degradation, 56
DNA Data Bank of Japan (DDBJ), 18
domain, 151, 415
dot notation, 42, 44
double-headed arrow, 25, 26, 27, 28–29, 346
double modulation, 157–158
double-tailed arrow, 27, 28–29, 87, 346
Drosophilia melanogaster, 16
drugs
  degradation of, 133
  inter-individual variability of response to, 9, 412

drugs (*cont.*)
  intervention simulation, 141, 357, 360–361
  metabolism, 133, 134, 137, 141
  resistance, 141
  screening/testing, 1, 7, 8, 327, 362, 399
  uricosuric, 367–368, 462
dynamic data, 143
  branched pathway, 176–179
  cascades, 179–184
  direct estimation from, 173–174
  parameter estimation from, 172–185
  regression, 173–174
  slopes instead of derivatives, 174–184
dynamic response, 80, 97, 112, 114, 142, 143, 182, 283, 325, 330, 364, 397
dynamical exploration, 342
dynamical solution, 84, 85, 119, 120, 122, 124, 126, 280, 353, 361, 448, 456
dynamics, 60, 61, 68, 78, 90, 94, 103, 128, 132, 140, 141, 144, 146, 155, 171, 172, 196, 198, 199, 200, 219, 221, 260, 262–264, 291, 292, 327, 330, 331, 333, 340, 342, 343, 346, 351, 357, 389, 398, 403, 406, 444, 445, 448, 449, 452, 453, 457, 461
  anaerobic fermentation pathway model, 282–283
  ATP, 275–277
  of biochemical systems, 41–49, 56, 282–283
  bolus experiments, 107–108
  closed vs. open systems, 93
  of components, 5
  computer simulation, 105–119, 125
  controlled comparisons, 114–118
  enzyme-catalyzed reactions, 43–44, 64
  feedback, 111–115
  modification of system characteristics, 109–110
  molecular level, 5, 38, 39
  of networks, 63
  of organisms, 5
  pathway, 38, 111–115
  persistant changes, 108–109
  of Petri nets, 64
  phase-plane plots, 110–111
  simple pathway with surprise, 115–119
  in S-systems, 54–55, 282–283
  steady state, 124, 130–131
  time scale and, 6
  of transient responses, 105–107

EC Enzyme Database, 15
EcoCyc, 16
ecosystems, 11, 400
educated guess, 144, 345
efficiency, 8, 19, 66, 67, 86, 137, 174, 216, 232, 314, 323, 402, 406, 419
efflux, 70, 73, 91, 133, 144, 164, 185, 194, 216, 217, 232, 267, 303, 305, 318, 319, 331, 445, 447
eigenvalues, 124–126, 209–213, 216, 280, 289, 307, 309–311, 319, 320, 324, 375, 376, 379, 380, 402, 448
  complex conjugate, 340
eigenvector, 209
Einstein's theory of relativity, 37
EKG, 125
elastase, 29–30
elasticities, 148, 153
electric circuit, 48
elemental chemical kinetics, 44, 46, 53, 55
EMP Enzymology Database, 15
enabling transitions, 64
end-product inhibition, 80–81, 111, 122
endonuclease, 346
energy
  chemical, 14, 28
  storage, 328, 365
enteropeptidase, 30, 31
environment, 5, 8, 24, 92, 103, 109, 117, 123, 222, 294, 401, 402, 407, 411, 412
enzyme(s), 7, 11, 16, 23, 25, 31, 32, 40–42, 49, 59, 62, 64, 68, 105, 125, 135, 138, 144, 146, 150, 155, 157, 159, 160, 163, 165, 168, 233, 246, 262, 264, 268, 274, 278, 284, 286, 289, 295–298, 300, 305, 312, 314, 315, 322–324, 327–329, 334, 340, 341, 343, 344, 348, 363, 368–370, 395, 396, 399, 405, 407, 421, 447
  activity, 17, 122, 139, 140, 148, 149, 170, 188, 222, 271, 273, 276, 287, 310, 311, 313, 320, 355–360, 360, 372, 373, 375, 443
  affinity, 109
  binding, 39
  bound form, 21, 90, 171
  concentration constraints, 287–288
  databases, 15
  deficiency, 356, 443
  dissociation, 39
  free form, 21, 90
  graphical representation, 20–21, 27
  in models, 38, 76–77
  overexpressed, 290
  polymorphisms, 9, 17
  scaffold, 171
  replacement, 362
  unbound form, 90, 171
enzyme-catalyzed processes, 11, 12, 25, 38, 43–44, 53, 56, 64, 298, 406
enzyme–enzyme interaction, 246
Enzyme Reaction Mechanism (ERM) database, 16
equation construction
  bimolecular reactions, 87, 88–89

SUBJECT INDEX

branch-point constraints, 82–86, 89
branched pathways, 77, 81–86, 89
cascades, 86–87
closed systems, 93–94
conserved masses, 90–91
constraint derivation, 87–91
dependent variables, 48, 52, 54, 77–78
end-product inhibition, 80–81
enumeration of variables, 76–77
examples, 78–87
exercises, 94–96
linear pathways, 78–81
open systems, 93–94
precursor–product constraints, 79–80
purine metabolism model, 332–338
reversible pathways, 91–93
S-system representations, 70, 74, 76, 77, 78–80, 82–84, 92, 94, 99–103, 277–278, 338–340, 375
stoichiometric constraints, 90
tricarboxylic acid cycle, 298–306
equilibrium, 163, 262, 263, 264, 271, 272, 434
thermodynamic, 3, 92, 163, 405, 462
equilibrium constant, 22, 162, 169, 323, 325, 367, 368, 372, 373, 374, 396, 398
equivalence, 79, 90, 232, 246, 316, 317, 379, 407, 408, 461
external, 114–115
equivalent, 43, 57, 66, 73, 79, 88, 89, 91, 93, 94, 130, 137, 147, 148, 184, 185, 204, 210, 225, 229, 248, 255, 267, 277, 306, 319, 371, 378, 379, 382, 389, 391, 397, 401, 404, 410, 422, 447, 449, 461
systems, 115
error structure, 148
Escherichia coli, 3, 16, 62, 293, 400, 401
essence of the system, 40
essentially irreversible reaction, 25, 28, 91, 168, 298, 300, 304, 329, 331, 367
ESSYNS, 67, 407
estimation, 92, 222, 265, 268, 269, 291, 307, 328, 333, 340, 351, 365, 391, 393, 395, 396, 397, 398, 404, 421, 426, 427 (*see also* parameter estimation)
ethanol, 260, 261, 262, 263, 267, 273, 275, 281, 285, 286, 288, 289, 290, 291
ethics, 8
eukaryotes, 8
Euler's algorithm, 66–67, 74
European Bioinformatics Institute Macromolecular Structure Database (EBI/MSD), 18
European Molecular Biology Laboratory (EMBL) database, 18
evolutionary, 6, 160, 222

exact analogues, 137–138
exact model, 50
exact representation, 418
Excel®, 228
exhaustive exploration, 139, 174
exogenous
bolus, 103, 108, 109, 122, 377, 397
influences, 119, 125, 126
products, 285
substrate, 41, 93, 94
supply, 93, 94, 109, 122, 187, 354
exonuclease, 346
expansion, 54, 131, 202, 203, 226, 284, 296, 366
of determinants, 211, 214, 432, 434, 436, 442, 467–468
ExPASy Molecular Biology Server, 16, 19
exponential function, 44, 53, 65–66, 66, 134, 137, 195, 381, 439, 466
expression profile, 9, 18
external equivalence, 114–115, 316, 319, 379, 407, 408
external variables, 22
extrapolation, 126, 149, 171, 226, 403

factorial, 418
feedback
activation, 198, 206, 220
branched pathways with, 111–115, 125, 126, 202
cascade with, 86
controls, 13
inhibition, 3, 4, 14, 32, 57, 80–82, 113, 115, 126, 157, 176, 208, 220, 243, 256, 279, 329, 334, 405
linear pathway with, 212–216
regulation, 86, 329, 405
feedforward activation, 279
fermentation, 2, 260, 261, 273, 275, 285, 290, 292, 399, 408 (*see* anaerobic fermentation pathway)
F-factor, 210, 211, 214
final time, 45, 80, 84, 103, 104, 105, 117, 378, 458
finish time, 44, 103
firing transitions, 64
first-order kinetic processes, 44, 53, 65, 145, 176, 276, 295
first-order rate constant, 308
fitness, 174, 401, 408
flow
arrows, 14, 20, 23, 25, 27, 28–29, 31, 93, 443
carbon, 14
of currents, 48
of information, 14
of material, 12–14, 20, 23, 24, 31, 38, 40, 82, 86, 145, 304

flux(es), 151–154, 158, 160, 164, 165, 170, 171, 183, 223, 233, 246, 248, 250, 260, 261, 267, 281, 282, 285, 290, 291, 303, 304, 310, 311, 322, 336, 340–344, 346, 354–357, 361–363, 368, 373, 376, 379, 397, 408, 445, 447 (*see also* effluxes; influxes)
  aggregation of, 82, 83, 87, 91–92, 119, 288
  arrows, 25, 26, 27
  back, 25
  at branch points, 17, 82, 119, 176, 184, 288, 314, 315, 323, 369, 391, 393
  changes in, 40
  conservation of, 79, 82, 391
  cycle, 314, 320
  degradation, 85, 191
  equivalence, 73, 130
  feedback inhibition and, 115, 149
  GMA models, 253–256
  glycolytic–glycogenolytic pathway, 385–388
  kinetic orders and, 82, 145, 148, 255–256
  logarithmic, 71, 146, 147, 148, 156, 190, 253, 325, 390
  log gains, 229–232, 253–254, 385–388
  maximization of, 286–289
  metabolite changes and, 62, 70, 74
  net, 25, 91, 92, 162, 163, 168, 169, 184, 185, 315, 316, 317, 457, 458
  parameter changes and, 244–245
  purine metabolism, 337–338
  rate constants, 78–79, 82, 254–255
  rates, 42, 78–79, 254–255
  relative changes in, 73–75, 147, 148
  reversal, 25, 92, 405
  secondary reactions and, 129
  sensitivity, 229–232, 244–245, 253–256
  size, 17
  smooth functions, 70
  steady state, 73, 119, 130, 146, 147, 287, 337–338, 370
  stoichiometry of, 83, 191, 288
  as transformations, 84–85
FlyNets, 16
forest, 400, 403
formal language approach, 59, 62–64
forward flux, 25
forward reaction, 96, 162, 451
fractals, 129
fractal kinetic orders, 55
fractal kinetics, 406
free energy, 27
frequency, 110, 117, 125, 409, 449
frequency-modulated Rössler oscillator, 136
fructose diphosphate, 90
fructose-1,6-diphosphatase, 262
fructose-1,6-diphosphate, 273–275
fructose-2,6-biphosphate, 367

fructose-6-phosphate (F6P), 28, 100, 101, 109, 122, 125, 167, 168, 170, 187, 262, 263, 264, 271, 272, 273–275, 290, 367, 369, 370, 374, 376, 384, 391, 398
fruiting body, 293
fudge factors, 58
fumarase, 300, 323, 325
fumarate (FUM), 188, 298, 299, 304, 307, 308, 309, 310, 319, 320, 324, 451
function(s) (*see also* power-law functions)
  algebraic, 2, 42
  change, 42
  classification of, 77, 406, 410
  continuous, 417, 439
  defined, 42
  degradation, 59
  discrete, 416, 439
  explicit, 417
  exponential, 44
  of functions, 47
  general features of, 414–417, 421–424
  linear, 46
  monotone, 57, 417, 439
  rational, 38
  of several variables, 225, 421–424, 474
  of single variable, 414–421
futile cycle, 262

gain, 37, 97, 125, 205, 215, 262, 282, 291, 315, 387, 391 (*see also* logarithmic gains)
  profile, 229, 313, 321, 323, 324, 328, 403
  types, 223
game theory, 5
gamma (Γ, mass action ratio), 162, 169, 372, 373, 374, 398
Gauss®, 200
Gear method, 200, 444, 445
GenBank, 18
gene
  circuits, 4, 158–159, 400, 401–402
  expression, 15, 17–18, 63, 145, 149, 400, 401, 402
  regulation, 4, 18, 62, 86, 158–159, 179, 400–402, 407
  sequences, 15–16
general system theory, 5
General mass action model and GMA system, 76, 85, 94, 95, 97, 105, 112, 126–128, 144, 159, 165, 188, 194, 200, 219, 241, 260, 272, 275, 277–281, 288, 290, 291, 324, 325, 328, 333–335, 338, 339, 342, 344, 346, 347, 350, 351, 353–361, 363, 364, 405, 407, 410, 444, 445, 455, 459, 461
  anaerobic fermentation pathway, 265–268
  branch-point constraints, 82–84
  canonical, 65

SUBJECT INDEX

computer simulations, 103
concentrations, 252–253
constrained systems, 82–84, 257–258
degradation in, 265, 267, 355, 362, 448
dependent variables, 77–78
equation specification, 76, 77–78, 82–83, 265–268
examples, 256–257
exercises, 258–259
fluxes, 253–256
kinetic orders, 255–256
logarithmic gains, 252–254
mathematical structure, 59
metabolite sensitivity, 254–255
power-law terms, 77
rate constants, 254–255
sensitivity analyses in, 251–259
terminology, 251–252
uses, 60
genetic
algorithm, 174
grammar, 62
language, 62
manipulation, 8, 62, 148
network, 63
polymorphisms, 17
rules, 62
genome sequences, 8
genomics
comparative, 8–9
functional, 8–9
GenProtEC, 16
global properties, 140
global stability, 208
glucagon, 151, 152
glucokinase, 28, 29, 100, 367, 369, 370, 373, 390, 396
gluconeogenesis, 150, 152
glucose, 6, 24, 28, 29, 40, 100, 108, 120, 122, 167, 168, 187, 188, 261, 262, 264, 268–270, 280, 282, 285, 294, 296, 299, 365, 366, 367, 369, 370, 378–380, 386–389, 393, 397, 398, 415, 421, 422, 462
formation, 151, 152
glucose-1-phosphate (G1P), 28, 100, 101, 103, 107–109, 122, 295, 366, 367, 369, 370, 372, 373, 377, 378, 384, 393, 397
glucose-6-phosphate (G6P), 17, 24, 28, 100, 101, 125, 126, 162, 167–171, 187, 188, 261–264, 267–269, 290, 365, 366, 368–370, 373, 374, 379, 380, 384, 385, 393, 397
degradation, 110, 167, 271–273, 278, 391
metabolism, 98–99, 103–105, 108–109
parameter sensitivities, 122–123
steady-state values, 120, 121–122
glucose-6-phosphate dehydrogenase, 17
glutamate (GLU), 32, 188, 299, 300, 309, 310, 315, 319, 325, 457
glutamate dehydrogenase, 300
glutamate semialdehyde, 32
glutamic acid, 12
glutamine synthase, 86
glutamine synthetase, 38–39
glyceraldehyde 3-phosphate dehydrogenase (GAPD), 264, 274, 275, 291
glycerol, 249, 261, 262, 263, 264, 273, 274, 285, 286, 290
glycogen, 28, 29, 100, 261, 262, 264, 267, 272, 295, 367, 368, 369, 370, 372, 396, 398
glycogen phosphorolysis, 366
glycogen synthetase, 187, 271, 272–273, 291
glycogenolysis, 365, 366, 390
glycolysis, 22, 28, 29, 90, 187, 228, 229, 238, 243, 365, 366, 372
glycolytic–glycogenolytic pathway
algebraic analysis, 365–398
branch point constraints, 393–395
changes in dependent variables, 377–378
changes in independent variables, 378–379
consistency analysis, 375–380
constraints, 370–372
exercises, 396–398
fluxes, 385–388
kinetic orders, 372–375, 395
log gains, 376–377, 384–388
map construction, 365–368
metabolites, 384–385
parameter values, 368–375, 379–380
perturbations, 377–380
precursor–product relationships, 391–393
rate constants, 375, 390–395
robustness of model, 375–380, 379–380
sensitivities, 376–377, 390–395
S-system equations, 375
steady-state data, 368–370
steady-state equations, 380–383
steady-state solution, 375–376
symbolic steady-state analysis, 380–395
transition times, 389–390
unconstrained parameters, 391
GMA model, 60, 76, 82, 83, 94, 103, 127, 128, 144, 165, 188, 251, 267, 268, 272, 277, 278, 280, 288, 291, 325, 335, 338, 342, 351, 353–357, 359–363, 405, 444, 461
GMA system, 59, 60, 77, 78, 84, 95, 103, 105, 112, 126, 194, 200, 219, 241, 281, 350, 354, 362, 405, 407, 444, 448, 455, 459
GMP reductase (GMPR), 334, 348
GMP synthetase (GMPS), 348
Gompertz function, 220

Gompertz law, 410
gout, 327
gradient-based search, 174
graphical representation, 6, 11, 311, 416
    chemical structures, 12–13
    databases, 15–16
    directed graph, 61, 63
    enzymes, 20–21, 27
    exercises, 31–36
    maps, 12–19, 23–31
    parameters, 21–23
    PLAS solutions, 107, 108
    of rate laws, 16
    three-dimensional biomolecular structures, 18–19
    variables, 19–23
grammar, 62
growth, 6, 55, 56, 114, 141, 144, 159, 172, 174, 189, 220, 294, 406, 410, 411, 417
    canonical modeling of, 402–404
    characteristics, 7
    logistic, 123, 124
guanase, 330
guanine (Gua), 329, 331, 333, 334, 335, 336, 340, 341, 344, 345, 347, 351, 353, 356, 347, 348, 363
guanine hydrolase (GUA), 334, 348
guanosine (Guo), 328, 331, 353
guanosine diphosphate (GDP), 330, 333, 336, 346, 347, 352
guanosine monophosphate (GMP), 333, 347
guanosine triphosphate (GTP), 330, 333, 336, 346, 347, 348, 352
guanylates, 56, 346
guideline, 24, 97, 316, 326
Guidelines for Exposure Assessment, 140

Haemophilus influenzae, 8
half-life, 361
half-systems, 60
harmonic oscillations, 110–111, 125, 219
Heisenberg uncertainty principle, 2–3
hepatocytes, 415, 421, 422
heterozygote, 149
hexokinase, 167, 187, 269–271, 276, 280, 291
hexose–pentose shunt, 296
hide if absolute value is < ..., 122
hierarchical structure, 6
hierarchy of organization, 2, 4, 6
higher-order terms, 69
Hill coefficient, 163, 186
Hill equation, 74, 137
Hill function, 137, 138, 142
Hill parameter, 273
Hill process, 186

Hill rate law, 142, 273
histidine, 362, 461
holistic approach, 1
homocysteine, 32
homogenate, 148, 369
homogeneous form, 90
homogeneous media, 171
homogeneous structure, 50, 58
homogeneous systems, 246
homoserine, 32
homoserine kinase, 32
homoserine succinylase, 32
homozygotes, 149
hormonal control, 86
Human and Ecological Risk Assessment journal, 140
Human Genome Project, 9, 18
hybridization, 18
Hybridoma cells, 62
hydrogen bomb, 139
hydrolysis, 25, 27
hydroxyadenine, 364
hyperbolic function, 74, 75, 145, 398
hyperlysinemia, 31
hyperuricemia, 327, 328, 341, 353, 356
hypoxanthine (HX), 328–331, 333, 335, 336, 340, 345–347, 351–357, 360, 361, 462
hypoxanthine-guanine phosphoribosyltransferase (HGPRT), 328, 329, 334, 341, 345, 348, 349, 353, 354, 356–360, 361

if–then scenario, 7, 20
ill-defined, 14
imaginary part, 124–125, 210, 216, 307, 376
imbalance, 333, 360, 366, 370, 462
immune response, 396, 400, 405
immunology, 86, 179
IMP dehydrogenase (IMPD), 334, 348
implementation, 45, 53, 62, 63, 79, 101, 103, 123, 130, 134, 135, 136, 220, 307, 357, 387, 394, 456, 457, 462
improper maps, 32
in vitro–in vivo extrapolation, 171
in vitro experiment, 6, 24, 25, 39, 68, 144, 159, 160, 161, 171, 172, 283, 295, 296, 297, 307, 323
inaccuracy, 307
inconsistency, 47, 46
independent variables, 51, 54, 56, 59, 68, 70, 78, 79, 90, 92–94, 99, 104, 105, 107, 111, 115, 117, 127, 141, 144, 155, 157, 158, 164, 165, 167, 168, 177, 187, 190, 192, 195–198, 200, 219, 222–233, 235, 237, 238, 241, 244, 247, 249, 250, 252–254, 258, 259, 262–265, 267, 268, 276, 279, 280–283, 285, 286, 288, 289,

SUBJECT INDEX

291, 292, 298, 301, 303, 312, 314, 315, 317, 320, 322–324, 335, 342, 346, 360, 367–369, 377, 381, 382, 384–386, 389, 390, 392, 396–398, 408, 415, 421, 423, 443, 445, 447, 449, 452, 456, 462
"&&" declaration, 121, 240, 376, 393, 394
biochemical vs. mathematical, 22
changes in, 108–109, 122, 378–379
defined, 21–22
dependent variables distinguished, 22, 23
enumeration of, 76–77
equation construction, 48, 49–50
in glycolysis, 28–29, 120, 370, 378–379
inhibitory effect, 52
mapping, 27, 28–29
notation, 47, 50
replacement with dependent variable, 128
and steady-state evaluation, 120, 121, 122, 201–208
S-systems with, 204–206
S-systems without, 201–204
system effects on, 22
value assignment, 102
in zymogen activation, 29–31
index notation, 19, 20, 23, 31, 42, 47, 48, 50, 54, 55, 59, 77, 78, 88, 99, 101, 103, 184, 200, 205, 243, 245, 247, 265, 338, 368, 413, 416, 427, 433
inducer, 149
inducible gene circuit, 4
induction, 17, 400, 402
infectious disease, 220
infinitesimally small, 122, 124, 148, 190, 208, 225, 307, 424
influx, 49, 70, 73, 91, 144, 164, 185, 194, 206, 231, 232, 303, 305, 314, 318, 319, 331, 338, 371, 444, 461
information, 3, 5, 7, 8, 12, 14–18, 25, 27, 39, 41, 42, 44, 47, 60, 68, 79, 83, 94, 99, 107, 119, 124, 128, 129, 139, 144, 145, 158, 159, 171, 179, 184, 206, 210, 216, 231, 260, 262, 268, 285, 295, 301, 307, 315, 317, 323, 327, 328, 335, 337, 338, 343–345, 351, 357, 358, 361, 364, 365, 368, 399, 403, 422, 435
information theory, 5
inhibition and inhibitors, 16, 26, 39–42, 55, 86, 112, 127, 138, 140, 145, 148, 149, 160, 162, 164, 179, 183, 184, 186, 193, 213, 215, 247, 248, 268, 269, 275, 308, 323, 333, 335, 340, 345, 351, 359, 360, 363, 364, 373, 379, 380, 400, 415, 425
activation compared, 4, 220
allosteric, 76
of AMPD, 355–356
branched pathways with, 81–82, 126

competitive, 24–25, 76, 24–25, 76, 154, 161, 274, 301, 302, 334
constants, 7
of degradation, 52, 195, 206, 220, 346, 355
end product, 80–81, 111, 122
equations, 86
feedback, 3, 4, 14, 32, 57, 80–82, 113, 115, 126, 157, 176, 208, 220, 243, 256, 279, 329, 334, 405
modulators, 27
noncompetitive, 152
parameter value, 113, 125
representation of, 12–13, 16, 26, 27, 55, 86, 349
inhibitory effect, 27, 52, 54, 80, 112, 145, 176, 247, 334, 341, 355, 379
initial conditions, 44, 66, 79–80, 95, 110, 125, 213, 251, 291
initial guess, 172, 173, 175, 176, 179, 182, 183, 289
initial time, 65, 79–80, 104
initial value, 44–46, 64–67, 95, 100, 102, 104, 105, 107, 111, 114, 116, 117, 119, 120, 123–128, 132, 133, 138, 141, 173, 179, 183, 195, 199, 200, 207, 283, 306, 325, 330, 340, 350, 362, 363, 444, 449, 453, 458, 461
inorganic phosphate, 12
inosine monophosphate (IMP), 328–331, 333–336, 340, 347–349, 351–357, 359, 361, 363, 364, 461, 462
input
changes, 116, 353–354
genetic networks, 63
modules, 133–134, 137, 361
signals, 22
strength of, 117–118
time-dependent, 133–134
variables, 22, 136, 141, 195, 223, 285
insensitive systems, 140
instability, 124, 215, 216, 409
Institute for Chemical Research, Kyoto University, 15
Integrated Genomics, Inc., 16
integrated model, 7, 39, 40–41, 327
integrated system, 3, 4–8, 39
intermediate metabolites, 15, 22, 76, 129, 327
internal equivalence, 408
internal representation, 103
internal variables, 22, 56, 109
intuitive analysis, 3, 4, 11, 13, 43, 50–51, 52, 81, 92, 98, 105, 107, 125, 126, 130, 133, 140, 164, 206, 219, 275, 327, 424, 425
inverse matrices, 203–204, 206, 227, 237, 252, 382, 384, 387, 414, 437–439, 441
irreducibility, 5

irregular S-system, 201, 206, 209
irreversible pathways, 91
irreversible reactions, 25, 47
irreversible strategy, 92, 163, 168, 447
isocitrate, 165–167, 308, 312, 314, 322
isocitrate dehydrogenase (ICDH), 165, 166, 167, 300, 314
isoleucine, 20, 32
isomorphism, 5
isopropylthiogalactoside, 149
iteration, 173, 284, 326
  of the modeling process, 328, 342
iterative process, 127, 326
IUPAC names, 19

Kalman filtering, 404
Kyoto Encyclopedia of Genes and Genomes (KEGG), 15, 62
kinetic analysis, 16
kinetic orders, 59, 76, 77, 79, 89, 90, 95, 105, 109, 122, 130, 140, 142–144, 165–168, 170, 171, 176–179, 183–188, 200, 214, 218, 222, 227, 228, 230, 231, 233, 239, 244–246, 248, 250, 252–254, 265, 268–276, 282, 287, 291, 299, 303, 305, 311, 314, 318, 321, 324, 325, 328, 333–335, 341, 343–347, 350, 351, 354, 356, 358, 362, 363, 368, 370, 371, 377, 381, 384, 385, 396–398, 405, 406, 407, 445–447, 449–452, 455, 456, 460–463, 465–467
  activation, 145
  adenine excretion, 149–150
  aggregated, 85, 87, 267, 338, 339, 342, 355, 357
  averaged, 84
  in biochemical system models, 37, 53–54, 55
  citrulline pathway, 152–155
  in computer simulations, 100–101
  in constraint equations, 89
  degradation, 55, 127, 169, 304, 339, 342, 448
  double modulation method, 157–158
  experimental measurement, 146–149
  for feedback regulation, 86
  fluxes and, 82, 145, 148, 255–256
  fractal, 55
  in glycolytic–glycogenolytic pathway model, 372–375, 395
  in GMA models, 255–256
  Kacser–Burns method, 190–191
  logarithmic gains, 155–158
  metabolite sensitivity and, 255
  parameter estimation, 53–54, 55, 146–164, 190–191, 241–243, 255–256, 338, 372–375, 395
  in purine metabolism, 338–340
  pyruvate carboxylate in rat liver, 151–152
  pyruvate kinase in rat liver, 150–151
  in S-systems, 53–54, 55, 74, 78, 79, 82, 88, 100–101
  sensitivity analyses and, 241–243, 255–256, 395
  from steady-state data, 146–158
  from traditional rate laws, 158–164
  weighted, 84
Kirchhoff's node equation, 48
K_M, 21, 43, 64, 142, 145, 159, 160, 165, 169, 188, 374, 418, 444

lac operon, 62
lac repressor, 149
lactate, 151, 152, 247, 249, 415, 416, 421, 422
language, formal, 59, 62
language, genetic, 62
language, metabolic, 62
Lawrence Berkeley National Laboratory, 18
Lesh–Nyhan syndrome, 341
level of organization, 11
life cycle, 293, 315
LIGAND, 15–16
limit cycle, 114, 118, 119, 125, 126, 219, 409, 448
linear
  differential equations, 51, 57, 59
  function, 46, 56, 69, 71
  models, 56, 57
  program/programming, 285–286
  regression, 148, 149, 150, 173, 176, 186, 191, 414, 419, 451
  system, 40
linear algebra, 173, 191, 209, 252, 391, 406, 407, 426
  determinants, 430–434
  inverse matrices, 437–439
  matrices, 427–430
  systems of linear equations, 434–437
  vectors, 427–430
linear pathways, 23, 95, 129, 130, 131, 136, 137, 141, 142, 157, 193, 204, 206, 210, 219, 220, 399, 405
  algebraic analysis, 84, 212–213
  computer simulation, 115–119
  with cross-activation, 213–216
  end-product inhibition, 80–81
  equation constriction, 78–81
  with feedback, 212–216
  S-system, 72, 78–80, 115–119, 212–213
  steady-state evaluation, 212–213
  surprise, 115–119
linear program/programming, 285, 286, 287, 289, 405, 408

SUBJECT INDEX

linear regression, 148, 149, 150, 173, 176, 186, 191, 414, 419, 451
linear system, 129, 142, 156, 158, 194, 195, 201, 209
linearization, 414, 417, 419, 420, 421, 425
link matrix, 257, 258
lipogenesis, 62, 284
liver, 133, 148, 150, 151, 153, 208, 297, 327, 335, 365, 369, 370, 374, 398
local minimum, 174, 179
local property, 140, 228
local stability, 124–125, 208–216, 220, 279, 280, 307, 310, 319, 340, 376
log gain (see logarithmic gain)
logarithmic amplification, 222
logarithmic coordinates, 88, 146, 147, 149, 153, 154, 160, 191, 202, 203, 206, 270, 278, 286, 287, 288, 291, 377, 384, 408, 419, 420, 425, 449
logarithmic derivative, 156, 191, 252, 450
logarithmic flux, 71, 146, 147, 148, 156, 190, 253, 325, 390
logarithmic gains (log gains), 122, 126, 144, 190, 191, 219, 222, 223, 242, 247, 249–251, 283, 292, 296, 307, 311–314, 320–325, 342, 353, 362, 380, 389, 390, 393, 395, 397, 398, 402, 405, 408, 456, 464
  anaerobic fermentation pathway, 280–282
  of concentrations, 235–241, 252–253
  defined, 121
  examples, 226–229
  experimental measurements of, 155–158
  of fluxes, 229–232, 253–254, 385–388
  in glucose-6-phosphate pathway, 123
  in glycolytic–glycogenolytic pathway model, 376–377, 384–388
  in GMA models, 252–254
  of metabolites, 224–229, 384–385
  sensitivity analysis with, 224–233, 252–254
  S-system models, 280–282
  transition time and, 232–233
logistic growth, 123, 410
lognormal, 139
lysine, 31, 32, 443

macro-mechanics, 5
macroscopic world, 11
magnesium, 12, 400, 403
malate, 299, 304, 308, 310, 311, 314, 315, 322, 323, 324, 456
malate dehydrogenase, 300, 310, 311, 313, 320, 322, 323, 325
malic enzyme, 310, 313, 322, 324
malonate, 153
maltose catabolic system, 401

map, see maps, biochemical
maps, biochemical, 19, 32, 37, 40, 41, 52, 61, 67, 70, 76, 79, 87, 93, 94, 97, 127, 135, 136, 199, 221, 239, 261, 265, 281, 301, 307, 315, 328, 330, 332, 339, 346, 362, 403
  arrows, 23, 25–27
  branched pathways, 12, 13, 448
  of cascaded mechanisms, 86
  components and pools of components, 23–24, 25
  constructing, 14–15, 17, 23–28
  database sources for, 15–18
  elements of, 14
  enzyme-catalyzed reactions, 23
  examples, 28–31, 54
  gene, 15
  glycolysis, 28–29
  glycolytic–glycogenolytic pathway, 365–368
  interactions, 24–25
  linear pathways, 115
  modulations, 24–25, 26
  nodes, 25
  purine biosynthesis, 13
  translation into equations, 49, 58 (see also equation construction)
  tricarboxylic acid cycle, 297–298
  zymogen activation, 29–31
Maple®, 200
marginally stable, 208, 209, 212, 216
mass action ratio (Γ), 162, 169, 372, 373, 374, 398
mass balance equations, 91, 92, 93, 94, 267, 371
mass conservation, 46, 91, 257, 265, 367
MathCad®, 200
Mathematica®, 167, 172, 200, 407
mathematical experiment, 40, 58, 135, 127–129
mathematical representation, 6 (see also maps, biochemical; index notation; notation)
mathematical rigor, 4, 98, 143, 406, 417
mathematical tractability, 7, 41, 59, 330
mathematically controlled experiment, 58
mathematically equivalent, 91, 137, 147
matrix, 156, 193, 207, 211, 232, 249, 257, 258, 285, 380, 385, 390, 391, 395, 398, 413, 433, 442, 452, 465
  equations, 61, 64, 200, 201, 202, 203, 204, 205, 218, 227, 231, 238, 250, 251, 253, 256, 284, 381, 386, 429, 430, 431, 434, 435, 436, 438, 441, 468
  identity, 203, 252, 254, 437, 438, 468
  inverse/inversion, 203–204, 206, 227, 237, 252, 382, 384, 387, 414, 437–439, 441
  linear algebra, 427–430
  multiplication, 386–387, 388, 428, 429–430, 432, 438
  notation, 200, 228, 231, 251, 253

matrix (*cont.*)
  stoichiometric, 16, 61, 251–252, 256, 284
  transpose, 427, 428, 434, 467
maximum velocity, 2, 162, 274, 374
MCA
measurable quantity, 40–41
mechanism, 1, 4, 12, 16, 25, 29, 32, 38, 39, 57, 76, 81, 86, 93, 136, 171, 179, 196, 208, 269, 323, 324, 328, 331, 400, 401, 405, 406, 462
mental retardation, 327, 328, 341
mercaptopicolinic acid, 152
merge from file option, 117
metabolic control analysis (MCA), 60, 83, 148, 245, 246, 249
metabolic disease, 345
metabolic language, 62
metabolic maps, 12, 15, 40; *see also* maps, biochemical
metabolic models, 7–8, 15, 16, 38, 59, 61
metabolic reconstruction, 16
metabolic rules, 59, 62
metabolism, 1, 4, 15, 56, 98, 149, 223, 249, 288, 294, 295, 313–315, 320, 322, 323, 326–330, 332, 335, 338, 341–343, 345, 356, 358, 364, 369, 399, 405, 407
  carbohydrate, 167, 168, 296, 297, 365
  drug, 133, 134, 137, 141
metabolite(s)
  concentration, 233–243, 288–289, 335–337
  databases, 15, 61
  degradation, 77, 78
  in GMA models, 254, 255
  graphic representation, 20, 21, 26
  intermediate, 15, 22, 76
  kinetic orders, 255
  logarithmic gain, 224–229
  in models, 38
  pooling of, 90
  rate constants, 254
  sensitivity analysis with, 224–229, 233–243, 254, 255
methionine, 32, 343
methionine adenosyltransferase (MAT), 342, 348
method of controlled mathematical comparisons, 114, 407
methylcobalamin (MeCbl), 32
methyltetrahydrofolate, 32
methyltransferase apoenzyme, 32
Michaelis constants, 2, 145, 161, 162, 169, 335, 368
Michaelis–Menten
  equation, 75
  rate laws, 37, 38, 42–43, 74, 75, 83, 144, 146, 159, 160, 246, 295, 296, 297, 361, 406, 418, 419
  reaction, 43, 76, 260, 273, 342

microarray, 17, 18, 145
microbial heat generation, 23
micro-mechanics, 5
microscopic world, 11
MIPS, 18
mitochondria, 148, 151–154, 187, 248, 297–298
mnemonic abbreviations, 19, 20
modal analysis, 129, 209
mode of regulation, 39, 401
models (*see also* biochemical system models)
  design, 16, 127–129, 262, 326, 327, 395
  structure, 40, 58, 146, 322, 326, 342, 396, 404–406, 407, 409
modifiers, 21, 39
modulated branch points, 87, 89
modulation/modulators, 11–13, 24–27, 31, 38, 39, 40, 50, 62, 81, 155, 157–158, 190, 285, 328, 366, 421
moiety, 274, 330, 331, 346, 349
mold, 167, 293
molecular biology, 5, 10, 399
molecular formulas, 19, 20
molecular trafficking, 171
Molecular Visualization Freeware, 18–19
Monte Carlo simulation, 138–140, 411
morphogenesis, 294
most likely situation, 138
motility, 187, 247, 248, 249
MPW Metabolic Pathway Database, 15
mRNA, 17, 63
mucopolysaccharide, 293
multi-enzymatic complexes, 171
multinomial system, 59
multivariate, 235, 424, 425, 426
muscle, 86, 365
mutagenesis, 284
mutant, 149
mutation, 17, 109, 222, 384, 401
Mycobacterium tuberculosis, 8
myxamoeba, 293

N-acetylglutamate, 152–153
NAD^+, 27, 262, 263, 264, 275, 283, 288, 366
NADH, 27, 135, 165, 166, 167, 262, 263, 264, 274, 283, 288, 300, 322, 324, 456
National Center for Biotechnology Information, 18
National Human Genome Research Institute, 18
National Library of Medicine, 8, 18
natural design, 115, 284
nematode sequence, 8
neobiosynthesis, 328
net change, 73, 194
net flux, 25, 91, 92, 162, 163, 168, 169, 184, 185, 315, 316, 317, 457, 458
neuron

SUBJECT INDEX

networks
  Boolean, 63
  metabolic, 59
Newton's laws of physics, 37
Newton–Raphson search method, 407
nitrogen flow, 14
nitrogen, dietary, 327
nodes, 25, 64
node equations, 48, 298, 299, 301, 304, 317
nomenclature, for biochemical systems, 19–23
nominal steady state, 146, 157, 159, 165, 208, 308, 312
non-competitive inhibitor, 152
nonlinear estimation, 173–174, 176, 179, 291
nonlinear regression, 172, 173, 174, 175, 176, 178, 189
nonlinearity, 3–4, 50–51, 56, 74, 131, 148, 151, 155, 179, 191, 202, 208, 209, 213, 226, 234, 291, 310, 343, 404, 408, 409, 417, 418, 420, 424, 426
nonoxidative pentose pathway, 262
non-trivial solution, 198
normalization, 257
norvaline, 154
notation (*see also* arrows)
  activation, 14, 27, 55, 349
  dot, 42, 44
  index, 19, 20, 23, 31, 42, 47, 48, 50, 54, 55, 59, 77, 78, 88, 99, 101, 103, 184, 200, 205, 243, 245, 247, 265, 338, 368, 413, 416, 427, 433
nucleic acid, 327, 328
Nucleic Acids Research journal, 15
5′(3′)-nucleotidase (3NUC), 334, 348
5′-nucleotidase (5NUC), 331, 334, 348
nucleotide, 9, 247, 331, 344, 363
numerical algorithms, 8, 44, 66–67, 68, 172, 173, 406–407
numerical implementation, 79
numerical steady-state solution, 85, 123, 219, 435

objective function, 174, 286–287
oil palm, 174
oleate, 151, 152
oligomeric, 135
open systems, 93–94
operating point, 60, 69, 83, 84, 88, 91, 96, 137, 140, 159, 160–162, 164, 165, 186, 189, 248, 269, 288, 297, 341, 355, 356, 361, 371, 408, 409, 417–420, 425, 426, 440, 447, 456, 466, 467
operations research, 62, 284
operon, 62
optimal solution, 285, 289, 409
optimality, criterion of, 285, 389
optimization, 59, 62, 261, 288, 396, 408–409

anaerobic fermentation pathway model, 284–286, 289–290
biotechnological, 83, 260, 285, 390, 397
linear programming, 285–286
orbit, 111, 114
order-of-magnitude model, 145–146
order-of-magnitude values, 149, 160, 311, 335
organization, 2, 5, 7, 11, 98, 129, 402
organizational level, 6, 11
organized complexity, 4, 5
ornithine, 32, 153, 154, 155
ornithine transcarbamoylase, 153, 154, 155
ornithine-δ-aminotransferase, 32
oscillations, 4, 45, 51, 57, 61, 63, 64, 110–111, 113, 117, 118, 124–125, 216, 224, 279, 283, 307, 376, 409, 410, 448, 449
  damped, 114, 116, 448
oscillatory behavior, 209, 216, 280, 340
O-succinylhomoserine, 32
output, 38, 45, 61, 63, 64, 93, 103, 105, 106, 111, 114, 122, 129, 134, 139, 140, 193, 216, 232, 280, 290, 326, 390, 396, 408, 412
outside the system, 22, 93, 94
overdetermined, 190
overexpression, 290, 327
overproduction, 340, 345
overshoot, 84, 308, 378, 417
oxalacetate (OAA), 151, 152, 153, 188, 298, 299, 300, 308, 309, 323, 324, 325, 450
oxidative pentose pathway, 262
oxidized form, 27
oxipurinol, 360, 361
oxygen consumption, 294
oxygen uptake, 248
oxypurines, 56, 364

pancreas, 24, 29–31
paper and pencil computation, 65, 97, 105, 119, 124, 143, 280, 406
paradigm, 3, 37, 417
paradigm shift, 4, 5–6
parameter estimation, 60, 143–145, 175, 328, 333, 335, 340, 351, 365, 369, 396, 405, 427
  adenine excretion, 149–150
  for branched pathway, 176–179
  for cascades, 179–184
  citrulline pathway, 152–155
  constraints used for, 191–192, 370–372
  by double modulation, 157–158
  from dynamic data, 172–185
  examples, 149–155, 165–171, 191–192
  exercises, 185–189
  experimental measurements, 146–149, 155–158
  flux stoichiometry, 191
  genetic algorithms, 174

parameter estimation (*cont.*)
  glucose-6-phosphate in *Aspergillus niger*, 167–171
  for *in vivo* systems, 171–172
  isocitrate in *Dictyostelium discoideum*, 165–167
  kinetic orders, 53–54, 55, 146–164, 190–191, 338, 372–375
  logarithmic gains, 121, 155–158
  purine metabolism, 335–338
  pyruvate carboxylate in rat liver, 151–152
  pyruvate kinase in rat liver, 150–151
  rate constants, 164–171, 375
  regression methods, 173–174
  scope of expected results, 145–146
  slopes replacing derivatives, 174–184
  S-system equations, 375
  from steady-state data, 146–158, 191–192, 335–338, 368–370
  from traditional rate laws, 158–164
  transient responses, 184–185
parameter values, 21, 27, 39, 52, 58, 60, 76, 79, 82, 85, 92, 95, 96, 115, 116, 121, 130, 135, 136, 138–140, 146, 157, 158, 168, 172–176, 179, 182–184, 186, 187–189, 191, 194, 206, 212, 220, 222–224, 234, 236, 259, 279, 280, 282, 283, 291, 307, 316, 317, 319, 322, 330, 335, 345, 346, 362, 363, 377, 383, 391, 393, 396, 403, 405, 407, 421, 431, 445, 447, 449, 451, 457
  anaerobic fermentation pathway, 268–277
  changes in, 109–110, 122–123, 379–380
  in computer simulations, 98, 99, 101
  and dependent variables, 122–123, 143, 144
  glucose-6-phosphate metabolism, 99
  glycolytic–glycogenolytic pathway, 368–375, 379–380
  inhibition, 113, 125
  tricarboxylic acid cycle, 298–306
parameterize, 174, 341
parameters, 4, 40, 65, 97, 103, 105, 107, 112, 125, 126, 127, 159, 193, 197, 210, 213, 233, 235, 239, 257, 272, 311, 314, 343, 344, 353, 395, 402, 404, 464, 466
  constrained, 71, 87, 191–192, 241, 243, 245, 287–288
  defined, 22
  graphical representation, 21, 22, 23
  rate law, 2, 45
  sensitivity, 122–123, 222, 244–245
  S-systems, 50, 51-55, 58, 60
  unconstrained, 390
partial derivatives, 88, 89, 166, 167, 170, 225, 230, 241, 304, 318, 394, 414, 424, 425, 440, 460, 461

partial differentiation, 89, 91, 147, 161, 165, 169, 268, 275, 277, 304, 371, 446, 447, 451, 456, 467
particle physics, 6
path computation tool, 16
PathComp, 16
pathological condition, 40
pathways, 47 (*see also* anaerobic fermentation pathway; branched pathways; linear pathways; superpathways)
  amphibolic, 25, 92
  citrulline, 152–155
  clamped, 92, 96
  converging, 77, 87, 95
  didactic, 115, 127, 176, 256, 405
  diverging, 95, 267, 288
  graphical representations, 15
  metabolic, 15–16
  physiological shortening of, 57
  polyamine, 342
  reversible, 91–93
  surprise, 115–119
pentose, 262, 284
pepsin, 160
permanent change, 222, 233, 462
permeabilized cells, 148
persistent changes, 108–109, 117–118, 155, 222, 231, 233, 378
perturbations, 17, 39, 103, 114, 117, 118, 125, 130, 143–145, 155–157, 172, 184–186, 208, 209, 222, 223, 231, 279–281, 283, 296, 297, 306–308, 310, 312, 319, 321, 325, 328, 340, 344, 345, 354–355, 357, 395, 397, 398, 402, 403, 405, 406, 408, 457, 458, 464 (*see also* changes)
  in dependent variable, 377–378
  in independent variable, 109, 378–379
  and local stability, 124
  log gains/sensitivities and, 123
Petri nets, 63–64
pH, 23, 24, 27, 148, 260, 270, 271, 272, 275, 281, 283, 286, 292, 415
pharmacogenomics, 9
phase plot, 250
phase plane, 110–111, 114, 116, 118, 119, 126, 250, 409, 448
phosphatase, 21, 22, 23, 25
phosphate (P_i), 12, 100, 153, 154, 155, 284, 346, 347, 353, 369, 385
phosphoenolpyruvate (PEP), 90, 150, 151, 152, 262, 264, 273, 274, 275, 280, 283
phosphoenolpyruvate carboxykinase (PEPCK), 152
phosphoenylpyruvate (PEP), 150

SUBJECT INDEX

phosphofructokinase (PFK), 28, 29, 100, 101, 109, 122, 125, 262, 264, 267, 271–272, 278, 281, 282, 283, 291, 367, 369, 370, 372, 374, 376, 390, 391, 398, 447, 455, 456
phosphoglucomutase, 28, 29, 100, 122, 366, 367, 370, 372, 384, 390
phosphoglucose isomerase, 28, 29, 100, 367, 370, 374, 377, 391, 398
3-phosphoglycerate (3PG), 20–21, 264
phosphoglycerate dehydrogenase, 20, 23
phosphohomoserine, 23, 27, 32
3-phosphohydroxyglycerate, 23
3-phosphohydroxypyruvate, 20, 22, 23
5-phosphoribosyl-α-1-pyrophosphate (PRPP), 12, 328, 334, 335, 349, 351, 362, 363, 461
phosphoribosyl pyrophosphate synthetase (PRPPS), 329, 334, 348, 355, 357, 361, 364
phosphoribosyl pyrophosphate synthetase superactivity, 340, 355, 357
phosphoribosyltransferase, 328, 329, 334, 341, 348, 349
phosphorylase, 28, 29, 86, 295, 331, 367, 368, 370, 372, 384, 396, 398
phosphorylase a, 100
phosphorylation, 63, 64, 148, 295, 379, 380
phosphorylization, 24, 249
phosphoserine, 20, 21, 22, 23
phosphoserine phosphatase, 20–21, 22, 23
phosphoserine transaminase, 20, 23
physiological variation, 57, 356
physiology, 1, 86, 458
physiomics, 9
P_i, 28, 345, 346, 347, 351, 353, 363, 369, 370, 372
ping pong, 301, 302, 303
places, 64
PLAS program (see computer simulations)
plasmids, 17, 149
plot, 43, 80, 108, 110, 111, 114, 116, 117, 118, 125, 126, 146, 147, 149, 175, 187, 311, 423, 448, 449, 450, 451
polyamine pathway, 331, 342, 343
polymerase, 348, 351, 363
polymerase chain reaction, 17, 363
polymorphisms, 9, 17
polynomials, 60, 67, 69, 210, 413, 418, 419, 439, 440, 466
  characteristic, 211, 212, 213, 453
polysaccharides, 262, 263, 264, 267, 271, 272, 276, 278, 282, 285, 291, 455, 456
pools/pooling, 14, 20, 21, 23–25, 27, 45, 46, 50, 56, 79, 83, 91, 83, 90, 119, 132, 135, 136, 144, 206, 221, 230, 263, 290, 297, 298, 299, 309, 314, 316, 322, 324, 328, 330, 331, 333, 335, 337, 340, 342–344, 346, 353, 354, 357, 361, 363, 403, 434, 444, 447, 449, 458 (see also aggregation)
  combined, 128, 296
  condensation of, 129–131
  sizes, 42, 49, 51, 296, 315
population model, 7
power-function representation, 41, 50, 53–54, 57, 59, 88
power-law approximation, 90, 144, 160, 164, 338, 414, 419–421, 425–426, 440, 447, 467
power-law equations, 50
power-law functions, 49–50, 55, 60, 74, 77, 102, 138, 142, 155, 164, 176, 192, 198, 251, 303, 305, 368, 389, 390, 398, 404, 406, 407, 408, 413, 420, 425, 426, 448, 452, 460, 467
  products of, 56–57, 59, 70, 73
power-law rate law, 167, 168, 451
power-law terms, 73, 77, 80, 86, 91, 95, 103, 137, 159, 160, 164, 165, 171, 177, 188, 269, 271, 277, 291, 334, 338, 347, 370, 375, 394, 444, 455, 465, 466
PowerPoint®, 106, 228
powers, 50
PP-ribose-P, 12
PP-ribose-P synthetase, 12
precursor, 67, 78, 79, 86, 87, 115, 119, 208, 285
precursor–product constraints, 79–80, 82, 83, 99
precursor–product relationships, 79, 84, 90, 99, 111, 115, 130, 132, 176, 184, 190, 193, 240, 266, 267, 271, 290, 322, 371, 391–393
prediction, 64, 65, 98, 117, 122, 126, 128, 140, 152, 226, 230, 234, 236, 243, 251, 262, 290, 295, 296, 297, 307, 323, 324, 325, 326, 341, 353, 356, 377, 384, 387, 390, 392, 399, 401, 402
prestalk cell, 293
procarboxypeptidase, 29–30
product notation, 51
production, 25, 26, 31, 47–49, 57, 59, 68, 72, 77, 78, 81, 86–109, 119, 126, 149, 153–155, 191, 193, 208, 236, 237, 247–249, 260, 262, 264, 267, 271–273, 276, 278, 281, 282, 284–286, 288–291, 294, 295, 304, 308, 316, 329, 344, 351, 360, 363, 366, 391, 408, 415, 421, 422, 448, 450, 451, 455, 456, 457, 461, 465
  function, 51
  parameters, 50, 53, 55
  term, 50, 51, 54, 55, 70, 94, 133, 265
productivity, 128, 131–133, 134, 141, 389, 396
proelastase, 30, 31
profile, expression, 9, 18
Prokaryotic Database, 18
proofreading, 223, 400

protein, 7, 9, 11, 15, 19, 63, 152, 168, 297, 298, 300, 304, 324, 325, 348, 369, 400, 457, 458
  catabolism, 294, 315, 317
  degradation, 294, 314, 315–316
  sequences, 16, 18
  in serum, 40
Protein Data Bank (PDB), 18
Protein Information Resource (PIR), 18
protein O-methyltransferase (MT), 348
proteomes, 9
proteomics, 9, 16
PRPP synthetase superactivity, 340
pseudoplasmodium, 293
pseudo-three-dimensional representation, 18, 123, 228, 237, 243, 249, 311, 320, 324, 325, 423
purine biosynthesis, 13, 328
purine metabolism, 12, 56, 149, 223, 326–328, 364, 399, 405
  activation of, 334, 349
  adenylate turnover, 343
  AMPD inhibition, 355–356
  biochemistry, 328
  computer simulations, 340–341, 353–361
  degradation, 330
  drug treatment simulation, 360–361
  enzyme activity changes, 355–360
  equation computation, 338–340
  equation construction, 332–338
  exercises, 362–364
  HGPRT deficiency, 356–360
  input changes, 353–354
  kinetic characteristics, 338
  models, 328–361
  modifications to model, 342–343
  parameter estimation, 335–338
  polyamine pathway, 342
  processes, 332–335, 346–353
  PRPP synthetase superactivity, 355
  sensitivity analysis, 342
  steady-state concentrations, 335–337
  steady-state fluxes, 337–338
  temporary perturbations, 354–355
purine nucleoside phosphorylase, 331
purine ring, 330, 331, 349
purine synthesis, 12, 328, 329, 334, 344, 359
purine synthesis *de novo*, 12, 329
pyridoxal phosphate, 32
pyrimidine, 344, 350, 362
pyrophosphate, 12, 328, 329
pyrroline-5-carboxylate, 32
pyruvate (Pyr), 22, 90, 150, 151, 153, 249, 262, 273–276, 280, 283, 287, 288, 297, 300, 308–310, 313–315, 320, 322–324, 456, 457, 461
pyruvate carboxylase, 150, 151–152, 153

pyruvate dehydrogenase, 300
pyruvate kinase, 150–151, 262, 273, 274, 275, 276, 280, 283, 287

qualitative behavior, 4, 283
qualitatively correct, 341, 362
qualitatively different, 74, 113, 118
quantitative laws, 7, 20
quantitative terminology, 19–20
quasi-steady-state assumption, 37

radioactive decay, 3
random number, 411
randomness, 3
rank correlation, 140, 247–249
rat liver, 148, 153, 297, 327
  glycolytic–glycogenolytic pathway in, 365–398
  pyruvate carboxylate in, 151–152
  pyruvate kinase in, 150–151
rate constants, 44–46, 59, 65–67, 76, 79, 81, 88, 90, 94, 95, 101, 103, 105, 110, 115, 120, 126, 127, 130, 134, 142, 143, 146, 147, 176, 177, 179, 183, 188, 191, 192, 194, 196–198, 200, 201, 205–207, 214, 217, 219, 220, 222, 227, 230, 233, 242, 244–247, 249, 252, 265, 268, 269, 271–274, 278, 287, 291, 298, 299, 303, 305, 308, 311–314, 318, 320, 324, 328, 338–340, 342, 350, 351, 353–355, 358, 363, 368, 370, 377, 379, 380, 396–398, 445, 446, 448, 449, 452, 455–457, 461–463, 466, 467
  branch point constraints, 393–395
  constrained systems, 239–241, 391–395
  degradation, 50, 123, 267
  estimation of, 164–171
  first-order, 308
  glucose-6-phosphate, 122–123, 167–171
  in glycolytic–glycogenolytic pathway model, 375, 390–395
  in GMA models, 77–78, 254–255
  indexed, 77
  isocitrate in *Dictyostelium discoideum*, 165–167
  in fluxes, 78–79, 82, 254–255
  in metabolite concentrations, 235–241, 254
  precursor–product relationships, 391–393
  sensitivities with respect to, 122–123, 235–241, 254–255, 375, 390–395
  in S-systems, 50, 51–52, 53, 70, 77, 78
  symbols, 50
  unconstrained parameters, 391
rate laws, 2, 6, 40, 47, 50, 58, 67, 142, 144–146, 150, 159, 160–164, 167, 168, 187, 189, 246, 248, 263, 270, 273, 274, 276, 295–297, 305, 330, 350, 361, 362, 406, 418, 419
  in enzyme-catalyzed reactions, 53

SUBJECT INDEX

Michaelis–Menten, 37, 38, 42–43, 74, 83
  power functions and, 55, 74
  representations of, 16
  number of terms, 39
  traditional, 37, 41, 60, 92, 165, 169, 186, 188, 260, 265, 268, 269, 272, 275, 277, 283, 291, 304, 338, 355
rational functions, 38, 50
reactions
  activation, 14, 27, 28, 86, 112–113
  occurring on surfaces, 53
real part, 124, 125, 126, 210, 212, 213, 216, 280, 307, 340, 376, 379, 448
realistic conditions, 40
recasting, 10, 137, 142, 409–410, 411
recombinant DNA, 287–288
reconstruction, 4–8, 16
reconstructionism, 5–6
recycling, 198, 206, 220, 331, 343, 356
red-blood-cell metabolism, 408
red-blood-cell model, 400
reduced form, 27
reduced set of dependent variables, 257, 258
reductionism, 2–4, 5–6
refinement, 298, 327, 366
regression, linear, 148, 149, 150, 173, 176, 186, 191, 414, 419, 451
regression, nonlinear, 172, 173–174, 175, 176, 178, 189
regular S-system, 201–206, 207
regulation, 12, 18, 39, 57, 62, 86, 160, 179, 208, 223, 261, 329, 400, 401, 405, 407
regulator, 58, 400, 402
regulatory mechanism, 4, 208, 400, 401
regulatory pattern, 7, 58
relative changes, 56, 57, 68, 71, 72, 73–75, 121, 149, 155, 190, 222, 225, 226, 227, 230, 235, 237, 241, 376, 377, 379, 390, 392, 464
relative concentrations, 146, 147, 148, 151
relative growth, 55–56
relative process rate, 146
relnodes, 64
report interval, 80, 103, 104, 111
repressible gene circuits, 4
repression, 400
Research Collaboratory for Structural Bioinformatics (RCSB), 18
residual, 4, 172, 174
resolution, 107
respiration, 2, 187, 247, 248, 249
response function, 147
responsiveness, 386, 402
results menu, 44, 106, 110, 111, 112, 114, 117, 130, 250, 444
reverse engineering, 63

reverse reaction, 32, 91, 162, 188, 317, 372, 462
reverse transcriptase, 17
reversible/reversibility, 32, 96, 155, 163, 164, 168, 170, 190, 298, 300, 334, 349, 447
  pathways, 91–93, 95
  reactions, 25, 26, 91, 92, 162, 167, 272, 304
  strategy, 92
ribonuclease, 346
ribonucleotides, 328, 329, 338
ribose ring, 12
ribose-5-phosphate (R5P), 12, 223, 346, 347, 353
Riccati systems, 60
risk assessment, 139, 411
RNA, 56, 327, 330, 346, 347, 348, 351, 352, 353, 363
RNAse, 348
robustness, 92, 123, 222, 223, 260, 279, 282, 283–284, 320, 321, 324, 345, 368, 372, 386, 395, 396, 408
  glycolytic–glycogenolytic pathway, 375–380, 379–380
  tricarboxylic acid cycle model, 306–315
Rössler, 136, 141
Routh, 210, 453
rule set, 62
run-time parameter, 103

Saccharomyces cerevisiae, 8, 330, 331
  anaerobic fermentation pathway, 260–292
saccharopine, 31, 443
S-adenosyl-L-methionine (SAM), 331, 333, 336, 342, 343, 344, 347, 352
S-adenosylmethionine, 32
S-adenosylmethionine decarboxylase (SAMD), 334, 348
salvage pathway, 129, 328, 329, 334, 335, 341, 349, 351, 354, 462
S-AMP (adenylosuccinate), 330, 331
sampling design, 139
Sarrus's rule, 208, 211, 213, 432, 433, 442, 468
SAS®, 178
saturation, 51, 57, 146, 160, 163, 186, 270, 273, 369, 374, 378
scaffold, 171
scaling, 161, 267
scenario, 47, 87, 107, 141, 148, 175, 201, 210, 222, 272, 310, 340, 377, 387, 448
screening tool, 8, 290, 327, 362
S-distribution, 164, 411
search algorithm, 353, 361
second-order approximation, 406
second-order reactions, 53
selectivity, 402
self-mutilation, 327, 328
self-thinning, 404

semi-quantitative, 359
sensitivity, 140, 291, 297, 307, 311, 314, 315,
    321, 323, 328, 336, 345, 346, 372, 375, 379,
    397, 398, 405, 426, 464
  anaerobic fermentation pathway model, 282
  aggregated, 353
  constrained, 312, 393, 465
  deaggregated, 363
  parameter, 122–123, 244–245
  profile, 324, 325, 342, 343, 351, 395, 396, 462
  types, 223
  unconstrained, 465
sensitivity analyses, 222
  with concentrations, 235–241, 252–253
  connectivity relationships, 246–249
  constrained systems, 239–241, 257–258
  examples, 226–229, 247–249, 256–257
  exercises, 249–250, 258–259
  with fluxes, 229–232, 244–245, 253–256
  of glycolytic–glycogenolytic pathway model,
    376–377, 390–395
  in GMA models, 251–259
  kinetic orders and, 241–243, 255–256, 395
  logarithmic gains and, 224–233, 252–254
  with metabolites, 224–229, 233–243, 254, 255
  purine metabolism model, 342
  ranking importance of processes, 247–249
  rate constants and, 235–241, 254–255,
    390–395
  summation relationships, 245–249
  terminology, 251–252
  trajectories, 250–251
  transition time and, 232–233
sequence, 12, 14, 32, 63, 329
sequence alignment, 18
serine, 20–21, 23, 25, 27
serine phosphatase, 25
serum, 40, 208, 357
signals, biochemical, 9, 14, 17, 47, 53, 63, 86, 93,
    136, 187, 248, 284, 314, 320, 330–331, 397,
    443, 444
simplification, 5, 6–7, 37, 57, 262, 325, 330, 366,
    373, 456
simulation, 41, 157, 158, 179, 230, 237, 243,
    297, 324, 325, 345, 356, 357, 360, 361, 363,
    368, 378, 387, 392, 403 (see also computer
    simulation)
  Monte Carlo, 138–140
sine waves, 110, 111, 125, 141, 449
single-displacement mechanism, 269
slime mold
slopes, 42, 67, 68, 74, 147–149, 151, 154,
    174–184, 233, 234, 235, 270, 278, 291, 377,
    417, 419, 420, 423, 449
slug, 293, 315

smoothing algorithm, 175
solution, 158, 437
  algebraic, 190, 383
  analytical, 65, 66, 75, 210, 350, 353, 406, 410
  dynamical, 84, 85, 119, 120, 122, 124, 126,
    280, 353, 361, 448, 456
  numerical, 44, 65–68, 75, 85, 108, 119, 123,
    132–133, 143, 174, 195, 205, 207, 216, 219,
    315, 357, 389, 406, 435
  parameter, 104, 107, 108
  points, 45, 444
  space, 207
sorocarp, 293, 296
spatial organization, 7
Spearman's rank correlation coefficient, 140
speed of response, 45, 52, 57, 75, 86, 164, 240,
    402
spermatozoa, 187, 247, 249
splitting reactions, 25, 26, 27, 90
spore, 293, 296
S-systems, 59, 60, 129, 174, 191, 193, 209, 210,
    217, 219, 251, 288, 307, 404, 405, 410, 411,
    452
  anaerobic fermentation pathway model,
    278–284
  allometric relationships in, 55–56, 70–73, 74
  analytical convenience, 58
  β-term aggregation, 277–278
  bimolecular reactions, 88–89
  branched pathways, 81–82, 83–84, 112–113
  canonical models, 65
  cascaded, 87
  computer simulations applied to, 97, 99–103
  constraints, 88–89
  degradation in, 265, 267, 277, 351, 362, 363,
    448, 461
  dependent variables, 77, 79
  dynamics, 54–55, 282–283
  equation specification, 70, 74, 76, 77, 78–80,
    82–84, 92, 94, 99–103, 277–278, 338–340,
    375
  examples, 52, 54–55
  fluxes, 83
  glucose-6-phosphate metabolism, 98–99
  in glycolytic–glycogenolytic pathway model,
    375
  independent variables absent, 201–204
  independent variables present, 204–206
  irregular, 206–208
  irreversible representations, 92
  kinetic orders, 53–54, 55, 74, 78, 79, 82
  linear pathway representation, 71, 78–80, 81,
    115–119
  log gains, 280–282
  local stability, 212

SUBJECT INDEX

parameters, 50, 51–55, 58, 60, 84
phase-plane plots, 110–111
power functions, 56–57, 74
precursor–product constraints, 79–80, 82, 83, 84
properties of models, 55–58, 279–284
purine metabolism, 338–340
rate constants, 50, 51–52, 71, 74, 79, 82
regular, 201–206
reversible representations, 92
robustness of models, 283–284
sensitivities, 282
steady-state, 119, 123–124, 201–208, 280, 407–409
surprise pathway, 115–119
two-variable, 212
telescopic property, 56–57, 64
theoretical justification, 57–58, 70–73
validity, 55–57, 73
stability, 98, 105, 127, 143, 145, 219, 220, 283, 292, 307, 310, 328, 344, 345, 357, 362, 364, 376, 380, 403, 407–409, 448, 453, 464
of branched pathways, 126
global, 208
local, 208–216
marginal, 208, 209, 212, 216
steady-state analysis, 123–125, 208–216
strength of inputs and, 117–118
two-variable S-system, 212
Stanford University, 18
starch, 149
start time, 44, 66
state variables, 22, 167
Statistica®, 178, 179, 180
statistical distributions, 139, 410, 411–412
statistical mechanics, 5
steady state, 52, 56, 61, 71, 73
anaerobic fermentation pathway, 280, 287
at branched pathways, 119, 213
characteristics, 80, 98, 127, 139, 157, 158, 328
concentrations, 108, 113, 126, 158, 168, 192, 193, 195, 196, 204, 208, 219, 224, 233, 309, 317, 324, 335–337, 108, 113, 126, 158, 168, 192, 193, 195, 196, 204, 208, 219, 224, 233, 309, 317, 324
constraints, 72, 191–192, 287
data, 143, 146–158
equations, 72, 191–192, 193–198, 200, 204–206, 211–213, 219, 380–383, 407–409
fluxes, 103, 130, 146, 147, 160, 165, 184, 230, 247, 268, 287, 308, 310, 311, 337–338, 341, 346, 356, 357, 362, 368, 370, 373, 447
kinetic orders estimated from, 146–158
nominal, 146, 157, 159, 165, 208, 308, 312
non-zero, 123–124

persistent changes and, 108
of S-system, 113, 119, 280
zero, 198–200
steady-state analysis (see also algebraic steady-state evaluation)
coefficients equations, 197–198, 200, 204, 205, 206, 211, 212, 213, 219
computer simulations, 105, 375–376
concentrations, 119–127
logarithmic gains, 120–122
parameter sensitivities, 122–123
stability, 123–125
step size, 44, 45, 67, 68, 75, 387
stiffness, 129
stochastic, 174, 404
stoichiometric
constraints, 83, 90, 191
matrix, 16, 61, 251–252, 256, 284
models, 63
network, 60–62, 284–285
proportions, 26, 27, 349
stoichiometry
at branch points, 184
deviation from, 83
of fluxes, 83, 184, 191, 288
of metabolic networks, 59
structural change, 109–110, 222, 324, 340, 355
structural stability, 409
structure of models, 40, 58, 146, 322, 326, 342, 396, 405, 407, 409
subscript notation, 19, 20, 23, 31, 42, 47, 48, 50, 54, 55, 59, 77, 78, 88, 99, 101, 103, 184, 200, 205, 243, 245, 247, 265, 338, 368, 413, 416, 427, 433
substitution, 56, 66, 90, 91, 128, 156, 171, 175, 177, 195, 200, 202, 203, 207, 208, 224, 272, 274, 275, 380, 385, 389, 392, 394, 426, 452
substrates, 7, 11, 12, 15, 16, 22, 25, 27, 31, 32, 37, 38, 40, 41, 45, 52, 54, 61, 62, 74, 76, 82, 93, 94, 115, 148, 151, 155, 157, 159, 160, 161, 163, 167–169, 171, 186, 188, 249, 263, 267, 269–271, 273, 275, 278, 293, 314, 327, 328, 333, 335, 338, 346, 350, 351, 366, 373, 402, 405, 418, 419, 421, 425, 444
change in concentration, 42–44, 47, 64
degradation, 46, 75, 90
subsystem, fast, 128–129
subsystem, slow, 128–129
succinate (SUC), 188, 298, 299, 305, 451
succinate dehydrogenase (SDH), 153, 188, 300, 322, 323, 325, 451
sucrose, 149
sugar transport, 268–269
summation relationship, 245–249
superactivity, 340, 355, 357

superpathways, 16
superposition principle, 48
surface, 209, 222, 295, 422
survival function, 411, 417
suspended cells, 260, 283
Swiss Institute of Bioinfomatics, 16, 19
SWISS-PROT, 18
symbolic
  algebra program, 167, 200, 395, 407
  evaluation, 143, 193, 205, 227, 268, 380–384, 386, 405, 407, 426, 439, 465
  form, 19, 20, 23, 28–31, 41, 50, 62, 82, 98, 99, 143, 147, 168, 177, 211, 220, 224, 236, 240, 242, 265, 285, 299, 317, 324, 334, 335, 347, 350, 351, 362, 364, 393–395, 408, 415
synergism, 4, 11, 51, 334
synthesis, 12, 20, 21, 27, 32, 50, 56, 72, 78–81, 86, 90, 111–115, 123, 126, 135, 145, 154, 167, 168, 170, 183, 315, 316, 325, 327–329, 333–335, 340, 343, 344, 346–348, 350, 351, 359, 362, 364, 368, 372, 385, 393, 458, 461
system(s)
  characteristics, modification of, 109–110
  component, 4, 14, 21, 23–24, 27, 39, 40, 58, 223
  description, 24, 48, 49, 56, 57, 59, 60, 70, 71, 73, 448
  of differential equations, 43, 51
  of linear equations, 434–437
  equations, 40, 48, 53, 72, 73, 76, 78, 79, 81, 93–94, 105, 128, 132, 172, 251, 298, 301, 330, 350, 379, 380, 434–437
  response, 4, 7, 48, 125, 139, 225, 292, 326, 341, 342, 362, 378, 380, 397, 406, 457
  variables, 22, 24, 63, 72, 134, 335, 340, 417, 423, 458, 461, 462

table option, 106, 110, 112, 130, 144
tangent vector, 209
Taylor approximation, 88, 406, 414, 417–419, 420
Taylor expansion, 226
Taylor polynomial, 60, 69, 418, 439
Taylor polynomial, truncated, 418, 440
Taylor series, 418, 440, 466
Taylor's theorem, 69, 70
TCA cycle, 294–297, 322, 323
teleology, 5
telescopic property, 56–57, 64, 129
temperature, 17, 19, 22–24, 27, 120, 128, 403
temporally dominant, 403, 411
terminology, 6, 7, 19–23, 41, 251–252
tetrahydrofolate, 32
thermodynamic equilibrium, 3, 92, 163, 405, 462
thiopurinol, 364

threonine, 32
threshold of stability, 4, 118, 126, 158, 210, 213, 215, 321, 402
threshold of toxicity, 285
throughput, 216, 232
tile command, 110, 112, 113
tiling, 109
time scales, 127–129, 118, 126, 158, 210, 213, 215, 321, 402, 457
  biochemical, 6, 209, 307, 403, 404
  biomolecular, 6
  hierarchy of, 6
time-dependent
  activation, 295
  inputs, 133–134
titration, 148
tokens, 64, 103, 131
topological structure, 59, 61, 144
total derivative, 252, 253, 254, 255
total differential, 155, 424
total enzyme concentration, 90
toxicity, 252, 253, 254, 255
tractability, 7, 41, 59, 330
traditional approach, 39, 81
traditional rate laws, 2, 6, 37, 39, 41, 50, 58, 60, 165, 169, 186, 188, 260, 265, 268, 269, 272, 275, 277, 283, 291, 304, 338, 355
  aggregation of fluxes and, 92
  kinetic order estimation from, 158–164
trajectories, 110, 111, 114, 116, 118, 209, 250–251
transaminase, 20, 23, 32, 300, 308
transcriptional unit, 400
transformations, 27, 28, 46, 84–85, 138, 148, 196, 201, 221, 287, 338, 354, 355, 419, 451
transient responses, 92, 103, 105–108, 113, 116, 128, 144, 172, 283, 307, 353, 355
  parameter estimation from, 184–185
transition time, 171, 216–217, 223, 249, 311, 320, 324, 396, 398, 409
  in glycolytic–glycogenolytic pathway, 389–390
  log gain and, 232–233
transitions, 64
transmethylation pathway, 343
transport process, 145, 159
treatment schedule, 141
tree, 189, 403, 404, 411
trehalose, 261, 262, 264, 267, 271, 272, 273, 294, 295
trend, 15, 43, 107, 148, 175, 295, 341, 402, 411, 419, 464
trial and error, 1, 172
tricarboxylic acid cycle, 129, 165, 171, 188, 223, 294–297, 306–310
  biochemical map, 297–298

SUBJECT INDEX

consistency analysis, 306–310, 319–322
in *Dictyostellum discoideum*, 293–325
equation construction, 298–315
exercises, 324–325
modification of model, 315–324
parameter values, 298–315
robustness, 310–315
truism, 5
trypsin, 29–30
trypsinogen, 30, 31
tryptophan, 19, 20, 362, 461
tubules, 7
turbidimetric method, 248
turnover, 11, 187, 247, 248, 249, 283, 309, 310, 315, 319, 343
adenylate, 343
ATP, 187, 247, 248, 249

UDPG phosphorylase, 295
unbound form of an enzyme, 90
uncertainty, 7, 76, 128, 139, 157, 183, 336, 346, 368, 375, 395, 396, 409
principle, 2–3
uncompetitive inhibition (*see* non-competitive inhibition)
unconstrained parameters, 76, 128, 139, 157, 183, 336, 346, 368, 375, 395, 396, 409
underdetermined system, 285
undershoot, 84, 112, 417
uni uni, 303
University of Minnesota Biocatalysis/Biodegradation Database (UMBBD), 16
University of Washington Resource Facility for Kinetic Analysis, 16
unstable, 118, 124, 126, 208, 209, 210, 212, 213, 215, 216, 307, 320, 380, 409, 448
uric acid (UA), 185, 327, 328, 330, 338, 340, 341, 345, 353, 354, 355, 356, 360, 361, 362, 363, 364, 462
excretion, 338, 362
uridine diphosphoglucose (UDPG), 187, 188, 264, 271, 272, 273, 295
U.S. Environmental Protection Agency, 140

validation, 328
validity, 39, 59, 144, 163, 171, 245, 246, 295, 307, 323, 355, 367, 408, 411, 413, 421
of S-system models, 55–57
vanadate, 248
variables (*see also* dependent variables; independent variables)
in anaerobic fermentation model, 262–265
auxiliary, 102, 138, 220, 370
in biochemical system models, 17, 24, 25, 38, 39, 46, 262–265
changes in, 39
control, 22
in fluxes, 82
indexed, 20, 23, 42, 47, 48
internal, 22, 56, 109
open vs. closed systems, 94
sensitivities
symbolic names, 19
vectors, 61, 193, 209, 233, 251–255, 257, 383, 386–388, 390, 392, 465, 427–430
column, 427–430
row, 200, 228, 231, 427, 428, 430
solution, 201, 202, 204, 205, 227, 429, 435, 436, 438
transpose, 427
validation, 328
validity, 39, 56, 57, 59, 144, 163, 171, 245, 246, 295, 307, 323, 355, 367, 408, 411, 413, 421
vanadate, 248
visual excitation, 86
velocities of transformation, 46

weight, 5, 64, 315, 365, 369, 370, 415, 464
body, 139, 150, 337
weighted average, 84, 371, 397
wholeness, 5
Windows menu, 407
WIT (What Is There?), 16
WormPD, 16
worst-case scenario

xanthine (Xa), 328–331, 333, 334, 336, 340, 341, 345–348, 352, 355, 360–364, 462
xanthine dehydrogenase (XD), 334, 345, 348, 363
xanthine monophosphate, 331
xanthine oxidase (XD), 334, 348, 360, 364
xanthinuria, 345
xanthosine monophosphate (XMP), 330, 331, 333, 336, 346, 347, 352
xenobiotic compounds, 16

yeast, 8, 18, 22, 260, 261, 290
Yeast Protein Database (YPD), 16
yield data, 174

zero steady state, 198–200
zooming, 64
zymogen, 29–31

Quick PLAS Syntax Reference
(for details see PLAS Help file)

1. GENERAL RULES
    - A system is defined in PLAS through **declarations** (variables, equations, constants, etc.) and **comments**.
    - **One declaration per line** of text.
    - With few exceptions, the **order of declarations** is arbitrary; therefore, declarations and comment lines can be freely intermixed.
    - **Comments** are prefaced by a double slash "//" and may follow a declaration on the same line. Also, any text line not in proper syntax is treated as a comment.

2. NAMES OF VARIABLES AND CONSTANTS
    - First character of a **name** must be a letter; following characters may be letters, numbers, or the underscore character "_". Maximum length is 255 characters. Spaces are not allowed in names.
    - PLAS is **case sensitive**: "X1" and "x1" are two different names.
    - Reserved (taboo) names: **t0, tf, hr,** and **_time**. PLAS reserves these names for the initial time, final time, report interval, and the current time value, respectively.

3. DECLARATION OF DIFFERENTIAL EQUATIONS AND CONSTANTS
    - **Format for differential equations** (apostrophe indicates the derivative):
      <dynamic variable name>′ = <mathematical expression>
      Example: $X1' = (2/25)X1^{\wedge}0.5 \, X2^{\wedge}-0.2 - \beta X1$
    - **Format for constants, initial values,** etc.: <constant name> = <mathematical expression>
    - **Initial values** must have the same name as a dynamic variable. Example: $X1 = 0.125$
    - Standard **operator notation**: $+ - * / \wedge$ (exponentiation). Also, a space indicates multiplication. Example: x*y = x y
    - Any level of nested **parentheses** allowed. ( ) only; do not use [ ] or { }.

4. DECLARATION OF TRANSFORMATION VARIABLES
    - A **transformation** is defined by a function of dynamic variables and/or other transformation variables. Example: $z = 3.2 \, X2^{\wedge}2 \, / \, \ln[X1/X3]$
    - **Predefined functions** in PLAS:

abs[ ]	absolute value	cos[ ]	cosine	ln[ ]	natural log	sin[ ]	sine
atan[ ]	arc tangent	exp[ ]	exponential	log[ ]	log base 10	sqrt[ ]	square root

5. OTHER DECLARATIONS, DIRECTIVES, BUTTONS, COMMANDS, OPTIONS

!!	!! X1 X2 Z	Show only X1, X2, and Z, but not other variables
&&	&& X5 X6	X5 and X6 are independent variables and not just constants
@	@ <time> <A> = <constant expression> @ 3.6 X1=ln[100]	At specified time (t0 ≤ time ≤ tf), set constant or variable A to the value of the specified constant expression
>>	>> Continue declaration on next line; next line should begin with >> (recommended)

- compute numerical solution (as function of time).
- compute steady-state features; show internal representation and values of variables and constants.
- Select **graph type** (time plot, 2D, or 3D phase plot, merge), or table representation under "Results." Right-click a graph to bring up the "Results" menu.
- Double-click graph to **change format** (axes, grid lines, line styles, colors, legend, etc.).
- To print or otherwise display, **cut, copy, paste** into word processors, spreadsheets, and presentation programs, using *Ctrl-X, Ctrl-C, Ctrl-Y.*
- PLAS offers two **numerical solvers**: a very efficient Taylor series method for power-law systems and a general purpose BDF stiff integrator ("Gear method"). Use "Options/Solver" to select.